AF598356

CYCLOSTRATIGRAPHY: APPROACHES AND CASE HISTORIES

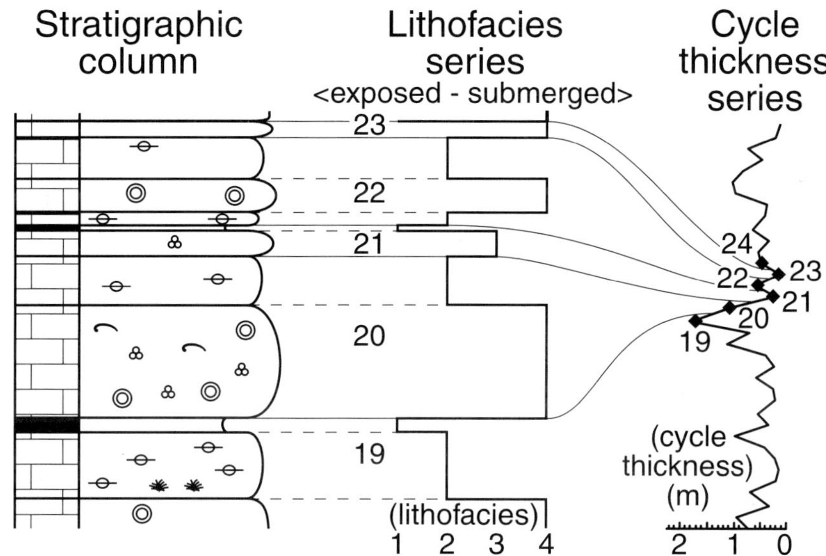

Edited by:

BRUNO D'ARGENIO

Istituto per l'Ambiente Marino Costiero, Geomare, National Research Council, Calata Porta di Massa, Porto di Napoli, 80133 Napoli, Italy

AND

Dipartimento di Scienze della Terra, Università "Federico II", Largo San Marcellino 10, 80138 Napoli, Italy

ALFRED G. FISCHER

Department of Earth Sciences, University of Southern California, Los Angeles, California 90089-0740

ISABELLA PREMOLI SILVA

Dipartimento di Scienze della Terra, Università di Milano, via Mangiagalli 34, 30133 Milano, Italy

HELMUT WEISSERT

Geological Institute ETH-Z, CH-8092 Zürich, Switzerland

AND

VITTORIA FERRERI

Dipartimento di Scienze della Terra, Università "Federico II", Largo San Marcellino 10, 80138 Napoli, Italy

Laura J. Crossey, Editor of Special Publications
SEPM Special Publication 81

Tulsa, Oklahoma, U.S.A. *October, 2004*

SEPM and the authors are grateful to the following
for their generous contribution to the cost of publishing

Cyclostratigraphy: Approaches and Case Histories

Institute for the Marine Coastal Environment (CNR), Naples, Italy

Department of Earth Sciences, University of Naples, Federico II, Naples, Italy

Department of Earth Sciences, University of Milan, Milan, Italy

Italian Ministry of Education, University and Research, Rome, Italy

Department of Geology, Temple University, U.S.A.

Ruprecht Karle University, Heidelberg, Germany

**Department of Earth and Life Sciences,
Free University Amsterdam, The Netherlands**

Contributions were applied to the cost of production, which reduced the
purchase price, making the volume available to a wide audience

---

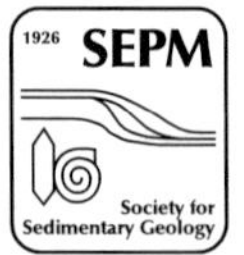

SEPM (Society for Sedimentary Geology) is an international not-for-profit Society based in Tulsa, Oklahoma. Through its network of international members, the Society is dedicated to the dissemination of scientific information on sedimentology, stratigraphy, paleontology, environmental sciences, marine geology, hydrogeology, and many additional related specialties.

The Society supports members in their professional objectives by publication of two major scientific journals, the Journal of Sedimentary Research (JSR) and PALAIOS, in addition to producing technical conferences, short courses, and Special Publications. Through SEPM's Continuing Education, Publications, Meetings, and other programs, members can both gain and exchange information pertinent to their geologic specialties.

For more information about SEPM, please visit **www.sepm.org.**

ISBN 1-56576-108-1

SEPM (Society for Sedimentary Geology)
6128 E. 38th Street, Suite 308
Tulsa, Oklahoma 74135-5814, U.S.A.

Printed in the United States of America

CYCLOSTRATIGRAPHY: APPROACHES AND CASE HISTORIES

Bruno D'Argenio, Alfred G. Fischer, Isabella Premoli Silva, Helmut Weissert, and Vittoria Ferreri, Editors

CONTENTS

Mixed and Detrital Marginal Facies

Lacustrine Facies

Appendix

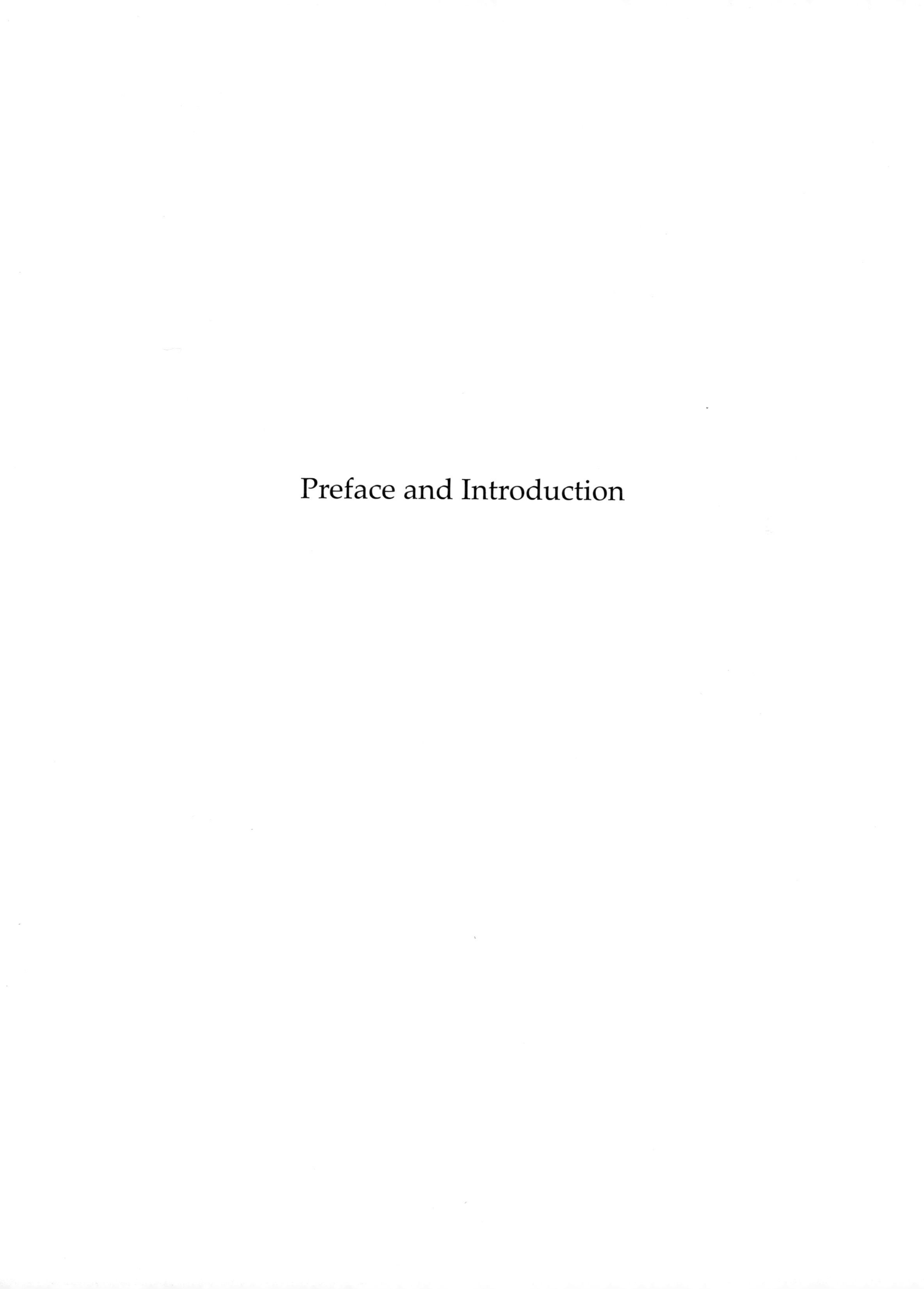

Preface and Introduction

CYCLOSTRATIGRAPHY: APPROACHES AND CASE HISTORIES

BRUNO D'ARGENIO
Istituto per l'Ambiente Marino Costiero, Geomare, National Research Council, Calata Porta di Massa Porto di Napoli, 80133 Napoli, Italy
e-mail: dargenio@gms01.geomare.na.cnr.it
AND
Dipartimento di Scienze della Terra, Università "Federico II", Largo San Marcellino 10, 80138 Napoli, Italy
ALFRED G. FISCHER
Department of Earth Sciences, University of Southern California, Los Angeles, California 90089-0740, U.S.A.
e-mail: fischer@earth.usc.edu
ISABELLA PREMOLI SILVA
Dipartimento di Scienze della Terra, Università di Milano, via Mangiagalli 34, 30133 Milano, Italy
e-mail: isabella.premoli@unimi.it
HELMUT WEISSERT
Geological Institute ETH-Z, CH-8092 Zurich, Switzerland
e-mail: helmi@erdw.ethz.ch
AND
VITTORIA FERRERI
Dipartimento di Scienze della Terra, Università "Federico II", Largo San Marcellino 10, 80138 Naples, Italy
e-mail: ciclisti@gms01.geomare.na.cnr.it

PREFACE

This volume is derived from an SEPM international workshop entitled *Multidisciplinary Approach to Cyclostratigraphy*, organized by the Editors in May 2001 and held in Sorrento (Naples, Italy). The Sorrento workshop was attended by some 60 people from nine European countries and from the United States, with a blend of senior scientists and research students.

Their topics explored cyclicity in strata ranging from Neogene–Quaternary back to Middle Triassic times, and from continental through paralic and shallow-marine to deep marine settings. Methodologies for recognizing orbital periodicities ranged from those based on visually apparent stratification and its hierarchical cyclic patterns to instrumentally detected oscillations (including stable isotopes and magnetic parameters). Tests for cyclicity and for the proper identification of orbital cycles included various types of numerical processing. Problems of formal stratigraphic definitions were also addressed. A field trip served to demonstrate the cyclic Cretaceous carbonate platform sequences of the southern Apennines.

In the Introduction we offer a brief history of how concepts of orbital cyclicity and its effects on the Earth evolved, an appraisal of the present state of research, and an overview of the papers in this volume. The main body of the volume consists of the contributed studies. These include a paper on conceptual and pragmatic approaches to stratification cycles by one of the pioneers of cyclostratigraphy, Walther Schwarzacher, who, in the 1940s, discovered the hierarchical expression of orbital cycles in rocks. The other contributions are specific studies of cyclic sequences, extending from the Quaternary back to the Triassic, covering the range from continental deposits to the deep sea, and employing a wide variety of techniques for extracting and processing the information.

As cyclostratigraphy comes to be an internationally practiced branch of geology, it is timely to standardize its language for purposes of communication. Toward that end the International Union of Geological Sciences and its Commission on Stratigraphic Nomenclature appointed a Working Group, consisting of Frits Hilgen, Walther Schwarzacher, and André Strasser. Their report, discussed and implemented at the Sorrento workshop, and here published, establishes some basic concepts and suggests the lines along which further codification may develop. We thank the authors and the IUGS–CSN for this useful contribution.

Finally we wish to acknowledge people and institutions that helped to organize and run the workshop and to prepare this volume.

Wolfgang Schlager, while serving as SEPM President in 1999–2000, feeling the need for a conference on cyclostratigraphic progress in Europe, suggested its organization to Bruno D'Argenio. Daniel Bernoulli, SEPM International Councilor, presented the plan to the SEPM Council. Sabrina Amodio, Francesco P. Buonocunto, Arturo Raspini, Rosaria Sandulli, and Patricia Sclafani worked tirelessly during the years 2000–2001 to organize and run the workshop. Financial contributions were received from the Consiglio Nazionale delle Ricerche, and from the Rectorial Authority and Earth Science Department of the University of Naples. We happily acknowledge all of this support.

As to this volume, we are much indebted to Sabrina Amodio and Patricia Sclafani for their patient and accurate work at all stages of the preparation of this SEPM Special Publication. Laura Crossey and Kris Farnsworth have been helpful and efficient throughout the process of organization and completion of the present publication. We also thank John Southard, SEPM Special Publication Copy Editor, for the manuscript's embellishment, and Robert T. Clarke for his patient desktop publishing work.

Cyclostratigraphy: Approaches and Case Histories
SEPM Special Publication No. 81, Copyright © 2004
SEPM (Society for Sedimentary Geology), ISBN 1-56576-108-1, p. 3–4.

Last but certainly not least we would like to thank sincerely the numerous referees who contributed notably to the quality and readability of the papers.

Bruno D'Argenio
Alfred G. Fischer
Isabella Premoli Silva
Helmut Weissert
Vittoria Ferreri

CYCLOSTRATIGRAPHIC APPROACH TO EARTH'S HISTORY: AN INTRODUCTION

ALFRED G. FISCHER
Department of Earth Sciences, University of Southern California, Los Angeles, California 90089-0740, U.S.A.
e-mail: fischer@earth.usc.edu
BRUNO D'ARGENIO
Istituto per l'Ambiente Marino Costiero, Geomare, National Research Council,
Calata Porta di Massa Porto di Napoli, 80133 Napoli, Italy
e-mail: dargenio@gms01.geomare.na.cnr.it
AND
Dipartimento di Scienze della Terra, Università "Federico II", Largo San Marcellino 10, 80138 Napoli, Italy
ISABELLA PREMOLI SILVA
Dipartimento di Scienze della Terra, Università di Milano, via Mangiagalli 34, 30133 Milano, Italy
e-mail: isabella.premoli@unimi.it
HELMUT WEISSERT
Geological Institute ETH-Z, CH-8092 Zurich, Switzerland
e-mail: helmi@erdw.ethz.ch
AND
VITTORIA FERRERI
Dipartimento di Scienze della Terra, Università "Federico II", Largo San Marcellino 10, 80138 Naples, Italy
e-mail: ciclisti@gms01.geomare.na.cnr.it

FROM EARLY INTUITIONS TO MODERN CYCLOSTRATIGRAPHY

The Orbital Theory of Ice Ages

Geological interest in Earth's orbital variations harks back to the discovery of the Pleistocene ice ages in the 1840 by Louis Agassiz, who convinced numerous prominent geologists that the "drift" that covered much of northern Europe was not a relict of the biblical deluge but of a great ice sheet (Imbrie and Imbrie, 1979). What had caused this to happen?

Adhemar—A First Attempt.—

The French tutor Joseph Alphonse Adhemar offered a theory in his *Revolutions des Mers* (1842). In 1754 the French astronomer–mathematician Jean le Ronde d'Alembert had published his analysis of the Earth's precessional cycle and its effects on insolation patterns. He had found that, with our closest approach to the sun (perihelion) in early January, the summer in the southern hemisphere is abridged by 168 hours of sunlight, while that of the northern hemisphere is correspondingly prolonged. To Adhemar this implied a loss of heat in the south and a gain of heat in the north, which explained to him the greater size of the southern ice cap. The precessional cycle implied that in 11,000 years the situation would be reversed. And thus he reasoned that ice caps would develop alternately in the northern and southern polar regions, each growing and melting away catastrophically on an 11 ky schedule. What he had neglected to note was that the greater proximity of Earth to the sun during perihelion offsets the loss of daylight duration by greater intensity, so that the total annual insolation received remains almost the same. Along with other unsupportable embellishments this first attempt turned out to be science fiction.

Croll—Climatic Effects of Orbital Rhythms.—

By 1843 the French astronomer Urbaine Leverrier had worked out solutions to the changes in Earth's orbital eccentricity, distorted by the gravitational links to our sister planets. He also discovered the oscillations in Earth's axial tilt or obliquity. Intrigued by Adhemar's hypothesis, the self-taught Scot James Croll (Fig. 1), using Leverrier's calculations, discovered that although the precessional effect on annual averages is insignificant, seasonal differences are noteworthy. Perihelion in midsummer brings hotter-than-normal summers and colder-than-normal winters, i.e., intensified seasonality, to the hemisphere in question. At the same time the other hemisphere's seasonality is correspondingly reduced, and these effects are directly proportional to the degree of eccentricity. Croll thus discovered the precession–eccentricity syndrome. Calculating eccentricity for the last 500,000 years, he showed how, now at ca. 3%, it had varied from about 2% to about 6%, and dramatically illustrated the 100 ky eccentricity cycle. He subsequently discovered the positive albedo feedback derived from the whiteness of the ice cap, the climatic effects of changes in axial obliquity, and the likelihood that cold poles imply a stronger zonal atmospheric circulation, strengthening the trade winds and thus involving oceanic circulation. By 1875 Croll had thus discovered the basic principles of orbital forcing (Fig. 2).

Croll supposed that ice caps would expand beyond present limits in response to cold winters, i.e., at high eccentricity and in the perihelial summer phase of the precessional cycle. He, like Adhemar, therefore viewed glaciations as ca. 10,000 year events, alternating between the northern and the southern hemispheres but absent during times of low eccentricity. Accordingly he concluded that the big ice sheets had retreated 80,000 years ago.

These concepts proved of great interest to the glaciologists of his time, who were just then discovering the evidence for mul-

Cyclostratigraphy: Approaches and Case Histories
SEPM Special Publication No. 81, Copyright © 2004
SEPM (Society for Sedimentary Geology), ISBN 1-56576-108-1, p. 5–13.

FIG. 1.—James Croll. (From Imbrie and Imbrie, 1979, modified.)

tiple glaciations. But that interest faded when anthropological and geological evidence showed this retreat to have occurred in the last 20,000 years.

Milankovitch—An Improved Glacial Theory.—

In the early 20th century Milutin Milankovitch (Fig. 3), a tireless Serbian scientist, picked up the research where Croll had left off (Milankovitch, 1920). Having calculated the seasonal fluctuations in insolation, latitude by latitude, he then turned to the effects of orbital variations and the problem of ice ages. Reasoning that the strong seasonality associated with high eccentricity would cause more summer melting than winter growth, he postulated ice advance during eccentricity lows, and retreat during eccentricity highs. Attributing less effect to the precession, he also visualized a somewhat slower pace of glacial successions, in a model that fit the geological data much better (Milankovitch, 1941). American geologists with a four-till model of the Pleistocene remained skeptical, but the Europeans were impressed with the fit to the Penck and Brückner model, based on river terraces. When the Penck–Brückner and the American models of the Pleistocene ice ages broke down in the early 1950s, the orbital baby was once again thrown out with the bathwater, but not for long.

Emiliani—The Marine Record and the Oxygen Isotope Proxy.—

As part of the post-war wave of isotope chemistry, Urey, Epstein, and Lowenstam had developed the use of oxygen isotope ratios in calcitic fossils to measure paleotemperatures in sea water. It seemed likely that ocean temperatures should have fluctuated with glaciations and that the calcitic foraminifera in pelagic sediments should have preserved a continuous record of such a history. Cesare Emiliani (1955), then a graduate student, found the expected variations in oxygen isotope ratios, and was able to develop a Pleistocene stratigraphy of "isotope stages". These recorded a more complex history than that of the four-till model, and, as his studies developed over the next twelve years, the timing of these fluctuations fell more and more into the schedule of the Milankovitch model (Emiliani, 1966, 1978). But something appeared to be wrong—the temperatures did not fit the foraminiferal associations being worked out by John Imbrie. By 1969 Nicholas Shackleton and Imbrie had concluded that the isotope variations had less to do with water temperature than with the transfer of ^{16}O from the ocean to the ice caps, and were thus primarily a measure of global ice volume—an even more direct measure of glaciation (Imbrie and Imbrie, 1979). A definitive paper by Hays, Imbrie, and Shackleton (1976), using refined and tuned time-series analyses of the oxygen isotope proxy in various cores, allayed all doubts about the link of global ice volume to orbital forcing and established a cyclochronology for that time: modern cyclostratigraphy was born.

Orbital Rhythmicity in Strata

De Geer—Annual Cyclicity in Strata.—

While one group of researchers thus carried Croll's principles forward to solve problems of glaciation, another group began to apply them to visual patterns of cyclicity in the stratigraphic record. The first to recognize orbital signatures in sedimentary sequences was Gerard De Geer (1910). During its retreat across Sweden, the last great ice sheet had left widespread muds deposited in ice-margin lakes. These muds were characterized by peculiar texture-graded laminations, "varves", which De Geer (1910) recognized as annual alternations of a silty summer phase and a thin clay winter phase. In a lifetime of work he and his students measured, correlated, and mapped these varves, to time the duration of this retreat across Sweden at 13,600 years (De Geer 1940). He thus became the first cyclostratigrapher (and the first to speak of "geochronology").

Gilbert—Precessional Cycles in Strata.—

In 1895, G.K. Gilbert called attention to the rhythmic occurrence of limestones in the marine Cretaceous of Colorado. Far too widely spaced to be comparable to De Geer's varves, Gilbert viewed them in the light of Croll's book of 20 years before, as precessional. Extrapolating them to the entirety of the Late Cretaceous succession, he estimated its duration at 20 million years, in vast disagreement with Lord Kelvin's concepts of geological time, and not far out of line with modern chronology.

Bradley—Precessional Signals Substantiated.—

Gilbert's paper was too visionary to elicit a rapid rush of stratigraphers into studies of cyclicity. It was 34 years before Wilmot H. Bradley (1929) took up the challenge, to study such cyclicity in more detail, in the Eocene lacustrine-playa complexes of the Green River Formation of the Rocky Mountains. This yielded good evidence for precessional (20 ky) spacing of dolomitic marlstones as measured by the number of varves in intervening oil shales, as well as evidence for sunspot cycles and 30-year cyclicity.

276 PRECESSION OF THE EQUINOXES. [CH. XIII.

is easy to see that, astronomically speaking, the northern winter in perihelion, as shown in fig. 16, ought to be less rigorous than winter of the same hemisphere in aphelion, as

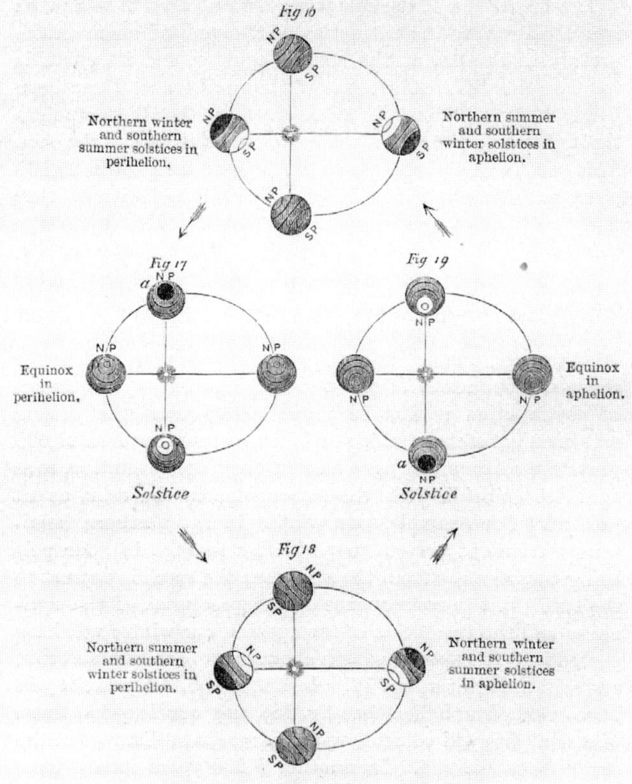

Diagram illustrative of the Precession of the Equinoxes.

In order to give a clear idea of the ellipse, it has been drawn as though the reader looked down upon it directly from above, while the figures of the globe are drawn as they would appear if placed rather more on a level with the eye.

The black patches represent winter; the white patches, summer.

in fig. 18, because the distance of the earth from the sun in the first case is less.

M. Adhémar, in a work entitled 'Les Révolutions de la

CH. XIII.] CLIMATAL EFFECTS OF EXCENTRICITY. 277

Mer,' published in 1840, suggested that a sensible effect has already resulted from the deviation which has occurred since the year 1248, of the position of the perihelion from the time of the winter solstice in the north. By this movement the nearest approach to the sun now occurs eleven days after the shortest day. M. Venetz had previously pointed out, as an historical fact, that before the tenth century the Swiss glaciers were larger than they are now, and that then, after retreating for four centuries, they advanced again, and have been slowly reacquiring their former dimensions. In other words, at that period when the sun was nearest the earth in midwinter or for two centuries before and two after 1248, there was the greatest melting of ice in the northern hemisphere. It may be questioned whether this slight astronomical change, which could hardly produce more than a difference of half a degree Fahrenheit between the cold of the present winter and that of 1248, would be appreciable in the course of 600 years; but the observation may help the reader to understand in what direction the precession of the equinoxes, if capable of producing a sensible change, would now be affecting climate.

It is obvious that this effect would be intensified whenever the ellipticity of the orbit is increased, for the difference of distance between aphelion and perihelion is now only 3,000,000 miles, but it would be at some periods as much as 14,000,000, causing, as Mr. Croll has pointed out, a difference of heat received at the two points amounting to about one-fifth of the entire heat received from the sun, because the amount of such heat varies inversely as the squares of the distance.*

Suggestions by Mr. Croll as to the effects of excentricity on climate.—Upon this difference of heat Mr. Croll has founded a theory which attempts to account for former changes of climate by the tendency which a maximum excentricity would have to exaggerate the cold in that hemisphere in which winter occurred in aphelion.†

* See Herschel's Astronomy, art. 368 *a*.

† Croll on the physical cause of change of climates during geological epochs. Phil. Mag., August 1864.

FIG. 2.—In search of climate-forcing mechanisms J. Adhémar (1842) and J. Croll (1864) suggested that changes in the precession of the equinoxes and in the eccentricity could have had an influence on past climate. This suggestion is reported in Lyell's classic *Principles of Geology* (1875) and shows that almost a century and half ago there had been an intuition on the orbital influence on climate.

Schwarzacher—The PES in Strata.—

Some 20 years later Walther Schwarzacher, writing a dissertation on the Upper Triassic Dachstein Limestone of the Northern Limestone Alps, discovered hierarchical rhythmicity in marine platform deposits. His teacher, Bruno Sander, had previously noted a regular alternation between massive limestones containing large clams and laminated dolomitic limestones with cheese-like textures, and had suggested a rhythmic oscillation in sea level as a possible cause. Schwarzacher (1947) not only documented the regularity of this "elementary rhythm" but also showed the grouping of its cycles into "bundles" of ca. five. Might not the elementary cycle represent the precession and the bundle the short eccentricity cycle, recording eustatic oscillations driven by the growth and decay of a small ice sheet on a Milankovitch schedule?

Subsequent Developments.—

Since then the number of cyclostratigraphic studies has grown at an ever-increasing pace. Cyclic sediments of purported orbital origin have been recognized in many facies and ages, though various examples have remained embroiled in controversy (note the Latemar controversy in this volume). The boots-and-hammer approach of the early workers, directed primarily at cyclicities visible in the field, has become supple-

FIG. 3.—Milutin Milankovitch in the years 1899–1901 (From M. Milankovitch, 1995).

mented by instrumental measurements, and numerous physical, chemical, and biological proxies have been found to carry an orbital imprint. Improved techniques for extracting such imprints made it possible to obtain ever larger data sets, and computers with dedicated software programs have eased their processing. But also it has become clear that dense sampling and large data sets are required if cyclostratigraphy is to become significant to geology.

Cyclochronologies, Anchored and Floating.—

On the one hand, the calculations of Berger (1978) and of Laskar (1999) have now extended astronomical solutions for precession, eccentricity, and obliquity back to 30 Ma. This provides a "target curve" to which the geological observations can be directly linked by magnetic reversals. It enabled Hilgen (1991) and associates, using outcrops of pelagic to lacustrine sediments, to push cyclochronology back through the Pliocene, achieving significant improvement over the radiometric chronology. Shackleton et al. (1999) used deep-sea drilling cores to carry the continuous records of cyclicity, expressible in years-before-present, back into the Oligocene.

Other cyclostratigraphers are working on older sequences to establish detailed cyclochronologies for stretches of geological time that remain "floating", dependent on radiometric estimates for their "before present" timing. With time the gaps will be filled, to provide a continuous cyclochronology.

Beyond Chronology.—

Chronology is not the only promise of cyclostratigraphy. It provides evidence of how components of the depositional system, at a given site and time, responded to the orbital variations. It is thus beginning to provide insight into ancient depositional settings, and into how the Earth's responses have changed with geography, in changing atmospheres and oceans.

Orbital Signatures

Milankovitch Frequency Band.—

For a recent review of orbital cyclicity and its record in sediments, see Hinnov (2000). The primary orbital parameters, such as the annual cycle, are subtly modulated by secondary ones, the orbital variations in the Milankovitch frequency band of 19–406 ky, with an additional ("resonance") frequency in the 2–2.8 My bracket. While these variations are engendered by gravitational interaction with our sister planets, they affect the Earth chiefly by varying the latitudinal and temporal distribution of solar energy, and thereby modifying the patterns of heat distribution on Earth. Above all, they vary the intensity of seasonal variations, i.e., the severity of climates. This in turn affects such factors as monsoons, seasonal precipitation patterns, and marine currents, and induces changes in biotic communities.

Precession–Eccentricity Syndrome.—

The variations fall into two categories, those related to the precession–eccentricity syndrome and those related to the obliquity cycle. The insolation effects of the precession–eccentricity syndrome are strongest in the middle and lower latitudes, those of the obliquity cycle in the polar regions. However, the complexities of global atmospheric and oceanic circulation imply interactions and responses over large distances. The eccentricity–precession syndrome (PES) consists of a basic carrier wave, the ≈ 20 ky precessional cycle relative to perihelion, modulated by the ≈ 100 ky and 400 ky eccentricity rhythms. The resulting pattern is that of the familiar "climatic precession" or "precession index" curve. This commonly finds stratigraphic expression in an "elementary cycle" commonly expressed as 20 ky bedding (couplets in the pelagic realm) , grouped into ca. 100 ky eccentricity "bundles", which in turn are grouped into ca. 400 ky eccentricity "superbundles".

Obliquity Rhythm.—

The obliquity cycles have ticked away in a relatively steady mode, though the changes in Earth rotation rate and lunar recession have dropped its main frequency from about 39 ky to about 41 ky in the last 100 Ma (Berger, 1978). In our experience the obliquity signal in the middle and lower latitudes appears and fades at irregular intervals, though it appears to have persisted steadily in some hemipelagic sequences.

Proxies

All of cyclostratigraphy depends upon discovering rhythmic stratigraphic variations in one or more parameters which can be considered as proxies for the long chain of effects that link back, through climate, to the variations in insolation patterns and the causal orbital variations.

Visual Phase Alternations.—

Initial cyclostratigraphic studies were based on visual recognition of sedimentary oscillations, in which quantification was limited to the stratigraphic spacing; examples are the silt–clay varving in glacial lakes (De Geer, 1940), varving and orbital periodicities in marine evaporites (Anderson, 1982), sedimentary oscillations in lake-playa complexes (Bradley, 1929), and depth oscillations in carbonate platforms (Schwarzacher, 1947). Wilkinson et al. (1996) and Wilkinson et al. (1998) have challenged some of the interpretations thus developed, pointing out that upward-shoaling cycles are not necessarily time-rhythmic. Also, not all rhythmic cycles are necessarily of orbital timing.

The alternations between marls and limestones may be due to variations in detrital influx, typical of hemipelagic settings, or of variations in skeletal productivity as in pelagic sediments (various papers, this volume) and may contain an overprint of cyclicities in redox potential on the seafloor. Cyclicity may extend to biotic differentiation (see below). Higher levels of cyclicity in such sediments may become apparent as modulations in the thickness of successive beds of bed couplets into bundles, or of bundles into superbundles.

Further Quantification: Rank Order.—

A first order of refinement in such studies is the introduction of rank series, in which visual judgement based on various criteria provides a semiquantitative estimate of such matters as physical agitation of the depositional interface, as an index to water depth (D'Argenio et al., this volume; Wissler et al., this volume; Ferreri et al., this volume). Such rank-order series lend themselves to formal time-series analysis (Olsen and Kent, 1999; Preto et al., this volume).

Instrumental Analyses of Bulk Properties.—

Further refinement is obtained by quantitative instrumental measurement of bulk properties, such as stratigraphic change in optical reflectivity (Grippo et al., this volume), neutron-density infrared reflectivity (Herbert and Mayer, 1991), carbonate content (Herbert and Fischer, 1986), or granularity (Sacchi and Müller, this volume), sonic impedence and gamma radiation (Fischer and Roberts, 1991).

Magnetics.—

Various magnetic parameters appear to reflect orbital cyclicity to some degree, but magnetic susceptibility stands out as a property easily measured in hand specimens or, continuously, in cores. Its resolution of orbital cyclicities has matched that of carbonate analyses in marly sequences (Ten Kate and Sprenger, 1993). But it also resolves orbital signals in the purer platform carbonates (Iorio et al., 1998). The role of generation of authigenic magnetite in sediments, by magnetotactic bacteria, remains unclear. A very fine record of orbital variations in the Pliocene Lake muds of Italy (Napoleone et al., this volume) is carried by the diagenetic iron sulfide greigite.

Paleontological Proxies.—

Not only did foraminiferal and coccolith production respond differently to orbital forcing, but the faunal composition of both fluctuated on orbital time scales (Erba, 1988, 1992; Premoli Silva et al., 1989a, 1989b; Herrle, 2002). The role of orbital forcing in biogeography and evolution is only beginning to be appreciated.

The trace-fossil record also shows rhythmic variations in sequences in which orbital forcing affected redox conditions on the sea floor (Erba and Premoli Silva, 1994; Grippo et al., this volume).

Isotope Proxies.—

As noted above, oxygen isotope ratios in foraminiferal tests extracted from deep-sea cores provided the critical evidence linking glaciations to orbital forcing. They have played a lesser role in the exploration of the more distant past, for several reasons. Chief amongst these are the susceptibility of carbonate to ionic exchange with surrounding waters, and the pervasive nature of carbonate mobilization and cementation. In carbonate sequences that suffered emergence on orbital schedules, oxygen isotope ratios have been useful in identifying horizons that suffered fresh-water diagenesis (Buonocunto et al., 2002).

Time Series

Sampling Density.—

Definitive identification of cycles requires first of all the extraction of time series, in which sampling density becomes an important factor. The theoretical treatment of frequencies assumes that the cycles are perfect sinusoids, and that the time dimension is perfectly well known. In stratigraphic cycles neither is strictly true. In the first place, the astronomic cycles range from the stable ca. 404 ka eccentricity cycle to the highly unstable precessional cycle whose mean length, now at ca. 21 ky, shows factor-of-two deviations and is split into a series of modes. Furthermore, the stratigraphic time axis is distorted by variations in accumulation rate, increasingly troublesome as our focus moves from lower to higher frequencies. As a result it becomes necessary to sample more densely than the Nyquist rules would suggest. In the case of the pelagic Albian (Grippo et al., this volume) the ca. 8 cm precessional cycle was missed by 2 cm (5,000 year) sampling in a 1.2–1.6 million year time series of carbonate values, while 6 mm (1,500 year) sampling in a gray-scale scan captured it with its two modes. In the shallow carbonate sequences a 2 cm sampling density (corresponding to ≈ 500 to 2,500 years) yields precessional to eccentricity periodicities and allows subtle stratigraphic gaps, common in this realm, to be individuated (D'Argenio et al., 1999; Ferreri et al., this volume)

Numerical Processing.—

Time-series analysis yielding power spectra and other forms of visualization have now become routine, with programs available as freeware (e.g., Paillard et al., 1996). The use of evolutive spectra of different sorts (cf. Grippo et al., this volume; Napoleone et al., this volume; Maurer et al., this volume) illustrates secular changes in sedimentary response. While multi-tapered Fourier spectra continue to find favor (Hinnov, 2000), Walsh spectra, less demanding of evenly spaced sampling in time series, have been used effectively (e.g., Weedon and Jenkyns, 1999) and the neural-net-aided detection (Brescia et al., 1996) as well as the wavelet (Pelosi and Amodio, 2001) approach may also turn out to be valuable. Of the imperfections introduced into time-series analysis by geological noise, loss of signals to bioturbation or erosional–redepositional events are irremediable, but those introduced by variations in accumulation rate can in many cases be ameliorated by tuning (cf. Grippo et al., this volume) to one of these frequencies. This involves equispacing the values that define cycle boundaries, and readjusting the time series accordingly.

CASE HISTORIES FROM THE PLEISTOCENE TO THE MESOZOIC

The variety of modern approaches to cyclostratigraphy is illustrated in the 16 case studies that form the heart of this volume. They span time from the Quaternary back to the Triassic. From the standpoint of geochronology they fall into two groups. Neogene studies are linked to the "anchored" astronomical target curve for the last 30 million years. These studies on the one hand gain significance from accurate placement in history, and at the same time add content to that history. The chronologies of the Cretaceous and older times remain floating, linked to the general geochronologic scale only by way of the loosely constrained radiometric estimates, in some cases refined by magnetostratigraphic correlation. In time they too will become linked to the "anchored" chronology.

In the meantime cyclostratigraphy is beginning to play a role in climatological and paleoceanographic reconstructions. That theme runs through much of this volume, and for this reason we shall review the papers in a facies context, from bathyal pelagic sediments through neritic, and estuarine–paralic sequences to lacustrine deposits.

Deep-Water Facies

The pelagic and hemipelagic facies of bathyal depths are attractive for various reasons. Depth and distance from the geological turmoil associated with lands generally imply continuity of accumulation, and limited fluctuations in facies and in accumulation rates. Pelagic faunas also offer optimal biostratigraphic control. Three papers deal with studies of Miocene pelagic sequences in the Mediterranean, where the isolation created in Middle Miocene time by the Suez isthmus originated a very special regime resonating with global signals.

Sprovieri, Bonanno, Barbieri, Bellanca, Neri, Patti, and Mazzola examine strontium variations in *Orbulina universa*, in the Late Miocene at Monte Gibliscemi (Sicily). The strontium isotope ratios fluctuated with amplitudes like those of the Pleistocene, but followed the 400 ky eccentricity cycle, which the authors attribute to oscillations in orbitally driven river discharge into the Mediterranean.

Iaccarino, Lirer, Bonomo, Caruso, Di Stefano, et al. examine the cyclostratigraphy of Miocene (Langhian–Tortonian) hemipelagic sediments in Malta, Tremiti Island (Adriatic), and Sicily. Working with lithic variations, calcium carbonate content, and abundance variations in selected species of planktonic foraminifera, they establish correlation with the astronomic target curve. Cyclicity is mainly of the precession–eccentricity syndrome, but obliquity appeared at times. Sapropels are judged to be products of the high seasonalities developed in the summer-perihelion phase at high eccentricity.

Sprovieri, Sgarrella, Russo, Bellanca, and Neri have traced benthic foraminiferal assemblages, processed by Q-mode analysis, through the cyclic hemipelagic Blue Clay of Malta, to reconstruct the nature of Mediterranean circulation in the Middle Miocene. Oxygen supply to the seafloor, indicating more vigorous circulation of the Mediterranean Intermediate Water, was tied to the 100 ky and 400 ky eccentricity maxima.

For earlier times, Grippo, Fischer, Hinnov, Herbert, and Premoli Silva readdress a Cretaceous deep-water section of rhythmic marls and limestones in the Marche Apennines (Central Italy). A drab facies, ranging from white limestones to black marls, represents deposition from a normally stratified water column with a cyclically magnified oxygen-minimum zone that resulted in a record punctuated by precessional anoxic pulsations (PAPs). A contrasting red facies is attributed to downwelling of warm saline waters in a Chamberlin or halothermal circulation. Orbital cyclicity pervades both, with consistent precession–eccentricity forcing and episodic obliquity. A combination of 100 ky and 400 ky cycles in gray-scale scans yields a refined chronology for the Albian.

Maurer, Schlager, and Hinnov describe the cyclicity of part of the Buchenstein Formation, a hemipelagic inter-reef deposit in the mid-Triassic of the Dolomites. After carbonate turbidites are extracted and discarded, time series of the pelagic background yield signatures very similar to those of the pelagic Albian. Middle Triassic (Anisian–Ladinian) stages appear to occupy more of Triassic time than traditional chronologies and zircon dating of tuffs would suggest.

As a general comment to the above papers we note that, despite the contrast between the great east–west expanse of early Cretaceous Tethys and the restricted nature of the Neogene Mediterranean, both settings produced pulses of carbon enrichment associated with the precessional (and at times obliquity) cycles. In both settings these mark precessions associated with high eccentricity. The Neogene sapropels resulted from the high productivity of the perihelial summer phase, whereas the precessional anoxic pulsations of the early Cretaceous seem to record enhanced preservation of organic matter in the stratified regimes of the perihelial winter phase. The hundreds of pulses recorded as centimeter-scale black marls or shales in the Italian Cretaceous extend from the Neocomian through the Cenomanian, a stretch of about 40 million years, and spread at least from the middle Mediterranean through the North Atlantic (Dean and Arthur, 1999). It is tempting to think of the oceanic anoxic events (OAE) of this time as cases in which such intensified and expanded oxygen minima maintained themselves for much longer periods and reached global dimensions.

Carbonate Platforms

Cycles in the carbonate-platform facies form a distinct contrast to pelagic deposits, primarily reflecting sea-level change. Rapid deepening was followed by progressive shoaling and, not uncommonly, brief emergence. Shoaling-upward cycles are typical of carbonate platforms, but the attribution of these cycles to orbital forcing (Schwarzacher, 1947; Goldhammer et al., 1990) has been challenged repeatedly (Wilkinson et al., 1996; Wilkinson et al., 1998) and even when workers agree on the existence of *bona fide* periodicities they may disagree on their identification.

D'Argenio, Ferreri, and students have long carried on studies of upward-shoaling cycles in Cretaceous platform limestones of the Apennines, Sicily, and Montenegro (southern Dinarids). Here, three papers consider these in a broader context.

D'Argenio, Ferreri, Weissert, Amodio, Buonocunto, and Wissler reconsider Aptian to earliest Albian platform sequences. In these, compound upward-shoaling cycles normally attain emergence. Elementary precession and obliquity cycles group into bundles and these into superbundles, representing the short and long cycles of eccentricity. Parts of longer transgressive–regressive cycles (T/RSTs) are comparable to widely recognized "sequences". Carbon-isotope profiles in the less frequently emerged sequences show good correlation with a carbon profile of Weissert et al. (1998) from the pelagic facies of the southern Alps. The combination with T/RSTs yields a broadly applicable cyclostratigraphy, spanning the ca. 8 My history from OAE-1a into the earliest Albian, at the level of the third-order sequences of Jacquin et al. (1998).

Wissler, Buonocunto, Weissert, Ferreri, and D'Argenio use carbon isotope stratigraphy to correlate Barremian–Aptian pelagic and shallow-water successions. This provides an opportunity to test the reliability of cyclostratigraphy established by two different research groups in two different paleoenvironmental settings, and serves to identify, in the carbonate-platform sequence, the equivalent of the *Selli Level*, a well known pelagic anoxic event (OAE-1a).

Ferreri, Amodio, Sandulli, and D'Argenio compare sections of the Valanginian–Hauterivian boundary near Naples and in Sicily, and find excellent correlation at the level of the 100 ky bundles. The same superbundle and T / RFT relationships prevail in both, and serve to propose a high-resolution global correlation at the level of the third-order sequences of Haq et al. (1987) and Jacquin et al. (1998).

Strasser, Hillgärtner, and Pasquier deal with the well-studied Berriasian carbonate platform in the Jura Mountains region. This platform, subsiding at lower rates than its Italian counterparts and sensitive to small changes in sea level, preserves much complexity in its elementary cycles, but still the model of precessional cycles grouped into eccentricity bundles serves best as an overall scheme, widely applicable.

A striking example of "shoaling-upward" cycles is that of the Triassic (Anisian–Ladinian) Latemar atoll in the Dolomites. A group of workers, related to the Johns Hopkins University and here represented by the paper of Preto, Hinnov, De Zanche, Mietto, and Hardie views the basic cycles of the lagoonal facies as precessional. This brings them into conflict with a group affiliated with the Eidgenössische Technische Hochschule in Zürich, whose radiometric zircon dates suggest a totally different chronology. Zuehlke, in a comprehensive look at the Latemar atoll, takes the latter's view. Assigning the elementary cycles to a sub-Milankovitch period of about 3000 years, he equates the eccentricity cycles of the Hopkins group with the precession. Preto et al. derive comfort from the quality of the spectra, which yield the full orbital range, and from the accumulation rates, very high in the Zuehlke model. Neither side has fully overcome the other, and the Latemar controversy (involving the integrity of tuff-derived zircon ages) will continue to bubble in the cauldron of controversies.

Mixed and Detrital Marginal Facies

Three papers deal with cyclicity in mixed to detrital sequences deposited at continental margins.

The paper by Aiello and Boudillon contributes data from entirely different observations: a set of sparker profiles across the Salento shelf in the southern Adriatic Sea (the south side of Italy's "heel"). These seismic profiles show the structure of prograding sedimentary wedges, viewed as formed during regressions forced by drops in sea level, reinforced by increased supply of bioclastic sediments. Sets of these progradational wedges are separated and truncated by three major ravinement surfaces, attributed to rapid sea-level rise in the last three major deglaciation events, tied to the 100 ky eccentricity schedule.

The paper by Iorio et al. contributes to the glacial history of the Antarctic in Pleistocene time, as recorded in ocean-drilling cores of the hemipelagic drifts of mud accumulated on the continental rise off the West Antarctic margin. Glacial pulses of mud are separated by diatom-rich episodes of detrital starvation. The interval analyzed extends back to 2.6 Ma. Parameters measured were bulk density, magnetic susceptibility, chromaticity (reflectivity), and granulometry. Spectral analysis shows that Antarctic glaciations followed a stable pattern throughout this time span, and exhibit all of the Milankovitch periodicities, including the precession and obliquity cycles, and thus demonstrates that the Antarctic ice sheet was not entrained by the ice regimes of the Northern Hemisphere but acted autonomously (as predicted in Fischer and Hinnov, 1997).

Anderson outlines the cyclic record of the earliest Cretaceous (Berriasian–lower Valanginian) Purbeck Group of southern England, a 7.7 My span. A lively history of transgressions and regressions is reflected in the varying levels of the cycle hierarchy, by interbedding of facies, and the occasional development of paleosols. Carbonate-rich strata with marine to brackish faunas are equated to times of sediment starvation when rising sea level trapped clays in fresh-water settings. They serve as boundaries of cycles or sequences. Correlations are provided by ostracode zones. The Purbeckian is viewed as consisting of four third-order sequences of about 2 My each. Details are provided for the third of these, the Middle Purbeck cycle. This is divisible into five fourth-order sequences corresponding to the 400 ky eccentricity cycle. Each of these is broken into three to four fifth-order sequences reflecting the 100 ky short eccentricity term, in turn showing the familiar bundling of strata in sixth-order sequences corresponding to the precessional cycle.

Lacustrine Facies

Lakes are excellent sediment traps. Most are small and transient, yet there always have been some large lakes that persist for some millions of years. Unlike shoal-water marine settings, they are largely independent of global eustasy, but they are very sensitive to changes in the precipitation–evaporation ratio within their drainage basins.

Napoleone et al. describe the late Gauss–early Matuyama cycle record, in the late Pliocene up to its very end, as magnetically recorded in rift-valley lakes of Italy. The signal is confined mostly to the clay facies, and is carried by the ferromagnetic mineral greigite, a product of very early diagenesis. This captured all of the magnetic-reversal events, while its susceptibility oscillations recorded the hierarchy of cycles. Very dense sampling reveals oscillations at the ca. 7,000 year level, but beyond this the periodicities are mainly those of the Milankovitch frequency band. Ongoing study of the pollen record in the same samples will provide further insights into the climatic responses.

Sacchi and Müller present the cyclostratigraphy of a Hungarian core from the upper Miocene of the Pannonian Basin. A thick sequence of sediments records the filling of a deep lake and the subsequent shoal-water history. A basic chronostratigraphic frame provided by magnetic reversals was refined by cyclostratigraphic studies based on granulometry. Mean grain size fluctuated with a remarkably persistent precessional rhythm, yielding 111 cycles with a mean thickness of 13 m. This demonstrates significant sensitivity of lake deposits to orbital forcing, presumably via climatic control of lacustrine hydrodynamics. Thus silt/sand ratio in the Pannonian lake and magnetic susceptibility in the Italian lakes would appear to be two faces of the same coin. Sacchi and Müller attribute the variations in silt deposition to variations in stream gradients in response to lake-level fluctuations with climate. We speculate that it may also have been influenced by changes in seasonal stratification patterns of the lake. A bundling of precessions into 100 ky eccentricity cycles is visible, with evidence for some 400 ky cycles.

REFERENCES

ADHÉMAR, J., 1842, Révolution des Mers: Déluges Périodiques, Publication privée, Paris, 184 p.

ANDERSON, R.Y., 1982, A long geoclimatic record from the Permian: Journal of Geophysical Research, v. 87, p. 7285–7294.

BERGER, A., 1978, Long-term variations of daily insolations and Quaternary climatic changes: Journal of Atmospheric Sciences, v. 35, p. 2362–2367.

BRADLEY, W.H., 1929, The varves and climate of the Green River epoch: U.S. Geological Survey, Professional Paper 158, p. 87–110.

BRESCIA, M., D'ARGENIO, B., FERRERI, V., PELOSI, N., RAMPONE, S., AND TAGLIAFERRI, R., 1996, Neural net aided detection of astronomical periodicities in geologic records: Earth and Planetary Science Letters, v. 139, p. 33–45.

BUONOCUNTO, F.P., SPROVIERI, M., BELLANCA, A., D'ARGENIO, B., FERRERI, V., NERI, R., AND FERRUZZA, G., 2002, Cyclostratigraphy and high-frequency carbon-isotope fluctuations in Upper Cretaceous shallow-water carbonates, southern Italy: Sedimentology, v. 49, p. 1331–1337.

CROLL, J., 1864, On the physical cause of the change of the climate during geological epochs: Philosophical Magazine, v. 28, p. 121–137.

CROLL, J., 1875, Climate and Time in Their Geological Relations: New York, Appleton & Co.

D'ARGENIO, B., AMODIO, S., FERRERI, F., AND PELOSI, N., 1997, Hierarchy of high-frequency orbital cycles in Cretaceous carbonate platform strata: Sedimentary Geology, v. 113, p. 169–193.

D'ARGENIO, B., FERRERI, V., RASPINI, A., AMODIO, S., AND BUONOCUNTO, F.P., 1999, Cyclostratigraphy of a carbonate platform as a tool for high-precision correlation: Tectonophysics, v. 315, p. 357–385.

DEAN, W.E., AND ARTHUR, M.A., 1999, Sensitivity of the North Atlantic Basin to cyclic climatic forcing during the Early Cretaceous: Journal of Foraminiferal Research, v. 29, p. 465–486.

DE GEER, G., 1912, A geochronology of the last 12,000 years: 11th International Geological Congress, Stockholm, 1910, Proceedings Report, v. 1, p. 241–253.

DE GEER, G., 1940, Geochronologia Suecica Principles: Atlas with Plates 54–90: Stockholm, Almquist and Wiksells.

EMILIANI, C., 1955, Pleistocene temperatures: Journal of Geology, v. 63, p. 538–578.

EMILIANI, C., 1966, Paleotemperature analysis of Caribbean cores P6304-6 and P6304-9 and a generalized temperature curve for the past 425,000 years: Journal of Geology, v. 74, p. 109–126.

EMILIANI, C., 1978, The cause of ice ages: Earth and Planetary Science Letters, v. 37, p. 349–352.

ERBA, E., 1988, Aptian–Albian calcareous nannofossils biostratigraphy of the Scisti a Fucoidi cored at Piobbico (central Italy): Rivista Italiana di Paleontologia e Stratigrafia, v. 94, p. 249–284.

ERBA, E., 1992, Calcareous nannofossil distribution in pelagic rhythmic sediments (Aptian–Albian Piobbico core, Central Italy): Rivista Italiana di Paleontologia e Stratigrafia, v. 97, p 455–484.

ERBA, E., AND PREMOLI SILVA, I., 1992, Orbitally driven cycles in trace-fossil distribution from Piobbico core (Late Albian, central Italy), *in* de Boer, P.L., and Smith, D.G., eds., Orbital Forcing and Cyclic Sequences: International Association of Sedimentologists, Special Publication 19, p. 211–225.

FISCHER, A.G., AND HINNOV, L.A., 1997, Orbitally forced glaciation in greenhouse and icehouse times, part 1: a global model (abstract): Geological Society of America, Annual Meeting, Salt Lake City, Abstracts with Program, p. A-211.

FISCHER, A.G., AND ROBERTS, L.I., 1991, Cyclicity in the Green River Formation, (lacustrine Eocene) of Wyoming: Journal of Sedimentary Petrology, v. 61, p. 1146–1154.

GILBERT G.K., 1895, Sedimentary measurement of geologic time: Journal of Geology, v. 3, p. 121–127.

GOLDHAMMER, R.K., DUNN, P.A., AND HARDIE, L.A., 1990, Depositional cycles, composite sea-level changes, cycle stacking-patterns, and the hierarchy of stratigraphic forcing: Example from Alpine Triassic platform carbonates: Geological Society of America, Bulletin, v. 102, p. 535–562.

HAQ, B.U., HARDENBOL, J., AND VAIL, P.R., 1987, Chronology of fluctuating sea levels since the Triassic: Science, v. 237, p. 1156–1167.

HAYS, J.D., IMBRIE, J., AND SHACKLETON, N.J., 1976, Variations in the Earth's orbit: pacemaker of the ice ages: Science, v. 194, p. 1121–1132.

HERBERT, T.D. AND FISCHER, A.G., 1986, Milankovitch climatic origin of mid-Cretaceous black shale rhythms in central Italy: Nature, v. 321, p. 739–743.

HERBERT, T.D. AND MAYER, L., 1991, Long climatic time series from DSDP / ODP physical property measurements: Journal of Sedimentary Petrology, v. 61, p. 1089–1108.

HERRLE, J.O., 2002, Paleoceanographic and paleoclimatic implications on mid-Cretaceous black shale formation in the Vocontian basin and the Atlantic: evidence from calcareous nannofossil and stable isotopes: Ph.D. thesis, Universität Tübingen, Institut and Museum für Geologie und Paläontologie, 114 p.

HILGEN, F.J., 1991, Extension of the astronomically calculated (polarity) time scale to the Miocene / Pliocene boundary: Earth and Planetary Science Letters, v. 107, p. 349–368.

HINNOV, L.A., 2000, New perspectives on orbitally forced stratigraphy: Annual Review of Earth and Planetary Sciences, v. 28, p. 419–475.

IMBRIE, J., AND IMBRIE, K.P., 1979, Ice Ages. Solving the Mystery: New York, Harvard University Press, 224 p.

IORIO M., TARLING D.H., AND D'ARGENIO B., 1998, The magnetic polarity stratigraphy of a Hauterivian / Barremian carbonate sequence from Southern Italy: Geophysical Journal International, v. 134, p. 13–24.

JACQUIN, T, RUSCIADELLI, G., AMEDRO, F., DE GRACIANSKY, P.C, AND MAGNIEZ-JANNIN, F., 1998, The North Atlantic cycle: an overview of 2nd order transgressive / regressive facies cycles in the Lower Cretaceous Western Europe, *in* de Graciansky, P.C., Hardenbol, J., Jacquin, T., Vail, P.R., and Farley, M.B., eds., Mesozoic and Cenozoic Sequence Stratigraphy of European Basins: SEPM, Special Publication 60, p. 397–409.

LASKAR, J., 1999, The limits of the Earth orbital calculations for geological time-scale use: Royal Society (London), Philosophical Transactions, Series A, v. 357, no. 1757, p. 1735–1759.

LEVERRIER, U., 1843, 1855, Connaissance des temps: Annales de l'Observatoire Imperial de Paris, v. II, 1855.

LYELL, C., 1875, Principles of Geology, Vol. I: London, John Murray, 655 p.

MILANKOVITCH, M., 1920, Théorie Mathématique des Phénomènes Termiques Produits par la Radiation Solaire: Académie Yougoslave des Sciences et des Arts de Zagreb, Gauthiers-Villiers, Paris.

MILANKOVITCH, M., 1941, Kanon der Erdbestrahlung und seine Anwendung auf das Eiszeitenproblem: Akademie Royale Serbe, v. 133, 633 p.

MILANKOVITCH, M., 1995, Milutin Milankovitch 1879–1958. (Autobiography with comments by V. Milankovitch and preface by A. Berger): European Geophysical Society, Kattlenburg–Lindau, Germany, XXIII, 181 p.

OLSEN, P.E., AND KENT, D.V., 1999, Long-term Milankovitch cycles from the Late Triassic and Early Jurassic of eastern North America and their implications for the calibration of the Early Mesozoic timescale and the long term behavior of the planets: Royal Society (London), Philosophical Transactions, Series A, v. 357, p. 1761–1788.

PAILLARD, D., LABEYRIE, L., AND YIOU, P., 1996, Macintosh program performs time-series analysis (abstract): Eos, Transactions, American Geophysical Union, v. 77, p. 379.

PELOSI, N., AND AMODIO, S., 2001, Wavelet application to cyclostratigraphy signals. The San Lorenzello section, Lower Cretaceous, Italy (abstract): SEPM, International Workshop on: "Multidisciplinary Approach to Cyclostratigraphy", Sorrento, Italy,. Abstract Volume, p. 39–40.

PREMOLI SILVA, I., ERBA, E., AND TORNAGHI, M-E, 1989a, Paleoenvironmental signals and changes in surface fertility in Mid-Cretaceous C-org-rich pelagic facies of the Fucoid Marls (Central Italy): Géobios, Mémoire Special, v. 11, p. 225–236.

PREMOLI SILVA, I., TORNAGHI, M-E., AND RIPEPE, M., 1989b, Planktonic foraminiferal distribution records productivity cycles: evidence from the Aptian–Albian Piobbico core (Central Italy): Terra Nova, v. 1, p. 443–448.

SCHWARZACHER, W., 1947, Über die Sedimentäre Rhythmik des Dachsteinkalkes von Lofer: Geologische Bundesanstalt, Verhandlungen, v. H10-12, p. 175–188.

SHACKLETON, N.J., CROWHURST, S.J., WEEDON, G.P., AND LASKAR, J., 1999, Astronomical calibration of Oligocene–Miocene time: Royal Society (London), Philosophical Transactions, Series A, v. 357, p. 1907–1930.

TEN KATE, W.G.H.Z., AND SPRENGER, A., 1993, Orbital cyclicities above and below the Cretaceous / Paleogene boundary (N Spain): Sedimentary Geology, v. 87, p. 69–101.

WEEDON, G-P., AND JENKYNS, H.C., 1999, Cyclostratigraphy and the Early Jurassic timescale: data from the Belemnite Marls, Dorset, southern England: Geological Society of America, Bulletin, v. 111, p. 1823–1840.

WEISSERT, H., LINI, A., FÖLLMI, K.B., AND KUHN, O., 1998, Correlation of Early Cretaceous carbon-isotope stratigraphy and platform drowning events: A possible link?: Palaeogeography Palaeoclimatology, Palaeoecology, v. 137, p. 189–203.

WILKINSON, B.H., DIEDRICH, N.W., AND DRUMMOND, C.N., 1996, Facies successions in peritidal carbonate sequences: Journal of Sedimentary Research, v. 66, p. 1065–1078.

WILKINSON, B.H., DIEDRICH, N.W., DRUMMOND, C.N., AND ROTHMAN, E.D., 1998, Michigan hockey, meteoric precipitation, and rhythmicity of accumulation on peritidal carbonate platforms: Geological Society of America, Bulletin, v. 110, p. 1075–1093.

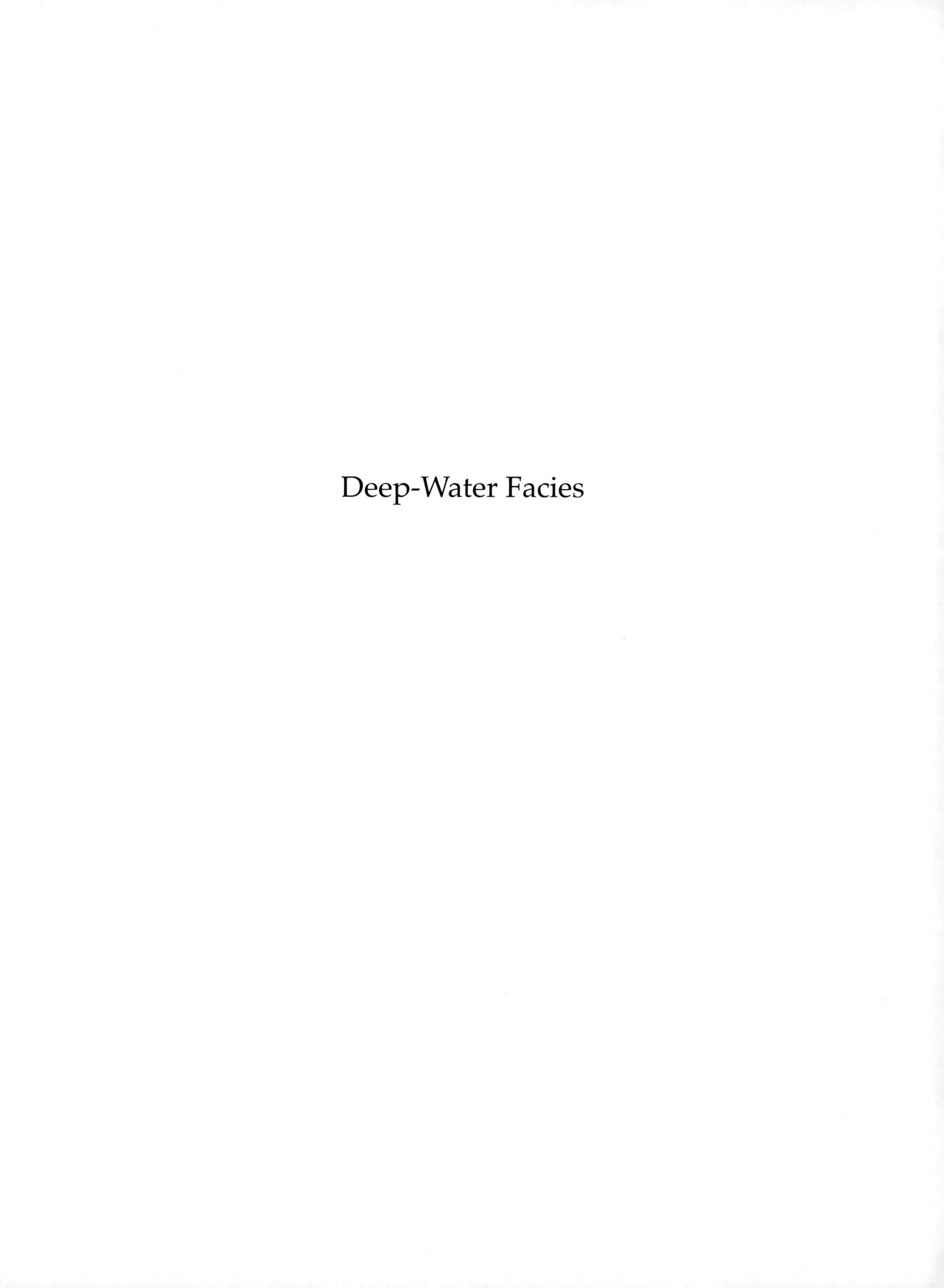

Deep-Water Facies

^{87}Sr/^{86}Sr VARIATION IN TORTONIAN MEDITERRANEAN SEDIMENTS: A RECORD OF MILANKOVITCH CYCLICITY

MARIO SPROVIERI
Istituto per l'Ambiente Marino Costiero, Geomare, National Research Council, Calata Porta di Massa, Porto di Napoli, 80133 Napoli, Italy
e-mail: sprovier@gms01.geomare.na.cnr.it
ANGELO BONANNO
Istituto per l'Ambiente Marino Costiero, IRMA, CNR, Via L. Vaccara 61, 91026 Mazara del Vallo, Italy
e-mail: bonanno@irma.pa.cnr.it
MARIO BARBIERI
Dipartimento di Scienze della Terra, Università "La Sapienza", Piazzale Aldo Moro 5, 00185 Roma, Italy
e-mail: mario.barbieri@uniroma1.it
ADRIANA BELLANCA AND RODOLFO NERI
Dipartimento di Chimica e Fisica della Terra (CFTA), Università di Palermo, Via Archirafi 36, 90123 Palermo, Italy
e-mail: bellanca@unipa.it; neri@unipa.it
AND
BERNARDO PATTI AND SALVATORE MAZZOLA
Istituto per l'Ambiente Marino Costiero, IRMA, CNR, Via L. Vaccara 61, 91026 Mazara del Vallo, Italy
e-mail: bpatti@irma.pa.cnr.it; mazzola@irma.pa.cnr.it

ABSTRACT: This work presents a detailed ^{87}Sr/^{86}Sr isotope curve for the 7.5–9.8 Ma time interval obtained by analyzing isotope compositions of the *Orbulina universa* planktonic foraminifer species from the Mediterranean Gibliscemi section (southern Sicily). The available astronomical tuning of the section provided the opportunity to assess a direct control of the Milankovitch climate cyclicity on the seawater Sr isotope changes. Results of spectral analysis suggest a linear forcing of the 400 ky eccentricity component on the ^{87}Sr/^{86}Sr ratios of the Mediterranean seawater. The recorded amplitude of Sr isotope 400 ky cycles ranges between $\pm$ 5.5 x 10^{-5} and $\pm$ 6 x 10^{-5} around the long-term trend for the Tortonian at global scale. Such a Δ^{87}Sr is of the same order of magnitude of that measured by Capo and DePaolo (1990) and Dia et al. (1991) for the Pleistocene 100 ky glacial–interglacial cycles and about two times larger than that reported for site 758 by Clemens et al. (1993) for the same periodic oscillations. Mass-balance calculations applied to our dataset suggest that periodic changes of about 100–150% in the riverine inputs can account for the amplitude oscillations of ^{87}Sr/^{86}Sr ratios recorded in the Mediterranean during the Tortonian, thus emphasizing the high potential of this basin as good recorder of climate-induced seawater Sr isotope changes.

INTRODUCTION

The seawater ^{87}Sr/^{86}Sr ratio has gradually increased during the last 40 My, probably because of enhanced weathering rates of continental rocks associated with the uplift of the Himalayas and Tibetan Plateau (Raymo et al., 1988; Hodell et al., 1990; Capo and DePaolo, 1990; Richter et al., 1992). Several papers have been published to refine the Sr isotope curve of the Neogene, thus improving the stratigraphic reliability of this tool for large-scale correlations (e.g., DePaolo and Ingram, 1985; DePaolo, 1986; Hess et al., 1986; Hodell et al., 1989; Ward et al., 1993; McArthur, 1994; Martin et al., 1999; Sprovieri et al., 2002). In recent years, some authors have combined the outcomes of detailed astronomical calibration of selected marine and land-based sections with high-resolution Sr isotope time series acquired from the same sedimentary sequences, thus overcoming the classic biostratigraphic constrain on the Sr isotope calibration and taking advantage of a most accurate age control (Martin et al., 1999; Sprovieri et al., 2002). Other authors (e.g., Clemens et al., 1993) tried to detect the short-term Milankovitch cyclicities in the Sr isotope record but any proof of a direct link between seawater ^{87}Sr/^{86}Sr variations and astronomical forcing remains only a controversial working hypothesis (Henderson et al., 1994). The small-amplitude oscillations of seawater ^{87}Sr/^{86}Sr ratio directly related to the classic astronomical forcing of precession, obliquity, and eccentricity fall within the order of magnitude of the analytical error, limiting the reliability of the cycle interpretation often judged as a simple technical artifact (e.g., Hodell et al., 1990; Henderson et al., 1994). Moreover, the estimated long-time residence of Sr in the ocean (~ 2.5 My; Hodell et al., 1990) strongly limits the possibility of revealing detectable short-term climate changes in the seawater Sr isotope curve.

In this work, we present a detailed Tortonian Sr isotope record from a Mediterranean astronomically calibrated land-based section.

Owing to the relatively limited geographic extent of the Mediterranean Sea and to the important contribution of river inputs to the basin, the residence time of strontium in seawater is lower than that estimated for the global ocean, ranging nowadays between 300 and 400 years (for details see Table 1). This feature, added to the characteristic strong link of the basin with the close continents, directly influenced by different climatic regimes, could represent an important advantage to verify reliably if the Milankovitch forcing works on the ^{87}Sr/^{86}Sr seawater changes. The shorter residence time of Sr in the Mediterranean basin, compared to that of the global ocean, makes this basin an optimal site to study high-frequency ^{87}Sr/^{86}Sr seawater changes modulated by the astronomical forcing. The calibrated ^{87}Sr/^{86}Sr time

Cyclostratigraphy: Approaches and Case Histories
SEPM Special Publication No. 81, Copyright © 2004
SEPM (Society for Sedimentary Geology), ISBN 1-56576-108-1, p. 17–26.

TABLE 1.—Hydrographic parameters used for the solution of the mass-balance differential equation with relative source bibliography.

Influxes	J_r (km³/year)	Sr (μmol/l)	$^{87}Sr/^{86}Sr$
Bosporus influx	200 (8)	2.759 (2)	0.7093 (2)
Nile	90 (4)	2.7 (1)	0.7060 (3)
Rhone	55 (1)	5.936 (1)	0.7087 (1)
Po	52 (5)	1.58 (5)	0.7097 (3)
Other rivers	53 (8)	0.92 (1)	No data
Total	450 (8)		
Estimated average			0.7085 (10)
Present Mediterranean volume		36.14 x 10^5 km³ (6)	
Present Atlantic influx		23 x 10^4 km³/year (7)	
Mediterranean Sr concentration		90 μmol/l (5)	

(1) Palmer and Edmond (1989); (2) Faure and Powell (1972); (3) Brass (1976); (4) Van Os and Rohling (1993); (5) Emelyanov and Shimkus (1986); (6) Benson et al. (1991); (7) Bryden et al. (1994); (8) Garrett et al. (1993); (9) Lane-Serff et al. (1997); (10) This paper

series is studied by spectral analysis. Mass-balance equations are proposed to independently evaluate the possible role of major hydrographic parameters in controlling changes in the Sr isotope seawater.

SECTION AND SAMPLING

The Gibliscemi section is located on the southern slope of Monte Gibliscemi in the south of Sicily (Fig. 1). The upper Miocene sequence is characterized by a cyclic alternation of homogeneous hemipelagic marls and sapropelic layers followed by the diatomites of the Tripoli Formation (Messinian) and capped by the "Calcare di Base" and evaporites (gypsum rocks) of the Gessoso-Solfifera Formation.

The studied sediments consist of open-marine sediments that show cyclic alternations of homogeneous marls and brownish laminated sapropelic layers. These sapropelic beds reveal characteristic patterns with both small-scale and large-scale clusters in addition to individual sapropels, which allows the section to be calibrated astronomically. Detailed analysis of the sedimentary cycle patterns of the studied sections and their correlation to the astronomical target curves $La90_{1,1}$ (Laskar et al., 1993) are reported in Hilgen et al. (1995).

The studied record covers the stratigraphic interval from 9.7 to 7.5 Ma. The lithologic features, the integrated plankton biostratigraphy (from Sprovieri et al., 1999), and the astronomical tuning to the insolation curve of Laskar et al. (1993), proposed by Hilgen et al. (1995), are shown in Figure 2.

Samples for Sr isotope measurements have been collected only from the homogeneous marls in the intermediate part of most sedimentary cycles, with an estimated average sampling periodicity of ~ 40 ky.

ANALYTICAL PROCEDURES

For the Sr isotope analyses, 300–500 mg of monospecific planktonic foraminifer shells (*Orbulina universa* species) were picked from the > 150 μm fraction. Each chamber was broken open and the pieces sonified in distilled water to remove any clay particles. The cleaned fragments were then dissolved in 2.5N HCl and evaporated to dryness. The residue was dissolved in a 2N HCl solution. The obtained solution was loaded onto Sr selective crown ether resin. Total blanks were < 2 ng of Sr; the sample contained > 5 ng of Sr. Concentrations of Rb were found to be too low to require corrections for radiogenic ^{87}Sr.

Sr isotope analyses were performed by using a Finnigan MAT 262 RPQ multicollector mass spectrometer and the software package described in Ludwig (1994). All samples were loaded with tantalum oxide onto tungsten filaments and analyzed in static mode. Sr analyses were normalized to $^{86}Sr/^{88}Sr = 0.1194$. More than 200 ratios were collected for a single analysis at 1.5 V. The errors of the analyzed sample are expressed as two standard errors of the mean (95% confidence interval). During the collection of data, the measured value for NIST 987 (45 measurements) was within 0.00002 of the value 0.710232, which, in our laboratory, corresponds to a value of 0.709175 for modern seawater from the Mediterranean Sea (MSW). All Sr isotope data are referred with respect to NIST 987 and MSW values.

FIG. 1.—Location map of the Gibliscemi section, southern Sicily.

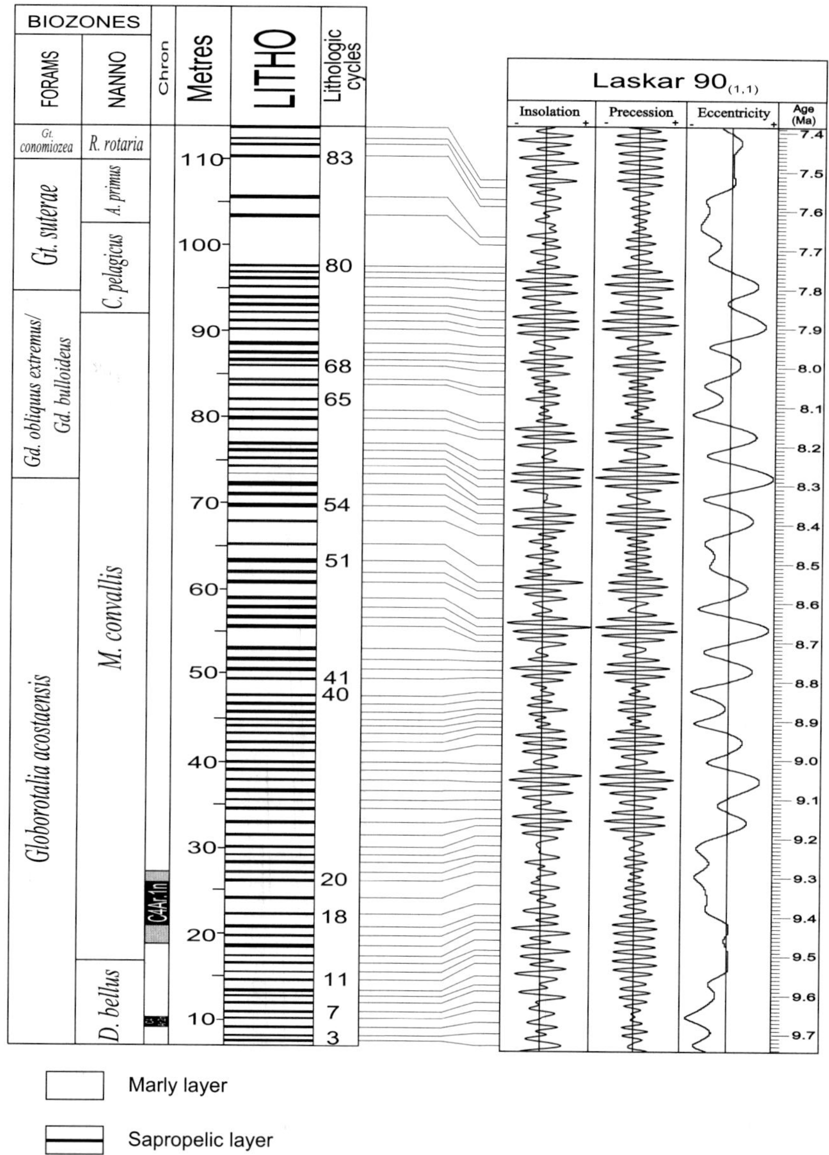

FIG. 2.—Lithology, magnetostratigrapy, integrated biostratigraphy, and astronomical calibration of the Gibliscemi section (modified from Hilgen et al., 1995, and Sprovieri et al., 1999).

RESULTS

A new set of 15 Sr isotope measurements was carried out on specimens of the planktonic foraminifer *Orbulina universa* from the Gibliscemi section to integrate the dataset presented by Sprovieri et al. (2002) and ensure the most homogeneous distribution of data along the sedimentary sequence.

The complete $^{87}Sr/^{86}Sr$ dataset obtained for the Gibliscemi section is reported versus time (Fig. 3) on the basis of the astronomical tuning of the same section presented by Hilgen et al. (1995). The Sr isotope curve shows an initial increasing trend from the base of the section (about 9.7 Ma) to ~ 8.5 Ma with an estimated $^{87}Sr/^{86}Sr$ change of about 7 x 10^{-5}. Then, a quasi-constant Sr isotope value (0.70892) is shown up to about 7.8 Ma, followed by a final slight negative excursion (down to about 0.70888) from 7.8 Ma up to the top of the succession (about 7.5 Ma). A detailed discussion of the long-term trends recorded in $^{87}Sr/^{86}Sr$ signal of Gibliscemi and a comparison with Atlantic and Pacific deep-sea records are reported in Sprovieri et al. (2002).

Superimposed on this long-term trend, shorter-term oscillations shift the $^{87}Sr/^{86}Sr$ ratios in a range of ± 2.5 x 10^{-5} / 0.2 Ma and ± 3 x 10^{-5} / 0.2 My. On the basis of SEM observations, the analyzed specimens of *Orbulina universa* appear to have avoided diagenetic alteration; therefore, their $^{87}Sr/^{86}Sr$ ratio can be considered reliable as a proxy for the seawater Sr isotope composition.

DISCUSSION

Spectral Analysis

The $^{87}Sr/^{86}Sr$ record was detrended using a sixth-order polynomial spline in order to subtract the long-term trend (Fig. 4A) and then interpolated to 40 ky constant steps using a Gaussian filter. Such a numerical approach ensures that in a given interval

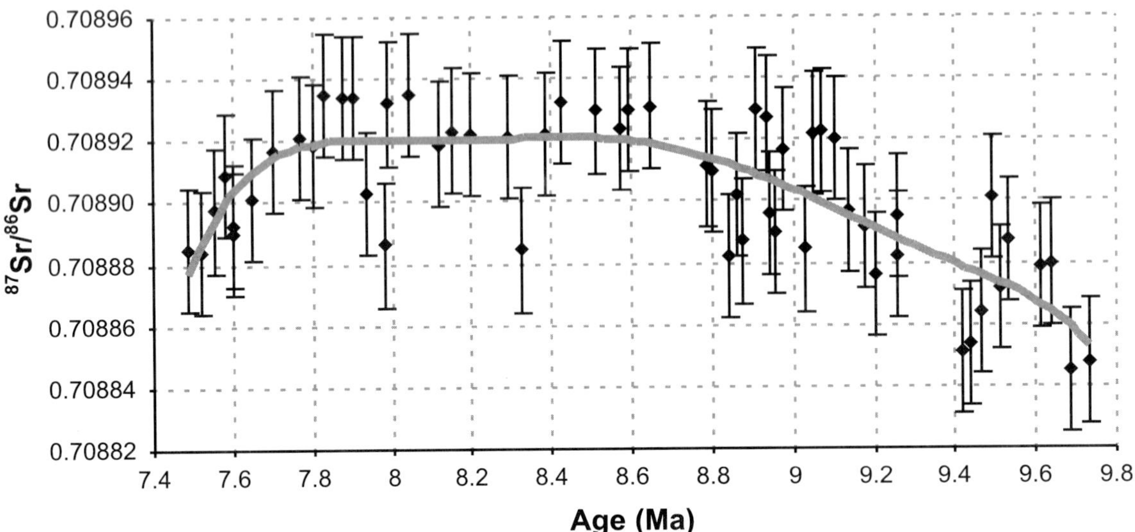

FIG. 3.—$^{87}Sr/^{86}Sr$ signal from the Gibliscemi section reported versus time on the basis of the astronomical calibration of the section. Thick gray line indicates the sixth-order polynomial fitting curve.

interpolated data more than original points are never present. The power spectrum of the $^{87}Sr/^{86}Sr$ signal was estimated (Fig. 4B) using the standard procedure reported in Jenkins and Watts (1968). The sampling rate adopted (about 1 sample/40 ky) does not allow us to reliably detect periodicity lower than 100 ky. The record shows variance concentration at 400 ky corresponding to the long-eccentricity periodicity band. The coherence and time-lag analysis between the insolation curve and the Sr isotope record quantify the degree to which amplitude variations are linearly related. Coherence values of ~ 0.8 in the 400 ky periodicity band suggest a direct link between astronomic forcing and $^{87}Sr/^{86}Sr$ response (Fig. 4C). The phase relationship between the two signals is ~ 180° in the same periodicity band (Fig. 4D). Therefore, the $^{87}Sr/^{86}Sr$ signal seems to fluctuate in the long-eccentricity frequency band, with lower values recorded during warmer semi-cycles and higher values during the colder ones. By using the Tukey–Hinnov band-pass filter (Jenkins and Watts, 1968), which preserves information about frequency and amplitude of selected harmonics, we estimated the range of isotope oscillation related to selected periodicity bands. The $\Delta^{87}Sr$ record filtered in the 400 ky periodicity band shows average amplitude of cycles of ~ 5.5–6 x 10^{-5} (Fig. 5), which is the same order of magnitude of the Pleistocene 100 ky glacial–interglacial cycles recorded by Hodell et al. (1989) and Dia et al. (1991). Although possible analytical artifacts cannot be ruled out definitively in the interpretation of the cycle pattern of Gibliscemi, variations of this amplitude are resolvable in our dataset, the estimated sample reproducibility being ± 20 x 10^{-6}. We think that the reliability of the recorded fine structure of the $^{87}Sr/^{86}Sr$ curve needs to be confirmed by new acquisition of isotope data from coeval astronomically tuned Mediterranean sections. However, we tried to estimate the magnitude of fluxes or isotopic changes that would be required to account for the recorded short-term variations in the Mediterranean Sr isotope ratios.

Origin of the 400 ky $^{87}Sr/^{86}Sr$ Oscillations

Mass-balance calculations (Brass, 1976; Hodell et al., 1990; Clemens et al., 1993) illustrate that the modern seawater $^{87}Sr/^{86}Sr$ ratio is driven by composite effects of riverine inputs, hydrothermal contributions, and diagenesis–dissolution of deep-sea carbonates. Therefore, comprehension of the oceanographic mechanisms that induced the 400 ky $^{87}Sr/^{86}Sr$ oscillations recorded during the Tortonian in the Mediterranean basin at least requires consideration of those three sources of variability of seawater Sr isotope.

The short-term nature of the $^{87}Sr/^{86}Sr$ changes recorded for the Gibliscemi section argues against the classically slow hydrothermal dominance of the isotope budgets (Hodell et al., 1990). Moreover, it should be difficult to explain climate-induced periodic changes of hydrothermal activity to account for coupled seawater $^{87}Sr/^{86}Sr$ oscillations. Thus, we neglected, as a first assumption, a direct hydrothermal control on the cyclic Sr isotope variations of the Tortonian Mediterranean basin.

On the other hand, dissolution effects of deep-sea carbonates in the Mediterranean can reasonably be neglected because of the relatively shallow depth of the basin (average 1429 m; Benson et al., 1991), which makes unlikely the development of a calcite compensation depth (CCD). Late Miocene paleogeographic reconstructions of the Mediterranean suggest unchanged cumulative average depth of the basin during the Tortonian (Rögl, 1998). Moreover, dissolution of deep-sea carbonates is an unlikely mechanism because of its small contribution to the Sr budget of the marine water and to $^{87}Sr/^{86}Sr$ ratios very similar to those of the overlying ocean.

Consequently, the 400 ky Sr isotope oscillations could be thought as controlled mainly by: (1) isotope variations of Atlantic seawater feeding into the Mediterranean, and/or (2) variations of riverine inputs into the basin. In order to verify each hypothesis or the combined effects of them, we used the following general mass-balance equation (Brass, 1976; Clemens et al., 1993) to quantify linear effects of a set of free parameters influencing the seawater $^{87}Sr/^{86}Sr$ ratios:

$$Sr_{SWM}\frac{dR_{SWM}}{dt} = J_A(R_{SWA} - R_{SWM}) + k\,J_F(R_F - R_{SWM}) \qquad (1)$$

In this equation (numerical details for the resolution of the differential equation are given in the appendix), Sr_{SWM} and R_{SWM} are the number of strontium moles (μmol Sr/l) and the $^{87}Sr/^{86}Sr$

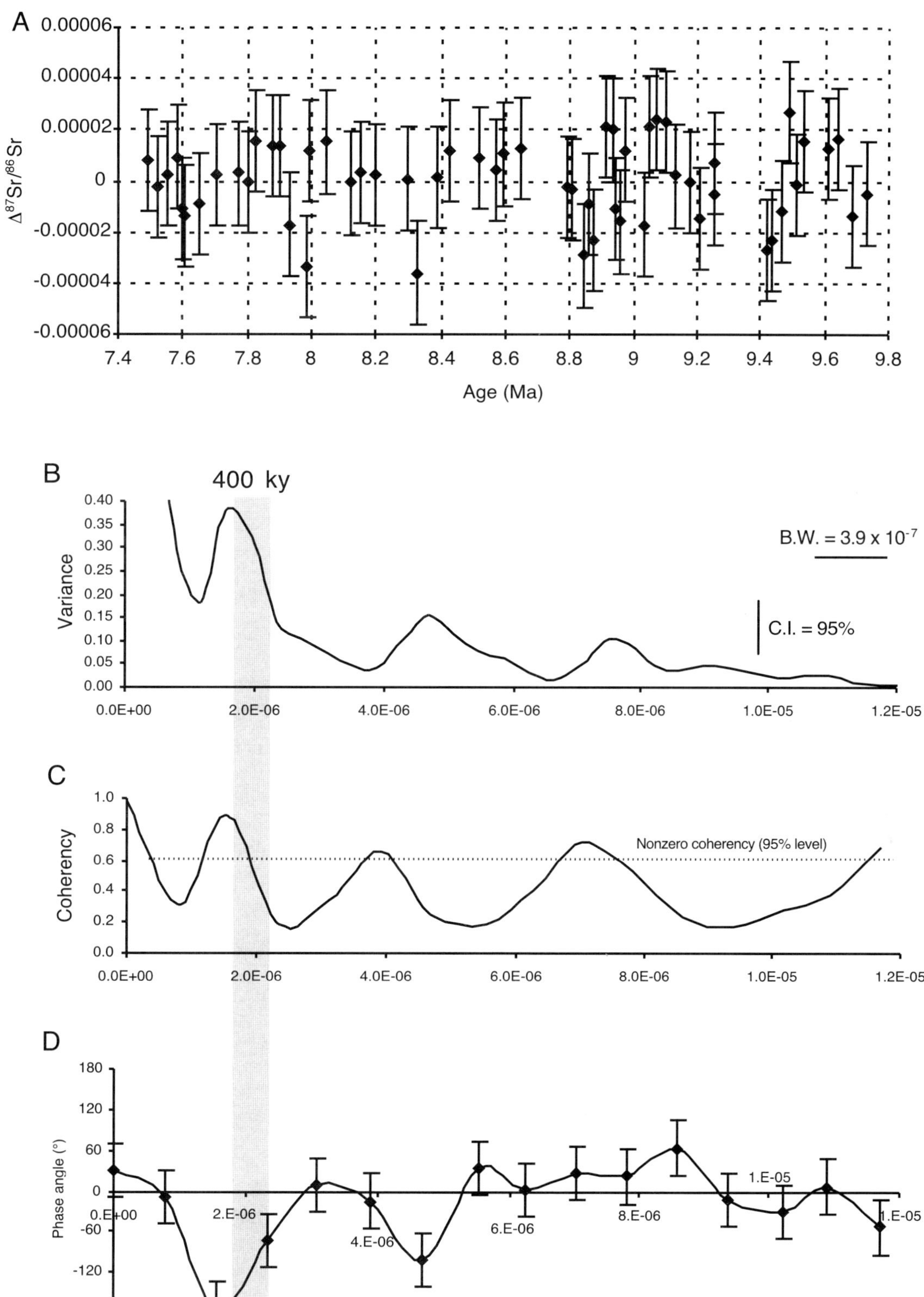

FIG. 4.—**A)** $^{87}Sr/^{86}Sr$ record detrended using a sixth-order polynomial spline. **B)** Power spectrum of the Sr isotope signal from the Gibliscemi section. **C)** Coherency analysis and **D)** phase-angle analysis between insolation forcing and Sr isotope curve. C.I., confidence interval; B.W, bandwidth. Spectral densities are plotted on a linear scale. The coherency spectrum is plotted on a hyperbolic arctangent scale. The dashed horizontal line indicates confidence at the 95% level.

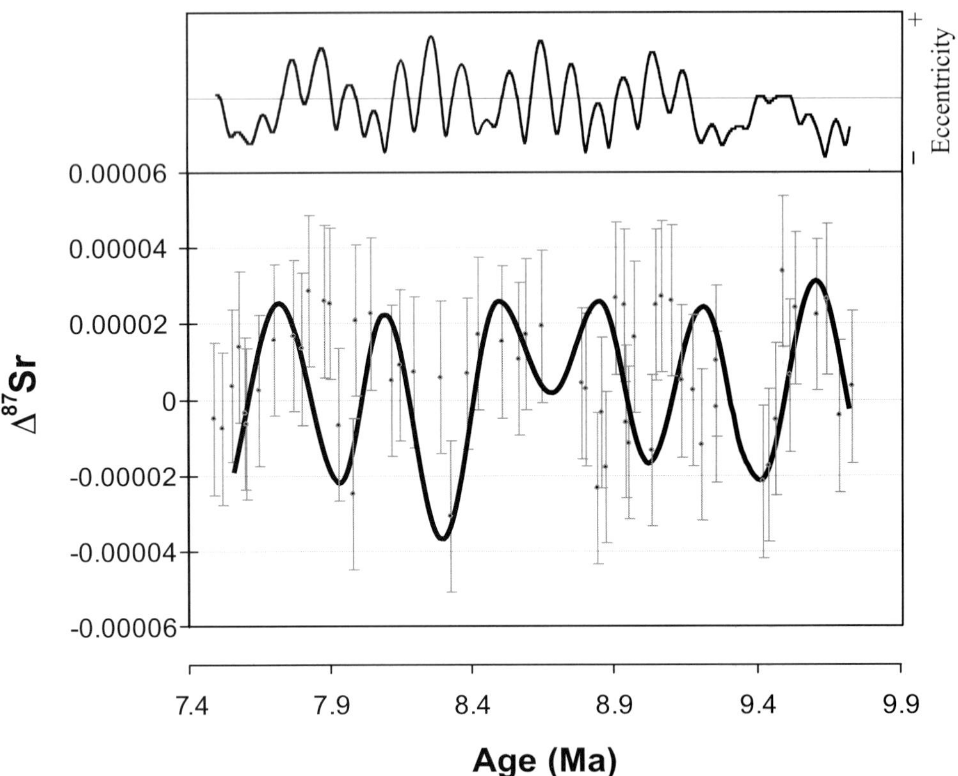

FIG. 5.—Comparison between original and 400 ky filtered (thick line) Sr isotope data at Gibliscemi. In the upper rectangle the eccentricity curve ($La90_{1,1}$) of Laskar et al. (1993) is reported for comparison. The filtering methodology used is a Tukey–Hinnov bandpass filter centered at the frequency of 0.0025 cycles / ky with a bandwidth of 0.055 cycles / ky corresponding to the periodicity cycles of 400 ky.

ratio of the Mediterranean seawater, respectively; J_A and R_{SWA} represent the flux (µmol Sr / Ma) and the $^{87}Sr/^{86}Sr$ ratio of Atlantic seawater inflowing the Mediterranean; J_R and R_R are the flux (µmol Sr / Ma) and the estimated average Sr isotope composition of the Mediterranean rivers; the parameter k takes into account the sign of the expression R_R-R_{SW} (see appendix for details).

A box model (Fig. 6) shows a simple scheme of the Mediterranean basin for the Tortonian interval where the main hydrographic parameters set up for the solution of the mass-balance equation are shown. All the numerical parameters used in the equation (with relative bibliographic sources) are reported in Table 1.

The average Sr isotope composition of rivers was estimated by calculating a weighted average of the most important Mediterranean tributaries (Table 1). Furthermore, to close up the total river input estimated by Garrett et al. (1993) for the Mediterranean region, we approximated a Sr concentration of the so-called "other rivers" similar to that calculated by Palmer and Edmond (1989) for the world rivers and used the same $^{87}Sr/^{86}Sr$ ratio calculated as a weighted average of the major Mediterranean rivers.

The value of Atlantic influx was estimated considering the twin channel Betic–Rifian corridors (see Benson et al., 1991, for details), both with an approximate sill depth of ~ 500 m. Considering the present value of Atlantic influx through the Gibraltar gateway, with a depth of sill at about 340 m, and according to the first-order estimate of Benson et al. (1991), we assumed for the Tortonian interval a 2.5–3 times larger influx of Atlantic water into the Mediterranean.

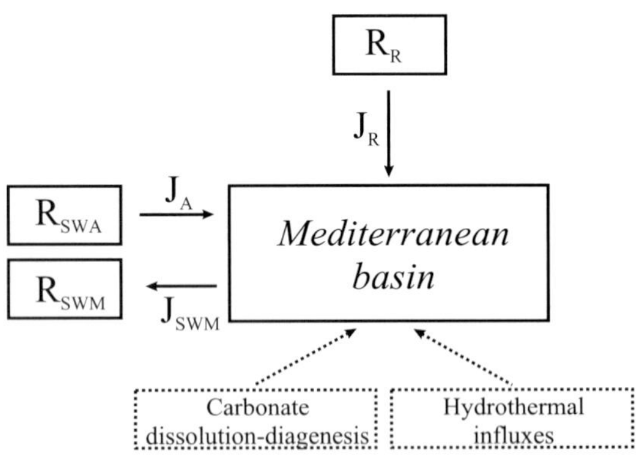

FIG. 6.—Layout of the model used to determine short-term basinwide $^{87}Sr/^{86}Sr$ changes. Hydrothermal influxes and carbonate dissolution effects are indicated with dotted lines to imply the limited capacity of these two forcings to affect the Sr isotope composition of the Mediterranean basin. Sr_{SWM} and R_{SWM} are the number of strontium moles (µmol Sr / l) and the $^{87}Sr/^{86}Sr$ ratio of the Mediterranean seawater, respectively; J_A and R_{SWA} represent the flux (µmol Sr / My) and the $^{87}Sr/^{86}Sr$ ratio of Atlantic seawater inflowing the Mediterranean; J_R and R_R are the flux (µmol Sr / My) and the estimated average Sr isotope composition of the Mediterranean rivers; J_{SWM} is the Sr flux from the Mediterranean basin into the Atlantic Sea [µmol / My].

We assumed constant values for some parameters of the differential equation (namely, Sr_{SWM}, J_A, and R_R), whereas the results of spectral analysis drove us in choosing a periodic structure for R_{SWA} and / or J_R to be used as forcing in the mass-balance equation. Thus, two different scenarios can explain the recorded 400 ky Sr isotope cyclicity during the Tortonian: (1) direct Sr isotope forcing by oceanic seawater flowing into the Mediterranean, and (2) variations of the Mediterranean river influxes.

The first hypothesis considers periodic Sr isotope fluctuations of Atlantic seawater, in a range of ~ $\pm 6 \times 10^{-5}$ / 0.2 Ma, as the direct cause of Mediterranean $^{87}Sr/^{86}Sr$ changes (Fig. 7). Owing to the three-orders-of-magnitude larger inflows of ocean water, compared to the budget of Mediterranean river influxes, the forcing signal (the Atlantic Sr isotope change) is approximately reproduced with a 1:1 ratio in the Mediterranean seawater Sr isotope response. Nonetheless, none of the detailed Sr isotope records from the Tortonian open-ocean deep-sea sediments (e.g., Hodell and Woodruff, 1994; Farrell et al., 1995; Martin et al., 1999) show evidence of cyclic patterns driven by Milankovitch forcing, thus making doubtful a direct control of the Atlantic seawater $^{87}Sr/^{86}Sr$ composition on the Mediterranean during the studied interval. Actually, because of the longer residence time of Sr in the global ocean, $^{87}Sr/^{86}Sr$ variations in the open ocean comparable to those recorded from the Tortonian Mediterranean sediments would require unrealistically large variations in the river influxes.

Conversely, the second scenario proposed to explain the Tortonian 400 ky $^{87}Sr/^{86}Sr$ oscillations is related to variations of the river influx into the Mediterranean basin, which directly could induce, with their relatively low $^{87}Sr/^{86}Sr$ ratios, periodic changes of the seawater isotope composition. In Figure 8, a set of sensitivity tests is reported showing the river budget variations (ΔF_r) needed to account for the $^{87}Sr/^{86}Sr$ variations recorded in the Tortonian dataset. The best agreement between the simulated data and the sedimentary record can be found by imposing the periodic river input oscillating in a range of ± 100–150% with respect to present conditions. For each test a variability range of ± 20% (around the average value of 2.5) estimated for the Atlantic influx is reported to take into account the combined effect of the various parameters influencing the seawater $^{87}Sr/^{86}Sr$ of the Mediterranean basin. The estimated seawater Sr residence time for this late Miocene Mediterranean oceanographic configuration is of the order of magnitude of about 1000 years, which already enables a reliable recording of the long-eccentricity cycles through the Gibliscemi succession.

General atmospheric circulation models suggest that the basic climate zonation in the Northern Hemisphere was established before the Late Miocene, following the definitive buildup of the Himalayas, which intensified the Indian monsoon system, and the stabilization of the Antarctic ice cap (e.g., Ruddiman and Kutzbach, 1989; Ruddiman et al., 1989). This would imply that the climatic conditions for the Europe–Mediterranean area were comparable to those recorded today, with the westerlies wind belt carrying Atlantic-sourced moisture promoting precipitation over most of Europe, including its eastern sector (e.g., Rozanski et al., 1993). The combined effects of periodic increase of continental runoff from the northeastern margin of the Mediterranean basin and from the southeastern Nile (e.g., Rossignol-Strick, 1983) could account for the recorded $^{87}Sr/^{86}Sr$ eccentricity cycles of the Tortonian.

Such a scenario has the potential to link Milankovitch forcing to changes in seawater $^{87}Sr/^{86}Sr$ induced by climate-sensitive riverine inputs. Moreover, it ensures an internal Mediterranean control on the marine isotope oscillations and gives the open ocean an auxiliary role to play in forcing of the seawater Sr isotope changes of the basin. Thus, variations of climate-sensitive riverine budget within the Mediterranean area represent a diluting autocyclic mechanism which drives short-term seawater Sr isotope changes into the basin.

CONCLUDING REMARKS

The recognized presence of 400 ky eccentricity cycles in the Sr isotope curve of the Tortonian Gibliscemi section highlights the

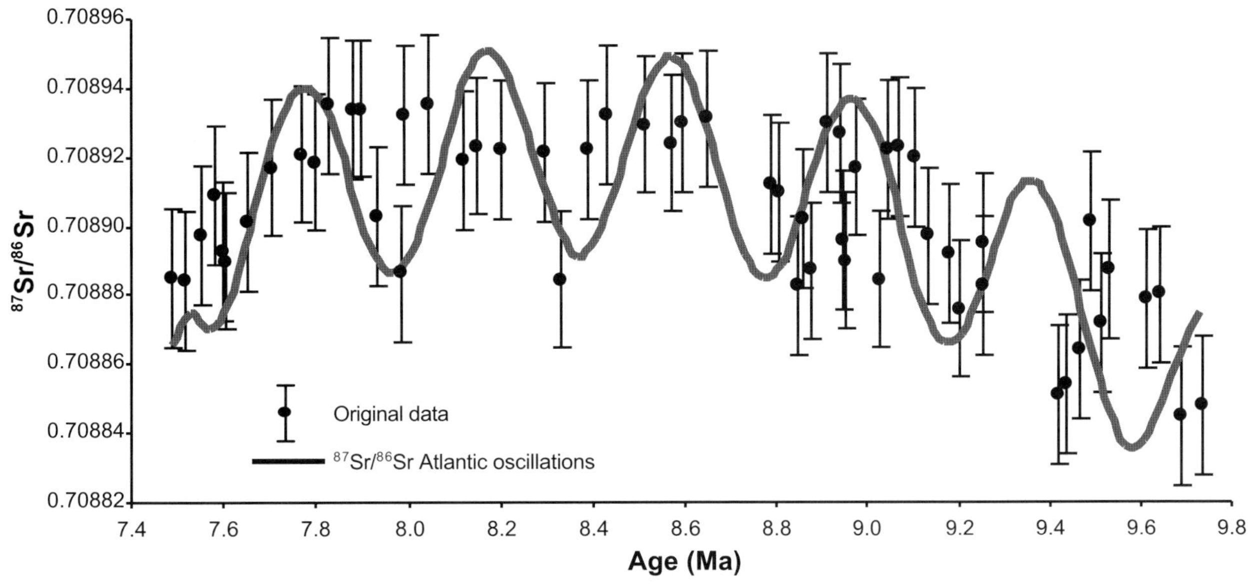

FIG. 7.—Simulated Sr isotope oscillations of the Mediterranean seawater (thick line) induced by phased Atlantic $^{87}Sr/^{86}Sr$ changes (in a range of ± 0.3 ppm / ky). The 400 ky Sr isotope cycles are superimposed on the sixth-order polynomial curve reported in Figure 3.

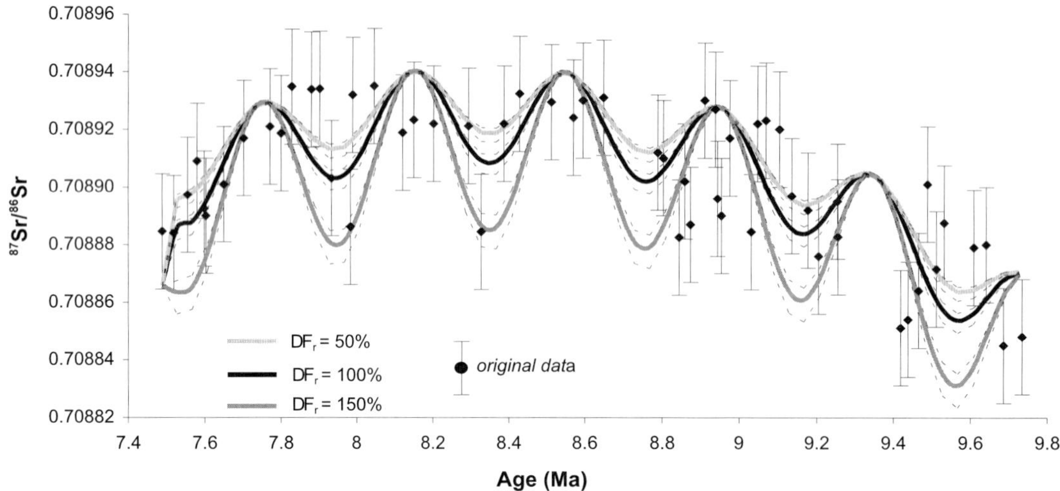

FIG. 8.—Simulated Sr isotope oscillations of the seawater Mediterranean basin induced by variations of the budget of river inputs (DF_r). Variations of 50% (light gray line), 100% (black line), and 150% (dark gray line) are reported. Dashed lines indicate variations of ± 20% of Atlantic influx around the estimated average input of 2.5 times with respect to present conditions. The 400 ky Sr isotope cycles are superimposed on the sixth-order polynomial curve reported in Figure 3.

high potential of the Mediterranean sedimentary record for astronomical calibration of global seawater $^{87}Sr/^{86}Sr$ signals. As proved by different forcing applied to the mass-balance equation for $^{87}Sr/^{86}Sr$ simulated changes, Milankovitch cyclicity in the Sr isotope curve can be induced in the Mediterranean basin by periodic variations of the river inputs. The limited geographic extent of the Mediterranean basin, combined with its strict climate control by the nearby continents, makes this area extremely suitable for detecting Milankovitch forcing in the Sr isotope record and for its fine time calibration to the insolation signal. Conversely, short-term Sr isotope oscillations are possibly masked in the open ocean by the buffering effect related to the huge amount of seawater and long Sr residence time. We think that the reliability of the recorded fine structure of the $^{87}Sr/^{86}Sr$ curve needs to be confirmed by new acquisition of isotope data from coeval astronomically tuned Mediterranean sections. Sampling of Mediterranean sections at the highest resolution possible could reveal the presence of the shorter Milankovitch periodicity bands (short-eccentricity, obliquity, and precession) in the Sr isotope record for a longer stratigraphic interval with the consequent possibility of improving the stratigraphic calibration of the $^{87}Sr/^{86}Sr$ curve and better understanding the climate–ocean interactions causing its short-term variability.

ACKNOWLEDGMENTS

Thanks are due to E. Di Biasio for help with chemical preparation of samples. Mass spectrometric analyses were performed at the Centro di Studio per il Quaternario e l'Evoluzione Ambientale (C.N.R. Rome). Thanks are given to F. Castorina for her laboratory assistance and helpful suggestions. Thanks are also due to H. Stoll and H. Weissert for their useful comments on an earlier version of the paper. This research has been financially supported by Murst Cofin 2000 to M.B. and A.B.

REFERENCES

BENSON, R.H., EL-BIED, K.R., AND BONADUCE, G., 1991, An important current reversal (influx) in the Rifian corridor (Morocco) at the Tortonian–Messinian boundary: the end of the Tethys ocean: Paleoceanography, v. 6, p. 164–192.

BRASS, G.W., 1976, The variation of the marine $^{87}Sr/^{86}Sr$ ratio during Phanerozoic time: interpretation using a flux model: Geochimica et Cosmochimica Acta, v. 40, p. 721–730.

BRYDEN, H.L., CANDELA, J., AND KINDER, T.H., 1994, Exchange through the Strait of Gibraltar: Progress in Oceanography, v. 33, p. 201–248.

CLEMENS, S.C., FARRELL, J.W., AND GROMET, L.P., 1993, Synchronous changes in seawater strontium isotope composition and global climate: Nature, v. 363, p. 607–610.

CAPO, R.C., AND DEPAOLO, D.J., 1990, Seawater strontium isotope variation from 2.5 million years ago to the present: Science, v. 249, p. 51–55.

DEPAOLO, D.J., 1986, Detailed record of the Neogene Sr isotopic evolution of seawater from DSDP Site 590B: Geology, v. 14, p. 103–106.

DEPAOLO, D.J., AND INGRAM, B., 1985, High-resolution stratigraphy with strontium isotopes: Science, v. 227, p. 938–941.

DIA, A.N., COHEN, A.S, O'NIONS, R.K., AND SHACKLETON, N.J., 1991, Seawater Sr-isotope variations over the last 300 ka and global climate change: Nature, v. 356, p. 786–789.

EMELYANOV, E.M., AND SHIMKUS, K.M., 1986, Geochemistry and Sedimentology of the Mediterranean Sea: Dordrecht, The Netherlands, Reidel Publishing Company, 553 p.

FARRELL, J.W., CLEMENS, S.C., AND GROMET, L.P., 1995, Improved chronostratigraphic reference curve of Late Neogene seawater $^{87}Sr/^{86}Sr$: Geology, v. 23, p. 403–406.

FAURE, G., AND POWELL, J.L., 1972, Strontium isotope geology, *in* Von Englhardt, E., and Roy, R., eds., Minerals, Rocks and Inorganic Materials, v. 5: Berlin, Springer-Verlag, 188 p.

GARRETT, C., OUTERBRIDGE, R., AND THOMPSON, K., 1993, Interannual variability in Mediterranean heat budget and buoyancy fluxes: Journal of Climatology, v. 6, p. 900–910.

HENDERSON, G.M., MARTEL, D.J., O'NIONS, R.K., AND SHACKLETON, N.J., 1994, Evolution of seawater $^{87}Sr/^{86}Sr$ over the last 400 ka: the absence of glacial/interglacial cycles: Earth and Planetary Science Letters, v. 128, p. 643–651.

HESS, J., BENDER, M.L., AND SHILLING, J.G., 1986, Evolution of the ratio of strontium-87 to strontium-86 in seawater from Cretaceous to present: Science, v. 231, p. 979–984.

HILGEN, F.J., KRIJGSMAN, W., LANGEREIS, C.G., LOURENS, L.J., SANTARELLI, A., AND ZACHARIASSE, W.J., 1995, Extending the astronomic (polarity) time scale into the Miocene: Earth and Planetary Science Letters, v. 136, p. 495–510.

HODELL, D.A., MUELLER, P.A., MCKENZIE, J.A., AND MEAD, G.A., 1989, Strontium isotope stratigraphy and geochemistry of the late Neogene ocean: Earth and Planetary Science Letters, v. 92, p. 165–178.

HODELL, D.A., MEAD, G.A., AND MUELLER, P.A., 1990, Variation in the strontium isotopic composition of seawater (8 Ma to present): Implications for chemical weathering rates and dissolved fluxes to the oceans: Chemical Geology (Isotope Geosciences Section), v. 80, p. 291–307.

HODELL, D.A., AND WOODRUFF, F., 1994, Variations in the strontium isotopic ratio of seawater during the Miocene: Stratigraphic and geochemical implications: Paleoceanography, v. 9, p. 405–426.

JENKINS, G.M., AND WATTS, D.G., 1968, Spectral Analysis and Its Applications: Oakland, California, Holden Day, 410 p.

LANE-SERFF, G.F., ROHLING, E.J., BRYDEN, H.L., AND CHANNOCK, H., 1997, Postglacial connection of the Black Sea to the Mediterranean and its relation to the timing of sapropel formation: Paleoceanography, v. 12, p. 169–174.

LASKAR, J., JOUTEL, F., AND BOUDIN, F., 1993, Orbital, precession and insolation quantities for the Earth from -20 Ma to +10 Ma: Astronomy and Astrophysics, v. 270, p. 522–533.

LUDWIG, K.R., 1994, Analyst: a Computer Program for Control of a Thermal-Ionization, Single Collector Mass-Spectrometer: U.S. Geological Survey, Open-File Report, p. 92–543

MARTIN, E.E., SHACKLETON, N.J., ZACHOS, J.C., AND FLOWER, B.P., 1999, Orbitally-tuned Sr isotope chemostratigraphy for the Late Middle to late Miocene: Paleoceanography, v. 14, p. 74–83.

MCARTHUR, J.M., 1994, Recent trends in Sr isotope stratigraphy: Terra Nova, v. 6, p. 331–358.

PALMER, M.R., AND EDMOND, J.M., 1989, The strontium isotope budget of the modern ocean: Earth and Planetary Science Letters, v. 92, p. 11–26.

RAYMO, M.E., RUDDIMAN, W.F., AND FROELICH, P.N., 1988, Influence of the late Cenozoic mountain building on ocean geochemical cycles: Geology, v. 16, p. 649–653.

RICHTER, F.M., ROWLEY, D.B., AND DEPAOLO, D.J., 1992, Sr isotope evolution of seawater: the role of tectonics: Earth and Planetary Science Letters, v. 109, p. 11–23.

RÖGL, F, 1998, Paleogeographic considerations for Mediterranean and Paratethys seaways (Oligocene and Miocene): Naturhistorisches Museum Wien, Annalen, v. 99A, p. 279–310.

ROSSIGNOL-STRICK, M., 1983, African monsoons, an immediate climate response to orbital forcing: Nature, v. 304, p. 46–49.

ROZANSKI, K., ARAGUAS-ARAGUAS, L., AND GONFIANTINI, R., 1993, Isotopic patterns in modern global precipitation, *in* Swart, P.K., Lohmann, K.C., McKenzie, J., and Savin, S., eds., Climate Change in Continental Isotopic Records: American Geophysical Union, Geophysical Monograph Series, v. 78, p. 1–36.

RUDDIMAN, W.F., AND KUTZBACH, J.E., 1989, Forcing of Late Cenozoic Northern Hemisphere climate by plateau uplift in southern Asia and the American West: Journal of Geophysical Research, v. 94, p. 18,409–18,427.

RUDDIMAN, W.F., PRELL, W.L., AND RAYMO, M.E., 1989, Late Cenozoic uplift in southern Asia and the American West: Rationale for general circulation modeling experiments: Journal of Geophysical Research, v. 94, p. 18,379–18,391.

SPROVIERI, M., BELLANCA, A., NERI, R., MAZZOLA, BONANNO, A., PATTI, B., AND SORGENTE, R., 1999, Astronomical calibration of Late Miocene stratigraphic events and analysis of precessionally driven paleoceanographic changes in the Mediterranean Basin: Società Geologica Italiana, Memorie, v. 54, p. 7–24.

SPROVIERI, M., BARBIERI, M., BELLANCA, A., AND NERI, R., 2002, Astronomical calibration of the $^{87}Sr/^{86}Sr$ curve for the upper Tortonian: Terra Nova, v. 15, p. 29–35.

VAN OS, B.J.H., AND ROHLING, H.J., 1993, Oxygen isotope depletions in eastern Mediterranean sapropels exclude antiestuarine circulation: Geologica Ultraiectina, v. 109, p. 1–12.

WARD, P.D., NELSON, B.K., AND MACLEOD, K.G., 1993, Marine isotope evolution: Nature, v. 358, p. 378.

APPENDIX

Equation 1 represents the Sr isotope mass-balance equation used to simulate the recorded Tortonian 400 ky cyclicity in the $^{87}Sr/^{86}Sr$ curve (Brass, 1976; Clemens et al., 1993):

$$Sr_{SWM}\frac{dR_{SWM}}{dt} = J_A(R_{SWA} - R_{SWM}) + k\,J_F(R_F - R_{SWM}) \qquad (1)$$

where

Sr_{SWM} is the number of Sr moles of the Mediterranean basin [μmol];
R_{SWM} is the Sr isotope ratio in the Mediterranean seawater;
J_A is the Sr flux from the Atlantic Ocean into the Mediterranean [μmol/My];
R_{SWA} is the Sr isotope ratio of the Atlantic Ocean;
J_F is the Sr flux from the rivers into the Mediterranean basin [μmol/My];
R_F is the mean Sr isotopic ratio of the Mediterranean rivers;
J_{SWM} is the Sr flux from the Mediterranean basin into the Atlantic Ocean [μmol/My].

The parameter k in Equation 1 was introduced in order to distinguish two different situations for the river influx in the Mediterranean basin. When R_F is greater than R_{SWA}, k is set to 1 and the rivers influx contributes to increase the Sr isotopic ratio (R_{SWM}) in the Mediterranean Sea. In the second condition R_F is lower than R_{SWA} and k is set to –1. In this case, the river outflow tends to reduce R_{SWM} in the Mediterranean:

$$J_F = J_{F0} + \Delta F\,\sin\,(\omega t + \varphi) \qquad R_{SWA} = R_{SWA0} + \Delta R_{SWA}\,\sin\,(\omega t + \varphi)$$

For the present study we considered a periodic structure both for R_{SWA} and for J_F, that is where J_{F0}, R_{SWA0} are the mean values for J_F and R_{SWA}, respectively, and ΔF, ΔR_{SWA} represent the fluctuations of the river outflow and the Sr isotopic ratio in the Atlantic Ocean.

According to this forcing structure, Equation 1 can be rearranged as

$$\begin{aligned} Sr_{SWM}\frac{dR_{SWM}}{dt} + (J_A + k\,J_F)R_{SWM} &= \\ J_A\left[R_{SWA0} + \Delta R_{SWA}\,\sin\,(\omega t + \varphi)\right] &+ \\ \left[J_{F0} + \Delta F\,\sin\,(\omega t + \varphi)\right] k\,R_F & \end{aligned} \qquad (2)$$

Taking into account that J_A is about 10,000 times the term $k\,J_F$, Equation 2 can be rewritten in a simpler form:

$$Sr_{SWM}\frac{dR_{SWM}}{dt} + J_A R_{SWM} = S + T\,\sin\,(\omega t + \varphi) \qquad (3)$$

where

$$S = J_A R_{SWA0} + J_{F0}R_F k \qquad \text{and} \qquad T = J_A \Delta R_{SWA} + R_F \Delta F\,k$$

The integral of the differential equation (3) is

$$\begin{aligned} R_{SWM}(t) &= c_1 \lambda^{-\frac{J_A t}{Sr_{SWM}}} + c_2 + a + bt + \\ &M\,\cos\,(\omega t + \varphi) + N\,\sin\,(\omega t + \varphi) \end{aligned} \qquad (4)$$

where

$$\begin{cases} b = 0 \\ a = S/J_A \\ N = \dfrac{T\,J_A}{J_A^2 + Sr_{SWM}^2\omega^2} \\ M = \dfrac{-T\,\omega\,Sr_{SWM}}{J_A^2 + Sr_{SWM}^2\omega^2} \end{cases}$$

and c_1, c_2 are two constant values to be evaluated by applying appropriate boundary conditions.

The first condition we considered is the starting point of the curve, i.e., for

$$t = t_1 \qquad R_{SWM}(t_1) = R_1 \qquad (5)$$

where t_1 is the starting time of the time series and R_1 is the related R_{SWM} value.

The second condition originates from budget considerations. In the proposed box model for the Mediterranean basin the Sr isotope sources are the Atlantic Ocean and all the rivers entering in the Mediterranean sea. For this reason the R_{SWM} cannot be greater than R_{SWA} or R_F.

We translated such a consideration into the following mathematical condition:

$$\text{Max}\left[R_{SWM}(t)\right] = \text{Max}\left[R_{SWA}; R_F\right] \qquad (6)$$

The conditions expressed by the constants c_1, c_2 can be evaluated from Equations 5 and 6.

ASTROCHRONOLOGY OF LATE MIDDLE MIOCENE MEDITERRANEAN SECTIONS

SILVIA MARIA IACCARINO
Dipartimento di Scienze della Terra, Università di Parma, Parco Area delle Scienze, 157 A, 43100 Parma, Italy
e-mail: iaccarin@unipr.it
FABRIZIO LIRER
Dipartimento di Scienze della Terra, Università di Parma, Parco Area delle Scienze, 157 A, 43100 Parma, Italy
AND
Istituto per l'Ambiente Marino Costiero, Geomare, National Research Council,
Calata Porta di Massa, Porto di Napoli, 80133 Napoli, Italy
SERGIO BONOMO AND ANTONIO CARUSO
Dipartimento di Geologia e Geodesia, Università di Palermo, Corso Tukory 131, 90134 Palermo, Italy
AGATA DI STEFANO
Dipartimento di Scienze della Terra, Università di Catania, Corso Italia 55, 55129 Catania, Italy
ENRICO DI STEFANO
Dipartimento di Geologia e Geodesia, Università di Palermo, Corso Tukory 131, 90134 Palermo, Italy
LUCA MARIA FORESI, ROBERTO MAZZEI, AND GIANFRANCO SALVATORINI
Dipartimento di Scienze della Terra, Università di Siena, Via Laterina 8, 53100 Siena, Italy
MARIO SPROVIERI
Istituto per l'Ambiente Marino Costiero, Geomare, National Research Council,
Calata Porta di Massa, Porto di Napoli, 80133 Napoli, Italy
RODOLFO SPROVIERI
Dipartimento di Geologia e Geodesia, Università di Palermo, Corso Tukory 131, 90134 Palermo, Italy
AND
ELENA TURCO
Dipartimento di Scienze della Terra, Università di Parma, Parco Area delle Scienze, 157 A, 43100 Parma, Italy

ABSTRACT: High-resolution cyclostratigraphy and calcareous plankton astrobiochronology have been obtained from the latest Langhian to the earliest Tortonian of the Mediterranean. The investigated areas (Malta, Tremiti, and Sicily) are located in different geological settings, and the three studied sections show different cyclicity. Direct correlation between the Laskar $90_{(1,1)}$ solution of the insolation curve and the sedimentary cycle pattern occurring in the investigated sections showed that all the sedimentary cycles are forced dominantly by Milankovitch periodicity. This forcing is also reflected in the climate-sensitive data ($CaCO_3$ content, and the relative abundance of the planktonic foraminifer *Globigerinoides*) as shown by the application of spectral and filtering analyses.

The calibration provided astronomical ages for all the sedimentary cycles and bioevents recorded in the sections. In particular, an age of 13.59 Ma was obtained for the extinction level of *Sphenolithus heteromorphus*, which is the best event approximating the Langhian–Serravallian boundary and an age of 10.55 Ma for the first regular occurrence of *Neogloboquadrina acostaensis*, the event that better approximates the Serravallian–Tortonian boundary in the Tortonian type section.

INTRODUCTION

Milankovitch cyclostratigraphy is the best tool to obtain high-resolution stratigraphy in different depositional environments (continental, marginal, and deep sea). The extraction of high-frequency cycles from deep-sea records and their calibration to the astronomical target curve, the most recent and accurate method, provides absolute ages for the cycles and consequently for all the bioevents. This approach improved the geological time scale, and more importantly, our knowledge of climate changes. In fact, the cyclic changes in the sedimentary record are related to climate variations that are ultimately controlled by variations in the shape of the Earth's orbit and the inclination of its rotation axis (Hilgen et al., 1997).

Orbitally forced climate oscillations are recorded in sediments through changes in sediment properties, fossil communities, and chemical and isotopic characteristics. Earth scientists can read these archives to reconstruct paleoclimate, while astronomers formulate models based on the mechanism of the solar–planetary system and Earth–Moon system to compute the past variations of precession, obliquity, and eccentricity of the Earth's orbit and rotation axis. Sedimentary cycles can be dated by matching patterns of paleoclimate variability with patterns of varying input of solar energy computed from the astronomical model solutions (Hilgen et al., 1999).

Application of cyclostratigraphy resulted in the construction of the astronomical time scale (ATS) for the Quaternary, the Pliocene, and the Late Miocene (Hilgen 1991a, 1991b; Lourens et al., 1996; Shackleton et al., 1995; Shackleton and Crowhurst, 1997; Hilgen et al., 1999; Hilgen et al., 2000a; Hilgen et al., 2000b) which is more accurate and has a much higher resolution than the previous geological time scales (Berggren et al., 1985; Berggren et al., 1995; Harland et al., 1990). Researchers are now working to extend the astronomical time scales to older intervals, even in continental records (Opdyke et al., 1997; Abdul Aziz et al., 2000; Agustì et al., 2001).

Astronomical calibration of sedimentary cycles enabled some authors (Lourens et al., 1996; Hilgen et al., 1995, Hilgen et al., 1999; Hilgen et al., 2000a; Hilgen et al., 2000b; Krijgsman, 1996; Krijgsman et al., 1995; Krijgsman et al., 1997; Krijgsman et al., 1999; Sierro et al., 2001) to extend the astrochronology of the Mediterranean records back into the Middle Miocene (to about 12 Ma). According to many authors (Lourens et al., 1992; Lourens et al., 1996; Versteegh, 1994; Van Vugt, 2000) the sediments from the Mediter-

Cyclostratigraphy: Approaches and Case Histories
SEPM Special Publication No. 81, Copyright © 2004
SEPM (Society for Sedimentary Geology), ISBN 1-56576-108-1, p. 27–44.

ranean are particularly sensitive in recording astronomically induced climate fluctuations, because of the paleo-latitudinal position of the Mediterranean in combination with its semi-enclosed, land-locked configuration. In the Mediterranean area, the extension of the astrochronological time scale back to 12.1 Ma is based on various Miocene sections outcropping in Metochia (Greece), Faneromeni (Greece), and Monte Gibliscemi (Sicily) (Ten Veen and Postma, 1993; Krijgsman et al., 1995; Postma and Ten Veen, 1999; Hilgen et al., 2000b).

Hilgen et al. (1995), Hilgen et al. (2000b), and Lourens et al. (1996) demonstrated that the La $90_{(1,1)}$ solution represents the best tuning between geological and astronomical records for the last 12 Ma, and in this paper we tuned the uppermost Langhian–lower Tortonian sedimentary record with the astronomical insolation curve of Laskar et al. (1993).

Geochemical fluctuation and faunal fluctuation in relative abundance identify the Milankovitch cyclicity in the absence of other sedimentary expression. Sprovieri (1992, 1993), Lourens et al. (1992), Lourens et al. (1996), Lourens (1994), Shackleton et al. (1995), Sprovieri, M., et al. (1999); Sprovieri, M., et al. (2002), Shackleton and Crowhurst (1997), and Turco et al. (2001) proved that fluctuations in relative abundance of planktonic foraminifera (essentially in the *Globigerinoides* population), calcium carbonate content, physical properties, and oxygen isotopes are forced by the precession or obliquity astronomical cycles.

In this paper we present the results of the cyclostratigraphic study and the high-resolution biostratigraphy of calcareous plankton of three Mediterranean sections belonging to different geodynamic and sedimentary environments.

The time interval investigated includes the Langhian–Serravallian and Serravallian–Tortonian boundaries, which at present are recognized to be approximated, respectively, by the LO (last occurrence) of *Sphenolithus heteromorphus* (Fornaciari et al., 1996; Rio et al., 1997) and the FCO (first common occurrence) or FRO (first regular occurrence) of *Neogloboquadrina acostaensis* in the stratotype sections (Foresi et al., 1998; Hilgen et al., 2000b).

The sections investigated are located on Malta and Sicily (central Mediterranean) and the Tremiti Islands (Adriatic Sea) (Fig. 1). Detailed descriptions of the methodologies and the results of the cyclostratigraphy and high-resolution biostratigraphy of the three sections are discussed in separate papers (Caruso et al., 2002; Di Stefano et al., 2002; Foresi et al., 2002a, Foresi et al., 2002b; Lirer et al., 2002; Sprovieri, R., et al., 2002).

GEOLOGICAL SETTING AND SECTIONS

Malta

The Malta Archipelago (the islands of Malta and Gozo) is located about 100 km south of Sicily and belongs to a graben system that was formed by two groups of faults: one with NW–SE orientation and the other with ENE–WSW orientation. This graben lies within the African Plate, in the foreland of the Sicilian Apennine–Maghrebian fold and thrust belt (Hyde, 1955; Felix, 1973; Pedley et al., 1976, 1978; Catalano et al., 2000; Foresi et al., 2002a). The geologic and tectonic evolution of the Archipelago is widely documented in numerous works (Cello et al., 1984; Finetti, 1984; Cello, 1987; Argnani, 1990; Dart et al., 1993; Robertson and Grasso, 1995). The sedimentary sequences cropping out on Malta and Gozo encompassing the Oligocene–Miocene time interval are characterized by four formations: (1) Lower Coralline Limestone, (2) Globigerina Limestone, (3) Blue Clay, and (4) Upper Coralline Limestone, described in detail by Giannelli and Salvatorini (1975).

The investigated Ras il-Pellegrin section is located on the northern side of Fomm ir-Rih Bay 20 km west of the town of Valletta (Fig. 1). The sequence, 60 m thick, belongs to the Blue Clay Formation (Fig. 2) and consists of clayey marls intercalated with whitish, calcareous layers, which are more evident after rain (Figs. 2, 3A).

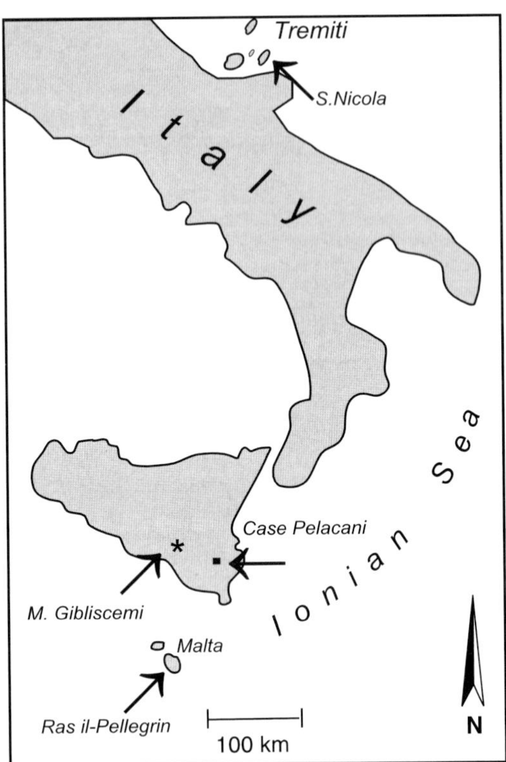

FIG. 1.—Location map of the investigated sections; * indicates the M. Gibliscemi section of Hilgen et al. (2000b).

Tremiti Islands

The Tremiti Islands (S. Domino, S. Nicola, Cretaccio, and Caprara), located in the southern part of Adriatic sea, belong to the Adria microplate, which is bounded by the Hellenides to the west and the Apennines to the east (Channell et al., 1979; Platt et al., 1989; Gambini and Tozzi, 1996). The sedimentary sequences cropping out on the Tremiti Islands span from the late Paleocene to the Quaternary (Selli, 1971; Iaccarino et al., 2001, among others). The Neogene sequence, consisting of the Cretaccio and San Nicola Formations, is transgressive on the pre-Neogene sediments (Selli, 1971; Pampaloni, 1988; Iaccarino et al., 2001). It crops out mainly on the islands of Cretaccio and S. Nicola and to a lesser extent on San Domino Island. The sections investigated are made up of deep-marine cyclically bedded hemipelagic sediments of Serravallian to early Tortonian age belonging to the Cretaccio Formation. Two sections have been sampled: the Cemetery (42° 07' 41.6" N; 15° 30' 47.9" E), 19 m thick, and the Castle (42° 07' 18.3" N; 15° 30' 26" E), 26.03 m thick, which are located in the NW and SE parts of S. Nicola Island, respectively (Figs. 1, 2). The detailed logging, which followed different trajectories because of the presence of small faults, provides a continuous record for the entire investigated sequence. Sedimentary cycles are not equally distinct and visible in the two sections because of lateral changes, different weathering, and poor exposure. Correlation between the two sections was established on the basis of calcareous plankton bioevents.

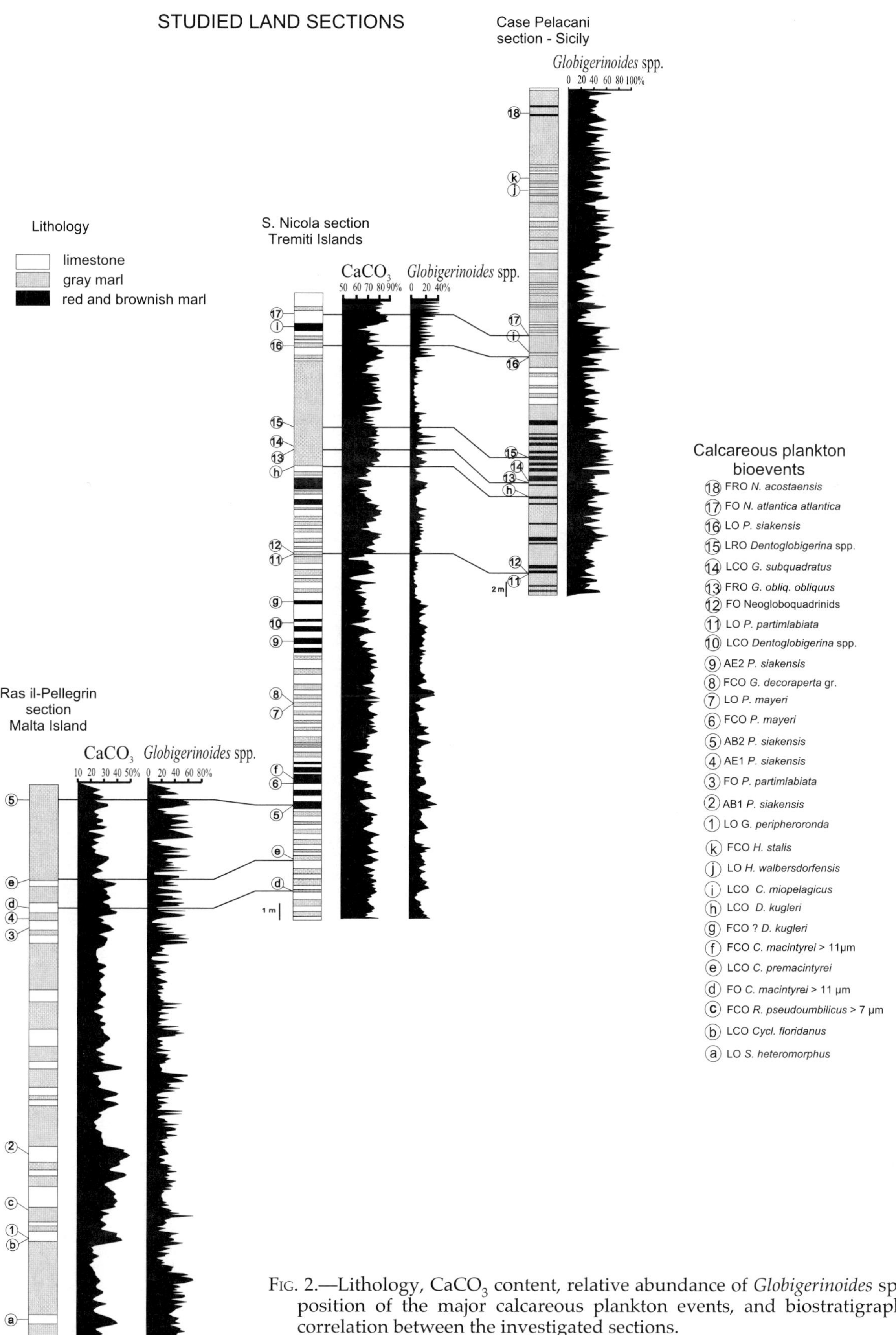

FIG. 2.—Lithology, $CaCO_3$ content, relative abundance of *Globigerinoides* spp., position of the major calcareous plankton events, and biostratigraphic correlation between the investigated sections.

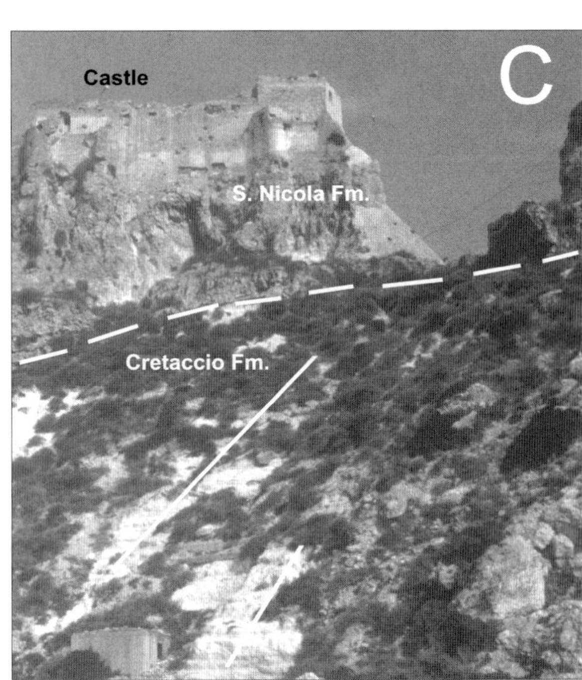

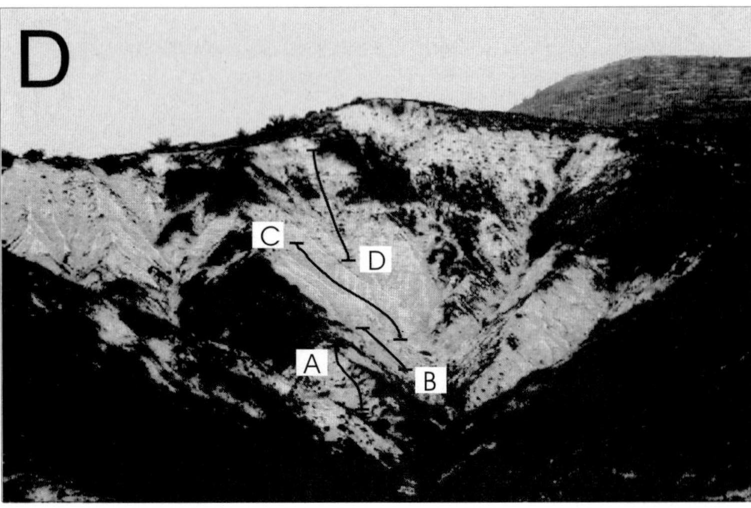

FIG. 3.—Sampling trajectories: **A)** Ras il-Pellegrin section; S. Nicola section (the dashed lines indicate the boundary between Cretaccio and S. Nicola Fms.): **B)** lower part (Cemetery segment), **C)** upper part (Castle segment); **D)** Case Pelacani section.

The two partially overlapping sections were combined into the S. Nicola composite section, whose thickness is 38 m (Figs. 2, 3B, 3C).

Sicily

The Case Pelacani section crops out along the Bianco River, 4 km west of the village of Palazzolo Acreide, in the Iblean Plateau (SE Sicily; Figs. 1, 2). The succession is characterized by Middle–Upper Miocene hemipelagic sediments, belonging to the Tellaro Formation, overlying the calcarenites and marls of the Ragusa Formation and underlying the calcarenites of the Palazzolo Formation (Rigo and Barbieri, 1959).

The whole succession 66.55 m thick, and is well exposed in a series of gullies along the southern slope of Mt. Cozzo Mastica (Lat. 37° 02' 54" N, Long. 14° 53' 00" E) (Figs. 2, 3D). It is composed of four segments (Pelacani A/B/C/D), showing different cyclicities from the base to the top, which have been correlated by means of the lithological alternations and biostratigraphic events.

ASTRONOMICAL CALIBRATION OF THE INVESTIGATED SECTIONS

The investigated sections have been calibrated astronomically through the tuning of the lithological cycles and periodic fluctuations of climate-sensitive records ($CaCO_3$ content and *Globigerinoides* spp.) to the astronomical target curves.

The astronomical tuning has basically been obtained through four successive steps.

1. We correlated the Case Pelacani and S. Nicola sections with the partially overlapping Monte Gibliscemi section, recently calibrated astronomically by Hilgen et al. (2000b). The correlation is based on the comparison of the abundance curves of selected planktonic foraminiferal species (*Globigerinoides subquadratus*, *Paragloborotalia siakensis*, and *Paragloborotalia partimlabiata*), (Fig. 4). The strong similarity of the curves allowed us to consider as starting point for our tuning the following bioevents: the acme end (AE) of *P. siakensis*, corresponding, in the M. Gibliscemi section, to the strong decrease of *P. mayeri* in cycle -87, calibrated at 12.006 Ma; the LO of *P. partimlabiata*, corresponding to the abrupt decrease of *P. partimlabiata* in cycle -80, calibrated at 11.78 Ma, and the last common occurrence (LCO) of *G. subquadratus* in cycle -72, dated at 10.539 Ma (Fig. 4). The stratigraphic correlation between the S. Nicola and Ras il-Pellegrin sections (Sprovieri, M., et al., 2002) is based on the FCO of *Paragloborotalia mayeri* and the LCO of *Calcidiscus premacintyrei*, calibrated respectively at 12.34 Ma and 12.51 Ma by Lirer et al. (2002).

2. We directly tuned the lithological cycles with the precession and eccentricity curves using the Laskar $90_{(1,1)}$ solution (Laskar et al., 1993). The tuning was based on the phase relation proposed by Hilgen (1991a): sapropels (and their equivalents) are correlated to precession minima–summer insolation maxima in the Northern Hemisphere, and small-scale and large-scale sapropel clusters to 100 ky and 400 ky eccentricity maxima, respectively, whereas the maxima in carbonate content are correlated to eccentricity minima. Alternating thin–thick sapropels, which reflect precession–obliquity interference, develop preferentially during the eccentricity minima when the precession influence is reduced (e.g., Lourens et al., 1996; Hilgen et al., 2000b; Sierro et al., 2001).

3. Cross-spectral techniques were used to identify in $CaCO_3$ and *Globigerinoides* spp. records the cyclic fluctuations and their phase relation with the precession and eccentricity curves.

4. The direct correlation of the lithological pattern to the astronomical curves of Laskar et al. (1993), combined with spectral and filtering techniques applied to $CaCO_3$ content and *Globigerinoides* spp. records, produced a good astronomical tuning of our sedimentary succession.

From the top of the Case Pelacani down to the base of Malta section, 160 precession-related sedimentary and/or carbonate and biotic cycles, numbered successively from 1 to 160, were recognized (Fig. 5).

Calibration of Sedimentary Cycles

The sedimentary cycles of the Case Pelacani and S. Nicola sections are defined by the cyclic occurrence of dark-gray organic-rich and gray marls (Case Pelacani) or red or gray layers (S. Nicola) alternating with $CaCO_3$-rich layers. These colored marly layers (characterized by high abundance of *Globigerinoides* spp.; Caruso et al., 2002; Lirer et al., 2002) are sapropel-equivalents as in other Mediterranean Neogene marine successions (Hilgen et al., 2000b; Sierro et al., 2001; Sprovieri et al., 1999). Therefore they are tuned to precession minima–summer insolation maxima, whereas the carbonate layers are related to precession maxima–insolation minima (Caruso et al., 2002; Lirer et al., 2002).

In the Case Pelacani section, the cyclic pattern changes throughout the sedimentary sequence and shows clusters of cycles of different types (Fig. 5). In the lower part of the section, between cycles 69 to 55, the occurrence of thick homogeneous whitish marls separating couplets of thin distinct dark-gray organic-rich marl layers suggests an additional effect of the obliquity on the insolation (Fig. 5). The alternating thin (or absent)–thick sapropel pattern, corresponding to prolonged intervals of interference in the insolation target, is centered around the 400 ky eccentricity minima at 11.70 Ma (Fig. 5) (Caruso et al., 2002). In the overlying interval between cycles 54 and 45, the 10 distinct dark-gray organic-rich marl layers (characterized by high abundance of the warm oligotrophic planktonic foraminifer *Globigerinoides* spp.; Caruso et al., 2002) have to be modulated by high fluctuations of the precession index and their possible best fitting in the precession–insolation curve is between 11.58 and 11.38 Ma (Fig. 5). Upwards, we can observe a progressive change in the sedimentary setting. The occurrence of biogenic turbiditic deposits indicates a new type of elementary cycle, in which the dark-gray organic-rich marls are replaced by gray marls and the previous gray marls are replaced by prominent whitish carbonate beds (Fig. 5). Cycles 44 to 37 are characterized by the thickest whitish carbonate layers, whose best fit is with the anomalously low amplitude of the eccentricity curve between 11.36 and 11.20 Ma (Fig. 5). This interval coincides with a period of low-amplitude variation in obliquity related to the 1.2 My cycle recorded by Turco et al. (2001). Upwards, cycles 36 to 9 show the modulation of the 100 ky eccentricity minima. Between cycles 36 and 34, 32 and 30, 18 and 15, and 12 and 10, the thick homogeneous marls correspond to 100 ky eccentricity minima centered around 10.70 Ma, 10.80 Ma, 11.00 Ma, and 11.20 Ma (Fig. 5). These homogeneous marls occur rhythmically below distinct lithologic clusters, characterized by couplets of white prominent carbonate beds and thin gray marls, interpreted to have been controlled by high fluctuation in the insolation curve. The homogeneous marls represent inter-

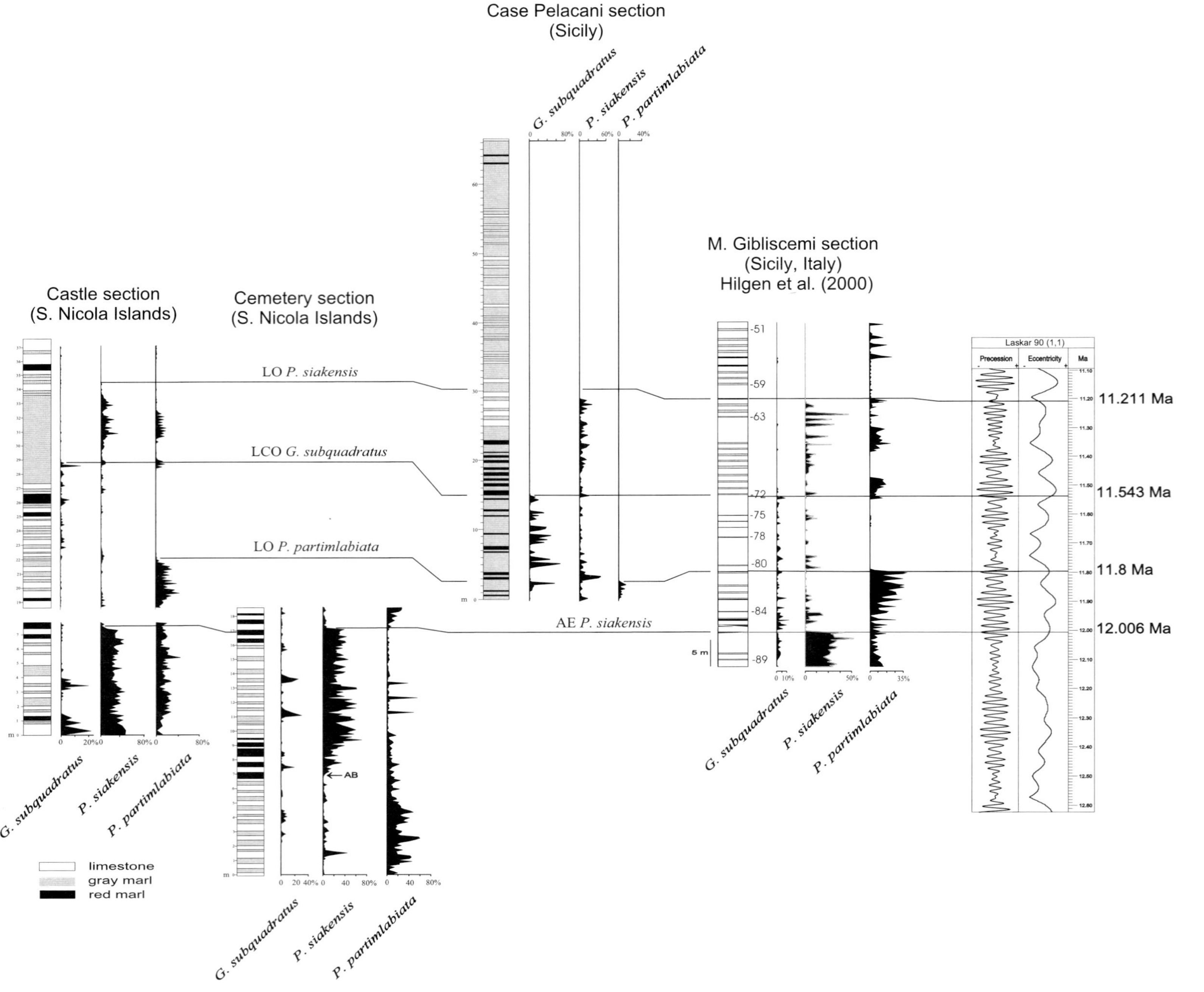

FIG. 4.—Correlation between investigated sections and Hilgen, et al. (2000b) Gibliscemi section through the abundance curves and bioevents of selected planktonic foraminifera.

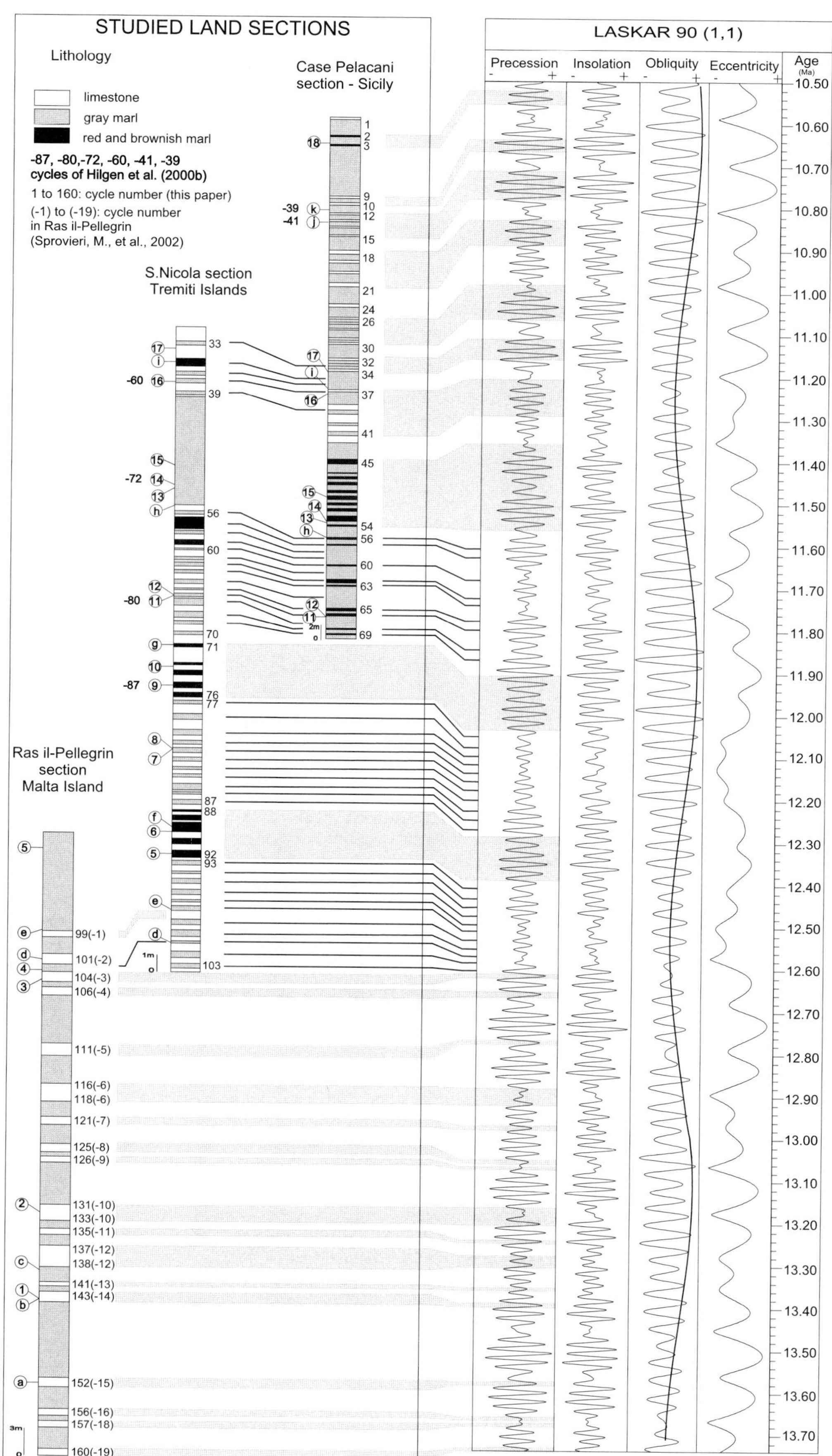

FIG. 5.—Tuning of the sedimentary cycles of the investigated sections to the astronomic target (precession, insolation, obliquity, and eccentricity). The black line in the obliquity curve corresponds to about 1.2 My obliquity cycle; the light gray bands only facilitate the correlation between the sedimentary cycles and the astronomic curves. The main planktonic foraminifera bioevents are indicated by the following numbers: 1, LO *G. peripheroronda*; 2, AB1 *P. siakensis*; 3, FO *P. partimlabiata*; 4, AE1 *P. siakensis*; 5, AB2 *P. siakensis*; 6, FCO *P. mayeri*; 7, LO *P. mayeri*; 8, FCO *G. decoraperta* gr.; 9, AE2 *P. siakensis*; 10, LCO *Dentoglobigerina* spp.; 11, LO *P. partimlabiata*; 12, FO Neogloboquadrinids; 13, FRO *G. obliq. obliquus*; 14, LCO *G. subquadratus*; 15, LRO *Dentoglobigerina* spp.; 16, LO *P. siakensis*; 17, FO *N. atlantica atlantica*; 18, FRO *N. acostaensis*. The main calcareous nannofossil bioevents are indicated by the following letters, a, LO *S. heteromorphus*; b, LCO *Cycl. floridanus*; c, FCO *R. pseudoumbilicus* > 7 mm; d, FO *C. macintyrei* ≥ 11 µm; e, LCO *C. premacintyrei*; f,, FCO *C. macintyrei* ≥ 11 µm; g, FCO ? *D. kugleri*; h, LCO *D. kugleri*; i, LCO *C. miopelagicus*; j, LO *H. walbersdorfensis*; k, FCO *H. stalis*.

vals in which the calcareous layers linked to times of very low-amplitude peaks in the precession–insolation curve related to the 100 ky eccentricity cycle modulation are missing. This phase relation between eccentricity minima and cycle patterns is not recorded in cycles 24–21 and 9–3.

The S. Nicola composite section shows a quasi-regular rhythmic alternation of red and / or gray, less indurated, $CaCO_3$-poor marly beds, and whitish $CaCO_3$-rich marly limestones. The cyclic pattern consists of an evident alternation of clusters, which are distinguishable by the color of the marly beds (Lirer et al., 2002) (Fig. 5). The upper part of the S. Nicola composite section, where the sedimentary cyclicity is not evident (cycles 55 to 39), was tuned to the astronomical target by fluctuations in abundances of *Globigerinoides* spp. and content of CaCO3, processed by spectral and filtering procedures (Lirer et al., 2002). In this section, clear sapropels are not recorded, but the distinct red layers (characterized by high abundance of *Globigerinoides* spp.) are considered to be sapropel equivalents (Lirer et al., 2002). These red layers, forming two clusters (cycles 76–71 and cycles 92–88), are well constrained between the time intervals 11.91–12.03 Ma and 12.29–12.38 Ma, when the precession fluctuations are at maximum (Fig. 5). Clusters (cycles 87–77 and 103–93) dominated by gray layers are controlled mainly by long periods of eccentricity minima, corresponding to the low influence of the precession (12.05–12.27 Ma and 12.41–12.60 Ma). Between cycles 70 and 60 the cyclicity is more evident than in the Case Pelacani section (Fig. 5), and all the sedimentary cycles were recognized.

In the Ras il-Pellegrin section, where cycles are less distinct and less frequent, and sapropel equivalents are not evident, Sprovieri, M., et al. (2002) obtained an astronomical tuning mainly through spectral analysis and filtering of abundance fluctuations of *Globigerinoides* spp. abundance and $CaCO_3$ content (Fig. 2). This tuning is based on the phase relation between eccentricity maxima–$CaCO_3$ minima (Shackleton et al., 1995) and eccentricity maxima–*Globigerinoides* spp. maxima (Sprovieri, M., et al., 1999).

The astronomical tuning of the Ras il-Pellegrin section led us to interpret the white carbonate levels, interbedded in the marly sequence and originally labeled from -1 to -19 (Fig. 5), as being controlled by 100 ky eccentricity minima modulation, where the influence of precession is very low (Fig. 5). Only the eccentricity minimum at 13.45 Ma has no sedimentary expression. At present, this missing carbonate layer is an unresolved problem.

Tuning of Climate-Sensitive Proxy Records

As already stated, cycles induced by Milankovitch periodicity were also identified by processing, through spectral analyses, the analytical data of $CaCO_3$ content, and relative abundance of *Globigerinoides* spp.

Power spectra (Fig. 6) show the presence in the sedimentary records of the precession, long-eccentricity, and short-eccentricity astronomical periodicities as the dominant forcing. The Tukey–Hinnov band-pass filter (Jenkins and Watts, 1968) was used to extract, from the original signals, selected frequency bands in order to compare them with the same periodicity of the insolation curve.

The tuning strategy adopted to correlate the insolation curve and the sedimentary records in all the Milankovitch frequencies associates lower values of Northern Hemisphere summer insolation with higher carbonate content (as suggested by Hilgen, 1991a) and with lower percentages of the *Globigerinoides* spp. (as suggested by Sprovieri et al., 1999).

In Figure 7 the comparison between the eccentricity component of the insolation and the 400 ky and 100 ky cycles in the climate-sensitive records (carbonate and *Globigerinoides* spp. signals) shows a good correspondence, emphasizing the appropriateness of the first-order tuning of the geological records.

At the level of 400 ky, it seems likely that the tuning is correct over the whole composite section. The filtered curves of $CaCO_3$ and *Globigerinoides* spp. show minima and maxima, respectively, in phase with eccentricity maxima.

At the level of 100 ky, the $CaCO_3$ and *Globigerinoides* spp. filtered curves show a good fit with the eccentricity, except for the overlapping intervals between the subsections. Moreover, within each subsection there is a slight mismatch, probably due to small changes in sedimentation rate.

The 19–23 ky filtered records were compared with the precession curve La $90_{(1,1)}$ to refine the tuning to the precessional scale.

Coherency estimated between the insolation curve and the various records (Fig. 6) shows the highest values at the eccentricity and precession Milankovitch periodicities, thus confirming, in those periodicity bands, a direct relationship between astronomical forcing and sedimentary records.

CALCAREOUS PLANKTON BIOSTRATIGRAPHY AND BIOCHRONOLOGY

The composite section consists of part of the Ras il-Pellegrin section (from cycle 160 up to cycle 104), part of the S. Nicola section (from cycle 103 up to cycle 56), and part of the Case Pelacani section (from the top of cycle 56 to cycle 1) (Figs. 8, 9). The abundance curves of the major taxa, the bioevents, and the biostratigraphy are plotted versus time in the composite section (Fig. 9).

Planktonic Foraminifera

The quantitative distribution patterns of selected foraminifera having biostratigraphic significance are plotted in Figure 9. These patterns were obtained by counting about 300 specimens of planktonic foraminifera from the > 125 μm fraction splits. Quantitative analysis was carried out on about 1200 samples. The sampling and collecting data are described in Di Stefano et al. (2002), Foresi et al. (2002a); and Foresi et al. (2002b). The high number of bioevents recorded and calibrated in the investigated sections gave us the opportunity to propose a new high-resolution biostratigraphy for the planktonic foraminifera (Sprovieri, R., et al., 2002). The major events are listed below, and their astrochronological calibration is reported in Table 1.

Globorotalia peripheroronda (Fig. 9: G. periph.).—

This taxon occurs only in the lowest part of the composite section (Ras il-Pellegrin). Its LO falls in cycle 143.

Globigerinoides subquadratus (Fig. 9: G. subq.).—

Very rare at the base of the sequence, it becomes more common upwards, reaching values between 30 and 40% at different levels of its range. Its LCO is a well detectable event, and it occurs in the white marly limestone of cycle 53.

Globigerinoides obliquus obliquus (Fig. 9: G. o. obliq.).—

This taxon is generally rare or randomly distributed in the lower to middle part of the sequence except for a short interval (cycles 81 to 77) in which it is slightly more common. It becomes common and regularly distributed (FRO) from cycle 53, coinciding with the LCO of *G. subquadratus* up to the top.

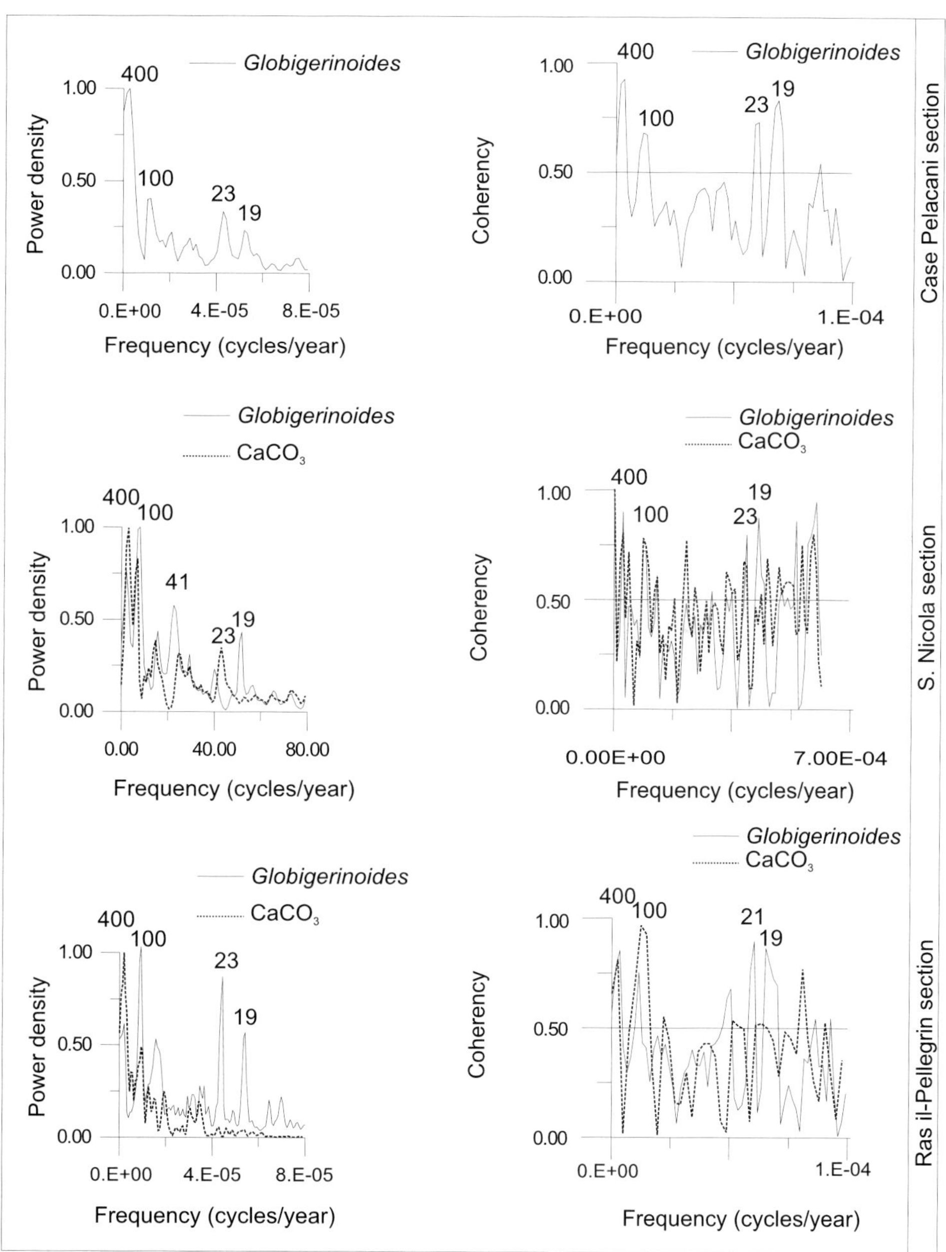

FIG. 6.—Spectral and coherence analysis performed on *Globigerinoides* and $CaCO_3$ curves.

Paragloborotalia siakensis (Fig. 9: P. siak.).—

We distinguished two acme abundance intervals, the lowest between cycles 132 and 131 and the youngest between cycles 92 and 75. The LO of *P. siakensis* occurs in cycle 37. However, in the uppermost part of its range it is represented by atypical and small-sized specimens.

Paragloborotalia partimlabiata (Fig. 9: P. part.).—

This taxon first appears in the white carbonate layer of cycle 104 in the Ras il-Pellegrin section, just below the top of the AE1 (acme end) of *P. siakensis*. The crossover between the distribution of *P. partimlabiata* and that of *P. siakensis* is clearly observable. The LO of this taxon occurs in cycle 66.

Paragloborotalia mayeri (Fig. 9: P. may.).—

This taxon (*sensu* Blow, 1969) has a very short distribution. Its FO (first occurrence) is not well detectable because it is very rare in its lowest distribution, but its FCO in cycle 90 is well recognizable. The LO falls within cycle 81.

Neogloboquadrinids (Fig. 9: Neogl.).—

N. atlantica praeatlantica is the first representative of the neogloboquadrinids, and its FO is recorded in cycle 66. The FO of *N. acostaensis* is recorded in the same cycle but not in the same sample. *N. atlantica praeatlantica* is common from the level of its first occurrence upwards and shows two distinct influxes, whereas *N. acostaensis* is very rare until cycle 3, where it becomes

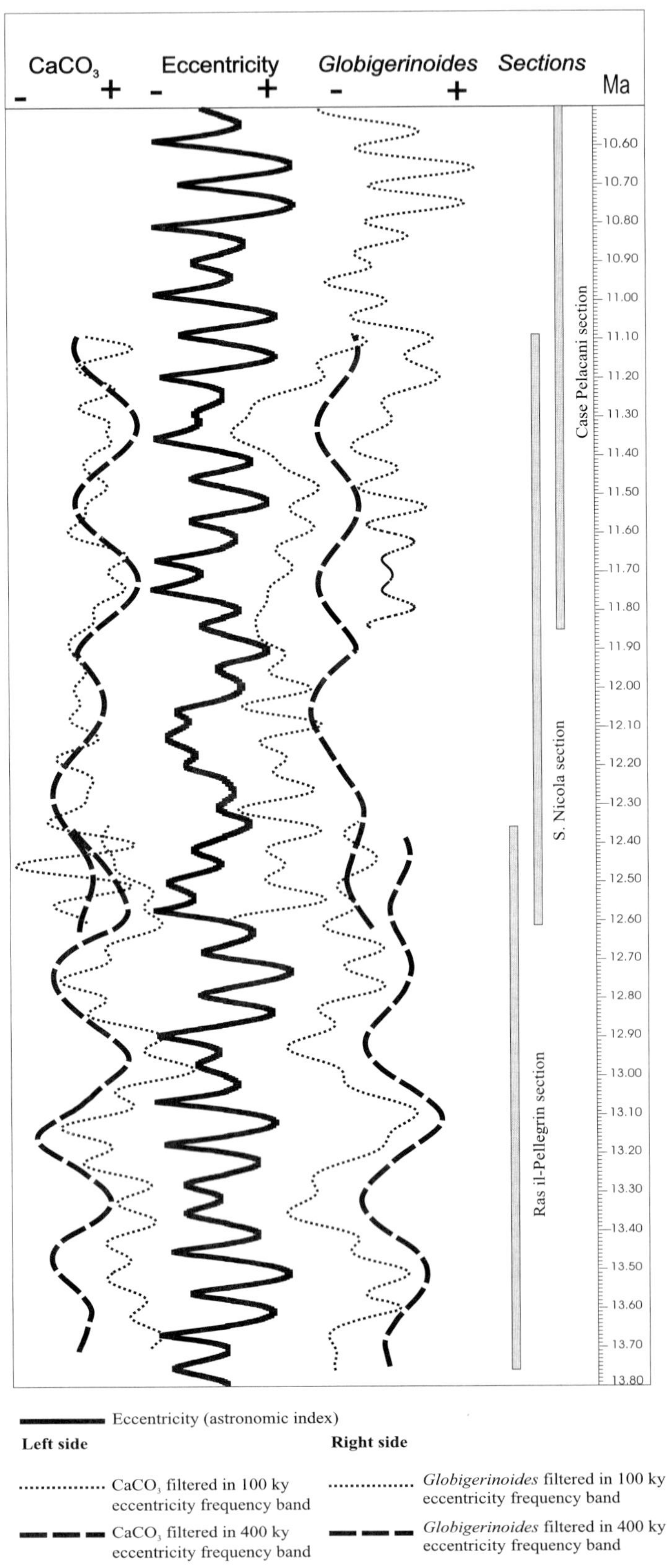

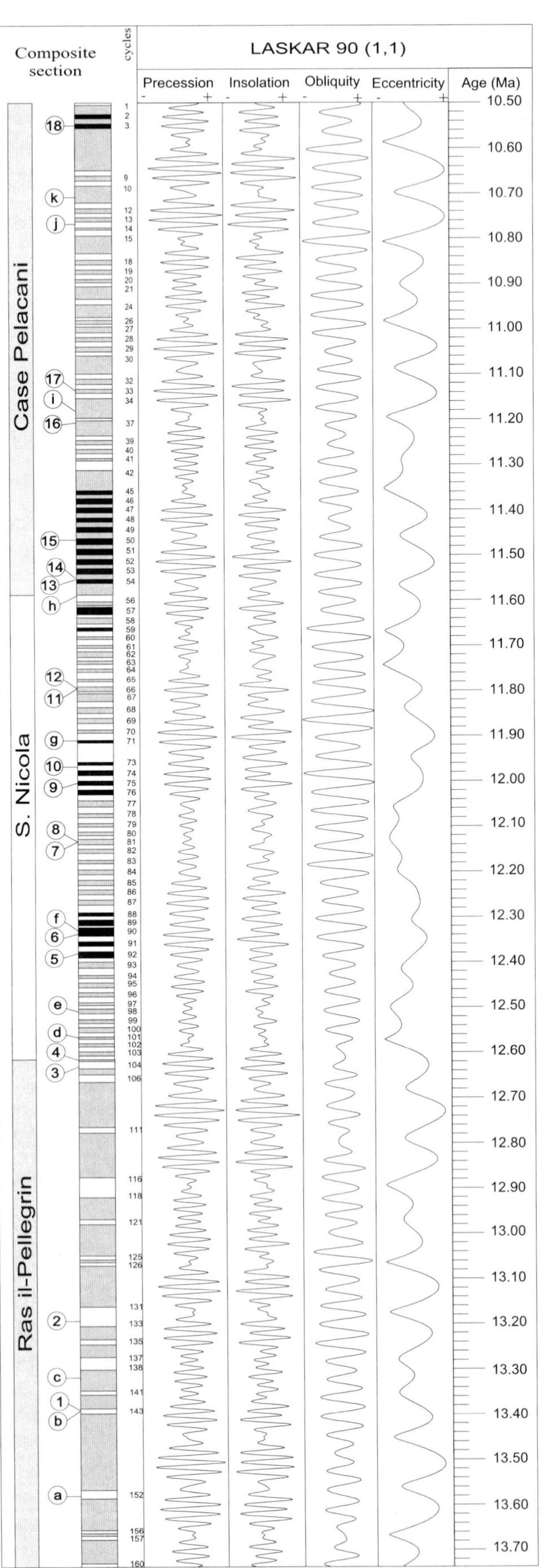

FIG. 7.—Comparison between *Globigerinoides* and $CaCO_3$ content filtered data in the 400 ky and 100 ky eccentricity bands with the astronomical curve of the eccentricity.

regularly present (FRO). The third representative of the group is *N. atlantica atlantica,* which occurs between cycles 34 and 24.

The arrival of the neogloquadrinids into the Mediterranean is a synchronous event, but it is diachronous with respect to the tropical area (Hilgen et al., 2000b; Turco et al., 2002).

Dentoglobigerina spp. *(Fig. 9: Dent. spp.).—*

Dentoglobigerina spp., which includes mainly *D. altispira altispira* and *D. altispira globosa,* occurs from the base of the composite section and shows its LCO and LRO in cycles 73 and 50, respectively.

Globorturborotalita decoraperta gr. (Fig. 9: G. dec. gr.).—

This taxonomic unit, including *G. decoraperta* and *G. woodi,* starts to be common (FCO) in cycle 81, where *P. mayeri* disappears.

Calcareous Nannofossils

To obtain the distribution patterns of selected calcareous nannofossil taxa, light-microscope analyses were performed (transmitted light and crossed nicols) at about 1000 magnification. Smear slides were prepared from unprocessed sediments following standard techniques.

Abundance data were collected using methodology described by Backman and Shackleton (1983) and Rio et al. (1990) and used extensively in Mediterranean and extra-Mediterranean quantitative biostratigraphic studies of Neogene marine record (ODP sequences and land sections) (Raffi and Flores, 1995; Raffi et al., 1995; Fornaciari et al., 1996; Backman and Raffi, 1997; Di Stefano, 1998; Hilgen et al., 2000a; Hilgen et al., 2000b).

The distribution pattern of the following taxa, having biostratigraphic significance, *Sphenolithus heteromorphus, Ciclycargolithus floridanus, Reticulofenestra pseudoumbilicus* > 7 μm, *Calcidiscus macintyrei* ≥ 11 μm, *C. premacintyrei, Coccolithus miopelagicus, Discoaster kugleri, Helicosphaera walbersdorfensis,* and *H. stalis,* are plotted in Figure 9.

The less detailed distribution pattern of *C. macintyrei, C. premacintyrei, C. miopelagicus, D. kugleri, H. walbersdorfensis,* and *H. stalis* from cycle 71 up to 56 is due to a smaller number of analyzed samples, and the jump in the abundance values of *C. macintyrei* from cycles 72 and 71 is due to better preservation of the calcareous nannofossils.

←

FIG. 8.—Composite section plotted versus time and astronomical parameters. The main planktonic foraminifera bioevents are indicated by the following numbers: 1, LO *G. peripheroronda;* 2, AB1 *P. siakensis;* 3, FO *P. partimlabiata;* 4, AE1 *P. siakensis;* 5, AB2 *P. siakensis;* 6, FCO *P. mayeri;* 7, LO *P. mayeri;* 8, FCO *G. decoraperta* gr.; 9, AE2 *P. siakensis;* 10, LCO *Dentoglobigerina* spp.; 11, LO *P. partimlabiata;* 12, FO Neogloboquadrinids; 13, FRO *G. obliq. obliquus;* 14, LCO *G. subquadratus;* 15, LRO *Dentoglobigerina* spp.; 16, LO *P. siakensis;* 17, FO *N. atlantica atlantica;* 18, FRO *N. acostaensis*. The main calcareous nannofossil bioevents are indicated by the following letters, a, LO *S. heteromorphus;* b, LCO *Cycl. floridanus;* c, FCO *R. pseudoumbilicus* > 7 μm; d, FO *C. macintyrei* ≥ 11 μm; e, LCO *C. premacintyrei;* f, FCO *C. macintyrei* ≥ 11 μm; g, FCO ? *D. kugleri;* h, LCO *D. kugleri;* i, LCO *C. miopelagicus;* j, LO *H. walbersdorfensis;* k, FCO *H. stalis*.

Sphenolithus heteromorphus (Fig. 9: S. hete.).—

This taxon occurs only in the lowest part of the composite section (Ras il-Pellegrin), and its LO in cycle 152 predates the LO of *G. peripheroronda,* in contrast to the tropical ocean (Ceara Rise, ODP Leg 154, Site 926), where it postdates the latter event. Its astrochronological calibration at 13.59 Ma slightly predates that obtained in Site 926 by Backman and Raffi (1997) at 13.523 Ma and more recently by Turco et al. (2002) at 13.51 Ma.

Ciclycargolithus floridanus (Fig. 9: Cycl. flori.).—

This form is present only in the lowest part (Ras il-Pellegrin) of the composite section, and its LCO in cycle 143 coincides with the LO of *G. peripheroronda.*

Reticulofenestra pseudoumbilicus > 7 μm (Fig. 9: R. pseud. > 7).—

Only morphotypes > 7 μm were considered. They occur discontinuously, with very low abundance values from the base of the composite section up to cycle 138, where the FCO of this taxon was recognized.

Calcidiscus macintyrei ≥ 11 μm (Fig. 9: C. mac. ≥ 11).—

The distribution pattern of the morphotypes ≥ 11 μm is not well detectable before the FCO, which falls in cycle 90 at 12.34 Ma. In fact, this taxon is very rare and scattered in the basal part of its range, and only by a very closely spaced analysis was its FO recognized in cycle 101, three precessional cycles below the LCO of *C. premacintyrei.*

Calcidiscus premacintyrei (Fig. 9: C. premac.).—

The distribution pattern of this taxon is characterized by high percentage values and is well recorded in the Ras il-Pellegrin section up to the lower part of the S. Nicola section, where the LCO in cycle 98 is clearly detectable. A short paracme is recorded in the lower part of the composite section within cycles 111 and 116.

Discoaster kugleri (Fig. 9).—

In the composite section, *D. kugleri* has been reported from cycle 71 to cycle 56. Indeed, this taxon occurs, in the Castle section, in the underlying levels down to cycle 86, whereas is totally absent in the Cemetery section (Foresi et al., 2002b). We tentatively placed its FCO in cycle 71 at 11.9 Ma, following the calibration of Hilgen et al. (2000b). *D. kugleri* shows fluctuating abundance, reaching a maximum density of 8 specimens per square millimeter in cycle 67. Its LCO was identified in cycle 56 and dated at 11.60 Ma. Above this interval it is extremely rare and was found in only three samples.

Coccolithus miopelagicus (Fig. 9: C. miop.).—

This taxon is commonly and continuously present in the analyzed intervals up to cycle 35 (Case Pelacani section), where its LCO has been placed.

Helicosphaera walbersdorfensis (Fig. 9: H. walb.).—

This taxon, showing wide fluctuations, is continuously present in the analyzed intervals of the composite section up to

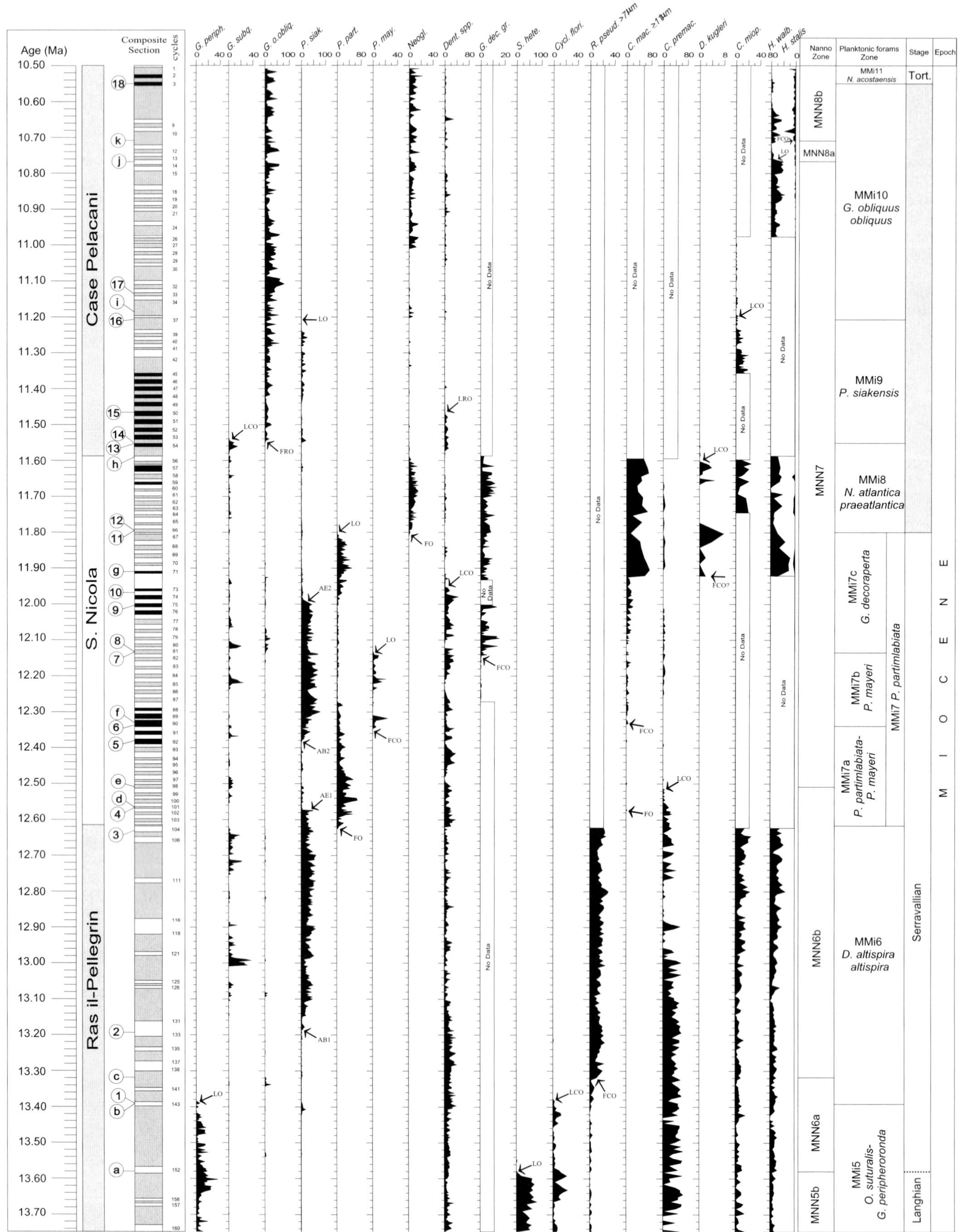
Age (Ma)
Composite Section
cycles
Case Pelacani
S. Nicola
Ras il-Pellegrin
G. periph.
G. subq.
G. o.obliq.
P. siak.
P. part.
P. may.
Neogl.
Dent. spp.
G. dec. gr.
S. hete.
Cycl. flori.
R. pseud. >7µm
C. mac. >11µm
C. premac.
D. kugleri
C. miop.
H. walb.
H. stalis
No Data
Nanno Zone
Planktonic forams Zone
Stage
Epoch
MMi11 N. acostaensis
Tort.
MNN8b
MNN8a
MMi10 G. obliquus obliquus
MMi9 P. siakensis
MMi8 N. atlantica praeatlantica
MNN7
MMi7c G. decoraperta
MMi7b P. mayeri
MMi7a P. partimlabiata-P. mayeri
MMi7 P. partimlabiata
MMi6 D. altispira altispira
MNN6b
MNN6a
MMi5 O. suturalis-G. peripheroronda
MNN5b
Serravallian
Langhian
M I O C E N E

TABLE 1.—Astronomical ages of the main calcareous plankton bioevents.

Bioevents	Pelacani Section		S. Nicola Section		Ras il-Pellegrin Section	
	Age (Ma)	Precessional Cycles	Age (Ma)	Precessional Cycles	Age (Ma)	Precessional Cycles
FRO *N. acostaensis*	10.55	3				
FCO *H. stalis*	10.71	11/12				
LO *H. walbersdorfensis*	10.76	14/15				
FO *N. atlantica atlantica*	11.16	34	11.16	34		
LCO *C. miopelagicus*	11.19	35/36	11.18	35		
LO *P . siakensis*	11.21	37	11.21	37		
FRO *G. obliquus obliquus*	11.54	54	11.54	53		
LCO *G. subquadratus*	11.54	54	11.54	53		
LCO *D. kugleri*	11.60	56	11.60	56		
FO *N. atlantica praeatlantica*	11.80	66	11.80	66		
FO *N. acostaensis*	11.80	66	11.80	66		
LO *P. partimlabiata*	11.80	66	11.80	66		
FCO? *D. kugleri*			11.90	71		
AE2 *P. siakensis*			12.00	75		
FCO *G. decoraperta* gr.			12.14	81		
LO *P. mayeri*			12.14	81		
FCO *C. macintyrei* ≥ 11 μm			12.34	90		
FCO *P. mayeri*			12.34	90	12.34	90
AB2 *P. siakensis*			12.39	92	12.38	91/92
LCO *C. premacintyrei*			12.51	98	12.51	98/99
FO *C. macintyrei* ≥ 11 μm			12.57	101	12.57	101
AE1 *P. siakensis*					12.58	101
FO *P. partimlabiata*					12.62	104
AB1 *P. siakensis*					13.22	131
FCO *R. pseudoumbilicus* > 7 μm					13.32	138
LO *G. peripheroronda*					13.39	143
LCO *C. floridanus*					13.39	143
LO *S. heteromorphus*					13.59	152

cycle 14, where a sharp decrease in abundance occurs. Di Stefano et al. (2002) considered the common, but badly preserved, specimens that occur above this level as reworked, and interpreted the above-mentioned decrease of *H. walbersdorfensis* as its LO (cycle 14). This horizon falls shortly below the FCO of *H. stalis*.

Helicosphaera stalis (Fig. 9).—

The distribution pattern of *H. stalis* shows an abundance increase (FCO) from the interval between cycles 12 and 11 up to the top of the composite. The age of this event, dated at 10.70 Ma, agrees well with that of Hilgen et al. (2000b), confirming that it is a reliable event for intra-Mediterranean correlations.

←

FIG. 9 (opposite page).—Quantitative distribution pattern of the major calcareous plankton marker species through time and position of major bioevents through the composite section. On the right side is the new planktonic foraminiferal zonal scheme proposed by Sprovieri, R., et al. (2002) for the Mediterranean late Middle Miocene and calcareous nannofossil zonation of Fornaciari et al. (1996). The circled numbers and letters correspond to the events listed in Figure 8.

CHRONOSTRATIGRAPHIC AND CHRONOLOGIC IMPLICATIONS

Langhian–Serravallian boundary (L–S)

The Langhian and Serravallian stages were erected by Pareto (1865). Following Pareto's indications, Cita and Premoli Silva (1960) designated the type section of the Langhian, which is represented mainly by the Pteropods Marls (Cessole Fm.) near the village of Cessole at the Bricco della Croce section, whereas Vervloet (1966) designated the type section of the Serravallian, which was equated to the Serravalle Sandstone Fm., near the village of Serravalle Scrivia along the Scrivia River.

The definition, geological setting, and timing of the two stages are widely documented by Rio et al. (1997), who highlighted a time gap between the top of the Langhian and the base of the Serravallian stratotypes. The FO of *Orbulina universa* and the FCO of *H. walbersdorfensis*, which are recorded at the top of the Langhian stratotype, occur some 70 m and 40 m, respectively, below the base of the Serravallian stratotype. The best planktonic foraminiferal events approximating the Langhian–Serravallian boundary are the FO of *G. praemenardii* and the LO of *G. peripheroronda*, which, however, are very rare in the Piedmont area. Fornaciari et al. (1996) documented that the LO of *S. heteromorphus* is remarkably close to the base of the Serravallian type section. Rio et al. (1997) proposed that this event represents the best tool for recognizing the Langhian–Serravallian boundary and therefore it must be the driving criterion to follow in designating the GSSP of the Serravallian. The LO of *S. heteromorphus* has been calibrated to Chron C5Abr (Backman et al., 1990; Olafsson, 1991) and is associated with the paleoclimatic–paleoceanographic mid-Miocene events of Miller et al. (1991a) and Miller et al. (1991b).

All the mentioned bioevents occur in the Ras il-Pellegrin section. The oldest events (*O. universa* FO and *H. walbersdorfensis* FCO) have been detected in the upper part of the Upper Globigerina Limestone, close to the boundary with the overlying Blue Clay Fm. (Giannelli and Salvatorini, 1975; Mazzei, 1985). Following the recommendations of Rio et al. (1997) and taking into consideration that this event is almost synchronous between the Mediterranean (13.59 Ma) and low latitudes (13.51 Ma), the LO of *S. heteromorphus* recorded in cycle 152 (8.8 m above the base of the section) could be a potential candidate for the GSSP of the Serravallian whereas *G. peripheroronda* LO is clearly diachronous (Fig. 10). *C. floridanus* LCO could be an additional event because its delay in low latitudes is 70 ky (Fig. 10). Magnetostratigraphic data, which could confirm the extinction of *S. heteromorphus* within Chron C5ABr of the geomagnetic polarity time scale of Cande and Kent (1992, 1995) and hence the synchroneity of the event, were unfortunately unreliable. Even the oxygen stable-isotope analysis carried out on the Ras il-Pellegrin section (Bellanca et al., 2002) does not show a clear increase in benthic foraminiferal $\delta^{18}O$ in correspondence with the *S. heteromorphus* LO.

Serravallian–Tortonian Boundary

The Serravallian–Tortonian (S–T) boundary is not yet formally defined by a GSSP but is historically placed at or close to the FO of *N. acostaensis* (Cita and Blow, 1969) because this bioevent was supposed to occur from the basal part of the Tortonian

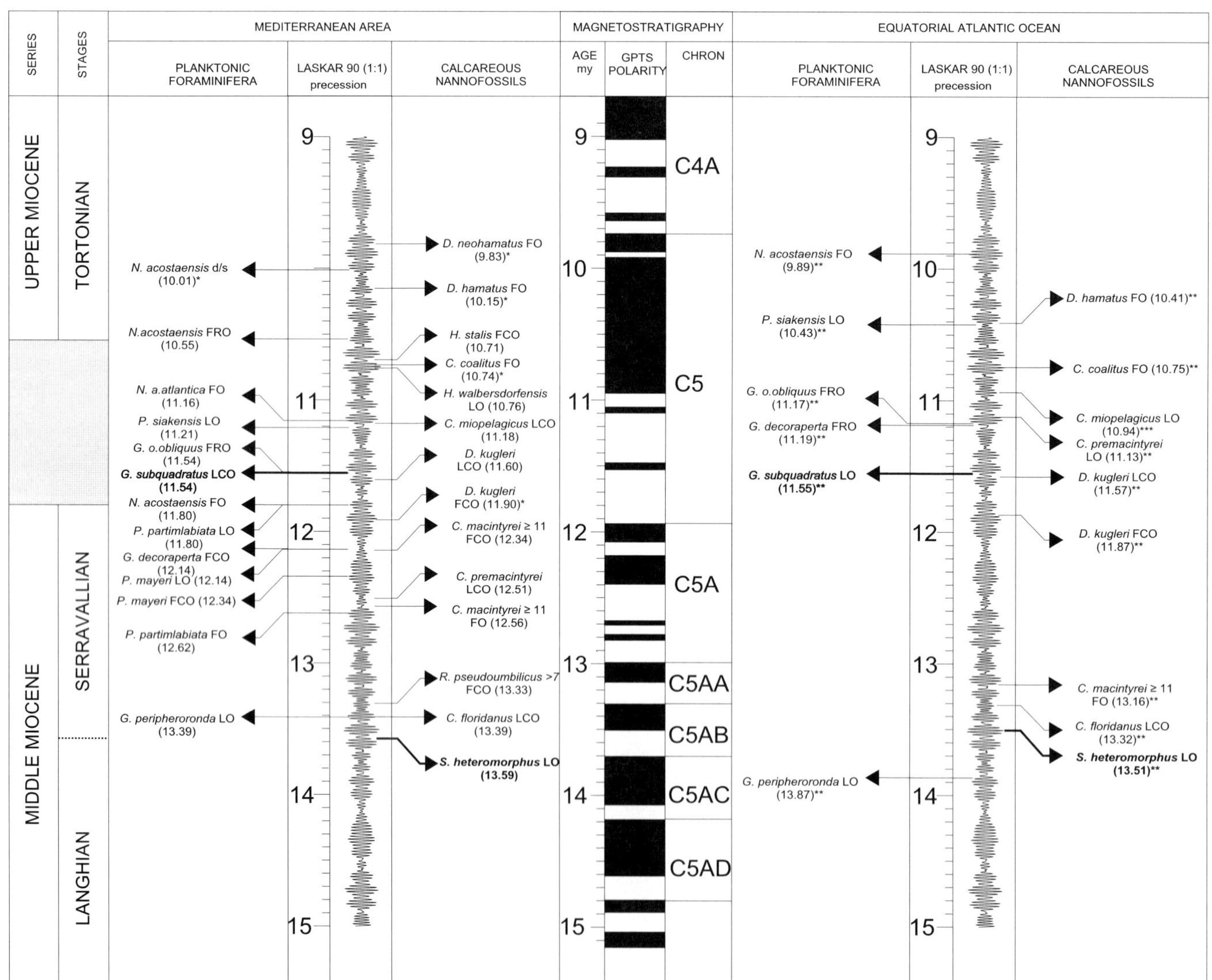

FIG. 10.—Correlation between Mediterranean and equatorial Atlantic bioevents calibrated in the precession curve. The GPTS is from Cande and Kent (1995); the asterisked events are: * from Hilgen et al. (2000 b), ** from Turco et al. (2002), and *** from Backman and Raffi (1997). The gray area in the chronostratigraphic column represents the interval within which the GSSP boundary of the Tortonian might be placed.

stratotype section of Rio Mazzapiedi–Castellania as defined by Cita and Premoli Silva (1968). In the low-latitude zonal scheme of Blow (1969), this bioevent defines the N15–N16 boundary and occurs later than the LO of *P. siakensis*, which defines the N14–N15 boundary. At middle latitudes (ODP Site 608), Miller et al. (1985) and Miller et al. (1991a) recorded these two events very close to each other. In the Mediterranean (Cita et al., 1978; Mazza, 1985; Coccioni et al., 1992) and the North Atlantic (Poore, 1979; Huddleston, 1984), partial overlaps in the range of the two taxa are recorded. In particular, Foresi (1993), Miculan (1994, 1997), and Foresi et al. (1998) recorded some rare specimens of *N. acostaensis* from the uppermost part of the Serravalle Sandstone Fm.

Zachariasse (1992) inferred that the FO of *N. acostaensis* in tropical areas resulted from migration from high latitudes. Following the astronomical tuning of Shackleton and Crowhurst (1997), the arrival of *N. acostaensis* at the equatorial Ceara Rise (ODP Leg 154) is dated at 9.82 Ma in Site 925 by Chaisson and Pearson (1997) and at 9.89 Ma in Site 926 by Turco et al. (2002), (Fig. 10). Foresi et al. (1998) estimated for the FO of *N. acostaensis* and LO of *P. siakensis* from Mediterranean sections, an age of 11.5 Ma and 10.5 Ma, respectively. Hilgen et al. (2000b), using astronomical tuning, dated the entry of the neogloboquadrinids at 11.78 Ma and the LO of *P. mayeri* (= *P. siakensis* of this paper) at 11.21 Ma in the M. Gibliscemi section, establishing that in this area the overlap between the two taxa lasts about 600 ky (Fig. 10). The arrival of *N. acostaensis* is delayed by almost 2 My in low latitudes, compared with the Mediterranean. In contrast, *P. siakensis* LO (dated at 10.43 Ma) shows a delay of 780 ky with respect to the Mediterranean (LO at 11.21 Ma), (Fig. 10).

The Serravallian–Tortonian boundary in the Tortonian stratotype based on calcareous nannofossils is in conflict with those of the planktonic foraminifera. In fact, the boundary occurs within zone NN9 in Mazzei (1977) or NN8 in Rio et al. (1997), whereas according to Foresi et al. (1998), Hilgen et al. (2000b), Foresi et al. (2002b), and Di Stefano et al. (2002) the *N. acostaensis* FO falls within NN7. It follows that the base of the Tortonian stratotype postdates the *N. acostaensis* FO and approximates its FRO at 10.55 Ma. Therefore, the GSSP of the Tortonian should be placed within the FO and FRO of *N. acostaensis*. The LCO of *G. subquadratus* at 11.54 Ma could be a potential candidate for the GSSP of the Tortonian for its synchroneity between Mediterranean and low latitude oceans as reported by Turco et al. (2002). Also the LCO of *D. kugleri* closely predates this event in both areas (Fig. 10). In terms of cycles, the LCO of *G. subquadratus* occurs just below the cluster of 10 thin sapropels both in Mediterranean (Case Pelacani and M. Gibliscemi sections) and equatorial areas (Ceara Rise, Leg 154, Site 926).

ACKNOWLEDGMENTS

We are grateful to W. Schwarzacher and F.J. Sierro for their critical review, which substantially improved the manuscript. This research is granted by Consiglio Nazionale delle Ricerche 1999.

REFERENCES

ABDUL AZIZ, H., HILGEN, F.J., KRIJGSMAN, W., SANZ, E., AND CALVO, J.P., 2000, Astronomical forcing of sedimentary cycles in the middle to late Miocene continental Calatayud Basin (NE Spain): Earth and Planetary Science Letters, v. 177, p. 9–22.

AGUSTÌ, J., CABRERA, L., GARGÉS, M., KRIJGSMAN, W., OMS, O., AND PARÉS, J.M., 2001, A calibrated mammal scale for Neogene of Western Europe: State of the art: Earth-Science Reviews, v. 52, p. 247–260.

ARGNANI, A., 1990, The Strait of Sicily rift zone: Foreland deformation related to the evolution of a back-arc basin: Journal of Geodynamics, v. 12, p. 311–331.

BACKMAN, J., AND RAFFI, I., 1997, Calibration of Miocene nannofossil events to orbitally tuned cyclostratigraphies from Ceara Rise: Scientific Results of the Ocean Drilling Program, v. 154, p. 83–99.

BACKMAN, J., AND SHACKLETON, N.J., 1983, Quantitative biochronology of Pliocene and early Pleistocene calcareous nannofossils from the Atlantic, Indian and Pacific oceans: Marine Micropaleontology, v. 8, p. 141–170.

BACKMAN, J., SCHNEIDER, D.A., RIO, D., AND OKADA, H., 1990, Neogene low-latitude magnetostratigraphy from Site 710 and revised age estimates of Miocene nannofossil datum events: Scientific Results of the Ocean Drilling Program, v. 115, p. 271–276.

BELLANCA, A., SGARRELLA, F., NERI, R., RUSSO, B., SPROVIERI, M., BONADUCE, G., AND ROCCA, D., 2002, Evolution of the Mediterranean basin during the late Langhian–early Serravallian: an integrated paleoceanographic approach, *in* Iaccarino, S.M., ed., Integrated Stratigraphy and Paleoceanography of the Mediterranean Middle Miocene: Rivista Italiana di Paleontologia e Stratigrafia, v. 108, p. 223–240.

BERGGREN, W.A, KENT, D.V., AND VAN COUVERING, J.A., 1985, Neogene geochronology and chronostratigraphy, *in* Shelling, N.J., ed., The Chronology of the Geological Record: Geological Society of America, Memoir 10, p. 211–259.

BERGGREN, W.A, KENT, D.V., SWISHER, C.C., AND AUBRY, M.P., 1995, A Revised Cenozoic Geochronology and Chronostratigraphy: SEPM, Special Publication 54, p. 129–212.

BLOW, W.H., 1969, Late Middle Eocene to Recent planktonic foraminiferal biostratigraphy, *in* Bronnimann, P., and Renz, H.H., eds., First International Conference on Planktonic Microforaminifers, Geneva 1967, Proceeding s, vol. 1, Brill E.J., p. 199–421.

CANDE, S.C., AND KENT, D.V., 1992, A new geomagnetic polarity time scale for the Late Cretaceous and Cenozoic: Journal of Geophysical Research, v. 97, p. 13,917–13,951.

CANDE, S.C., AND KENT, D.V. 1995, Revised calibration of the geomagnetic polarity timescale for the Late Cretaceous and Cenozoic: Journal of Geophysical Research, v. 100, p. 6093–6095.

CARUSO, A., SPROVIERI, M., BONANNO, A., AND SPROVIERI, R., 2002, Astronomical calibration of the Serravallian–Tortonian Case Pelacani section (Sicily, Italy), *in* Iaccarino, S.M., ed., Integrated Stratigraphy and Paleoceanography of the Mediterranean Middle Miocene: Rivista Italiana di Paleontologia e Stratigrafia, v. 108, p. 297–306.

CATALANO, R., FRANCHINO, A., MERLINI, S., AND SULLI, A., 2000, Central western Sicily structural setting interpreted from seismic reflection profiles, *in* Catalano, R., and Lo Cicero, G., eds., Sicily, A Natural Laboratory in the Mediterranean Area Structures, Seas, Resources and Hazards: Società Geologica Italiana, 79ª Riunione Estiva, Palermo, 21–23 September 1998, Proceedings: Società Geologica Italiana, Memorie, v. 55, p. 5–16.

CELLO, G., 1987, Structure and deformation processes in the Strait of Sicily "rift zone": Tectonophysics, v. 141, p. 237–247.

CELLO, G., TORTORICI, L., AND TURCO, E., 1984, Evidenze di processi deformativi continui e di tettonica trascorrente nel Canale di Sicilia: le Isole Maltesi: Società Geologica Italiana, Bollettino, v. 103, p. 591–600.

CHAISSON, W.P., AND PEARSON, P.N., 1997, Planktonic foraminifer biostratigraphy at Site 925: Middle Miocene–Pleistocene: Scientific Results of the Ocean Drilling Program, v. 154, p. 3–31.

CHANNELL, J.E.T., D'ARGENIO, B., AND HORWARTH, F., 1979, Adria, the African promontory, in Mesozoic Mediterranean paleogeography: Earth-Science Reviews, v. 15, p. 213–292.

CITA, B.M., AND BLOW, W.H., 1969, The biostratigraphy of the Langhian, Serravallian and Tortonian stages in the type-sections in Italy: Rivista Italiana di Paleontologia e Stratigrafia, v. 75, p. 549–603.

CITA, M.B., AND PREMOLI SILVA, I., 1960, Pelagic foraminifera from the type Langhian: Proceedings of the International Paleontological Union, Norden, v. 22, p. 39–50.

CITA, M.B., AND PREMOLI SILVA, I., 1968, Evolution of the planktonic foraminiferal assemblages in the stratigraphical interval between the

Type-Langhian and the Type-Tortonian and biozonation of the Miocene of Piedmont: Giornale di Geologia, v. 25 (3), p. 1–28.

CITA, M.B., PREMOLI SILVA, I., AND ROSSI, R., 1965, Foraminiferi planctonici del Tortoniano tipo: Rivista Italiana di Paleontologia e Stratigrafia, v. 71, v. 217–308.

CITA, M.B., COLALONGO, M.L., D'ONOFRIO, S., IACCARINO, S., AND SALVATORINI, G., 1978, Biostratigraphy of Miocene deep-sea sediments (Sites 372 and 375), with special reference to the Messinian/pre-Messinian interval: Initial Reports of the Deep Sea Drilling Project, v. 42, p. 671–685.

COCCIONI, R., DI LEO, C., AND GALEOTTI, S., 1992, Planktonic foraminiferal biostratigraphy of the upper Serravallian –lower Tortonian Monte dei Corvi Section (Northeastern Apennines, Italy), *in* Montanari, A., Odin, G.S., and Coccioni R., eds., Conferenza interdisciplinare di geologia sull'epoca miocenica con enfasi sulla sequenza umbro-marchigiana, Ancona 1992: Miocene Columbus Project, International Union of Geological Sciences, abstracts and field trips, p. 53–56.

DART, C.J., BOSENCE, W.J., AND MCCLAY, K.R., 1993, Stratigraphy and structure of the Maltese graben system: Geological Society of London, Journal, v. 150, p. 1153–1166.

DI STEFANO, E., 1998, Calcareous nannofossils quantitative biostratigraphy of Holes 969E and 963B (Eastern Mediterranean): Scientific Results of the Ocean Drilling Program, v. 160, p. 99–112.

DI STEFANO, E., BONOMO, S., CARUSO, A., DINARÉS-TURELL, J., FORESI, L., SALVATORINI, AND SPROVIERI, R., 2002, Calcareous plankton bio-events in the Miocene Case Pelacani section (Southeastern Sicily, Italy), *in* Iaccarino, S.M., ed., Integrated Stratigraphy and Paleoceanography of the Mediterranean Middle Miocene: Rivista Italiana di Paleontologia e Stratigrafia, v. 108, p. 307–324.

FELIX, R., 1973, Oligo-Miocene stratigraphy of Malta and Gozo: Ph.D. thesis University of Utrecht, Utrecht, The Netherlands, 104 p.

FINETTI, I., 1984, Structure, stratigraphy and evolution of the central Mediterranean: Bollettino di Geofisica Teorica ed Applicata, v. 24, p. 247–312.

FORESI, L.M., 1993, Biostratigrafia a foraminiferi planctonici del Miocene medio del Mediterraneo e delle basse latitudini con considerazioni cronostratigrafiche: Ph.D. Dissertation, Parma University, 143 p.

FORESI, L.M., IACCARINO, S., MAZZEI, R., AND SALVATORINI, G., 1998, New data on middle to late Miocene calcareous plankton biostratigraphy in the Mediterranean area: Rivista Italiana di Paleontologia e Stratigrafia, v. 104, p. 95–114.

FORESI, L.M., BONOMO, S., CARUSO, A., DI STEFANO, E., SALVATORINI, G., AND SPROVIERI, R., 2002a, Calcareous plankton high resolution biostratigraphy (foraminifera and nannofossils) of the uppermost Langhian–lower Serravallian Ras Il-Pellegrin section (Malta), *in* Iaccarino, S.M., ed., Integrated Stratigraphy and Paleoceanography of the Mediterranean Middle Miocene: Rivista Italiana di Paleontologia e Stratigrafia, v. 108, p. 195–210.

FORESI, L.M., BONOMO, S., CARUSO, A., DI STEFANO, A., DI STEFANO, E., IACCARINO, S.M., LIRER, F., MAZZEI, R., SALVATORINI, G., AND SPROVIERI, R., 2002b, High resolution calcareous plankton biostratigraphy of the Serravallian succession of the Tremiti Islands (Adriatic Sea, Italy), *in* Iaccarino, S.M., ed., Integrated Stratigraphy and Paleoceanography of the Mediterranean Middle Miocene: Rivista Italiana di Paleontologia e Stratigrafia, v. 108, p. 257–274.

FORNACIARI, E., DI STEFANO, A., RIO, D., AND NEGRI, A., 1996, Middle Miocene quantitative calcareous nannofossil biostratigraphy in the Mediterranean region: Micropaleontology, v. 42, p. 37–63.

GAMBINI, R., AND TOZZI, M., 1996, Tertiary geodynamic evolution of the Southern Adria microplate: Terra Nova, v. 8, p. 593–602.

GIANNELLI, L., AND SALVATORINI, G., 1975, I foraminiferi planctonici dei sedimenti terziari dell' arcipelago maltese. Biostratigrafia di "Blue Clay", "Green Sands" e "Upper Globigerina Limestone": Società Toscana di Scienze Naturali, Atti, Memorie, Serie A, v. 79, p. 49–74.

HARLAND, W.B., AMSTRONG, R.L., COX, A.V., GRAIG, L.E., SMITH, A.G., AND SMITH, D.G., 1990, A Geological Time Scale 1989: Cambridge, U.K., Cambridge University Press, 263 p.

HILGEN, F.J., 1991a, Astronomical calibration of Gauss to Matuyama sapropels in the Mediterranean and implication for Geomagnetic Polarity Time Scale: Earth and Planetary Science Letters, v. 104, p. 226–244.

HILGEN, F.J., 1991b, Extension of the astronomically calibrated (polarity) time scale to the Miocene/Pliocene boundary: Earth and Planetary Science Letters, v. 107, p. 349–368.

HILGEN, F.J., ABDUL AZIZ, H., KRIJGSMAN, W., LANGEREIS, C.G., LOURENS, L.J., MEULENKAMP, J.E., RAFFI, I., STEENBRINK, J., TURCO, E., VAN VUGT, N., WIJBRANS, J.R., AND ZACHARIASSE, W.J., 1999, Present status of the astronomical (polarity) time-scale for the Mediterranean Late Neogene: Royal Society (London), Philosophical Transactions, A, v. 357, p. 1931–1947.

HILGEN, F.J., BISSOLI, L., IACCARINO, S., KRIJGSMAN, W., MEIJER, R., NEGRI, A., AND VILLA, G., 2000a, Integrated stratigraphy and astrochronology of the Messinian GSSP at Oued Akrech (Atlantic Morocco): Earth and Planetary Science Letters, v. 182, p. 237–251.

HILGEN, F.J., KRIJGSMAN, W., LANGEREIS, C.G., AND LOURENS, L.J., 1997, Breakthrough made in dating of the geological record: EOS, Transactions, American Geophysical Union, v. 78, p. 288–289.

HILGEN, F.J., KRIJGSMAN, W., LANGEREIS, C.G., LOURENS, L.J., SANTARELLI, A., AND ZACHARIASSE, W.J., 1995, Extending the astronomical (polarity) time scale into the Miocene: Earth and Planetary Science Letters, v. 136, p. 495–510.

HILGEN, F.J., KRIJGSMAN, W., RAFFI, I., TURCO, E., AND ZACHARIASSE, W.J., 2000b, Integrated stratigraphy and astronomical calibration of the Serravallian–Tortonian boundary section at Monte Gibliscemi (Sicily, Italy): Marine Micropaleontology, v. 38, p. 181–211.

HUDDLESTON, P.F., 1984, Planktonic foraminiferal biostratigraphy, Deep Sea Drilling Project Leg 81: Initial Reports of the Deep Sea Drilling Project, v. 81, p. 429–438.

HYDE, H.P.T., 1955, Geology of the Maltese Islands: Malta, Lux Press, 135 p.

Iaccarino, S., Foresi, L.M., Mazzei, R., and Salvatorini, G., 2001, Calcareous plankton biostratigraphy of the Miocene sediments of the Tremiti Islands (southern Italy): Rivista Española de Micropaleontología, v. 33, p. 237–248.

JENKINS, G.M., AND WATTS, D.G., 1968, Spectral Analysis and its Applications: Oakland, California, Holden Day, 410 p.

KRIJGSMAN, W., 1996, Miocene magnetostratigraphy and cyclostratigraphy in the Mediterranean: extension of the astronomical polarity time scale: Geologica Ultraiectina, v. 141, 207 p.

KRIJGSMAN, W., HILGEN, F.J., LANGEREIS, C.G., LOURENS, L.J., SANTARELLI A., AND ZACHARIASSE W.J., 1995, Late Miocene magnetostratigraphy, biostratigraphy and cyclostratigraphy in the Mediterranean: Earth and Planetary Science Letters, v. 136, p. 475–494.

KRIJGSMAN, W., HILGEN, F.J., NEGRI, A., WIBRANS, J.R., AND ZACHARIASSE, W.J., 1997, The Monte del Casino section (Northern Apennines, Italy); a potential Tortonian/Messinian boundary stratotype?: Palaeogeography, Palaeoclimatology, Palaeoecology, v. 133, p. 27–47.

KRIJGSMAN, W., HILGEN, F.J., RAFFI, I., SIERRO, F.J., AND WILSON, D.S., 1999, Messinian astrochronology and its implications for salinity crisis: Nature, v. 400, p. 652–655.

LASKAR, J., JOUTEL, F., AND BOUDIN, F., 1993, Orbital, precessional, and insolation quantities for the Earth from –20 Myr to +10 Myr: Astronomy and Astrophysics, v. 270, p. 522–533.

LIRER, F., CARUSO, A., FORESI, L.M., SPROVIERI, M., BONOMO, S., DI STEFANO, A., DI STEFANO, E., IACCARINO, S.M., SALVATORINI, G., SPROVIERI, R., AND MAZZOLA, S., 2002, Astrochronological calibration of the upper Serravallian/lower Tortonian sedimentary sequence at Tremiti Islands (Adriatic Sea, Southern Italy), *in* Iaccarino S.M., ed., Integrated Stratigraphy and Paleoceanography of the Mediterranean Middle Miocene: Rivista Italiana di Paleontologia e Stratigrafia, v. 108, p. 241–256.

LOURENS, L.J., 1994, Astronomical forcing of Mediterranean climate during the last 5.3 million years: Ph.D. Dissertation, Utrecht University, 247 p.

LOURENS, L.J., HILGEN, F.J., GUDJONSSON, L., AND ZACHARIASSE, W.J., 1992, Late Pliocene to Early Pleistocene astronomically forced sea surface productivity and temperature variations in the Mediterranean: Marine Micropaleontology, v. 19, p. 49–78.

LOURENS, L.J., HILGEN, F.J., ZACHARIASSE, W.J., VAN HOOF, A.A.M., ANTONARAKOU, A., AND VERGNAUD-GRAZZINI, C., 1996, Evaluation of Plio-Pleistocene astronomical time scale: Paleoceanography, v. 11, p. 391–413.

MAZZA, P., 1985, On the occurrence of *Neogloboquadrina acostaensis* in the upper Serravallian sediments of Sicily: Rivista Italiana di Paleontologia e Stratigrafia, v. 91, p. 513–518.

MAZZEI R., 1977, Biostratigraphy of the Rio Mazzapiedi–Castellania section (type-section of the Tortonian) based on calcareous nannoplankton: Società Toscana di Scienze Naturali, Atti, Memorie, Serie A, v. 84, p; 15–24.

MAZZEI, R., 1985, The Miocene Sequence of the Maltese Island: Biostratigraphic and chronostratigraphic references based on nannofossils: Società Toscana di Scienze Naturali, Atti, Memorie, Serie A, v. 92, p. 165–197.

MICULAN, P., 1994, Planktonic foraminiferal biostratigraphy of the middle Miocene in Italy: Società Paleontologica Italiana, Bollettino, v. 33, p. 299–339.

MICULAN, P., 1997, Planktonic foraminiferal biostratigraphy of the Tortonian historical stratotype Rio Mazzapiedi–Castellania section, Northwestern Italy, *in* Montanari, A., Odin, G.S., and Coccioni, R., eds., Miocene Stratigraphy; An Integrated Approach: Amsterdam, Elsevier, Developments in Paleontology and Stratigraphy, no. 15, p. 97–106.

MILLER, K.G., AUBRY, M.P., KHAN, M.J., MELILLO, A.J., KENT, D.V., AND BERGGREN, W.A., 1985, Oligocene–Miocene biostratigraphy, magnetostratigraphy and isotopic stratigraphy of western North Atlantic: Geology, v. 13, p. 257–261.

MILLER, K.G., FEIGENSON, M.D., WRIGHT, J.D., AND CLEMENT, B.M., 1991a, Miocene isotope reference section, Deep Sea Drilling Project Site 608: an evaluation of isotope biostratigraphic resolution: Paleoceanography, v. 6, p. 33–52.

MILLER, K.G., WRIGHT, J.D., AND FAIRBANKS, R.G., 1991b, Unlocking the ice house: Oligocene–Miocene oxygen isotopes, eustasy and marginal erosion: Journal of Geophysical Research, v. 96, p. 6829–6848.

OLAFSSON, G., 1991, Quantitative calcareous nannofossil biostratigraphy and biochronology of early through late Miocene sediment from DSDP Hole 608: Stockholm University, Institute of Geology and Geochemistry, v. 203, 28 p.

OPDYKE, N., MEIN, P., LINDSAY, E., PEREZ-GONZALES, A., MOISSENET, E., AND NORTON, V.L., 1997, Continental deposits, magnetostratigraphy and vertebrate paleontology, late Neogene of Eastern Spain: Palaeogeography, Palaeoclimatology, Palaeoecology, v. 133, p. 129–148.

PAMPALONI, M.L., 1988, Il Paleogene–Neogene delle Isole Tremiti (Puglia, Italia Meridionale): Stratigrafia ed analisi paleoambientale: Ph.D. Dissertation, Università degli Studi di Roma, 65 p.

PARETO, L., 1865, Note sur les subdivisions que l'on pourrait établir dans les terrains tertiaires de l'Apennin septentrional: Societé Géologique de France, Bulletin, ser. 2, v. 22, p. 210–217.

PEDLEY, H.M., HOUSE, M.R., AND WAUGH, B., 1976, The geology of Malta and Gozo: Proceedings of the Geologists' Association, v. 87, p. 325–341.

PEDLEY, H.M., HOUSE, M.R., AND WAUGH, B., 1978, The geology of the Pelagian block: the Maltese Islands, *in* Nairn, A.E.M., Kanes, W.H., and Stehli, F.G., eds., The Ocean Basins and Margins: The Western Mediterranean: New York, Plenum Press, 417–433 p.

PLATT, J., BEHERMANN, H., CUNNINGHAM, P.C., DEWEY, J.F., HELMAN, H., PARISH, M., SHEPLEY, M.G., WALLIS, S., AND WESTON, P.J., 1989, Kinematics of the Alpine arc and the motion history of Adria: Nature, v. 337, p. 158–161.

POORE, R.Z., 1979, Oligocene through Quaternary planktonic foraminiferal biostratigraphy of the North Atlantic: DSDP Leg 49: Initial Reports of the Deep Sea Drilling Project, v. 49, p. 447–517.

POSTMA, G., AND TEN VEEN, J.H., 1999, Astronomically and tectonically linked variations in gamma-ray intensity in Late Miocene hemipelagic succession of the Eastern Mediterranean basin: Sedimentary Geology, v. 128, p. 1–12.

RAFFI, I., AND FLORES, J.A., 1995, Pleistocene through Miocene calcareous nannofossils from Eastern Equatorial Pacific Ocean (LEG 138): Scientific Results of the Ocean Drilling Program, v. 138, p. 233–286.

RAFFI, I., RIO, D., D'ATRI, A., FORNACIARI, E., AND ROCCHETTI, S., 1995, Quantitative distribution patterns and biomagnetostratigraphy of middle and late Miocene calcareous nannofossils from equatorial Indian and Pacific oceans (Legs 115, 130 and 138): Scientific Results of the Ocean Drilling Program, v. 138, p. 479–502.

RIGO, M., AND BARBIERI, F., 1959, Stratigrafia pratica applicata in Sicilia: Servizio Geologico Italiano, Bollettino, v. 80, p. 351–441.

RIO, D., RAFFI, I., AND VILLA, G., 1990, Pliocene–Pleistocene calcareous nannofossil distribution patterns in the Western Mediterranean: Scientific Results of the Ocean Drilling Program, v. 107, p. 513–533.

RIO, D., CITA, M.B., IACCARINO, S., GELATI, R., AND GNACCOLINI, M., 1997, Langhian, Serravallian, and Tortonian historical stratotypes, *in* Montanari, A., Odin, G.S., and Coccioni, R., eds., Miocene Stratigraphy: An Integrated Approach: Amsterdam, Elsevier, Developments in Paleontology and Stratigraphy, no. 15, p. 57–87.

ROBERTSON, A.H.F., AND GRASSO, M., 1995, Overview of the Late Tertiary–Recent tectonic and paleo-environmental development of the Mediterranean region: Terra Nova, v. 7, p. 114–127.

SELLI, R., 1971, Isole Tremiti e Pianosa, *in* Cremonini, G., Elmi, C., and Selli, R., eds., Foglio 156: S. Marco in Lamis: Note illustrative alla Carta Geologica d'Italia: Servizio Geologico d'Italia, scala 1:100.000, Rome.

SHACKLETON, N.J., BALDAUF, J.G., FLORES, J.A., MOORE, T.C., JR., RAFFI, I., AND VINCENT, E., 1995, Biostratigraphic Summary for Leg 138: Scientific Results of the Ocean Drilling Program, v. 138, p. 517–533.

SHACKLETON, N.J., AND CROWHURST, S., 1997, Sediment fluxes based on orbital tuned time scale 5 Ma to 14 Ma, Site 926: Scientific Results of the Ocean Drilling Program, v. 154, p. 69–82.

SIERRO, F.J, HILGEN, F.J., KRIJGSMAN, W., AND FLORES, J.A., 2001, The Abad composite (SE Spain): a Messinian reference section for the Mediterranean and the APTS: Palaeogeography, Palaeoclimatology, Palaeoecology, v. 168, p. 141–169.

SPROVIERI, M., BELLANCA, A., NERI, R., MAZZOLA, S., BONANNO A., BERNARDO, P., AND SORGENT, R., 1999, Astronomical calibration of Late Miocene stratigraphic events and analysis of precessionally driven paleoceanographic changes in the Mediterranean Basin: Società Geologica Italiana, Memorie, v. 54, p. 7–24.

SPROVIERI, M., CARUSO, A., FORESI, L.M., BELLANCA, A., NERI, R., MAZZOLA, S., AND SPROVIERI, R., 2002, Astronomical calibration of the upper Langhian/lower Serravallian record of Ras- Il Pellegrin section (Malta Island, Central Mediterranean), *in* Iaccarino, S.M., ed., Integrated Stratigraphy and Paleoceanography of the Mediterranean Middle Miocene: Rivista Italiana di Paleontologia e Stratigrafia, v. 108, p. 183–194.

SPROVIERI, R., 1992, Mediterranean Pliocene biochronology: a high resolution record based on quantitative planktonic foraminiferal distribution: Rivista Italiana di Paleontologia e Stratigrafia, v. 98, p. 61–100.

SPROVIERI, R., 1993, Pliocene–early Pleistocene astronomically forced planktonic foraminiferal abundance fluctuations and chronology of calcareous plankton bio-events: Rivista Italiana di Paleontologia e Stratigrafia, v. 99, p. 371–414.

SPROVIERI, R., BONOMO, S., CARUSO, A., DI STEFANO, A., DI STEFANO, E., FORESI, L.M., IACCARINO, S.M., LIRER, F., MAZZEI, R., AND SALVATORINI, G., 2002, An integrated calcareous plankton biostratigraphic scheme and biochronology for the Mediterranean Middle Miocene, *in* Iaccarino, S.M., ed., Integrated Stratigraphy and Paleoceanography of the Mediterranean Middle Miocene: Rivista Italiana di Paleontologia e Stratigrafia, v. 108, p. 337–353.

TEN VEEN, J.H., AND POSTMA, G., 1993, Astronomically forced variations in gamma-ray intensity: Late Miocene hemipelagic succession in the eastern Mediterranean basin as test case: Geology, v. 24 (1), p. 15–18.

TURCO, E., BAMBINI, A.M., FORESI, L.M., IACCARINO, S.M., LIRER, F., MAZZEI, R., AND SALVATORINI, G., 2002, Middle Miocene high-resolution calcareous plankton biostratigraphy at Site (Leg 154, equatorial Atlantic Ocean): palaeoecological and palaeobiogeographical implications: Géobios, v. 35, Mémoire Special 24, p. 257–276.

Turco, E., HILGEN, F.J, LOURENS, L.J., SHACKLETON, N.J., AND ZACHARIASSE, W.J., 2001, Punctuated evolution of global climate cooling during the late Middle to early Late Miocene: High-resolution planktonic foraminiferal and oxygen isotope records from the Mediterranean: Paleoceanography, v. 16, p. 405–423.

VAN VUGT, N., 2000, Orbital forcing in late Neogene lacustrine basins from the Mediterranean: Geologica Ultraiectina, v. 189, 167 p.

VERSTEEGH, G.J.M., 1994, Recognition of cyclic and non-cyclic environmental changes in the Mediterranean Pliocene: a palynological approach: Marine Micropaleontology, v. 23, p. 147–183.

VERVLOET, C.C., 1966, Stratigraphical and Micropaleontological Data on the Tertiary of Southern Piedmont (Northern Italy): Utrecht, Schotanus, 88 p.

ZACHARIASSE, W.J., 1992, Neogene planktonic foraminifers from Sites 761 and 762 off Northwest Australia: Scientific Results of the Ocean Drilling Program, v. 122, p. 665–675.

A MILANKOVITCH CLIMATE CONTROL ON THE MIDDLE MIOCENE MEDITERRANEAN INTERMEDIATE WATER

MARIO SPROVIERI
Istituto per l'Ambiente Marino Costiero, Geomare, National Research Council, Calata Porta di Massa, Porto di Napoli, 80133 Napoli, Italy
e-mail: sprovier@gms01.geomare.na.cnr.it
FRANCA SGARRELLA
Dipartimento di Scienze della Terra, Università Federico II, Largo San Marcellino 10, 80138 Napoli, Italy
e-mail: garella@unina.it
BIANCA RUSSO
Dipartimento di Scienze della Terra, Università Federico II, Largo San Marcellino 10, 80138 Napoli, Italy
e-mail: brusso@unina.it
ADRIANA BELLANCA
Dipartimento di Chimica e Fisica della Terra (CFTA), Via Archirafi 36, 90123 Palermo, Italy
e-mail: bellanca@unipa.it
AND
RODOLFO NERI
Dipartimento di Chimica e Fisica della Terra (CFTA), Via Archirafi 36, 90123 Palermo, Italy
e-mail: neri@unipa.it

ABSTRACT: Methodology of cyclostratigraphic analysis applied to benthic foraminifera is verified utilizing a faunal and geochemical dataset, from the Middle Miocene Ras il-Pellegrin composite section (Malta Island, central Mediterranean). Benthic data were elaborated by Q-mode varimax principal factor analysis. In this paper, spectral analysis is carried out only on two factors, which have a clear paleoecological significance: Factor 1 (loaded by *Cibicidoides ungerianus* and *Siphonina reticulata*), indicative of oxic bottom waters, and Factor 2 (loaded by *Bulimina elongata* group), indicative of oxygen-stressed conditions. Results of these analyses show that Factor 1 and Factor 2 curves are, respectively, in and out of phase with maxima of the eccentricity (100 and 400 ky).

We utilize the 3D paleocenographic model of Bellanca et al. (2002) as reference for the Middle Miocene Mediterranean circulation and focus our attention on the Mediterranean Intermediate water, characterized by hydrographic and hydrodynamic features similar to those presently recorded in the Levantine Intermediate Water (LIW). Consequently, we suppose that Middle Miocene Mediterranean Intermediate Water, here defined as proto-MIW, played a role similar to that of present Mediterranean Intermediate Water (MIW).

Following this hypothesis, Factor 1, which is indicative of oxic bottom waters, is interpreted as a tracer of high production of proto-MIW, during periods of high eccentricity and, probably, precession minima, characterized by coldest winter seasons.

These results point out a direct link between selected benthic species, long-term astronomical forcing, and deep-water response and provide a useful tool for astronomical calibration of geologic time and for paleoceanographic reconstructions.

INTRODUCTION

The use of geochemical and/or faunal tracers in recording of Milankovitch astronomical forcing, and then as helpful tools for astronomical tuning of marine sequences, is based on their sensitivity to the periodic climatic or paleoceanographic changes induced by the variation of precession, obliquity, and eccentricity of the Earth's orbit. This is well explained if we consider that the predominant climatic effects of the above variations are changes in seasonality, as pointed out by Berger (1980, 1988).

Presently, cyclostratigraphic reconstruction of marine sedimentary sequences is carried out by means of integrated analyses of rhythmic variations of lithology and of geochemical and faunal tracers (Hilgen, 1987, 1991; Thunell et al., 1991; Hilgen et al., 1999; Hilgen et al., 2000). In sediments characterized by homogeneous lithologies this reconstruction is based on the study of periodic oscillation of selected physical or chemical parameters of water masses, reflected in the coupled variation of planktonic foraminiferal species and their oxygen and carbon isotopes as well as in carbonate contents, which are considered good proxies for surface-water changes, directly or secondarily linked to astronomically induced insolation variability (Sprovieri, 1992; Lourens et al., 1996; Sgarrella et al., 1999; Sprovieri et al., 2002).

As to the utilization of faunal tracers, less developed is the application of cyclostratigraphic methodologies to selected species and/or groups of species of benthic foraminifera for the astronomical tuning of marine sedimentary sequences. This is due to the complex structure of the benthic assemblage, from which it is generally difficult to infer straightforward relationships between environmental forcing and the response of the different species. Only few authors (e.g., Streeter, 1973; Lutze, 1979; Schnitker, 1979, 1984; Verhallen, 1991; Coccioni and Galeotti, 1993; Sgarrella et al., 1999) have previously pointed out cyclic changes in benthic foraminifera reflecting modifications in paleoecological and/or paleoceanographic bottom conditions.

In particular, Schnitker (1984) pointed out benthic foraminiferal fluctuations related to insolation changes, occurring at about 20 ky, 40 ky, 100 ky, and 400 ky intervals, in the North Atlantic DSDP Hole 552A sequence, from the Late Miocene to the Quaternary, utilizing Q-mode varimax component and power-spectrum analyses.

Cyclostratigraphy: Approaches and Case Histories
SEPM Special Publication No. 81, Copyright © 2004
SEPM (Society for Sedimentary Geology), ISBN 1-56576-108-1, p. 45–55.

Coccioni and Galeotti (1993) grouped benthic foraminiferal assemblages of the Late Albian "Amadeus Segment" of Scisti a Fucoidi Formation (Umbria–Marche region, Central Italy) into morphogroups and reported the faunal changes as related to short-eccentricity (93–123 ky), obliquity, and precession astronomical record by means of spectral analysis. However, they pointed out that distributions of species, genera, and morphogroups were discontinuous and difficult to evaluate.

PREVIOUS WORK AND AIM OF STUDY

In this paper we present the results of a cyclostratigraphic study of data on benthic foraminifera, reported by Bellanca et al. (2002), from the marly sediments of the Blue Clay Formation in the upper part of the Middle Miocene Ras il-Pellegrin composite section (Malta Island, central Mediterranean), (Fig. 1).

The astronomical calibration of the whole section (Fig. 2) using the astronomical solution of Laskar et al. (1993), reported by Sprovieri et al. (2002), indicates for the deposition of analyzed sediments a time interval ranging between 13.75 and 12.32 Ma.

In our opinion, this time interval is useful for investigating the oceanographic evolution of the (paleo)Mediterranean after the interruption of communications between the Mediterranean and Indo-Pacific areas. This important paleogeographic event was estimated at about 16 Ma by Johnson (1985) and at about 14.5 Ma by Woodruff and Savin (1991) and represented the first step of a progressive oceanographic evolution of the Tethys water masses towards present Mediterranean conditions.

Results of integrated analyses of benthic microfauna (foraminifera and ostracoda) and planktonic and benthic oxygen isotopes in the Ras il-Pellegrin composite section, combined with data from other Mediterranean sedimentary sequences, allowed Bellanca et al. (2002) and Russo et al. (2002) to reconstruct a thermohaline circulation system for the upper Langhian –lower Serravallian Mediterranean, characterized by three main water masses: (1) surface Atlantic water inflowing the Mediterranean; (2) intermediate outflowing Mediterranean water (in this paper reported as proto-MIW), originating in the surface eastern zone; and (3) Atlantic psychrospheric bottom water.

In particular, the same authors suggested that the hydrographic and hydrodynamic features of the Mediterranean Intermediate Water, during the Middle Miocene, were similar to those presently recorded in the Levantine Intermediate Water (LIW), which is the major constituent of Mediterranean Intermediate Water (MIW) (Wüst, 1961). At present the LIW is the "endemic" Mediterranean water mass for antonomasia, representing the master engine ensuring the ventilation and the oxygenation at the bottom of the Mediterranean (POEM group, 1992). Its formation is a seasonal event occurring mainly in the Rhodes cyclonic gyre, where, during winter, predominantly in March, dry and cold continental air masses cross the eastern Mediterranean Sea, evaporation and mixing are enhanced by strong winds, the seasonal thermocline is eradicated, and the Modified Atlantic Waters (MAW) are well mixed and do not impede convection (Wüst, 1961; Ozsoy, 1981; Ovchinnikov and Plakhin, 1984; POEM group, 1992). The combination of these events produces intense loss of buoyancy of the surface waters and their sinking down to a depth of 300–400 m (MIW) whence they spread throughout the Mediterranean Sea at extremely slow rate.

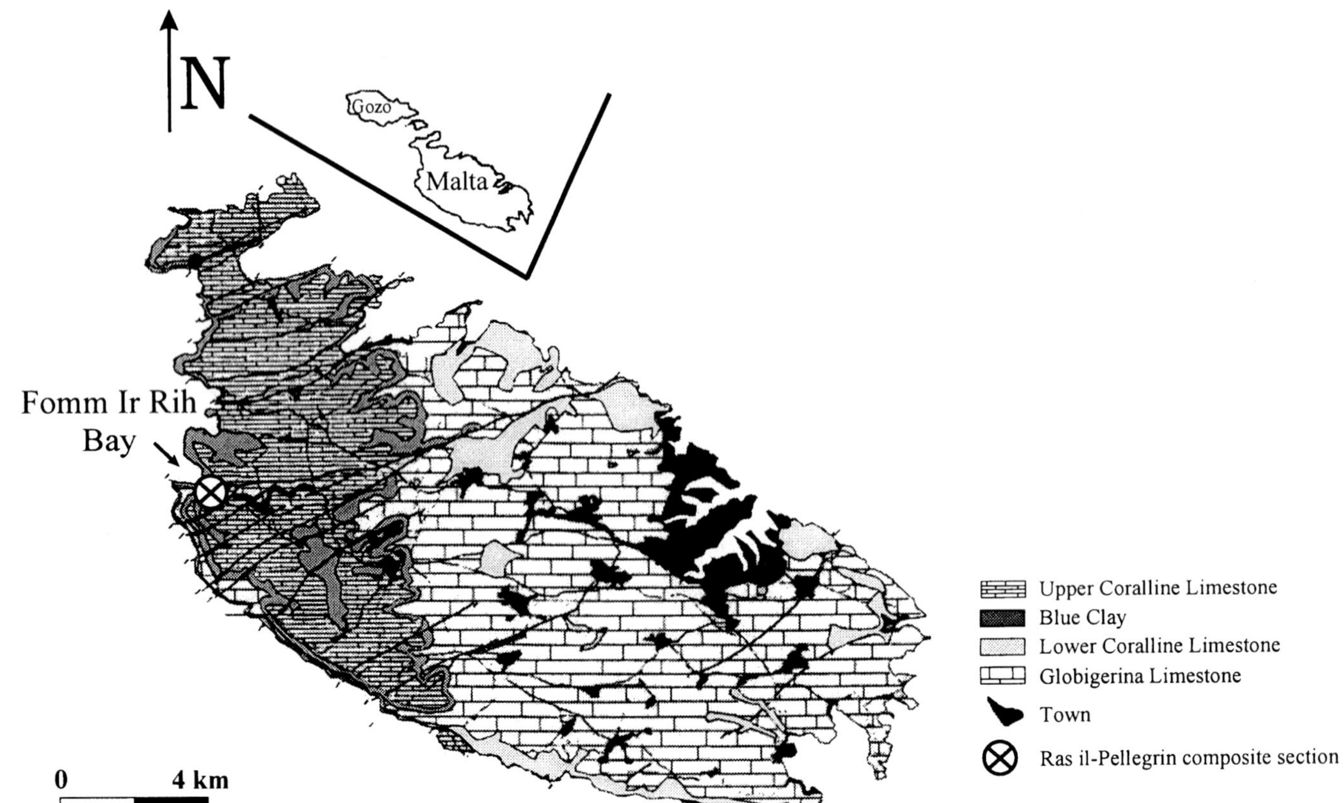

FIG 1.—Location map of the Ras il-Pellegrin composite section. Lithostratigraphic units are reported by Foresi et al. (2002).

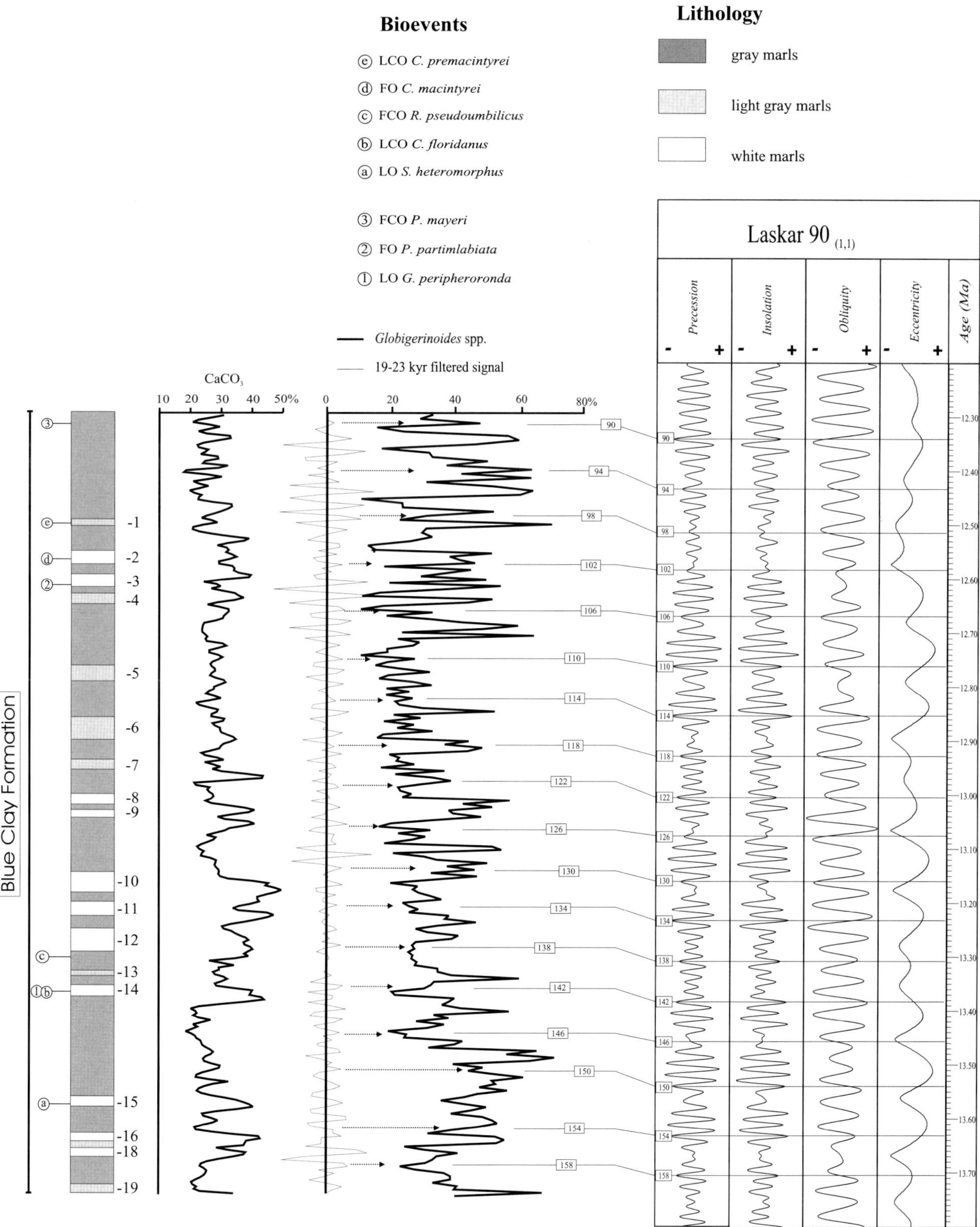

FIG 2.—Lithology, main calcareous plankton events, carbonate-content curve, *Globigerinoides* spp. curve, and astronomical tuning of the Blue Clay Formation in the Ras il-Pellegrin composite section (as reported in Sprovieri et al., 2002). White layers coincide with more carbonate-rich intervals. Numbers to the right of the lithologic column mark thick white layers in the sequence. Labeling of the precession cycles is reported for the *Globigerinoides* spp. peaks and the correspondent minima in the precession component of the insolation curve. Correlation between the lithologic cycles and the astronomical solution of Laskar et al. (1993) and the present-day values for tidal dissipation and dynamical ellipticity of the Earth are reported.

Periodic variations of LIW formation could be ascribed to changes of hydrographic parameters in the key area of the eastern basin occurring during the winter seasons. In particular, changes of sea-surface temperature and of temperature gradient in the land–sea system, triggering northwesterly wind circulation, could stimulate more or less intense LIW production.

A direct consequence of the hypothesized similarity of present MIW and Middle Miocene Intermediate Waters is that some climate–ocean interaction structures in the Mediterranean area were similar to present conditions, possibly since the Middle Miocene, and the proto-MIW was a suitable mechanism of oxygenation for the bottom waters of the Mediterranean basin.

This suggests that the intermediate water mass was a constant feature of the (paleo)Mediterranean since the Middle Miocene, although at that time the basin was characterized by different physiography and by a circulation of water masses generally different from that of the water masses of the present Mediterranean. According to Hsü and Bernoulli (1978), during the Early and Middle Miocene the Mediterranean basin was broken up and changed into a number of deep basins. Paleobathymetric analyses of Benson (1978), Wright (1978), and Cita et al. (1978) yielded convincing evidence that the pre-Messinian basins were very deep. In particular, Benson (1973, 1978) pointed out the occurrence of psycrospheric ostracodes in the central and western Mediterranean, which indicated a paleo-depth of 1500 m or more, and suggested that the Middle Miocene Atlantic threshold was deep enough to allow those lower-slope species to enter into the Mediterranean.

In this paper, we speculate on hydrographic and hydrodynamic changes that could have induced variations of proto-MIW formation during the time of deposition of the Blue Clay Formation.

We focus our attention on those benthic species which can be considered the best recorders of variation of proto-MIW production. The results should offer the opportunity to use the benthic fauna as a valuable instrument for cyclostratigraphic analysis of Mediterranean marine sedimentary sequences.

Materials

The sedimentary interval studied belongs to part of the Langhian–Tortonian Blue Clay Formation (Fig. 2) and was referred to the upper Langhian–lower Serravallian (Jacobs et al., 1996; Foresi et al., 2002). It crops out in the northwestern margin of the Malta Island in Fomm Ir Rih Bay about 20 km west of the town of Valletta (Fig. 1) and is about 68 m thick. It consists of blue pelagic marls intercalated by thick pale layers (light-gray marls and/or white calcareous marls). These sediments conformably overlie the Globigerina Limestone Formation (Giannelli and Salvatorini, 1975; Fornaciari et al. 1996). Biostratigraphic results (Giannelli and Salvatorini, 1975; Mazzei, 1985; Fornaciari et al., 1996) suggest a Langhian–Tortonian age for the deposition of the whole Blue Clay Formation.

The Blue Clay Formation is unconformably overlain by the Coralline Limestone Formation. The latter consists mainly of limestones, wackestones, and packstones with abundant macrofossils. Coralline algae are the most common organogenic remains. Quaternary deposits, consisting of sands and conglomerates interbedded with paleosols and remains of continental fauna, are present at the top of the sedimentary succession of the island of Malta.

From a tectonic point of view, the studied sediments constitute part of the Maltese sedimentary system (Dart et al., 1993) characterized by a graben structure within the African Plate.

Faunal Data, Paleoecology, and Q-mode Principal Component Analysis

Quantitative analysis of the benthic foraminiferal assemblage (> 125 μm fraction) was carried out by Bellanca et al. (2002) on 221 samples, collected with a mean sampling interval of 30 cm. In this previous paper taxa abundance, expressed as percentage of the total fauna, was calculated, and some species, such as *Gyroidina* spp., *Lenticulina* spp., *Uvigerina rutila–barbatula* goup, *Bulimina elongata* group, and nodosariids were lumped together. These authors recognized 88 species and groups of species.

The resulting benthic foraminiferal assemblages are characterized by dominance of *Cibicidoides ungerianus*, *Cibicidoides subhaidingerii*, *Uvigerina rutila–barbatula* group, and *Bulimina elongata* group, mostly represented by *B. lappa*, as common species. *Siphonina reticulata*, *Bulimina costata*, *Uvigerina peregrina*, *Gyroidina* spp. (mostly *G. soldanii*), *Uvigerina pygmaea*, *Pullenia bulloides*, *Bolivina reticulata*, *Bolivina dilatata*, *Uvigerina proboscidea*, and *Spiroplectinella carinata* are additional but significant species. The distributions of these species through the composite section are reported in Figure 3. Moreover, percentage values of *Angulogerina anulosa*, *Anomalinoides helicinus*, *Astrononion umbilicatulum*, *Bigenerina nodosaria*, *Bolivina arta*, *Bolivina* sp., *Bolivina spathulata*, *Bulimina exilis*, *Cassidulina carinata*, *Cassidulina laevigata*, *Cibicidoides dutemplei*, *Globobulimina* spp., *Globocassidulina subglobosa*, *Hanzawaia boueana*, *Lenticulina* spp., *Martinottiella* spp., *Melonis pompilioides*, nodosarids, *Oridorsalis umbonatus*, *Pullenia quinqueloba*, *Sphaeroidina bulloides*, and *Trifarina bradyi* show a significant distribution all through the composite section.

Bellanca et al. (2002) suggested a paleo–water depth of 500–600 m and a persistent slightly under-oxygenated bottom paleoenvironment during the deposition of the studied Blue Clay sediments.

Because of the generally high complexity of the benthic assemblage and of the difficulty to unequivocally interpreting variations of singular species as tracers of well-defined environmental parameter changes, Bellanca et al. (2002) reduced the dimensionality of the original benthic assemblage by application of Q-mode varimax principal-component analysis. Such a numerical procedure consists in linear combinations of the original benthic species for the construction of a reduced number of new variables (factors), which are interpreted on the basis of the benthic species present with the highest loadings (Imbrie and Kipp, 1971; Klovan and Imbrie, 1971; Davis, 1986). The linear combination of the different benthic species in a limited number of factors statistically reduces the role of random variability present in the total benthic assemblage and allows one a selection, based on "ecologically analogous" benthic species, of the major environmental variables that influenced the bottom seawater.

Principal-component analysis was performed on the benthic dataset utilizing abundant or common species, excluding those with percentage values less than 2% occurring only in few widely spaced samples. Four factors are pointed out (Fig. 4), explaining a total variance of 91%. Factor 1 (F1, variance 27%), loaded by *C. ungerianus* and *S. reticulata*, was interpreted as a tracer of oxic bottom waters on the basis of ecological information reported by van der Zwaan (1982) and Sgarrella et al. (1999) for the two species. Factor 2 (F2, variance 25%), loaded by the *B. elongata* group, was considered to be a recorder of stressed oxygen bottom conditions, according to Murray (1991), Pujos (1976), Bizon and Bizon (1984), and Sgarrella and Moncharmont Zei (1993). Factor 3 (F3, variance 21%), dominated by *C. subhaidingerii*, *U. rutila–U. barbatula* group, and *B. costata*, was interpreted to be indicative of moderate oxygen deficiency (Corliss, 1991; Kaiho, 1994, van der Zwaan, 1982, Borsetti et al., 1986; Jonkers, 1984). Factor 4 (F4,

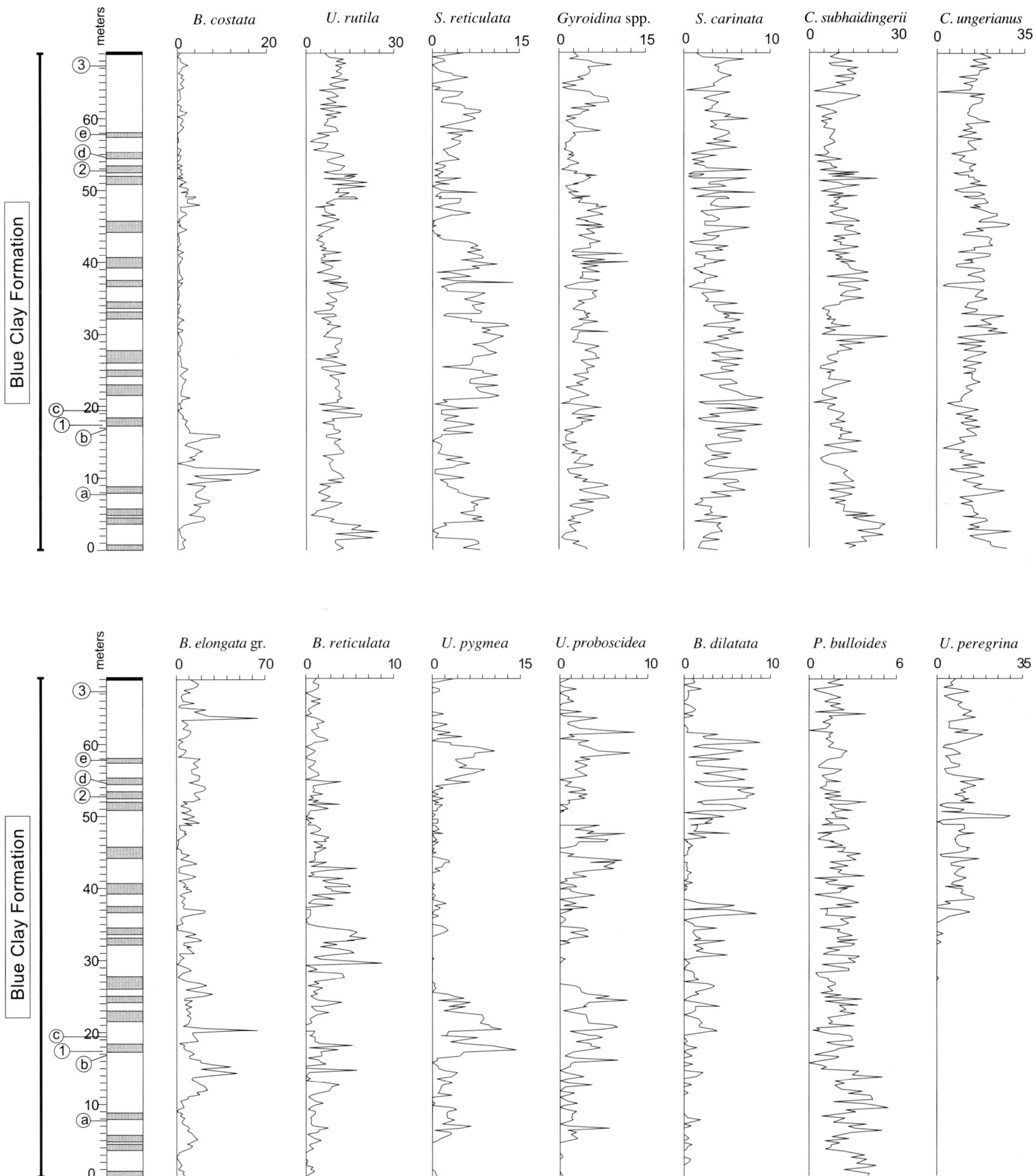

FIG 3.—Distribution (percentage values) of the dominant, common, and significant benthic foraminiferal species throughout the Blue Clay Formation in the Ras il-Pellegrin composite section. Lithology and biostratigraphic events are reported as in Figure 2.

variance 17%), loaded by *Uvigerina peregrina*, was considered to be a good tracer of high bottom organic-matter content (Miller and Lohmann, 1982; Lutze and Coulbourn, 1984; Lutze, 1986; van Leeuwen, 1986; van der Zwaan et al., 1986; Miao and Thunell, 1993) and/or low bottom-water oxygen content (Pflum and Frerichs, 1976; Lohmann, 1978; Streeter and Shackleton, 1979; Schnitker, 1979, 1980, 1994).

The Middle Miocene Surface-To-Bottom Oxygen Isotope Gradient: Clues for Proto-MIW

Results of integrated analyses of benthic microfauna (foraminifera and ostracoda) and planktonic and benthic oxygen isotopes in the Ras il-Pellegrin composite section, combined with data from western Mediterranean (Wright, 1978; Benson, 1978) and eastern Mediterranean (Hsü and Bernoulli, 1978; Cita and Grignani, 1982; Russo et al., 2002) sequences allowed reconstruction of the previously reported 3D thermohaline circulation system for the upper Langhian–lower Serravallian Mediterranean (Bellanca et al., 2002). In particular, these authors suggested, by comparing long-term planktonic and benthic $\delta^{18}O$ trends, that the proto-MIW had hydrographic features similar to those of the present LIW. In fact these two signals showed a parallel trend and a quasi-constant isotope difference of about 1–1.2‰ throughout the succession (Fig. 5). Therefore, assuming a constant temperature/isotope ratio of about 4°C/δ‰, the temperature gradient between surface waters and deep waters was inferred to have remained quite constant throughout the succession in a range of about 5–6°C, the same value presently recorded in the Mediterranean.

Such a result is confirmed by the ostracod data from the Ras il-Pellegrin composite section, indicating the absence of psycrospheric environment and a stable thermospheric system (Bonaduce and Barra, 2002).

According to our previous paleoceanographic reconstruction (Bellanca et al., 2002), we believe that the high-resolution analysis of the benthic microfauna carried out on the Blue Clay Formation can offer an opportunity to monitor periodic interactions between climate changes and proto-MIW production, assuming that highly efficient convective mechanism of rapid transmission of hydrographic parameters between surface and subsurface waters has operated since the Middle Miocene.

TRACERS AND CLIMATE FORCING OF MEDITERRANEAN INTERMEDIATE WATER (MIW)

Spectral Analysis and Milankovitch Control on Formation of Proto-MIW

An improvement on the preliminary results of spectral analysis presented by Bellanca et al. (2002) is here discussed. In our previous paper we selected for spectral analysis only two benthic species (*C. ungerianus* and *C. subhadingerii*), because they showed

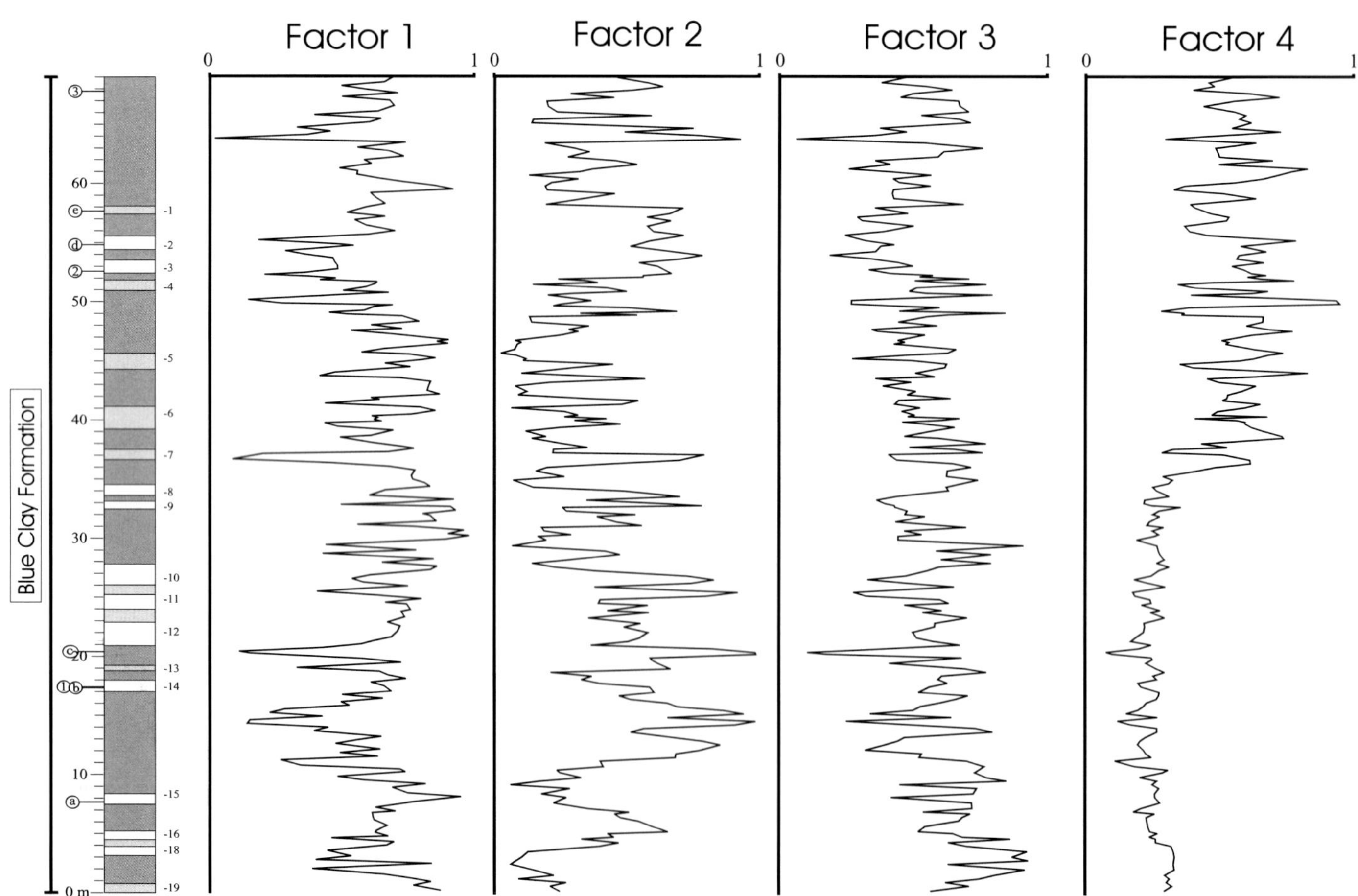

FIG 4.—Factor scores obtained from the benthic foraminiferal assemblage Q-mode factor analysis of the Blue Clay Formation in the Ras il-Pellegrin composite section. Lithology and biostratigraphic events are reported as in Figure 2.

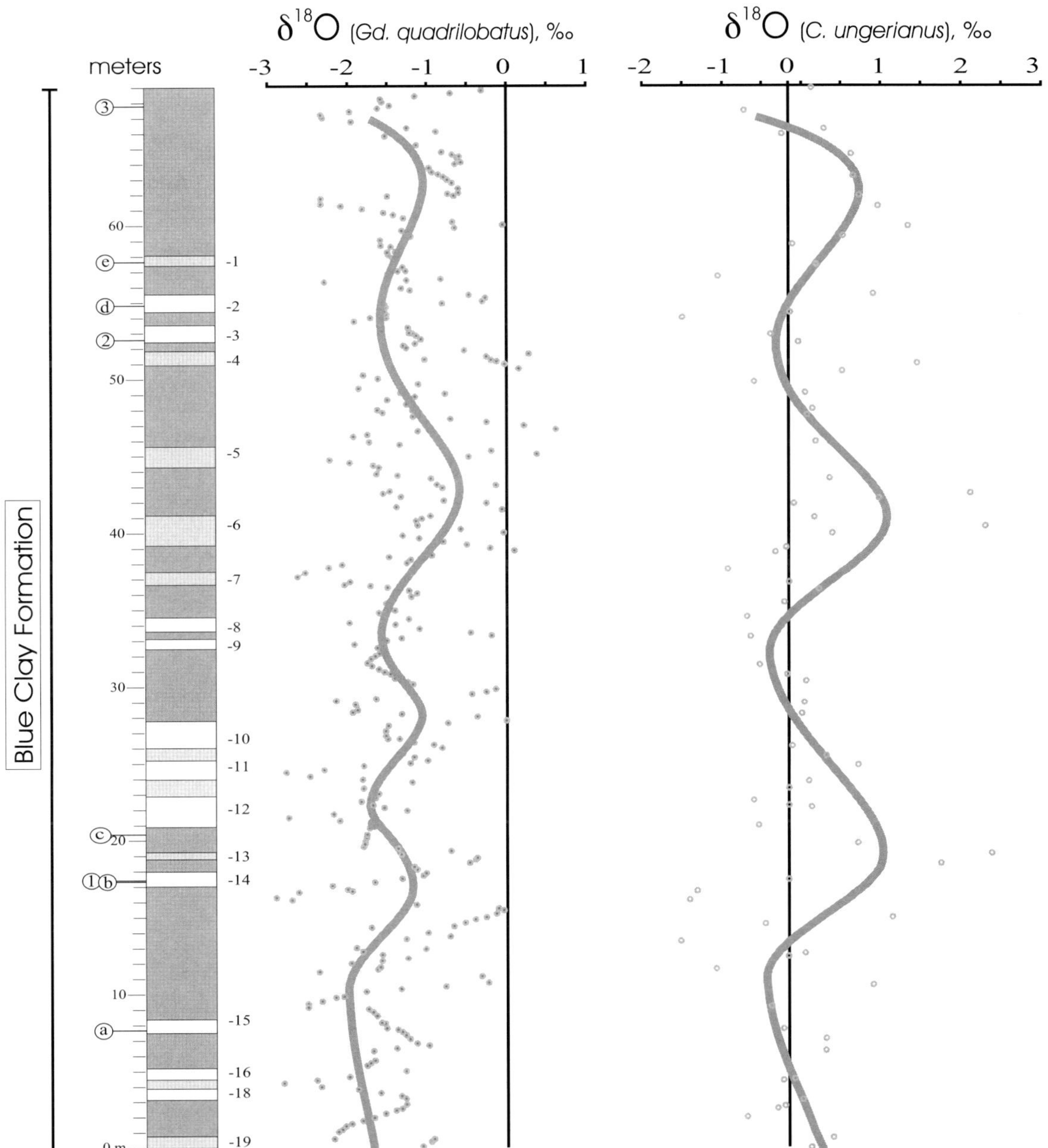

FIG 5.—Oxygen isotopes of the selected planktonic (*Gd. quadrilobatus*) and benthic (*C. ungerianus*) foraminiferal species in the Blue Clay Formation of the Ras il-Pellegrin composite section (Bellanca et al. 2002, modified). Gray lines represent two sixth-order polynomial curves highlighting long-term trends in the two isotope records. Lithology and biostratigraphic events are reported as in Figure 2.

the most nearly continuous distribution with the highest percentages throughout the section. In the present paper, we tested another methodological approach, performing spectral analysis on the factor signals, because in our opinion these signals could reflect the fluctuations in paleoenvironmental parameters better than the single species.

Spectral analysis was performed only on Factor 1 (loaded by *C. ungerianus* and *S. reticulata*) and Factor 2 (loaded by the *B. elongata* group), because both are clearly significant from a paleoecological point of view. Ecological information carried by the benthic foraminifera led us to interpret Factor 1 as a tracer of good ventilation of the basin bottom waters and, consequently, directly associated with high production of proto-MIW. Conversely, stressed conditions represented by Factor 2 allow us to associate this factor with reduced formation of proto-MIW.

Factors 3 and 4 were discarded because Factor 3 corresponds to ecological conditions intermediate between Factors 1 and 2, and Factor 4, recorded only in the upper part of the section, indicates very particular and locally controlled paleoceanographic conditions.

The F1 and F2 signals were first detrended using a Gaussian filter in order to subtract the long-term trend and then interpolated to 9 ky constant steps. Such a numerical approach ensures that more interpolated data in a given interval than original points are never present. Power spectra estimated for F1 and F2 signals (Fig. 6) show the presence of the two dominant periodicity bands of long-eccentricity and short-eccentricity components. This result suggests a direct control of the astronomic forcing on the modulation of the two factors.

The two time series were filtered in the 400 ky and 100 ky periodicity bands and compared directly with the astronomical eccentricity curve (Fig. 7). The comparison highlights a good match (in-phase and out-of-phase response of F1 and F2 signals, respectively) between the astronomic eccentricity curve and the filtered time series. Cross-spectral analysis of the insolation curve and the different time series shows high values of coherence in the 400 ky and 100 ky periodicity bands (Fig. 8). This confirms a linear link between the astronomical forcing in the eccentricity frequency bands and the faunal response.

Bellanca et al. (2002) pointed out an opposite behavior of the *C. ungerianus* curve with respect to the insolation curve. In this paper, instead the different time lags for the Milankovitch periodicity bands measured between the F1 signal (loaded by *C. ungerianus* and *S. reticulata*) and the insolation curve are interpreted as due to a slightly different response of the two *C. ungerianus* and F1 time series throughout the Blue Clay Formation in the Ras il-Pellegrin composite section as well as to a more adequate application of spectral analysis techniques to the available time series.

Assuming Factor 1, extracted from the total benthic assemblage of the Blue Clay Formation, to be a reliable tracer of high production of proto-MIW, we found periodic enhanced production of proto-MIW phased to maxima of the short and long eccentricity cycles (Fig. 7). During these time intervals, the Earth's orbital configuration corresponds to periods of largest excursion of the 19–23 ky insolation cyclicity. Thus, we can speculate that intervals of eccentricity maxima represent the periods of maximum production of proto-MIW in the Mediterranean basin. Furthermore, the coldest winter seasons, occurring during periods of precession minima and high eccentricity (in the 100 ky and 400 ky periodicity bands), should favor formation of proto-MIW.

Factor 2, indicative of stressed oxygen bottom conditions, should reflect times of maxima of precession cycles characterized by warmer winter seasons and consequently less intense production of proto-MIW.

Unfortunately, the sampling rate adopted for the analysis of benthic assemblage (~ 30 cm corresponding to ~ 9 ky) did not allow a reliable record of the 19–23 ky cyclicity from the Blue Clay Formation, and thus the mechanism for short-term forcing of proto-MIW remains at present to be further investigated.

CONCLUSION

In this paper we focus our attention on the Middle Miocene Mediterranean Intermediate Water, here defined as proto-MIW, pointed out in the 3D paleoceanographic model proposed by Bellanca et al. (2002). These authors suggested for this water mass hydrographic and hydrodynamic features similar to those presently recorded in the Levantine Intermediate Water (LIW), which is the major constituent of the Mediterranean Intermediate Water (MIW). Presently, MIW is considered to be a suitable mechanism to oxygenate the deep Mediterranean bottom waters, and therefore we speculate that proto-MIW played a similar role during the Middle Miocene.

The results of the cyclostratigraphic analysis of benthic foraminifera from the Blue Clay Formation, cropping out through the Ras il-Pellegrin composite section, suggest a link between selected benthic species and rate of the proto-MIW formation during the late Langhian–early Serravallian.

Principal-component analysis, performed on benthic data (Bellanca et al., 2002), singled out two significant factors, which are related to bottom-water circulation: Factor 1 (loaded by *C. ungerianus* and *S. reticulata*), interpreted as a tracer of oxic bottom waters, and Factor 2 (loaded by *B. elongata* group), interpreted as a recorder of stressed oxygen conditions.

Spectral analysis of these two factors indicated that signals are modulated by short and long eccentricity. In particular, the Factor 1 and Factor 2 curves are in phase and out of phase with maxima of the eccentricity, respectively.

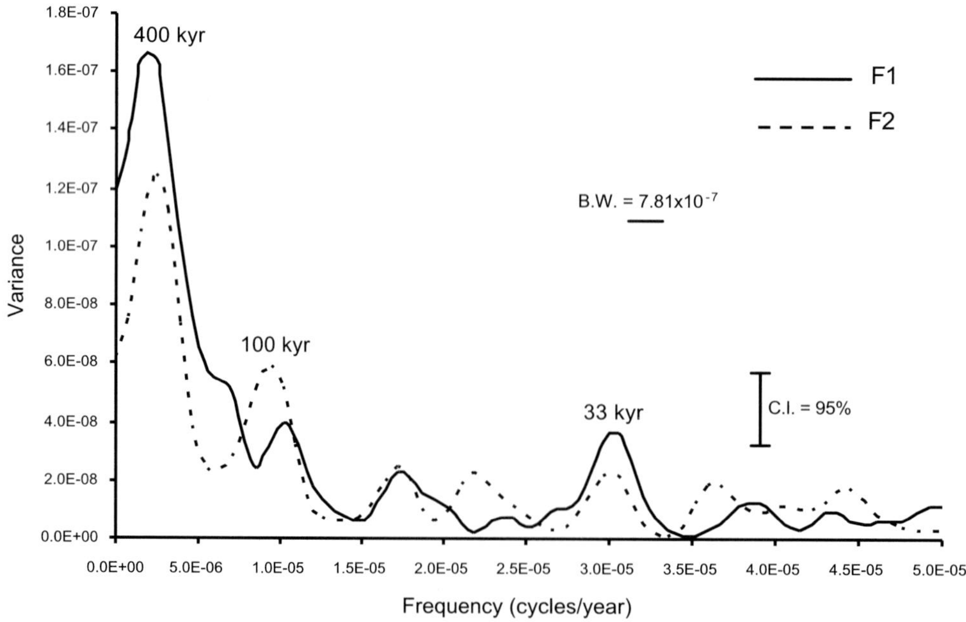

FIG 6.—Power spectra of the F1 (solid line) and F2 (dashed line) signals. Confidence interval (C.I.) and bandwidth (B.W.) are reported.

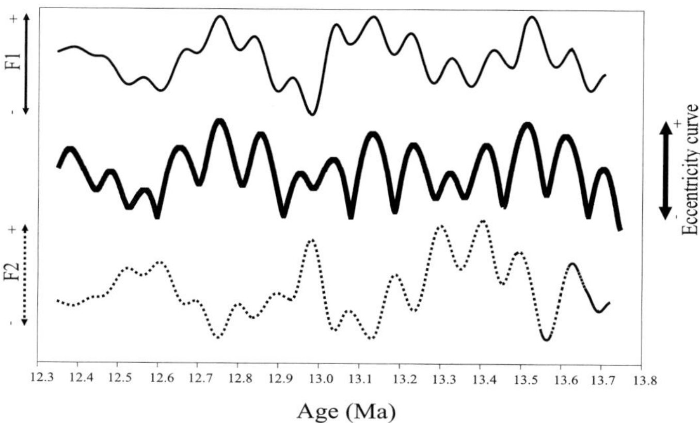

FIG 7.—Comparison between the 100 ky and 400 ky periodicitiy bands recognized in the eccentricity curve (thick line) and in the F1 (solid line) and F2 (dashed line) signals.

Because Factor 1 is indicative of oxic bottom waters, we assume that it is a reliable tracer of high production of proto-MIW during periods of high eccentricity, and probably, precession minima, characterized by the coldest winter seasons. Therefore, we infer a direct climate control, driven by the classic Milankovitch astronomical forcing, on the proto-MIW seasonal formation during the Middle Miocene. If such a climate–ocean interaction is confirmed by results of analysis of other Mediterranean Neogene sedimentary sequences, it should represent the direct oceanographic link between long-term astronomical forcing and deep-water response.

Our preliminary results offer a stimulating perspective for applying cyclostratigraphic methodology to benthic assembages. If the hypothesized oceanographic mechanism, triggering the abundance fluctuations of selected benthic species, responds directly to changes of climate-sensitive deep water masses, it should be possible to correlate the Milankovitch forcing to the periodicity of the benthic foraminifera signals. This methodological approach may provide a new tool to improve the astronomical calibration of geological time scales, integrating data on surface water masses (planktonic $\delta^{18}O$, carbonate abundance, planktonic foraminifera data, etc.) with information from the deep-water environment.

ACKNOWLEDGMENTS

Thanks are due to M. Mutti and M. Howell for their useful comments on an earlier version of the paper.

REFERENCES

BELLANCA, A., SGARRELLA, F., NERI, R., RUSSO, B., SPROVIERI, M., BONADUCE, G., AND ROCCA, D., 2002, Evolution of the Mediterranean Basin during the Late Langhian–Early Serravallian: an integrated paleoceanographic approach, *in* Iaccarino, S., ed., Integrated Stratigraphy and Paleoceanography of the Mediterranean Middle Miocene: Rivista Italiana di Paleontologia e Stratigrafia, v. 108, p. 223–239.

BENSON, R.H., 1973, An ostracodal view of the Messinian salinity crisis, *in* Drooger C.W., ed., Messinian Events in the Mediterranean: Amsterdam, North-Holland Publishing Company, p. 235–242.

BENSON, R.H., 1978, The paleoecology of the ostracodes of DSDP Leg 42A: Initial Reports of the Deep Sea Drilling Project, v. 42, p. 777–783.

BERGER, A., 1980, Milankovitch astronomical theory of paleoclimates: a modern review: Vistas in Astronomy, v. 24, p. 103–122.

BERGER, A., 1988, Milankovitch theory and climate: Reviews of Geophysics, v. 26, p. 624–657.

BIZON, G., AND BIZON, J.J., 1984, Ecologie des foraminifères en Méditerranée nord-occidentale. Les foraminifères des sédiments profond, *in* Ecologie des microorganismes en Méditerranée occidentale: "Ecomed", Pétrole et Techniques, v. 301, p. 104-139.

BONADUCE, G., AND BARRA, D., 2002, The ostracods in the paleoenvironmental interpretation of the Langhian–Serravallian section of Ras il-Pellegrin (Malta), *in* Iaccarino, S., ed., Integrated Stratigraphy and Paleoceanography of the Mediterranean Middle Miocene: Rivista Italiana di Paleontologia e Stratigrafia, v. 108, p. 211–222.

BORSETTI, A.M., IACCARINO, S., JORISSEN, F.J., POIGNANT, A., SZTRAKOS, K., VAN DER ZWAAN, G.J, AND VERHALLEN, P.J.J.M., 1986, The Neogene development of *Uvigerina* in the Mediterranean: Utrecht Micropaleontological Bulletin, v. 35, p. 183-236.

CITA, M.B., WRIGHT, R.C., RYAN, W.B.F., AND LONGINELLI, A., 1978, Initial Reports of the Deep Sea Drilling Project, v. 42, p. 1003–1035.

CITA, M.B., AND GRIGNANI, D., 1982, Nature and origin of Late Neogene Mediterranean sapropels, *in* Schlanger, S.O., and Cita, M.B., eds., Nature and Origin of Cretaceous Carbon-Rich Facies: London, Academic Press, p. 165–196.

COCCIONI, R., AND GALEOTTI, S., 1993, Orbitally induced cycles in benthonic foraminiferal morphogroups and trophic structure distribution pat-

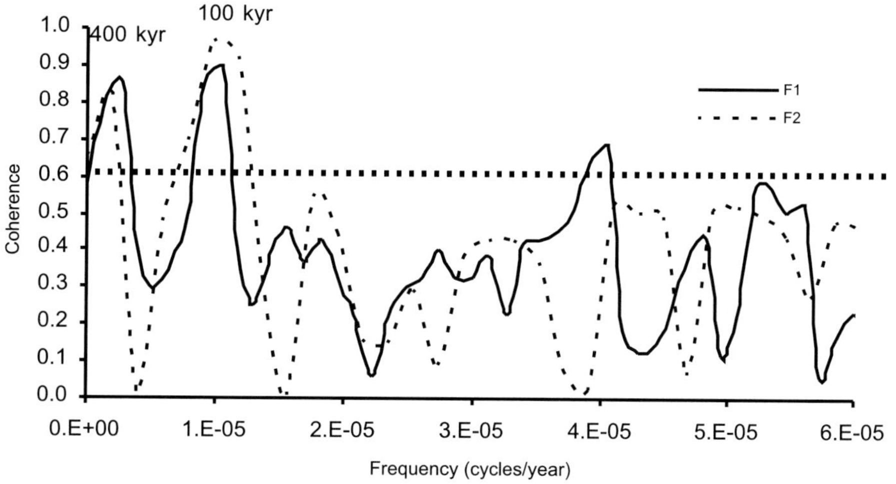

FIG 8.—Coherence values between F1 (solid line) and F2 (dashed line) signals and the eccentricity curve La90$_{(1,1)}$.

terns from the Late Albian "Amadeus Segment" (Central Italy): Journal of Micropaleontology, v. 12, p. 227–239.

Corliss, B.H., 1991, Morphology and microhabitat preferences of benthic foraminifera from the Northwest Atlantic Ocean: Marine Micropaleontology, v. 17, p. 195–236.

Dart, C.J., Bosence, W.J., and McClay, K.R., 1993, Stratigraphy and structures of the Maltese graben system: Geological Society of London, Journal, v. 150, p. 1153–1166.

Davis, J.C., 1986, Statistics and Data Analysis in Geology, Second Edition: New York, Wiley, 646 p.

Foresi, L., Bonomo, S., Caruso, A., Di Stefano, E., Salvatorini, G., and Sprovieri, R., 2002, Calcareous plankton high-resolution biostratigraphy (planktonic foraminifera and calcareous nannofossils) of the uppermost Langhian–Lower Serravallian Ras il-Pellegrin section (Malta), *in* Iaccarino S., ed., Integrated Stratigraphy and Paleoceanography of the Mediterranean Middle Miocene: Rivista Italiana di Paleontologia e Stratigrafia, v. 108, p. 195–210.

Fornaciari, E., Di Stefano, A., Rio, D., and Negri, A., 1996, Middle Miocene quantitative calcareous nannofossil biostratigraphy in the Mediterranean region: Micropaleontology, v. 42, p. 37–63.

Giannelli, L., and Salvatorini, G., 1975, I Foraminiferi planctonici dei sedimenti terziari dell'Arcipelago maltese. I. Biostratigrafia di "Blue Clay", "Green Sands" e "Upper Globigerina Limestone": Società Toscana di Scienze Naturali, Atti, Memorie, Serie A, v. 79, p. 49–74.

Hilgen, F.J., 1987, Sedimentary rhythms and high-resolution chronostratigraphic correlations in the Mediterranean Pliocene: Newsletters on Stratigraphy, v. 17, p. 109–127.

Hilgen, F.J., 1991, Extension of the astronomically calibrated (polarity) time scale to the Miocene/Pliocene Boundary: Earth and Planetary Science Letters, v. 107, p. 349–368.

Hilgen, F.J., Abdul Aziz, H., Krijgsman, W., Langereis, C.G., Lourens, L.J., Meulenkamp, J.E., Raffi, I., Steenbrink, J., Turco, E, van Vugt, N., Wijbrans, J.R., and Zachariasse, W.J., 1999, Present status of the astronomical (polarity) time-scale for the Mediterranean Late Neogene: Royal Society (London), Philosophical Transactions, A, v. 357, p. 1931–1947.

Hilgen, F.J., Krijgsman, W., Raffi, I., Turco, E., and Zachariasse, W.J., 2000, Integrated stratigraphy and astronomical calibration of the Serravallian/Tortonian boundary section at Monte Gibliscemi (Sicily, Italy): Marine Micropaleontology, v. 38, p. 181–211.

Hsü, K.J., and Bernoulli, D., 1978, Genesis of the Tethys and the Mediterranean: Initial Reports of the Deep Sea Drilling Project, v. 42, p. 943–949.

Imbrie, J., and Kipp, N.L., 1971, A new micropaleontological method for quantitative paleoclimatology: application to a Late Pleistocene Caribbean core, *in* Turekian, K.K., ed., The Late Cenozoic Glacial Ages: New Haven, Connecticut, Yale University Press, p. 78–181.

Jacobs, E., Weissert, H., and Shields, G., 1996, The Monterey event in the Mediterranean: a record from shelf sediments of Malta: Paleoceanography, v. 11, p. 717–728.

Johnson, D., 1985, Abyssal teleconnections II. Initation of Antarctic Bottom Water flow in the southwestern Atlantic, *in* Hsü, K., and Weissert, H., eds., South Atlantic Paleoceanography: Cambridge, U.K., Cambridge University Press, p. 283–325.

Jonkers, H.A., 1984, Pliocene benthonic foraminifera from homogeneous and laminated marls on Crete: Utrecht Micropaleontological Bulletin, v. 31, 179 p.

Kaiho, K., 1994, Benthic foraminiferal dissolved-oxygen index and dissolved-oxygen levels in the modern ocean: Geology, v. 22, p. 719–722.

Klovan, J.E., and Imbrie, J., 1971, An algorithm and Fortran-IV program for large scale Q-mode factor analysis and calculation of factor scores: Mathematical Geology, v. 3, p. 61–77.

Laskar, J., Joutel, F., and Boudin, F., 1993, Orbital precession and insolation quantities for the Earth from –20 Myr to +10 Myr: Astronomy and Astrophysics, v. 270, p. 522–533.

Lohmann, G.P., 1978, Abyssal benthonic foraminifera as hydrographic indicators in the western South Atlantic Ocean: Journal of Foraminiferal Research, v. 7, p. 7–34.

Lourens, L.J., Antonarakou, A., Hilgen, F.J., van Hoof, A.A.M., Vergnaud Grazzini, C., and Zachariasse, W.J., 1996, Evaluation of the Pliocene to early Pleistocene astronomical time scale: Paleoceanography, v. 11, p. 391–413.

Lutze, G.F., 1979, Benthic foraminifers of Site 397: faunal fluctuations and ranges in the Quaternary: Initial Reports of the Deep Sea Drilling Project, v. 47, Pt. 1, p. 419–431.

Lutze, G.F., 1986, *Uvigerina* species of the Eastern North Atlantic: Utrecht Micropaleontological Bulletin, v. 35, p. 21–46.

Lutze, G.F., and Coulbourn, W.T., 1984, Recent benthic foraminifera from the continental margin of the Northwest Africa: community structure and distribution: Marine Micropaleontology, v. 8, p. 361–401.

Mazzei, R., 1985, The Miocene sequence of the Maltese Islands: Biostratigraphic and chronostratigraphic references based on nannofossils: Società Toscana di Scienze Naturali, Atti, Memorie, Serie A, v. 92, p. 165–197.

Miao, Q., and Thunell, R.C., 1993, Recent deep-sea benthic foraminiferal distributions in the South China and Sulu Seas: Marine Micropaleontology, v. 22, p. 1–32.

Miller, K.G., and Lohmann, G.P., 1982, Environmental distribution of Recent benthic foraminifera on the northeastern United States continental slope: Geological Society of America, Bulletin, v. 93, p. 200–206.

Murray, J.W., 1991, Ecology and Palaeoecology of Benthic Foraminifera: New York, John Wiley & Sons, 397 p.

Ovchinnikov, I.M., and Plakhin, Y.A., 1984, Formation of the intermediate waters in the Mediterranean Sea: Oceanology, v. 24, p. 317–319.

Ozsoy, E., 1981, On the atmospheric factors affecting the Levantine Sea: European center for Medium Range Weather Forecasts, Technical Report 25, 29 p.

Pflum, C.E., and Frerichs, W.E., 1976, Gulf of Mexico Deep-Water Foraminifers: Cushman Foundation for Foraminiferal Research, Special Publication 14, 124 p.

POEM Group, 1992, General circulation of the Eastern Mediterranean: Earth-Science Reviews, v. 32, p. 285–309.

Pujos, M., 1976, Ecologie des foraminifères benthiques et des thécamoebiens de la Gironde et au plateau continental Sud-Gascogne. Application à la connaissance du Quaternaire terminal de la région Ouest-Gironde: Institut de Géologie du Bassin d'Aquitaine, Mémoires, no. 8, 274 p.

Russo, B., Sgarrella, F., and Gaboardi, S., 2002, Benthic foraminifera as indicators of paleoecological bottom conditions in the Serravallian Tremiti sections (Eastern Mediterranean, Italy), *in* Iaccarino, S., ed., Integrated Stratigraphy and Paleoceanography of the Mediterranean Middle Miocene: Rivista Italiana di Paleontologia e Stratigrafia, v. 108, p. 275–287.

Schnitker, D., 1979, The deep-water of the western North Atlantic during the past 24,000 years and re-initiation of the western boundary undercurrent: Marine Micropaleontology, v. 4, p. 265–280.

Schnitker, D., 1980, Quaternary deep-sea benthic foraminifers and bottom water masses: Annual Review of Earth and Planetary Sciences, v. 8, p. 343–370.

Schnitker, D., 1984, High resolution records of benthic foraminifers in the Late Neogene of the northeastern Atlantic: Initial Reports of the Deep Sea Drilling Project, v. 81, p. 611–622.

Schnitker, D., 1994, Deep-sea benthic foraminifers: food and bottom water masses, *in* Zahn, R., ed., Carbon Cycling in the Glacial Ocean: Constraints on the Ocean's Role in Global Change: NATO ASI series, v. 117, p. 539–554.

Sgarrella, F., and Moncharmont Zei, M., 1993, Benthic foraminifera of the Gulf of Naples (Italy): systematics and autoecology: Società Paleontologica Italiana, Bollettino, v. 32, p. 145–264.

SGARRELLA, F., SPROVIERI, R., DI STEFANO, E., CARUSO, A., SPROVIERI, M., AND BONADUCE, G., 1999, The Capo Rossello bore-hole (Agrigento, Sicily) cyclostratigraphic and paleoceanographic reconstructions from quantitative analyses of the Zanclean foraminiferal assemblages: Rivista Italiana di Paleontologia e Stratigrafia, v. 105, p. 303–322.

SPROVIERI, M., CARUSO, A., FORESI, L., BELLANCA, A., NERI, R., MAZZOLA, S., AND SPROVIERI, R., 2002, Astronomical calibration of a upper langhian / lower serravallian record from the Ras il-Pellegrin section (Malta island, central Mediterranean), *in* Iaccarino, S., ed., Integrated Stratigraphy and Paleoceanography of the Mediterranean Middle Miocene: Rivista Italiana di Paleontologia e Stratigrafia, v. 108, p. 183–193.

SPROVIERI, R., 1992, Mediterranean Pliocene biochronology: a high resolution record based on quantitative planktonic foraminifera distribution: Rivista Italiana di Paleontologia e Stratigrafia, v. 98, p. 61–100.

STREETER, S.S., 1973, Bottom water and benthonic foraminifera in the North Atlantic Glacial–Interglacial contrasts: Quaternary Research, v. 3, p. 131–141.

STREETER, S.S., AND SHACKLETON, N.J., 1979, Paleocirculation in the deep north Atlantic: 150,000 yr. record of benthic foraminifera and oxygen-18: Science, v. 203, p. 168–170.

THUNELL, R., RIO, D., SPROVIERI, R., AND RAFFI, I., 1991, Limestone–marl couplets: origin of the early Pliocene Trubi marls in Calabria, southern Italy: Journal of Sedimentary Petrology, v. 61, p. 1109–1122.

VAN DER ZWAAN, G.J., 1982, Paleoecology of Late Miocene foraminifera: Utrecht Micropaleontological Bulletin, v. 25, 201 p.

VAN DER ZWAAN, G.J., JORISSEN, F.J., VERHALLEN, P.J.J.M., AND VON DANIELS, C.H., 1986, *Uvigerina* from the Eastern Atlantic, North Sea Basin, Paratethys and Mediterranean: Utrecht Micropaleontological Bulletin, v. 35, p. 7–20.

VAN LEEUWEN, R.J.W., 1986, The distribution of *Uvigerina* in the Late Quaternary sediments of the deep eastern south Atlantic: Utrecht Micropaleontological Bulletin, v. 35, p. 47–50.

VERHALLEN, P.J.J.M., 1991, Late Pliocene to early Pleistocene mediterranean mud-dwelling foraminifera: influence of a changing environment on community structure and evolution: Utrecht Micropaleontological Bulletin, v. 40, 219 p.

WOODRUFF, F., AND SAVIN, S.M., 1991, Mid-Miocene isotope stratigraphy in the deep-sea: high resolution correlations, paleoclimatic cycles, and sediment preservation: Palaeoceanography, v. 6, p. 755–806.

WRIGHT, R., 1978, Neogene benthic foraminifers from DSDP Leg 42 A, Mediterranean Sea: Initial Reports of the Deep Sea Drilling Project, v. 42, p. 709–726.

WÜST, G., 1961, On the vertical circulation of the Mediterranean Sea: Journal of Geophysical Research, v. 66, p. 3261–3271.

CYCLOSTRATIGRAPHY AND CHRONOLOGY OF THE ALBIAN STAGE (PIOBBICO CORE, ITALY)

ALESSANDRO GRIPPO
Department of Earth Sciences, University of Southern California, Los Angeles, California 90089-0740, U.S.A.
e-mail: grippo@earth.usc.edu
ALFRED G. FISCHER
Department of Earth Sciences, University of Southern California, Los Angeles, California 90089-0740, U.S.A.
LINDA A. HINNOV
Department of Earth and Planetary Sciences, Johns Hopkins University, Baltimore, Maryland 21218, U.S.A.
TIMOTHY D. HERBERT
Department of Geological Sciences, Box 1846, Brown University, Providence, Rhode Island 02912, U.S.A.
AND
ISABELLA PREMOLI SILVA
Dipartimento di Scienze della Terra, Università di Milano, via Mangiagalli 34, 30133 Milano, MI, Italy

ABSTRACT: The mid-Cretaceous (Albian) deep-water sediments (coccolith–globigerinacean marls) of the Umbria–Marche Apennines show complex rhythmic bedding. We integrated earlier work with a time-series study of a digitized and image-processed photographic log of the Piobbico core. A drab facies is viewed as recording normal stratified conditions, and a red facies as the product of downwelling warm saline (halothermal) waters. Both are pervaded by orbital (Milankovitch) rhythms. These reflect fluctuations in the composition and abundance of the calcareous plankton in the upper waters. The drab facies is overprinted by redox oscillations on the bottom, including episodic precessional anaerobic pulses (PAPs). Contrasts between the individual beds representing the alternate phases of the precessional rhythm rose and fell with orbital eccentricity, in the classical pattern of Berger's climatic precession or precession index curve, varyingly complicated by the obliquity rhythm. We conclude that greenhouse oceans in general, and perhaps this area in particular, were very sensitive to orbital forcing. Our count of 29 406-ky eccentricity cycles yields an Albian duration of 11.8 ± 0.4 My.

INTRODUCTION

The Umbria–Marche arc of the Apennines, in the region west of Ancona, contains a continuous history of pelagic–hemipelagic sedimentation extending from Early Jurassic times into the Miocene (Cresta et al., 1989). It is here (at Gubbio) that the impact nature of the Cretaceous–Tertiary boundary was discovered, and that continuous stratigraphic profiles of Cretaceous–Paleogene magnetic zonation were developed. The Cretaceous biostratigraphy of the region is summarized in Premoli Silva and Sliter (1995).

The Albian Stage is here represented by 50–60 m of pelagic sediment, initially coccolith–globigerinid ooze and marl, now compacted into rhythmically alternating beds of marlstone and limestone that form the Scisti a Fucoidi (Fucoid Marls) and the basal beds of the succeeding Scaglia Bianca Limestone.

History of Study

The rhythmicity was first studied in discontinuous outcrops by de Boer (1982,1983) and de Boer and Wonders (1984). A stratigraphic thickness of 50–60 m, representing a stage with an estimated duration of ca. 12 million years, implies a mean accumulation rate of 4–5 Bubnoff units (mm/ky, m/My). That made the ca. 8 cm bedding couplets likely candidates for the precessional cycle, and their grouping into bundles of five a likely expression of the short-eccentricity cycle. Schwarzacher and Fischer (1982) had reached similar conclusions about cyclic patterns in the underlying Barremian and overlying Cenomanian limestones.

Continuity of sequence was obtained in the Piobbico core, drilled by Premoli Silva, Napoleone, and Fischer through the Scisti a Fucoidi at the Le Brecce farm northwest of Piobbico (Fig. 1). Pratt and King (1986) described the composition of the organic matter. Detailed studies mainly directed at an 8 m core segment (cycles 8–12, Fig. 16) yielded calcium carbonate profiles and a gray-scale scan by densitometry of diapositives (Herbert and Fischer, 1986, Herbert et al., 1986). These yielded new insights into rhythmicity and sedimentation, and the gray-scale scan showed reflectivity to be an excellent proxy for calcium carbonate content, save for a step function from gray to black associated with black marlstone beds (the PAPs of this paper). Spectral studies by Park and Herbert (1987), Premoli Silva et al. (1989a), and Premoli Silva et al. (1989b) confirmed the assignment of cyclicities and discovered the presence of an obliquity signal. The stratigraphy of the core was described by Erba (1988) and Tornaghi et al. (1989).

The chemical and gray-scale analyses employed had proved too tedious for large-scale application to cyclostratigraphic studies. The visual count of the short-eccentricity bundles in core and outcrop by Herbert et al. (1995) yielded an Albian duration of 11.9 My, but it remained undocumented.

By then, digitization of photographs and programs for time-series analysis had brought the entire core into reach of study and documentation. Fischer and Grippo, with the aid of Hinnov, undertook a restudy of the core photographs (for details, see Appendix). By means of a photolog, gray-scale log, and spectral analyses we were able to document the cyclicity in the ca. ten million years of history recovered by the core, and we extended this to the full Albian by extrapolation to the surface data of Herbert et al. (1995).

While these studies were in progress, Fiet et al. (2001) measured a surface section of the Albian on Monte Petrano, some 13

Cyclostratigraphy: Approaches and Case Histories
SEPM Special Publication No. 81, Copyright © 2004
SEPM (Society for Sedimentary Geology), ISBN 1-56576-108-1, p. 57–81.

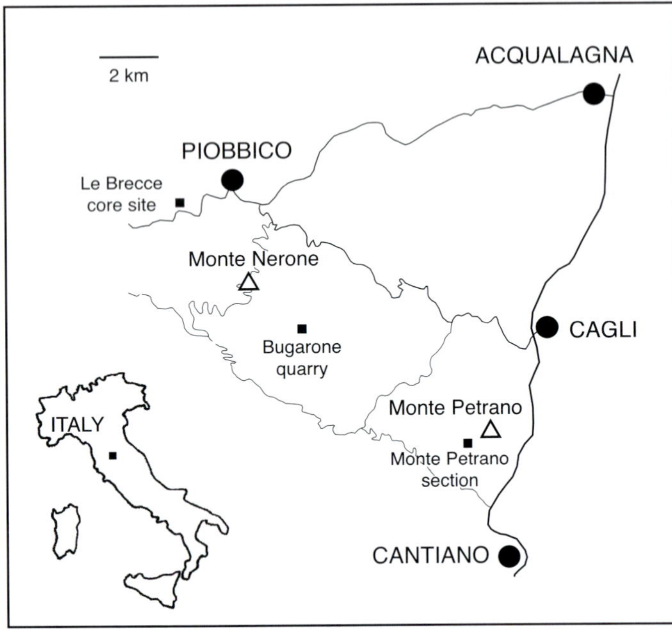

FIG. 1.—Index map.

km to the southeast of our drill site. Although some distinctive lithic and faunal markers provide a general correlation, an overall cycle-to-cycle correlation is not achieved, though in the end their assignment of 11.4 My to the Albian is only one 406 ky cycle short of ours.

Conventions

We here use the terms *cyclicity* or *rhythm* for the general process that, plotted through time, appears as a wave train, and we reserve *cycle* for the individual wave.

The need to refer to individual cycles made it necessary to devise a system of numeration. Superbundles corresponding to 406 ky eccentricity E-cycles are numbered from the top of the Albian downward (Figs. 4, 15, 16), and bundles for each superbundle are lettered a, b, c, d in upward progression. The core begins in cycle 3. Spectral coverage extends from the top of cycle 8 downward, and the base of the Albian lies somewhere between the upper part of cycle 29 and the lower part of 31, and is tentatively placed in cycle 30 (Fig. 15).

We give radiometric age estimates in kiloyears and million years as ka and Ma, and stratigraphic ones, including those derived with astronomic help, as ky and My.

SCISTI A FUCOIDI

The Scisti a Fucoidi represent coccolith–foraminiferal marls, comparable to modern marly globigerinid ooze, deposited at a depth estimated at ca. 2 km (Premoli Silva and Sliter, 1995).

The mean accumulation rate (compacted) was on the order of 4 Bubnoff units (mm/ky, m/My). Restriction of calcareous fossils to those initially calcitic indicates deposition below aragonite compensation level but generally above the calcite lysocline.

All of the Tethyan foraminiferal and nannofossil zones of the Albian are represented, and deposition is thought to have been essentially continuous (Premoli Silva and Sliter, 1995), extending from its base in the overlap between the nannofossil *Predicosphaera columnata* and the foraminiferal *Hedbergella planispira* zones, to its top at the base of the *Rotalipora brotzeni* zone.

Sedimentation

Our understanding of sedimentation goes back to the observations of de Boer (1982, 1983) and the study of an 8 m segment (our cycles 8–12) by Herbert and Fischer (1986) and Herbert et al. (1986), supplemented by other observations in outcrop and on the core (e.g., Tornaghi et al., 1989).

The initial sediment consisted mainly of coccoliths, planktonic foraminifera, a minor phase of biogenic silica partly or wholly attributable to radiolarians, and terrigenous siliciclastic matter (probably eolian dust) (de Boer, 1982, 1983; de Boer and Wonders, 1984; Herbert and Fischer, 1986; Herbert et al., 1986). The dust supply appears to have been comparatively constant, whereas the supply of skeletal calcite fluctuated, in rhythmic fashion, to yield stratification. Beds are planar, and bedding planes are generally not sharply defined but gradational. The more calcareous beds contain, at odd intervals, centimeter-scale lenses of cross-bedded radiolarite, representing ripple-drifted radiolarian ooze. In these the tests have largely been calcified while the silica moved to fill the interior as chalcedony. (Fischer, personal observations). Excess silica in the limestones suggests that biogenic silica, generally not visibly preserved, fluctuated along with carbonate (Herbert et al., 1986).

These findings, based on detailed study of parts of the sequence, appear applicable to all of it, but are in no sense complete, and our interpretations thus remain generalized. Isotopic records are strongly overprinted by pervasive calcite cement.

While thus relatively simple in basic constituents, the stratigraphic sequence is dramatically differentiated in color, ranging from greenish gray to white, to black, and to red (Figs. 2, 4), in response to wide fluctuations in depositional redox conditions.

Drab Facies.—

Most of the sequence takes the drab form (Fig. 2). Greenish-gray beds of marlstone grade on the one hand into limestones, whitish by virtue of the reflectance of calcite. On the other hand, the greenish marls grade into thin black marlstones colored by organic matter and iron sulfide, and representing precessional anoxic pulsations (PAPs).

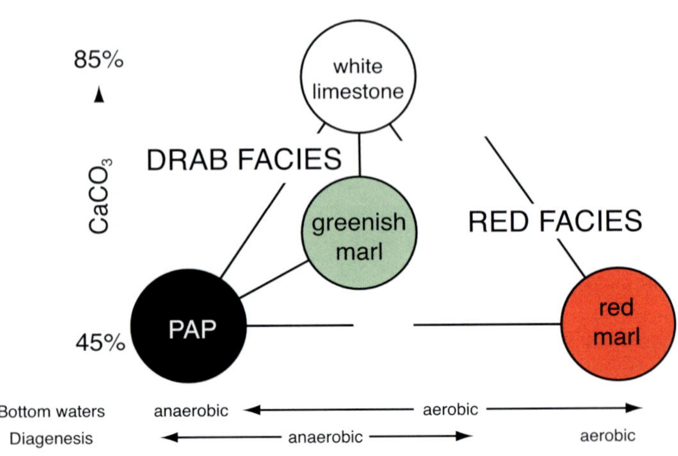

FIG. 2.—Lithic phases and inferred aeration states.

Carbonate Oscillations.—The initial calcite constituents were of two main sorts: coccoliths, and planktonic foraminifera. The purest limestones are composed mainly of coccoliths, planktonic foraminifera being suppressed in numbers and restricted to smaller, simpler forms (hedbergellids). Planktonic foraminifera reach greatest abundance and diversity in the gray-green marls and PAPs, where they attain the full complement of the Tethyan faunas, including the large and more highly ornamented species (Premoli Silva et al., 1989a; Premoli Silva et al., 1989b).

Bioturbation.—Our information on bioturbation is based on observations by Erba and Premoli Silva (1994), confined largely to cycles 8–12, but appear applicable to the entire sequence by general observations in the field and on the core. The ichnofauna lacks the largest bioturbators such as *Thalassinoides*, probably because of water depth, but it contains *Chondrites* (the "fucoids" from which the formation derives its name), as well as *Teichichnus* and *Zoophycos*. Varied in the more calcareous beds, it becomes restricted to *Chondrites* in the more clay-rich members and tends to dwindle to small *Chondrites* at the boundary with PAPs.

The variation in ichnofauna thus parallels that found by Savrda and Bottjer (1994) in the Niobrara Formation in Colorado. There, diversity of ichnofauna, size of burrows, and diversity of shelly benthos are associated with oxygenated seafloors of the carbonate facies. They decline to *Chondrites* in the dysaerobic shales and are lacking in the anaerobically deposited black shales. It would appear that the limestones in the Scisti a Fucoidi were deposited on moderately well aerated bottoms, which turned dysaerobic with the deposition of the greenish marls.

Precessional Anoxic Pulsations (PAPs).—Episodic intensification of this redox oscillation produced anoxia or near-anoxia as recorded in black marls, which may occupy part or all of the marly bed. These PAPs have generally been referred to as black shales, but, lacking fissility and retaining considerable carbonate, are marlstones. Their carbonate content averages lower than that of the greenish-gray marls (Herbert and Fischer, 1986), owing to a reduced content of coccoliths (Erba, 1988, 1992). The organic-matter content generally lies in the 1–2% range, and the intensity of pigmentation suggests presence of elemental carbon, supplemented by iron sulfides (de Boer, 1983; Herbert and Fischer, 1986). The average hydrogen content of the organic matter is generally low, and the abundance of wax-derived n-alkanes in the extractable compounds (Pratt and King, 1986) suggests that soot may constitute up to 50% of the organic fraction. Lack of fine laminations implies mixing on a microscopic scale, implying dysaerobic episodes when small nektonic animals may have disturbed the bottoms or when very small benthonic ones obtained temporary footholds.

Urbino Anoxic Event.—Distinct from the PAPs is the Urbino marlstone bed in our cycle 27 (Fig. 16). In this, persistence of anaerobic conditions formed a 40 cm bed of thinly laminated black marlstone. In persistence of anoxia and in its high organic and hydrogen content (Pratt and King, 1986) it resembles the Selli (OAE1a) unit of the lower Aptian and the Bonarelli (OAE2) unit of the uppermost Cenomanian, but unlike these it contains considerable calcite in the form of recrystallized laminae of planktonic foraminifera, and it is not cherty. It appears to represent an appreciable fraction of a 406 ky eccentricity cycle, an episode of prolonged anaerobic conditions. Its timing is roughly that of OAE1b, but its identity with this remains in doubt.

Red Facies.—

The red facies also consists of coccolith–globigerinid marls, showing a similar rhythmicity in oscillating carbonate contents and foraminifer/coccolith ratios, but its carbonate values remain below those attained in the drab facies. The redox cycle is generally lacking; these sediments were deposited in well-aerated water, in which all of the reactive iron minerals were converted to the ferric form. Furthermore, little or no reactive organic carbon was buried, so that the iron remained ferric through diagenesis. The degree of redness varies, there being a gradation from greenish gray marls through units tinged by pink and lavender, as in cycle 13 (Fig. 4, 16), to bright red marls and (rare) decalcified maroon mudstones. The amplitude of carbonate variation is reduced from that shown in the drab facies, and PAPs are missing except for a few in the transitional facies.

The red facies is also set apart by lack of bioturbation. In the drab facies the vigor and diversity of bioturbators grew with oxygen supply. One might thus expect the red facies to show an even more diverse ichnofauna, with conspicuous large burrows, but this is not the case. No conspicuous burrows were noted in field and core (Fischer, personal observations). The sediments are not conspicuously laminated, suggesting that mixing occurred on a fine scale, but the common presence of individual fine foraminiferal laminae with centimeter and sub-centimeter spacing implies little large-scale burrow mixing. We conclude that macroburrowers, abundant at this depth in the flysch facies, were scarce on these bottoms, limited not by oxygen but by scarcity of food.

Condensed Mudstones.—A variant of the red facies is represented by maroon mudstones, condensed zones of carbonate dissolution in the lower part of the succession. They lack microfossils and appear to be dissolution zones, stratigraphically condensed.

The red–drab alternations seem unrelated to the pattern of orbital variations.

NEW CORE SCAN

The Piobbico core (Fig. 1, 16) was drilled in 1982 on the Le Brecce farm some 2 km west of Piobbico on the road to Apecchio (Tornaghi et al., 1989). It yielded 44 m of Albian strata, missing the uppermost 6 m. It obtained excellent recoveries except for gaps in its uppermost 5.5 m, and missed 2 m of the middle Albian cut out by a fault (Herbert et al., 1995) (Fig. 16).

Soon after drilling, the core was split by diamond saw and divided into < 30 cm segments. Those of the archive set were embedded in plastic and smoothed with 600 mesh carborundum. Acetate replication of this surface after a light etch by hydrochloric acid provided microscopic access to the entire core, allowing precise delimitation of biostratigraphic zones. Each of these segments was photographed on Ektachrome film, with sub-millimeter resolution. These pictures served as the basis for the new analysis.

Scans and Logs

Our return to the Piobbico core was motivated by the desire to test more rapid yet quantitative techniques for extracting and processing cycle information. Digitization of pictures, and software for rapid gray-scale scanning and time-series processing, allowed us to undertake a reanalysis (for details, see Appendix).

Photolog.—

We digitized the pictures and cleared a path across each of nonstratigraphic information. We concatenated these paths into a single stratigraphically continuous scanning path, the consecutive pixels of which form, in theory, a historical series of observations (samples). This, the colored series, was corrected for dip to true stratigraphic depth and printed as a *photolog* (Figs. 4, 16A), with a pixel density equivalent to 6 mm of rock.

Gray-Scale Series.—

We then converted the colored series into a gray-scale series that served as the basis for subsequent calculations. Small gaps due to incomplete recovery render the upper 5.5 m unsuitable for detailed analysis, but from this point down the entire record was analyzed, with exception of the two meters excised by the fault.

Gray-Scale Log.—

The gray-scale series can be printed out as a gray-scale log, which, for greater visual impact, we present in mirrored form (Figs. 4, 16B). The convention we chose was to mirror the plot on black, so as to spread the limestones and contract the PAPs, which thus come to segment the image into the ca. 40 cm bundles. The cyclicity can be pictured in quite the opposite way (Fig. 13B), with white in the center, contracting limestones and expanding the PAPs into spikes.

Power Spectra

In order to obtain a more objective view of the cyclicities we turn to the frequency domain of power spectra. We employed tapered Thomson spectra, well adapted for study of irregularly time-sampled series.

The spectrum of the entire scan (Fig. 3A) shows groups of peaks, at certain stratigraphic spacings. Peaks concentrated at the extreme (low frequency) left have little significance ("red noise" enhanced by the logarithmic scale). The group in the 30–60 cm bracket corresponds to the stratigraphic bundle, identified by de Boer (1982), Herbert and Fischer (1986), and Park and Herbert (1987) with the 95 ky eccentricity cycle. The prominent peak at ca. 20 cm corresponds to no stratigraphic feature observed in field or core, but to the obliquity cycle found spectrally by Park and Herbert (1987) and Premoli Silva et al. (1989b). A scattering of peaks in the 9–14 cm region corresponds to the couplets identified by the same authors with the two precessional modes. The main periodicities in this spectrum are thus consonant with those previously found and identified as the major elements of the orbital variations,

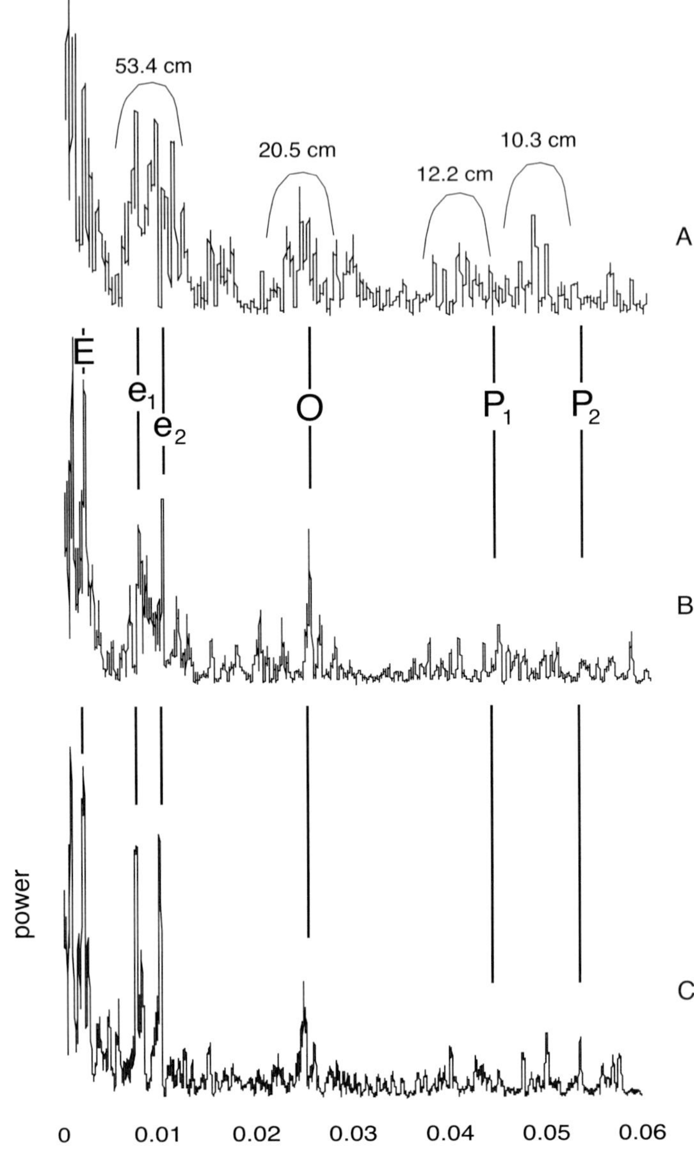

FIG. 3.—Power spectra of full time series. **A)** Untuned spectrum, referred to stratigraphic space. **B)** Spectrum tuned to 95 ky bundle. **C)** Spectrum tuned to 406 ky superbundle or its equivalent. E, predicted 406 ky (long) eccentricity rhythm; e_1 and e_2, predicted modes of the short (ca. 95 ky) eccentricity rhythm; O, predicted obliquity rhythm; P_1 and P_2, predicted modes of the ca. 20 ky precessional rhythm.

Tuning.—

The multiplicity of peaks in Figure 3A derives mainly from variations in accumulation rate, i.e., from deviations of the stratigraphic space dimension from time. "Tuning" involves the choice of some consistently recorded cyclicity, whose individual cycles in the data series can be stretched or condensed to the same length, in essence moving them out of the space domain into the time domain. The eccentricity values were taken to be the modern ones, as computed by Laskar for the last 19 My (1999). For obliquity and precession we use the Berger et al. (1989) estimates for 95 Ma, 38.9 ky for obliquity, and 22.3 and 19.5 ky for the precessional mode.

Success or failure of tuning can be judged by whether tuning to any one cyclicity helps to align the others or scatters them even more widely, and by whether the resulting alignment of peaks corresponds to the astronomic predictions. Choice of the periodicity to tune to is obviously limited to one that is consistently present and shows well-defined cycle boundaries in the gray-scale log. It should be unimodal, because choice of a bimodal cyclicity would introduce a pseudobimodality into the others, particularly those of higher frequency.

The precessional rhythm fails to qualify. The astronomical signal is strongly bimodal, and its preservation in the series is quite irregular (Figs. 4, 5), with cycles apparently lost to bioturbation and in other cases added by double beats.

The obliquity cycle is very stable in period but is commonly lacking, and its cycle boundaries are masked by interference from the precession–eccentricity syndrome.

The short-eccentricity rhythm is consistently recorded in our series and is thus practical to tune to. The thus-tuned spectrum (Fig. 3B) shows considerable simplification. The peak for the long eccentricity emerges clearly, as do the two modes of the short-eccentricity rhythm. The 20 cm peak is now an excellent match for the obliquity cycle. Couplet peaks remain scattered but show a semblance of two groups corresponding to P_1 and P_2. Some distortion, however, has been introduced by tuning to a bimodal cyclicity.

The long-eccentricity rhythm is very stable at 406 ky, but its sporadic appearance precludes tuning to it directly. But it sets a frame into which the short-eccentricity signals can be fitted. In a complete series, every fourth (occasionally fifth) e-cycle boundary should come close to matching an E-cycle boundary, and this made it possible to extrapolate E-cycle boundaries for the entire sequence, to yield the E-cycle sequence shown in Figures 15 and 16, numbered from the top of the Albian down. This exercise necessitated additions of four previously unrecognized 95 ky cycles, obscured by condensed sedimentation. The resulting spectrum (Fig. 3C) shows further simplification. Peaks for E, e_1, e_2, and O remain aligned with predicted values. The couplet peaks are much simplified from Figure 3B, presumably because no longer doubled by tuning to the bimodal 95 ky cycle, but their fit to the precession estimates is not improved.

Precessional Rhythms

The "precession of the equinoxes" derives from the gyration of Earth's rotational axis. This describes a cone with a period of ca. 25.671 ky relative to the stars. Its influence on climates depends on its interaction with the ellipticity (technically eccentricity) of Earth's (planet 3) orbit. The axis of that orbit, as defined by perihelion (closest approach to the Sun) moves counter to the precessional rotation, cutting the length of the precessional cycle to 19.104 ky. Its orbit is further distorted by the gravitational attractions of Mercury (1), Venus (2), Mars (4), and Jupiter (5), each of which has its own period or frequency, expressed as the sine of its longitude relative to the moving perihelion. Jupiter and Mercury combine to provide a strong precessional periodicity at ca. 23 ky (the P_1 mode) while Earth and Mars combine to form one at ca. 19 ky (P_2). Bracketing the effect of Venus, these modes are commonly thought of as combined into a precessional cyclicity at ca. 21 ky.

As the orbital eccentricity waxes and wanes, it brings changes of insolation that find expression mainly in seasonality. The more circular orbits produce moderate summers and winters. Higher eccentricity brings more intense summer heat and winter cold in the half of the cycle in which perihelion in summer, while during the other (winter-perihelial) half it warms winters and cools summers beyond those of a circular orbit. When perihelion occurs in the summer (and aphelion in the winter), the seasons in that hemisphere are intensified. At the same time those in the other hemisphere are damped. In the succeeding phase these relationships are reversed. The intensity or amplitude of these climatic deviations is proportional to the amount of eccentricity, which varies as described below.

de Boer (1982) and Herbert and Fischer (1986) reckoned that if this Albian sequence of ca. 50 m had been continuously deposited for an estimated 12 My of Albian time, for an accumulation rate of ca. 4 Bubnoff units (mm/ky), then the 6–8 cm couplets would approximately match the precession, and the ca. 30–50 cm bundles would match the ca. 100 ka short cycle of eccentricity. As shown below, the inference is now supported by close comparisons of the spectral signatures.

The "elementary cycle" is one of ca. 8 cm *couplets* that record a pulsation in the rate of carbonate production, attributable to variations in the vigor of coccolithophoracean algae in the photic zone. Not only did the vigor of coccolith blooms vary, but so did the composition of the flora, with species thought to be diagnostic of higher fertility (upwelling) confined to the more calcareous beds (Erba, 1992).

This cyclicity pervades both the drab facies and the red facies. In the drab facies it is joined by a variation in redox conditions on the sea floor. The more calcareous beds show strong bioturbation in which *Zoophycos* and *Planolites* play a role, evidence of moderately good aeration. In the marly members the ichnofauna became restricted to *Chondrites*, indicating dysaerobic conditions, and at times reached anoxia in precessional anoxic pulsations (PAPs) evidenced by black coloration.

In the gray-scale log the precessional couplets appear as high-frequency serrations (Figs. 4, 5, 16). Strength and regularity of the couplets vary, with PAPs deeply segmenting the gray-scale curve. Some couplets have been amalgamated by bioturbation (Fig. 5). In the red facies the couplets show a lower-amplitude variation than in the drab, corresponding to a lesser contrast in carbonate content. In decalcified condensed maroon clays (in cycles 27, 29, and 30) the precessional couplets are wiped out. On the other hand, the transitionally red marls of cycles 13 and 14 preserve the number of precessional couplets most faithfully, owing to the smaller degree of bioturbation (Fig. 5).

Occasional couplets or groups of couplets have a bifid signature, suggesting a precessional double beat previously noted in the precessional cycles at low latitudes, where the counterphased precessional cycles of the two hemispheres interact (Park et al., 1993).

As described above, the decrease in deep-water oxygenation at times became intensified to the point of anoxia, so that part or all of this marl bed turns black, with elevated organic content and iron sulfide. These thin precessional anoxic pulsations (PAPs) segment the stratal sequence in an eye-catching manner (Figs. 4, 16). Their cyclic distribution (Fig. 13) is discussed below. Ash-gray (rather than greenish-gray) beds, shown in Figures 4 and 16, appear to represent relicts of such PAPs. Seasonal anoxia may have produced an alternation of anoxic and aerated times that, with gentle bioturbation in the aerated intervals, yielded a mixed product. But even black marls one to two centimeters thick were threatened by subsequent bioturbation, as shown in outcrop, where, in the Monte Petrano section, a PAP was laterally terminated by dense *Chondrites* burrowing from above (Fischer, personal observations). Either process could readily have left enough carbon particles in the resulting sediment to give it a distinctive ash-gray color. We speak of such beds as inferred PAPs.

The tuned spectrum (Fig. 3C) does indeed show a group of peaks in the estimated precessional spectral band, but these are scattered as compared to the peaks for eccentricity and obliquity. When examined in smaller time slices (Fig. 6), about half of the spectra show two well-defined peaks, in some characteristically paired into the two modes, but slightly displaced toward higher and lower frequencies. The tuning did not, of course, correct for distortions by variations in accumulation rate at frequencies beyond the 406 ky level. Existence of such uncorrected variations is demonstrated in Figure 8. Here individual 95 ky eccentricity bundles show time distortions of a factor of two, much larger than

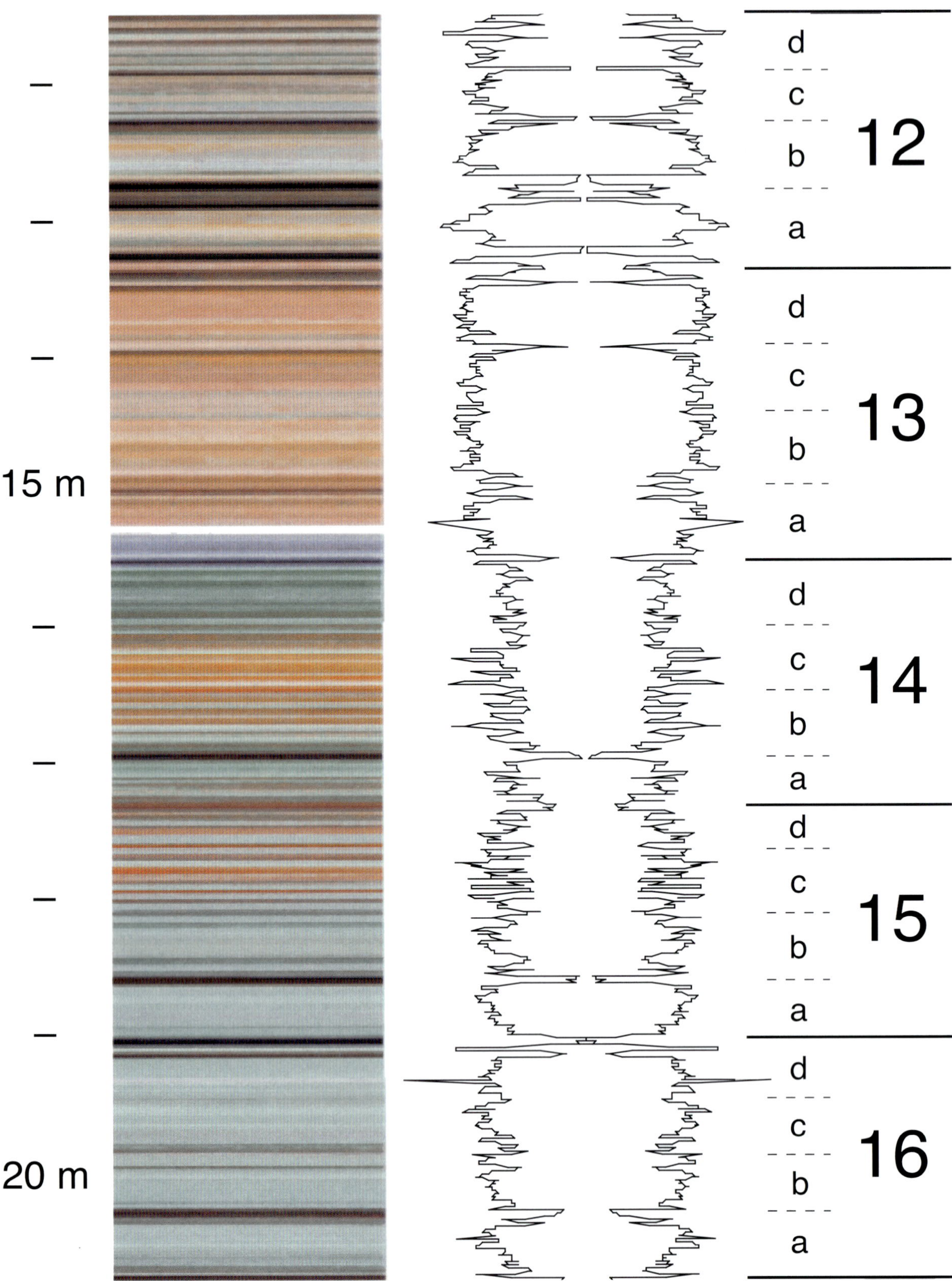

FIG. 4.—Photolog and gray-scale log of a ca. 2 My section of core, comprising superbundles 12–16. Gray-scale scan shown as butterfly plot, from black in center to white (limestone) on margins. Numbers refer to superbundles (406 ky eccentricity cycles), numbered from top of Albian downward. The constituent bundles (95 ky cycles), lettered, a–d, are punctuated by single or groups of precessional black pulses (PAPs). Bundles b and c have become wholly confluent in cycle 13.

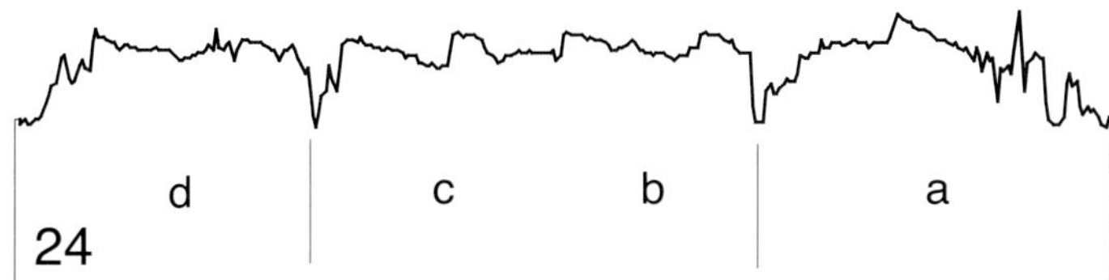

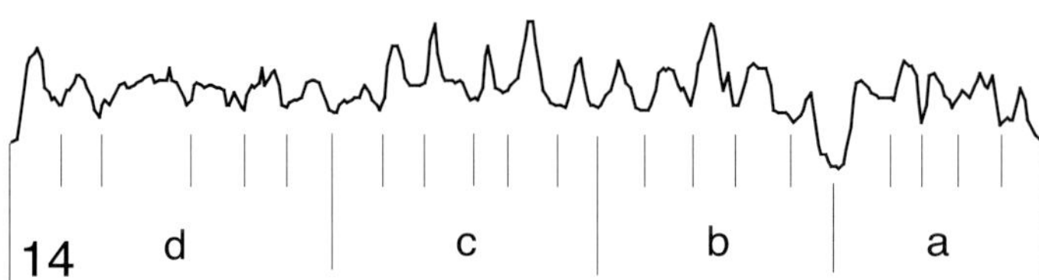

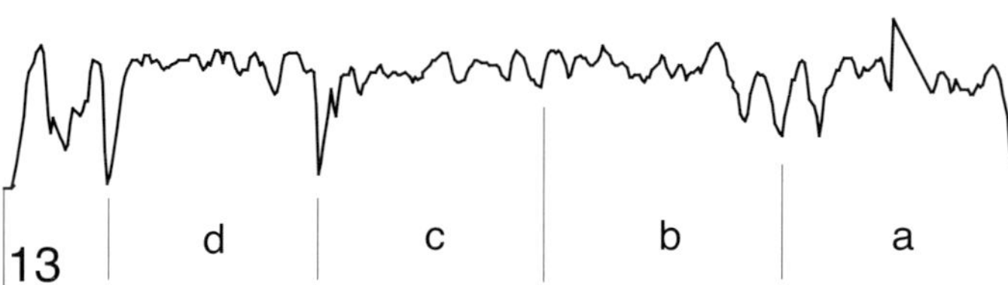

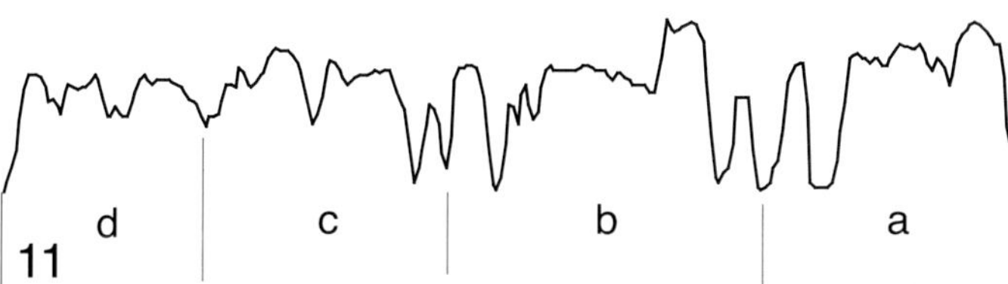

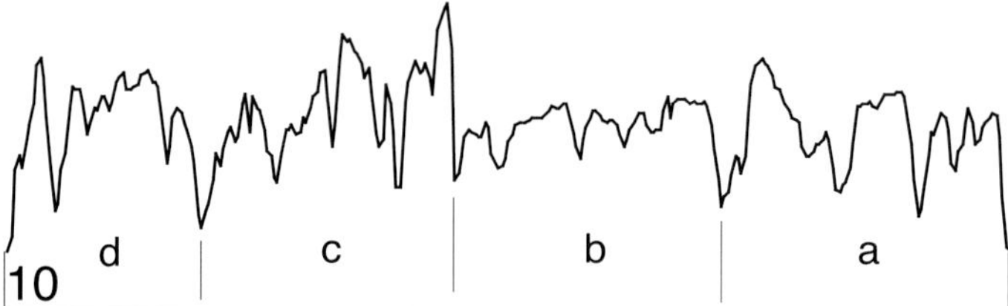

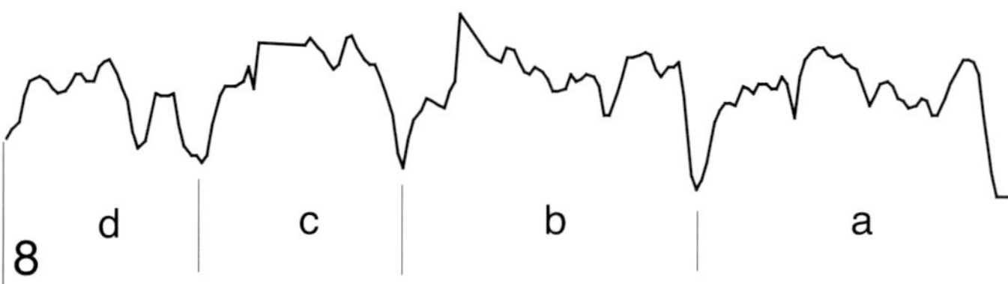

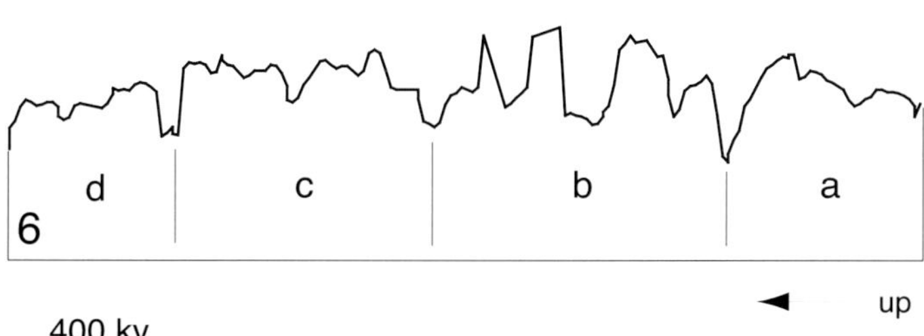

the actual variation in their period. Variations below the 95 ky level could only add to this distorting factor. It nevertheless seemed desirable to test the assignment of the couplets to precessional forcing, in another way.

Precession Confirmed.—

The 10 My duration of the Piobbico series provides an opportunity to investigate the orbital modulations during the mid-Cretaceous. Below we assess the evidence as follows (procedures after Hinnov et al., 2002). (1) isolation of the individually recorded orbital parameters by bandpass filtering; (2) estimation of amplitude modulations of the filtered series using quadrature signal analysis; and (3) spectral analysis of the amplitude modulations and comparison with those of the predicted orbital parameters.

To establish the presence of precessional forcing in the Piobbico series, we filtered the precession band of the tuned series and looked for amplitude modulations matching Earth's orbital eccentricity (following Shackleton et al., 1999; Hinnov, 2000). Because the eccentricity is intact in the tuned series, the amplitude modulations of any existing precession signal should also be intact, despite rate-induced time misalignments in the underlying precession cycles.

The filtered series is shown in Figure 7A. Remarkably, the spectrum of the AM series (Fig. 7B) reveals the clear presence of E, and doublets e_1 and e_2 at frequencies comparable to those observed in the eccentricity band of the tuned series (cf. Fig. 7D). As with the short eccentricity of the tuned series, the AM components interpreted as e_1 and e_2 are centered on frequencies that are slightly red-shifted relative to those of theoretical eccentricity (compare Figs. 7B, D; Table 1B). They match a similar shift in the eccentricity periods measured by us (Fig. 10). In sum, these results confirm that precession forcing played a decisive role in the cyclic sedimentation of the Piobbico series, as has long been hypothesized.

Most of the precession power (Fig. 7A) is concentrated in the PAPs-rich intervals (cycles 8–12 in Figs. 4 and 16). This irregular spacing of amplitude anomalies may preclude detection of the long-period eccentricity patterns sought for. This may explain why the AM series fails to mimic eccentricity periodicities longer than 10^6 years (compare black curves, Fig. 7A, C).

Eccentricity Rhythms

The climatic effect of the precessional cycles is modulated as the four eccentricity rhythms introduced by gravitational interactions with Mercury, Venus, Mars, and Jupiter go into and out of phase. In any one case, the largest effect is obtained at the frequency corresponding to the "difference tone" or "combination tone" between them as shown in Table 1B.

←

FIG. 5 (opposite column).—Fine structure of selected 406 ky superbundles, tuned to same length. a–b–c–d, 95 ky cycles in ascending order. In superbundles 13, and 14, transitional to red facies, show precessional signals of low amplitude but high fidelity, and confluence of central bundles. Amplitudes are high but fidelity of preservation is low in cycles 8 and 10, reflecting a strong obliquity signal. In cycle 24 the precessional signal has largely been wiped out by bioturbation. Confluence of the b and c bundles is shown in cycles 13, 14, and 24.

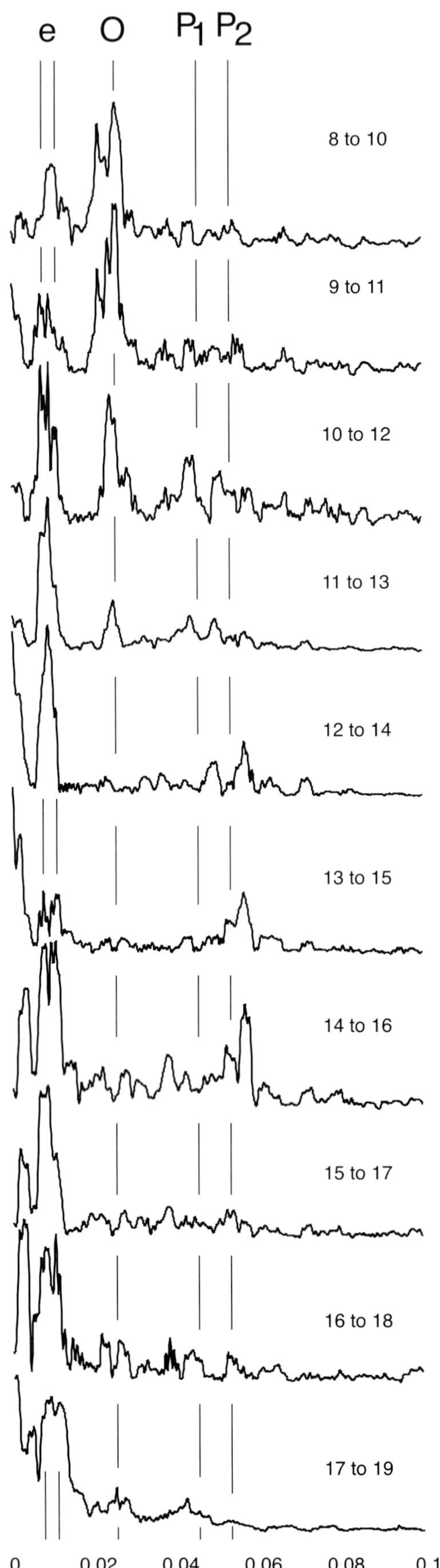

Short-Eccentricity Rhythms.—

The close proximity of the frequencies for the Jupiter–Mars relationship and the Jupiter–Earth relationship combines these in the period for e_1. That between Mars–Venus and Earth–Venus defines e_1. These are the modes of the short-eccentricity rhythm, to which we ascribe the "bundles" of couplets in the Scisti a Fucoidi.

These bundles, generally 30–50 cm thick, conspicuous in outcrop as well as in our logs, are defined by two features. The central couplets in such a bundle are generally more calcareous and more bioturbated, and therefore less sharply differentiated than are those at the boundaries. Also, in the drab facies they are punctuated by PAPs and inferred PAPs, as shown in Figures 4 and 9.

A statistical distribution of PAPs and gray layers (inferred PAPs) (Fig. 9C, D) in a 10 My series is plotted for the 20 precessions of the 406 ky eccentricity cycle. It shows their preferred incidence at bundle boundaries and peak incidence at superbundle boundaries. This distribution of PAPs seems to be largely a primary feature, but the greater proportion of inferred ones in the mid-bundle precessions suggests that the pattern has become somewhat enhanced by the stronger bioturbation in the better-aerated conditions of the mid-bundle region.

The scattering of peaks in the short-eccentricity band of the raw spectrum (Fig. 3A) is logically attributable to "stratigraphic distortion" by varied accumulation rates. That they should be pulled together into a single one by tuning is to be expected, but the period of that peak is a very good match to the predicted one (Fig. 3B, C).

Long-Eccentricity Rhythm.—

The phase relationship between Earth and Jupiter (Table 1B) yields the difference tone of 404.100 ky, the long-eccentricity rhythm, which is very stable. This is the frequency to which we ascribe the superbundle in the Scisti a Fucoidi.

At intervals the bundles of the gray-scale log are clearly grouped into sets of four, the 406-ky-long eccentricity cycle (cycles 10, 13, 16, 24, and 30 in 4 and 8, and in Fig. 16). These superbundles are a larger version of the bundles. They show a similar punctuation by PAPs, a similar confluence of central bundles, and a tendency to be separated by more marly bundles. The intervening bundles do not reveal such grouping, but, as explained above, we found it possible to project the 406 ky cycle through these intervals by assigning every fourth bundle boundary to the superbundle schedule. Three superbundles (7; 27 with the Urbino black marlstone; and 30 with a thick decalcified maroon clay) contain condensed intervals that require 100 ky additions to meet the next superbundle in phase.

Red Shift.—

In the spectra tuned to the 406 ky cycle the e_1 and e_2 eccentricity modes, presently at 94.8 and 123.9 ky (Laskar, 1999, his

←

FIG. 6 (opposite column).—Spectra of 1,200 ky segments (three 406 ky cycle equivalents) as recognized by 406 ky tuning, overlapped. Note the great rise of obliquity power in upper part of core, and the red and blue shifts in the precessional modes. e, modes of short (ca. 95 ky) eccentricity rhythm with which the spectra have been aligned; O, P, astronomically predicted modes of obliquity and precessional rhythms.

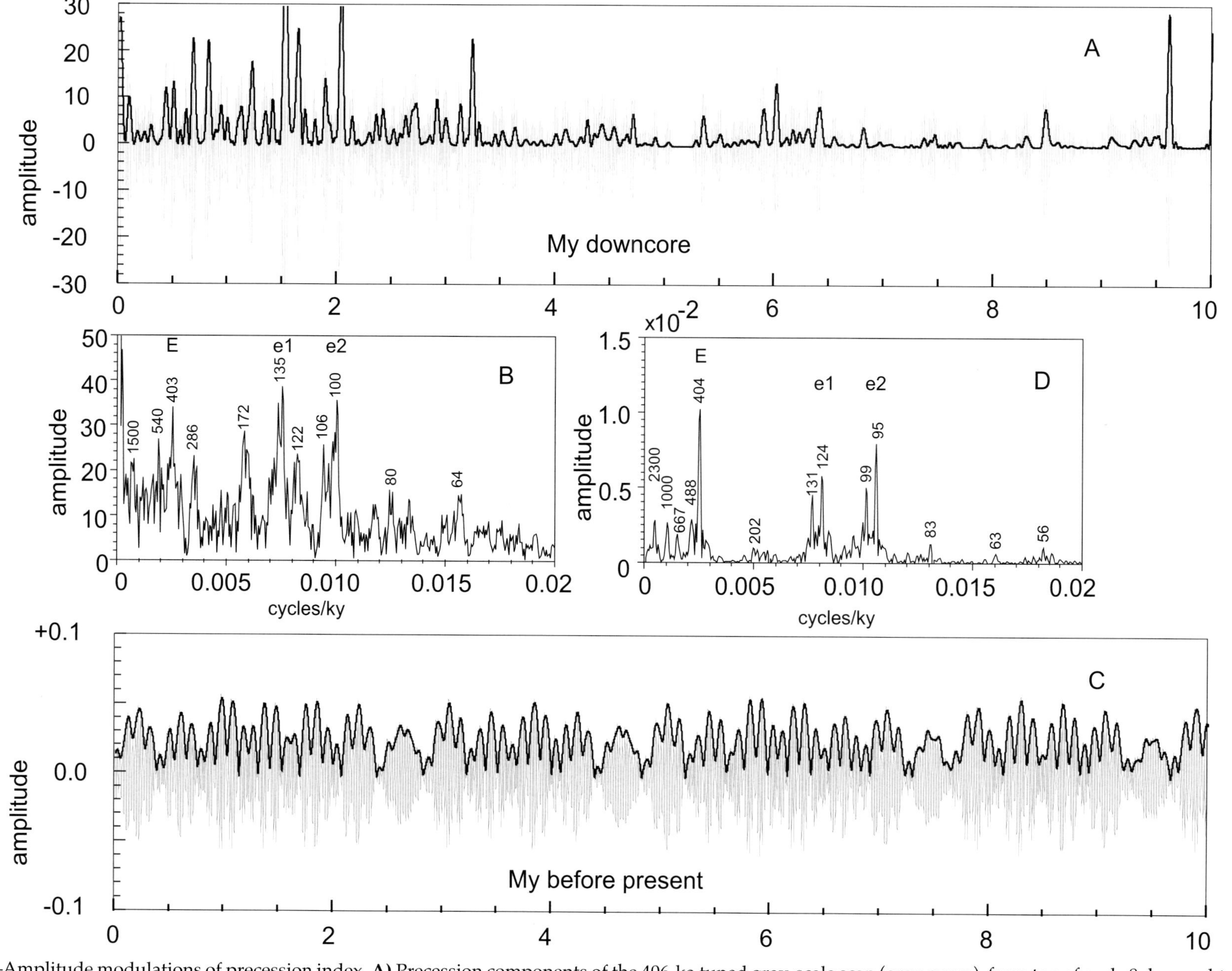

FIG. 7.—Amplitude modulations of precession index. **A)** Precession components of the 406-ka tuned gray-scale scan (gray curve), from top of cycle 8 down, obtained by Taner bandpass filtering with a lower cutoff frequency of 0.035 cycles/ky, and upper cutoff frequency of 0.060 cycles/ky, with a cutoff slope of 15 db/octave. The amplitude modulations were obtained by Hilbert transformation (black curve). **B)** Amplitude spectrum of the amplitude modulation series shown in Part A. **C)** Theoretical precession index over the past 10 million years (gray curve), and eccentricity series (black curve). **D)** Amplitude spectrum of the eccentricity series shown in Part C. Labels indicate periodicity in ky.

table 6), appear significantly "red-shifted" toward lower frequencies (Fig. 10). An overlay of the spectra for the stacked quarter-series, the stacked half-series, and the full series (Fig. 10E) demonstrates the gain in resolution obtained from the longer series, shows this red-shift in all, and establishes the e_1 and e_2 modes at 100.2 and 132.9 ky. Despite the shift, E has remained essentially invariant at 407.2 ky (Fig. 10E), indistinguishable from the present value of 406 ky. On the other hand, it is only spottily clear in the time series, and strong only in the spectrum of the second quarter, where it also appears to be displaced slightly toward lower frequencies.

The explanation for the red-shift in e_1 and e_2 lies in tuning errors. We demonstrate this as follows. In Figure 11A, the "true eccentricity series" (top curve) contains the 406 ky component (middle curve) that we seek to use as a uniform "metronome", through identification of successive 406 ky minima in the gray-scale curve. In tuning the Piobbico series, we identified the 406 ky component primarily through visual identification of bundle boundaries (Figs. 4, 5, 8). However, as shown in Figure 11A (arrows), the minima of the "true" eccentricity series do not occur at precise 406 ky intervals because of interference with the short-eccentricity cycles. In fact, over a 10-million-year simulation, 75% of the local minima of the "true" eccentricity series occur at intervals slightly less than 406 ky. This means that (incorrectly) treating these minima as uniform 406 ky increments, and tuning the series accordingly, effectively "stretches" e_1 and e_2 within the assumed 406 ky increments to slightly longer periodicities (compare the bottom gray and black curves). This is confirmed in the spectra of the original "true" and mistuned "distorted" eccentricity series shown in Figure 11B (top and middle curves).

The "distorted" eccentricity spectrum closely resembles the eccentricity band of the Piobbico series (Fig 10E, middle and bottom curves). Not only is power diverted into virtually the same (lower) frequencies, but additionally, the doublet structure of e_1 is lost, and that of e_2 is broader and less distinct. This nearly identical fine-scale structure in e_1 and e_2 of the Piobbico series and the experimentally distorted eccentricity spectra suggest that the red shift is due to a small systematic error in picking cycle boundaries. Finally, the absence of a sharp 406 ky spectral peak in the tuned Piobbico series suggests that the minima selected in the tuning deviated significantly from those of the true E component in the series.

Longer Periodicities.—

Although the Piobbico series is theoretically long enough to resolve the very long-period AM components at 2.35 My, 967 ky, and 697 ky (Table 1B, Fig. 12D), neither the eccentricity band (Fig.

→

TABLE 1 (opposite column).—The Earth's orbital parameters in terms of the planetary fundamental frequencies. The Earth's axial precession rate is estimated at $k = 50.4712''$/year, which gives a frequency of revolution of 3.8944 x 10^{-2} cycles/ky (i.e., a period of revolution of 25,678 years). Subscript numbers identify the planet: 1 = Mercury, 2 = Venus, 3 = Earth, 4 = Mars, 5 = Jupiter, and 6 = Saturn; for each of these planets, the fundamental frequencies g_i indicate changes in orbital eccentricity and longitude of perihelion, and s_i indicate changes in orbital inclination and longitude of the ascending node, for the *i*th planet (Laskar 1990). Labels P1, P2, E, e_1, e_2, and O indicate major orbital components labeled in Figures 6, 7, and 8.

A. Precession index

	Fundamental frequency	frequency (cycles/kyr)	period (years)
P_1	$k+g_5$	4.22×10^{-2}	23684
P_1	$k+g_1$	4.33×10^{-2}	23115
	$k+g_2$	4.47×10^{-2}	22373
P_2	$k+g_3$	5.23×10^{-2}	19104
P_2	$k+g_4$	5.28×10^{-2}	18952

B. Amplitude modulations of the precession index

	Fundamental frequency	frequency (cycles/kyr)	period (years)
	g_4-g_3	4.250×10^{-4}	2352900
	g_1-g_5	1.040×10^{-3}	961700
	g_2-g_1	1.435×10^{-3}	697000
E	g_2-g_5	2.475×10^{-3}	404100
e_1	g_3-g_2	7.646×10^{-3}	130800
e_1	g_4-g_2	8.071×10^{-3}	123900
e_2	g_3-g_5	1.012×10^{-2}	98800
e_2	g_4-g_5	1.055×10^{-2}	94800

C. Obliquity oscillations

	Fundamental frequency	frequency (cycles/year)	period (years)
O	$k+s_3$	2.44×10^{-2}	40996
	$k+s_4$	2.52×10^{-2}	39657
	$k+s_6$	1.86×10^{-2}	53714
	$k+s_1$	3.46×10^{-2}	28889

D. Amplitude modulations of the obliquity

Fundamental frequency	frequency (cycles/kyr)	period (years)
s_4-s_3	8.24×10^{-4}	1214174
s_3-s_6	5.78×10^{-3}	173145
s_4-s_6	6.60×10^{-3}	151536
s_1-s_4	9.40×10^{-3}	106394
s_1-s_3	1.02×10^{-2}	97822
s_1-s_6	1.60×10^{-2}	62507

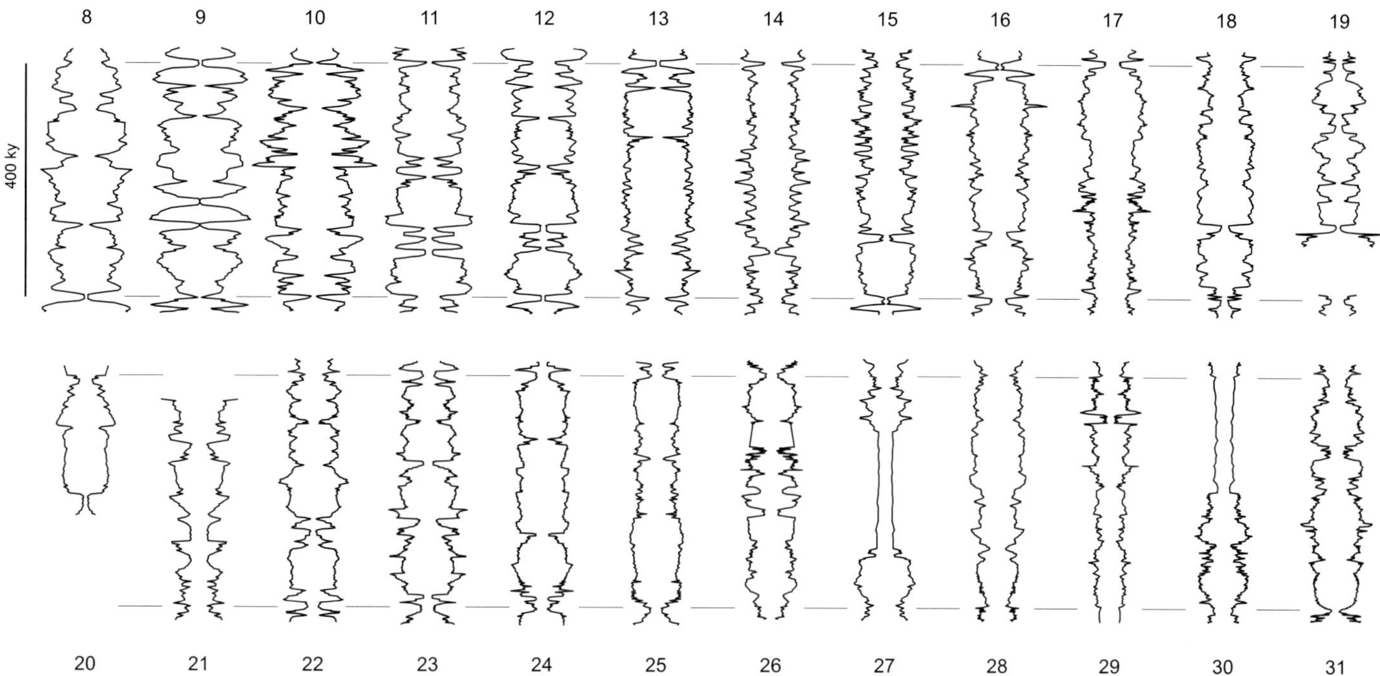

FIG. 8.—Comparison of the successive superbundle signatures in gray-scale log, tuned to 406 ky cycle and numbered from top Albian downward. One 95 ky cycle has been added to the Urbino black shale (superbundle 27), and two to the maroon shale of cycle 30.

3) nor the AM series of the filtered precession (Fig. 7B) show evidence for these periodicities, although the latter indicates a spectral peak at 1.5 My.

An alternative way to search for these AM components is through quadrature analysis of the short eccentricity (Table 1B): e_1 and e_2 are in reality doublets, each of which through time undergoes an amplitude modulation 2.35 My long; the four components comprising e_1 and e_2 together also interact to produce a shorter 406 ky modulation. The filtered short eccentricity of the tuned series is shown along with its AM series in Figure 12A, and their filtered theoretical equivalents in Figure 12C. The AM spectrum of the data (Fig. 12B) reveals major modulation periodicities in the 400 to 580 ky range. Of these, the peak at 400 ky is very sharply defined, and another, broader periodic component occurs at 1.5 My. The peaks at 500–580 ky may reflect tuning errors, i.e., errors in the choice of some of the 406 ky segments that caused one or more (true) 406 ky amplitude modulations to mistakenly be distributed over more than one 406 ky (tuned) period at a time.

Most noteworthy is the absence of the predicted major 2.35 My amplitude modulation (Fig. 12D). This modulation is caused by the interaction between the moving orbital perihelia and eccentricities of Earth and Mars, g4–g3, and has been observed in the short eccentricity in Oligocene–Miocene sediment (Zachos et al. 2001). Here, however, the Piobbico series shows instead a modulation at 1.5 My; this is also present in the AM series of the filtered precession index band (Fig. 7B), presumably the direct stratigraphic expression of the same component. Whether this 1.5 My component is the result of geological distortion to what was originally a 2.35 My modulation, i.e., disruptions by episodic PAPs, or is evidence for a real and significantly different g4–g3 during the mid-Cretaceous will require further investigation. There has been discussion, however, that during the Late Triassic and Early Jurassic, g4–g3 may have had a ca. 1.6 to 1.7 My periodicity (Olsen and Kent, 1999; Hinnov and Park, 1999). This would suggest that the Piobbico 1.5 My component may be a true expression of Cretaceous g4–g3, and that a shift in g4–g3 of astrodynamical significance occurred sometime between late Mesozoic and middle Cenozoic times.

Obliquity Rhythm

Our nearest sister planets affect not only the eccentricity of Earth's orbit but also its axial inclination because of the pull exerted on our equatorial bulge (Table 1C) by planets with an ecliptic notably different from ours. The main period of the self-induced oscillation of 41 ky, while Saturn induces a lesser one at ca. 54 ky. The closely spaced frequencies of the obliquity rhythm also produce amplitude modulations, shown in Table 1D.

The obliquity rhythm comes and goes through the core, as seen in spectra (Fig. 6) and in an extraction of the obliquity signal (the 0.018 to 0.038 ky pass-band) from the gray-scale series (Fig. 13). This inconstancy is attributable not to modulations of the obliquity cycle itself but to the Earth's response. Largely generated in the polar regions, its episodic strength in the mid-latitudes may be largely dependent upon its transmission.

Like the cycles of the precession–eccentricity syndrome, those of the obliquity drove variations in carbonate production and in bottom redox conditions. Thicker and abundant PAPs and relatively pure limestones coincide with the presence of strong obliquity signals in cycles 8–12, suggesting that primary productivity was increased at these times. The changing phase relations between the precession index and the obliquity signal largely mask the visual impression of a 40 ky regularity (Fig. 5).

Interactions among the varying inclinations and ascending nodes of the planetary orbits produce amplitude modulations in the Earth's obliquity variation listed in Table 1D. Because orbital inclination is the most unsteady of the planetary motions, the Earth's obliquity variation is not expected to have remained stable over remote geologic times (Laskar, 1999). Nonetheless,

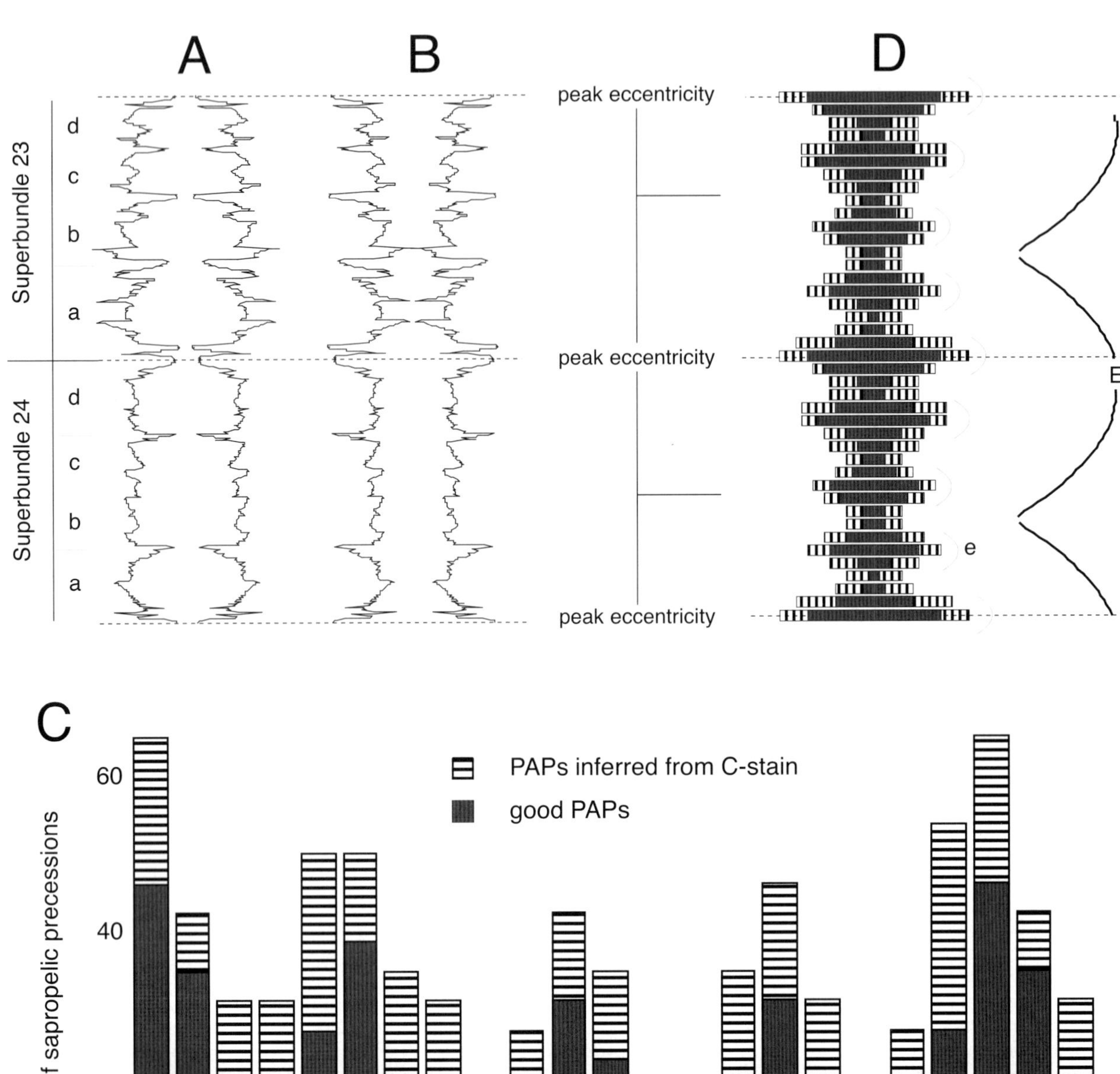

FIG. 9.—PAPs and the phasing of cycles. **A)** Gray-scale curve mirrored on black, yielding an image of carbonate cycles bounded by precessional black pulses (PAPs). **B)** The alternative, mirrored on white, yielding an image of cycles centered on PAPs. **C)** Statistical distribution of PAPs and grayed (PAP-inferred) couplets (precessions) in a stack of 26 406-ky supercycles. Incidence of PAPs in the first precession of each 95 ky cycle would be near 100% but for their absence in the red facies and for a reduction of peaks by a running three-point average. **D)** Patterns of statistical PAP incidence, projected to 800 ky, and suggested phase match to eccentricity cycles.

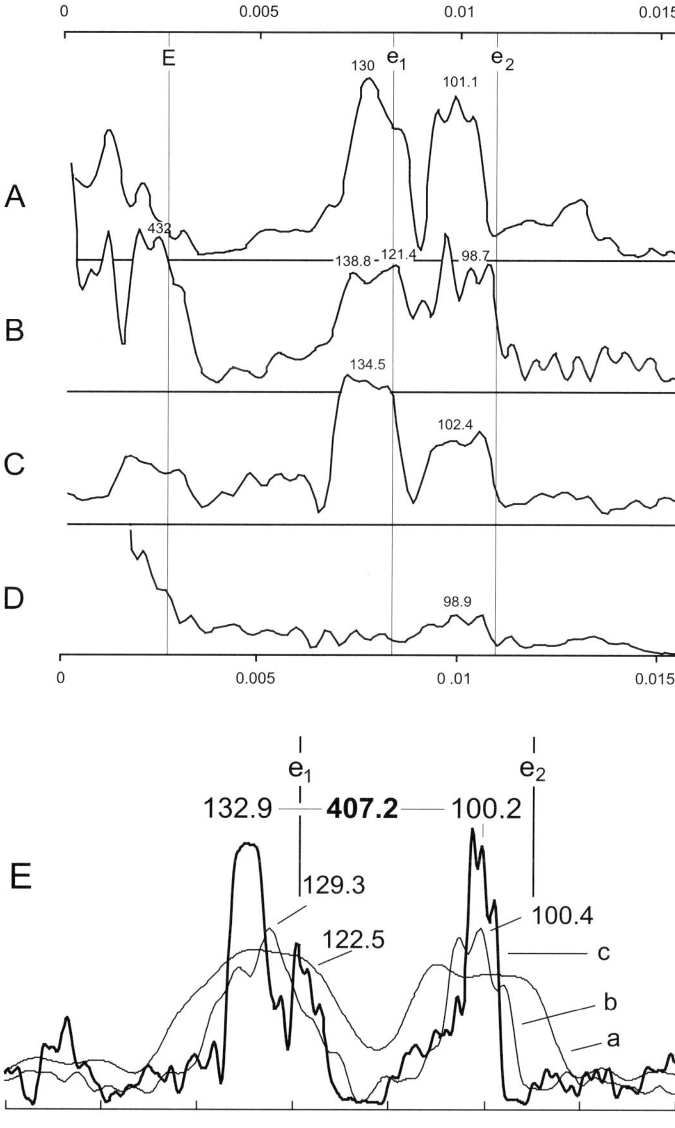

FIG. 10.—Red shift in frequency modes of short eccentricity cycles (e_1, e_2) tuned to 406 ky cycle. E, the 406 ky cycle of eccentricity, is not well resolved in such short series. **A–D)** Power spectra for the four "quarters" of the 10 My core series, tuned to the 406 ky cycle. (cycles 0–13, 14–19, 21–26, 27–31). (Ea) the sum of the above spectra; **E)** comparison of (a) the sum of the four quarter series; (b), the sum of the two half-series spectra; (c), the spectrum for the full series. The frequency modes of the short-eccentricity (e) rhythm have been displaced to 100.2 and 132.9 ky, respectively, but E, their difference tone, remains close to present value of 406 ky.

Figure 13C shows that, in the tuned Piobbico series, spectral power is concentrated at the two main predicted obliquity frequencies, k + s3 = 0.0244 cycles / ky and k + s4 = 0.0252 cycles / ky (see Table 1C), suggesting that s3 and s4 may have been invariant over the past 100 My.

The Piobbico filtered obliquity series (Fig. 13A, B) lacks certain modulations to be expected. There is no dominating 1.2 My modulation that would indicate the presence of predicted s4–s3, nor are there modulations at the 100 ky and 170 ky time scales, also predicted for the obliquity (Figs. 13C, D; Table 1D). This may be attributable to degradation of signals during atmospheric or oceanic transmission of signals from distant sites of generation.. But, if taken at face value, it would suggest that the two obliquity frequencies in Figure 3C cannot be ascribed to s3 and s4 as given in current models (i.e., Laskar, 1999) and that the other s_i are probably different as well.

Instead, broad peaks centered on 242 ky and 133 ky periods (Fig. 13D) characterize the AM spectrum. Whether these peaks reflect true interactions s_i–s_j for the Cretaceous, or irregular amplitude patterns not related to orbital forcing, will require further investigation. Evidence suggests that the s_i during the Jurassic Period were different from those predicted for the past 20 My (Hinnov and Park, 1999). By Oligocene–Miocene times, however, s4–s3 with a 1.2 My variation in the periodicity is clearly visible in obliquity-forced sedimentation (Shackleton et al., 1999; Zachos et al., 2001).

STRATA AND PHASE

How do the stratigraphic cycles—the eccentricity bundles and the precessional couplets—relate to the precession index of Berger and others? The alternatives are shown in Figure 9A, our conventional gray-scale log, and Figure 9B, its inverse. Do PAPs, segmentations in A and spikes in B, represent the lows or the highs of the eccentricity cycles?

In the precession index curve, low eccentricity implies moderate seasonality with little contrast between the perihelial summer phase and the perihelial winter phase of the precession. This pattern finds its match in the damped gray-scale variations characteristic of middle of the bundles. As eccentricity grows, the seasons become more damped in the perihelial winter phase and more extreme in the perihelial summer phase. We take this to be expressed in the higher amplitudes of the precessional signals at bundle boundaries. Accordingly, PAPs were preferentially associated with eccentricity peaks, at the level of both the short and the long eccentricity rhythms. The statistics for their incidence are shown in Figure 9C and their inferred relationship to the eccentricity cycles is shown in Figure 9D. The Quaternary sapropels of the Mediterranean—also PAPs of a sort—are also associated with high eccentricity (Rossignol-Strick, 1985).

Turning to the precessional cycle, we ask whether the PAPs were formed in the phase of perihelial summers or that of perihelial winters. The former, marked by hotter-than-normal summers and colder-than-normal winters, leads to high monsoonality, likely to be expressed in high productivity. The perihelial winter phase, on the other hand, reduces seasonality and monsoons, and presumably favored stable stratification.

This poses the question of whether the PAPs resulted from organic hyperproduction or from exceptional preservation of organic carbon. de Boer (1982, 1983) and de Boer and Wonders (1984) took carbonate productivity to be a function of primary productivity. Herbert et al. (1986) presented evidence for more organic silica in the limestones (versus PAPs and other marls), suggesting higher primary production in the carbonate-rich phase. The microbiota offers further support: limestones denote the times when coccolithophoraceans, part of the phytoplankton, flourished and contained the species generally identified with upwelling and fertility (Erba, 1988, 1992). The richest assemblages of planktonic foraminifera, associated with the marls and PAPs, would seem to have resulted from depth tiering, suggesting a preference for clearer oligotrophic waters (Premoli Silva et al., 1989b).

These observations support the de Boer model, that PAPs represent not productivity but episodes of carbon preservation.

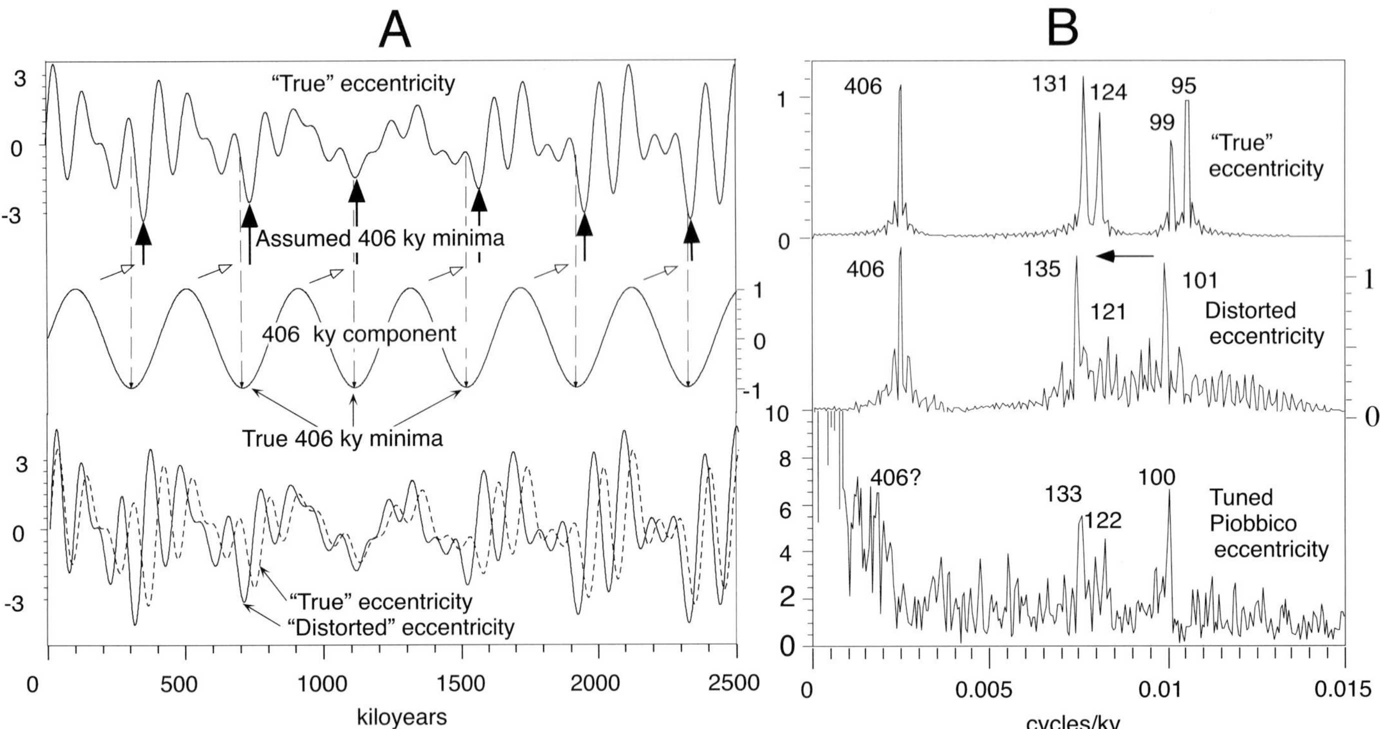

FIG. 11.—**A)** An approximation of the "true" eccentricity, $x(t) = \sin(2\pi t/406) + 0.9\sin(2\pi t/131) + 0.7\sin(2\pi t/124) + 0.6\sin(2\pi t/99) + 0.9\sin(2\pi t/95)$, with minima indicated every ca. 400 ky by heavy black arrows (top curve); the 406 ky component of the "true" eccentricity, i.e., $\sin(2\pi t/406)$ (middle curve), with minima indicated by dashed arrows; and the "true" eccentricity replotted (bottom gray curve) and compared to a "distorted" eccentricity (bottom black curve) obtained by tuning the "true" eccentricity minima to constant 406 ky increments. **B)** The amplitude spectrum of a "true" eccentricity series 10 million years long (top curve), computed as in Part A, of a 10-million-year long "distorted" eccentricity (middle curve) computed as in Part A, and of the eccentricity band of the tuned Piobbico series 9.6 million years long (bottom curve).

Accordingly we assign the PAPs to the perihelial winter phase of the precessional cycle, when seasonality was damped, monsoons were suppressed, hydrodynamics reduced and stratification enhanced. This would seem to differentiate them from the Quaternary sapropels, which originated in the very different hydrodynamic setting of a blind-alley Mediterranean and are thought to represent the high productivity in the perihelial summer phase (Rossignol-Strick, 1985).

PALEOCLIMATIC–PALEOCEANOGRAPHIC INTERPRETATIONS

We think of the greenhouse world as a warmer one in which latitudinal climatic gradients were so low as to have kept permanent ice from reaching sea level and in which ocean temperatures, now near 3°C, hovered around what is now Earth's average surface temperature of ca. 15°C. That is the temperature obtained for Cretaceous ocean bottoms (Douglas and Savin, 1975; papers in Barrera and Johnson, 1999; Huber, Macleod and Wing, 2000).

The ocean is density-stratified. In the present mode, bottom waters are supplied by the cold, though not highly saline, waters of the high latitudes. Chamberlin (1906) reasoned that in greenhouse times the production of such waters would be diminished, while the increased production of dense (saline though warm) waters in the paratropical dry belts would play a competitive role.

As argued by Hay and DeConto (1999), a global "halothermal" circulation of this sort could not have become permanent, lest sequestration of salt in the deep ocean turn the surface brackish. Yet episodic or localized subtropical downwelling of saline waters seems likely. Figure 14, Herrle's (2002) model of this in the Mediterranean Tethys, embodies such downwelling and the very processes that our interpretation of the Scisti a Fucoidi call for.

While global changes in atmospheric and oceanic behavior set the stage for a strong response to orbital forcing, that response is likely to have been heightened by the local geography (Fig. 14). The opening of Tethys to the west, in Jurassic–Cretaceous times, provided an avenue for east–west water transport. It seems likely that trade winds drove warm, saline, nutrient-depleted waters into the westward-narrowing funnel of Tethys, as suggested in Figure 14, and that at times and places such waters of the mixed layer foundered to bottom.

As pointed out by Herrle (2002), the still comparatively aggregated pattern of lands must have left this part of the world particularly liable to orbitally modulated monsoons, which presumably played a large part in local ocean dynamics and elicited strong biotic response.

Drab Facies

The drab facies represents the more prevalent condition, and records sedimentation in a stratified water column, which we

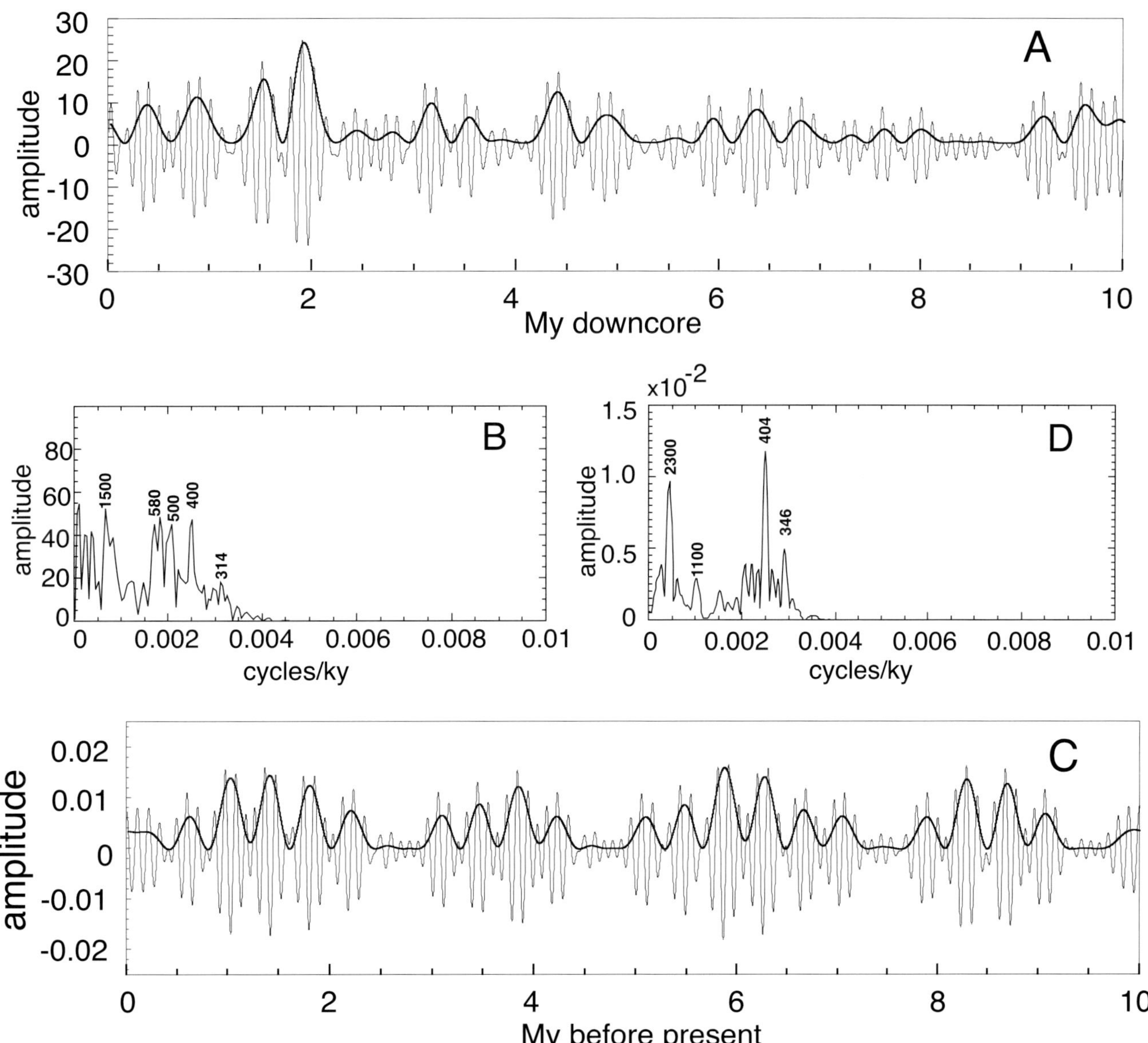

FIG. 12.—Amplitude modulations of eccentricity. **A)** Short-eccentricity components (gray curve) of the 406 ky tuned gray-scale scan, from top of cycle 8 down, obtained by Taner bandpassing with a lower cutoff frequency of 0.007 cycles/ky and an upper cutoff frequency 0.0112 cycles/ky, with roll-off slope of 15 db/octave. Also shown is the amplitude modulation series (black curve), obtained by Hilbert transformation. **B)** Amplitude spectrum of the amplitude modulation series shown in Part A. **C)** Theoretical short-eccentricity components (gray curve) obtained by the same Taner bandpass applied in Part A, and the amplitude modulation series (black curve), obtained by Hilbert transformation. **D)** Amplitude spectrum of the amplitude modulation series in Part C. Labels indicate periodicities in ky.

visualize something like that of the North Pacific. There the oxygen content of surface waters, from 4–6 cm^3/l, drops to a minimum of < 1 cm^3/l at about 1 km, and generally remains below 2 cm^3/l to depths of ca. 2 km, gradually rising to 4–4.5 cm^3/l on the deep bottoms (Dietrich and Ulrich, 1968). In the Albian Tethys, temperatures were higher and shortened reaction rates, while at times orbitally modulated monsoons brought more oxygen to the profile.

We postulate that times of moderate to strong seasonality, associated with the summer perihelion phase of the precessional cycle, generated monsoons, invigorated oceanic circulation, and increased nutrient supply to the surface. This favored coccoliths in the photic zone, yet continued to supply oxygen to the deep bottoms, maintaining a diverse ichnofauna.

The perihelial winter phase, with its reduced seasonality, would have decreased monsoonality and ocean dynamics. This would have brought more oligotrophic conditions to the surface waters, with the clarity that allowed planktonic foraminifera to thrive in depth-zoned communities. It reduced oxygen replenishment, bringing dysaerobic conditions to the bottoms and reducing the ichnofaunas to *Chondrites*. At such times anoxia may well have occurred in an oxygen minimum at lesser depths, but

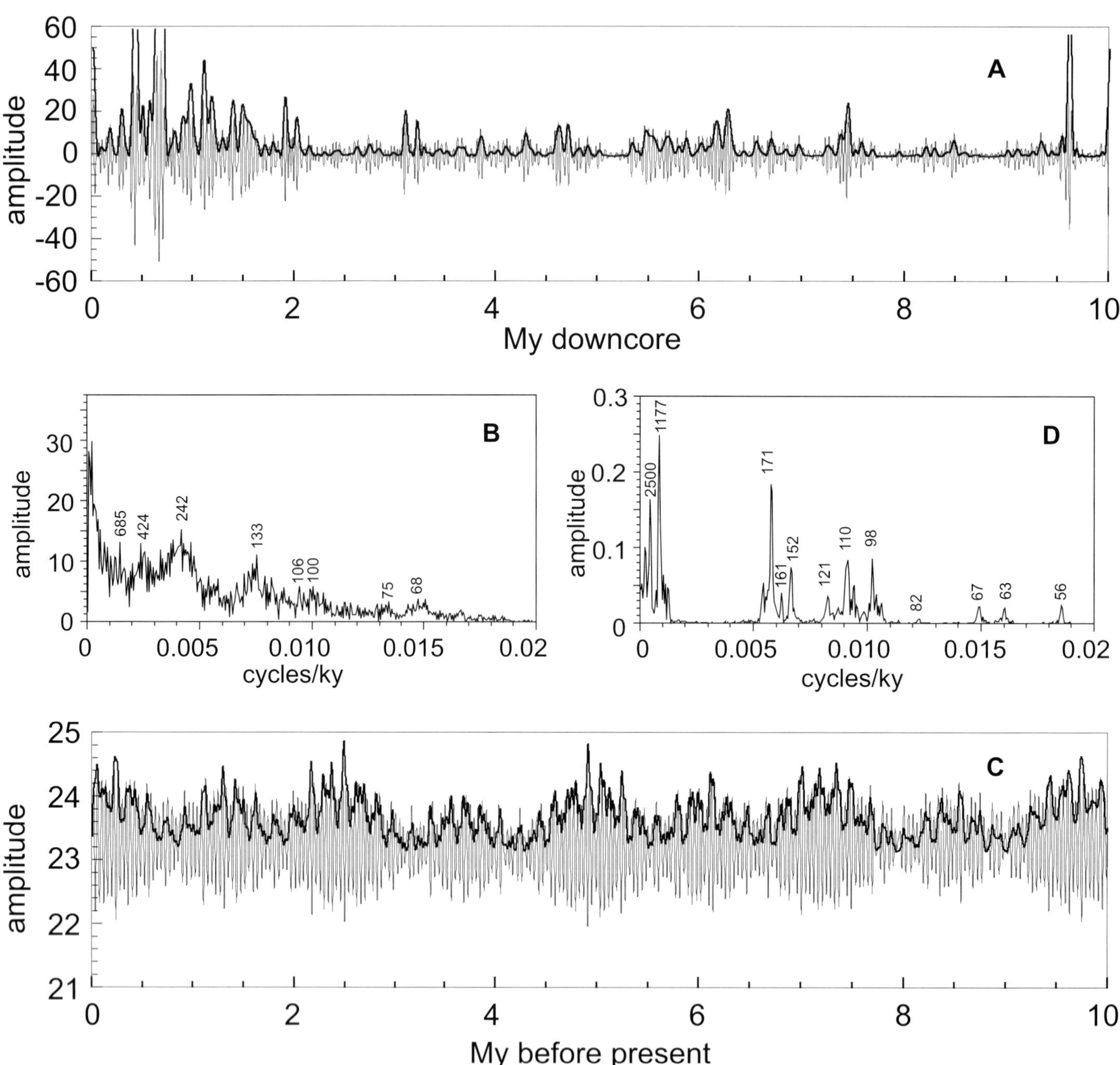

FIG. 13.—Amplitude modulations of obliquity rhythm. **A)** Obliquity components of the gray-scale scan (gray curve), from top of cycle 8 down, obtained by Taner filtering with a lower cutoff frequency of 0.018 cycles/ky and an upper cutoff frequency of 0.038 cycles/ky, with 15 db/octave slope at the cutoffs, and the amplitude modulations (black curve) obtained by Hilbert transformation. **B)** Amplitude spectrum of the amplitude modulation series estimated in Part A. **C)** Theoretical obliquity (gray curve) and its amplitude modulation series (black curve) obtained by Hilbert transformation. **D)** Amplitude spectrum of the theoretical amplitude modulations of the obliquity given in Part C. Labels indicate periodicity in ky.

when magnified by high eccentricity the anoxic zone became extended to the bottom, to form the PAPs.

Red Facies

The persistence of the carbonate cycle in the red facies implies that the upper water regime continued as outlined above, though at somewhat lower carbonate levels (Herbert and Fischer, 1986), implying lower fertility. Deposition at depth occurred in very different, well-aerated waters. These we interpret as Chamberlin's downwelling saline, nutrient-depleted tropical waters, which brought with them their high temperature and relatively high oxygen content. Supply of organic matter, relatively low to begin with, was now further reduced owing to the acceleration of decomposition in the 30°C regime, leaving the deep bottoms depleted in food, while oxidation of sediments would have proceeded rapidly at those temperatures and would have persisted into early diagenesis occurring within the sediment.

GEOCHRONOLOGY

The Albian Stage is defined, by ammonites, as extending from the base of the *schrammeni* zone (in Tethys, the *tardifurcata*) zone to the end of the *perinflata* zone. In the absence of ammonites we depend on foraminiferal and nannofossil zonation. This does not present a problem at the top of the Albian, where the ammonite-defined boundary nearly coincides with the beginning of the *Rotalipora brotzeni* (= *R. globotruncanoides*) foraminiferal zone. It does present some uncertainty for the base of the Albian, where the ammonite-defined base falls between the first occurrence (FO) of *Predicosphaera columnata* and the FO of *Ticinella primula* (Tornaghi et al., 1989) (Figs. 15, 16).

As noted above, our gray-scale scan allowed us to combine the short and long eccentricity records into an extrapolated count of the more stable 406 ky cycles, which also revealed the need to add a total of 400 ky to condensed zones. This count places the base of the *P. columnata* zone at 30.6 E-cycles for an Albian duration of 12.4 My, 500 ky longer than the calculation by Herbert et al. On the other hand, the appearance of *P. columnata* preceded the beginning of Albian time. In an attempt to move as close to the ammonite-defined base as possible we have placed the boundary midway between the first appearance of *P. columnata* and that of *T. primula* (Figs. 15, 16), at the base of cycle 29 or 11.8 My before the end of Albian time. By chance this closely approximates the 11.9 My figure initially suggested by Herbert et al. (1995). We cannot exclude the possibility that one or two 406 ky cycles are missing.

BROADER IMPLICATIONS

The patterns of orbital cyclicity in the Aptian–Albian Scisti a Fucoidi began in the Tithonian–Neocomian Maiolica Limestone (Herbert, 1992) and lasted, in somewhat degraded form, through the Cenomanian Scaglia Bianca Limestone (Schwarzacher, 1994). In those limestones, carbonate productivity was doubled, which generally doubles the thickness of the couplets and reduces the marly member to a thin interbed, which may or may not be developed as a PAP. Black radiolarian cherts suggest complications.

Precessional redox pulses recorded in organic-matter content, associated with fertility fluctuations in the coccolith flora, have been well documented by Herrle (2002) in the Aptian–Albian hemipelagic deposits of the Vocontian basin of southern France.

Thin black shales and marls, showing remarkable similarity to the PAPs, occur throughout the Early and mid-Cretaceous of the North Atlantic (Dean and Arthur, 1999 and references therein). It thus appears that these precessional fluctuations in planktonic productivity and in bottom redox conditions, modulated by the precession–eccentricity syndrome, were widespread in Tethys.

We suggest that the greenhouse state sensitized the oceans to orbital forcing. Although halothermal circulation at times brought more oxygen into the lower waters of the tropics and subtropics, the lower carrying capacity of the warm waters implies a diminution in the amount of oxygen thus taken down globally, per unit volume of water. It is difficult to judge whether the mean turnover rate of the oceans was higher or lower than the present one, and how it may have varied with orbital forcing, but conceivably

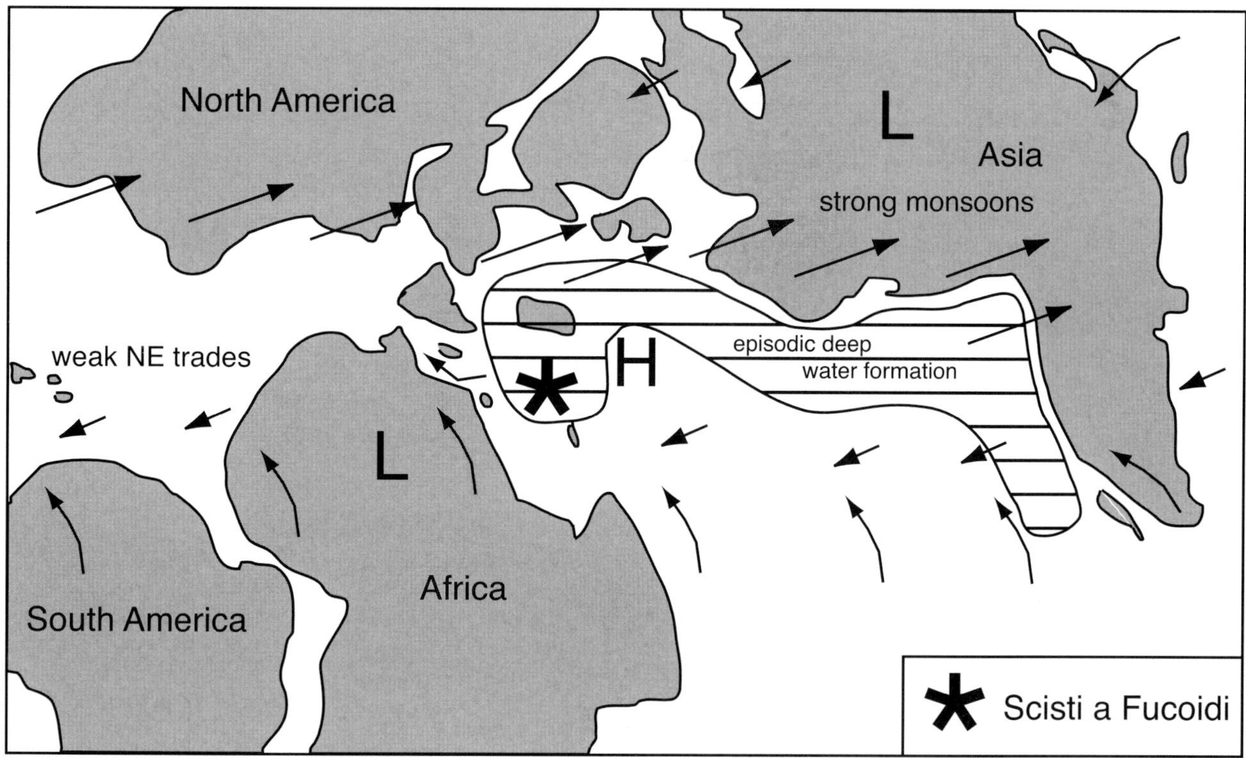

FIG. 14.—Paleogeographic reconstruction for Albian time, showing suggested directions of monsoonal winds and of downwelling warm saline waters. Modified from Herrle (2002), with permission. H, high-pressure area; L, low-pressure area. The star shows hypothetical location of studied sequence.

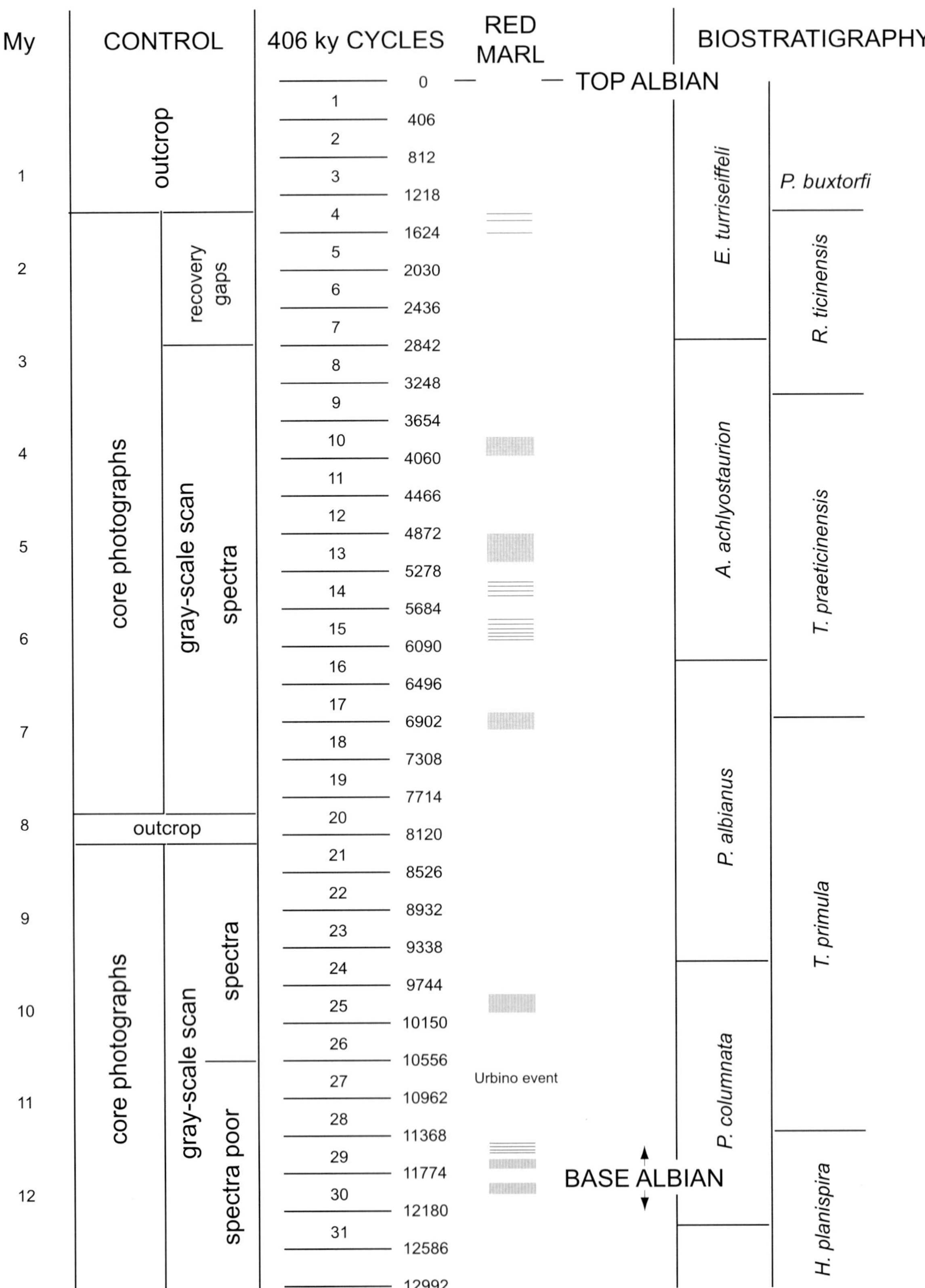

FIG. 15.—Albian cyclochronology as deduced from the Scisti a Fucoidi–Scaglia Bianca sequence in the Apennines of Umbria and Marche (Italy). Observations mainly on Piobbico core, supplemented by data from outcrop (Herbert et al., 1995). Paleontology after Premoli Silva (1977) and Erba (1986, 1988, 1992) and Tornaghi et al. (1989). Duration of the Albian is estimated at 11.9 ± 0.5 My.

the greater role of salinity in circulation slowed the return of some of the deeper waters. All this would have tended to deplete the oxygen content of the mid-water masses.

The high temperatures, implying more rapid decomposition of sinking organic matter, must have intensified the oxygen minimum and at times extended it downward. Consumption of more organic matter in the water column left less for benthic life and for burial the deep-sea sediments.

As pointed out by de Boer (1983), a number of observations suggest an accelerated rate of carbon burial in Early Cretaceous time. These include the episodic formation of large and persistent anoxic water masses (OAEs), the widespread formation of black shales in epeiric seas, and the extraordinary role which this time played in the generation of petroleum (Irving et al., 1974; Tissot, 1979). A corollary to carbon burial is the release of oxygen. Atmospheric oxygen levels at this time may have been higher than present ones, leading to a greater incidence of wildfires, possibly reflected in the high proportion of soot in the Scisti a Fucoidi (Pratt and King, 1986). The possible relation of wildfires to orbital forcing remains to be explored.

As carbon dioxide emissions now move us toward a greenhouse state, we may wonder what lessons the Albian world holds for us. So long as we continue to have extensive ice in the polar regions we shall not come to feel the full brunt of a greenhouse such as the one that developed in Cretaceous time. Furthermore, we are living at a prolonged time of low eccentricity, in which orbital forcing remains moderate. Yet it seems inevitable that some changes in circulation will occur and that the development of higher temperatures in the mid-water masses is bound to accelerate the decomposition of organic matter settling through the water column. This may in places drive the oxygen minimum to critically low levels that could interfere with diurnal migration of organisms which, in the Neogene icehouse, lost the means of coping with such oxygen deficits.

SUMMARY AND CONCLUSIONS

(1) Albian coccolith–globigerinacean marls and limestones of the Umbria–Marche belt (Italian Apennines), deposited at a depth of ca. 2 km, are remarkably cyclic in structure. Building on earlier studies we reanalyzed the Piobbico core with an image-processed photolog and a variety of time series studies designed to follow orbital forcing through ten million years of deep-water sedimentation.

(2) Of two basic facies, we attribute the drab one (white–greenish gray–black) to deposition in a stratified water column, the red one to deposition in downwelling saline (halothermal) waters.

(3) Both reflect oscillations in the composition and vigor of planktonic carbonate producers in the upper waters. These are expressed in ca. 8 cm marl–limestone *couplets*. Abundance and diversity of coccoliths and planktonic foraminifera are antithetic. Coccolith production, with indicators of high fertility, dominated the purer limestones. Abundance and diversity of planktonic foraminifera is greatest in the less-calcareous marls, and implies a preference for oligotrophic conditions.

(4) In the drab facies, this record of oscillations in the upper waters received the overprint of redox oscillations on the bottom. A diverse ichnofauna in the limestones reflects aerated conditions, whereas restriction to *Chondrites* in the marly member of the couplet suggests oxygen deficiency. Episodically this dropped to anoxia, resulting in black marlstones (PAPs, precessional anoxic pulsations).

(5) Such couplets are grouped into ca. 40 cm bundles punctuated by more marly couplets containing PAPs.

(6) The red facies shows similar carbonate fluctuations of somewhat lower carbonate content but lacks the redox cycle except in transitional conditions. Despite deposition in highly aerated settings, it lacks extensive bioturbation, suggesting that supply of organic matter to the bottom was minimal.

(7) To the eye, the Scisti a Fucoidi as seen in outcrop, core, photolog, and gray-scale log reveal cyclicity at two major levels, that of the elementary couplet oscillation, and its ca. 5:1 grouping into PAP-punctuated bundles.

(8) Digitized pictures, converted to a photolog stripped of nonstratigraphic information, served to construct a gray-scale series and gray-scale log. This, by way of spectra tuned to the 406 ky eccentricity cycle, yielded confirmation of the precessional, obliquity, and short and long eccentricity rhythms.

(9) Regarding precessional cyclicity, the elementary doublet cycle, impressive in the field, yielded only marginal spectral results. Low and scattered peaks, though in some intervals appropriately spaced for the two precessional modes, do not come to coincide precisely with the predicted values for the precession. This is due in large part to variations in accumulation rate, below the tuning level of 406 ky. The couplet record also shows extensive damage from bioturbation and, in places, dissolution. But amplitude-modulation analysis of these signals, filtered out of the series, shows the precise eccentricity modulations to be expected, namely the eccentricity frequencies observed, and leaves no doubt as to the precessional origin of the couplets

(10) Signals of the obliquity rhythm are strong in cycles 8–12, where they increased the incidence and thickness of PAPs. They are weaker and not consistently present in the rest of the core.

(11) The short (ca. 95 ky) eccentricity cycles are recorded in the ca. 40 cm bundles of couplets. The low-amplitude gray-scale variations of the middle couplets of such a bundle suggest low eccentricity, whereas the high amplitudes (PAPs) that punctuate bundles would appear to represent eccentricity peaks. The bundles as plotted are thus inverted expressions of the 95 ky cycle.

(12) Such bundles in turn are intermittently grouped into superbundles of four, inverted expressions of the long (406 ky) eccentricity cycles. These are differentiated in the manner of the bundles.

(13) Interpreting the couplets in terms of precessional phases suggests that the summer perihelial phase brought monsoons, fertility that triggered maximal production of coccoliths but limited abundance and variety in planktonic foraminifera. In the drab facies bottoms were moderately well aerated, as attested by bioturbation by a varied ichnofauna. The precessional winter-perihelial phase, of damped seasonality, brought a lessened production of coccoliths but a more abundant and varied foraminiferal fauna, suggesting olig-

otrophic conditions. In the drab facies, bottoms became stressed by lack of oxygen, as evidenced by restriction of bioturbation to *Chondrites*. When this phase coincided with high seasonality, extreme damping of seasonality brought anoxia to the bottoms and resulted in deposition of black marlstones, precessional anoxic pulsations (PAPs).

(14) To derive a chronology for the Albian, we turned to the gray-scale log, where the 95 ky bundles provide a good but not unambiguous record of the short-eccentricity rhythm. The intermittent appearance of the 406 ky cycle sets a framework. Fitting the 95 ky cycles into that frame necessitated the addition of a total of 400 ky, to three intervals of condensed accumulation.

(15) Extrapolating the 406 ky cycle counts in the core to the entire stage, we estimate the length of the Albian at 11.9 ± 0.5 My, short of the radiometric estimate of 13.3 My by Gradstein et al., (1995) but within their confidence limits of 1.7 My. we cannot, however, exclude the possibility that one or two 406 ky cycles are missing.

(16) The sensitivity of these sediments to orbital forcing may reflect in part the complexities of oceanic circulation in a greenhouse world, and in part the increased rate of chemical and bacterial processes at high ocean temperatures.

(17) Factors contributing to orbital sensitivity may have included the location of the region, in a tropical seaway narrowing westward, thus admitting flux of oceanic waters. Proximity to large continental masses implies confrontation of such nutrient-depleted waters with nutrient-rich ones, fluctuating with weathering and runoff. Those factors as well as upwelling and mixing of water masses fluctuated with the orbitally driven variations in monsoonality, highest in the precessional summer phases at times of high eccentricity.

(18) The time seems to have been one of exceptional rates of carbon burial, which implies an exceptional flux of oxygen to the atmosphere. The high proportion of soot in organic matter suggests that this may have found expression in extensive wildfires.

(19) Although the now impending greenhouse state is not likely to reach the strength of the mid-Cretaceous one, it is likely to intensify the oxygen-minimum zone, and may in places come to interfere with the accustomed patterns of diurnal migration.

(20) The photoscan method proved to be an effective means of exploring patterns of cyclicity in a sedimentary sequence.

APPENDIX: THE PHOTOSCAN APPROACH

Many sequences of stratified rocks recorded climatic variations by a variety of parameters, and thereby incorporated into themselves a record of a hierarchy of cycles forced by the Earth's orbital variations. Extracting this record depends on finding efficient ways to measure these variations through the sequence of strata.

The photoscan method measures variations in a bulk property of the sediment—light reflectivity—as a proxy for one or more oscillating compositional variations.

In the Piobbico core the high reflectivity of calcite renders the most calcareous rocks in shades of white (Herbert and Fischer, 1986), whereas the clay minerals with included or adsorbed iron compounds color the marls gray-green, or, when thoroughly oxidized, red. In addition, certain strata are colored black or neutral gray by elemental carbon (lamp black) or organic compounds, and possibly iron sulfides.

The strata recorded in these color variations are planar, and devoid of sedimentary structures save for occasional ripple-drifted and cross-laminated accumulations of radiolarian tests (particles of extreme mobility).

Photolog

To facilitate magnetic studies the core had been drilled in rocks with a dip of 23°. It was split in the dip direction by diamond saw and cut into segments < 30 cm long. Smoothing with 600 grade carborundum and a light acid etch prepared these for photography as 35 mm Ektachrome diapositives (slides). Shortness of the core segments left the images free of the "edge bias" (the lighting gradients so common in core pictures), which readily produces pseudocyclicity in gray-scale scans (Herbert et al., 1999; Thurow and Nederbragt, 1999).

The Ektachrome slides were initially digitized at a pixel density corresponding to 0.8 mm of rock. These files could be concatenated into a single long log-strip, but one that would be full of nonstratigraphic distractions, from coarse bioturbation and the ca. 23° dip to fractures and chipping of the core induced during and after drilling and in extraction and preparation.

In the beginning we corrected the dip in each picture by means of the skew function in Adobe Photoshop®. But time was saved by running the optical scans parallel to the core axis and correcting drilling depth to stratigraphic depth by the factor cos 23° = 0.92.

Digitization allowed us to remove blemishes by image processing. This we initially applied to the full width of the pictures, but clearing a narrow path sufficed to capture the desired information: a time series of successive color pixels representing stratigraphic features only. Along this path we replaced the images of cracks, shadows, pits, and coarse bioturbation with the color of the beds they crossed. The scans of individual core segments were then overlapped wherever possible to concatenate them into a continuous series, broken here and there where core loss left gaps as estimated at the time of drilling. The construction of this colored image was done in Adobe Photoshop®. Once completed, the image was transferred to Adobe Illustrator® for better-quality imaging and compatibility with other files (i.e., the gray-scale time-series file) that had to be successively included. The time series could then be printed as a noise-free Photolog, to any desired format (Fig. 16).

Gray-Scale Series and Log

Such a time series contains two sets of parameters: the stratigraphic depth range of each stratum, and its color, a multidimensional quality. For normal two-dimensional time-series processing we reduced the color to its gray-scale components, though cognizant of the loss of valuable information—the difference between red and gray, an expression of oxidation state.

For conversion of the colored profiles into their gray-scale equivalents we used NIH Image, software provided by NIH as freeware (http://rsb.info.nih.gov/nih-image/index.html). To convert the color series of the individual pictures, using a standard window of 20 cm length, the edited page is selected, copied, and pasted into NIH Image. The graph command then automatically converts color values into a graph of gray-scale values ranging from 0 (white) to 255 (black), pixel by pixel.

Recording and Smoothing

Pasting the gray-scale graph into Kaleidagraph® generates pixel values in the Data window. The "create series" option in the Functions menu serves to provide the relative depth for each data point. For example, given a scan segment extending from depth centimeters 300 to 310, these are entered into the "initial value" boxes of the Create Series panel. 200 pixels in this interval implies a "pixel density" of 0.5 mm / pixel, a figure to be entered into the "linear increment" box. The time series can then be smoothed by choosing the most appropriate pixel size, as explained below.

The dip in the core generally provided stratigraphic overlap between successive segments, helping to concatenate the segments accurately.

Conversion resulted in a gray-scale time series of ca. 56,000 pixels, each equivalent to 0.6 mm of rock. In terms of time, at a mean accumulation rate of ca. 4 Bubnoffs (mm / ky) this implies a mean sample spacing of 200 years.

The definition of geologically distorted responses to simple sinusoidal forcing requires denser sampling than the Nyquist rule would suggest. However, the retention of oscillations at frequencies beyond those sought for is costly in computer time and complicates the graph of the gray scale by a clutter of sub-Milankovitch oscillations, a "furriness" that obscures the larger pattern. We therefore changed pixel density to one equivalent to 6 mm of rock, thereby smoothing the gray-scale log. Figures 4 and 16 show the photolog and, alongside, the gray-scale log, in mirrored form, with "black" at the center, in order to increase visual impact.

Spectra and Tuning.—

Subsequent extraction of power spectra, tuning of the time series to given frequencies, etc., has been made possible by the use of the Analyseries© (Paillard et al., 1996). Cyclicities were apparent from visual inspection of the photolog and the corresponding gray-scale log, and were further explored by means of Fourier frequency spectra of the Thompson multi-tapered type. For tuning the time series to specific frequencies we selected the cycle boundaries and located them in the time series. We then expanded or shrank successive sets of the cycles, by adjusting them to a template with an equal number of equispaced boundaries. This new, tuned graphic series was then resampled at equispaced intervals, in an approach toward the Fourier demand of sampling at equal time intervals. This new series, theoretically closer to time than the initial one, provided improved spectra (Figs. 3, 6, 11). Further operations are described in the text.

ACKNOWLEDGMENTS

Our early work on cyclicity in the Italian Cretaceous, including the drilling of the core, was supported by the National Science Foundation. This and the Italian Consiglio Nazionale delle Ricerche supported the drilling and processing of the Piobbico core. Early studies of the core were aided by the American Chemical Society's Petroleum Research Fund and the International Office of the National Science Foundation. The reanalysis of the Piobbico core was supported by the ENI Petroleum Company's AGIP Division. We are most grateful for the support of these organizations.

Lisa Pratt supervised the coring by the skillful CNR drillers and the beginning stages of core description. Core preparation was performed by several students of Premoli Silva at Universitá di Milano. Elisabetta Erba and Maria Emilia Tornaghi contributed a lithologic study of the core, Maurizio Ripepe the computer files and early computer studies. Discussions with Giovanni Napoleone and Andrea Albianelli engaged in studies of the core's magnetic structure, with Elisabetta Erba, studying the nannoflora, and with Maurizio Orlando and Luciano Gorla of AGIP have proved most rewarding. We thank them all for their participation and aid.

REFERENCES

BARRERA, E., AND JOHNSON, C.C., EDS., 1999, Evolution of Cretaceous Ocean–Climate Systems: Geological Society of America, Special Paper 332, 445 p.

BEAUDOIN, B., M'BAN, E.P., MONTANARI, A., AND PINAULT, M., 1996, Lithostratigraphie haute résolution (< 20 ka) dans le Cénomanien du bassin d'Ombrie-Marches (Italie): Académie des Sciences (Paris), Comptes Rendus, v. 323, no. 2a, p. 689–696.

BERGER, A., 1989, Pre-Quaternary Milankovitch frequencies: Nature, v. 342, p. 133

BERGER, A., LOUTRE, M.F., AND DEHANT, V., 1989, Influence of the changing lunar orbit on the astronomical frequencies of Pre-Quaternary insolation patterns: Paleoceanography, v. 4, p. 555–564.

BROMLEY, R.G., AND EKDALE, A.A., 1984, *Chondrites*, a trace-fossil indicator of anoxia in sediments: Science, v. 224, p. 872–874.

CHAMBERLIN, T.C., 1906, On a possible reversal of deep-sea circulation and its influence on geologic climates: Journal of Geology, v. 14, p. 363–373.

CRESTA, S., MONECHI, S., AND PARISI, G., 1989, Stratigrafia del Mesozoico e Cenozoico nell'area Umbro-Marchigiana: Memorie Descrittive della Carta Geologica d'Italia, v. XXXIX, 185 p.

DEAN, W.E., AND ARTHUR, M.A., 1999, Sensitivity of the North Atlantic Basin to cyclic climatic forcing during the Early Cretaceous: Journal of Foraminiferal Research, v. 29, p. 465–486.

DE BOER, P.L., 1982. Cyclicity and storage of organic matter in Middle Cretaceous pelagic sediment, *in* Einsele, G., and Seilacher, A., eds., Cyclic and Event Stratification: New York, Springer Verlag, p. 456–474.

DE BOER, P.L., 1983, Aspects of Middle Cretaceous pelagic sedimentation in southern Europe: production and storage of organic matter, stable isotopes and astronomic influences: Geologica Ultraiectina, Utrecht, v. 31, 112 p.

DE BOER, P.L., AND WONDERS, A.A.H., 1984, Astronomically induced rhythmic bedding in Cretaceous pelagic sediments near Moria (Italy), *in* Berger, A.L., Imbrie, J., Hays, J., Kukla, G., and Saltzmann, B., eds., Milankovitch and Climate: Dordrecht, The Netherlands, D. Reidel, p. 177–190.

DECONTO, R.M., BRADY, E.C., BERGENGREN, J., AND HAY, W.W., 2000, Late Cretaceous climate, vegetation, and ocean interactions, *in* Huber, B.T., Macleod, K.G., and Wing, S., eds., Warm Climates in Earth History: Cambridge, U.K., Cambridge University Press, p. 276–296.

DIETRICH, G., AND ULRICH, J., 1968, Atlas zur Ozeanographie: Bibliographisches Institut Mannheim, Germany, p. 40.

DOUGLAS, R.G., AND SAVIN, S.M., 1975, Oxygen and carbon isotope analyses of Tertiary and Cretaceous microfossils from Shatsky Rise and other sites in the North Pacific Ocean, *in* Gardner, J.V., ed., U.S. Government Printing Office, Washington, D.C., Initial Reports of the Deep Sea Drilling Project, v. 32, p. 509–520.

ERBA, E., 1986, I nannofossili calcarei nell'Aptiano–Albiano (Cretacico inferiore): Biostratigrafia, paleoceanografia e diagenesi degli Scisti a Fucoidi del pozzo Piobbico (Marche): Ph.D. Dissertation, Università degli Studi di Milano, 313 p.

ERBA, E., 1988, Aptian–Albian calcareous nannofossils biostratigraphy of the Scisti a Fucoidi cored at Piobbico (central Italy): Rivista Italiana di Paleontologia e Stratigrafia, v. 94, p. 249–284.

ERBA, E., 1992, Calcareous nannofossil distribution in pelagic rhythmic sediments (Aptian–Albian Piobbico core, Central Italy): Rivista Italiana di Paleontologia e Stratigrafia. v. 97, p. 455–484.

ERBA, E., AND PREMOLI SILVA, I., 1994, Orbitally driven cycles in trace-fossil distribution from the Piobbico core (Late Albian, central Italy), *in* de Boer, P.L., and Smith, D.G., eds., Orbital Forcing and Cyclic Sequences: International Association of Sedimentologists, Special Publication 19, p. 211–226.

FIET, N., BEAUDOIN, B., AND PARIZE, O., 2001, Lithostratigraphic analysis of Milankovitch cyclicity in pelagic Albian deposits of central Italy: implications for the duration of the stage and substages: Cretaceous Research, v. 22, p. 265–275

FISCHER, A.G., HERBERT, T.D., NAPOLEONE, G., PREMOLI-SILVA, I., AND RIPEPE, M., 1991, Albian pelagic rhythms (Piobbico Core): Journal of Sedimentary Petrology, v. 61, no. 7 (Special Issue), p. 1164–1172.

GRADSTEIN, F.M., AGTERBERG, F.P., OGG, J.G., HARDENBOL, J., VAN VEEN, P., THIERRY, J., AND HUANG, Z., 1995, A Triassic, Jurassic and Cretaceous Time Scale, *in* Berggren, W.A., Kent, D.V., Aubry, M.-P., and Hardenbol, J., eds., Geochronology, Time Scales and Global Stratigraphic Correlation: SEPM, Special Publication 54, p. 95–126.

HAY, W.W., AND DECONTO, R., 1999, Comparison of modern and Late Cretaceous meridional energy transport and oceanology, *in* Barrera, E., and Johnson, C.C., eds., 1999, Evolution of Cretaceous Ocean–Climate Systems: Geological Society of America, Special Paper 332, p. 283–300.

HERBERT, T.D., 1992, Paleomagnetic calibration of Milankovitch cyclicity in Lower Cretaceous sediments: Earth and Planetary Science Letters, v. 112, p. 15–28.

HERBERT, T.D., AND FISCHER, A.G., 1986, Milankovitch climatic origin of mid-Cretaceous black shale rhythms in central Italy: Nature, v. 321, p. 739–743.

HERBERT, T.D., GEE, J., AND DIDONNA, S., 1999, Precessional cycles in Upper Cretaceous pelagic sediment of the South Atlantic: Long-term patterns from high frequency climate variations, *in* Barrera, E., and Johnson, C.C., eds., Evolution of the Cretaceous Ocean–Climate System: Geological Society of America, Special Paper 332, p. 105–220.

HERBERT, T.D., PREMOLI SILVA, I., ERBA, E., AND FISCHER, A.G., 1995, Orbital chronology of Cretaceous–Paleocene marine sediments, *in* Berggren, W.A., Kent, D.V., Aubry, M.-P., and Hardenbol, J., eds., Geochronology, Time Scales and Global Stratigraphic Correlation: SEPM, Special Publication 54, p. 81–94.

HERBERT. T.D., STALLARD, R.F., AND FISCHER, A.G., 1986, Anoxic events, productivity rhythms and the orbital signature in a mid-Cretaceous deep-sea sequence from Central Italy: Paleoceanography, v. 1, p. 495–506.

HERRLE, J.O., 2002, Paleoceanographic and paleoclimatic implications on mid-Cretaceous black shale formation in the Vocontian basin and the Atlantic: evidence from calcareous nannofossil and stable isotopes: Tübinger Mikropaläontologische Mitteilungen, v. 27, 114 p.

HINNOV, L.A., 2000, New perspectives on orbitally forced stratigraphy: Annual Review of Earth and Planetary Sciences, v. 28, p. 419–475.

HINNOV, L.A., AND PARK, J., 1999, Strategies for assessing Early–Middle (Pliensbachian–Aalenian) Jurassic cyclochronologies: Royal Society (London), Philosophical Transactions, Series A, v. 357, p. 1831–1859.

HINNOV, L.A., SCHULZ, M., AND YIOU, P., 2002, Interhemispheric space–time attributes of the Dansgaard–Oeschger oscillations between 0–100 ka: Quaternary Science Reviews, v. 21, p. 1213–1228.

HUBER, B.T., MACLEOD, K.G., AND WING, S. L., EDS., 2000, Warm Climates in Earth History: Cambridge, U.K., Cambridge University Press, 462 p.

IRVING, E., NORTH, F.K., AND COUILLARD, F., 1974, Oil, climate and tectonics: Canadian Journal of Earth Sciences, v. 11, p. 1–15.

JENKYNS, H.C., 1980, Cretaceous anoxic events: from continents to oceans: Geological Society of London, Journal, v. 137, p. 275–291.

LASKAR, J., 1999, The limits of Earth orbital calculations for geological time-scale use: Royal Society (London), Philosophical Transactions, Series A, v. 357, p. 1735–1759

MANN, M.E., AND LEES, J.M., 1996, Robust estimation of background noise in signal detection in climatic time series: Climatic Change, v. 35, p. 409–455.

OLSEN, P.E., AND KENT, D.V., 1999, Long-period Milankovitch cycles from the Late Triassic and Early Jurassic of eastern North America and their implications for the calibration of the Early Mesozoic timescale and the long term behavior of the planets: Royal Society (London), Philosophical Transactions, Series A, v. 357, p. 1761–1786.

PAILLARD, D., LABEYRIE, L., AND YIOU, P., 1996, Macintosh program performs time-series analysis: Internet program referenced to Eos, Transactions, American Geophysical Union, v. 77, p. 379.

PARK, J., D'HONDT, S.L., KING, J.W., AND GIBSON, G., 1993, Late Cretaceous precessional cycles in double-time: a warm-Earth Milankovitch response: Science, v. 261, p. 1331–1334.

PARK, J., AND HERBERT, T.D., 1987, Hunting for paleoclimatic periodicities in a geologic time series with an uncertain scale: Journal of Geophysical Research , v. 92, p. 14,027–14,040.

PRATT, L.M., AND KING, J.D., 1986, Variable marine productivity and high eolian input recorded by rhythmic black shales in mid-Cretaceous pelagic deposits from central Italy: Paleoceanography, v. 1, p. 507–522.

PREMOLI SILVA, I., 1977, Upper Cretaceous–Paleocene magnetic stratigraphy at Gubbio, Italy: II—Biostratigraphy: Geological Society of America, Bulletin, v. 88, p. 367–389.

PREMOLI SILVA, I., ERBA, E., AND TORNAGHI, M.E., 1989a, Paleoenvironmental signals and changes in surface fertility in Mid-Cretaceous C-org-rich pelagic facies of the Fucoid Marls (central Italy): Géobios, Mémoire Spécial, v. 11, p. 225–236, 6 figs., 3 tables, Lyon.

PREMOLI SILVA, I., TORNAGHI, M.E., AND RIPEPE, M., 1989b, Planktonic foraminiferal distribution records productivity cycles: evidence from the Aptian–Albian Piobbico core (Central Italy): Terra Nova, v. 1, p. 443–448.

PREMOLI SILVA, I., AND SLITER, W.V., 1995, Cretaceous planktonic foraminiferal biostratigraphy and evolutionary trends from the Bottaccione section, Gubbio, Italy: Palaeontographica Italica, v. 82, p. 1–89.

PRETO, N., HINNOV, L.A., HARDIE, L.A., AND DE ZANCHE, V., 2001, A Middle Triassic orbital signature recorded in the shallow-marine Latemar carbonate buildup (Dolomites, Italy): Geology, v. 29, p. 1123–1126.

RIPEPE, M., AND FISCHER, A.G., 1991, Stratigraphic rhythms synthesized from orbital variations, *in* Franseen, E.K., Watney, W.L., Kendall, C.G.St.C., and Ross, W., eds., Sedimentary Modeling: Computer Simulations and Methods for Improved Parameter Definition: Kansas Geological Survey, Bulletin, 233, p. 335–344.

ROSSIGNOL-STRICK, M., 1985, Mediterranean Quaternary sapropels, and immediate response of the African monsoon to variation of insolation: Palaeogeography, Palaeoclimatology, Palaeoecology, v. 49, p. 237–263.

SAVRDA, C.E., AND BOTTJER, D.J., 1994, Ichnofossils and ichnofabrics in rhythmically bedded pelagic–hemipelagic carbonates: recognition and evaluation of benthic redox and scour cycles, *in* de Boer, P.L., and Smith, D.G., eds., Orbital Forcing and Cyclic Sequences: International Association of Sedimentologists, Special Publication 19, p. 195–210.

SCHWARZACHER, W., 1953, Cyclostratigraphy and the Milankovitch Theory: Amsterdam, Elsevier, 225 p.

SCHWARZACHER, W., 1994, Cyclostratigraphy of the Cenomanian in the Gubbio district, Italy, a field study, *in* de Boer, P.L., and Smith, D.G., eds., Orbital Forcing and Cyclic Sequences: International Association of Sedimentologists, Special Publication, 19, p. 87–98.

SCHWARZACHER, W., AND FISCHER, A.G., 1982, Limestone–shale bedding and perturbations of the earth's orbit, *in* Einsele, G., and Seilacher, A., eds., Cyclic and Event Stratification: New York, Springer-Verlag, p. 72–95.

Shackleton, N.J., Crowhurst, S.J., Weedon, G.P., and Laskar, J., 1999, Astronomical calibration of Oligocene–Miocene time: Royal Society (London), Philosophical Transactions, Series A, v. 357, p. 1907–1930.

Thurow, J., and Nederbragt, A.J., 1999, Cretaceous laminated sediments as climate archive (abstract): European Union of Geosciences 10, Strasbourg, France, Abstract volume, p. 225.

Tissot, B., 1979, Effects of prolific petroleum source rocks and major coal deposits caused by sea-level changes: Nature, v. 277, p. 463–465.

Tornaghi, M.E., Premoli Silva, I., and Ripepe, M., 1989, Lithostratigraphy and planktonic foraminiferal biostratigraphy of the Aptian–Albian "Scisti a Fucoidi", Piobbico core, Marche, Italy: background for cyclostratigraphy: Rivista Italiana di Paleontologia e Stratigrafia, v. 95, p. 223–264.

Zachos, J.C., Shackleton, N.J., Revenaugh, J.S., Pälike, H., and Flowers, B.P., 2001, Climate response to orbital forcing across the Oligocene–Miocene boundary: Science, v. 292, p. 474–478.

STATISTICAL TIME-SERIES ANALYSIS AND SEDIMENTOLOGICAL TUNING OF BEDDING RHYTHMS IN A TRIASSIC BASINAL SUCCESSION (SOUTHERN ALPS, ITALY)

FLORIAN MAURER
Department of Earth and Life Sciences, Free University, De Boelelaan 1085, 1081 HV Amsterdam, The Netherlands
present address: The Petroleum Institute, P.O. Box 2533, Abu Dhabi, U.A.E.
e-mail: fmaurer@pi.ac.ae
LINDA HINNOV
Morton K. Blaustein Department of Earth and Planetary Sciences, Johns Hopkins University,
Baltimore, Maryland 21218, U.S.A.
e-mail: hinnov@jhu.edu
AND
WOLFGANG SCHLAGER
Department of Earth and Life Sciences, Free University, De Boelelaan 1085, 1081 HV Amsterdam, The Netherlands
e-mail: wolfgang.schlager@falw.vu.nl

ABSTRACT: Zircon dates and orbital interpretation of bedding rhythms have yielded very different estimates on the duration of Middle Triassic stages. Recently, a core was drilled in Middle Triassic basin sediments at Seceda (western Dolomites) to directly compare cyclostratigraphy with geochronologic data. Detailed study of facies, sediment sources, and transport mechanisms formed the basis of the statistical analysis of bedding rhythms that are based on a grayscale scan and a gamma-ray well log. Amplitude spectrograms reveal strong frequency components at $f = 0.025$ cycles / cm in the main nodular limestone interval (92–64 m core depth), corresponding to the dominant 40 cm bedding thickness. Significant spectral differences were found between the grayscale and gamma-ray bedding proxies, placing doubt on the appropriateness of the use of the latter as an effective tool in cyclostratigraphy. In the uppermost part of the succession (59–45 m core depth) calciturbidites constitute more than 50% of the rock volume. If turbidites and tuffs are removed from the rock column, the spectrogram in this interval becomes much smoother and significant peaks appear at higher frequencies. The signals of this pelagic background sedimentation were extracted by bandpass filtering and show strong similarities to Milankovitch cycles in the Quaternary. According to this cyclostratigraphic interpretation, the dominant 40 cm bedding rhythm was produced by eccentricity, and the average sedimentation rate results in ~ 3.6 mm / ky. This estimate is in contrast to zircon data from volcaniclastic layers that bracket this core interval and suggest a sedimentation rate of 13.5 mm / ky. As it currently stands, neither of the two interpretations is yet fully satisfactory. Although the presence of orbital variations in the Triassic analogous to those predicted for the last 20 My remains questionable owing the presumed chaotic behavior of the planets, the zircon age data have uncertainties related to their origin that remain unaccounted for and require further investigation.

INTRODUCTION

The recognition of orbitally forced paleoclimatic signals in marine sedimentary rocks has gained increased attention over recent years as investigations have advanced further into the remote geological past. Orbital signals recovered from the ocean sedimentary record have already been used to tune and improve Plio-Pleistocene chronology. The detection of orbital variations over longer and increasingly older time scales, and their concatenation into an astronomical time scale, has even led to significant revisions in the ages of some major stratigraphic boundaries (Shackleton et al., 1990; Shackleton et al., 1999). Correlations of deep-sea records with bedding patterns observed in onshore outcrops in the Mediterranean area has enabled the construction of a tuned astronomical polarity time scale for the whole Neogene (Hilgen et al., 1999). The extension of such a time scale and its use parallel with conventional radiometric age dating is tempting, because the accuracy of radiometric age data decreases back in time. An astronomically based calibration technique seems even more attractive now that Milankovitch cycles have been shown to be present in the Mesozoic (e.g., Herbert et al., 1995, for the Cretaceous; Olsen and Kent, 1999, for the Triassic).

Shallow-water cycles in Alpine Triassic carbonate platforms have long been regarded as allocyclic and of Milankovitch origin (e.g., Fischer, 1964; Schwarzacher and Haas, 1986). The latest example of orbital control on the cyclicity in shallow-water carbonates was reported from the Middle Triassic Latemar platform in the Southern Alps (Goldhammer et al., 1987, 1990). Hinnov and Goldhammer (1991) found evidence for 5:1 thickness bundling of the Latemar cycles consistent with Milankovitch forcing. They postulated a precession-scale timing for most of the 600 platform cycles and estimated a period of approximately 12 My for the deposition of the Latemar cyclic succession. Subsequently, Mundil et al. (1996) dated single zircons of tuff layers in age-equivalent basinal limestones (Buchenstein Fm.) and obtained a time interval of less than 5 My for their deposition. Because these results were highly controversial, a heated debate ensued over the duration of the Middle Triassic, raising questions both on the accuracy of radiometric age data and on the allocyclic nature of peritidal carbonate cycles (Brack et al., 1996, 1997; Hardie and Hinnov, 1997). New studies carried out on the Latemar succession lent new support for the Milankovitch forcing hypothesis (Preto et al., 2001), whilst others proposed alternative hypotheses involving sub-Milankovitch forcing to resolve the conflict (Mundil et al., 2003; Zühlke et al., 2003).

In an attempt to solve the dispute, scientists of both parties joined in the Seceda coring project to drill a core in the basinal Buchenstein Fm. near Seceda (Western Dolomites) to analyze the

Cyclostratigraphy: Approaches and Case Histories
SEPM Special Publication No. 81, Copyright © 2004
SEPM (Society for Sedimentary Geology), ISBN 1-56576-108-1, p. 83–99.

bedding rhythms of these deposits (Brack et al., 2000). If strong orbital rhythms were to be found, so the argument, then direct comparison of the zircon ages with the orbital chronology would be possible. Furthermore, the basinal equivalent is suggested to have a more continuous record than the frequently exposed platform top, as demonstrated for recent sedimentary systems (Haak and Schlager, 1989). A complicating attribute of the Buchenstein Fm. is that deposition was fed from at least three different sediment sources: a) detritus from surrounding carbonate platforms, b) planktic material, consisting of organic matter, and siliceous and possibly also calcareous tests of microorganisms, and c) volcanic detritus. Because different sedimentation rates can be expected for the three lithosomes, time-series analyses will be meaningful only after sedimentological studies have succeeded in separating them and quantifying the variations of input through time.

The present contribution is a first attempt to statistically analyze bedding rhythms in the Buchenstein Fm. Standard time-series and spectral-analysis techniques as well as statistical significance tests are used to identify the significant frequency components in the succession. To decide whether orbital forcing is present, the spectra created are compared with those of theoretical Milankovitch cycles. Thus, the approach used in this study is a test that is independent of the accuracy of the zircon dating and the controversial Milankovitch interpretation of the (partially) coeval Latemar platform. The present statistical analysis builds on sedimentological studies on the Seceda core and surrounding outcrops that documented depositional facies, sources of sediment, and changes in sediment input through time (Maurer and Schlager, 2003; Maurer et al., 2003). The information on the sedimentology of the Buchenstein basin should help to improve the statistical evaluations and recognize the limits in time series and spectral analysis.

CHARACTERISTICS OF THE BUCHENSTEIN BASIN FILL

The Middle Triassic Buchenstein Fm. was deposited in interplatform basins up to 1000 m deep throughout the Southern Alps (Brack and Rieber, 1993). In the Dolomites these hemipelagic sediments extend over an area of 500 km^2 and consist, where complete, of a succession, 50–70 m thick, of limestones and marls with volcaniclastic intercalations (Fig. 1; Viel, 1979; Bosellini and Ferri, 1980). The biostratigraphic age of the formation ranges from the late Anisian to late Ladinian, includes five ammonoid zones, and covers two normal-magnetic-polarity and one reversed-magnetic-polarity zones (Fig. 2; Brack and Rieber, 1993; Brack and Muttoni, 2000). The Buchenstein Fm. is developed in

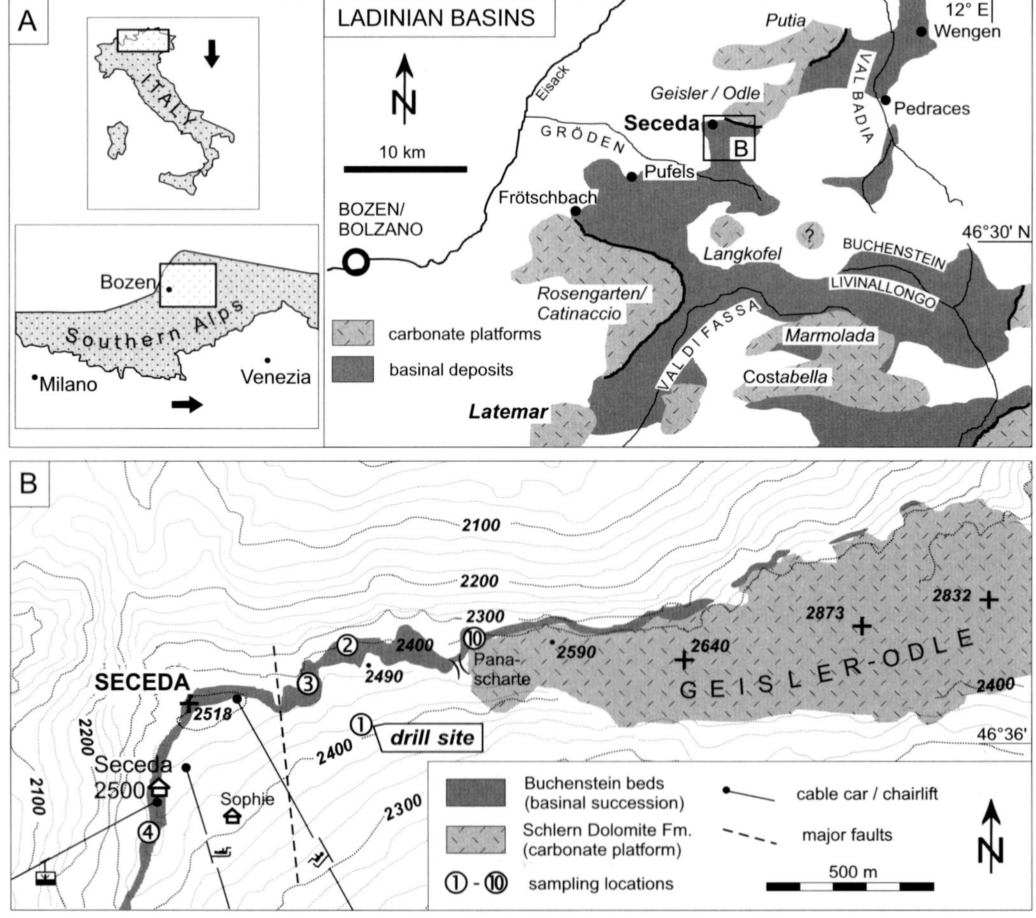

FIG. 1.—**A)** Distribution of Middle Triassic carbonate platforms and basins in the Dolomites of northern Italy during the Middle Triassic (Ladinian). Black lines mark approximate slope–basin transitions. **B)** Close-up of the study area around Seceda showing main outcrops of Buchenstein Fm. and age-equivalent platform carbonates. The encircled numbers indicate the location of the drill site (1) and studied outcrop sections.

both laminated ("Lower Plattenkalke", "Bänderkalke") and bioturbated ("Knollenkalke") facies and locally interfingers with the slopes of high-rising carbonate platforms (Bosellini, 1984). The laminated facies yields filament-radiolarian wackestones and clay and is rich in organic matter. The bioturbated intervals are made up of centimeter- to decimeter-bedded siliceous nodular limestones with centimeter-thick intercalations of marls that are less resistant to surface weathering. Most of these beds, such as marker beds 1–6 (Fig. 2), can be correlated over more than 30 km in the Buchenstein basin of the Dolomites (Brack and Muttoni, 2000). A comparison with the laminated facies shows that at least part of the bedding is caused by different stages of cementation during early diagenesis (Maurer and Schlager, 2003).

The Buchenstein basin was fed from different sediment sources, an attribute that complicates the detection of any cyclicity potentially present in the background sediment of the succession. Background sedimentation was interrupted by fallout of volcaniclastics and frequent deposition of turbidites shed from adjoining carbonate platform slopes. Both turbidites and volcaniclastics change laterally in amount and thickness, in respect to their source area. Therefore it is likely that removing these sediments from the succession will decrease the noise in the dataset and improve the accuracy of time series and spectral analysis. The volcaniclastic fraction is concentrated in three intervals and was used to establish a detailed tephrastratigraphy in the Buchenstein basin (e.g., the T_c–T_e tuffs in Fig. 2; Brack and Rieber, 1993). These greenish volcaniclastic sandstones and siltstones are easily distinguished from the carbonate fraction and were removed prior to running analyses. The input of turbidites and its relation to the bedding patterns and the background sediment was quantified by observations in the Seceda core and surrounding outcrops (Maurer and Schlager, 2003; Maurer et al., 2003) and is summarized as follows:

- Calciturbidites and breccias make up approximately one-third of rock volume in the Seceda core, increasing upward from 5% to 60%, in step with the progradation of the adjacent platform slopes towards the well site.

- Calciturbidites are well identifiable in the core and occur in various dimensions, from millimeter- to decimeter-thick layers of carbonate sandstones to mudstones. Over 70% of the turbidites are less than 1 cm thick, but these contribute only about 3.5% to the total turbidite input.

- Comparison of variations in thickness and sediment composition in outcrop sections and single beds indicates a close relationship between the amount of shallow-water limestones present in the succession and the distance to a carbonate platform source. Turbidity currents are identified as a main mechanism for transporting shallow-water carbonate debris into the basin; apart from the carbonate sand fraction, they carry large amounts of micrite, suggesting a prolific shallow-water source of lime mud.

In the Buchenstein Fm. around Seceda a lateral transition from laminated to bioturbated facies is observed (Maurer and Schlager, 2003). A comparison shows that decimeter-thick bedding is preserved in all facies types, while millimeter-thick to centimeter-thick layers are destroyed by bioturbation. This observation shows how bioturbation affects bedding and marks the limit in the detection of signals in the nodular limestone facies. In contrast to the background sediment, with a considerably low sedimentation rate, turbidites are deposited in time intervals from some hours to days. These abrupt changes in sedimentation rate are likely to disturb the continuous signature of any potential cyclicity, and their removal may therefore help to improve the detection of sedimentary rhythms. Furthermore, core-to-outcrop correlation reveals that around the 60 m level in the core approximately 4 m of the succession present in the outcrop is missing because of local tectonic unconformities (Fig. 2; fig. 3 in Maurer and Rettori, 2002). This gap has to be considered for the statistical evaluation of the succession.

DATA AND METHODS

Data Acquisition and Description

To capture the sedimentary rhythmicity in the Seceda core, two high-resolution data series were developed in a format that could be submitted to statistical time series analysis: (1) a 1-mm-resolution grayscale scan based upon photographs of the slabbed Seceda core and (2) a 1-cm-resolution gamma-ray well log acquired during logging operations. A description of the two datasets, their acquisition procedure, and limitations for time-series analysis follows. The techniques of spectral estimation and filtering, and the associated hypothesis testing performed on the two datasets, are described subsequently.

Grayscale Scan.—

High-resolution color digital photographs of the slabbed Seceda core were used as basis for grayscale analysis. For a 1-m-long core archive box, three photographs, each covering a vertical length of 43 cm, were acquired. Subsequently, for each core meter, three images were stitched together with graphics software Corel Draw. Because the overlap between the photos is 25%, it was possible to minimize optical distortion during this process. On average, each single image covers 33 cm of core, with deviation between 31 and 36 cm. Brightness differences between individual photographs were corrected by calibration to a Kodak color and grayscale bar present on each image. However, slight differences in brightness are still present when observing the contact between two photographs (Fig. 3). The resulting photographs were exported as grayscale images with a resolution of 75 dpi (or 1.67 pixels per mm core). The software Scion Image version 3 for Windows (www.scioncorp.com) was used to convert the pixels into grayscale values between 0 (= white) and 255 (= black). Scanning was performed on a single-pixel transect approximately in the center of the core (Fig. 3). Only the hard parts of the core were scanned; tuff layers and loose core intervals were not included. In an MS Excel data sheet, the intervals composed of sandy turbidite layers were marked on a second column next to the grayscale log in order to be able to later remove these values from the dataset easily. The resulting grayscale log covers 92% of the succession in the processed interval (45–92 m core depth).

As illustrated in Figure 3, the grayscale scan records the bedding pattern in the core in high detail. Marly intercalations result in much higher values than light carbonate beds (see, for instance, core interval 73.2–73.3 m in Fig. 3). The grayscale scan thus represents mainly a proxy for carbonate content, with lighter values for higher and darker values for lower concentrations. Within the limestone beds, lighter gray intervals can be distinguished from darker gray chert patches and nodules (e.g., 73.50–73.55 m in Fig. 3). The artificial shadow effect on top and at the base of each core box, resulting from the positions of spotlights during image acquisition, was removed from the grayscale scan by subtracting the obfuscation caused by the shadow (Fig. 3). To accomplish this, the degree of obfuscation was first determined by observing the light–shadow transition in cores with unique lithology (white

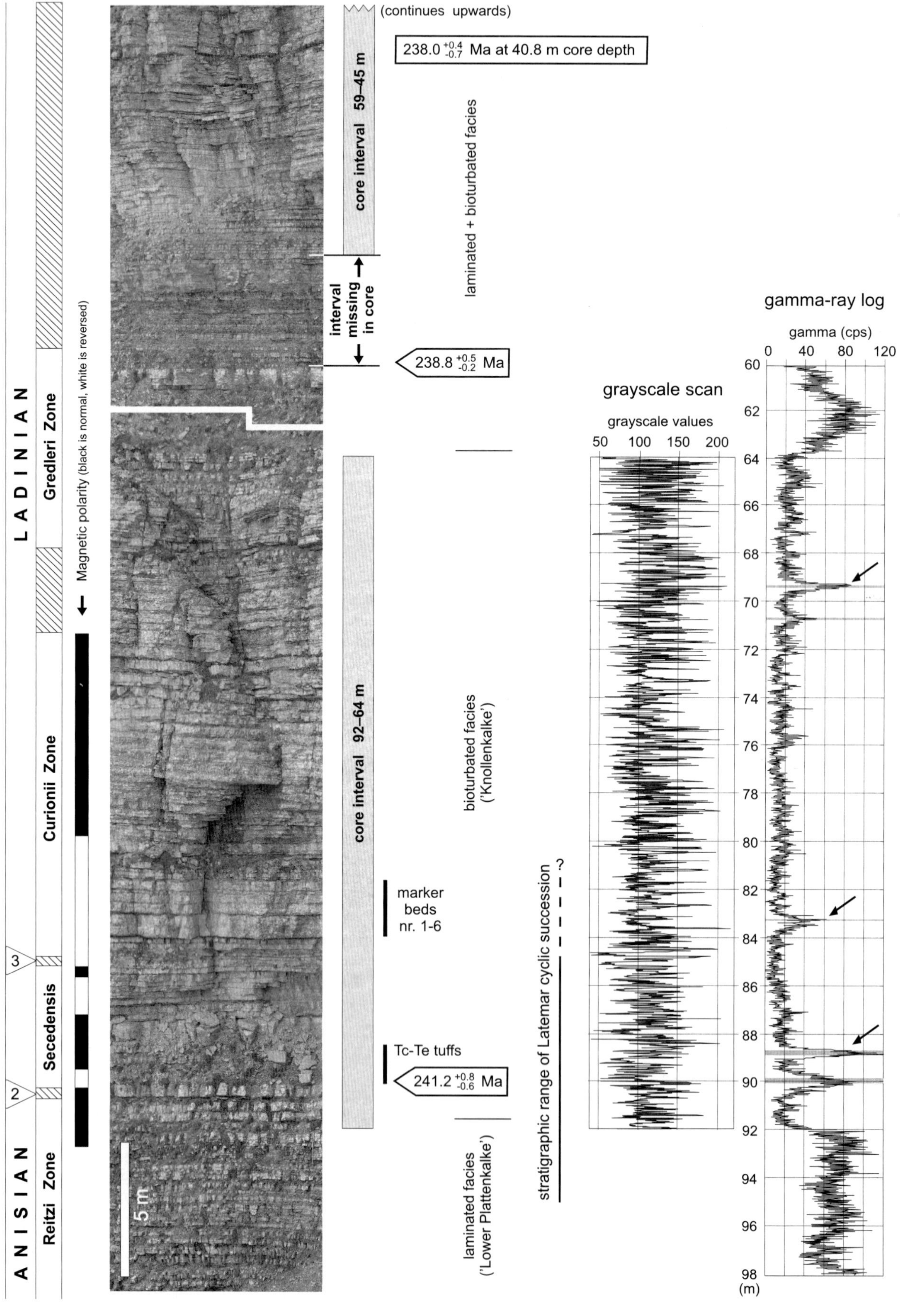
ANISIAN
LADINIAN
Reitzi Zone
Secedensis
Curionii Zone
Gredleri Zone
3
2
Magnetic polarity (black is normal, white is reversed)
5 m
(continues upwards)
238.0 +0.4 -0.7 Ma at 40.8 m core depth
core interval 59–45 m
interval missing in core
238.8 +0.5 -0.2 Ma
core interval 92–64 m
marker beds nr. 1-6
Tc-Te tuffs
241.2 +0.8 -0.6 Ma
laminated + bioturbated facies
bioturbated facies ('Knollenkalke')
laminated facies ('Lower Plattenkalke')
stratigraphic range of Latemar cyclic succession ?
grayscale scan
grayscale values
50 100 150 200
gamma-ray log
gamma (cps)
0 40 80 120
60 62 64 66 68 70 72 74 76 78 80 82 84 86 88 90 92 94 96 98
(m)

dolomites of Contrin Fm. in cores from 102 to 109 m core depth). These reveal that the shadow adds to the lithology around 90 points to real grayscale value (in the spectrum from 0 to 255) and that the transition from the fully lighted part to the shadowed part of the core is more or less a straight line. With this information, the shadow artifact was eliminated from the affected core intervals (Fig. 3). Finally, the grayscale log was resampled to a final resolution of 1 mm (from a previous of 0.6 mm or 1.67 pixels per mm).

Natural Gamma Ray.—

A gamma-ray log was acquired during well logging in the open borehole following completion of drilling operations (Brack et al., 2000). Natural gamma radiation was measured with a sodium iodide crystal 22.22 mm in diameter and 76.2 mm long. Logging was performed from bottom to top with an average velocity of 1 m / min and a sampling frequency of 0.6 s, resulting in a sampling resolution of 1 cm. The crystal receives all the radiation that enters from the side, thus the maximum resolution of the tool is limited to its length (7.62 cm). Signal variations at wavelengths shorter than the length of the crystal are considered to be noise. For variations with wavelengths in the 8–12 cm range, slightly "smoothed" amplitudes in the power spectra are expected. This assumption is based on the fact that in the borehole the strata have an average dip of 25°. With a borehole diameter of 11 cm, the crystal therefore includes radiation of layers from 2.5 cm above and below the crystal (Fig. 4).

Intervals characterized by rather pure limestone beds show the lowest radiation (0–30 counts per second), followed by intercalations of siliceous marls (20–50 cps) and laminated limestones and marls rich in organic matter (40–100 cps). Tuff layers reach the highest values, with sometimes over 100 cps (Fig. 2). The high radiation spikes caused by the tuffs produce the strongest signal in the very low frequencies of the estimated spectra, obscuring adequate detection of possible significant higher frequencies. In order to visualize the frequencies governing the bedding patterns, frequencies between $f = 0$ and $f = 0.016$, caused by the high gamma radiation of tuff layers, were removed from the original gamma-ray log by bandpass filtering. The filter method is described below.

Techniques of Time-Series Analysis

Amplitude Spectrogram.—

The amplitude spectrogram consists of the fast Fourier transformation of overlapping segments of the data series at regular steps along the series, with the output amplitude estimates mapped as a function of time frequency. This technique allows visualization of timewise shifts in frequencies that can result from unstable sedimentation rates (e.g., Meyers et al., 2001). Unless specified otherwise, segments with 4 m length were analyzed; the offset (step) between two adjacent segments is 20 cm. The different colors of the spectrogram reflect amplitude magnitude (red = high values, blue = low values). Because the color scheme of the spectrogram is relative, the comparably "weak" amplitudes in the higher frequencies can come to drown in dark blue. For the calculation of the amplitude spectrograms, the software Matlab release 12 for Windows (www.mathworks.com) was used, with an original script developed for increased windowing flexibility, which is available upon request.

Lomb–Scargle Periodogram.—

This simple spectral technique is based on the algorithms of Lomb (1976) and Scargle (1982) and is usually applied to time series with unequally spaced data. Here it is used for high-resolution periodogram estimation and preliminary evaluation of observed peak significance in terms of periodogram probability. The Lomb–Scargle periodogram follows the Chi-squared distribution with two degrees of freedom (Percival and Walden, 1993, p. 221), allowing specific probability levels to be drawn directly on the power axis (Press et al., 1996). By this measure, if a data spectral peak exceeds the 95% probability level, it may constitute evidence for a significant sinusoid. Lomb–Scargle periodograms were computed in Matlab using the function described in Muller and MacDonald (2000, p. 290).

Multitaper Spectral Analysis.—

The multitaper method (Thomson, 1982, 1990) has proven to be particularly suitable for the statistical time-series analysis of short, highly colored (quasi-periodic) geological data (Hinnov, 2000). The method estimates the power spectral density of a time series through application of a set of orthogonal tapers, called $m\pi$ tapers, the sum of which approximates a rectangular observation time window that is the length of the data. The frequency resolution of the output spectral density is defined by the averaging bandwidth $\beta = 2\,m / T$, where T is the length of the time series, and the parameter m, or time-bandwidth product, is selected, usually with a value between 2 and 10. The number of multitapers to apply follows the general rule $k = 2m$-1. A "jackknife" procedure is used to determine the statistical uncertainties of the output spectral estimates; these are reported in terms of ± 95% confidence limits. A significance test for the presence of harmonic lines (the F-test) is also provided, and is described subsequently. Multitaper analysis and the harmonic F-test were calculated using the software Analyseries for Macintosh (Paillard et al., 1996).

Hypothesis Testing and Noise Modeling

To investigate the presence of significant nonrandom frequency components in the data, hypothesis tests were performed on the estimated data spectra. We considered several approaches, including whether the data conform to a continuous random process, and, if not, whether the indicated nonrandomness can be explained by specific sinusoidal components in the data.

Autoregressive Noise Tests.—

To decide whether the data conform generally to a random process ("noise"), the estimated data spectrum can be evaluated

←

FIG. 2 (opposite page).—Outcrop of the Buchenstein Fm. at the northern slope of Seceda (location 10 in Fig. 1) with indication of intervals used for time-series analysis. The grayscale scan and the gamma-ray well log used for cyclostratigraphic analysis in the lower part of the section are illustrated on the right. Arrows in the gamma-ray log mark the intervals rich in volcaniclastics. Biostratigraphic and magnetostratigraphic data on the left are after Brack and Rieber (1993), Brack and Muttoni (2000), and Brack et al. (2001); radiometric age data are after Mundil et al. (1996). The numbers in the triangles in the biostratigraphic subdivision mark candidates 2 (after Krystyn, 1983) and 3 (after Brack and Rieber, 1993) currently evaluated by the Subcommission for Triassic Stratigraphy for the definition of the base of the Ladinian stage. The stratigraphic range of the equivalent platform interior succession at Latemar, where orbital cycles have been reported, is indicated.

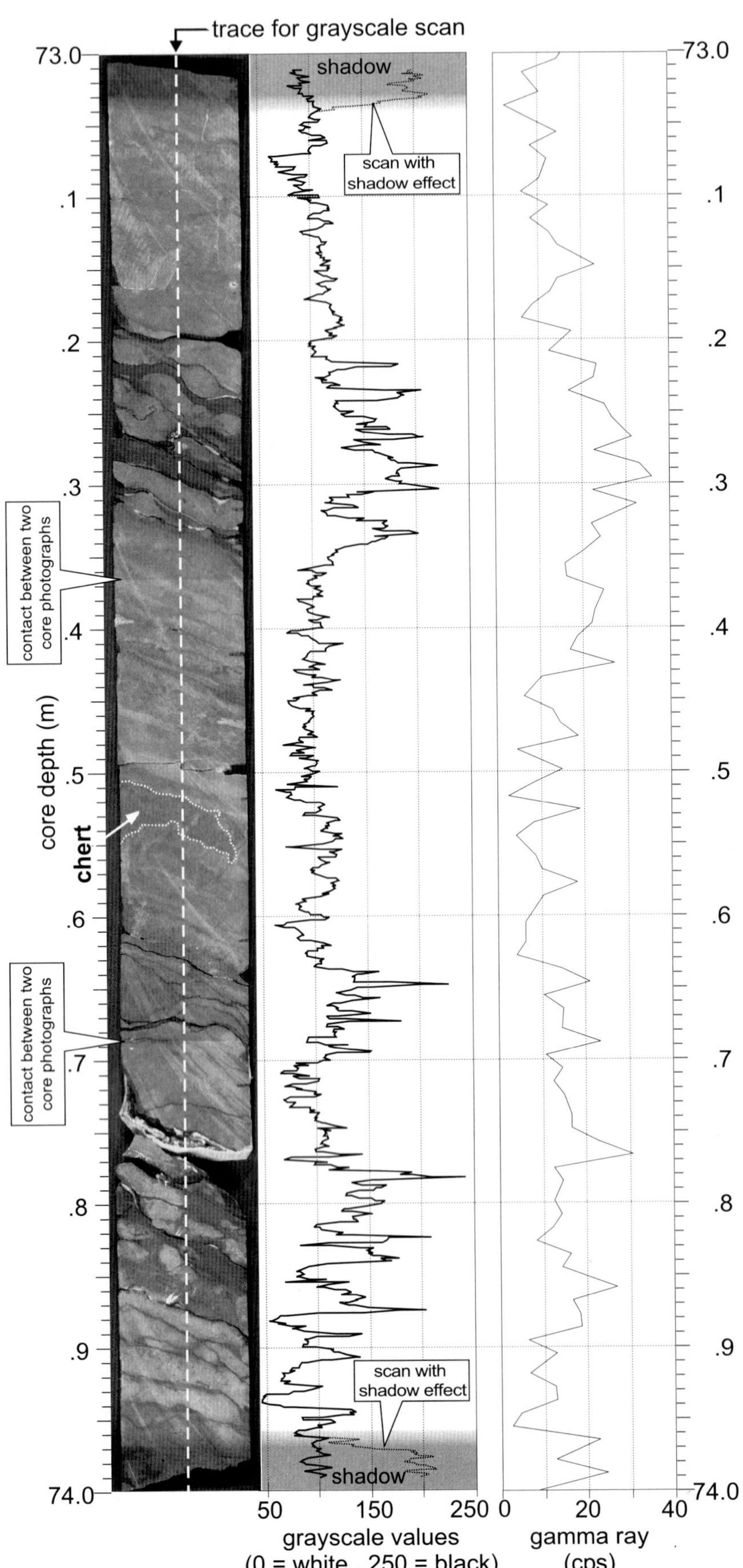

FIG. 3.—73–74 m interval of slabbed Seceda core with corresponding grayscale scan and gamma-ray log. The white dotted line marks the transect where the pixels of the core photographs were converted to grayscale values. The grayscale scan records the bedding pattern in high detail as well as the artificial shadow effect at both ends of the core and the contrast in brightness between the single core photographs visible at their borders. In contrast, the gamma-ray log has a much lower resolution.

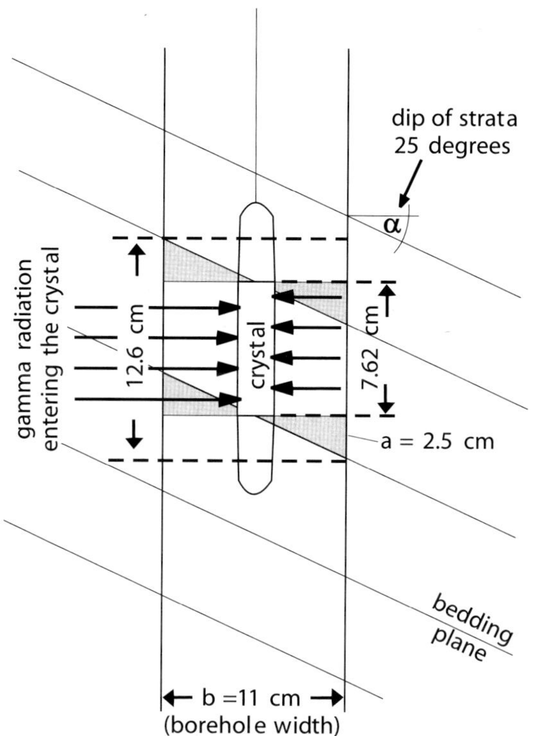

FIG. 4.—Illustration of well logging of natural gamma radiation. The crystal measures all radioactivity entering from the side. Because of the 25° dip of strata the measurement includes $a =$ 2.5 cm of strata above and below the receiver (gray intervals). The length of a is calculated with the equation $a = b/2\tan\alpha$. Both dip of strata and length of the logging crystal contribute to the low resolution of the gamma-ray log.

statistically against that of a prescribed noise model with an equivalent total power (variance). If the data spectrum deviates significantly from that of the noise over a specific frequency band, it can be concluded that the data are nonrandom, and that they contain a "signal" at one or more frequencies within that band. The prevailing thinking is that time-dependent random phenomena tend to conform to what is described as a first-order autoregressive process, or a random process with a spectral power distribution that is weighted more heavily at the lowest frequencies ("red noise"). Here we compute red noise spectra on the basis of this model and "white" noise counterparts with power distributed at all frequencies equally, using autocovariance estimates of the data (Box and Jenkins, 1970, p. 60). Frequencies where the lower 95% confidence limit of the estimated data spectrum exceeds the level of the noise spectra will be considered to be statistical evidence for nonrandomness. It should be noted that because these noise models are based upon the data themselves, which may contain elements of both signal and noise, they will have power levels that are in excess of the true noise in the data. As such, these tests should be considered conservative.

Harmonic F-Test.—

Thomson's harmonic F-test (Thomson, 1982, 1990) is used to identify phase-coherent sinusoids, or harmonics, in data. Here it is used to seek evidence for individual harmonics within nonrandom frequency bands that may be related to specific orbital frequencies. Using the multitapered, Fourier-transformed data, the test sets up a "signal-to-noise" ratio, in which the numerator expresses the spectral power estimated at a central frequency within a bandwidth β, and the denominator expresses the integrated power over the same bandwidth minus that of the central frequency (which is in the numerator). This ratio is equivalent to an F-statistic following a Fisher distribution with 2 and $2m$-2 degrees of freedom. The Fisher distribution is integrated out to the ratio value to determine the significance of the test relative to the ratio of the variance of two random processes; the result is reported as the "F-test," with values close to 1 considered evidence for a significant harmonic at the center of the averaging bandwidth β. All frequencies resolved by the multitapered Fourier-transformed data can be assessed by the test, performed here using Analyseries for Macintosh (Paillard et al., 1996).

Bandpass Filter

The traditional way to visualize the time–frequency behavior of individual frequency bands involves isolating them from the entire spectrum of frequencies by bandpass filtering. In its simplest form, the filtering is applied as follows (code given by Muller and MacDonald, 2000, p. 294): (a) the data are Fourier transformed; (b) power at frequencies outside the range of interest are set to zero; and (3) the result is inverse Fourier transformed. This technique was also used to eliminate strong low-frequency components in the original gamma-ray series (caused by the high gamma radiation of the tuffs; Fig. 2) as follows: all frequency components between $f = 0$ and $f = 0.016$ cycles/cm were bandpassed; the output was then subtracted from the original gamma-ray series to obtain a "high-pass" representation of the data.

RESULTS

Bedding Frequencies Revealed by the Spectrograms

Both amplitude spectrograms for the grayscale and gamma-ray data are composed of several overlapping, unsmoothed periodograms, generated by FFT of zero-padded, segments 4 m long, with an offset of 20 cm between them. They cover the main nodular limestone interval (92–64 m core depth), which was selected for this analysis for the following reasons: (1) the core recovery in this interval, at 98%, is very high, and there are no indications of major gaps in the series; (2) the bedding patterns in the nodular limestone facies are clearly recognizable in core and outcrop, and are laterally consistent over several kilometers in the basin (Brack and Muttoni, 2000; Maurer and Schlager, 2003); (3) the proportion of turbidites is only 10–20%, and single calciturbidite beds are only 2–5 cm thick, and thus much smaller than the average bed thickness observed in the outcrop (Maurer et al., 2003); (4) the input of volcaniclastics is small and limited mainly to three levels, and (5) the "Knollenkalke" interval also covers the time of deposition of the Latemar cyclic succession (Fig. 2; Preto et al., 2001; Zühlke et al., 2003). In the spectrograms the display is limited to the range between $f = 0$ and $f = 0.08$ cycles/cm. Thus, only decimeter-scale bedding is considered. This is the bedding pattern that is preserved in both the laminated and the bioturbated facies.

Because of the average 25° dip of strata, the bed thickness in the core is approximately 10% larger than in the outcrop. The bed thickness in the outcrop is calculated by multiplication of the bed thickness in the borehole times the cosine of the dip angle. Thus, the calculated accumulation rates for the core are also 10% higher than those corrected for dip. If not further specified, all values of bed thicknesses, frequencies, and accumulation rates are related to the core.

Spectrogram of Grayscale Data.—

The grayscale spectrogram reveals the highest amplitudes in the frequency range between $f = 0.02$ and $f = 0.034$ cycles/cm (Fig. 5B). The high power at $f = 0.022$ cycles/cm, between 84 and 81 m core depth, reflect the bedding thickness of 40–50 cm in this interval. The same holds for the high amplitudes located around $f = 0.024$ cycles/cm in the interval 75–71 m. In the low-frequency range ($f = 0.002$ to $f = 0.08$ cycles/cm) high amplitudes occur in the intervals 86–81 m, 79–75 m, and 70–66 m. They correlate with packages of beds of 2–4 m thickness that are more resistant against outcrop weathering in nearby outcrops (Fig. 5A). At 75–66 m core depth, amplitudes around $f = 0.018$ follow the path of the main frequency around $f = 0.024$ cycles/cm and probably reflect partial bundling of beds.

As previously noted, two artifacts exist in the grayscale scan: the shadow effect every meter ($f = 0.01$ cycles/cm) and the brightness differences between the single images approximately every third of a meter ($f = 0.033$ cycles/cm). The latter is most probably responsible for the signal around $f = 0.032$ cycles/cm. In contrast to the previously described components, it is present over the whole interval and is quite stable, whereas the other components have modulating frequencies. Around $f = 0.01$ cycles/cm a weak signal is present only sporadically, confirming that the method to eliminate the shadow effect was successful. The attempt to correct the brightness differences between the single core photographs was less effective.

Spectrogram of Gamma-Ray Log.—

The amplitude spectrogram of the original gamma-ray series is dominated by the high gamma radiation of the tuff layers. As a result, the highest power is present in frequencies lower than $f = 0.016$ cycles/cm and all frequencies higher up are masked by this effect. In order to visualize the power of the higher frequencies in the spectrogram, the low frequencies between $f = 0$ and $f = 0.016$ cycles/cm were removed by high-pass filtering (described above) over the intervals 90–88 m, 84.5–82.5 m, and 69–70 m. In the spectrogram generated from this filtered gamma ray log (Fig. 6B), these same intervals are characterized by loss of power in the frequency range between $f = 0$ and $f = 0.016$ cycles/cm. The previously masked, higher-frequency components occur in the frequency range from $f = 0.0025$ to $f = 0.025$ cycles/cm. High power between $f = 0.02$ and $f = 0.025$ cycles/cm in the intervals 83–80.5 m and 76–71 m reflect the bedding thicknesses of 35–50 cm. The most consistent signal occurs near $f = 0.013$ cycles/cm and extends over the core depth 78–66 m; it largely follows the bedding thickness around $f = 0.022$ cycles/cm and probably reflects a two-bed bundling, partly visible in the outcrop. In the low-frequency range ($f = 0.002$ to $f = 0.004$ cycles/cm) high amplitudes occur in the intervals 82.5–80 m and 78–76 m. They correlate with packages of beds of 2–4 m thickness that are more resistant against weathering (Fig. 6A).

In comparing the two spectra, the following observations can be made: (1) in both spectrograms dominant bedding thicknesses are present as high amplitudes in some intervals, (2) bundling of beds recognized in the outcrop appear as frequency components moving in concordance with the main bedding signal, and (3) the suggested artifact in the grayscale spectrum caused by variations in brightness is not present in the gamma-ray frequencies. This illustrates the importance of using independently sampled datasets in time-series analysis. The bedding thickness in the interval 83-80 m occurs in the grayscale spectrogram in slightly lower frequencies ($f = 0.02$ to $f = 0.023$ cycles/cm) and over a longer part of the section than in the gamma-ray spectrogram ($f = 0.022$ to $f = 0.024$ cycles/cm). The two datasets thus record the bedding rhythms slightly differently; the reasons for that are addressed in the discussion. In both spectrograms, persistent signal components at higher frequencies have so far not been recognized.

Significance Tests

Selected intervals in the two datasets are used to calculate the Lomb–Scargle periodogram and power spectral density via the multitaper method with F-test in order to assess the statistical significance of the various frequency components. The intervals are the same as those in the unsmoothed periodograms illustrated in Figures 5C and 6C. The same intervals were chosen in both datasets. Hence, direct comparison between the grayscale and gamma ray data is possible.

In the Lomb–Scargle periodograms (not shown), for most of the intervals the main bedding rhythm around $f = 0.025$ cycles/cm is represented by spectral peaks exceeding the 99% probability. Because FFT and Lomb–Scargle calculations produce periodograms whose data have only two degrees of freedom, the power spectral density was calculated for the selected intervals using the multitaper method. In doing this, 2π and 4π tapers were applied, which give 6 and 12 degrees of freedom, respectively, allowing calculation of the harmonic F-test. Signal components with over 90% significance, which have been identified by F-test computed in combination with 2π multitaper analysis, are indicated together with the averaging bandwidth β in the periodograms in Figures 5C and 6C. These F-tests show in most spectra highly significant frequency components caused by the bed thicknesses. Thereby, the dominant 40 cm bedding rhythm reaches the highest significance.

DISCUSSION

Comparison of Datasets

As noted previously, comparison of the results of the signal analysis between the two datasets reveals some differences. In the spectrograms one can observe that the main bedding thickness in the interval 83–80 m occurs in the grayscale spectrogram at slightly lower frequencies ($f = 0.02$ to $f = 0.023$ cycles/cm; Fig. 5B) and over a longer part of the section than in the gamma-ray spectrogram ($f = 0.022$ to $f = 0.024$ cycles/cm; Fig. 6B). One reason for these dissimilarities lies in the manner of data acquisition in relation to the dip of strata. The grayscale scan was created using a single-pixel transect down the center of the core (Fig. 3), and so the dip of strata influences the data only in the resulting length of the recorded signals. In contrast, the gamma-ray log is a heavily averaged transect, whose measurement is influenced by multiple dipping strata in the borehole (Fig. 4). This shows that the grayscale scan is by far the more accurate of the two bedding proxies and is therefore the only dataset considered in further analyses presented below.

The differences were further evaluated in a coherency analysis (not shown) of the datasets over the core intervals 81.7–83.3 m and 72.5–74.5 m. These two segments were chosen because they show distinct bedding and because already a good visual correlation between the gamma ray log and grayscale scan is present. However, the analysis reveals that the two datasets are significantly coherent only at a single frequency centered at $f = 0.025$ cycles/cm (the dominant bedding thickness), and not down to

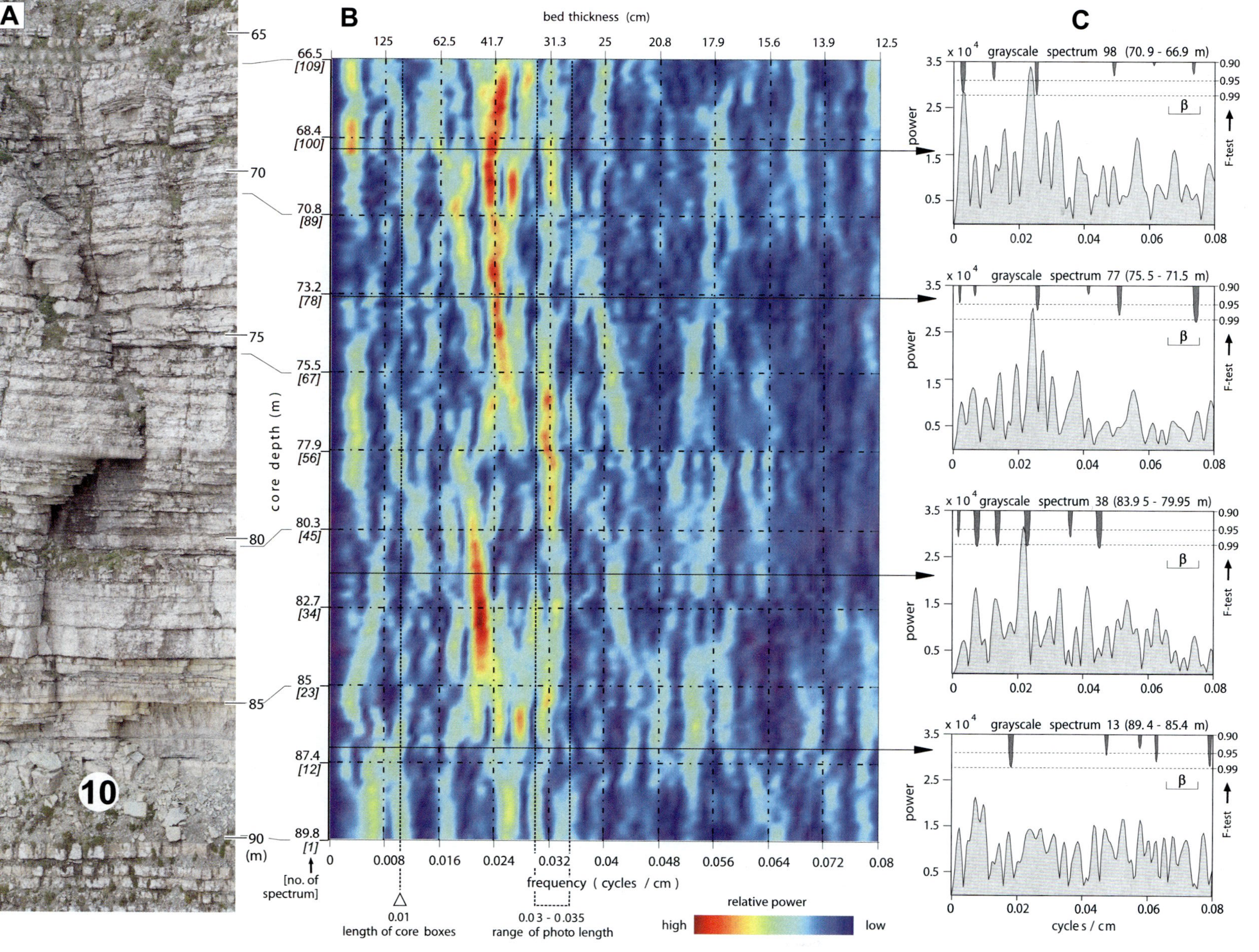

FIG. 5.—**A)** Photo of outcrop section 10 (for location see Fig. 1), showing main nodular limestone interval in the Buchenstein Fm. **B)** Amplitude spectrogram of grayscale scan in core interval 92–64 m, corresponding to interval illustrated in Part A. The strongest amplitudes occur around $f = 0.024$ cycles/cm and reflect the bedding thicknesses observed in the outcrop. **C)** Selection of four single power spectra out of the spectrogram, created by FFT of segments 4 m long data. Significance of harmonics, as calculated by 2π multitaper analysis with F-test, are indicated along with the averaging bandwidth b. Associated multitaper spectra and Lomb–Scargle periodograms are illustrated in the first author's Ph.D. thesis.

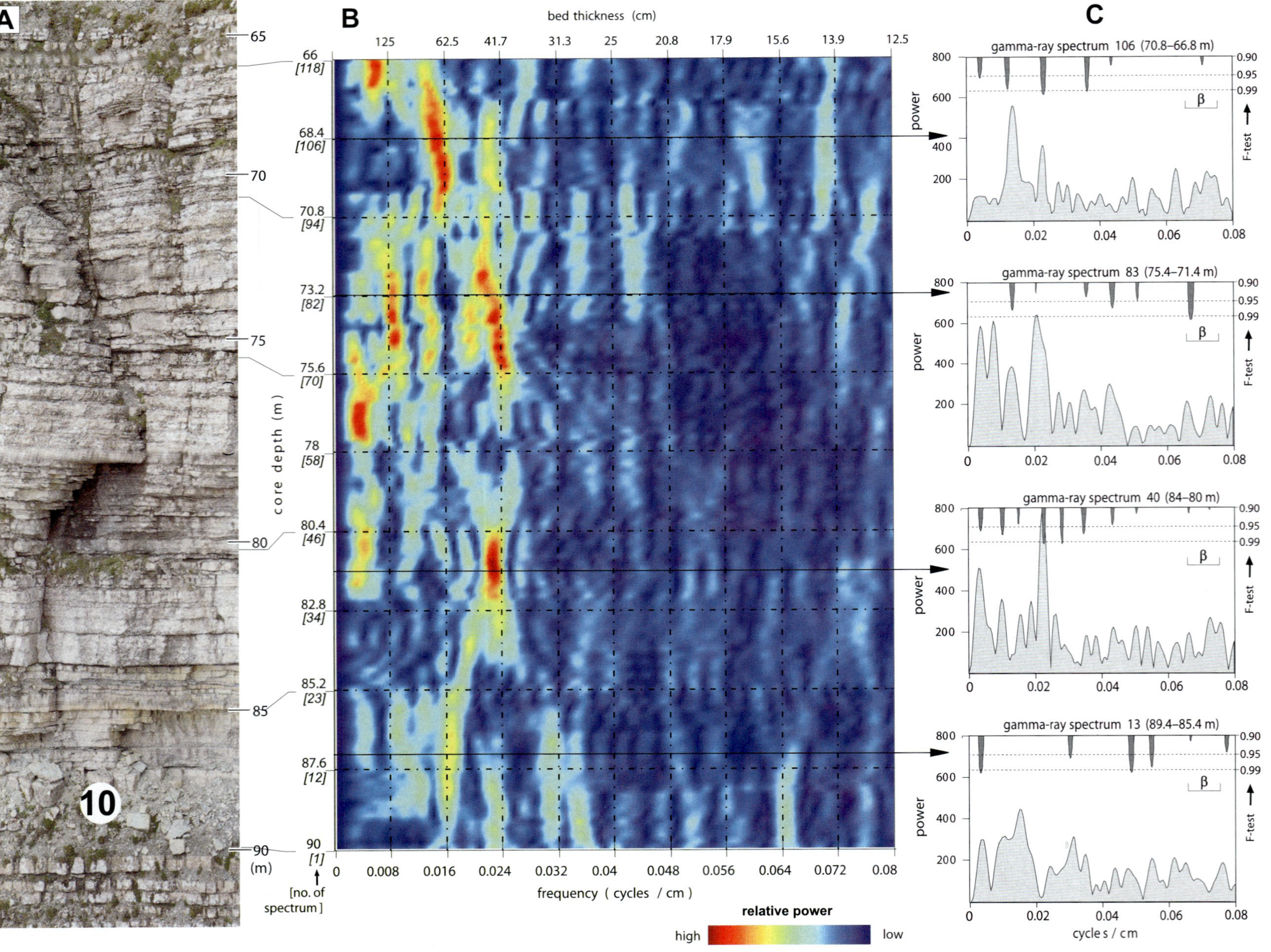

FIG. 6.—**A)** Photo of outcrop section 10 (same as in Fig. 5A). **B)** Amplitude spectrogram of gamma-ray log in core interval 92–64 m, where strongest amplitudes occur in the lower frequencies up to $f = 0.024$ cycles / cm. **C)** Selection of single power spectra out of the spectrogram, taken from the same intervals as the ones in Figure 5C. Significance of harmonics, as calculated by 2π multitaper analysis with F-test, are indicated along with the averaging bandwidth b. Associated multitaper spectra and Lomb–Scargle periodograms are illustrated in the first author's Ph.D. thesis.

the gamma-ray resolution limit of $f = 0.13$ cycles / cm, as expected. The reason most probably lies in the acquisition of the gamma-ray log. The speed with which the logging tool was moved in the borehole varied between 0.9 and 1.1 m / min in the analyzed core interval. This probably caused the gamma-ray series to lag in some parts of the succession and lead in others relative to the core, resulting in disappointingly low coherency between it and the grayscale scan at most frequencies.

These factors illustrate that the use of well-log data alone may not lead to a successful detection of quasiperiodic signals potentially present in a bedding sequence, particularly in basinal successions with bedding rhythms at the centimeter to decimeter scale.

Signal Modulation and Effect of Turbidites

Meyers et al. (2001) explain how a time-frequency signal analysis, such as the spectrogram used here, can reveal modulation in signal frequencies associated with changes in accumulation rate. If the accumulation rate increases, signal components move to lower frequencies (i.e., cyclic bedding thicknesses increase). Lower accumulation rates produce thinner bedding, and therefore signals move to higher frequencies. If the accumulation rate is constant, signals remain stable over the observed interval. Episodic changes in bed thickness in the Buchenstein Fm. are clearly visible in the outcrop (Fig. 2). In the spectrograms (Figs. 5B, 6B) modulations of signal components can be observed as well, and the question arises whether these movements are the result of changes in accumulation rate.

Because sandy turbidites are well recognizable in the core and certainly disturb sedimentation rates, their removal eliminates one source of variability of sedimentation rates. According to Maurer et al. (2003) the percentage of turbidites in the analyzed core interval (92–64 m) varies between 8 and 20%. Although these changes are rather small, modulation of frequencies that can be related to their presence can still be observed. For example, in the grayscale spectrogram (Fig. 5B) the signal component caused by bedding thicknesses near $f = 0.024$ cycles / cm and most of the other prominent components vary slowly in frequency through the section. Only the amplitudes at $f = 0.032$ cycles / cm, which are caused by the brightness differences between the single photographs, remain stable.

The grayscale interval between 59 and 45 m core depth (for stratigraphic position see Fig. 2) is now observed in more detail, because there the changes in accumulation rate are stronger and the effect of frequency modulation is likely to be more eye-catching in the spectrogram. In fact, in this interval the percentage of turbidites increases from 20% at the base to over 50% at the top and goes in step with the progradation of the nearby Geisler carbonate platform towards the well site (Maurer et al., 2003). Furthermore, only 4 m of this interval is bioturbated, and for the most part the original bedding is preserved. This attribute increases the separation of turbidites from the background sediment, in contrast to the bioturbated facies (core interval 92–64 m), where due to heavy nodularity this task is more difficult to accomplish. The amplitude spectrogram of this interval reveals a shift of the amplitudes towards lower frequencies, in tune with an increase in accumulation rate (Fig. 7B, indicated by arrows). Only the amplitudes at $f = 0.032$ cycles / cm, caused by the artifact, are stable over the entire spectrogram. As a consequence of the shifting frequencies, the 2π multitaper power spectrum above the spectrogram is very complex, with high-amplitude peaks occurring throughout the illustrated frequency range. Some of these spectral peaks are generated from the same signal component, but they appear to be unrelated in the averaged spectrum. In contrast to the amplitude spectrogram, modulatory frequencies are not recognizable in a single power spectrum. This illustrates that the use of a single power spectrum over an interval with changing accumulation rates could lead to misinterpretations of the estimated signal frequencies.

As an experiment, the turbidites were removed from the core interval 59–45 m in order to observe changes in the spectrogram and in the multitaper power spectral density. To eliminate the turbidites, all values corresponding to turbidite deposits in the grayscale scan were removed, and the intervening spaces were collapsed. In doing this, changes in accumulation rate are minimized. This procedure is tantamount to a "tuning", but one that is based solely upon sedimentological criteria.

The results of this experiment are shown in Figure 7C. The amplitude spectrogram looks stretched in comparison to the one on the left. This is because both spectrograms were computed with power spectra of data segments 4 m long, and the cleaned interval on the right is with approximately 7 m, only half the length of the raw series on the left. In the cleaned spectrogram most amplitudes are stable over the observed interval, which strongly suggests that turbidites are indeed largely responsible for changes in accumulation rate. The strongest amplitudes occur at $f = 0.025$ cycles / cm, in agreement with the dominant bedding rhythm observed in the interval 92–64 m. In the corresponding 2π multitaper spectrum, strong power is concentrated in three intervals at $f = 0.005$, $f = 0.025$, and $f = 0.063$ cycles / cm. The dotted line shows the amount of power derived from the artifact that contributes to the overall power spectral density. It was calculated by first bandpassing the raw series at $f = 0.033$ cycles / cm, then removing those sampled positions corresponding to the turbidites, and finally performing the multitaper analysis on that dataset. Note that the artifact does not contribute any power to the three major peaks in the spectrum.

Time Calibration of Bedding Rhythms

Triassic time scales differ significantly in the estimate of the duration of single stages (e.g., Gradstein et al., 1994, fig. 3). These discrepancies arise from uncertainties in radiometric age data, which sometimes have analytical errors of millions of years. According to Gradstein et al. (1994, fig. 9) the time span that corresponds to the deposition of the Buchenstein Fm. would have encompassed approximately 5 My. Mundil et al. (1996) dated U–Pb decay in single zircons from three ash layers in the Buchenstein Fm. At Seceda, the "Tc"-tuff (at 90 m core depth) gives an age of 241.2 +0.8 / -0.6 Ma, and another tuff, at 40 m core depth, 238.0 +0.4 / -0.7 Ma (see also core–outcrop correlation in fig. 4 in Brack et al., 2000). The third tuff layer, dated in Bagolino (Lombardian Alps), has age of 238.8 +0.5 / -0.2 Ma and corresponds approximately to the lapilli layer in the Seceda section (core interval 59–60 m; Maurer and Rettori, 2002).

The zircon data indicate a duration of 2.4 My for the deposition of the nodular limestone interval between 90 and 60 m core depth. With a thickness of 25 m in outcrop, this interval has an average accumulation rate of approximately 10.4 mm / ky. According to these values, the main bedding rhythm, which in the outcrop approximates 36 cm, would correspond to an ~ 34.6 ky cycle and thus matches the period of the Earth's obliquity estimated for the Early Permian (Berger and Loutre, 1994). Assuming the shortest possible duration of 1.3 My according to the analytical errors, the accumulation rate could be as high as 19.2 mm / ky, calibrating the main bedding rhythm at 18.7 ky, in tune with the estimated periodicity for climatic precession (Berger and Loutre, 1994). Taking the longest possible duration of 3.4 My into account, the accumulation rate would result in 7.3

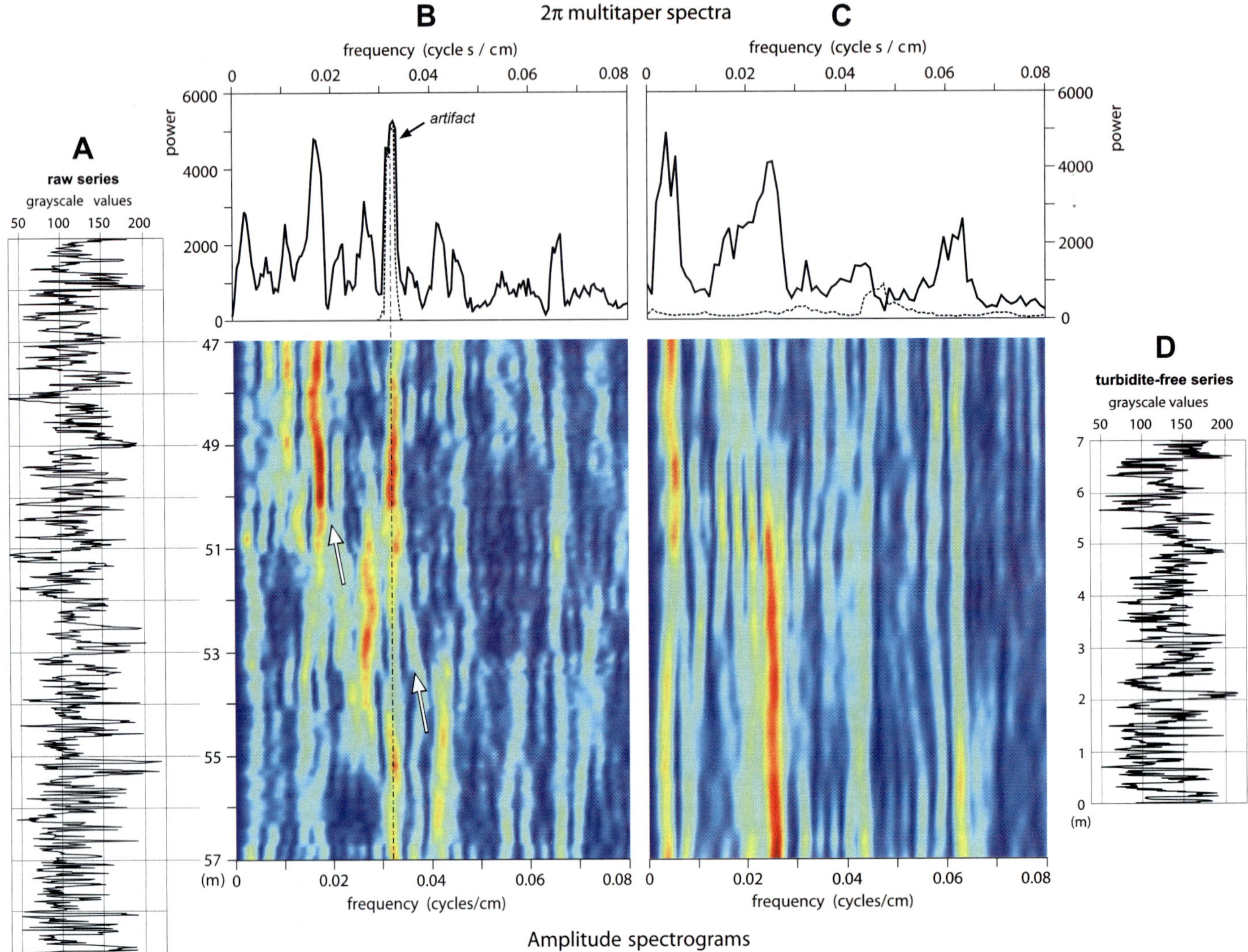

FIG. 7.—Analysis of the grayscale data in core interval 59–45 m, showing the sedimentological tuning. **A)** Raw grayscale series. **B)** The increase in sedimentation rate in this interval is manifested in the amplitude spectrogram by a shift of strong amplitudes towards lower frequencies (arrows). **C)** By removing the turbidites, the signals in the spectrogram become stable. This goes along with a reduction in the number of spectral peaks in the corresponding 2π multitaper spectrum illustrated above. The dashed lines in the power spectra give the amount of power that the artifact caused by brightness differences of the core photos contributes to the overall power spectral density. **D)** Grayscale input series for analyses in Part C. It covers the same interval as in Part A, but here all data points corresponding to turbidites have been removed.

mm/ky, in which case the bedding rhythm is not related to a prominent orbital cycle. These considerations make clear that at present geochronology is not sufficiently precise to argue for one orbital parameter over another. However, arguments for one or the other hypothesis may be found if the signal component responsible for the bedding rhythm is extracted from the dataset and compared with the time–frequency behavior of theoretical orbital signals.

For such an analysis the core interval 59–45 m cleaned of turbidites is most suitable, because the effect of variable accumulation rate caused by the turbidites was removed, and stable signal components can be effectively isolated using bandpass filtering. In Figure 8 this interval is illustrated on the left and compared with theoretical Milankovitch cycles computed for the Pleistocene on the right (data from Laskar et al., 1993). The core interval is now examined out to a frequency of $f = 0.16$ cycles/cm, in order to be able to detect any additional important signal components also in the higher frequencies.

In Figure 8, 2π multitaper analysis (A), amplitude spectrogram (B), and filtered signals (C) all suggest the presence of Milankovitch cycles in the cleaned core interval. First considering the multitaper power spectral density, that of the Seceda core (left) reveals strong spectral peaks that are distributed in frequency similarly to those in the theoretical spectrum (right); proposed long (E) and short (e_1, e_2) eccentricity, obliquity (O), and precession (P_1, P_2) components are indicated. In the Seceda core interval a relation of 5:1 is observed between the signal component at $f = 0.025$ and the one at $f = 0.13$ cycles/cm. The component at $f = 0.063$ cycles/cm appears in the same relation to its neighboring ones as the obliquity to eccentricity and precession in the Laskar spectrum on the right. Assuming a short-eccentricity origin for the component at $f = 0.025$ cycles/cm (= 40 cm bedding rhythm in the core), the duration of deposition of this 6.9-m-long interval is 1730 ky.

Figure 8B shows the amplitude spectrograms of the Seceda data and the theoretical model. The spectrogram on the left was calculated with power spectra of data segments 2 m long and an offset of 10 cm between them. It depicts the same interval as that in Figure 9C, which was computed with power spectra of segments 4 m long and plotted only to a frequency of $f = 0.08$ cycles/cm. The spectrogram on the right illustrates the theoretical orbital variations over the last 1730 ky. It was computed using data segments of 500 ky and an offset of 25 ky, thus with the same relation in length of data segments and offset as the spectrogram on the left. In the Laskar model both short eccentricity (e) and long precession (P_1) appear as signals with two parallel lines of amplitudes, merging to nodes every 400 ky. The obliquity (O) is represented as a straight line of amplitudes with maximum power every 1200 ky. In the spectrogram created with the Seceda grayscale data a straight line of amplitudes is present in the presumed frequency range for obliquity at $f = 0.063$ cycles/cm. The nodal features of the short eccentricity and long precession are not clearly identifiable in this display. However, in Figure 7C, where the windowing is twice as long, in the interval 51–57 m the signal component centered at $f = 0.024$ cycles/cm has a "doublet" structure with a stronger line on the right and a weaker one on the left, which is a feature consistent with the time–frequency behavior of short eccentricity.

Figure 8C illustrates the filtered signal components. In the Laskar model on the right the filtered short eccentricity displays maxima every 400 ky, and the filtered obliquity every 1200 ky (asterisks). The precession, which is modulated by the eccentricity, has maxima every five cycles, corresponding to the length of one short-eccentricity cycle. The maximum of every eccentricity cycle coincides with the maximum in a bundle of five precession cycles. The filtered signals in the Seceda grayscale data show similar features. In the interval between 4 and 7 m the relation between short eccentricity and precession is exactly 5:1, and the amplitude maxima in the precession bundles correspond to the maxima in the eccentricity cycles. The distance between two maxima in the obliquity (asterisks) is 32 cycles long, and thus slightly longer than in the theoretical obliquity signal with 28 cycles.

Figure 9 presents the significance tests for the presumed Milankovitch signal in the cleaned core interval 59–45 m. In the 6π multitaper estimate, all presumed orbital components exceed 95% significance in the F-test, and obliquity and precession even surpass the 99% significance level (Fig. 9A). Also in the Lomb–Scargle periodogram all orbital signals exceed the 95% probability level (Fig. 9B). Another way to determine the significance of spectral peaks is to compare the relation of the power of spectral peaks with the overall noise of the dataset (e.g., Hinnov and Goldhammer, 1991). If the lower 95% confidence interval of normalized power exceeds all noise levels, the associated frequency component is interpreted to be significantly different from the noise. In Figure 9C the 2π multitaper power spectral estimate normalized to the sum of power is plotted together with the noise levels and shows that only peaks associated with orbital forcing exceed the levels of white and red noise.

Assuming a duration of 98 ky for the short eccentricity, the duration of one obliquity cycle at $f = 0.063$ cycles/cm results in 38.9 ky and the one of the precession signal at $f = 0.128$ becomes 19.1 ky (Fig. 8A). The durations for obliquity and precession are slightly shorter than the recent Milankovitch cycles, but they still lie in the range estimated by Berger and Loutre (1994) for the late Paleozoic. According to this cyclostratigraphic interpretation, the accumulation rate in the observed core interval results in 4.08 mm/ky, or approximately 3.6 mm/ky if corrected for dip. This estimate can be compared with the zircon data of Mundil et al. (1996), measured at the bottom and the top of the investigated interval (Fig. 10). They indicate an average sedimentation rate of 13.5 mm/ky (corrected for dip) for the core interval ~ 59.3–40.8 m cleaned of turbidites, including the outcrop interval missing in the core due to local tectonic unconformities (Maurer and Rettori 2002, Fig. 3). Thus, the proposed radiometric data suggest a sedimentation rate that is more than three times higher than the cyclostratigraphic analysis. However, the analytical error of the zircon data (i.e., the maximum possible time of 2 My for deposition of the interval versus the proposed one of 800 ky) is 140%. Assuming the maximum time estimate of 2 My, the sedimentation rate results in 5.4 mm/ky and would therefore be much closer to the cyclostratigraphic interpretation.

All timing of geologic bedding rhythms depends on the accuracy with which one can reconstruct the orbital variations of the distant past. The degree of insolation on Earth depends on the planet's position in space and its orientation relative to the Sun. From general physical considerations, shorter frequencies are expected for obliquity and precession in the distant geological past. Berger and Loutre (1994) estimate a period of approximately 35 ky for the main obliquity signal and periods of 21 and 17.6 ky for the long and short precession during the Early Permian. In the 2π multitaper spectrum of the cleaned interval the strongest peak in the obliquity range occurs at $f = 0.063$ cycles/cm (= 38.9 ky), and in the precession range the strongest frequency component is located at $f = 0.128$ cycles/cm (= 19.1 ky). Both periods are longer than the estimates of Berger and Loutre (1994), but distinctly shorter than at the present, in agreement with the predicted trend. Very high accuracy of orbital reconstructions may be difficult to achieve for the distant past because of the strongly chaotic behavior of the inner Solar System (Laskar, 1999).

CONCLUSIONS

In basins with frequent interruption of background sedimentation the detection of quasiperiodic signals in the sedimentary record is difficult to accomplish. Changes in accumulation rate due to variations in the input of calciturbidites and extraordinarily high gamma radiation linked to episodic fallout of volcaniclastics were identified as the most disturbing factors for cyclostratigraphic analyses in the Buchenstein basin. Once these disruptive layers are cleaned from the sedimentary succession, time-frequency analysis reveals stabilization of the background sedimentary signal. Thus, sedimentological studies are crucial prior to performing time-series analysis in successions like the Buchenstein Fm. with many event beds. Furthermore, comparison between the 1-mm-resolution grayscale scan and the 1-cm-resolution gamma-ray log shows that the use of well-log data alone without sedimentological control may not lead to a recognition of Milankovitch cycles in ancient sedimentary rocks.

ACKNOWLEDGMENTS

This contribution is part of the first author's Ph.D. thesis, which is supported by the Austrian Academy of Sciences and the VU Industrial Associates Program in Sedimentology. LH was supported by National Science Foundation Grant EAR-

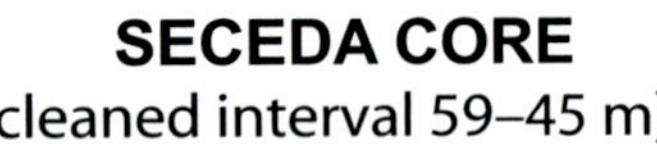

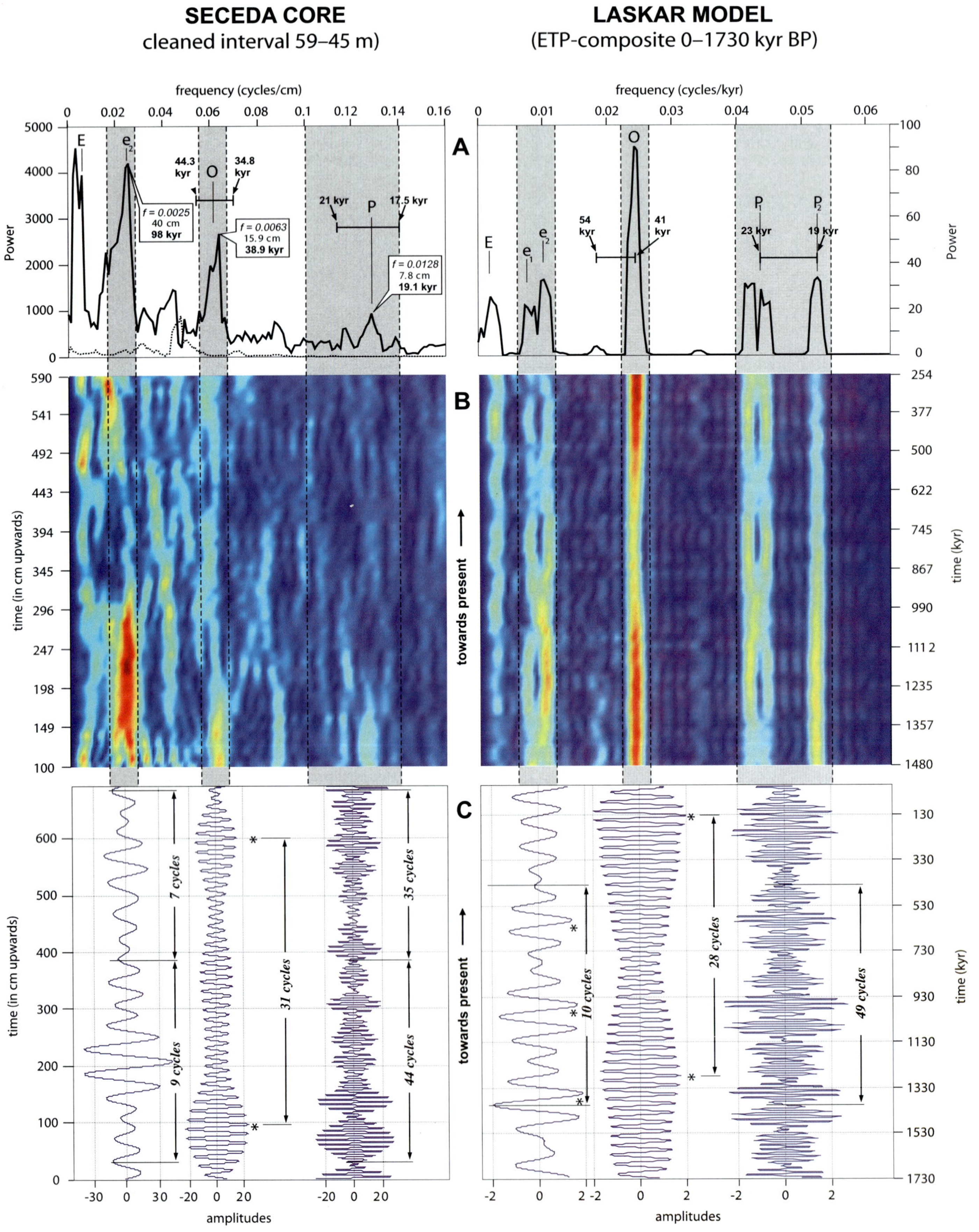
SECEDA CORE
cleaned interval 59–45 m)
LASKAR MODEL
(ETP-composite 0–1730 kyr BP)
frequency (cycles/cm)
frequency (cycles/kyr)
Power
A
B
C
E
e2
e1
O
P
P1
P2
44.3 kyr
34.8 kyr
21 kyr
17.5 kyr
f = 0.0025
40 cm
98 kyr
f = 0.0063
15.9 cm
38.9 kyr
f = 0.0128
7.8 cm
19.1 kyr
54 kyr
41 kyr
23 kyr
19 kyr
time (in cm upwards)
time (kyr)
towards present
7 cycles
9 cycles
31 cycles
35 cycles
44 cycles
10 cycles
28 cycles
49 cycles
amplitudes

9909528. We thank Jeroen Kenter and Michel Groen (Vrije Universiteit Amsterdam) for acquisition of well logs and discussion on the gamma-ray dataset. FM also thanks Lawrence Hardie for invitation to the Faculty of Earth Sciences at Johns Hopkins University (Baltimore). Al Fischer and an anonymous reviewer provided useful suggestions for improvement of the manuscript.

REFERENCES

BERGER, A., AND LOUTRE, M.F., 1994, Astronomical forcing through geological time, *in* de Boer, P.L., and Smith, D.G., eds., Orbital Forcing and Cyclic Sequences: International Association of Sedimentologists, Special Publication 19, p. 15-24.

BOSELLINI, A., 1984, Progradation geometries of carbonate platforms: examples from the Triassic of the Dolomites, northern Italy: Sedimentology, v. 31, p. 1-24.

BOSELLINI, A., AND FERRI, R., 1980, La Formazione di Livinallongo (Buchenstein) nella valle di S. Lucano (Ladinico inferiore, Dolomiti Bellunesi): Universitá di Ferrara, Annali, sezione IX, v. 6, p. 63-89.

BOX, G.E.P., AND JENKINS, G.M., 1970, Time Series Analysis; Forecasting and Control: San Francisco, Holden-Day, 575 p.

BRACK, P., AND MUTTONI, G., 2000, High-resolution magnetostratigraphic and lithostratigraphic correlations in Middle Triassic pelagic carbonates from the Dolomites (northern Italy): Palaeogeography, Palaeoclimatology, Palaeoecology, v. 161, p. 361-380.

BRACK, P., AND RIEBER, H., 1993, Towards a better definition of the Anisian / Ladinian boundary: new biostratigraphic data and correlations of boundary sections from the Southern Alps: Eclogae Geologicae Helvetiae, v. 86, p. 415-527.

BRACK, P., MUTTONI, G., AND RIEBER, H., 2001, Comment on: 'Magnetostratigraphy and biostratigraphy of the Middle Triassic Margon section (Southern Alps, Italy)': Earth and Planetary Science Letters, v. 193, p. 253-255.

←

FIG. 8 (opposite page).—Comparison between the core interval 59–45 m cleaned of turbidites and the theoretical orbital variations over the past 1730 ky (data from Laskar et al., 1993). **A)** 2π multitaper power spectra with indication of Milankovitch periodicities (long (E) and short (e) eccentricity, obliquity (O), and precession (P)). The dotted line in the spectrum created with the Seceda grayscale data shows the amount of power that the artifact signal contributes to the overall power. The gray background marks the frequency ranges that were selected to filter the signal components plotted in Part C. Horizontal bars indicate the frequency range of obliquity and precession for the late Paleozoic (left spectrum) and the Quaternary (right) according to Berger and Loutre (1994). **B)** Amplitude spectrograms created using the same relation between size of data segments and offset. The spectrogram of the Seceda core differs from that in Figure 7C in that it is now calculated with 2 m data segments and extended to $f = 0.16$ cycles / cm. **C)** On the left are illustrated the signals recovered by bandpass filtering the grayscale data in the frequency ranges $f = 0.016 - f = 0.025$ (e), $f = 0.058$ to $f = 0.067$ (O), and $f = 0.1$ to $f = 0.14$ (P) cycles / cm, corresponding to areas with gray background in Part A. These signals show similarities with the characteristics of Milankovitch cycles in the Pleistocene plotted on the right, in which the contribution of each orbital signal was standardized to unit variance separately, then summed together to yield an "ETP" model.

BRACK, P., MUNDIL, R., OBERLI, F., MEIER, M., AND RIEBER, H., 1996, Biostratigraphic and radiometric age data question the Milankovitch characteristics of the Latemar cycles (Southern Alps, Italy): Geology, v. 24, p. 371-375.

BRACK, P., MUNDIL, R., OBERLI, F., MEIER, M., AND RIEBER, H., 1997, Biostratigraphic and radiometric age data question the Milankovitch characteristics of the Latemar cycles (Southern Alps, Italy). Reply: Geology, v. 25, p. 471-472.

BRACK, P., SCHLAGER, W., STEFANI, M., MAURER, F., AND KENTER, J., 2000, The Seceda drill hole in the Middle Triassic Buchenstein beds (Livinallongo Formation, Dolomites, Northern Italy) a progress report: Rivista Italiana di Paleontologia e Stratigrafia, v. 106, p. 283-292.

FISCHER, A.G., 1964, The Lofer cyclothems of the Alpine Triassic, *in* Merriam, D.F., Symposium on Cyclic Sedimentation: Kansas Geological Survey, Bulletin 169, p. 107-149.

GOLDHAMMER, R.K., DUNN, P.A., AND HARDIE, L.A., 1987, High frequency glacio-eustatic sea-level oscillations with Milankovitch characteristics recorded in Middle Triassic platform carbonates in Northern Italy: American Journal of Science, v. 287, p. 853-892.

GOLDHAMMER, R.K., DUNN, P.A., AND HARDIE, L.A., 1990, Depositional cycles, composite sea-level changes, cycle stacking patterns, and the hierarchy of stratigraphic forcing: examples form Alpine Triassic platform carbonates: Geological Society of America, Bulletin, v. 102, p. 535-562.

GRADSTEIN, F.M., AGTERBERG, F.P., OGG, J.G., HARDENBOL, J., VAN VEEN, P., THIERRY, J., AND HUANG, Z., 1994, A Mesozoic time scale: Journal of Geophysical Research, v. 99B, p. 24,051-24,074.

HAAK, A., AND SCHLAGER, W., 1989, Compositional variations in calciturbidites due to sea-level fluctuations, late Quaternary, Bahamas: Geologische Rundschau, v. 78, p. 477-486.

HARDIE, L.A., AND HINNOV, L.A., 1997, Biostratigraphic and radiometric age data question the Milankovitch characteristics of the Latemar cycles (Southern Alps, Italy). Comment: Geology, v. 25, p. 470-471.

HERBERT, T.D., PREMOLI SILVA, I., ERBA, E., AND FISCHER, A., 1995, Orbital chronology of Cretaceous–Paleocene marine sediments, *in* Berggren, W.A., Kent, D.V., Aubry, M.-P., and Hardenbol, J., eds., Geochronology, Time Scales and Global Stratigraphic Correlation: SEPM, Special Publication 54, p. 81-94.

HILGEN, F.J., ABDUL AZIZ, H., KRIJGSMAN, W., LANGEREIS, C.G., LOURENS, L.J., MEULENKAMP, J.E., RAFFI, I., STEENBRINK, J., TURCO, E., VAN VUGT, N., WIJBRANDS, J.R., AND ZACHARIASSE, W.J., 1999, Present status of the astronomical (polarity) time-scale for the Mediterranean Late Neogene: Royal Society (London), Philosophical Transactions, v. A 357, p. 1931-1947.

HINNOV, L.A., 2000, New perspectives on orbitally forced stratigraphy: Annual Review of Earth and Planetary Sciences, v. 28, p. 419-475.

HINNOV, L.A., AND GOLDHAMMER, R.K., 1991, Spectral analysis of the Middle Triassic Latemar Limestone: Journal of Sedimentary Petrology, v. 61, p. 1173-1193.

KRYSTYN, L., 1983, Das Epidaurus-Profil (Griechenland)—ein Beitrag zur Conodonten-Standardzonierung des tethyalen Ladin und Unterkarn, *in* Zapfe, H., ed., Neue Beiträge zur Biostratigraphie der Tethys-Trias: Österreichische Akademie der Wissenschaften, Erdwissenschaftliche Kommission, Schriftenreihe, v. 5, p. 231-258.

LASKAR, J., 1999, The limits of Earth orbital calculations for geological time-scale use: Royal Society (London), Philosophical Transactions, v. A 357, p. 1735-1759.

LASKAR, J., JOUTEL, F., AND BOUDIN, F., 1993, Orbital, precessional, and insolation quantities for the Earth from -20 Myr to +10 Myr: Astronomy and Astrophysics, v. 270, p. 522-533.

LOMB, N.R., 1976, Least squares frequency analysis of unequally spaced data: Astrophysics and Space Science, v. 39, p. 447-462.

MAURER, F., AND RETTORI, R., 2002, Middle Triassic foraminifera from the Seceda core (Dolomites, northern Italy): Rivista Italiana di Paleontologia e Stratigrafia, v. 108, p. 391-398.

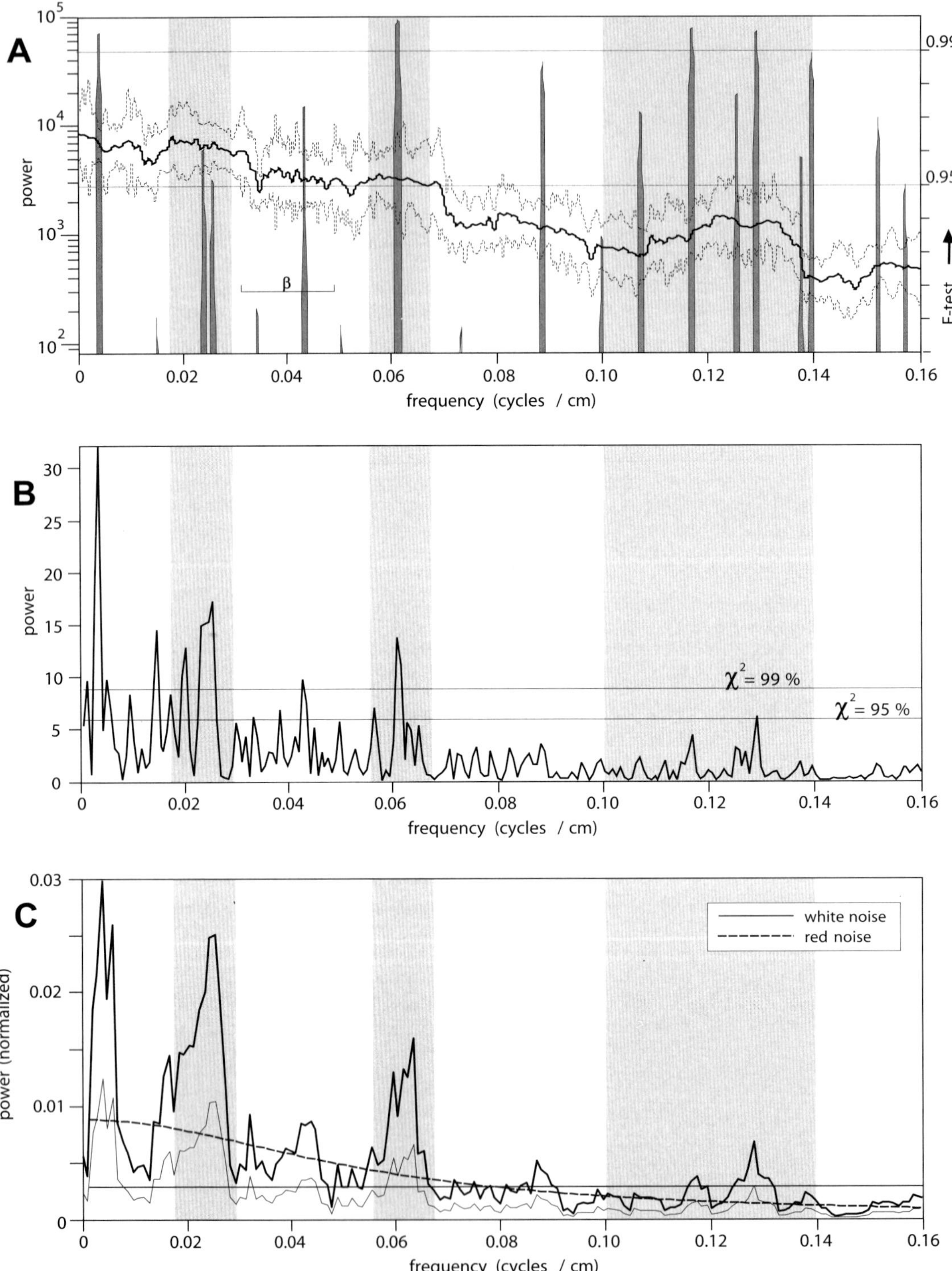

FIG. 9.—Multitaper analysis and Lomb–Scargle periodograms of grayscale scan of the cleaned core interval 59–45 m and associated significance tests. The gray background marks frequency range of presumed Milankovitch signals. **A)** 6π multitaper power spectrum with logarithmic *y* axes and F-tests. Thick solid lines give power spectral density, thin dotted lines represent upper and lower 99% confidence levels, and dark gray spikes indicate significant frequency components. All presumed orbital signal components exceed the 95% significance level of the F-test. **B)** Lomb–Scargle periodogram, in which all amplitudes related to the presumed Milankovitch signals exceed the 95% chi-squared probability distribution. **C)** Same 2π multitaper power spectrum as in Figure 8A but with power normalized to sum of power. The thick solid line gives the power spectral density, and the thin solid line the associated lower 95% confidence level. White and red noise levels are indicated by dashed–dotted lines. Note that only in the frequency ranges of presumed Milankovitch signals does the lower 95% confidence interval exceed both noise levels.

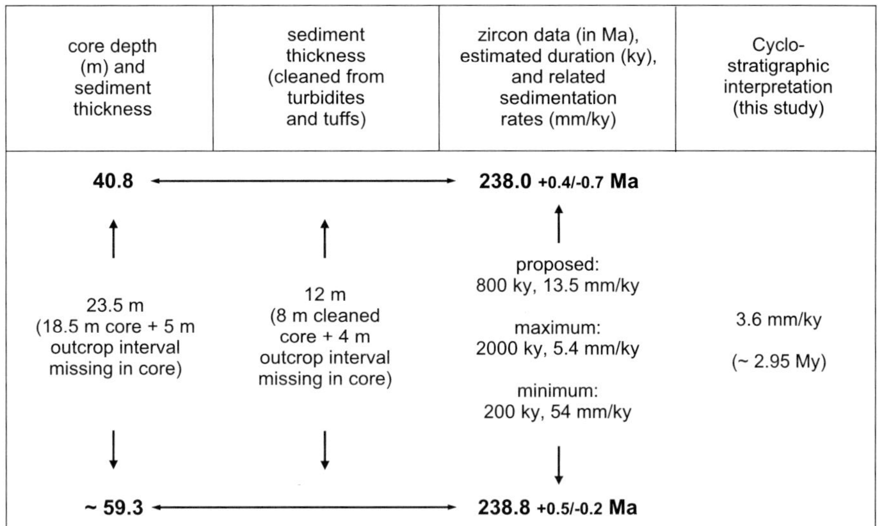

FIG. 10.—Duration of deposition of core interval 59–45 m and related sedimentation rates (corrected for dip) as estimated with zircon data and cyclostratigraphic interpretation. In both estimates an interval missing in the core, but present in the outcrop, is incorporated. For details on core–outcrop correlation in this interval, see Figure 3 in Maurer and Rettori (2002).

MAURER, F., AND SCHLAGER, W., 2003, Lateral variations in sediment composition and bedding in Middle Triassic interplatform basins (Buchenstein Formation, Southern Alps, Italy): Sedimentology, v. 50, p. 1-22.

MAURER, F., REIJMER, J.J.G., AND SCHLAGER, W., 2003, Quantification of input and compositional variations of calciturbidites in a Middle Triassic basinal succession (Seceda, Dolomites, Southern Alps): International Journal of Earth Sciences, in press.

MEYERS, S.R., BRADLEY, B., AND HINNOV, L.A., 2001, Integrated quantitative stratigraphy of the Cenomanian–Turonian Bridge Creek limestone member using evolutive harmonic analysis and stratigraphic modelling: Journal of Sedimentary Research, v. 71, p. 628-644.

MULLER, R.A., AND MACDONALD, G.J., 2000, Ice Ages and Astronomical Causes; Data, Spectral Analysis and Mechanisms: Berlin, Springer-Verlag, Springer-Praxis Books in Environmental Sciences, 318 p.

MUNDIL, R., BRACK, P., MEIER, M., OBERLI, F., AND RIEBER, H., 1996, High-resolution U–Pb dating of Middle Triassic volcaniclastics: time-scale calibration and verification of tuning parameters for carbonate sedimentation: Earth and Planetary Science Letters, v. 141, p. 137-151.

MUNDIL, R., ZÜHLKE, R., BECHSTÄDT, T., BRACK, P., EGENHOFF, P., MEIER, M., OBERLI, F., PETERHÄNSEL, A., AND RIEBER, H., 2003, Cyclicities in Triassic platform carbonates: synchronizing radio-isotopic and orbital clock: Terra Nova, v. 15, p. 81–87.

OLSEN, P.E., AND KENT, D.V., 1999, Long-period Milankovitch cycles from the Late Triassic and Early Jurassic of eastern North America and their implications for the calibration of the Early Mesozoic time-scale and the long-term behaviour of the planets: Royal Society (London), Philosophical Transactions, v. A 357, p. 1761-1786.

PAILLARD D., LABEYRIE, L., AND YIOU, P., 1996, Macintosh program performs time-series analysis (abstract): Eos, Transactions, American Geophysical Union, v. 77, p. 379.

PERCIVAL, D.B., AND WALDEN, A.T., 1993, Spectral Analysis for Physical Applications: Cambridge, U.K., Cambridge University Press, 583 p.

PRESS, W.H., FLANNERY, B., TEUKOLSKY, S., AND VETTERLING, W., 1996, Numerical Recipes, Third Edition: Cambridge, U.K., Cambridge University Press, 992 p.

PRETO, N., HINNOV, L.A., HARDIE, L.A., AND DE ZANCHE, V., 2001, Middle Triassic orbital signature recorded in the shallow marine Latemar carbonate buildup (Dolomites, Italy): Geology, v. 29, p. 1123-1126.

SCARGLE, J.D., 1982, Studies in astronomical time series analysis II. Statistical aspects of spectral analysis of unevenly spaced data: Astrophysical Journal, v. 263, p. 835-853.

SCHWARZACHER, W., AND HAAS, J., 1986, Comparative statistical analysis of some Hungarian and Austrian Upper Triassic peritidal carbonate sequences: Acta Geologica Hungarica, v. 29, p. 175-196.

SHACKLETON, N.J., BERGER, A., AND PELTIER, W.A., 1990, An alternative astronomical calibration of the lower Pleistocene timescale based on ODP Site 677: Royal Society of Edinburgh, Transactions, Earth Science, v. 81, p. 251-261.

SHACKLETON, N.J., CROWHURST, S.J., WEEDON, G.P, AND LASKAR, J., 1999, Astronomical calibration of Oligocene–Miocene time: Royal Society (London), Philosophical Transactions, v. A 357, p. 1907-1930.

THOMSON, D.J., 1982, Spectrum estimation and harmonic analysis: Institute of Electrical and Electronics Engineers, Proceedings, v. 10, p. 1055-1096.

THOMSON, D.J., 1990, Time series analysis of Holocene climate data: Royal Society (London), Philosophical Transactions, v. A 330, p. 601-616.

VIEL, G., 1979, Litostratigrafia Ladinica: una revisione. Ricostruzione paleogeografica e paleostrutturale dell'area Dolomitico-Cadorina (Alpi Meridionali): Rivista Italiana di Paleontologia e Stratigrafia, v. 85, p. 85-125.

ZÜHLKE, R., BECHSTÄDT, T., AND MUNDIL, R., 2003, Submilankovitch and Milankovitch forcing on a model Mesozoic carbonate platform—The Latemar (Middle Triassic, Italy): Terra Nova, in press.

Carbonate Platforms

A MULTIDISCIPLINARY APPROACH TO GLOBAL CORRELATION AND GEOCHRONOLOGY. THE CRETACEOUS SHALLOW-WATER CARBONATES OF SOUTHERN APENNINES, ITALY

BRUNO D'ARGENIO
Istituto per l'Ambiente Marino Costiero, Geomare, National Research Council,
Calata Porta di Massa, Porto di Napoli, 80133 Napoli, Italy
e-mail: dargenio@gms01.geomare.na.cnr.it
AND
Dipartimento di Scienze della Terra, Università "Federico II", Largo San Marcellino 10, 80138 Napoli, Italy
VITTORIA FERRERI
Dipartimento di Scienze della Terra, Università "Federico II", Largo San Marcellino 10, 80138 Napoli, Italy
e-mail: ciclisti@gms01.geomare.na.cnr.it
HELMUT WEISSERT
Geological Institute ETH-Z, CH-8092 Zürich, Switzerland
e-mail: helmi@erdw.ethz.ch
SABRINA AMODIO
Istituto per l'Ambiente Marino Costiero, Geomare, National Research Council,
Calata Porta di Massa, Porto di Napoli, 80133 Napoli, Italy
e-mail: amodio@gms01.geomare.na.cnr.it
AND
Dipartimento di Scienze della Terra, Università "Federico II", Largo San Marcellino 10, 80138 Napoli, Italy
FRANCESCO P. BUONOCUNTO
Istituto per l'Ambiente Marino Costiero, Geomare, National Research Council,
Calata Porta di Massa, Porto di Napoli, 80133 Napoli, Italy
e-mail: ciclisti@gms01.geomare.na.cnr.it
AND
LUKAS WISSLER
Geological Institute ETH-Z, CH-8092 Zürich, Switzerland
e-mail: helmi@erdw.ethz.ch

ABSTRACT: Detailed sedimentological and carbon-isotope data have allowed us to propose a high-resolution regional correlation between four Lower Cretaceous carbonate-platform successions, Early Aptian to Early Albian in age, cropping out in the southern Apennines. These successions, formed in open to restricted lagoonal and peritidal–supratidal settings, reveal a high-frequency cyclic recurrence of depositional and early meteoric (karstic and/or pedogenetic) features. The latter are normally superimposed on subtidal deposits, suggesting that the above cyclicity may be linked to sea-level changes. In the studied sections, elementary cycles are grouped into bundles, which in turn are grouped into superbundles. Although bundles and superbundles appear to be related to the Earth's orbital short-eccentricity and long-eccentricity signal, respectively, elementary cycles seem to record either the precession or a combination of the precession and obliquity periodicity. Moreover, the stacking pattern of the orbitally controlled cycles suggests that they are superimposed on lower-frequency sea-level fluctuations (transgressive–regressive facies trends).

The $\delta^{13}C$ curves, established throughout the sections, show the same carbon-isotope pattern as the time-equivalent pelagic strata (two positive carbon-isotope episodes separated by an interval with lower carbon-isotope values). On the basis of this correspondence, and integrating the cyclostratigraphy and the carbon-isotope stratigraphy with the sequence stratigraphy, we propose a high-precision regional correlation and a chronostratigraphic chart and suggest a duration of 7.8 My for the studied interval. Moreover, on the basis of sequence stratigraphy and isotope geochemical criteria, and using our orbital chronostratigraphy as a reference frame, a correlation with current global scales is here proposed.

INTRODUCTION

There is considerable interest in the understanding the detailed behavior of the ocean–atmosphere system through time as the main driving force of the Earth's global changes whose primary evidence is embedded in sedimentary sequences. Moreover, high-precision long-distance correlation among stratal successions may be crucial in the restoration of ancient sedimentary systems for exploration purposes. To this aim, cyclic sedimentation controlled by orbital forcing has become a very useful tool in the analysis of pelagic as well as shallow-water strata, particularly if such successions consist of carbonate rocks (Herbert and Fischer, 1986; Fischer et al., 1991; Pasquier and Strasser, 1997; D'Argenio et al., 1998; Strasser et al., 2001; see

Cyclostratigraphy: Approaches and Case Histories
SEPM Special Publication No. 81, Copyright © 2004
SEPM (Society for Sedimentary Geology), ISBN 1-56576-108-1, p. 103–122.

also Einsele et al., 1991; de Boer and Smith, 1994). Moreover, in shallow-water domains like carbonate platforms, water chemistry and temperature, ecology of carbonate-producing organisms, and early marine and meteoric diagenesis are amplified by the shallowness of the depositional sites, inducing a more distinct partitioning of the above sedimentary parameters. This is the case for the Mesozoic carbonate platforms of southern Italy, which had a high rate of sedimentation, commonly exceeding 25–30 m / My (D'Argenio and Alvarez, 1980). The strata of these platforms reveal a hierarchical cyclic organization, like those of the pelagic realms that have been traditionally used for high-resolution studies. It has been shown elsewhere that this type of cycle hierarchy can be traced laterally and correlated at regional (D'Argenio et al., 1999b) to global scale (Ferreri et al., 2001).

Using an integrated approach, we have revisited four Lower Cretaceous (Lower Aptian to Lower Albian) carbonate-platform successions of southern Italy previously analyzed (Buonocunto et al., 1994; Longo et al., 1994; Buonocunto, 1998; Raspini, 1998, 2001), adding carbon-isotope data to the sedimentological, biostratigraphic, and cyclostratigraphic studies. The aim of this multidisciplinary work is to show that positive shifts in $\delta^{13}C$ and low-frequency relative sea-level fluctuations (transgressive–regressive facies trends: T / RFTs) recorded in the studied sections can be related to oceanic changes of the carbon cycle and to eustatic fluctuations, respectively. By combining cyclostratigraphy with sequence stratigraphy and with carbon-isotope stratigraphy, we propose an orbital chronostratigraphy for the studied Early Cretaceous time interval and assess the role of eustasy in controlling high-frequency and lower-frequency environmental oscillations. On these bases we demonstrate that major changes in carbon reservoirs are reflected similarly in the carbon-isotope stratigraphy of both pelagic and shallow-water deposits.

APTIAN–ALBIAN HIGH-RESOLUTION CYCLOSTRATIGRAPHY

A total of about 515 m of Aptian–Albian carbonate-platform deposits from the southern Apennines (Italy) were studied at centimeter scale in four well-exposed sections at Serra Sbregavitelli, Monte Raggeto, Monte Faito, and Monte Tobenna (Fig. 1). The main results of previously published sedimentological and cyclostratigraphic studies are summarized below (Buonocunto et al., 1994; Longo et al., 1994; Ferreri et al., 1997; Buonocunto, 1998; Raspini, 1998, 2001).

Sedimentology

Serra Sbregavitelli.—

The Serra Sbregavitelli (SS) succession has a total thickness of about 165 m (Buonocunto et al., 1994; D'Argenio et al., 1999b) and crops out along the margins of a large tectonic depression, at the core of the Matese Mountains (Servizio Geologico d'Italia, 1971, map no. 161, Isernia). The lower–middle part of the studied section is Late Aptian in age, and its uppermost part reaches the Aptian–Albian transition, as suggested by the first occurrence of the benthic foraminifer *Ovalveolina reicheli* (De Castro). According to Chiocchini et al. (1994), this form indicates the uppermost Aptian, whereas according to De Castro (1991) and Bravi and De Castro (1995) it suggests the base of the Albian.

Monte Raggeto.—

The Monte Raggeto (MR) sequence crops out some 35 km North of Naples, in the southern part of the Monte Maggiore Range (Servizio Geologico d'Italia, 1971, map no. 172, Caserta). We measured approximately 280 m of well-bedded limestones

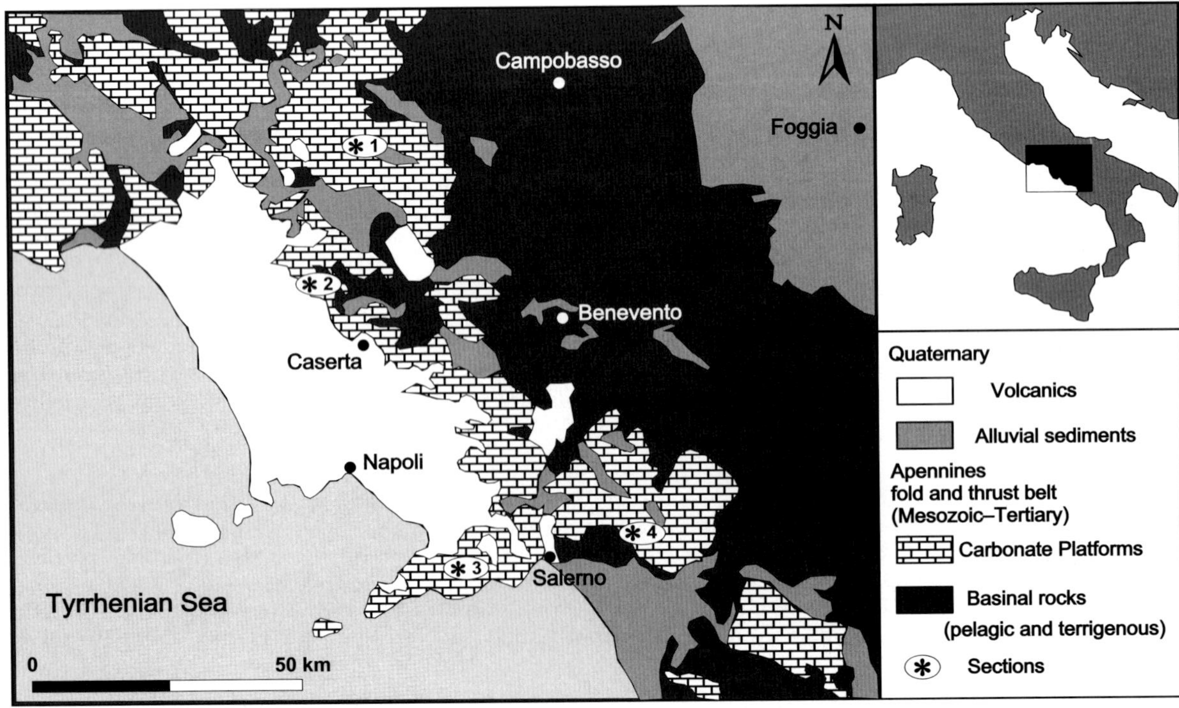

Fig. 1.—Location of the studied sections: 1, Monte Raggeto; 2, Serra Sbregavitelli; 3, Monte Faito; 4, Monte Tobenna.

and dolomitic limestones, white to light-brown or gray in color, ranging in age from Barremian to Early Albian (e.g., Ferreri et al., 1997). For the purposes of the present work, only the uppermost 130 m are discussed. This interval is characterized by the disappearance of *Palorbitolina lenticularis* (Blumenbach) and by the occurrence of *Salpingoporella dinarica* (Radoicic) associated with *Debarina* sp. and *Praechrysalidina infracretacea* (Luperto-Sinni). The transition to the Albian is suggested by the first occurrence of *Valdanchella* sp. and *Neoiraquia* sp.

Monte Tobenna.—

The Monte Tobenna (MT) section crops out in the Picentini Mountains (central part of the Campanian Apennines; Servizio Geologico d'Italia, 1969, map no. 185, Salerno) and is about 33 m thick. It is exposed in an old quarry on the southern slope of Monte Tobenna, near the village of S. Mango Piemonte (Raspini, 1998; D'Argenio et al., 1999b). Its middle part encloses the *Orbitolina* level Auct., a well-known Upper Aptian (Gargasian) biostratigraphic marker of the Campanian Apennines (Cherchi et al., 1978; Bravi and De Castro, 1995).

Monte Faito.—

The Monte Faito (MF) section is located near the town of Vico Equense in the Monti Lattari (Servizio Geologico d'Italia, 1959, map no. 196, Sorrento). The studied interval is part of a succession that crops out along the road from Vico Equense to the Monte Faito and reaches a total thickness of about 400 m (D'Argenio et al., 1999b; Raspini, 2001). Its age spans from the Late Hauterivian to the Early Albian (Robson, 1987). We have analyzed some 53 m of this succession, just above the *Orbitolina* level; the first occurrence of *Ovalveolina reicheli* (De Castro) was observed in the upper part of the section, where it marks the Aptian–Albian boundary (see previous discussion for Serra Sbregavitelli).

Lithofacies and Their Associations.—

For each of the four studied sections, the detailed analysis of textures and sedimentary features has allowed us to recognize several lithofacies (Fig. 2), which have been combined into seven *lithofacies associations* and two *clans* (groups of lithofacies associations; see Buonocunto et al., 1994; D'Argenio et al., 1999b). Their environmental interpretation suggests a shallow carbonate platform system, characterized by a depositional regime oscillating between peritidal and shallow-subtidal conditions.

In Figure 2 lithofacies are arranged from more open (normal-marine circulation) to more restricted (peritidal) settings. We observe that the Serra Sbregavitelli and the Monte Raggeto successions suggest, on the whole, more open lagoonal environments, but the Monte Faito and Monte Tobenna sections indicate a prevalence of peritidal conditions.

According to some authors (D'Argenio et al., 1975; Laubscher and Bernoulli, 1977; Channell et al., 1979), the Serra Sbregavitelli and Monte Raggeto sections and the Monte Faito and Monte Tobenna sections were originally situated on two different carbonate platforms separated by a wide basinal area. According to another interpretation (Marsella et al., 1995) the four studied successions were part of a single and large carbonate-platform domain with intervening shallow basins. Despite discussions in the literature on the paleogeography of the southern Apennines, it is worth noting that neither the analytical procedure nor the stratigraphic conclusions and correlations between the distant sequences are biased by these contrasting paleogeographic views (D'Argenio et al., 1999b).

Cyclostratigraphy

The vertical organization of lithofacies shows a cyclic recurrence, which can be analyzed in terms of depositional as well as early diagenetic features. The stacking pattern of the lithofacies and their grouping (lithofacies association and clans) as well as the variation of the early diagenetic features along the sections have been interpreted as the result of the prevailingly aggradational growth of a carbonate-platform system, forced by high-frequency eustatic oscillations. The overprint of early meteoric diagenesis, normally directly superimposed on subtidal deposits, supports this contention (see: *diagenetic cyclothems* as originally defined by D'Argenio, 1976, p. 151, and later by Hardie et al., 1986; see also Strasser, 1991; Buonocunto et al., 1994; Longo et al., 1994; D'Argenio et al., 1997; D'Argenio et al., 1999a). Two types of cycle boundaries were recognized in the present study. In the first type, the upper surface of the cycle shows weak pedogenesis and/or microkarstic features directly superimposed on marine sediments (em1-type emersion surface, as indicated in Figs. 3–6). Here the solution networks are of millimeter size and the resulting cavities are occluded by geopetal crystal silt; this suggests that an early diagenetic fabric developed under a meteoric regime in the upper supratidal zone. In the second type of cycle boundary, the meteoric overprint is enhanced and penetrates deeply, locally even into the underlying cycle, suggesting a more prolonged exposure and a more accentuated drop of sea level. In this case paleosols, locally marked by centimeter-thick clayey horizons and/or centimeter- to decimeter-size karstic cavities, also develop (em2-type emersion surface, as indicated in Figs. 3–6).

Considering the variation of the cycle thickness, as well as the variable penetration depth of early meteoric diagenetic overprint recorded at the cycle boundaries, a stacking pattern emerges (Fig. 7), where the elementary cycles are organized in bundles (groups of 2–5 elementary cycles) and the bundles in superbundles (groups of 2–4 bundles). The cycle hierarchy as well as the vadose caps directly superimposed on subtidal deposits have suggested a Milankovitch periodicity (Buonocunto et al., 1994; Buonocunto et al., 1999; D'Argenio et al., 1997; Raspini, 1998). This interpretation was also confirmed by mathematical treatment of the sedimentary data (lithofacies and related early meteoric fabric thickness) carried out at centimeter scale on numerous other Cretaceous carbonate-platform sequences in Italy (Longo et al., 1994; Brescia et al., 1996; Tagliaferri et al., 2001), as well as by the analysis of the magnetostratigraphic parameters (remanent inclination and declination) measured in the Hauterivian–Barremian of the Monte Raggeto section (Iorio et al., 1996). Cycle duration was estimated by comparing relative ratio sets (RRSs) for the recurrence of the sedimentary features, expressed in centimeters, with RRSs of the Earth's orbital parameters (precession, obliquity, and eccentricity) expressed in years, according the figures given for the corresponding geological time interval (Berger et al., 1992). The two ratio sets show a very good linear correlation ($r \geq 0.99$), suggesting that the elementary cycles were controlled by precession or obliquity (or a combination of both), whereas bundles and superbundles reflect a regulation by the short and long eccentricity cycles, respectively.

Moreover, the vertical variation of superbundle thicknesses suggests that the orbital cycles are superimposed on long-term transgressive–regressive facies trends (T/RFTs). A transgressive facies trend (TFT) is singled out on the basis of: (a) upward increase in thickness of elementary cycles and bundles; (b) lithofacies associations, evolving towards more open-marine conditions; and (c) development of cyclic units showing an upward decrease (or even lack) of early meteoric diagenetic overprint.

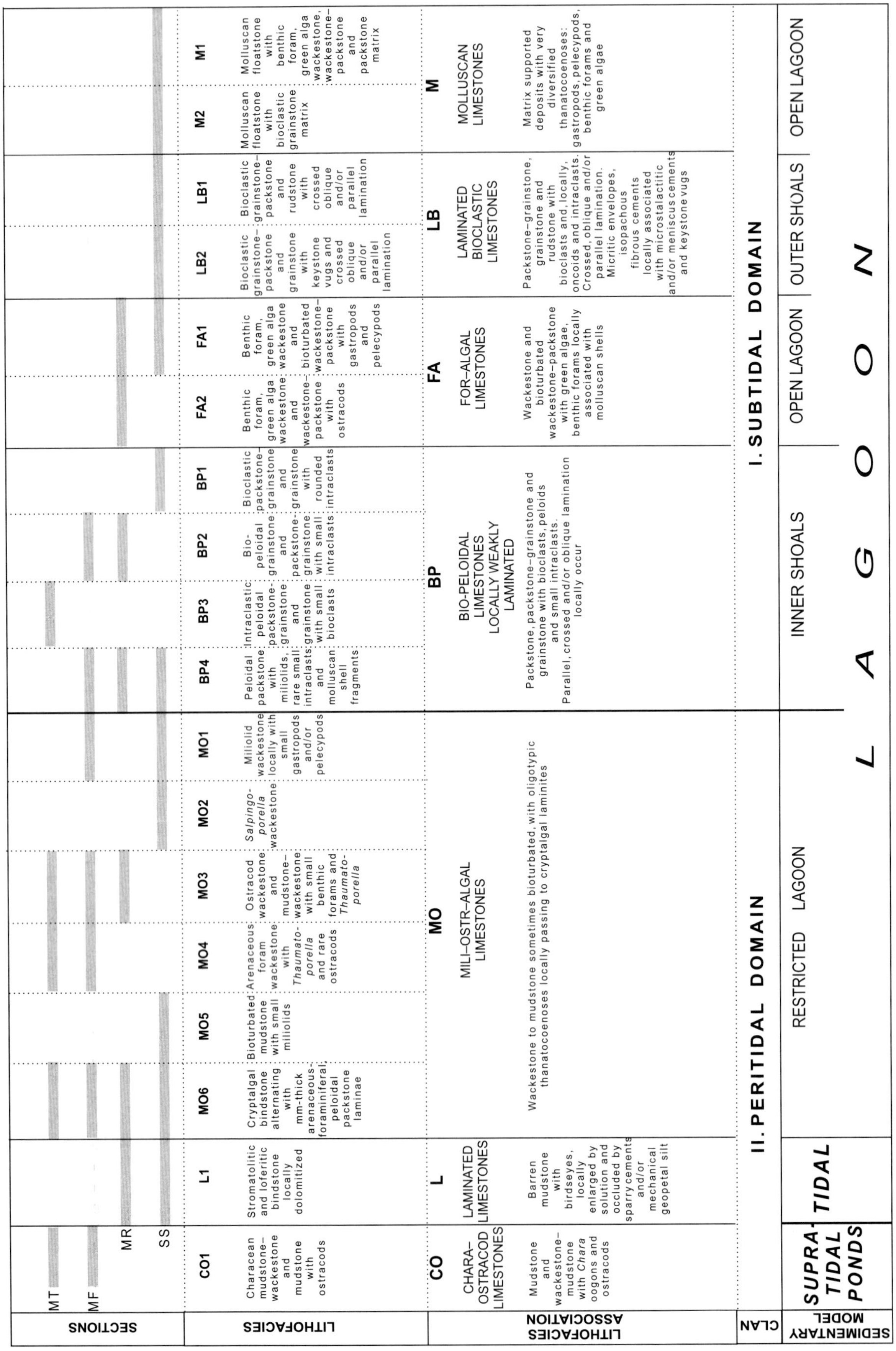

SECTIONS	MT MF MR SS																	
LITHOFACIES	CO1	L1	MO6	MO5	MO4	MO3	MO2	MO1	BP4	BP3	BP2	BP1	FA2	FA1	LB2	LB1	M2	M1
	Characean mudstone–wackestone and mudstone with ostracods	Stromatolitic and loferitic bindstone locally dolomitized	Cryptalgal bindstone alternating with mm-thick arenaceous-foraminiferal peloidal packstone laminae	Bioturbated mudstone with small miliolids	Arenaceous foram wackestone with *Thaumato-porella* and rare ostracods	Ostracod wackestone and mudstone–wackestone with small benthic forams and *Thaumato-porella*	*Salpingo-porella* wackestone	Miliolid wackestone locally with small gastropods and/or pelecypods	Peloidal packstone with miliolids, rare small intraclasts and molluscan shell fragments	Intraclastic peloidal packstone-grainstone and grainstone with small bioclasts	Bio-peloidal grainstone and packstone-grainstone with small intraclasts	Bioclastic packstone-grainstone and grainstone with rounded intraclasts	Benthic foram, green alga wackestone and wackestone–packstone with ostracods	Benthic foram, green alga wackestone and bioturbated wackestone–packstone with gastropods and pelecypods	Bioclastic grainstone–packstone and grainstone with keystone vugs and crossed oblique and/or parallel lamination	Bioclastic grainstone–packstone and rudstone with crossed oblique and/or parallel lamination	Molluscan floatstone with bioclastic grainstone matrix	Molluscan floatstone with benthic foram, green alga wackestone, wackestone–packstone and packstone matrix
LITHOFACIES ASSOCIATION	**CO** CHARA–OSTRACOD LIMESTONES	**L** LAMINATED LIMESTONES	**MO** MILI–OSTR–ALGAL LIMESTONES						**BP** BIO-PELOIDAL LIMESTONES LOCALLY WEAKLY LAMINATED				**FA** FOR–ALGAL LIMESTONES		**LB** LAMINATED BIOCLASTIC LIMESTONES		**M** MOLLUSCAN LIMESTONES	
	Mudstone and wackestone–mudstone with *Chara* oogons and ostracods	Barren mudstone with birdseyes, locally enlarged by solution and occluded by sparry cements and/or mechanical geopetal silt	Wackestone to mudstone sometimes bioturbated, with oligotypic thanatocoenoses locally passing to cryptalgal laminites						Packstone, packstone–grainstone and grainstone with bioclasts, peloids and small intraclasts. Parallel, crossed and/or oblique lamination locally occur				Wackestone and bioturbated wackestone–packstone with green algae, benthic forams locally associated with molluscan shells		Packstone–grainstone, grainstone and rudstone with bioclasts and, locally, oncoids and intraclasts. Crossed, oblique and/or parallel lamination. Micritic envelopes, isopachous fibrous cements locally associated with microstalactitic and/or meniscus cements and keystone vugs		Matrix supported deposits with very diversified thanatocoenoses: gastropods, pelecypods, benthic forams and green algae	
CLAN	II. PERITIDAL DOMAIN								I. SUBTIDAL DOMAIN									
SEDIMENTARY MODEL	SUPRA-TIDAL PONDS	TIDAL	RESTRICTED LAGOON						INNER SHOALS				OPEN LAGOON		OUTER SHOALS		OPEN LAGOON	
			L A G O O N															

FIG. 2.— Chart of lithofacies and lithofacies association of the analyzed sequences and related environmental interpretation. The shaded bars in the top row indicate the facies distribution in the studied sections. MT, Monte Tobenna; MF, Monte Faito; MR, Monte Raggeto; SS, Serra Sbregavitelli. Note that from Serra Sbregavitelli to Monte Tobenna the environmental conditions become increasingly restricted.

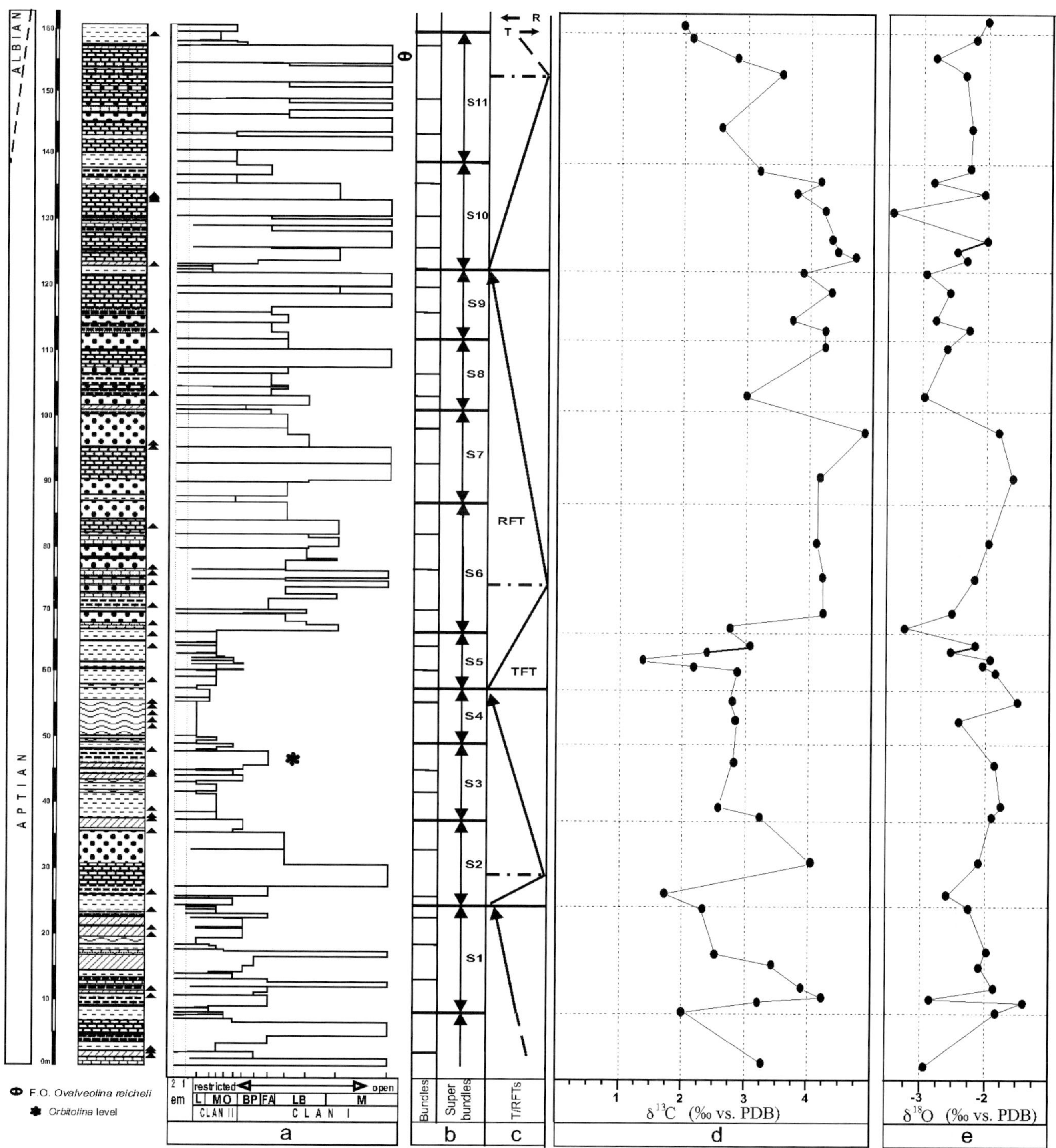

FIG. 3.—Facies organization and isotope stratigraphy recorded in the Serra Sbregavitelli section. Biostratigraphic constraints: *Orbitolina* (*Mesorbitolina*) *texana* and *Orbitolina* (*Mesorbitolina*) *parva* (Middle Gargasian = Late Aptian) in the central part (superbundle S3) and first occurrence of *Ovalveolina reicheli* (Aptian–Albian transition) in the uppermost part (superbundle S11). Scale in meters on the vertical axis. Key to the columnar logs: (a) Elementary cycles; lithofacies, lithofacies associations, clans, and cycle boundaries are indicated on the horizontal axis; Lithofacies associations: L, Laminate limestones; MO, mili–ostr–algal limestones; BP, bio-peloidal limestones; FA, for-algal limestones; LB, laminated-bioclastic limestones; M, molluscan limestones, (CO, chara–ostracod limestones only in Figs. 5 and 6); from the em1 cycle boundary (microkarst) to em2 boundary (thicker vadose caps, pervasive paleokarstic features) the early meteoric diagenetic overprint is more and more penetrative; em2 type cycle boundaries are locally marked by greenish to reddish, thin clayey levels; (b) High-frequency hierarchical organization of the elementary cycles into bundles (on the left) and superbundles (on the right). (c) transgressive–regressive facies trends (T/RFTs). (d) Carbon-isotope stratigraphy. (e) Oxygen-isotope stratigraphy. See Figure 2 for further information.

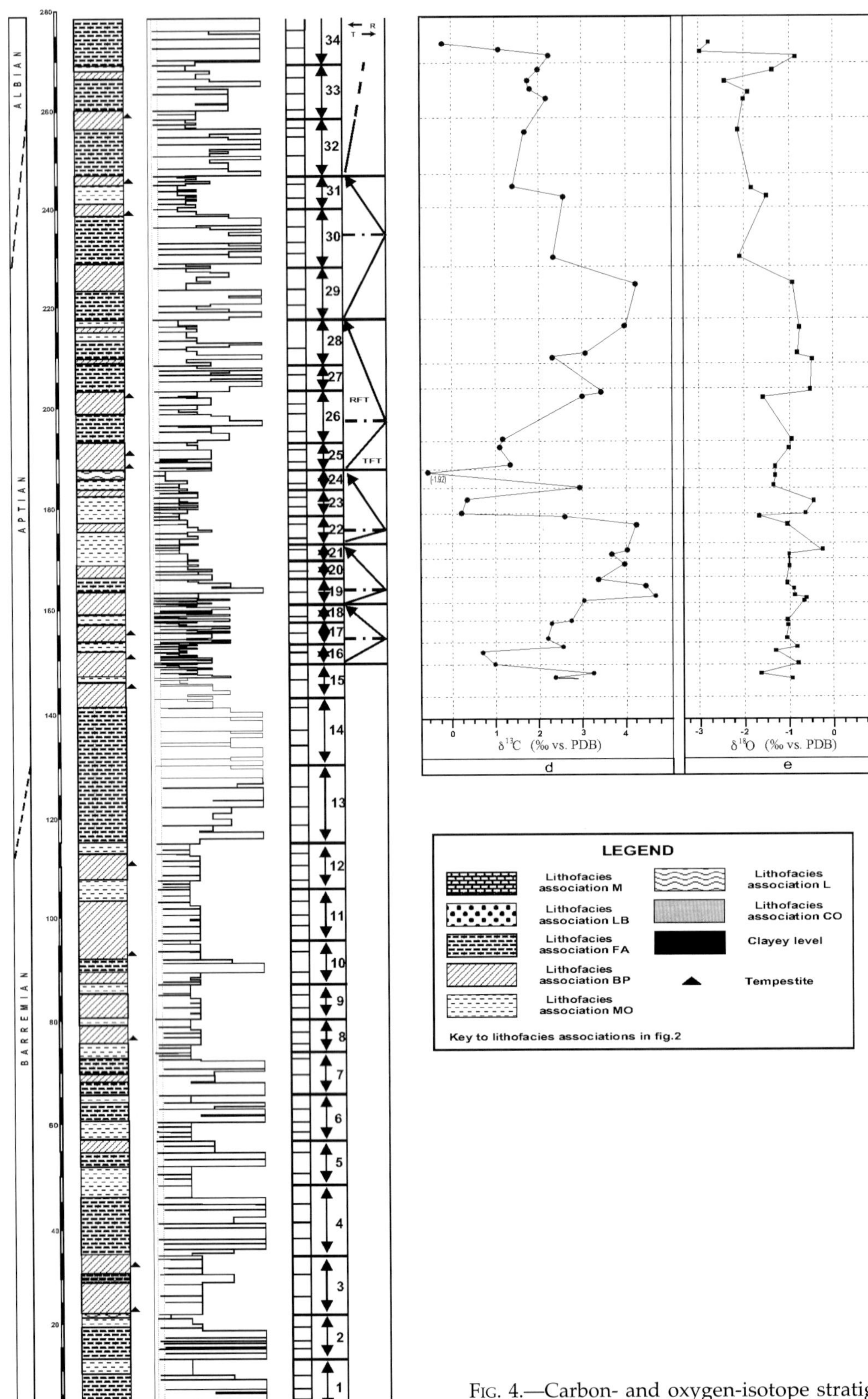

FIG. 4.—Carbon- and oxygen-isotope stratigraphy, and facies organization recorded in the Monte Raggeto section. The Aptian–Albian transition is suggested in the upper part of this section by the first occurrence of orbitolinids as *Valdanchella* sp. and *Neoiraquia* sp. Key to the columnar logs as in Figure 3. See also Figure 2 for further information.

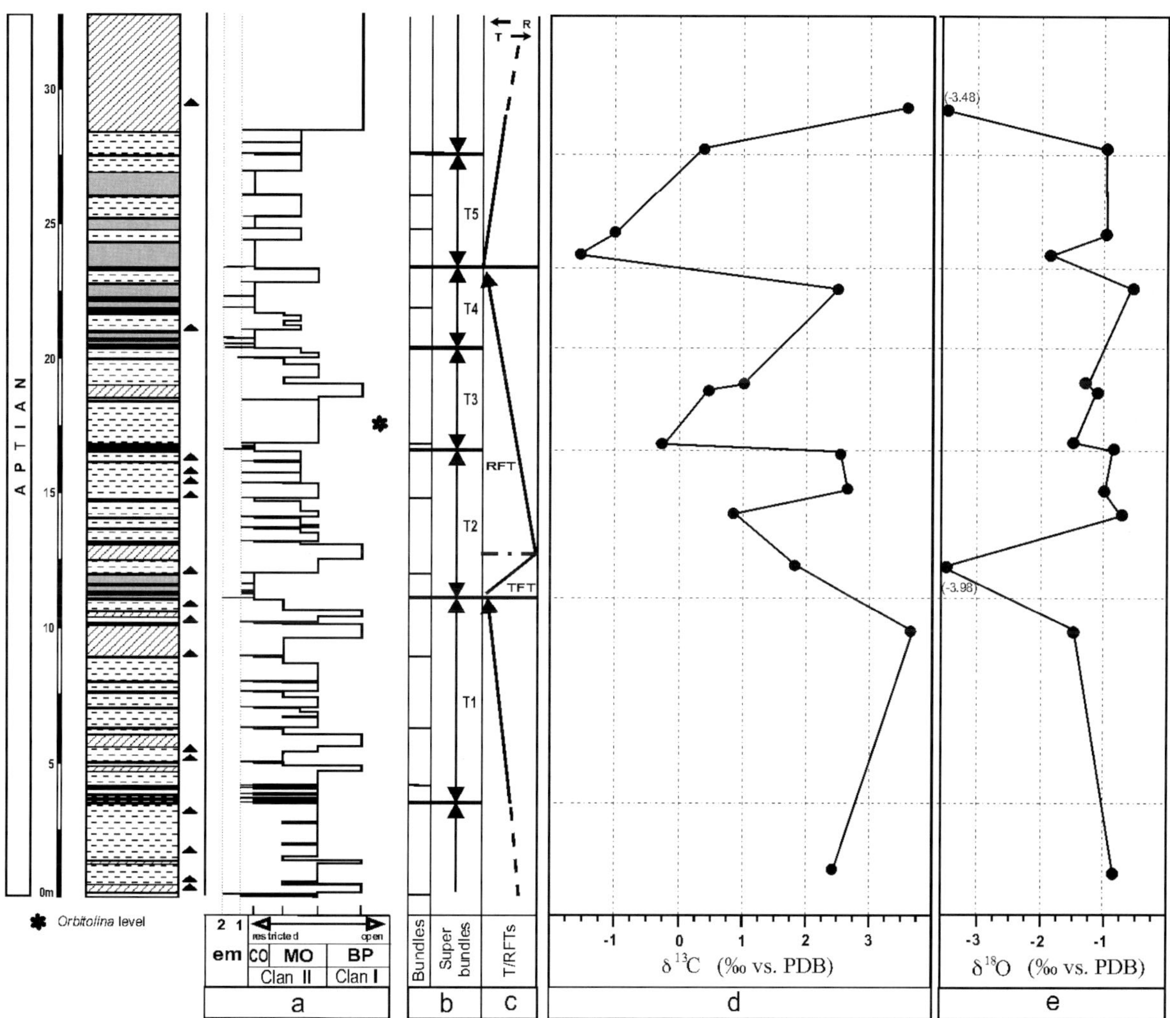

FIG. 5.—Carbon- and oxygen-isotope stratigraphy, and sedimentary evolution recorded in the Monte Tobenna section. The *Orbitolina* level Auct. (*Orbitolina texana* and *Orbitolina parva*, Middle Gargasian, Late Aptian) corresponds to superbundle T3. Columnar logs as in Figure 3. See also Figures 2 and 4 for the legend, and text for additional comments. Source of cyclostratigraphic data: Raspini (1998).

The transgressive peak of each TFT is located where the superbundle is thickest and formed of the relatively more open-marine lithofacies associations. On the contrary, regressive facies trends (RFTs) show: (a) upward-decreasing thickness of elementary cycles, bundles, and superbundles; (b) predominance of lithofacies associations suggesting increasingly restricted (peritidal) conditions; and (c) upward development of cycles characterized by boundaries showing progressively enhanced subaerial exposure features. The regressive peak of each RFT is located where the superbundle is thinnest and is formed of more restricted, inner-shelf lithofacies associations.

CARBON-ISOTOPE STRATIGRAPHY

Methods

Samples were collected after careful inspection on the outcrop in order to select intervals without (or with negligible) evidence of meteoric overprint, and these conditions were further confirmed under the microscope using polished sections and thin sections.

The isotopic composition of the bulk carbonate was determined by analysis of CO_2 evolved by reaction of bulk samples with 100% phosphoric acid at 50° C (McCrea, 1950). The $\delta^{18}O$ and $\delta^{13}C$ values of the released CO_2 gas were analyzed on a triple-collecting VG Micromass-903 mass spectrometer or on a fully automated VG Prism mass spectrometer in the stable-isotope laboratory at ETH in Zürich, whose internal standard is Carrara marble. The results were reported in the usual per mil δ notation relative to the V-PDB standard. Replicate analysis of selected samples showed a reproducibility of ± 0.1‰ for $\delta^{13}C$ and ± 0.2‰ for $\delta^{18}O$. The bulk carbonate $\delta^{18}O$ and $\delta^{13}C$ values for the studied sections are shown in Figures 3–6, plotted in their stratigraphic context.

Cycle Stacking Pattern and Isotope Record

Serra Sbregavitelli Succession.—

The succession is composed of 91 shallowing-up elementary cycles ranging from 0.30 to 5.55 m in thickness (column *a*

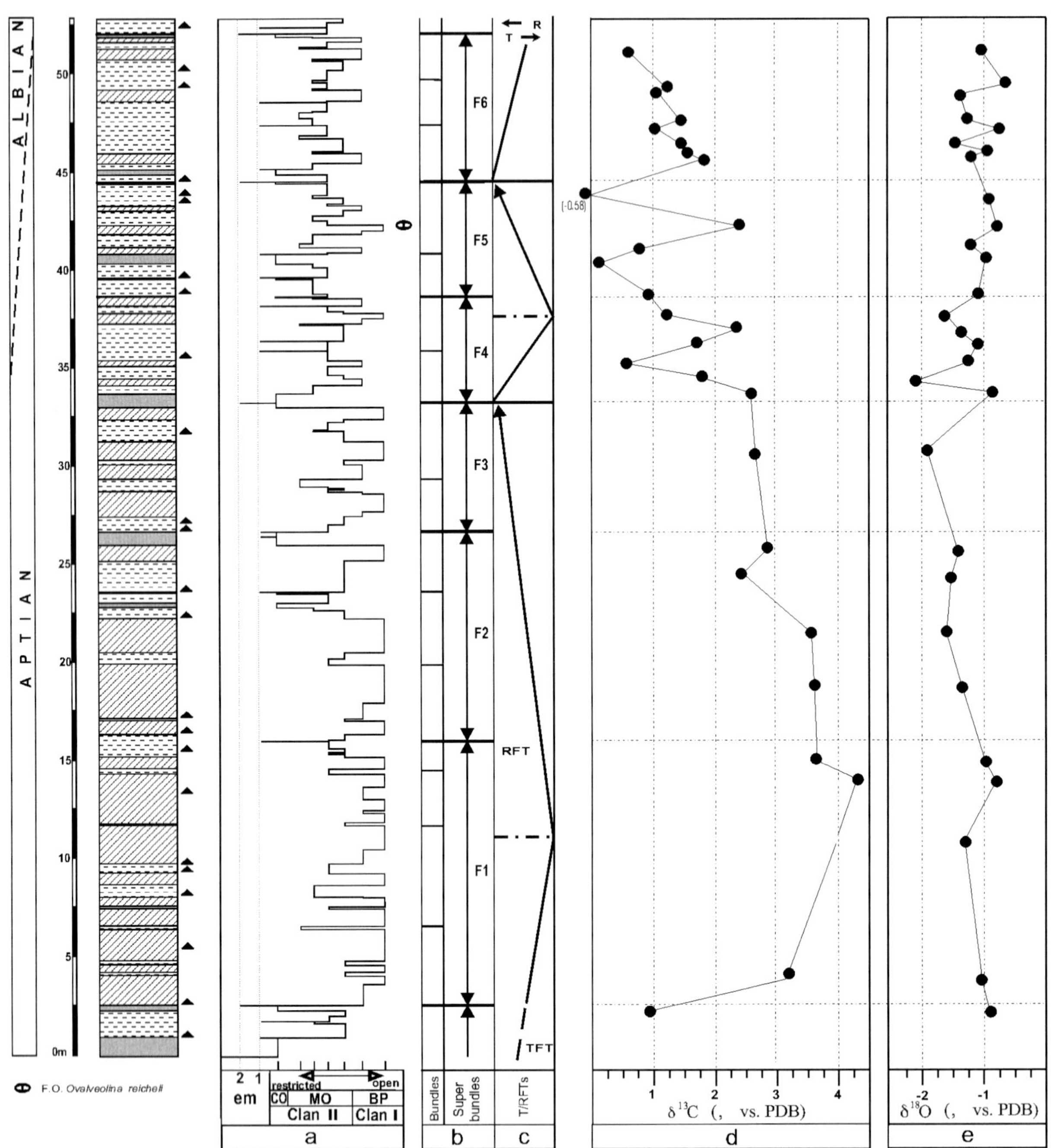

FIG. 6.—Carbon- and oxygen-isotope stratigraphy and sedimentary evolution of the Monte Faito section. Here the Aptian–Albian transition was recognized in the stratigraphic interval corresponding to the superbundles F4 and F5. Columnar logs as in Figure 3. See also Figures 2 and 4 for the legend, and text for additional comments. Source of cyclostratigraphic data: D'Argenio et al. (1999b); Raspini (2001).

in Fig. 3), hierarchically ordered in 36 bundles (each from 1 to 8.10 m in thickness) and 11 superbundles (from 8.05 to 20.63 m), plus two lowermost bundles forming an incomplete basal superbundle (Fig. 3, column *b*; Fig. 7). Moreover, these high-frequency cycles are organized in two complete T/RFTs (superbundles S2 to S4 and S5 to S9) plus a RFT (superbundle S1) and a TFT (superbundles S10 and S11, column *c* in Fig. 3). Throughout the section peritidal deposits are well represented from superbundles S1 to S5, whereas open lagoonal deposits prevail from superbundle S6 to S11, except the lower part of the superbundle S10.

In the 165-m-thick Serra Sbregavitelli section we measured carbon-isotope values that range between +1‰ and +5‰ (Table 1, Fig. 3, column *d*). In the lowermost 10 m, values fluctuate between +2‰ and +4‰. Between 10 m and 25 m (superbundle S1), values decrease from +4.2‰ to +1.2‰. Above, in the lower

SERRA SBREGAVITELLI

Bundles		Superbundles	
Number	Thickness (m)	Number	Thickness (m)
4 3 2 1	2.17 8.10 5.35 4.30	S11	19.92
4 3 2 1	3.25 5.00 4.80 3.40	S10	16.45
3 2 1	2.60 3.90 4.10	S9	10.60
3 2 1	5.30 3.80 2.12	S8	11.22
3 2 1	2.80 5.40 5.90	S7	14.10
4 3 2 1	4.90 5.55 6.63 3.55	S6	20.63
3 2 1	2.30 3.85 3.00	S5	9.15
2 1	2.15 5.90	S4	8.05
3 2 1	3.99 3.15 4.30	S3	11.44
3 2 1	4.50 7.05 1.00	S2	12.55
4 3 2 1	1.25 4.80 4.40 5.85	S1	16.30
1	6.08		
Average:	**4.23**		**13.67**

MONTE RAGGETO

Bundles		Superbundles	
Number	Thickness (m)	Number	Thickness (m)
2 1	3.55 3.45	R34	
4 3 2 1	2.70 3.10 3.30 1.85	R33	10.95
4 3 2 1	1.90 3.10 2.40 4.10	R32	11.50
3 2 1	1.80 2.80 2.05	R31	6.65
4 3 2 1	3.70 3.20 2.20 2.85	R30	11.95
3 2 1	4.50 3.20 2.80	R29	10.50
3 2 1	2.40 3.50 3.40	R28	9.30
2 1	1.90 3.30	R27	5.20
4 3 2 1	2.70 2.00 3.50 2.15	R26	10.35
3 2 1	1.50 2.55 1.70	R25	5.75
3 2 1	1.00 1.60 1.35	R24	3.95
3 2 1	1.50 1.90 1.85	R23	5.25
3 2 1	1.65 3.00 1.15	R22	5.80
2 1	1.05 2.20	R21	3.25
2 1	2.10 1.65	R20	3.75
2 1	2.80 2.61	R19	5.41
2 1	2.42 1.21	R18	3.63
2 1	2.03 1.14	R17	4.17
2 1	1.69 2.25	R16	3.94
Average:	**2.42**		**6.74**

MONTE FAITO

Bundles		Superbundles	
Number	Thickness (m)	Number	Thickness (m)
3 2 1	2.37 2.28 2.87	F6	7.52
2 1	3.68 2.19	F5	5.87
2 1	2.77 2.63	F4	5.40
2 1	3.94 2.72	F3	6.66
3 2 1	3.07 3.43 4.19	F2	10.69
4 3 2 1	1.69 2.45 5.29 4.01	F1	13.44
Average:	**3.10**		**8.26**

MONTE TOBENNA

Bundles		Superbundles	
3 2 1	1.49 1.25 1.49	T5	4.23
2 1	1.66 1.31	T4	2.97
2 1	3.42 0.30	T3	3.72
3 2 1	1.84 2.70 9.4	T2	5.48
4 3 2 1	2.13 2.62 2.13 0.64	T1	7.52
1	3.37		
Average:	**1.82**		**4.78**

FIG. 7.—Thicknesses (m) of bundles and superbundles in the analyzed sequences. Note that the average thicknesses of bundles and superbundles are proportional to the original environmental characteristics (see also inset of Fig. 8).

part of superbundle S2 (TFT), $\delta^{13}C$ increases rapidly to +4.0‰, then decreases again within 15 m and forms a plateau around +2.7‰ between 40 m and 70 m. Between 60 and 65 m above the base of the section, the plateau is interrupted by a narrow negative spike within the superbundle S5 (Fig. 3, column *d*). From 70 to 135 m relatively high carbon-isotope values between +3.7‰ and +5.0‰ are recorded, except one measurement of +3.0‰ close to 100 m. We note that the carbon-isotope signature records a positive pattern also close to the regressive peak, at S9 superbundle boundary (Fig. 3, columns *c* and *d*). Finally, above 135 m (Aptian–Albian transition interval), $\delta^{13}C$ values decrease to +2.5‰ within 10 m. At 150 m a last increase to +3.5‰ occurs, then the values decrease to +2.0‰ within the uppermost 10 m of the profile (Fig. 3, column *d*).

Monte Raggeto Succession.—

The stratigraphic interval here considered starts from about 148 m from the base (superbundle R16) up to 278 m (superbundle R34, Fig. 4). The succession is composed of 117 shallowing-up elementary cycles (from 0.15 to 2.50 m thick), hierarchically ordered in 53 bundles (from 1 to 4.50 m thick) and 18 superbundles (from 3.25 to 11.95 m thick), plus two uppermost bundles forming an incomplete superbundle (Fig. 4, columns *a* and *b*; Fig. 7).

TABLE 1.—C and O isotope values measured in the Serra Sbregavitelli succession.

Sample	$\delta^{13}C$ ‰ vs. PDB	$\delta^{18}O$ ‰ vs. PDB	Lithofacies	Distance from the base of the section (m)
82	2.00	-2.02	MO2	159.00
81	2.13	-2.19	MO5	157.50
82b	2.84	-2.80	M1	155.02
80	3.53	-2.36	M1	153.00
78(2)	2.60	-2.25	M1	144.22
77	3.20	-2.28	BP1	137.62
76(2)	4.15	-2.83	MO2	136.00
75(1)	3.78	-2.09	M2	132.00
75	4.22	-3.47	M1	131.00
74	4.37	-2.03	M1	127.50
73(2)	4.45	-2.46	FA1	125.47
73(1)	4.70	-2.34	M2	123.97
72	3.90	-2.94	MO6	121.97
70	4.23	-2.59	M2	119.22
68	3.73	-2.81	LB2	114.87
67	4.23	-2.30	FA1	113.37
66(4)	4.25	-2.62	LB2	111.26
65(3)	3.02	-2.98	FA1	103.87
60(3)	4.89	-1.81	LB2	100.96
58(3)	4.15	-1.59	M1	91.30
54(2)	4.14	-1.97	M2	79.80
52(2)	4.23	-2.19	M1	74.40
49(2)	4.21	-2.54	LB1	69.88
48(2)	2.77	-3.25	M1	66.00
46(1)	3.09	-2.17	MO5	63.45
45(3)	2.39	-2.52	MO5	62.78
43(1)	1.37	-1.95	MO1	61.00
42(2)	2.19	-2.04	BP4	60.50
41(1)	2.86	-1.84	BP4	59.55
37(1)	2.76	-1.50	MO6	54.80
36(2)	2.86	-2.41	L1	52.00
32(2)	2.81	-1.86	FA1	45.88
28(1)	2.59	-1.76	MO5	40.20
27(1)	3.22	-1.90	MO5	37.30
25(1)	3.96	-2.10	LB2	31.00
23(2)	1.74	-2.60	FA1	25.90
19(1)	2.30	-2.25	MO5	24.00
13(3)	2.58	-1.98	M1	17.53
12(2)	3.41	-2.08	BP1	14.63
9(1)	3.91	-1.87	M1	12.33
8(1)	4.20	-2.87	BP1	10.80
7(1)	3.19	-1.43	FA1	10.10
6(2)	1.97	-1.83	MO6	8.60
1a	3.24	-2.92	M1	0.10

TABLE 2.—C and O isotope values measured in the Monte Raggeto succession.

Samples	$\delta^{13}C$ ‰ vs. PDB	$\delta^{18}O$ ‰ vs. PDB	Lithofacies	Distance from the base of the section (m)
T3	-0.20	-2.81	FA1	275.85
T4	1.02	-3.07	FA1	271.70
T6	2.27	-0.90	FA1	270.55
31	1.88	-1.41	BP1	268.15
33	1.67	-2.45	BP1	266.65
34	1.86	-1.95	FA2	264.35
35	2.20	-2.03	FA2	262.80
38	1.67	-2.16	BP1	255.40
40	1.42	-1.86	BP2	244.40
43	2.61	-1.54	BP2	243.20
47	2.32	-2.16	FA1	230.80
50	4.08	-0.96	BP1	225.90
51	3.98	-0.80	BP2	216.20
52	3.04	-0.86	FA1	210.30
77	2.33	-0.53	MO1	209.80
80	3.50	-0.57	BP2	203.40
82	2.97	-1.61	BP1	201.90
86	1.17	-0.99	BP2	193.55
88	1.08	-1.05	BP2	192.15
90	1.27	-1.35	BP2	188.00
Ar	-1.92	-1.35	L1	186.70
93	2.83	-1.40	MO3	184.30
96	-0.10	-0.49	MO1	181.20
98	0.24	-0.68	MO3	177.70
100	2.59	-1.70	MO1	176.90
101	4.24	-1.09	BP2	175.70
103	4.02	-0.32	MO1	171.10
104	3.68	-1.07	MO1	170.30
E	3.96	-1.07	BP2	168.30
L	3.37	-1.11	FA2	164.10
C	4.49	-0.97	FA1	163.10
44c	4.64	-0.90	FA1	161.60
38c	2.93	-0.64	BP1	160.70
30c	2.76	-1.02	BP2	157.20
29c	2.32	-1.00	MO1	156.80
22c	2.20	-1.07	MO1	153.70
47f	2.56	-0.81	BP2	152.20
12c	0.68	-1.28	L1	151.30
20f	1.00	-0.80	L1	148.80
8f	3.27	-1.70	BP2	147.10
3f	2.38	-0.98	MO2	145.80

Moreover, five T/RFTs plus one TFT (the uppermost 32 m) were recognized, (Fig. 4, column *c*).

Peritidal conditions (lithofacies associations MO and L, Fig. 2) are well represented from superbundles R21 to superbundle R24, whereas more open, subtidal conditions (lithofacies associations FA and BP, Fig. 2) prevail in the lowermost part and in the middle–upper part of the section (Fig. 4, column *a*). In the latter cycle, boundaries range from em1 type (microkarst, weak pedogenesis) to em2 type (paleokarst and/or paleosols), whereas in the former em2 type boundaries prevail. Here the emersion-related features are increasingly penetrative. The carbon-isotope values vary between –2‰ and +4.5‰ (Table 2; Fig. 4, column *d*). From 148 to 161 m (R16–R18) a positive $\delta^{13}C$ shift can be observed, ranging from +0.6‰ to +3‰. The carbon-isotope values remain close to +4.0‰ from 161 to 176 m. Within the following 10 m, at 187 m from the base, just below a paleosol marking the regressive peak of the R24 superbundle boundary, measurements fluctuate between +3‰ and –2‰. Above this paleosol, between 188 m and 226 m, the values increase from +1‰ to +4‰, then slowly decrease to +1.42‰ at 245 m from the base. We note that the Monte Raggeto section, like that of Serra Sbregavitelli, records a negative trend from superbundle R29 to R31, close to the Aptian–Albian transition. Finally, in superbundles R32 and R33, corresponding to the uppermost TFT, the carbon-isotope profile shows values close to +2‰, then (uppermost 5 m) the $\delta^{13}C$ values decrease rapidly to –1‰ (Fig. 4, columns *c* and *d*).

Monte Tobenna Succession.—

The succession is composed of 53 shallowing-up elementary cycles (from 0.07 to 1.35 m thick), organized in 15 bundles and

5 superbundles, plus a lowermost bundle forming an incomplete superbundle (Fig. 5, columns *a* and *b*; Fig. 7). At the upper limit of superbundle T1, a regressive peak was identified. Above this limit the cyclic stacking pattern includes (Fig. 5, column *c*) a T/RFT (superbundles T2 to T4) plus a TFT segment (superbundle T5).

Lithofacies associations MO and CO (Fig. 2) are widespread throughout the section. Inner-shoal deposits (lithofacies association BP, Fig. 2), suggesting more open conditions, are rare and occur in some elementary cycles at times of relative high sea level. Emersion-related features characterize the cycle limits; they are revealed by microkarstic cavities, filled by geopetal crystal silt or by pedogenetic features (em1 or em2 type boundaries), both superimposed directly on subtidal deposits. Commonly the cycle limits are characterized by centimeter-thick, green clayey horizons that are thicker and more diffuse in the upper part of the section.

The Monte Tobenna carbon-isotope profile shows relatively stable values between +2.5‰ and +3.5‰ (Table 3, Fig. 5, column *d*). Three negative spikes are recorded at 15, 17, and 24 m from the base of the section, reaching values as depleted as –1.5‰ at the regressive peak located at the upper boundary of T4.

Monte Faito Succession.—

This section has been already analyzed in the past few years (D'Argenio et al., 1999b; Raspini, 2001). Here we add some refinements. The Monte Faito succession (Fig. 6, columns *a* and *b*; Fig. 7) is composed of 50 shallowing-up elementary cycles (from 0.24 to 3.03 m thick), organized in 16 bundles (from 1.69 to 5.29 m thick) and 6 superbundles (from to 5.40 to 13.44 m thick). Moreover, as can be seen in Figure 6, column *c*, high-frequency cycles are superimposed on two T/RFTs (superbundles F1 to F3 and F4 to F5, respectively) plus the base of a TFT in the uppermost 7 m (superbundle F6).

Subaerial exposure is recorded mostly in the upper part of the section, where vadose caps (em1 and em2 type boundaries), locally marked by centimeter-thick, green clayey horizons, define the tops of many cyclic units. No deposits suggesting deposition in lagoonal environments with normal-marine circulation (see environmental interpretation of lithofacies associations in Fig. 2) were observed. Relatively open conditions are testified only by inner-shoal deposits (lithofacies association BP, Fig. 2); they are widespread in the lower part of the section and decrease in superbundle F3.

Within the 53-m-thick Monte Faito succession, carbon-isotope values range between 0 and +4.3‰ (Table 4, Fig. 6, column *d*). In the basal 16 m (superbundle F1), including the first transgressive peak, the carbon-isotope profile shows an increase from +1‰ to +4.3‰ (Fig. 6, columns *b* and *d*). Within the overlying 17 m, values slowly decrease to +2.56‰, immediately above the first regressive peak, between superbundles F3 and F4. From 33 to 45 m, during the second T/RFT, this general trend is interrupted by three negative spikes, the last and most negative value (–0.6‰) occurring immediately below the regressive peak at F5 superbundle limit (Fig. 6, columns *c* and *d*). Finally, in the last 7 m, carbon-isotope values fluctuate between +1.04‰ and +1.79‰, reaching +0.64‰ below the F6 upper boundary.

Considering the $\delta^{18}O$ data, we note that in all the samples examined they fluctuate between –0.32‰ and –3.98‰ (Figs. 3–6, column *e*; Tables 1–4). These data indicate that the analyzed shallow-water carbonate of the studied sections was lithified at an early diagenetic stage, under marine conditions. This finding aids interpretation of the measured $\delta^{13}C$ values as primary Cretaceous signatures recording fluctuations of the marine carbon-isotope reservoir.

TABLE 3.—C and O isotope values measured in the Monte Tobenna succession.

Samples	$\delta^{13}C$ ‰ vs. PDB	$\delta^{18}O$ ‰ vs. PDB	Lithofacies	Distance from the base of the section (m)
44e	3.56	-3.48	BP3	29.26
42a	0.37	-0.99	MO4	27.78
37b	-1.02	-0.97	MO4	24.58
36g	-1.57	-1.85	CO1	23.81
34	2.50	-0.58	CO1	22.55
27a	1.01	-1.27	BP3	19.00
2b	0.45	-1.14	BP3	18.80
24d	-0.23	-1.49	CO1	16.78
23top	2.55	-0.87	MO4	16.50
20c	2.66	-1.00	MO3	15.06
19c	0.88	-0.74	MO3	14.19
15c	1.84	-3.98	MO3	12.28
12c	3.69	-1.49	BP3	9.84
6	2.43	-0.87	MO3	0.86

TABLE 4.—C and O isotope values measured in the Monte Faito succession.

Samples	$\delta^{13}C$ ‰ vs. PDB	$\delta^{18}O$ ‰ vs. PDB	Lithofacies	Distance from the base of the section (m)
72e	0.64	-1.04	BP4	51.15
71d	1.19	-0.67	MO3	49.50
70c	1.08	-1.39	BP4	49.01
68a	1.45	-1.26	MO4	47.70
67a	1.04	-0.77	MO3	47.25
65top	1.49	-1.46	MO1	46.50
64	1.58	-0.96	MO4	46.12
63d	1.79	-1.21	BP4	45.83
58b	-0.58	-0.93	MO1	43.65
54f	2.40	-0.79	BP2	42.27
53top	0.76	-1.23	MO6	41.40
52d	0.13	-0.98	CO1	40.70
48a	0.94	-1.10	MO3	38.83
45a	1.20	-1.61	BP2	37.77
43b	2.32	-1.37	MO1	36.90
42a	1.71	-1.12	MO3	36.25
41e	0.59	-1.27	BP4	35.37
40b	1.76	-2.11	BP4	34.41
39b	2.56	-0.87	CO1	33.86
35b	2.64	-1.92	BP2	30.90
24b	2.85	-1.42	BP2	25.69
23a	2.42	-1.53	MO1	24.51
17b	3.56	-1.61	BP2	21.61
12b	3.61	-1.36	BP2	18.83
7e	3.66	-0.98	BP2	14.99
7b	4.30	-0.82	BP2	14.07
4b	3.97	-1.39	BP2	11.07
-7(2)	3.18	-1.04	BP2	3.91
-11(3)	0.97	-0.92	MO1	2.33

DISCUSSION

Regional Cyclostratigraphic Correlation

In a previous paper (D'Argenio et al., 1999b), we presented a centimeter-scale study of three of the four sections discussed here: Serra Sbregavitelli, Monte Tobenna, and Monte Faito. In that paper we suggested a regional correlation at superbundle level, showing that the sedimentary record can be considered more continuous at Serra Sbregavitelli, whereas it is lacking one superbundle at Monte Faito. This correlation was based on biostratigraphic constraints and sequence-stratigraphic criteria. As to the biostratigraphy, we considered the Upper Aptian litho-bio-horizon "*Orbitolina* level" as the regional datum (see, e.g., Cherchi et al., 1978; Bravi and De Castro, 1995, for additional information on this biostratigraphic marker of southern Apennines Lower Cretaceous). We have also considered the first occurrence of *Ovalveolina reicheli* (De Castro) to be an evidence of the Aptian–Albian transition and hence we correlated superbundle S11 of Serra Sbregavitelli with superbundle F5 of Monte Faito (Figs. 9 and 11 in D'Argenio et al., 1999b).

As to the sequence stratigraphy, we considered the superbundles as equivalent to depositional sequences (D'Argenio et al., 1997; D'Argenio et al., 1999b). This was suggested by a higher probability for superbundles to be represented in the stratigraphic record, even if lacking some of their elementary cycles (and/or bundles). The latter are more likely to be missing because of low accommodation. According to this interpretation, we locate the sequence boundaries at the superbundle limits, and the maximum flooding surfaces at the occurrence of the most open-marine lithofacies (normally developed within the thickest elementary cycle).

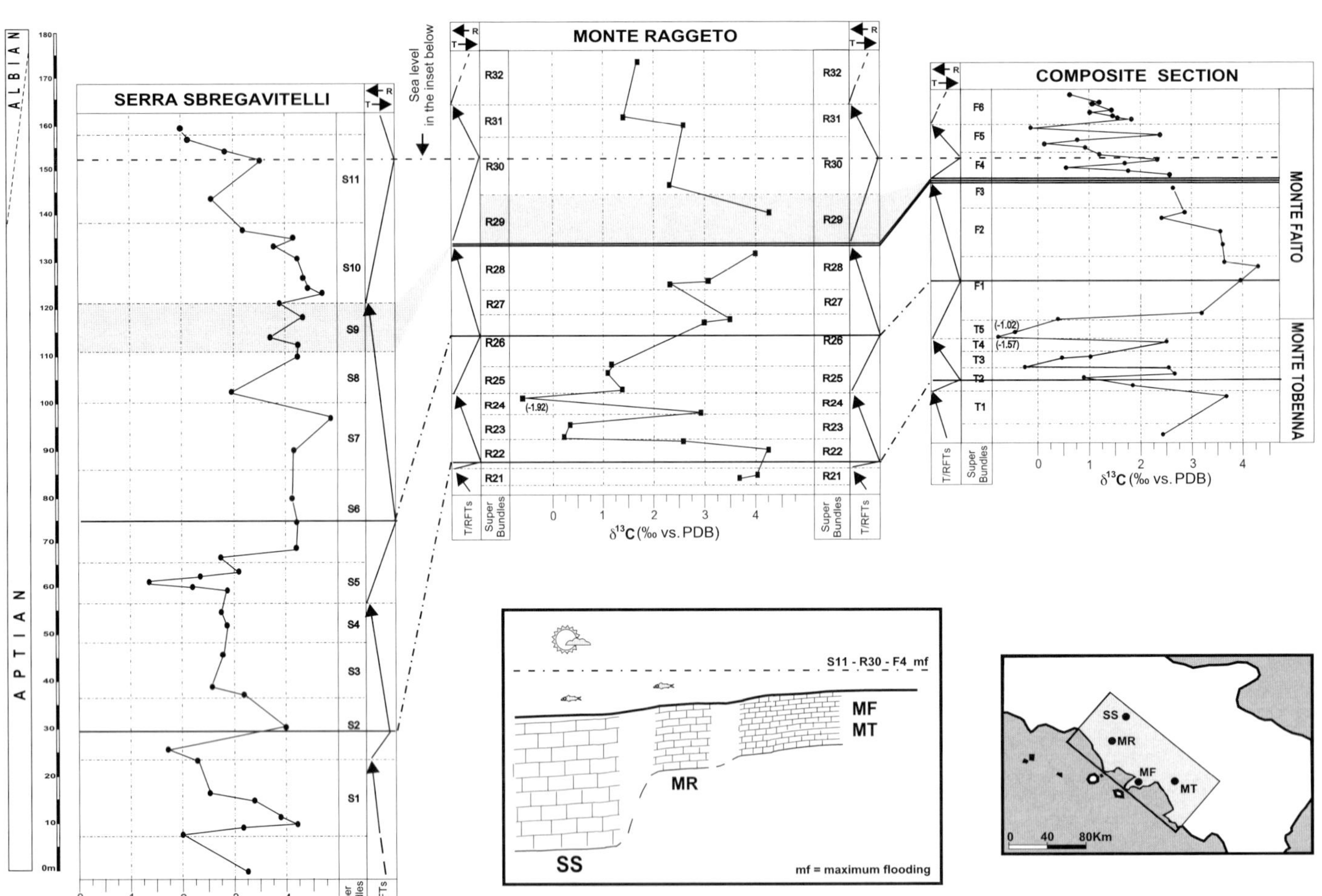

FIG. 8.—High-resolution regional correlation among the studied sequences. This chart is based on the following biostratigraphic constraints: the *Orbitolina* level, recorded within superbundles S3 and T3 (Figs. 3 and 5, respectively) and the Aptian–Albian transition, recognized as a stratigraphic interval in superbundles R30–R32 (M. Raggeto) as well as in superbundle S11 (Serra Sbregavitelli) and F4–F6 (M. Faito) at the first occurrence of *Ovalveolina reicheli* (Figs. 3, 6). Also the correspondence among the transgressive–regressive facies trends recognized in the studied sections was taken into account (dash-and-dot lines connecting the transgressive peaks are considered as time lines). Gaps are shown in gray. The inset at the center of the figure shows a restored profile across an idealized carbonate platform. SS, Serra Sbregavitelli; MR, Monte Raggeto; MT, Monte Tobenna; MF, Monte Faito. Note that some discrepancies among corresponding tracts of the $\delta^{13}C$ curves in the compared sections may be a consequence of both the different sampling intervals and the variable size of the gaps in the stacked superbundles. Moreover, in the Monte Tobenna–Monte Faito composite section the $\delta^{13}C$ signal appears to be noisier.

In the present paper, we enlarge the regional frame of the high-resolution correlation integrating also the much thicker section of Monte Raggeto. Furthermore, we apply carbon-isotope geochemistry as an additional tool to the cyclostratigraphic correlation. The new data have allowed us to define an improved fit, which refines the previous correlation, as well as to determine the equivalence of the superbundles at regional scale (Fig. 8). On the basis of depositional and early-diagenetic features, we assume that at TFT peaks there is the minimum probability for cycles to be missing, because of their more open lithofacies associations (and hence highest accommodation), as confirmed also by increasing marine cementation and decreasing meteoric overprint. On this assumption, and considering the Aptian–Albian boundary as a biostratigraphic time line, we use the maximum-flooding surfaces at our transgressive peaks to correlate stratigraphic intervals characterized by analogous transgressive–regressive facies trends and carbon-isotope trends. According to our previous correlation (see D'Argenio et al., 1999b), the sedimentary record of Serra Sbregavitelli section appears to be the most continuous; on the contrary, karst, penetrating also more than a cycle, allows gaps to be recognized between superbundles R28 and R29 at Monte Raggeto and between F3 and F4 at Monte Faito; in particular, the equivalent of superbundle S9 is not recorded at Monte Raggeto, whereas at Monte Faito the equivalents of superbundles S9 and S10 (the latter corresponding to R29 of Monte Raggeto) are missing. The inset in Figure 8 depicts a restored profile across a schematic platform where strata, facies, and thickness are correlative among them, from a more "external" position (most complete and thickest section) to a more "internal" position (less complete and thinner). It is noticeable that more restricted, and hence internal, settings of a carbonate platform, *per se* prone to omissions (Monte Tobenna and Monte Faito sections), find a good correlation with the more open Monte Raggeto and Serra Sbregavitelli sections, notwithstanding the widening of the regional frame.

Orbital Chronostratigraphy

On the basis of the cyclostratigraphic correlation shown in Figure 8, it was also possible to assemble a chronostratigraphic chart based on the bundles and superbundles of the analyzed sections (Fig. 9). In each diagram the precise location of maximum flooding surfaces and of missing bundles is suggested by the high-resolution sequence stratigraphic interpretation of both the superbundles and the transgressive–regressive facies trends, as well as by the correlation of the stacking patterns of each section. Figure 9 shows that in each section both location and time duration of the missing intervals are variable; moreover, longer gaps can be found in correspondence of RFT boundaries, marked by more deeply penetrating subaerial features (paleokarst, paleosols). As shown elsewhere (D'Argenio et al., 1999b), accommodation space as well as bed thickness and facies of time-equivalent cycles are controlled predominantly by regional changes in subsidence and carbonate productivity. Therefore a minor number of missing bundles is recorded in successions formed in more open carbonate-platform environments (hence producing higher sediment volumes), and *vice versa*. This might account for the Serra Sbregavitelli sedimentary record, which appears to be thicker and more continuous, at least at the superbundle level. In contrast, gaps of 500 and even 1000 ky can be recognized in the Upper Aptian of Monte Raggeto and Monte Tobenna–Monte Faito sections, respectively (Fig. 9). On these bases, a composite Serra Sbregavitelli–Monte Raggeto orbital chronostratigraphy was assembled (superbundles 1–20 in Fig. 9), suggesting that the minimum duration of the T/RFTs ranges from 1.2 to 2 My and that the total time represented in the studied stratigraphic interval is 7.8 My.

Carbon-Isotope Stratigraphy and Regional Correlation

Looking at the carbon-isotope stratigraphy, it is worth noting that the time interval selected for our investigation is characterized in pelagic sediments by two major positive carbon-isotope anomalies, interrupted by an interval marked by lower carbonate carbon-isotope values (see Weissert and Lini, 1991, Weissert et al., 1998, Bralower et al., 1999; see also Fig. 10, column *D*). In addition, in the pelagic record a pronounced negative carbon-isotope anomaly occurs immediately below the first Late Aptian positive carbon-isotope shift (*Selli Level* Auct. or equivalent intervals), which has been found in deep-water and shallow-water carbonates that contain both marine and terrestrial organic matter (Erba and Larson, 1998; Menegatti et al., 1998; Gröcke et al., 1999; Ando et al., 2002).

The carbon-isotope stratigraphy established in pelagic successions has been calibrated by biostratigraphic and magnetostratigraphic methods, and the pattern of the carbon-isotope curve has also been identified in hemipelagic and even shallow-water carbonate successions (Weissert and Bréhéret, 1991; Jenkyns, 1995; Ferreri et al., 1997; Buonocunto et al., 2002). The carbon-isotope composition of shallow-water carbonates, however, may be an unreliable tracer of marine carbon-isotope compositions. Variations in the chemistry of water covering shallow platforms (Patterson and Walter, 1994) and alteration of marine carbon-isotope composition during early meteoric diagenesis may have altered average oceanic carbon-isotope values considerably (e.g., Jenkyns, 1995; Ferreri et al., 1997; Grötsch et al., 1998; Davey and Jenkyns, 1999; Immenhauser et al., 2002). This may also explain why carbon-isotope curves established in shallow-water environments are sometimes more "noisy" than curves established in pelagic successions.

Despite these potential sources of error, three carbonate carbon-isotope curves were established through the sections studied for cyclostratigraphy. The most complete Aptian succession at Serra Sbregavitelli indeed preserves the well known carbon-isotope pattern of the Aptian stage (Fig. 8; Fig. 10, column *D*). Two episodes with positive carbon-isotope values near +4‰ are interrupted by an interval with lower carbon-isotope values. In this part of the section (superbundles S3 to S5) the higher frequency of em2 type boundaries (deeper early meteoric diagenetic overprint) suggests pronounced subaerial exposure, which is not recorded in the $\delta^{13}C$ values (Fig. 3). The Monte Raggeto succession contains the transition from the first positive carbon-isotope anomaly to the second one (Fig. 4), and the combined Monte Tobenna–Monte Faito succession, where the $\delta^{13}C$ signature appears to be noisy, reveals both carbon anomalies as well (Figs. 5, 6, 8). Because of the rather low resolution of the carbon-isotope record, we attempt to trace only the major anomalies recognized in the studied sections. It is worth noting that some discrepancies among corresponding tracts of the $\delta^{13}C$ curves in the compared sections may be a consequence of both the different sampling intervals and the variable size of the gaps in the stacked superbundles. Figure 8 suggests that, although the correlation of superbundles S1 and S2 respectively with R21 and R22 and with T1 and T2 seems trustworthy, carbon-isotope data do not clearly fit in the correlation of superbundles S3 to S5 with R23 to R25 and with T3 to T5. This may be because the compared stratigraphic intervals correspond to a time of more restricted marine circulation, as suggested by the type of lithofacies associations and by cycle boundaries more deeply overprinted by early meteoric diagenesis (Figs. 3–5). Intertidal to supratidal conditions and

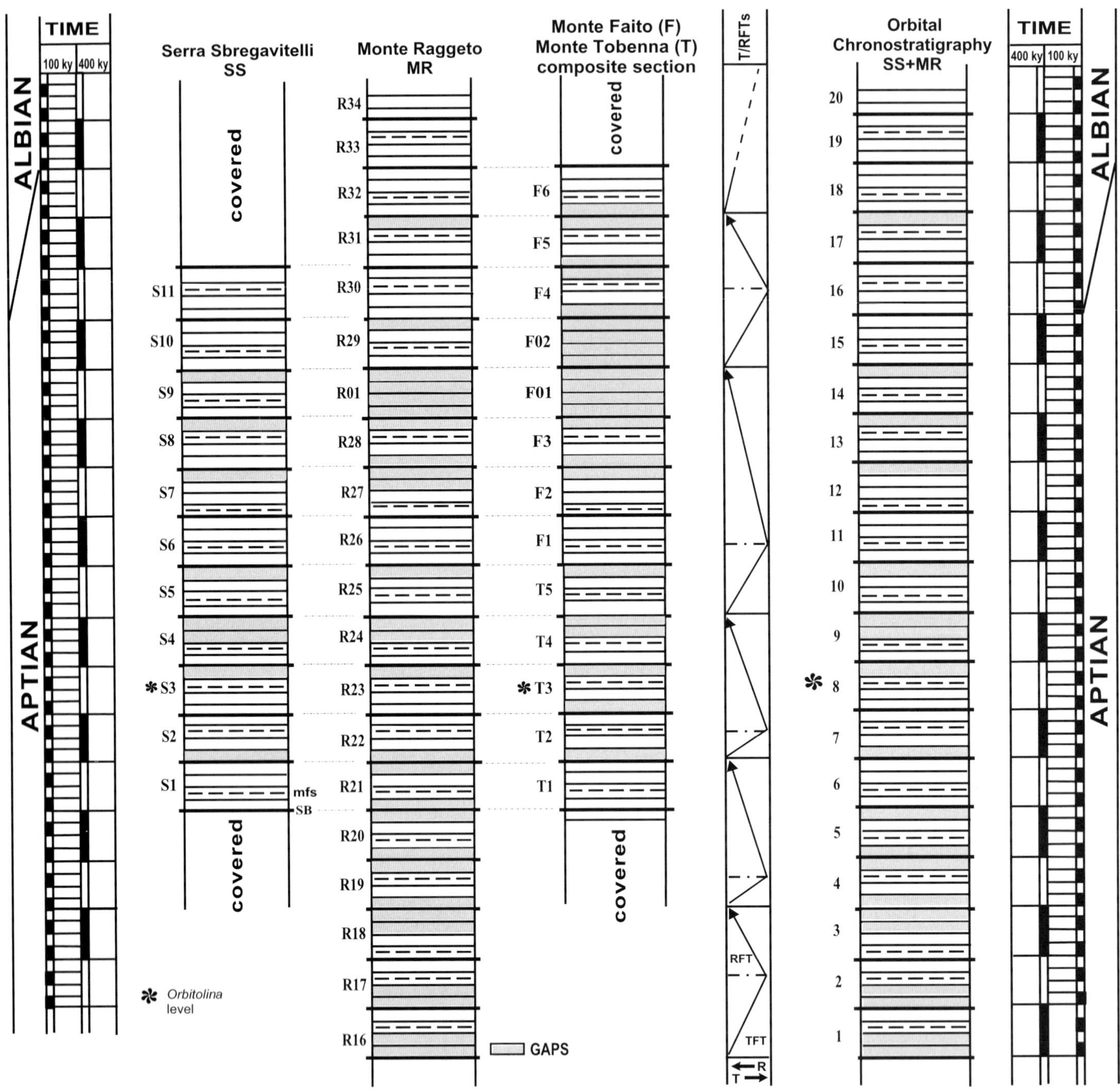

FIG. 9.—High-resolution chronostratigraphic correlation of the studied sections. Time scale is based on bundles (~ 100 ky cycles) and superbundles (~ 400 ky cycles). Missing cycles are shown in gray. T/RFTs, transgressive–regressive facies trends. The composite orbital chronostratigraphy of the studied interval (last column on the right, superbundles 1–20) was obtained using the Serra Sbregavitelli and the Monte Raggeto sections, which appear to preserve a more continuous sedimentary record, at least at the superbundle level.

influence of soil gas may explain the negative carbon-isotope values measured in the Monte Raggeto and Monte Tobenna sections.

We use the rapid transition to the second positive carbon-isotope anomaly recorded in the Serra Sbregavitelli sequence as the next marker level, and we correlate superbundles S6 to S8, respectively, with R26 to R28 and with F1 to F3. Moreover, the carbon-isotope signal recorded in superbundle S9 does not find an equivalent at Monte Raggeto and at Monte Faito, where, according to our cyclostratigraphic correlation, the corresponding superbundle is missing (Fig. 8).

A $\delta^{13}C$ trend to lower values was recognized in the uppermost part of the studied sections, from superbundles S10, R29, and F4 up to the biostratigraphic Aptian–Albian boundary (a

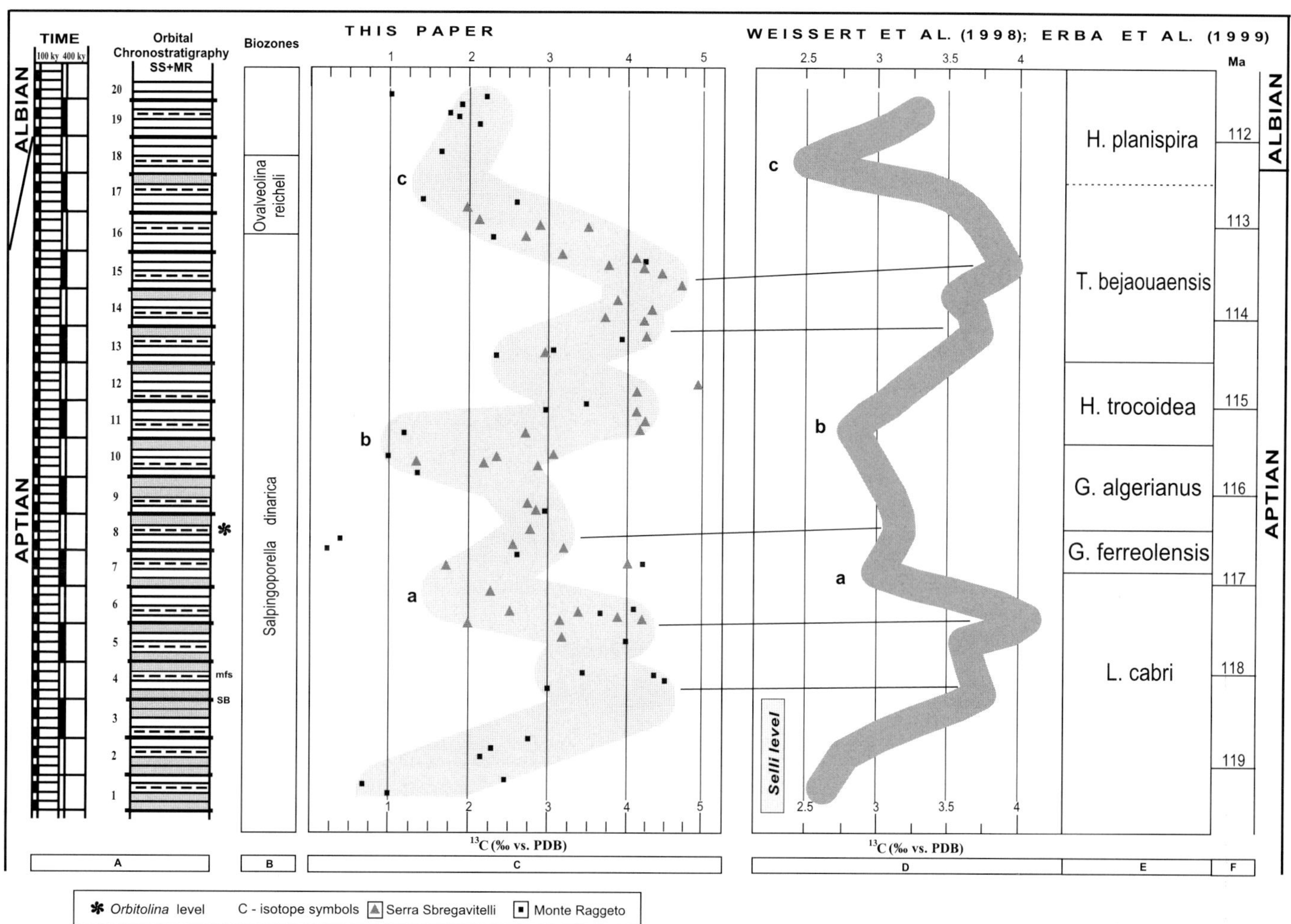

FIG. 10.—Global-scale correlation based on carbon-isotope stratigraphy. **A)** Orbital chronostratigraphy of the analyzed interval as in Figure 9. **B)** Carbonate platform biozones according to Chiocchini et al. (1994). **C)** $\delta^{13}C$ values plotted against orbital chronostratigraphy of the studied interval. The gray envelope curve shows the $\delta^{13}C$ long-term trends registered in the Aptian–Lower Albian of southern Italy as suggested by Serra Sbregavitelli and Monte Raggeto carbon-isotope stratigraphy (Figs. 3, 4). **D)** pelagic $\delta^{13}C$ curve (from Weissert et al., 1998), calibrated against biostratigraphy (as in column E) and enlarged to fit the orbital time scale (column A). The letters (a) to (c) on the curves of the columns C and D suggest synchronous turning points in the global ocean–biosphere system. Note that the $\delta^{13}C$ positive shift occurring within superbundle 11 does not have an equivalent in the pelagic curve of Weissert et al. (1998). **E)** Planktonic foraminifer biozones from Weissert et al. (1998) and Erba et al. (1999). **F)** Numerical ages from Channell et al. (1995). See text for discussion.

stratigraphic interval about 15–20 m thick; see also Figs. 3, 4, 6). This trend seems to correspond to the generally negative carbon-isotope values in the pelagic realm (Weissert et al., 1998; see also Fig. 10). On these bases we correlate superbundle S10 with R29, whereas the corresponding interval seems to be missing between superbundles F3 and F4 of the Monte Faito section.

Finally, we tentatively correlate superbundle S11 with R30 and F4, and superbundles R31 and R32 are correlated with F5 and F6, according to their well comparable carbon-isotope trends. Superbundles F4 and F5 (Monte Faito) show a broader dispersion of carbon-isotope values. It is also worth noting that at the three negative spikes of this section there is no evidence of meteoric diagenesis (Fig. 6). These values might hence indicate particular episodes of isolated lagoonal conditions, when the exchange with the open ocean was reduced and the shallow platform waters became enriched in the light carbon isotope because of organic respiration (Holmden et al., 1998; Patterson and Walter, 1994).

Global-Scale Correlation

To attempt a global-scale correlation (Figs. 10, 11) we use our composite orbital chronostratigraphic scheme shown in Figure 9 as a reference frame to plot the isotopic data and the transgressive–regressive facies trends of this study against both the pelagic isotopic curve of Weissert et al. (1998) and the third-order stratigraphic cycles of Jacquin et al. (1998).

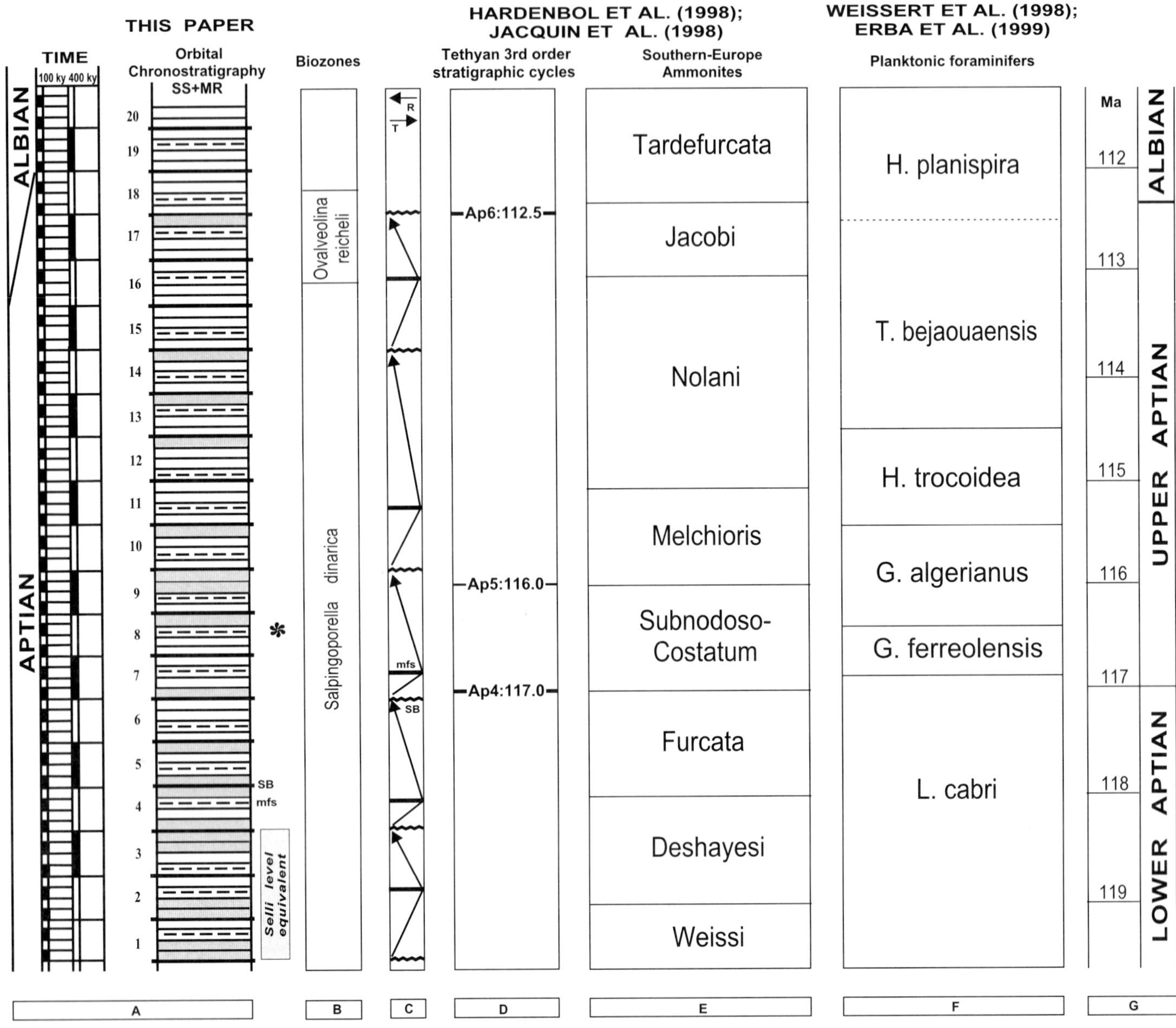

FIG. 11.—Global-scale correlation based on sequence-stratigraphy criteria. **A)** Orbital chronostratigraphy of the analyzed interval. This is a composite section from Monte Raggeto and Serra Sbregavitelli successions (see Fig. 9). **B)** Carbonate platform biozones according to Chiocchini et al. (1994). **C)** Transgressive–regressive facies trends recognized in the studied successions and calibrated against orbital chronostratigraphy. **D, E)** Tethyan third-order stratigraphic cycles of Jacquin et al. (1998), calibrated against southern Europe ammonite biozones (From Hardenbol et al., 1998). **F)** Planktonic foraminifer biozones from Weissert et al. (1998) and Erba et al. (1999). **G)** Numerical ages from Gradstein et al. (1995). Note that the superbundles 1 to 3 correspond also in facies characteristics to the pelagic *Selli Level* Auct. (Menegatti et al., 1998; Wissler et al., this volume).

Carbon-Isotope Stratigraphy.—

The pelagic $\delta^{13}C$ curve shown in Figure 10 starts from the oceanic anoxic event (OAE 1a) known as the "Cismon event" or the "Selli Level" (see Weissert et al., 1985; Weissert et al., 1998; Erba and Larson, 1998) and includes, in its uppermost part, the Aptian–Albian transition. At Monte Raggeto the equivalent of the OAE 1a is located in correspondence of superbundles R16–R18 (i.e., superbundles 1–3 of our composite chronostratigraphy; Figures 9–11) as shown by Wissler et al. (this volume). This interval is about 12 m thick and is formed of thin-bedded, dark-colored limestones showing the same Early Aptian carbon-isotope positive shift registered in the pelagic Selli Level. The carbon-isotope curve from the studied sections (Fig. 10) shows three intervals defined by low $\delta^{13}C$ values ($< +2‰$), labeled with a to c, and five positive peaks with $\delta^{13}C$ values higher than $+3‰$. A duration ranging from about 1 to about 2 million years is suggested by our orbital time scale for each of these positive–

negative carbon-isotope trends. If we compare our shallow-water carbon-isotope curve, calibrated on the basis of our orbital chronostratigraphy, with the pelagic curve of Weissert et al. (1998) calibrated against the Channell et al. (1995) time scale, we observe higher-amplitude fluctuations in the platform record but we also recognize a clear correspondence between the trends. The positive peak registered in our sections within superbundle 11 (Fig. 10, columns *A, C*) does not have an equivalent in the pelagic carbonates studied in southern Alps (Weissert et al., 1985; Weissert et al., 1998). We think that the enhancement towards more negative $\delta^{13}C$ values is due to depositional and early meteoric (diagenetic) influence (e.g., Holmden et al., 1998; Patterson and Walter, 1994). Moreover, differences in the compared Upper Aptian $\delta^{13}C_{carbonate}$ profile can be attributed to local changes in the carbon reservoir of carbonates that belong to different geological and paleoenvironmental settings of the Tethyan region (Menegatti et al., 1998; Jenkyns and Wilson, 1999; Ando et al., 2002).

Sea-Level Changes.—

We propose here a global-scale correlation of our transgressive–regressive facies trends, calibrated against orbital chronostratigraphy, with the third-order stratigraphic cycles of Jacquin et al. (1998) recognized in the Late Aptian–Early Albian time interval and calibrated against the Gradstein et al. (1995) time scale. It is worth noting that also Jacquin et al. (1998) reach similar results in correlating their third-order cycles with the depositional sequences of Haq et al. (1987).

On the basis of biostratigraphic and cyclostratigraphic criteria, and considering that Channell et al. (1995) and Gradstein et al. (1995) present identical numerical ages for the Aptian stage, we plotted the Upper Aptian Tethyan ammonite biozones and the third-order cycles of Jacquin et al. (1998) against our T/RFTs and orbital chronostratigraphy (Fig. 11). We also used the Upper Aptian planktonic foraminifer biozones (Figs. 10, 11) in order to establish a more precise chronostratigraphic relationship between the third-order sequence boundaries (Ap4–Ap6) of Jacquin et al. (1998) and our regressive facies trend (RFT) peaks (see also Hardenbol et al., 1998, Chart 4). By using these constraints we correlated our regressive peak at the upper boundary of superbundle 6 with the sequence boundary Ap4 (at 117 Ma) and our regressive peaks at the upper boundaries of superbundles 9 and 17 with the sequence boundaries Ap-5 (at 116 Ma) and Ap-6 (at 112.5 Ma), respectively. We note that neither of the RFT peaks at the upper boundaries of superbundles 3 and 14 have equivalents in the third-order Tethyan stratigraphic cycles of Jacquin et al. (1998). We suggest that a combination of regional factors such as different rate of subsidence and sedimentation as well as patterns of water circulation may have enhanced the regressive peaks occurring in the sections studied in southern Apennines (Fig. 8).

In conclusion, the global-scale correlation shown in Figure 11 allows us to propose that most of our superbundle boundaries and related regressive peaks, recorded in the Aptian–Lower Albian of southern Italy, are synchronous and can be globally correlated to the third-order cycles of Jacquin et al. (1998) and Haq et al. (1987). This supports the idea of Gale et al. (2002) that during pre-Quaternary times the 400 ky long-eccentricity cycle was a primary control on sea-level changes at the third-order (0.5–3,0 My) sequence scale (Vail et al., 1991).

Global Correlations and Geochronology.—

If we consider the carbon-isotope correlation shown in Figure 10, we note that our orbital chronostratigraphy suggests a duration of 7.2 My from the base of superbundle 1 to the top of superbundle 18. For the compared stratigraphic interval about 7.5 My is estimated on the basis of the Channell et al. (1995) time scale. Moreover, orbital chronostratigraphy suggests a duration of 4.4 My from the base of superbundle 7 to the top of superbundle 17, which is a figure well comparable to that of Jacquin et al. (1998), who estimated a duration of 4.5 My for the equivalent stratigraphic interval (from Ap4 to Ap6 sequence boundaries; Fig. 11). The correspondence in duration of the discussed intervals, as suggested by the related time scales, thus appears to be good. Negligible discrepancies in the correlation shown in Figure 11 probably result from the different approaches: cyclostratigraphy in our case and sequence stratigraphy in Jacquin et al. (1998). In our discussion we have used the superbundle stacking pattern to characterize the T/RFTs on which the orbital cycles are superimposed. Moreover, we have estimated the duration of the T/RFTs (Fig. 9) by considering the superbundles as time units (400 ky), regardless of their completeness. Instead, in the sequence-stratigraphic approach of Jacquin et al. (1998) the third-order sequences were calibrated against biostratigraphy, which in turn was tied to radiometric ages obtained for the corresponding stage boundaries.

It is worth noting that the correlation shown in Figure 11 is the result of a long-lasting process of increasing precision, based on various physical and biostratigraphic data. For these reasons, it is very difficult to obtain a more precise tuning (e.g., < 100 ky) in correlating very distant sections, because of the different morphology of the original depositional environments, the variable subsidence rates influenced by local tectonics, and the variable carbonate productivity (not to mention the intrinsic differences due to analytical approach and scale).

In conclusion, from the above discussion we note that it is possible to reach a quite precise global correlation also in carbonate platform domains, *per se* prone to frequent gaps in sedimentation. The good correlation of our carbonate strata with the pelagic sequences hence allows an orbital time scale to be proposed for the studied Lower Cretaceous interval (Figs. 9–11).

CONCLUSIONS

On the basis of a multidisciplinary (sedimentologic, biostratigraphic, cyclostratigraphic, and isotopic) approach, four successions of shallow-water carbonate strata from Lower Cretaceous of southern Italy were analyzed at a centimeter scale for a total thickness of about 300 m. The collected data have allowed us: (a) to recognize a hierarchy of cyclic oscillations of the depositional and diagenetic environments; (b) to constrain these oscillations in definite time intervals, typical of Milankovitch periodicities (from 20 to 400 ky); (c) to point out the relationships between sedimentary cyclicity and carbon-isotope stratigraphic variations recorded in the studied sections (Figs. 3–6); (d) to recognize lower-frequency relative sea-level oscillations (transgressive–regressive facies trends) on which the Milankovitch cycles are superimposed; and (e) to correlate with high precision the above sequences at regional scale (Fig. 8), and assemble chronostratigraphic diagrams showing the location of the sedimentary gaps that punctuate the stratigraphic record of the studied shallow-water strata (Fig. 9).

Regional correlation and orbital chronostratigraphy allow us to put into evidence that the Serra Sbregavitelli section, which developed in a more open carbonate platform sector, preserves an almost complete sedimentary record, whereas only two gaps longer than 400 ky are found at Monte Raggeto and at Monte Faito (Figs. 8, 9). Carbon-isotope data, combined with the detailed facies analysis, provide an additional control on the regional cyclostratigraphic correlation, confirming the position of the above gaps recorded in the platform stratigraphy (Fig. 8). Moreover, the oceanic $\delta^{13}C$ values appear to have been less overprinted by early meteoric diagenesis when moving from Monte Tobenna and Monte Faito, displaying more restricted facies, to Monte Raggeto and Serra Sbregavitelli successions, characterized by more open-marine facies (Figs. 3–6, 8).

Finally, a composite orbital chronostratigraphy (superbundles 1–20 in Fig. 9) was used to extend our regional correlation to global scale. We correlated (Figs. 10, 11) the shallow-water carbon-isotope signal with the coeval pelagic curves of Weissert et al. (1998) and our transgressive–regressive facies trends with the third-order depositional sequences of Jacquin et al. (1998).

The proposed global-scale correlations allow us (a) to propose an Early Aptian to Early Albian orbital time scale (Figs. 10, 11), which in the studied interval appears to fit well the time scales of Channell et al. (1995) and Gradstein et al. (1995); (b) to demonstrate that the sedimentary record of the analyzed carbonate-platform strata can be considered quasi-continuous, at least at the superbundle level; (c) to show that shallow-water carbonates preserve the global signature of $\delta^{13}C$ changes like their coeval pelagic deposits; and, finally, (d) to demonstrate that the lower-frequency relative sea-level oscillations (T/RFTs) recorded in the Aptian–Lower Albian of southern Apennines were eustatically controlled and driven almost entirely by the long-eccentricity cycle (400 ky).

ACKNOWLEDGMENTS

This work was carried out in the frame of the High-Resolution Stratigraphy Program of the Istituto per l'Ambiente Marino Costiero (Geomare), National Research Council, Napoli, and it was supported by the ETH Zürich and by the Swiss Science Foundation. We wish to thank the Geomare for space and laboratory facilities. Special thanks are due to R. Radoičić and R. Schroeder for their valuable suggestions on biostratigraphic problems. We are grateful to A. Raspini, who diligently assembled the samples of Monte Tobenna and Monte Faito for the isotopic analysis, plotting them against the cyclostratigraphic logs shown in Figures 5 and 6. We also thank H. Jenkyns and W. Schlager for the critical review of an earlier draft of the manuscript and for their helpful suggestions. Finally, would also like to acknowledge our reviewers A. Immenhauser and A. Strasser for their much appreciated suggestions and comments.

REFERENCES

Ando, A., Kakegawa, T., Takashima, R., and Saito T., 2002, New perspective on Aptian carbon-isotope stratigraphy: Data from $\delta^{13}C$ records of terrestrial organic matter: Geology, v. 30, p. 227–230.

Berger, A., Loutre, M.F., and Laskar, J., 1992, Stability of the astronomical frequencies over the Earth's history for paleoclimate studies: Science, v. 255, p. 560–565.

Bralower, T.J., Cobabe, E., Clement, W., Sliter, W.V., Osburn, C.L., and Longoria, J., 1999, The record of global change in mid-Cretaceous (Barremian–Albian) sections from the Sierra Madre, NE Mexico: Journal of Foraminiferal Research, v. 29, p. 418–437.

Bravi, S., and De Castro, P., 1995, The Cretaceous fossil fish level of Capo d'Orlando, near Castellammare di Stabia (NA): biostratigraphy and depositional environment: Memorie di Scienze Geologiche, v. 47, p. 45–72.

Brescia, M., D'Argenio, B., Ferreri, V., Pelosi, N., Rampone, S, and Tagliaferri, R., 1996, Neural net aided detection of astronomical periodicities in geologic records: Earth and Planetary Science Letters, v. 139, p. 33–45.

Buonocunto, F.P., 1998, Litostratigrafia di alta risoluzione nel Cretacico di piattaforma Carbonatica dell'Appennino Campano: Ph.D. Dissertation, Università Federico II, Napoli, 184 p.

Buonocunto, F.P., D'Argenio, B., Ferreri, V., and Raspini, A., 1994, Microstratigraphy of highly organized carbonate platform deposits of Cretaceous age. The case of Serra Sbregavitelli (Matese, Central Apennines): Giornale di Geologia, v. 56, p. 179–192.

Buonocunto, F.P., D'Argenio, B., Ferreri, V., and Sandulli, R., 1999, Orbital cyclostratigraphy and sequence stratigraphy of Upper Cretaceous platform carbonates at Monte Sant'Erasmo, southern Apennines—Italy: Cretaceous Research, v. 20, p. 81–95.

Buonocunto, F.P., Sprovieri, M., Bellanca, A., D'Argenio, B., Ferreri, V., Neri, R., and Ferruzza, G., 2002, Cyclostratigraphy and high-frequency carbon-isotope fluctuations in Upper Cretaceous shallow-water carbonates, southern Italy: Sedimentology, v. 49, p. 1331–1337.

Channell, J.E.T., D'Argenio, B., and Horvath, F., 1979, Adria, the African promontory in Mesozoic Mediterranean palaeogeography: Earth-Science Reviews, v. 15, p. 213–292.

Channell, J.E.T., Erba, E., Nakanishi, M., and Tamaki, K., 1995, Late Jurassic–Early Cretaceous time scale and oceanic magnetic anomaly block, *in* Berggren, W.A., Kent, D.V., Aubry, M.P., and Hardenbol, J., eds., Geochronology, Time Scales and Global Stratigraphic Correlation: SEPM, Special Publication 54, p. 51–64.

Cherchi, A., De Castro, P., and Schroeder, R., 1978, Sull'età dei livelli a Orbitolinidi della Campania e delle Murge Baresi (Italia meridionale): Società dei Naturalisti in Napoli, Bollettino, v. 87, p. 363–385.

Chiocchini, M., Farinacci, A., Mancinelli, A., Molinari, V., and Potetti, M., 1994, Biostratigrafia a foraminiferi, dasicladali e calpionelle delle successioni carbonatiche mesozoiche dell'Appennino centrale (Italia), *in* Mancinelli, A., ed., Biostratigrafia dell'Italia Centrale: Università di Camerino, Studi Geologici Camerti, Volume Speciale, p. 9–128.

D'Argenio, B., 1976, Le piattaforme carbonatiche periadriatiche. Una rassegna di problemi nel quadro geodinamico mesozoico dell'area mediterranea: Società Geologica Italiana, Memorie, v. 13, p. 137–160.

D'Argenio, B., Amodio, S., Ferreri, F., and Pelosi, N., 1997, Hierarchy of high-frequency orbital cycles in Cretaceous carbonate platform strata: Sedimentary Geology, v. 113, p. 169–193.

D'Argenio, B., and Alvarez, W., 1980, Stratigraphic evidence for crustal thickness changes on the southern Tethyan margin during the Alpine cycle: Geological Society of America, Bulletin, v. 91, part I, p. 681–689 and part II, p. 2558–2587.

D'Argenio, B., De Castro, P., Emiliani, C., and Simone, L., 1975, Geologic Notes, Bahamian and Apenninic limestones of identical lithofacies and age: American Association of Petroleum Geologists, Bulletin, v. 59, p. 524–530.

D'Argenio, B., Ferreri, V., Iorio, M., Raspini, A., and Tarling, D.H., 1999a, Diagenesis and remanence acquisition in the Cretaceous carbonates of Monte Raggeto, Southern Italy, *in* Tarling, D.H., and Turner, P., eds., Palaeomagnetism and Diagenesis in Sediments: Geological Society of London, Special Publication 151, p. 147–156.

D'Argenio, B., Ferreri, V., Raspini, A., Amodio, S., and Buonocunto, F.P., 1999b, Cyclostratigraphy of a carbonate platform as a tool for high-precision correlation: Tectonophysics, v. 315, p. 357–385.

D'Argenio, B., Fischer, A.G., Richter, G.M., Longo, G., Pelosi, N., Molisso, F., and Duarte Morais, M.L., 1998, Orbital cyclicity in the Eocene of Angola: visual and image-time-series analysis compared: Earth and Planetary Science Letters, v. 169, p. 147–161.

Davey, S.D., and Jenkyns, H.C., 1999, Carbon-isotope stratigraphy of shallow-water limestones and implications for the timing of Late Cretaceous sea-level rise and anoxic events (Cenomanian–Turonian of the peri-Adriatic carbonate platform, Croatia): Eclogae Geologicae Helvetiae, v. 92, p. 163–178.

deBoer, P.L., and Smith, D.G. eds., 1994, Orbital Forcing and Cyclic Sequences: International Association of Sedimentologists, Special Publication 19, 559 p.

De Castro, P., 1991, Mesozoic, *in* Barattolo, F., De Castro, P., and Parente, M., eds., 5th International Symposium on Fossil Algae: Field Trip Guide-Book, Capri, p. 21–38.

Einsele, G., Ricken, W., and Seilacher, A., eds., 1991, Cycles and Events in Stratigraphy: Berlin, Springer-Verlag, 955 p.

Erba, E., and Larson, R.L., 1998, The Cismon Apticore (Southern Alps Italy): A "reference section" for the Lower Cretaceous at low latitudes: Rivista Italiana di Paleontologia e Stratigrafia, v. 104, p. 181–192.

Erba, E., Channell, J.E.T., Claps, M., Jones, C., Larson, R., Opdyke, B., Premoli Silva, I., Riva, A., Salvini, G., and Torricelli, S., 1999, Integrated stratigraphy of the Cismon Apticore (Southern Alps, Italy): a "Reference Section" for the Barremian–Aptian interval at low latitude: Journal of Foraminiferal Research, v. 29, p. 371–391.

Ferreri, V., Amodio, S., D'Argenio, B., and Sandulli, R., 2001, High-resolution correlation and orbital chronostratigraphy at the Valanginian–Hauterivian boundary (abstract), *in* Amodio, S., Buonocunto F.P., and Sandulli R., eds., Multidisciplinary Approach to Cyclostratigraphy: SEPM, International Workshop, Sorrento, Abstract Volume, p. 27–28.

Ferreri, V., Weissert, H., D'Argenio B., and Buonocunto F.P., 1997, Carbon-isotope stratigraphy: a tool for basin to carbonate platform correlation: Terra Nova, v. 9, p. 57–61.

Fischer, A.G., Herbert, T.D., Napoleone, G., Premoli Silva, I., and Ripepe, R., 1991, Albian pelagic rhythms (Piobbico core): Journal of Sedimentary Petrology, v. 61, p. 1164–1172.

Gale, A.S., Hardenbol, J., Hathway, B., Kennedy, W.J., Young, J.R., and Phansalkar, V., 2002, Global correlation of Cenomanian (Upper Cretaceous) sequences: Evidence for Milankovitch control on sea level: Geology, v. 30, p. 291–294.

Gradstein, F.M., Agterberg, F.P., Ogg, J.G., Hardenbol, J., Van Veen, P., Thierry, J., and Huang, Z., 1995, A Triassic, Jurassic and Cretaceous time scale, *in* Berggren, W.A., Kent, D.V., Aubry, M.P., and Hardenbol, J., eds., Geochronology, Time Scales and Global Stratigraphic Correlation: SEPM, Special Publication 54, p. 95–126.

Gröcke, D.R., Hesselbo, S., and Jenkyns, H.C., 1999, Carbon-isotope composition of Lower Cretaceous fossil wood: ocean–atmosphere chemistry and relation to sea-level change: Geology, v. 27, p. 155–158.

Grötsch, J., Billing, J., and Vahrenkamp, V., 1998, Carbon-isotope stratigraphy in shallow-water carbonates: implications for Cretaceous black-shale deposition: Sedimentology, v. 45, p. 623–634.

Haq, B.U., Hardenbol, J., and Vail, P.R., 1987, Chronology of fluctuating sea levels since the Triassic: Science, v. 237, p. 1156–1167.

Hardenbol, J., Thierry, J., Farley, M.B., Jacquin, T., de Graciansky, P.C., and Vail, P.R., 1998, Cretaceous sequence chronostratigraphy, *in* de Graciansky, P.C., Hardenbol, J., Jacquin, T., Vail, P.R., and Farley, M.B., eds., Mesozoic and Cenozoic Sequence Stratigraphy of European Basins: SEPM, Special Publication 60, Charts.

Hardie, L.A., Bosellini, A., and Goldhammer, R.K., 1986, Repeated subaerial exposure of subtidal carbonate platforms, Triassic, northern Italy: evidence for high-frequency sea-level oscillations on 10^4 year scale: Paleoceanography, v. 1, p. 447–457.

Herbert, T., and Fischer, A.G., 1986, Milankovitch climatic origin of mid-Cretaceous black shale rhythms in central Italy: Nature, v. 321, p. 739–743.

Holmden, C., Creaser, R.A., Muehlenbachs, K., Leslie, S.A., and Bergström, S.M., 1998, Isotopic evidence for geochemical decoupling between ancient epeiric seas and bordering oceans: Implications for secular curves: Geology, v. 26, p. 567–570.

Immenhauser, A., Kenter, J.A.M., Ganssen, G., Bahamonde, J.R., Van Vliet, A., and Saher, M.A., 2002, Origin and significance of isotope shifts in Pennsylvanian carbonates (Asturias, NW Spain): Journal of Sedimentary Research, v. 72, p. 82–94.

Iorio, M, Tarling, D.H., D'Argenio, B., and Nardi, G., 1996, Ultra-fine magnetostratigraphy of Cretaceous shallow-water carbonates, Monte Raggeto, Southern Italy, *in* Morris, A., and Tarling, D.H., eds., Paleomagnetism and Tectonics of the Mediterranean Region: Geological Society of London, Special Publication 105, p. 195–203.

Jacquin, T., Rusciadelli, G., Amedro, F., de Graciansky, P.C., and Magniez-Jannin F., 1998, The North Atlantic cycle: an overview of 2nd order transgressive / regressive facies cycles in the Lower Cretaceous Western Europe, *in* de Graciansky, P.C., Hardenbol, J., Jacquin, T., Vail, P.R., and Farley, M.B., eds., Mesozoic and Cenozoic Sequence Stratigraphy of European Basins: SEPM, Special Publication 60, p. 397–409.

Jenkyns, H.C., 1995, Carbon-isotope stratigraphy and paleoceanographic significance of the Lower Cretaceous shallow-water carbonates of Resolution Guyot, Mid Pacific Mountains: Proceedings of the Oceanic Drilling Program, Scientific Results, v. 143, p. 99–104.

Jenkyns, H.C., and Wilson, P.A., 1999, Stratigraphy, paleoceanography, and evolution of Cretaceous Pacific guyots: Relics from a green-house Earth: American Journal of Science, v. 299, p. 341–392.

Laubscher, H.P., and Bernoulli, D., 1977, Mediterranean and Tethys, *in* Nairn, A.E.M., Kanes, W.H., and Stehli, F.G., eds., The Ocean Basins and Margins: London, Plenum Press, vol. 4, p. 1–28.

Longo, G., D'Argenio, B., Ferreri, V., and Iorio, M., 1994, Fourier evidence for high-frequency astronomical cycles recorded in Early Cretaceous carbonate platform strata, Monte Maggiore, southern Apennines, Italy, *in* de Boer, P.L., and Smith, D.G., eds., Orbital Forcing and Cyclic Sequences: International Association of Sedimentologists, Special Publication 19, p. 77–85.

Marsella, E., Bally, A.W., Cippitelli, G., D'Argenio, B., and Pappone, G., 1995, Tectonic history of the Lagonegro Domain and Southern Apennine thrust belt evolution: Tectonophysics, v. 252, p. 307–330.

McCrea, J.M., 1950, The isotopic chemistry of carbonates and a paleotemperature scale: Journal of Chemical Physics, v. 18, p. 849–857.

Menegatti, A.P., Weissert, H., Brown, R., Tyson, R.V., Fairmond, P., Strasser, A., and Caron, M., 1998, High-resolution δ^{13}C-stratigraphy through the Early Aptian "Livello Selli Equivalent" of the Alpine Tethys: Paleoceanography, v. 13, p. 530–545.

Pasquier, J.-B., and Strasser, A., 1997, Platform-to-basin correlation by high-resolution sequence stratigraphy and cyclostratigraphy (Berriasian, Switzerland and France): Sedimentology, v. 44, p. 1071–1092.

Patterson, W.P., and Walter, L.M., 1994, Depletion of ^{13}C in seawater CO_2 on modern carbonate platforms: Significance for the carbon isotopic record of carbonates: Geology, v. 22, p. 885–888.

Raspini, A., 1998, Microfacies analysis of shallow-water carbonates and evidence of hierarchically organized cycles. Aptian of Monte Tobenna, southern Apennines, Italy: Cretaceous Research, v. 19, p. 197–223.

Raspini, A., 2001, Stacking pattern of cyclic carbonate platform strata: Lower Cretaceous of southern Apennines, Italy: Geological Society of London, Journal, v. 158, p. 353–366.

Robson, J., 1987, Depositional models for some Cretaceous carbonates from the Sorrento Peninsula: Società Geologica Italiana, Memorie, v. 40, p. 251–257.

Servizio Geologico d'Italia, 1959, Carta Geologica d'Italia scala 1:100.000, Foglio n. 196, Sorrento.

Servizio Geologico d'Italia, 1969, Carta Geologica d'Italia scala 1:100.000, Foglio n. 185, Salerno.

Servizio Geologico d'Italia, 1971, Carta Geologica d'Italia scala 1:100.000, Foglio n. 161, Isernia.

Servizio Geologico d'Italia, 1971, Carta Geologica d'Italia scala 1:100.000, Foglio n. 172, Caserta.

Strasser, A., 1991, Lagoonal–peritidal sequences in carbonate environments: autocyclic and allocyclic processes, *in* Einsele, G., Ricken, W., and Seilacher, A., eds., Cycles and Events in Stratigraphy: Berlin, Springer-Verlag, p. 709–721.

Strasser, A., Caron, M., and Gjermeni, M., 2001, The Aptian, Albian and Cenomanian of Róter Sattel, Romandes Prealps, Switzerland: a high-resolution record of oceanographic changes: Cretaceous Research, v. 22, p. 173–199.

Tagliaferri, R., Pelosi, N., Ciaramella, A., Longo, N., Milano, M., and Barone, F., 2001, Soft computing methodologies for spectral analysis in cyclostratigraphy: Computers & Geosciences, v. 27, p. 535–548.

Vail, P.R., Audemard, F., Bowman, S.A., Eisner, P.N., and Perez-Cruz, C., 1991, The stratigraphic signatures of tectonics, eustasy, and sedimentology—an overview, *in* Einsele, G., Ricken, W., and Seilacher, A., eds., Cycles and Events in Stratigraphy: Berlin, Springer-Verlag, p. 617–659.

Weissert, H., and Bréhéret, J.G., 1991, A carbonate carbon-isotope record from Aptian–Albian sediments of the Vocontian trough (SE France): Société Géologique de France, Bulletin, v. 162, p. 1133–1140.

Weissert, H., and Lini, A., 1991, Ice age interludes during the time of Cretaceous greenhouse climate, *in* Müller, D.W., McKenzie, J.A., and Weissert, H., eds., Controversies in Modern Geology: London, Academic Press, p. 173–191.

Weissert, H., Lini, A., Föllmi, K.B., and Kuhn, O., 1998, Correlation of Early Cretaceous carbon-isotope stratigraphy and platform drowning events: A possible link?: Palaeogeography, Palaeoclimatology, Palaeoecology, v. 137, p. 189–203.

Weissert, H., McKenzie, J.A., and Channell, J.E.T., 1985, Natural variations in the carbon cycle during the Early Cretaceous, *in* Sundquist, E.T., and Broeker, W.S., eds., The Carbon Cycle and Atmospheric CO_2: Natural Variations Archean to the Present: American Geophysical Union, Geophysical Monographs, v. 32, p. 531–545.

CALIBRATION OF THE EARLY CRETACEOUS TIME SCALE: A COMBINED CHEMOSTRATIGRAPHIC AND CYCLOSTRATIGRAPHIC APPROACH TO THE BARREMIAN–APTIAN INTERVAL, CAMPANIA APENNINES AND SOUTHERN ALPS (ITALY)

LUKAS WISSLER AND HELMUT WEISSERT
Geological Institute ETH-Z, CH-8092 Zürich, Switzerland
e-mail: helmi@erdw.ethz.ch
FRANCESCO P. BUONOCUNTO
Istituto per l'Ambiente Marino Costiero, Geomare, National Research Council,
Calata Porta di Massa, Porto di Napoli, 80133 Napoli, Italy
e-mail: buonocun@gms01.geomare.na.cnr.it
VITTORIA FERRERI
Dipartimento di Scienze della Terra, Università "Federico II", Largo San Marcellino 10, 80138 Naples, Italy
e-mail: ciclisti@gms01.geomare.na.cnr.it
AND
BRUNO D'ARGENIO
Istituto per l'Ambiente Marino Costiero, Geomare, National Research Council,
Calata Porta di Massa, Porto di Napoli, 80133 Napoli, Italy
e-mail: dargenio@gms01.geomare.na.cnr.it
AND
Dipartimento di Scienze della Terra, Università "Federico II", Largo San Marcellino 10, 80138 Napoli, Italy

ABSTRACT: A detailed carbon-isotope stratigraphy has been generated from Barremian to Lower Aptian shallow-water carbonate sections in the Campania Apennines (Monte Raggeto, southern Italy). The new isotope curve is correlated with the magnetostratigraphically and biostratigraphically dated pelagic carbon-isotope stratigraphy from the Cismon locality (Southern Alps, northern Italy). All the major positive and negative carbon-isotope excursions that characterize the Barremian and Early Aptian carbon-isotope stratigraphy can be recognized in the shallow-water curve. Cyclostratigraphy, which was established earlier at the Monte Raggeto section, is used as an age calibration tool for the Barremian and Early Aptian isotope stratigraphy. The duration of the isotopically calibrated stratigraphic interval between the top of Chron M3 and the base of Chron M0 is estimated as 4 My. These time calculations are in good agreement with cyclostratigraphic data from the Cismon locality but differ from estimates based on a magnetic anomaly block model for the interval between M3 and M0 that yield only 3 My. We have also calculated that the Selli Level Equivalent (SLE) at the Monte Raggetto locality was deposited within 1.2 My. Our results demonstrate that the combination of chemostratigraphic and cyclostratigraphic studies can contribute significantly to the calibration of the Mesozoic time scale.

INTRODUCTION

Over the last decade, orbital cyclostratigraphy has developed into a highly valuable age calibration tool in Neogene chronostratigraphy. Pre-Neogene sedimentary successions in pelagic and shallow marine settings also preserve a record of sedimentary cycles, which register orbitally induced changes in ocean and climate behavior (e.g., Herbert and Fischer, 1986; D'Argenio et al., 1997; D'Argenio et al., 1999b). Cyclostratigraphy established in these successions can be used for high-resolution age calibration between radiometric tie points. In this study we focus on the Early Cretaceous time scale, which is based on few radiometric ages: 121 Ma for the base of magnetic Chronozone M0, 137 Ma for the CM16–CM17 interval, and 154 Ma for the top of CM25 (Channell et al., 1995). Age estimates for the magnetic chrons between the datum levels rely on interpolations that are based on oceanic anomaly block models, as outlined by Channell et al. (1995). Cyclostratigraphic analysis of successions with magnetostratigraphy provides a potential means of testing these estimations.

In this study, we combine cyclostratigraphic analyses of shallow-water and deep-water sections with chemostratigraphy and magnetostratigraphy. This approach allows us to test the reliability of cyclostratigraphy established in different paleoenvironmental settings and to correlate deep-water with shallow-water successions with a resolution of ten to twenty thousand years.

We concentrate on the Barremian–Aptian transition interval and correlate cyclostratigraphies established in deep-water (Herbert, 1992) and shallow-water (Ferreri et al., 1997; Buonocunto, 1998; D'Argenio et al., this volume) sequences. Using carbon-isotope stratigraphy as an independent correlation tool, we first test whether the obtained cyclostratigraphic curves can be correlated to each other. Second, we use cyclostratigraphy to estimate the duration of the interval between Chron M3 and the top of Chron M0 and of the oceanic anoxic event 1a, represented by the Selli Level and its equivalents (e.g., Coccioni et al. 1987).

Cyclostratigraphy: Approaches and Case Histories
SEPM Special Publication No. 81,
SEPM (Society for Sedimentary Geology), ISBN 1-56576-108-1, p. 123–133.

SECTIONS STUDIED

The Monte Raggeto section is located in the southern sectors of the Monte Maggiore Mountains (Campania Apennines), about 35 km north of Naples (Fig. 1). The outcropping section is 280 m thick and consists of well-bedded limestones and dolomitic limestones, ranging in age from Early Barremian to Early Albian (D'Argenio et al., 1993; Iorio, 1995; Iorio et al., 1995; Iorio et al., 1998; Ferreri et al., 1997; Buonocunto, 1998). We discuss here only the Late Barremian to Early Aptian time interval. The Barremian is marked by the occurrence of green algae like *Salpingoporella muhelbergi* and *Salpingoporella melitae*, associated with foraminifers like *Cuneolina laurenti* and *Debarina* sp.; the transition to the Early Aptian is indicated by the first appearance of *Salpingoporella dinarica* associated with *Praechrysalidina infracretacea* and *Debarina* sp.

Microstratigraphic (centimeter-scale) analysis of textures and sedimentary features has allowed 7 lithofacies, grouped into four lithofacies associations (Fig. 2) to be recognized (Ferreri et al., 1997; Buonocunto, 1998). Lithofacies and their grouping suggest a carbonate-platform system oscillating from relatively open lagoon to more restricted (peritidal) conditions. Vertical variation of depositional textures (lithofacies) and early diagenetic features (from marine to meteoric) results in a hierarchy of cycles (Fig. 3, columns a and b) formed of elementary cycles (normally corresponding to single beds), bundles (groups of 2–5 elementary cycles), and superbundles (groups of 2–4 bundles). High-frequency sea-level fluctuation best explains this cyclic organization, as well as the predominance of cycles showing emersion-related features directly superimposed on subtidal deposits (D'Argenio et al., 1997; D'Argenio et al., 1999a; D'Argenio et al., 1999b; Buonocunto, 1998; Raspini, 1998, 2001). Sedimentological interpretation of the cycle hierarchy is supported by a quantitative analysis of the data, including neural network and image analysis (Longo et al., 1994; Brescia et al., 1996; Tagliaferri et al., 2001). Spectral analysis based on the so-called modified Scargle algorithm (see, e.g., Longo et al., 1994) was used to search for thickness frequencies and possible Milankovitch periodicities in the sedimentary signal (textures and bed thickness measured at centimeter scale) of the Monte Raggeto section (Fig. 4). Duration of the cycles has been calculated for a total of about 270 m of succession, Barremian to Albian in age, by comparing relative ratio sets (RRSs) of the recurrence of sedimentary features (expressed in centimeters) with the relative ratio sets (RRSt) of orbital parameters based on Berger et al. (1992) for the Cretaceous (expressed in years). RRSs and RRSt show a very good linear correlation ($r > 0.99$), suggesting that the Monte Raggeto Barremian–Albian strata have a dominant allocyclic organization linked to Milankovitch periodicities (Table 1). The spectral periodicities (precession, obliquity, and eccentricity) appear to be well comparable with the stratigraphic periodicities expressed as average thickness of elementary cycles, bundles, and superbundles (Table 2). So, it may be assumed that elementary cycles record the obliquity (about 40 ky), and bundles and superbundles correspond to the short (about 100 ky) and long eccentricity (about 400 ky), respectively. Orbital periodicities are also suggested by spectral analysis of textures and paleomagnetic data (declination, intensity, and inclination) carried out on a core 64 m long drilled in the Hauterivian–Barremian of the same Monte Raggeto (Fig. 5; Table 1), partly overlapping the outcropping section (Iorio et al., 1995; Brescia et al., 1996).

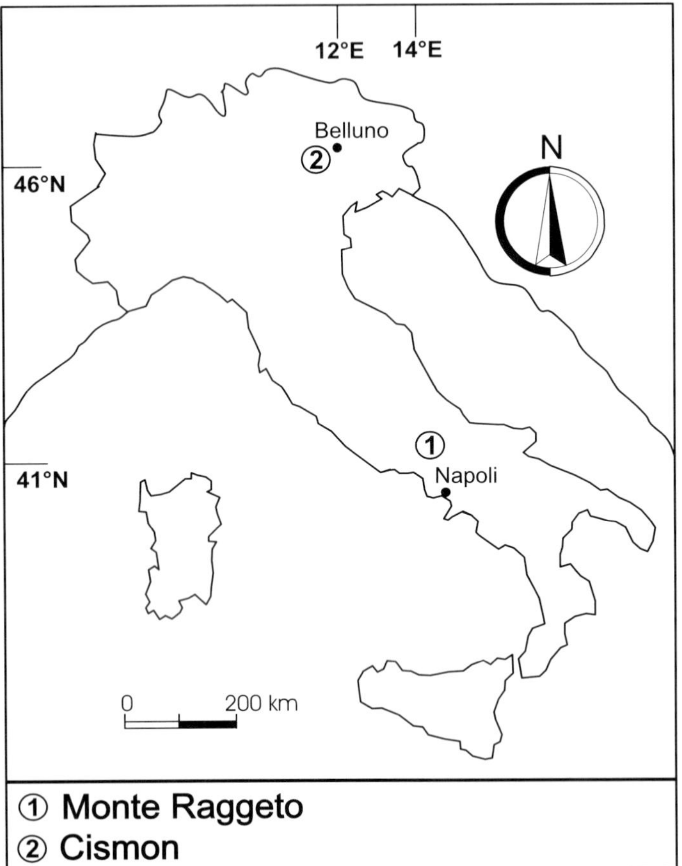

FIG. 1.—Location of the studied sections. The Monte Raggeto section is located at 10 km from the Caserta Nord exit, of Autostrada del Sole highway (coordinates: 41° 10' N, 14° 10' E, "Structural Model of Italy" 1:500.000, sheet 4, Bigi et al., 1992). The Cismon section is located at 17 km from the main road Strada Statale 50 del Monte Grappa in the valley of the Cismon River (coordinates: 46° 03' N, 11° 45' E, "Structural Model of Italy" 1:500.000, sheet 1, Bigi et al., 1992).

In the Monte Raggeto we have integrated the isotopic study of Ferreri et al. (1997) with more dense sampling and extending the measurements into the Lower Barremian. In total we have analyzed an interval of about 190 m (Fig. 3; Table 3). Moreover, it was possible to trace these strata into a borehole 64.4 m thick, which was drilled in the same locality at about 200 m from the sampled profile (Fig. 5). Paleomagnetic analysis allowed magnetozone M3 to be identified in the core (Iorio, 1995; Iorio et al., 1995; Iorio et al., 1998). This magnetozone was traced into our profile with centimeter precision on a lithostratigraphic basis (Fig. 5).

The pelagic carbonates of the Cismon section were drilled as part of the APTICORE project, close to the Cismon River in the Southern Alps of Northern Italy (Erba et al., 1999). The detailed lithology, biostratigraphy, magnetostratigraphy, and stable-carbon-isotope stratigraphy of this drill core is described in Erba et al. (1999). A high-resolution carbon-isotope stratigraphy established in the core and nearby outcrop serves as informal global reference section for the Barremian and Aptian interval (Menegatti et al., 1998; Erba et al., 1999). Herbert (1992) chose selected intervals of the Cismon section for an investigation of the thickness of limestone–marl couplets and documented in them a Milankovitch cyclicity. Frequencies identified by spectral analysis are consistent with mechanisms of astronomically forced climate variations.

	L1	MO2	MO1	BP2	BP1	FA2	FA1
LITHOFACIES	Stromatolitic and loferitic bindstone locally dolomitized	Cryptalgal bindstone alternating with mm-thick grain supported peloidal foraminiferal laminae	Ostracods wackestone and mud./wackestone with small benthonic foraminifers and *Thaumatoporella*	Locally laminated peloidal packstone with miliolids, rare small intraclasts and molluscan shell fragments	Bio-peloidal grainstone and pack/grainstone with small intraclasts	Benthonic foraminifers, green algae wackestone and wack./packstone with ostracods	Benthonic foraminifers, green algae wackestone and bioturbated wack./packstone with gastropods and pelecypods
LITHOFACIES ASSOCIATION	**L** **LAMINATED LIMESTONES** Barren mudstone with birdseyes, locally enlarged by solution and occluded by sparry cements and/or mechanical geopetal silt	**MO** **MILI–OSTR–ALGAL LIMESTONES** Wackestone to mudstone with oligotypic thanatocoenoses locally passing to cryptalgal laminites		**BP** **BIO-PELOIDAL LIMESTONES LOCALLY WEAKLY LAMINATED** Packstone, pack/grainstone and grainstone with bioclasts, peloids and small intraclasts. Parallel, crossed and/or oblique lamination may be locally found		**FA** **FOR–ALGAL LIMESTONES** Wackestone and bioturbated wack./packstone with green algae, benthonic foraminifers locally associated with molluscan shells	
ENVIRONMENTAL INTERPRETATION	TIDAL TO SUPRATIDAL	RESTRICTED LAGOON		INNER SHOALS		OPEN LAGOON	
		L A G O O N					

FIG. 2.—Lithofacies and lithofacies associations of the Monte Raggeto section and their environmental interpretation. Lithofacies associations BP and FA suggest more open carbonate platform settings characterized by normal marine circulation (subtidal domain). Lithofacies associations L and MO show sedimentary features of a more restricted carbonate-platform sector and suggest a peritidal domain.

METHODS

Carbon and oxygen isotope composition of calcite was measured on bulk rock samples from limestone beds. Carbonate powder was extracted with a micro drill to avoid diagenetic calcite from veins, microkarst, and macroscopic fossils. The sampled material was reacted with 100% phosphoric acid, and the carbon and oxygen isotope composition of the evolved CO_2 gas was analyzed with a PRISM dual-inlet mass spectrometer. The results are expressed in the common δ notation in per mil relative to the VPDB standard (VPDB is Vienna Peedee Belemnite). Precision is ± 0.1‰ for $\delta^{13}C$ and ± 0.2‰ for $\delta^{18}O$.

RESULTS

The results of oxygen and carbon isotope analyses carried out at Monte Raggeto are presented in Table 3 (outcrop) and Table 4 (bore-core S1). Figure 3 shows the carbon-isotope stratigraphy of the outcropping section plotted against sedimentological and cyclostratigraphic data. Figure 5 shows the carbon-isotope correlation between the outcropping section and the bore-core S1. In the outcropping section the oxygen-isotope values vary between +0.7‰ and -2.8‰, and carbon-isotope values fluctuate between -1.92‰ and +4.6‰ (Table 3). In the lowermost 50 m of the section, carbon-isotope values increase from +1.0‰ to +2.0‰. At 55 m from the base, between superbundles 5 and 6, a short negative excursion occurs with an amplitude of 2‰. Values then increase from 0‰ to +2.3‰ in the next 20 m of the profile. At 100 m, between superbundles 10 and 11, the $\delta^{13}C$ values shift to +3.0‰ and then decrease continuously in the overlying 35 m down to +1‰ (superbundle 14). In the following 10 m, values increase up to +3.4‰ (superbundle 15). In superbundle 16 the $\delta^{13}C$ values decrease rapidly to +0.68‰ and in superbundles 17 and 18 they reach values near +3‰ (except for the low value of 1.2‰ at 153 m from the base); the carbon-isotope values then increase up to +4.64‰. Finally, at the top of the section, $\delta^{13}C$ decreases rapidly, reaching -1.92‰ at 186.7 m from the base (Table 3, Figs. 3, 5, 6).

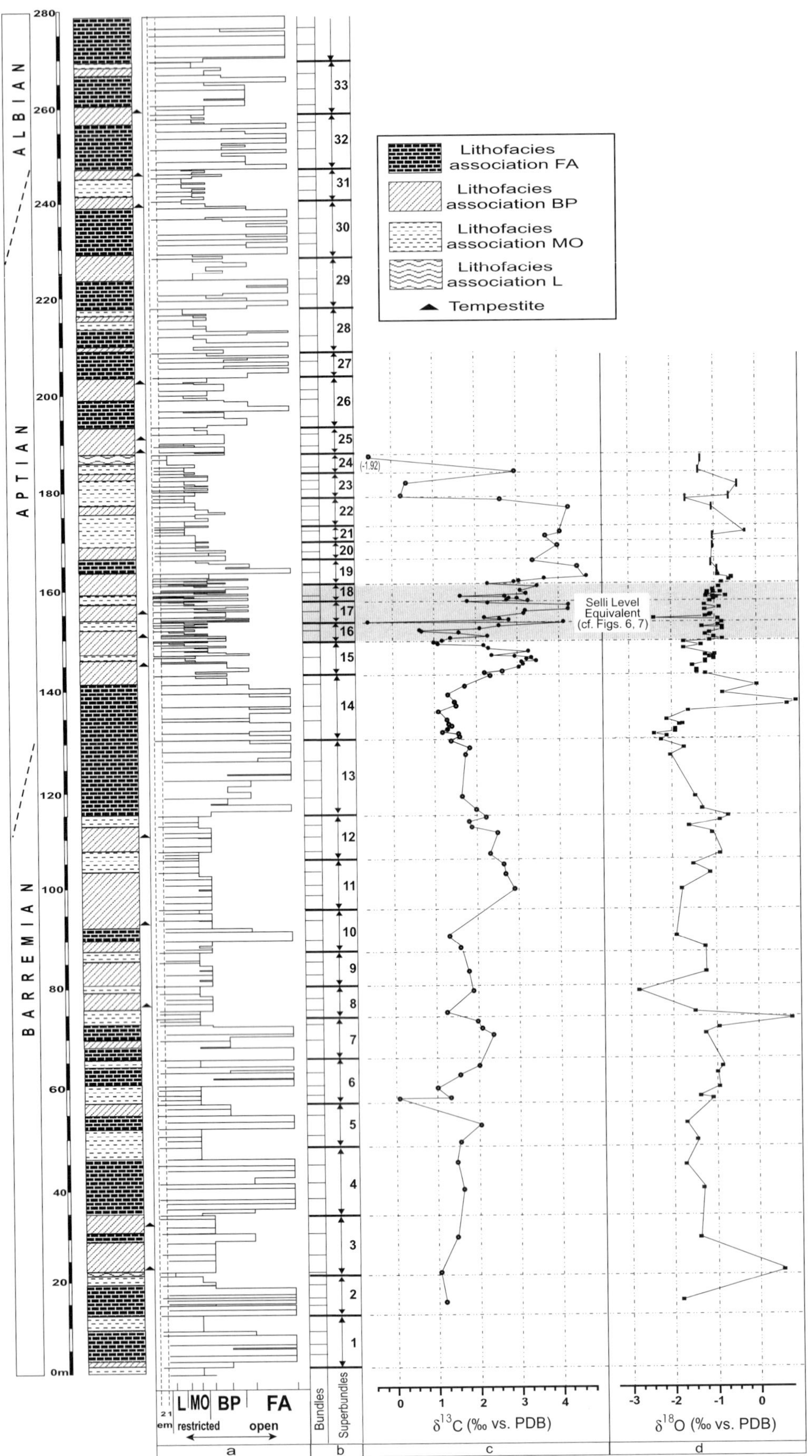

FIG. 3.—Barremian to Lower Aptian carbon and oxygen isotope stratigraphy of Monte Raggeto (columns c, d), plotted against sedimentological (graphic log) and cyclostratigraphic data (columns a, b). em, emersion-type surfaces at the top of the cycles; em1, microkarst, pedogenesis; em2, paleokarst, paleosols. L to FA, Lithofacies associations as in Figure 2. carbon-isotope values below 0.5‰ are interpreted to mirror early meteoric overprint and restricted lagoonal conditions.

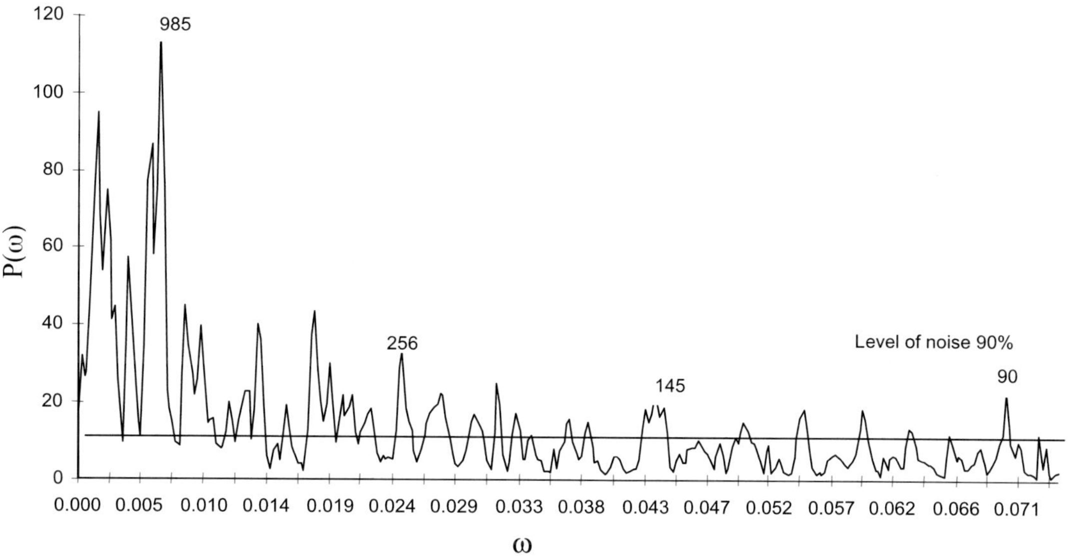

FIG. 4.—Barremian–Albian of Monte Raggeto. Power spectrum of the textural signal; periodicities are expressed in centimeters. From Brescia et al. (1996), modified. ω, frequency (cycles/cm); P(ω), power spectrum at frequency ω.

DISCUSSION

The shallow-water carbonates of the Monte Raggeto section were lithified at an early diagenetic stage and under marine conditions (e.g., D'Argenio et al., 1999a). This observation supports the interpretation of carbon-isotope values to be a marine signal that experienced little alteration during diagenesis. Several other potential factors may have influenced the bulk-rock carbon-isotope values. Platform-derived carbonates may have been formed from water masses covering isolated platforms (Holmden et al.,

TABLE 1.—Thickness and time periodicities of Monte Raggeto carbonate deposits (outcrop and core). From Brescia et al. (1996), modified.

Signal	Age	Signal Periodicities (cm)	(ratio) RRSs	Correction of Coefficiency	Orbital Ratios RRSt	Orbital Periodicities (years) Eccentricity	Obliquity	Precession	Average Accumulation Rate (cm/ky)
Textures (outcrop)	Albian	985	10.944		10.568	403800			
		256	2.844	0.9995	2.510	95800			2.60
		145	1.611		1.292		49370		
	Barremian	90	1.000		1.000		38210		
Bed thickness (outcrop)	Albian	985	10.707		10.568	403800			
		229	2.489	0.9997	2.510	95800			2.47
		130	1.413		1.292		49370		
	Barremian	92	1.000		1.000		38210		
	Barremian	1168	22.902		22.200	404200			
		266	5.216		5.211	94900			
Textures (core)		139	2.725	0.9998	2.617		47650		2.92
		119	2.333		2.084		37950		
		65	1.274		1.210			22000	
	Hauterivian	51	1.000		1.000			18210	
	Barremian	1073	19.870		22.200	404200			
Paleomagneti c inclination (core)		260	4.815		5.211	94900			
		142	2.630	0.9999	2.617		47650		2.93
		118	2.185		2.084		37950		
		69	1.278		1.210			22000	
	Hauterivian	54	1.000		1.000			18210	

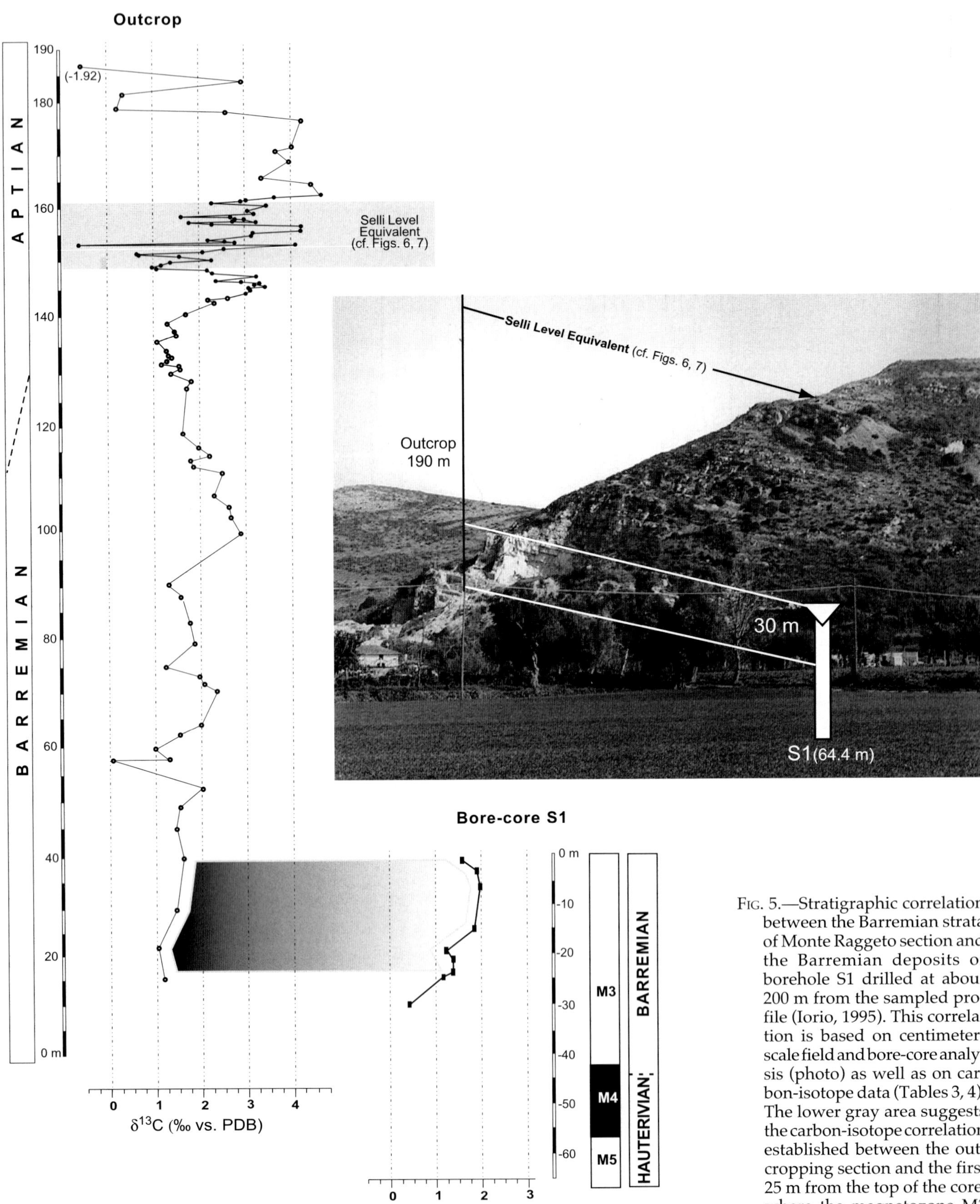

FIG. 5.—Stratigraphic correlation between the Barremian strata of Monte Raggeto section and the Barremian deposits of borehole S1 drilled at about 200 m from the sampled profile (Iorio, 1995). This correlation is based on centimeter-scale field and bore-core analysis (photo) as well as on carbon-isotope data (Tables 3, 4). The lower gray area suggests the carbon-isotope correlation established between the outcropping section and the first 25 m from the top of the core, where the magnetozone M3 occurs (Iorio et al., 1998).

TABLE 2.—Monte Raggeto (outcrop). Sedimentary and spectral periodicities (expressed in centimeters) and related long eccentricity and short eccentricity (expressed in ky) and obliquity (also expressed in ky).

Cycles		Periodicities (cm) Sedimentary Signal	Spectral Signal
Superbundles	(~ 400 ky)	810	985
Bundles	(~ 100 ky)	264	256
Elementary cycles	(~ 40 ky)	130	145–90

1998) depleted in carbon-13 because of enhanced organic-matter demineralization (e.g., Patterson and Walter, 1994). Moreover, Andrews et al. (1997) have shown that aragonitic precipitates are enriched in ^{13}C by about 1‰, suggesting that variations in carbonate mineralogy could have an impact on bulk carbonate signature. Previous studies (Ferreri et al., 1997; D'Argenio et al., this volume) as well as our results indicate that the main features of the studied shallow-water carbon-isotope profile are similar to those of the pelagic Cismon section (as shown by the three-point moving average) and therefore reflect the isotopic composition of the global carbon reservoir (Fig. 6; see also Bralower et al., 1994; Jenkyns, 1995; Weissert et al., 1998; Bralower et al., 1999). In the Cismon and Raggeto sections, a positive carbon-isotope shift across

TABLE 3.—Monte Raggeto (outcrop). Carbon and oxygen isotope data of the analyzed bulk sediment samples.

Sample	$\delta^{13}C$ (‰ vs. PDB)	$\delta^{18}O$ (‰ vs. PDB)	Lithofacies	Distance from the base (m)	Sample	$\delta^{13}C$ (‰ vs. PDB)	$\delta^{18}O$ (‰ vs. PDB)	Lithofacies	Distance from the base (m)
					5n	3.20	-1.10	MO1	145.2
Ar	-1.92	-1.35	L1	186.7	4n	3.40	-1.00	BP1	144.9
93	2.83	-1.40	MO2	184.3	3n	3.10	-1.20	BP1	144.6
96	-0.10	-0.49	MO1	181.2	2n	3.10	-1.20	BP1	144.2
98	0.24	-0.68	MO2	177.7	1n	3.00	-1.50	BP1	143.6
100	2.59	-1.70	MO1	176.9	10c2	2.60	-1.40	MO1	142.7
101	4.24	-1.09	BP2	175.7	10c1	2.20	-1.40	BP1	142.3
103	4.02	-0.32	MO1	171.1	9ctop	2.30	-1.20	MO1	142.0
104	3.68	-1.07	MO1	170.3	8c	1.70	0.00	FA1	139.7
E	3.96	-1.07	MO2	168.3	7c2	1.30	-0.80	FA2	138.1
L	3.37	-1.11	FA1	164.1	7c	1.47	0.91	FA1	136.5
C	4.49	-0.97	FA1	163.1	6c2	1.50	0.70	FA2	135.9
44c	4.64	-0.90	FA2	161.6	6c1	1.10	-1.60	FA2	134.6
107	3.03	-0.58	BP2	161.0	4c3	1.30	-2.10	FA1	132.9
38c	2.93	-0.64	BP1	160.7	4c	1.37	-1.73	FA1	132.0
104f	2.30	-0.82	MO2	160.1	4c2	1.40	-1.80	FA1	131.9
100f	3.47	-0.88	FA2	159.2	4c1	1.30	-1.90	FA2	131.0
92f	3.08	-1.03	BP1	158.6	3c3	1.20	-1.90	FA2	130.5
86f	3.21	-1.16	BP1	158.4	3c2	1.60	-2.40	FA1	130.0
82f2	1.63	-1.02	L1	158.3	3c1	1.60	-2.10	FA1	129.6
82f1	2.72	-0.84	MO1	158.2	2c	1.40	-2.23	FA2	128.8
108	3.65	-0.59	FA2	158.1	2c1	1.80	-1.70	FA1	127.3
78f	2.80	-1.17	BP1	157.7	1c	1.74	-2.02	FA1	127.0
76f	2.97	-0.73	MO2	157.4	1x	1.65	-1.42	FA2	119.0
75f	3.26	-1.18	BP2	157.3	15a	1.96	-1.31	FA1	115.5
30c	2.76	-1.02	BP2	157.2	15b	2.22	-0.64	MO1	112.8
71f	1.80	-0.94	MO2	157.1	15c	1.82	-0.85	MO1	112.5
29c	2.32	-1.00	MO1	156.8	14a	1.86	-1.61	MO1	111.0
69f	4.22	-1.09	FA2	156.4	14b	2.51	-1.05	BP2	110.0
67f	4.20	-1.23	FA2	155.6	13a	2.32	-0.82	MO1	106.0
65f	3.19	-0.87	MO1	155.1	13b	2.66	-1.51	MO1	104.0
64f	3.15	-1.21	BP1	154.6	13c	2.68	-1.06	MO1	101.5
22c	2.20	-1.07	MO1	153.7	6x	2.91	-1.71	BP2	99.5
59f	2.60	-1.10	BP1	153.5	36x	1.29	-1.91	FA1	90.4
57f	2.80	-1.20	FA2	153.4	34x	1.56	-1.16	BP2	87.6
54f	-1.20	-2.40	MO2	153.0	30x	1.76	-1.16	BP2	83.0
53f	4.10	-0.90	BP2	152.8	29x	1.88	-2.84	MO1	79.0
47f	2.56	-0.81	BP2	152.2	25x	1.23	-1.45	MO1	74.5
44f	2.10	-0.90	BP2	151.6	24x	2.01	0.81	MO1	72.8
12c	0.68	-1.28	L1	151.3	23x	2.11	-0.91	FA1	71.0
42f	0.70	-0.80	MO2	151.2	R2	2.35	-1.25	BP1	69.7
39f	1.50	-0.80	BP2	150.8	R5	2.01	-0.81	FA2	64.5
30f	2.30	-1.10	BP1	150.1	R7	1.54	-0.98	FA1	63.0
26f	1.40	-1.20	BP1	149.8	e	1.01	-0.86	MO1	59.4
22f	1.20	-1.00	MO2	149.3	c	1.31	-1.41	MO1	57.5
20f	1.00	-0.80	L1	148.8	a	0.06	-1.03	MO1	57.0
18f	1.10	-1.10	MO2	148.6	11x	2.06	-1.73	FA1	52.0
15f	2.20	-1.70	BP1	148.3	10x	1.56	-1.38	MO1	49.0
11f	2.30	-1.30	BP2	147.7	8x	1.45	-1.75	FA1	45.0
8f	3.27	-1.70	BP2	147.1	12m	1.61	-1.27	FA1	40.0
3f	2.38	-0.98	MO2	145.8	3m	1.44	-1.41	FA2	30.2
7n	2.90	-1.20	MO2	145.7	M4	1.04	0.64	L1	22.0
6n	3.30	-1.00	MO2	145.5	M10	1.16	-1.84	BP2	15.0

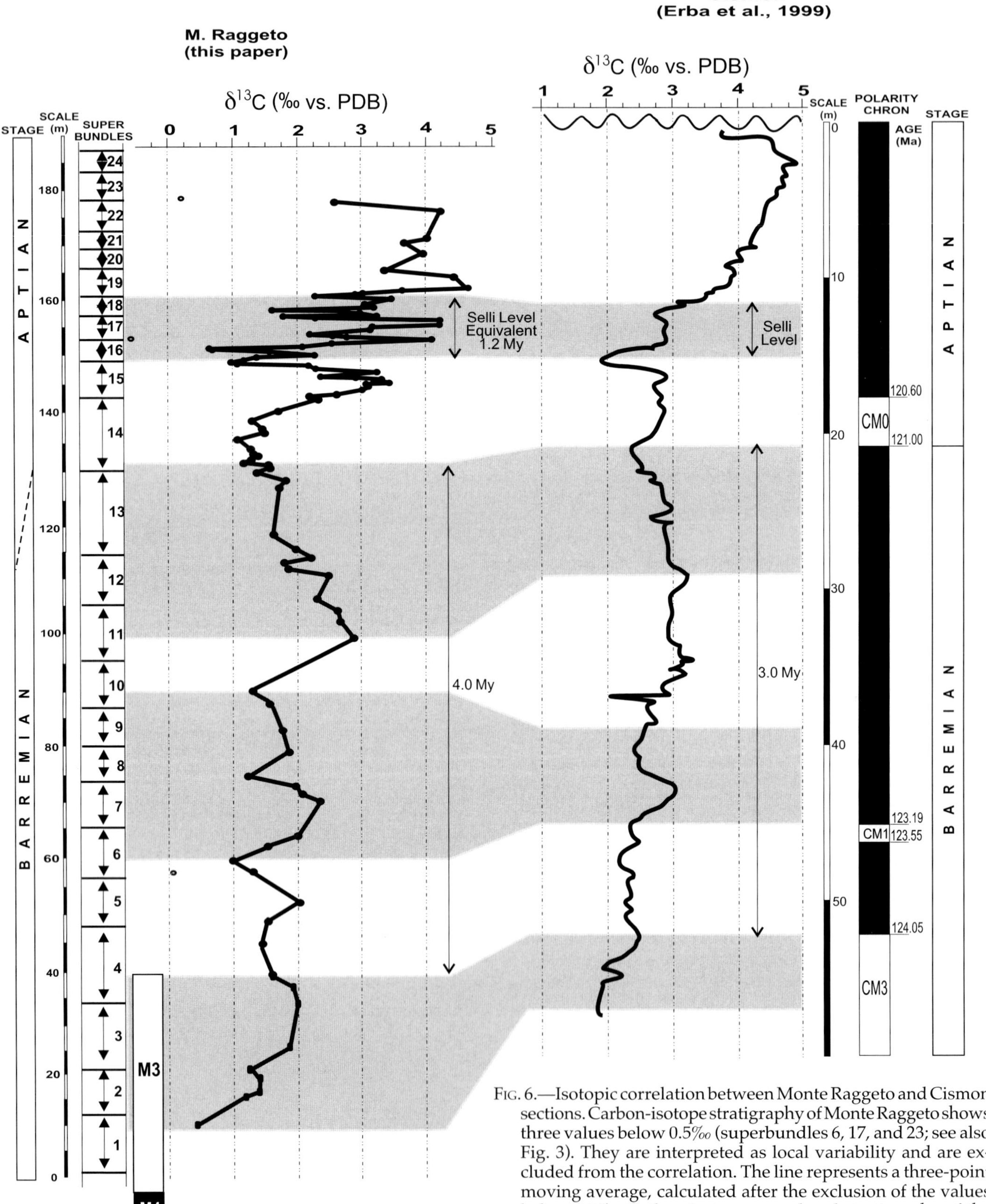

FIG. 6.—Isotopic correlation between Monte Raggeto and Cismon sections. Carbon-isotope stratigraphy of Monte Raggeto shows three values below 0.5‰ (superbundles 6, 17, and 23; see also Fig. 3). They are interpreted as local variability and are excluded from the correlation. The line represents a three-point moving average, calculated after the exclusion of the values below 0.5‰. Carbon-isotope record and stratigraphy of the Cismon section is after Erba et al. (1999). Magnetic reversal is from Channell et al. (1995).

TABLE 4.—Monte Raggeto (bore-core S1). Carbon and oxygen isotope data of the analyzed bulk sediment samples.

Sample	$\delta^{13}C$ (‰ vs. PDB)	$\delta^{18}O$ (‰ vs. PDB)	Distance from the base (m)
1bc	1.59	-1.28	-1
2bc	1.91	-1.62	-3
3bc	1.98	-1.44	-5.5
4bc	1.83	-0.78	-15
5bc	1.22	-1.80	-20
6bc	1.41	-0.71	-22
7bc	1.37	-1.42	-24
8bc	1.17	-2.68	-25
9bc	0.42	-1.58	-29

the top of magnetozone M3 is followed by an interval with relatively constant values. Increasing and decreasing carbon-isotope values within the overlying 30 m of the Monte Raggeto section (superbundles 6 to 10) most probably correspond to the small positive peak recorded in the Cismon profile at the base of magnetozone M1n. The subsequent shift to more positive values within M1n at Cismon occurs between superbundles 10 and 11 in the Monte Raggeto section. The declining $\delta^{13}C$ trend from superbundle 11 to superbundle 14 correlates with decreasing carbon-isotope values in the uppermost part of M1n at Cismon. Above this interval in both sections carbon-isotope values increase and stabilize around +3‰, showing a narrow negative peak followed by a plateau at +3‰; then they increase again, reaching values around +4.5‰. The four negative spikes in superbundles 6, 17, and 23 of the Monte Raggeto profile do not occur in the pelagic Cismon section (Fig. 6) and are therefore interpreted as a local signal. Because depleted $\delta^{13}C$ values occur within restricted facies, they may have resulted from isolated lagoonal conditions and/or meteoric overprint (Patterson and Walter, 1994). We have also measured $\delta^{13}C$ values in the Selli Level Equivalent (SLE) of Monte Raggeto. The highly negative carbon-isotope values in the SLE may reflect episodically restricted lagoonal conditions and/or early meteoric diagenesis, which is more common in this stratigraphic interval. However, because the shift to very positive values at the top of the SLE occurs in both open and restricted sediments, we interpret the general trend in this interval as a global signal. The low carbon-isotope values at the top of the section occur within, and immediately below, a deeply penetrating paleokarst horizon and hence they result from meteoric overprint (D'Argenio and Mindszenty, 1995).

Carbon Isotopes and Cyclostratigraphy

Microstratigraphic (centimeter-scale) analysis of textures and sedimentary features carried out on the Barremian–Aptian strata of Monte Raggeto section allows us to recognize a hierarchy of cycles (Fig. 3, columns a, b) that is attributed to the response of an aggradational carbonate platform to orbitally modulated, high-frequency sea-level changes (Longo et al., 1994; Brescia et al., 1996; Tagliaferri et al., 2001). Using the astronomically determined average rates of sediment accumulation (2.93 cm/ky in Table 1), the time represented by M3 in the Monte Raggeto core (Fig. 5) is about 1.5 My (Iorio et al., 1998). A comparable duration of 1.6 My is obtained from the sampled section, where the strata corresponding to the magnetozone M3 form about 4 superbundles (Fig. 6).

Carbon-isotope and magnetostratigraphic correlation between Raggeto and Cismon (Fig. 6) demonstrates that superbundle 4 at the Monte Raggeto section correlates with the top of magnetozone M3. On the basis of the carbon-isotope stratigraphy we correlate the base of superbundle 14 of Monte Raggeto with the base of M0, which at Cismon occurs at the top of the decreasing $\delta^{13}C$ trend that characterizes the upper part of magnetozone M1. The sediments recovered at Monte Raggeto between these two tie points feature 10 superbundles and represent roughly 4 My. This is in agreement with Herbert (1992), who estimated the duration of the same interval as 4 My on the basis of astronomical cycles recorded in pelagic sediments, but it is about 1 My longer than the estimation of Channell et al. (1995), which relies on a Hawaiian magnetic block model. Because the results derived from both cyclostratigraphic methods are similar, we believe that the oceanic anomaly block model of Channell et al. (1995) underestimates the time between Chrons M3 and M0. In fact, assuming an age of 121 Ma for the base of M0, cyclostratigraphy gives an age of about 125 Ma for the top of M3 (Fig. 6), instead of the 124.05 Ma suggested by Channell et al. (1995) and by Larson and Erba (1999).

Duration of the Selli Level

The Selli Level is the name given to a 5-m-thick organic-carbon-rich shale intercalated in Lower Aptian pelagic limestones of the Umbrian Apennines of Italy (e.g., Coccioni et al., 1987). A high-resolution carbon-isotope stratigraphy of the Selli Level was generated at Cismon, a locality in the southern Alps, by Menegatti et al. (1998). A pronounced negative peak in the carbon-isotope profile was identified at the base of this unit. Throughout the Selli Level carbon-isotope values remain around +3‰, whereas above the Selli Level the carbon-isotope curve shifts from +2.5‰ towards more positive values (Menegatti et al., 1998). At Monte Raggeto the Selli Level Equivalent (SLE), as identified on the basis of the carbon-isotope stratigraphy (Fig. 6), corresponds to a 12-m-thick interval of thin-bedded, dark-gray limestones that contain the superbundles 16, 17, and 18 (Figs. 6, 7). These superbundles represent a time interval of 1.2 My, which corresponds fully to the duration of the Selli Level in the pelagic section of Cismon as calculated by Herbert (1992) on the basis of the numbers of 100 ky cycles recognized in this interval.

CONCLUSIONS

The $\delta^{13}C$ signal measured in shallow-water carbonates of the southern Apennines mirrors the carbon isotope composition of the global carbon reservoir. The chemostratigraphic correlation between Monte Raggeto and Cismon shows complete agreement of Milankovitch cyclicity documented in shallow-water and pelagic sediments. Our results (Fig. 6) indicate that the top of magnetozone M3 is about 1 My older than estimated by the oceanic anomaly block model of Channell et al. (1995). Finally, the duration of 1.2 My estimated at Monte Raggeto for the SLE (Figs. 6, 7) fully agrees with the estimate of Herbert (1992).

We conclude that chemostratigraphy, combined with cyclostratigraphy, may provide a high-precision stratigraphic correlation tool between distant shallow-water and deep-water carbonate sections.

ACKNOWLEDGMENTS

This project was supported by the ETH Zürich and by the Swiss Science Foundation. The Italian work was financially sponsored by the Research Institute "Geomare sud", CNR, Napoli,

FIG. 7.—**A)** Lower Aptian part of the Monte Raggeto section, showing the vertical organization of the strata. As suggested by carbon-isotope correlation shown in Figure 6, the magnetozone M0 and the Selli Level Equivalent (SLE) occur in this stratigraphic interval. In correspondence of M0 (superbundle 14) the carbon-isotope signature suggests a shift towards more positive values recorded in thicker cycles. These cycles are formed of more open-marine lithofacies associations and are characterized by em1-type emersion limits (microkarst). The SLE (superbundles 16, 17, and 18) shows thinner cycles with em1- or em2-type emersion limits. These cycles are formed mainly of inner, dark-colored deposits suggesting, on the whole, a transgressive trend. In this interval the large scatter (around +3‰) of the $\delta^{13}C$ signal is most probably due to the recurrence of restricted lithofacies, overprinted more strongly by pedogenesis and / or dissolution processes, which may explain the most negative values. For key to the lithofacies and their grouping and for additional comments, see text and Figures 2 and 3. **B)** Detail of the Selli Level Equivalent.

which provided both laboratory and computer facilities. We wish to thank Raika Radoii from Belgrade for her help in biostratigraphic determination of benthonic foraminifers and green algae. Detailed reviews by T. Bralower and E. Erba helped to improve the manuscript.

REFERENCES

ANDREWS, J.E., CHRISTIDIS, S., AND DENNIS P.F., 1997, Assessing mineralogical and geochemical heterogeneity in the sub 63 micron size fraction of Holocene lime muds: Journal of Sedimentary Research, v. 67, p. 531–535.

BERGER, A.L., LOUTRE, M.F., AND LASKAR, J., 1992, Stability of the astronomical frequencies over the Earth's history for paleoclimate studies: Science, v. 255, p. 560–565

BIGI, G., COSENTINO, D., PAROTTO, M., SARTORI, R., AND SCANDONE, P., EDS., 1992, Structural Model of Italy: scale 1:500.000, 6 sheets, Consiglio Nazionale delle Ricerche (CNR), Progetto Finalizzato Geodinamica.

BRALOWER, T.J., ARTHUR, M.A., LECKIE, R.M., SLITER, W.V., ALLARD, D.J., AND SCHLANGER S.O., 1994, Timing and paleoceanography of oceanic dysoxia / anoxia in the Late Barremian to Early Aptian (Early Cretaceous): Palaios, v. 9, p. 335–369.

BRALOWER, T. J., COBABE, E., CLEMENT, B., SLITER, W.V., OSBURN, C.L., AND LONGORIA, J., 1999, The record of global change in mid-Cretaceous (Barremian–Albian) sections from the Sierra Madre, northeastern Mexico: Journal of Foraminiferal Research, v. 29, p. 418–437.

BRESCIA, M., D'ARGENIO, B., FERRERI, V., PELOSI, N., RAMPONE, S, AND TAGLIAFERRI, R., 1996, Neural net aided detection of astronomical periodicities in geologic records: Earth and Planetary Science Letters, v. 139, p. 33–45.

BUONOCUNTO, F.P., 1998, Litostratigrafia di alta risoluzione nel Cretacico di piattaforma Carbonatica dell'Appennino Campano: Ph.D. Dissertation, Università Federico II, Napoli, 184 p.

CHANNELL, J.E.T., ERBA, E., NAKANISHI, M., AND TAMAKI, K., 1995, Late Jurassic–Early Cretaceous time scale and oceanic magnetic anomaly block, *in* Berggren, W.A., Kent, D.V., Aubry, M.P., and Hardenbol, J., eds., Geochronology, Time Scales and Global Stratigraphic Correlation: SEPM, Special Publication 54, p. 51–64.

Coccioni, R., Nesci, O., Tramontana, M., Wezel, C.F., and Moretti, E., 1987, Descrizione di un livello guida "Radiolaritico-Bituminoso-Ittolitico" alla base delle Marne a Fucoidi nell' Appennino Umbro-Marchigiano: Società Geologica Italiana, Bollettino, v. 106, p. 183–192.

D'Argenio, B., Amodio, S., Ferreri, V., and Pelosi, N., 1997, Hierarchy of high-frequency orbital cycles in Cretaceous carbonate platform strata: Sedimentary Geology, v. 113, p. 169–193.

D'Argenio, B., Ferreri, V., Ardillo, F., and Buonocunto, F.P., 1993, Microstratigrafia sequenziale. Studi sui depositi di piattaforma carbonatica, Cretacico del Monte Maggiore (Appennino Meridionale): Società Geologica Italiana, Bollettino, v. 112, p. 739–749.

D'Argenio, B., Ferreri, V., Iorio, M., Raspini, A., and Tarling, D.H., 1999a, Diagenesis and remanence acquisition in the Cretaceous carbonates of Monte Raggeto, Southern Italy, *in* Tarling, D.H., and Turner, P., eds., Palaeomagnetism and Diagenesis in Sediments: Geological Society of London, Special Publication 151, p. 147–156.

D'Argenio, B., Ferreri, V., Raspini, A., Amodio, S., and Buonocunto, F.P., 1999b, Cyclostratigraphy of a carbonate platform as a tool for high-precision correlation: Tectonophysics, v. 315, p. 357–385.

D'Argenio, B., and Mindszenty, A., 1995, Bauxites and related paleokarst: Tectonic and climatic event markers at regional unconformities: Eclogae Geologicae Helvetiae, v. 88, p. 453–499.

Erba, E., Channell, J.E.T., Claps, M., Jones, C., Larson, R., Opdyke, B., Premoli Silva, I., Riva, A., Salvini, G., and Torricelli, S., 1999, Integrated stratigraphy of the Cismon Apticore (Southern Alps, Italy): a "Reference Section" for the Barremian–Aptian interval at low latitude: Journal of Foraminiferal Research, v. 29, p. 371–391.

Ferreri, V., Weissert, H., D'Argenio B., and Buonocunto F.P., 1997, Carbon-isotope stratigraphy: a tool for basin to carbonate platform correlation: Terra Nova, v. 9, p. 57–61.

Herbert, T., 1992, Paleomagnetic calibration of Milankovitch cyclicity in Lower Cretaceous sediments: Earth and Planetary Science Letters, v. 112, p. 15–28.

Herbert, T., and Fischer, A.G., 1986, Milankovitch climatic origin of mid-Cretaceous black shale rhythms in central Italy: Nature, v. 321, p. 739–743.

Holmden, C., Creaser, R.A., Muehlenbachs, K., Leslie, S.A., and Bergström, S.M., 1998, Isotopic evidence for geochemical decoupling between ancient epeiric seas and bordering oceans: Implications for secular curves: Geology, v. 26, p. 567–570.

Iorio, M., 1995, High resolution palaeomagnetic analyses of Cretaceous shallow water carbonates: Ph.D. Dissertation, University of Plymouth, U.K., 370 p.

Iorio, M., Tarling, D.H., D'Argenio, B., Nardi, G., and Hailwood, E.A., 1995, Milankovitch cyclicity of magnetic directions in Cretaceous shallow-water carbonate rocks, southern Italy: Bollettino di Geofisica Teorica ed Applicata, v. 37, p. 109–118.

Iorio, M., Tarling, D.H., and D'Argenio, B., 1998, The magnetic polarity stratigraphy of a Hauterivian/Barremian carbonate sequence from southern Italy: Geophysical Journal International, v. 134, p. 13–24.

Jenkyns, H.C., 1995, Carbon-isotope stratigraphy and paleoceanographic significance of the Lower Cretaceous shallow-water carbonates of Resolution Guyot, Mid Pacific Mountains: Proceedings of the Oceanic Drilling Program, Scientific Results, v. 143, p. 99–104.

Larson, R.L. and Erba, E., 1999, Onset of mid-Cretaceous greenhouse in the Barremian–Aptian. Igneous events and the biological, sedimentary, and geochemical responses: Paleoceanography, v. 14, p. 663–678.

Longo, G., D'Argenio, B., Ferreri, V., and Iorio, M., 1994, Fourier evidence for high-frequency astronomical cycles recorded in Early Cretaceous carbonate platform strata, Monte Maggiore, southern Apennines, Italy, *in* de Boer, P.L., and Smith, D.G., eds., Orbital Forcing and Cyclic Sequences: International Association of Sedimentologists, Special Publication 19, p. 77–85.

Menegatti, A.P., Weissert, H., Brown, R., Tyson, R.V., Farrimond, P., Strasser, A., and Caron, M., 1998, High-resolution $\delta^{13}C$-stratigraphy through the early Aptian "Selli Level Equivalent" of the Alpine Tethys: Paleoceanography, v. 13, p. 530–545.

Patterson, W.P., and Walter, L.M., 1994, Depletion of $\delta^{13}C$ in seawater ΣCO_2 on modern carbonate platforms: Significance for the carbon isotopic record of carbonates: Geology, v. 22, p. 885–888.

Raspini, A., 1998, Microfacies analysis of shallow-water carbonates and evidence of hierarchically organized cycles. Aptian of Monte Tobenna, southern Apennines, Italy: Cretaceous Research, v. 19, p. 197–223.

Raspini, A., 2001, Stacking pattern of cyclic carbonate platform strata: Lower Cretaceous of southern Apennines, Italy: Geological Society of London, Journal, v. 158, p. 353–366.

Tagliaferri, R., Pelosi, N., Ciaramella A., Longo, N., Milano, M., and Barone, F., 2001, Soft computing methodologies for spectral analysis in cyclostratigraphy: Computer & Geosciences, v. 27(5), p. 535–548.

Weissert, H., Lini, A., Föllmi, K., and Kuhn, O., 1998, Correlation of Early Cretaceous carbon-isotope stratigraphy and platform drowning events: A possible link?: Palaeogeography, Palaeoclimatology, Palaeoecology, v. 137, 189–203.

CYCLOSTRATIGRAPHIC TIMING OF SEDIMENTARY PROCESSES: AN EXAMPLE FROM THE BERRIASIAN OF THE SWISS AND FRENCH JURA MOUNTAINS

ANDRÉ STRASSER
Department of Geosciences, University of Fribourg, Pérolles, CH-1700 Fribourg, Switzerland
e-mail: andreas.strasser@unifr.ch
HEIKO HILLGÄRTNER
Shell International Exploration and Production BV, Volmerlaan 8, NL-2280 AB Rijswijk, The Netherlands
e-mail: heiko.hillgartner@shell.com
AND
JEAN-BRUNO PASQUIER
GéoVal Ingénieurs-Géologues SA, Rue de la Majorie 8, CH-1950 Sion, Switzerland
e-mail: jbpasquier@bluewin.ch

Abstract: The Berriasian Pierre-Châtel Formation in the Swiss and French Jura Mountains is dominated by shallow-marine carbonates that overlie lacustrine and marginal-marine sediments with a major transgressive surface. Detailed facies analysis of five sections allows the definition of elementary and small-scale depositional sequences, which commonly exhibit deepening–shallowing trends. Benthic foraminifera and rare ammonites on the platform, as well as a sequence-stratigraphic correlation with a well-dated deeper-water section, furnish the biostratigraphic framework. Thus, the large-scale sequence boundaries below and at the top of the Pierre-Châtel Formation can be correlated with dated boundaries in other European basins. This time constraint and the hierarchical stacking pattern on the platform as well as in the basin suggest that the sea-level fluctuations influencing the formation of the depositional sequences were controlled, at least partly, by Milankovitch cycles. The elementary sequences correspond to the 20 ky precession cycle, and the small-scale sequences to the 100 ky eccentricity cycle.

Uncertainties in the definition of sequences exist if facies contrasts are too low to develop clearly marked sequence boundaries or maximum-flooding intervals. Nevertheless, a best-fit solution for the correlation of the small-scale sequences between the studied sections can be proposed. The lowermost three small-scale sequences of the Pierre-Châtel Formation are analyzed in detail. They are decompacted and correlated on the level of the elementary sequences. Within this relatively precise time frame, the flooding of the Jura platform (following the early Berriasian sea-level lowstand) can be monitored. It is seen that the transgression occurred stepwise: every 20 ky, a transgressive pulse established marine facies farther towards the platform interior.

This study demonstrates that the cyclostratigraphical approach makes it possible to construct a narrow time frame, within which the rates of sedimentary, ecological, and diagenetic processes can be evaluated, phases of differential subsidence identified, and the durations of stratigraphic gaps estimated. The complex and dynamic evolution of an ancient carbonate platform can thus be studied with a time resolution of 20 to 100 ky.

INTRODUCTION

The correct evaluation of time is an everlasting quest in the geological sciences. In the Holocene, several dating methods (e.g., ^{14}C, dendrochronology) are available that allow reaching time resolutions of a few tens of years to a single year. Further back in the geologic past, the error margins of radiometric dating increase and reach values on the million-year scale (Berggren et al., 1995). In order to estimate the rates of sedimentological processes in the past (such as sediment production and accumulation, diagenesis, ecological changes), a much higher precision is needed. It is useless to just divide the thickness of the stratigraphic column by the corresponding time interval when calculating sedimentation rates: these rates are facies dependent, time may be condensed in hiatuses, sediment production rates may have been much higher than the final accumulation, and differential compaction may have distorted the sedimentary record.

Cyclostratigraphy allows constructing a relatively precise time scale in the geologic past even though radiometric dating gives large error margins (e.g., Schwarzacher, 1993; Lourens et al., 1996). If detailed analysis of the sedimentary record demonstrates that the formation of depositional sequences ("sedimentary cycles") is related to the quasi-periodic perturbations of the Earth's orbit (Milankovitch cycles), then a time resolution of 20–100 ky can potentially be reached. This time frame is comparable to that of the Pleistocene and Holocene, where the parameters controlling sedimentary processes are better known.

Detailed analyses of facies and stacking pattern in Berriasian (lowermost Cretaceous) shallow-water, carbonate-dominated sequences in Switzerland and France have shown that sedimentation was controlled at least partly by Milankovitch cycles (Pasquier, 1995; Pasquier and Strasser, 1997; Strasser and Hillgärtner, 1998; Hillgärtner, 1999). It appears that sea-level fluctuations were the most dominant parameter but that climatic changes controlling nutrient and siliciclastic fluxes, and synsedimentary tectonics shaping the substrate, also were important in determining facies distribution. Elementary sequences (the smallest units where facies evolution indicates an environmental cycle) formed in tune with the 20 ky precession cycle. Elementary sequences stack into small-scale and medium-scale sequences, representing the 100 ky and 400 ky eccentricity cycles, respectively. Sequences corresponding to the obliquity cycle of 40 ky could not be identified. Large-scale sequences are composed of several medium-scale sequences and mostly reflect tectono-eustatic changes.

Cyclostratigraphy: Approaches and Case Histories
SEPM Special Publication No. 81, Copyright © 2004
SEPM (Society for Sedimentary Geology), ISBN 1-56576-108-1, p. 135–151.

In the present paper, the focus is on the detailed analysis of the record of a large-scale transgression, which flooded a shallow platform. Within a biostratigraphic and sequence-stratigraphic framework, cyclostratigraphy is used to monitor this transgression with time steps corresponding to the 20 ky of the orbital precession cycle.

GEOGRAPHIC AND PALEOGEOGRAPHIC SETTING

Five shallow-water sections are situated in the Swiss and French Jura Mountains, and a deeper-water section used for biostratigraphic and cyclostratigraphic calibration is located in the Vocontian Basin in France (Fig. 1). In the Early Cretaceous, the Jura platform was part of the complexly structured northwestern margin of the Ligurian Tethys, and the Vocontian Basin represented a dead-end branch of this ocean (Fig. 2). The study area in the Jura was situated at a latitude of about 32 to 33° N (Dercourt et al., 2000). Paleoenvironmental conditions were subtropical and ideal for extensive carbonate production on the shallow platform. In the Vocontian Basin, hemipelagic and pelagic carbonate sedimentation was active but carbonate mud was also shed from the adjacent platforms (Hillgärtner, 1999).

During the latest Jurassic and earliest Cretaceous, a slowdown of sea-floor spreading in the western Tethys and accelerated rifting in the North Atlantic (Late Cimmerian phase; Sinclair et al., 1994) led to thermal doming and a long-term tectono-eustatic sea-level fall, which caused widespread emergence of the Armorican, Central, and Rhenish–Bohemian massifs (Fig. 2; Ziegler, 1988). This, together with episodically increased rainfall, furnished siliciclastics to platform and basin. Block faulting and differential subsidence affected the Jura platform and the Vocontian domain (Wildi et al., 1989; de Graciansky and Lemoine, 1988) and caused a very heterogeneous facies distribution, especially on the shallow platform.

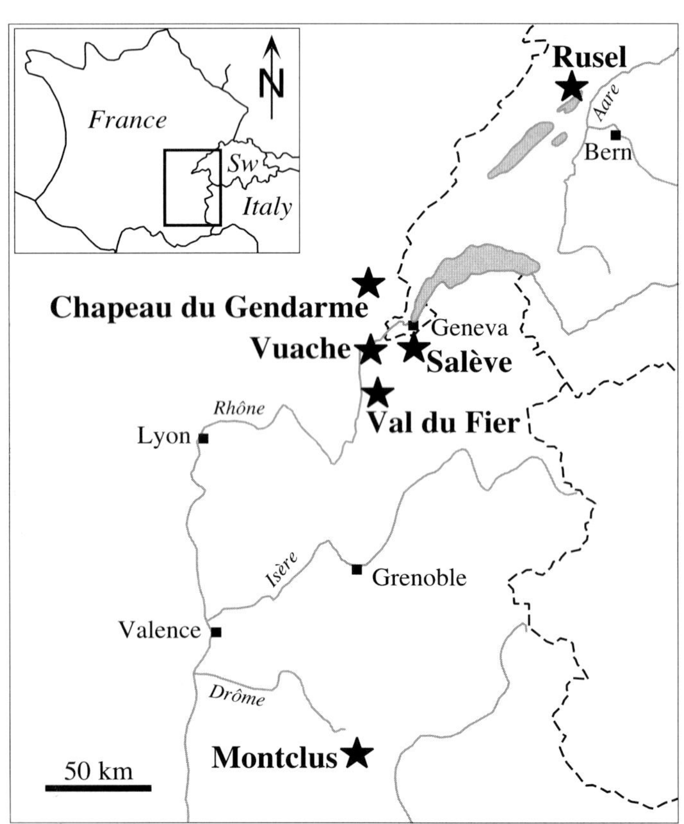

FIG. 1.—Location of the studied sections in Switzerland (Sw) and France.

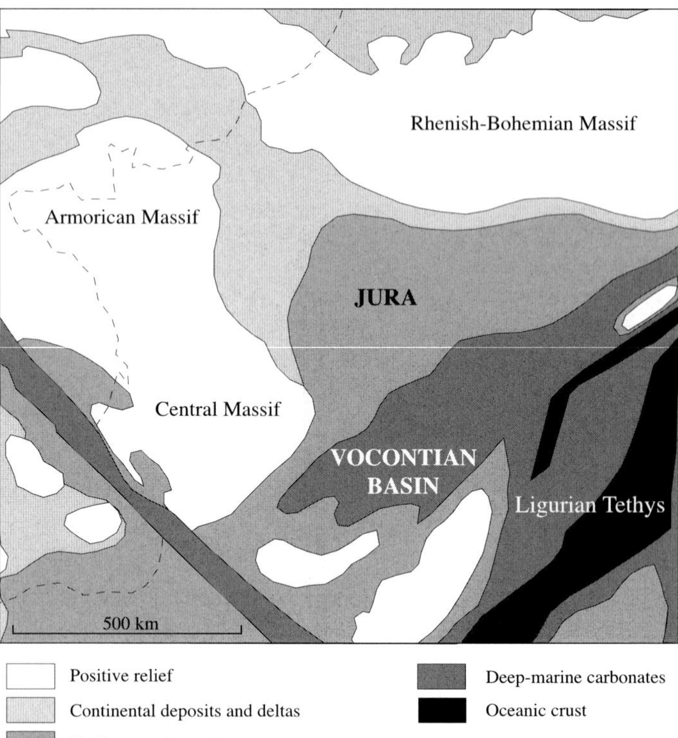

FIG. 2.—Paleogeography in the Early Cretaceous, including the shallow Jura platform and the Vocontian Basin (based on Ziegler, 1988).

Following the Alpine orogeny, the Jura Mountains were faulted and folded, the main tectonic activity going on during the late Miocene–early Pliocene. Overburden was a few hundred meters in the north and about 2000 m in the southern French Jura (Trümpy, 1980). For the present study, no palinspastic reconstructions were undertaken. The distances between the studied sections would have to be stretched by a few kilometers in a north–south direction, and some sections would be offset by a few kilometers by north–south-trending faults (Meyer, 2000). However, the general order on the transect from inner platform in the north (Rusel section; Fig. 1), central and outer platform (Chapeau-du-Gendarme, Vuache, Val-du-Fier, and Salève sections), and basin (Montclus section) in the south has not changed. The Salève section contains massive ooid grainstones that represent high-energy shoals typical of platform margins. On the transect, this section is therefore placed in the corresponding position, and the Val-du-Fier section in a more protected location. It is assumed that the platform margin was not a straight line but displayed promontories and bays.

LITHOSTRATIGRAPHIC AND BIOSTRATIGRAPHIC FRAMEWORK

In the platform sections, the studied interval comprises the top of the Goldberg Formation, the Pierre-Châtel Formation, and the base of the Vions Formation (Fig. 3). The Goldberg Formation (defined by Häfeli, 1966) displays to a large part the "Purbeckian"

formations	*biostratigraphy*	*ammonite zones*	*ammonite subzones*	*stage*
Vions	*Pseudotextulariella courtionensis* *Pavlovecina allobrogensis* *Picteticeras* aff. *moesica*	Boissieri	Picteti	upper Berriasian
			Parami-mounum	
Pierre-Châtel		Occitanica	Dalmasi	middle Berriasian
	M4 *Subalpinites* sp.		Privasensis	
Goldberg	*M2/M3* *Pseudosubplanites lorioli*		Subalpina	
	Pseudosubplanites combesi *Tirnovella* gr. *allobrogensis-suprajurensis* *M1b*	Jacobi-Grandis		lower Berriasian

FIG. 3.—Lithostratigraphy, biostratigraphy, and chronostratigraphy of the studied interval (biostratigraphy of ammonites and benthic foraminifera according to Clavel et al., 1986; charophyte–ostracod assemblages M1b, M2, M3, and M4 according to Détraz and Mojon, 1989).

facies: peritidal carbonates with charophytes, black pebbles, and, locally, evaporite pseudomorphs (Strasser, 1988). The top of this formation is locally eroded and condensed, which can be explained by differential subsidence and uplift due to enhanced tectonic activity during the Early Cretaceous (De Graciansky and Lemoine, 1988). The Pierre-Châtel Formation was defined by Steinhauser and Lombard (1969). Its base represents a rapid flooding of the partly emergent Purbeckian platform, and its bulk is made up of shallow but normal-marine carbonates (Pasquier, 1995). The overlying Vions Formation (Steinhauser and Lombard, 1969) displays mainly shallow-marine facies but also contains lacustrine and palustrine levels. Furthermore, it is characterized by abundant detrital quartz and clays (Hillgärtner, 1999).

Dating of these formations is difficult because biostratigraphically relevant fossils are rare. Clavel et al. (1986) described ammonites at the top of the Goldberg Formation, at the base of the Pierre-Châtel Formation, and at the base of the Vions Formation. The base of the Pierre-Châtel Formation can thus be attributed to the Subalpina Subzone and its top to the Paramimounum Subzone (Fig. 3). The foraminifer *Pavlovecina allobrogensis* (commonly associated with *Pseudotextulariella courtionensis*) typically appears in the Paramimounum Subzone (Clavel et al., 1986). Charophyte–ostracod assemblages have been defined by Détraz and Mojon (1989) and are calibrated on ammonite zones. Thus, even in the absence of ammonites, these associations allow constraining the studied interval. In the Montclus section of the Vocontian Basin, ammonites and calpionellids are abundant and a good biozonation is available (Le Hegarat, 1971).

MATERIAL AND METHODS

The platform sections were logged and sampled in considerable detail. The Rusel section is based on work by Pasquier (1995), and the Chapeau-du-Gendarme section has been analyzed by Waehry (1989) and Hillgärtner (1999). The Vuache section was first described by Blondel (1984) and was studied again by Hillgärtner (1999). The section of Val du Fier is based on work of Darsac (1983) and Hillgärtner (1999). The Salève outcrop was studied by Waehry (1989), Strasser and Hillgärtner (1998), and Hillgärtner (1999). The depositional environments and their evolution through time have been interpreted from the analysis of microfacies, sedimentary structures, and bedding surfaces. A high-resolution sequence-stratigraphic interpretation is made where facies evolution allows identification of accommodation changes. The sequence-stratigraphic terminology follows Vail et al. (1991).

An example of this approach is given in Figure 4. On the shallow platform, rapid loss of accommodation leading to the development of a sequence boundary is indicated by lacustrine or tidal-flat facies, and by birdseyes. Because of lack of accommodation, lowstand deposits are not developed or only thinly developed, or reworked within early transgressive deposits. The transgressive surface is commonly underlain by reworked material, including black pebbles indicative of subaerial exposure (Strasser and Davaud, 1983). Relatively deepest or most open-marine water is indicated by fauna such as echinoderms or brachiopods. Levels with increased bioturbation suggest a reduced sedimentation rate and, when combined with open-marine fauna, are interpreted as intervals of maximum flooding (maximum-flooding surfaces are not always developed). The marly levels can have different origins: clays may be washed into the system during relative sea-level lowstands and thus emphasize sequence boundaries, they can accumulate during maximum flooding when the seafloor is below wave base, or they can be related to increased rainfall in the hinterland (Strasser and Hillgärtner, 1998). The interpretation of the marls must therefore be based on their faunal and floral content, and on the context within the depositional sequence. Two orders of depositional sequences can be distinguished in the example of Figure 4: elementary sequences, consisting most commonly of one bed, and small-scale sequences, composed of three to five elementary sequences.

The deeper-water Montclus section was sampled in less detail because the facies consist mainly of limestone–marl alternations and are relatively homogeneous. The section presented here and its interpretation is based on work by Pasquier (1995), Pasquier and Strasser (1997), Hillgärtner (1999), and Strasser et al. (2000).

The methodology used for the sequence-stratigraphic and cyclostratigraphic interpretation of the studied interval follows several steps (Strasser et al., 1999):

1. Identification of elementary depositional sequences, i.e., of facies evolution through time corresponding to one cycle of environmental change (mainly a deepening–shallowing trend in the platform sections, and a limestone–marl couplet in the basin);
2. Identification of the stacking pattern of elementary sequences that compose small-scale, medium-scale, and large-scale depositional sequences, which again show characteristic facies evolutions (general deepening–shallowing trends on the platform, and upward thinning or upward thickening of the limestone beds in the basin);
3. Sequence-stratigraphic interpretation of the depositional sequences on all scales, independently for each section;
4. Comparison of this interpretation between the sections, and correlation of the sequence-stratigraphic elements to find a best-fit solution that is also compatible with the biostratigraphic framework;

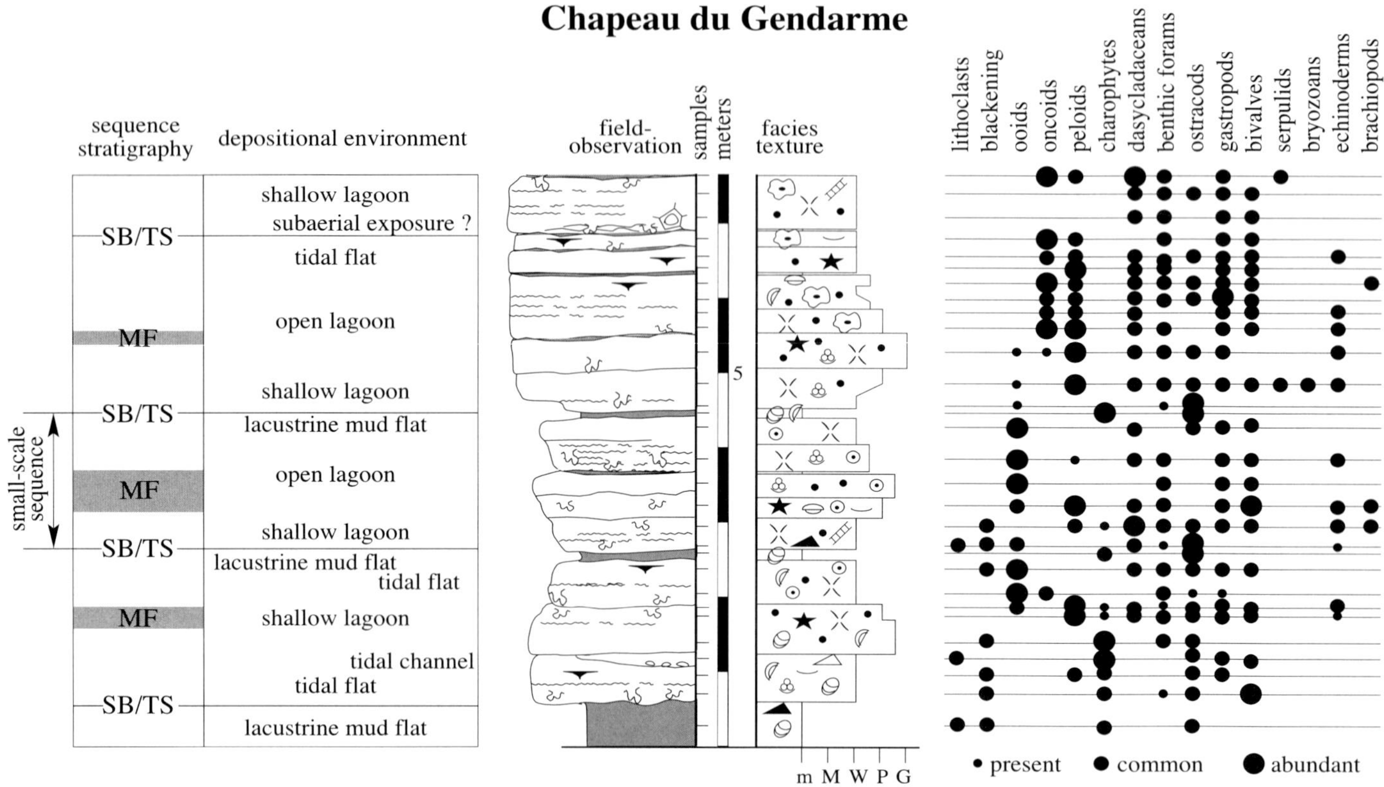

FIG. 4.—Example of detailed facies analysis, and interpretation of depositional environments and easily identifiable sequence-stratigraphic elements (base of Pierre-Châtel Formation, Chapeau-du-Gendarme section; microfacies analysis by Waehry, 1989). Symbols for facies and sedimentary structures as in Figure 5.

5. Comparison of large-scale sequence boundaries with boundaries identified in other sedimentary basins in the same biostratigraphic position;
6. Counting of the elementary and small-scale sequences between dated large-scale sequence boundaries and dated limits of biostratigraphic zones;
7. If the duration of these depositional sequences falls within the Milankovitch frequency band, a cyclostratigraphic time scale can be proposed.

CORRELATION

A best fit-solution of correlation of small-scale sequences between the studied sections is shown in Figure 5. First, the platform sections are discussed, then a correlation with the basinal section of Montclus is attempted.

Platform Sections

The top of the Goldberg Formation exhibits birdseyes, black pebbles, and pedogenetic brecciation (below the transgressive surface at the base of the sections in Fig. 5). The Subalpina Subzone is condensed (Clavel et al., 1986). This indicates prolonged emersion at this level, which is interpreted as a major sequence boundary. According to its biostratigraphic position it can be correlated with sequence boundary Be4 of Hardenbol et al. (1998), which has been recognized also in other European basins.

The base of the Pierre-Châtel Formation is marked by a sharp transgressive surface (TS in Fig. 5). Facies then indicate periodic deepening and shallowing of depositional environments, and / or suggest that conditions changed from restricted marine to open marine and back to restricted. The identification of depositional sequences is relatively easy in the sections representing the platform interior, where facies contrasts are well developed (Rusel, Chapeau du Gendarme; Fig. 5). In the Salève section, however, surfaces within bioclastic and ooid grainstones may well have formed through shoal migration and not through relative sea-level changes. The correlation of several small-scale sequences therefore is uncertain.

In a few cases, it is not clear where the limit of a small-scale sequence should be chosen, because two or more bedding surfaces of elementary sequences show the characteristics of a sequence boundary. This is interpreted to be due to the superposition of high-frequency sea-level fluctuations on a longer-term trend of sea-level evolution (Fig. 6). In this way, not only sequence boundaries but also transgressive and maximum-flooding surfaces may be repeated and define zones that correspond to the time of longer-term sea-level fall, transgression, or maximum flooding, respectively (Montañez and Osleger, 1993). Furthermore, changing amplitudes of the high-frequency fluctuations may attenuate or accentuate such surfaces and zones (Strasser et al., 1999). In Figure 5, sequence-boundary zones of small-scale and large-scale sequences are marked in gray.

In the sections of Rusel, Chapeau du Gendarme, Vuache, and Val du Fier, the small-scale sequences display a general thickening-upward trend. This trend, however, is interrupted by karst surfaces in the first two sections, by vadose cementation and circumgranular cracks indicating pedogenesis at Vuache, and by several levels of root traces in Val du Fier (sequences 15 to 18, Fig. 5). This apparently rapid loss of accommodation is interpreted as a major sequence boundary. Above, the appearance of the foraminifera *Pavlovecina allobrogensis* and *Pseudotextulariella courtionensis* indicate the Paramimounum Subzone (Fig. 3; Clavel et al., 1986). This major sequence boundary can therefore be correlated with Be5 of Hardenbol et al. (1998). The maximum-flooding interval of the large-scale sequence (defined between Be4 and Be5) is placed where the thickest beds suggest highest accommodation gain and keep-up of the carbonate system (small-scale sequence 16 at Chapeau du Gendarme), or where intense bioturbation points to reduced sedimentation rates because of a low carbonate production in deeper water (middle part of sequence 16 at Salève).

Between the transgressive surface at the base of the Pierre-Châtel Formation and sequence boundary Be5, 6 to 9 small-scale sequences are counted (Fig. 5). However, the correlation across the platform suggests that at Rusel and Chapeau du Gendarme the uppermost sequences are truncated. At Vuache, strong condensation at this level is indicated by the limited extension of *P. allobrogensis* (Hillgärtner, 1999). A fall of eustatic sea level alone cannot explain these features, and faulting of the Jura platform inducing differential subsidence has to be assumed at that time (De Graciansky and Lemoine, 1988; Hillgärtner, 1999).

Basinal Section

In order to better constrain the biostratigraphic guidelines and to better understand the sequential evolution of the platform sections, a correlation with the well-dated Montclus section in the Vocontian Basin was undertaken. There, a major sequence boundary is indicated at the base of an interval containing relatively thick, irregular, and locally channeled limestone beds, which are interpreted as lowstand deposits (Pasquier, 1995; Pasquier and Strasser, 1997). This boundary coincides with the Subalpina–Privasensis Subzone boundary (Le Hegarat, 1971) and can therefore be correlated with Be4 of Hardenbol et al. (1998). The lowstand deposits are overlain by an interval of thinner and more homogeneous limestone–marl alternations, which are thought to represent transgressive deposits. A rapid change to even thinner alternations is interpreted as a second transgressive pulse (TS2, Fig. 5). The marliest part of the section (partly covered in the outcrop) is attributed to a condensed interval (corresponding to the maximum flooding of this large-scale sequence), and the following alternations displaying a thickening-up trend are seen as highstand deposits. A major sequence boundary is placed at the base of the particularly thick limestone bed at meter 22.5, or at the base of a slumped interval (thus defining a sequence-boundary zone). This boundary is situated in the Paramimounum Subzone and corresponds to Be5 of Hardenbol et al. (1998).

The limestone–marl couplets group into bundles of 2 to 6 (Fig. 5). These bundles are interpreted as small-scale sequences, and one couplet as an elementary sequence (Pasquier and Strasser, 1997). The limestone beds are composed to a large part of nannoplankton (Cotillon et al., 1980; Strohmenger and Strasser, 1993). The limits of the small-scale sequences have been chosen at the bases of the thickest limestone beds, because these beds are thought to have formed during lowstand conditions when planktonic productivity was concentrated above the hemipelagic realm while the platform was at least partly exposed. With rising sea level, carbonate productivity was active also on the platform, less nannoplankton was produced in the open ocean, and marly facies formed. However, this oversimplified model was certainly complicated by phases of export of carbonate mud from platform to basin (Schlager et al., 1994; Pittet et al., 2000), by ocean currents, temperature, and nutrients influencing planktonic productivity (Einsele and Ricken, 1991), and by climate- and current-controlled input of clay minerals.

Basin-to-Platform Correlation

Eighteen to 19 small-scale sequences have been identified at Montclus between Be4 and Be5 (Fig. 5). Be4 in the basin correlates well with the top of the Goldberg Formation on the platform, where the Subalpina Subzone is condensed. It is tempting to correlate the transgressive surface TS1 at the top of the lowstand at Montclus with the base of the Pierre-Châtel Formation (TS at Salève; Fig. 5). However, when counting the small-scale sequences downward from Be5, the fit is better if sequence 10 at Montclus is correlated with the first small-scale sequence of the Pierre-Châtel Formation (Fig. 5). This solution implies that the platform already started being flooded while lowstand conditions continued in the basin (Pasquier and Strasser, 1997). In sequence 11, which in the basin is interpreted as following a transgressive pulse, the platform experienced an opening to more marine conditions, which allowed the growth of echinoderms and brachiopods even in the platform interior (Rusel and Chapeau-du-Gendarme sections). Also, it is evident from Figure 5 that nine small-scale sequences constituting the bulk of the lowstand in the basin are missing or strongly condensed on the platform.

It is interesting to note that the large-scale condensed section at Montclus (at the top of small-scale sequence 15) is not isochronous with the maximum flooding defined on the platform within small-scale sequence 16. Apparently, the sedimentary systems in the basin and on the platform reacted differently to the same eustatic sea-level change, or differential tectonic movements caused a shift of the time interval of fastest relative sea-level rise.

In lack of continuous outcrop or seismic profiles that allow tracing of sequences from the platform to the basin, sequence boundaries, transgressive surfaces, and maximum-flooding surfaces have to be identified on the basis of the observations in the individual sections. There, the best-developed surfaces will be chosen, which, however, may be offset by one or two high-frequency sea-level cycles when compared to the theoretical position on the long-term sea-level trend (Strasser et al., 1999). This offset may vary from one section to the other because of differences in basin morphology and/or sediment availability.

CYCLOSTRATIGRAPHY

Building a Time Scale

In the Montclus section, 71 to 104 limestone–marl couplets are counted between Be4 and Be5, depending on considering the thin beds of marly limestone as part of a couplet or as a separate couplet, and on excluding or including small-scale sequence 19 (Fig. 5). According to Hardenbol et al. (1998), Be4 is dated at 141.04 Ma, and Be5 at 139.33 Ma. The high precision of these numbers is of course unrealistic, considering that the Tithonian–Berriasian boundary is dated at 144.2 Ma with an error margin of ± 2.6 Ma (Gradstein et al., 1995). Nevertheless, the time between these two sequence boundaries can be estimated to be in the range of about 1.7 million years.

Assuming that one limestone–marl couplet represents an equal time increment, the duration of one couplet would vary between 24 and 16 ky. The small-scale sequences, on the other hand, would have durations of 90 to 95 ky. These numbers are close to the periodicities of the orbital precession cycle (20 ky in the Early Cretaceous; Berger et al., 1989) and the first eccentricity

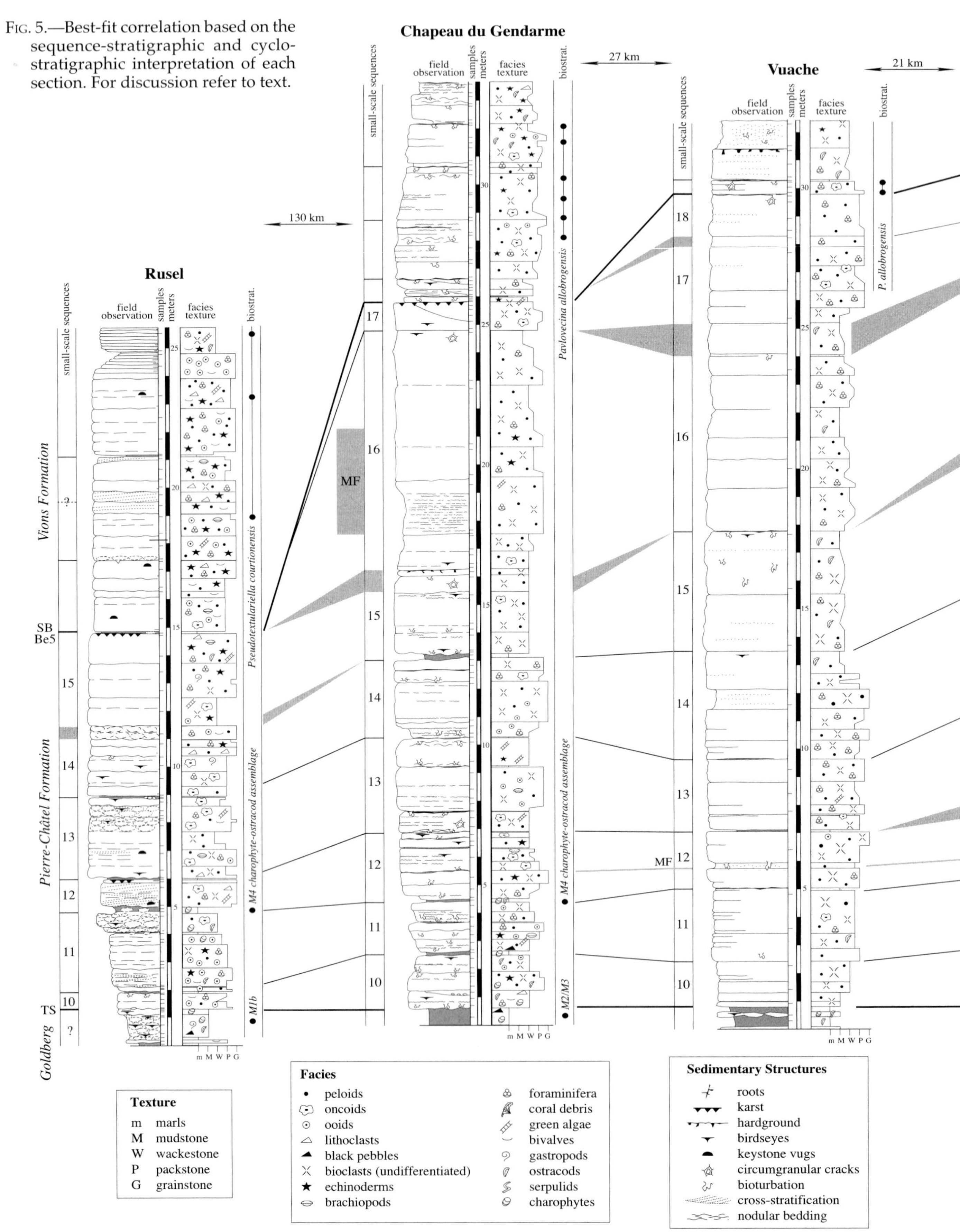

FIG. 5.—Best-fit correlation based on the sequence-stratigraphic and cyclostratigraphic interpretation of each section. For discussion refer to text.

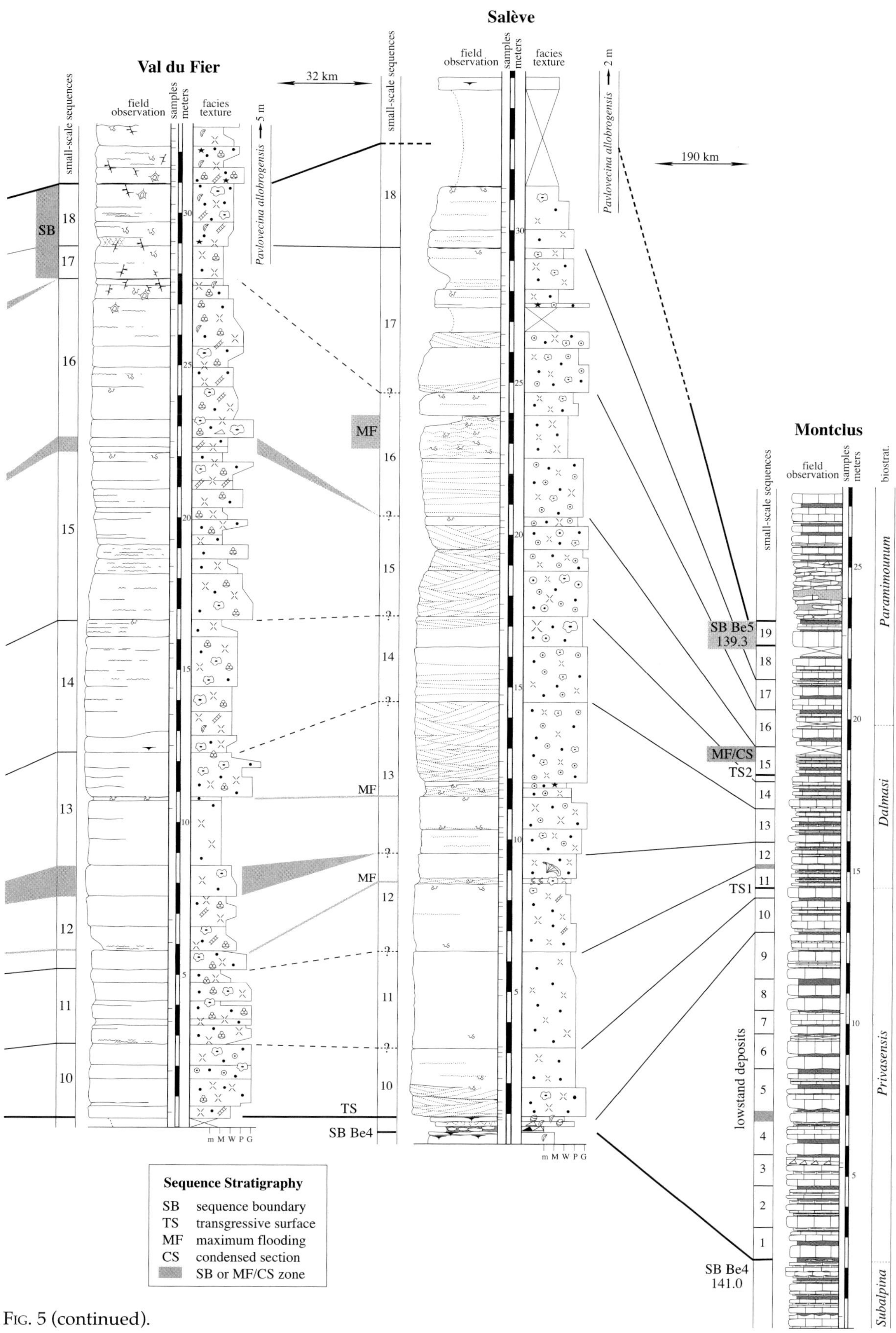

FIG. 5 (continued).

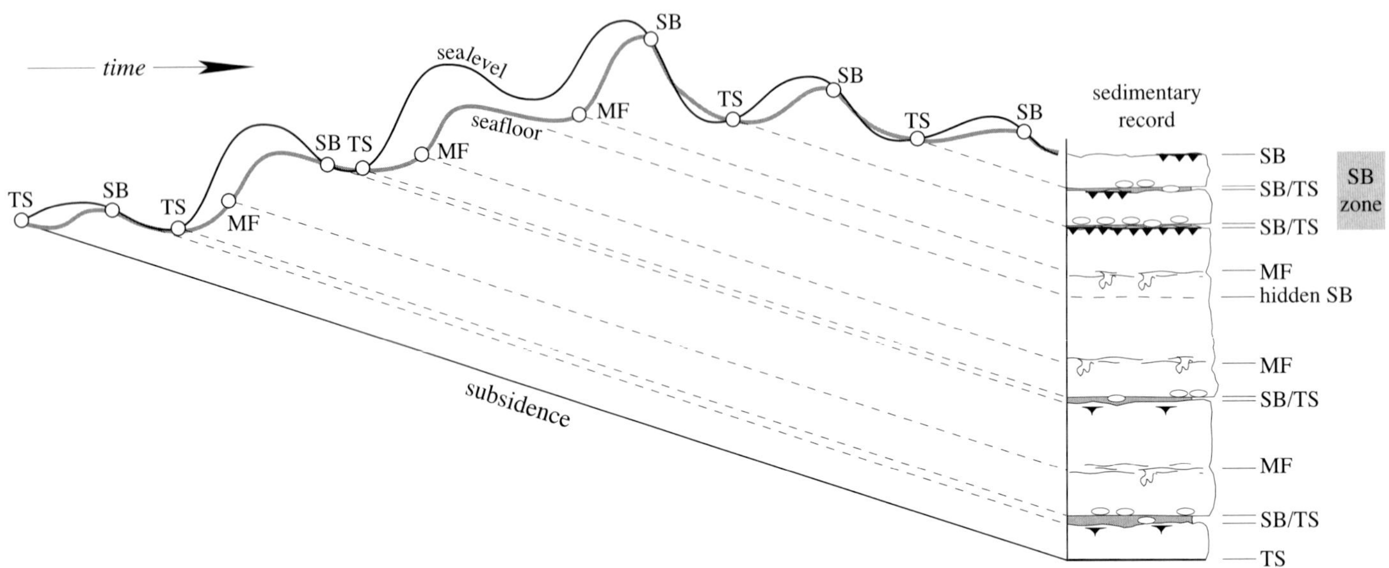

Fig. 6.—Hypothetical depositional sequences formed by high-frequency sea-level fluctuations superimposed on a longer-term rising and falling sea-level trend. Symbols are as in Figure 5. Note that stacked sequence boundaries form a sequence-boundary zone during the falling trend, and that a sequence boundary may not be expressed lithologically on the rising trend ("hidden SB").

cycle (100 ky), respectively. It is therefore implied that the sedimentary record at Montclus formed in tune with environmental changes controlled by insolation changes in the Milankovitch frequency band.

The counting of inferred 20 ky and 100 ky cycles between sequence-stratigraphic surfaces and limits of biozones now allows an improved estimate of time in the sections studied. Figure 7 shows that there are some discrepancies between the interpretation of the Montclus section and the chart published by Hardenbol et al. (1998). At Montclus, the Privasensis Subzone apparently lasted 950 ky, but only 500 ky are given in the chart. The duration of the Dalmasi Subzone, however, is consistent. Hardenbol et al. place the condensed interval of the large-scale sequence (Be4–Be5) at 139.7 Ma in the Paramimounum Subzone, whereas at Montclus it appears in the uppermost Dalmasi Subzone. The duration of the highstand, however, is consistent with 300 to 400 ky. The discrepancies may be due to problems with attributing absolute ages to biozones, and/or by the fact that the best-developed physical expression of a sequence boundary or a maximum-flooding interval does not necessarily occur at the same time in different paleogeographic positions (Jacquin and de Graciansky, 1998).

Hardenbol et al. (1998)					Montclus				platform sections		
stage	*ammonite subzones*	*duration*	*sequence stratigraphy*	*duration*	*ammonite subzones*	*duration*	*sequence stratigraphy*	*duration*	*bio-stratigraphy*	*sequence stratigraphy*	*duration*
Berriasian upper	Parami-mounum		**Be5** 139.33		Parami-mounum		SB		*P. allo-brogensis*	SB	
			CS 139.7	350 ky				300–400 ky		hiatus MF	150–250 ky
	140.05						CS				
Berriasian middle	Dalmasi	500 ky		1.35 My	Dalmasi	500 ky		500–600 ky			650 ky
	140.55						TS		*M4*	TS	
	Privasensis	500 ky			Privasensis	950 ky		900–1000 ky		hiatus	
	141.04		**Be4** 141.04				SB			SB	
	Subalpina				Subalpina				*M2/3*		

Fig. 7.—Chronostratigraphy of the studied interval according to Hardenbol et al. (1998), and comparison with the estimation of time based on the cyclostratigraphic analysis of the deeper-water Montclus section and the platform sections. For discussion refer to text.

Timing of the Platform Sections

If the correlation presented in Figure 5 is accepted as a working hypothesis, the small-scale sequences identified in the platform sections would correspond to 100 ky. These sequences are commonly composed of 2 to 6 beds, whereby a bed would correspond to an elementary sequence with a duration of 20 ky. However, in many cases it is difficult to define the limits of the elementary sequences because facies contrasts are too low. Bedding planes can also have formed through autocyclic processes such as shifting mudbanks (Pratt and James, 1986) or through clay input that was not related to sea-level change. In the case of sequence 12 at Rusel, only transgressive shoals are preserved. The karst surface that terminates this small-scale sequence may have formed through a drop of eustatic sea level (Fig. 6) and/or through tectonic uplift in this area.

HISTORY OF A TRANSGRESSION

Decompaction and High-Resolution Correlation

The relatively good time resolution obtained by cyclostratigraphy now permits detailed monitoring of the flooding of the Jura platform. For this purpose, the lowermost three small-scale sequences of the Pierre-Châtel Formation were chosen. The elementary sequences are relatively well defined (at least in the platform-interior sections), and a correlation with a time resolution of 20 ky can thus be proposed. To better understand the sea-level history, however, these sequences first have to be decompacted in order to evaluate the true accommodation changes (Fig. 8).

Mechanical reorganization of grains and dewatering in carbonate mud leads to a porosity loss of 10 to 30% after the first 100 m of burial (Moore, 1989). The experiments of Shinn and Robbin (1983) yielded values of 20 to 70% of volume loss through mostly mechanical and dewatering compaction. These values will be less if carbonate cementation sets in very early (e.g., Halley and Harris, 1979). With deeper burial, chemical compaction becomes important. Burial of over 2 km probably never occurred in the Jura Mountains (Trümpy, 1980) but pressure solution at grain contacts testifies to some dissolution processes in the studied sediments. Goldhammer (1997) proposes a compaction of slightly over 50% for 1 m of carbonate mud buried at 1000 m, and of about 15% for 1 m of carbonate sand at the same burial depth. According to Enos (1991), muddy terrigenous and muddy carbonate sediments do not have significantly different compaction curves. However, pressure solution along clay seams may of course enhance chemical compaction in carbonates (e.g., Bathurst, 1987). On the basis of these published values, the following decompaction factors have been chosen for the present study: 1.2 for grainstones and 2.5 for mudstones. For packstones and wackestones, the intermediate factors 1.5 and 2 are assumed, and for marls the factor 3. These factors are very rough estimates, but the decompacted sections nevertheless give a more realistic impression of the Early Cretaceous sediment accumulations than today's outcrops.

Figure 8 shows the effect of this decompaction and a best-fit solution of correlation on the 20 ky scale for the first three small-scale sequences. In some small-scale sequences, also the maximum-flooding intervals are identified and correlated.

Reconstruction of High-Frequency Eustatic Sea-Level Fluctuations

Reconstruction of the "true" eustatic sea-level changes that controlled sedimentation in the past is of course not possible. An approximation, however, can be proposed if the original sediment thickness, the range of water depth for each facies encountered, and the subsidence rate can be estimated. The first three small-scale sequences of the Chapeau-du-Gendarme section are well suited for such an approximation: they all end with intertidal or lacustrine facies, which sets mean sea level at zero. Maximum flooding of these sequences is indicated by subtidal, normal-marine, bioturbated, and low-energy facies. Water depth there is estimated at a minimum of one meter but could have been considerably more. The elementary sequences are well developed and give the timing of the sea-level fluctuations in 20 ky steps. Farther up in the section, the elementary sequences are less well defined, and intertidal facies are less common. The reconstruction of the sea-level curve therefore becomes very hypothetical (Fig. 9).

It is assumed that the sea-level fluctuations were more or less symmetrical. In the Early Cretaceous, ice in high latitudes and on mountains was probably present but volumes were small (Fairbridge, 1976; Frakes et al., 1992; Eyles, 1993; Price, 1999). Orbitally controlled climatic changes would have resulted in only minor glacio-eustatic sea-level fluctuations. However, insolation changes could also contribute to low-amplitude sea-level fluctuations through thermal expansion and contraction of the uppermost layer of ocean water (Gornitz et al., 1982), thermally induced volume changes in deep-water circulation (Schulz and Schäfer-Neth, 1998), and/or water retention and release in lakes and aquifers (Jacobs and Sahagian, 1993).

In order to accumulate 12 meters of sediment in the 300 ky corresponding to the first three small-scale sequences, accommodation gain must have been 4 m/100 ky on average. For small-scale sequences 13 to 15, 5 m/100 ky is assumed, and sequence 16 required 13 m of accommodation (Fig. 9). Part of the accommodation gain was due to long-term sea-level rise, which led to the general flooding of the Jura platform. However, the thickness variations of the small-scale sequences across the platform (Fig. 5) suggest that differential subsidence significantly influenced accommodation. During the deposition of sequence 16, a rapidly subsiding block was situated at Chapeau du Gendarme, while subsidence was attenuated or even uplift occurred at Rusel. According to Wildi et al. (1989), average subsidence rate on the Jura platform in the Late Jurassic and Early Cretaceous varied between 1 and 7 m/100 ky.

For the first three small-scale sequences of the Pierre-Châtel Formation, the long-term accommodation gain was subtracted from the reconstructed sea-level curve. This results in a high-frequency sea-level signal (Fig. 10). The wavelengths were adjusted slightly to fit the inferred 20 ky cyclicity. The amplitudes are minimum values because minimum estimates of water depth were used. If subsidence and long-term sea-level rise were regular, this signal theoretically reflects the eustatic sea-level changes created by the orbital precession and first eccentricity cycles.

Paleogeographic Evolution

On the basis of the high-resolution correlation of the elementary sequences (Fig. 8) and the inferred high-frequency sea-level curve (Fig. 10), the paleogeographic evolution of the platform can now be reconstructed with time steps of 20 ky (Fig. 11). Although large distances separate the studied sections and the correlation of the elementary sequences is uncertain in places, the image of a stepwise transgression is evident. While the platform interior (Rusel section) is still dominated by freshwater lakes, ooid bars form at the platform edge (Salève section) and shelter lagoons from high energy (Vuache section). Washovers occasionally carry marine material into the coastal lakes at Chapeau du Gendarme.

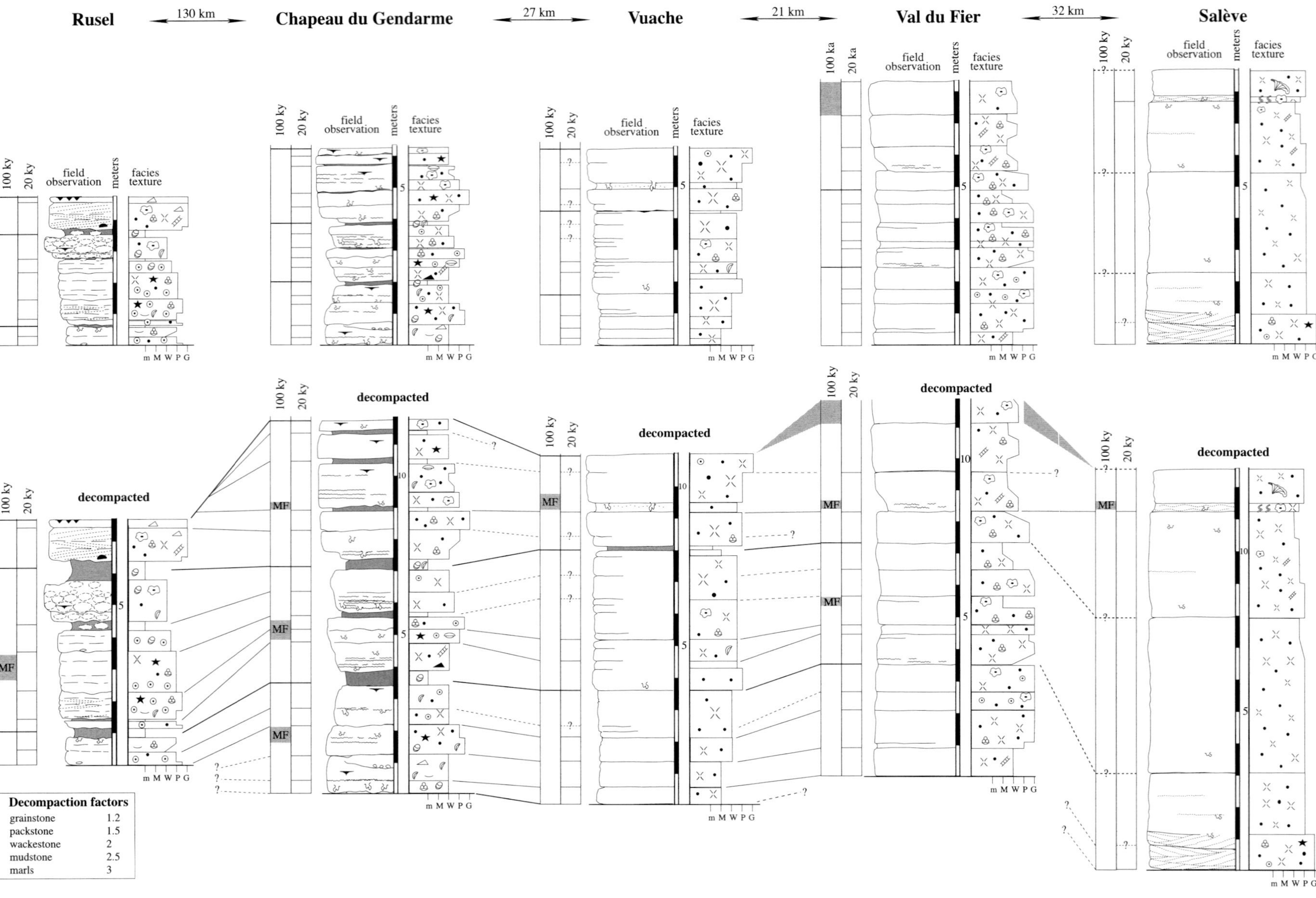

FIG. 8.—Decompaction of the lowermost three small-scale sequences of the Pierre-Châtel Formation, and best-fit correlation of the elementary sequences. The maximum-flooding interval of the third small-scale sequence is assumed to have been isochronous and is used as reference level. Symbols as in Figure 5. For discussion see text.

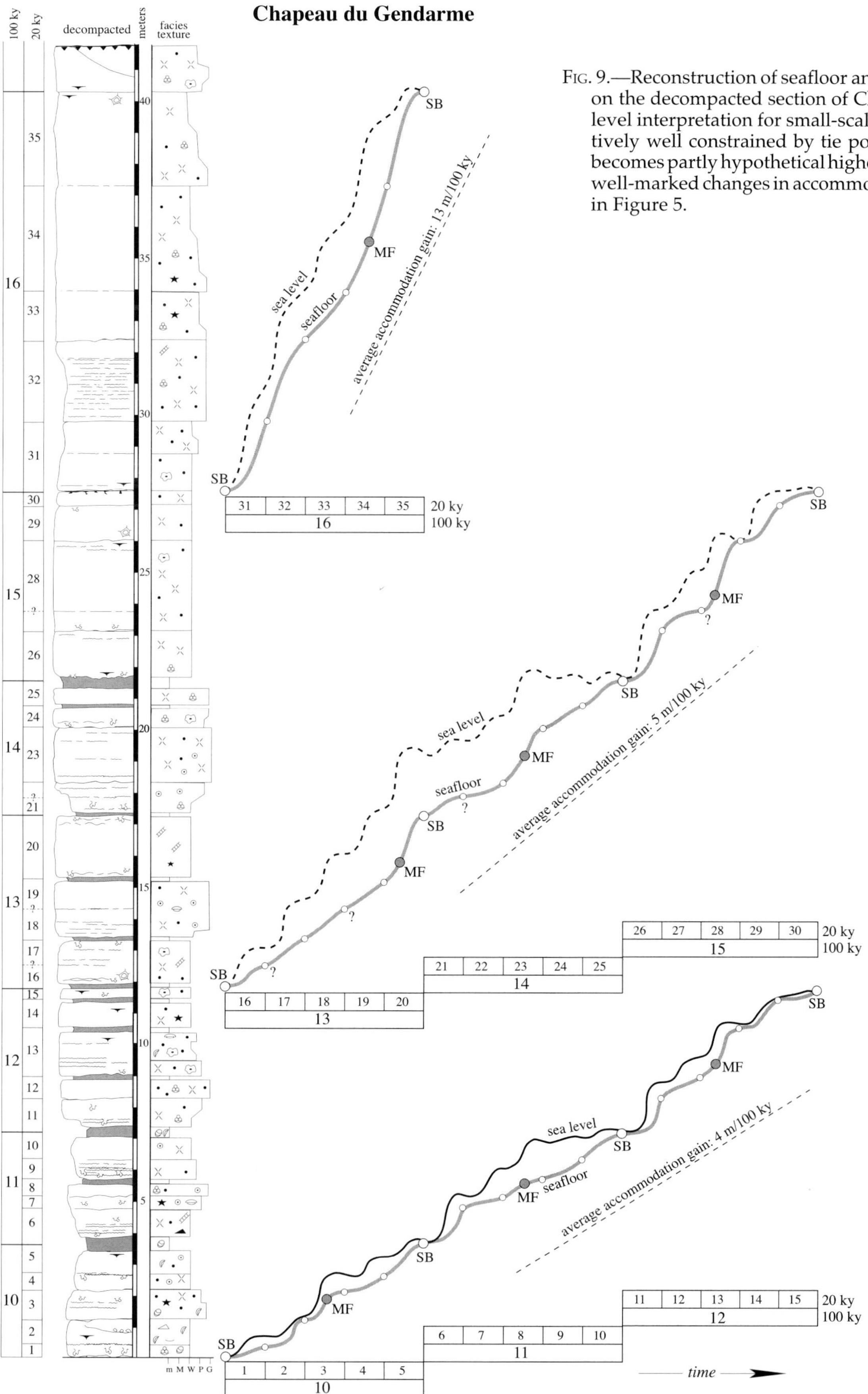

FIG. 9.—Reconstruction of seafloor and sea-level evolution based on the decompacted section of Chapeau du Gendarme. Sea-level interpretation for small-scale sequences 10 to 12 is relatively well constrained by tie points of intertidal facies but becomes partly hypothetical higher up in the section. Note the well-marked changes in accommodation gain. Symbols are as in Figure 5.

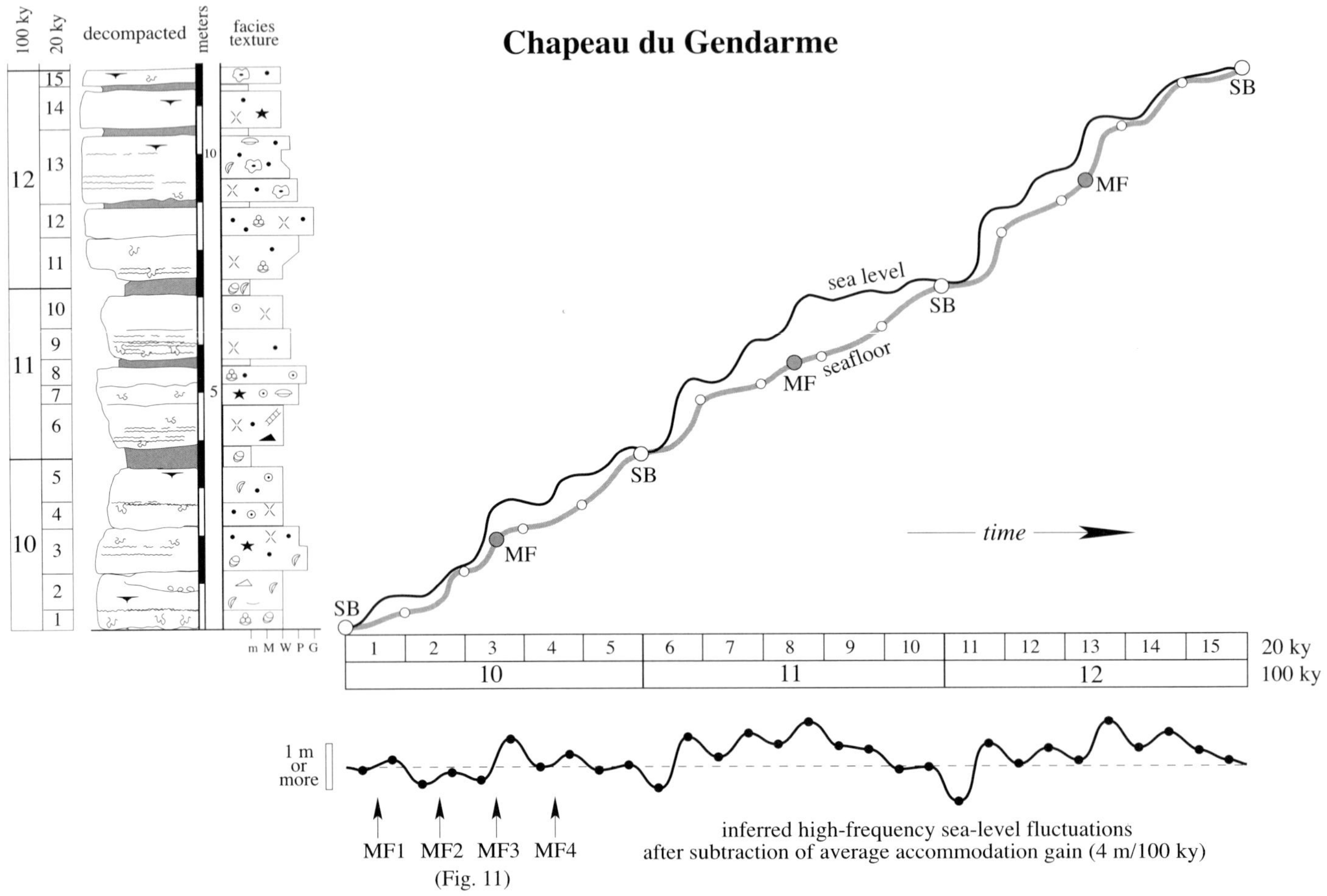

FIG. 10.—Reconstruction of a high-frequency eustatic sea-level curve from the decompacted lower part of the Chapeau-du-Gendarme section. For discussion refer to text. Symbols are as in Figure 5.

With the next transgressive pulse 20 ky later (which has a very low amplitude; Fig. 10), the situation does not change much except for a tidal flat with channels that develops at Chapeau du Gendarme. The following sea-level rise, however, is of higher amplitude and causes marine flooding of the platform as far as Chapeau du Gendarme. Rusel still keeps its lacustrine facies, and ooid shoals develop there only after the fourth transgressive pulse. Although the amplitude of this sea-level fluctuation is low, it is able to pass a threshold and establish marine conditions even in the platform interior.

DISCUSSION

Definition and Correlation of Depositional Sequences

A depositional sequence is defined by its facies evolution, which translates environmental change through time. In the studied sections, deepening–shallowing facies trends imply that accommodation changes are an important factor for sequence development. Supratidal features directly superimposed on subtidal facies cannot be produced by simple progradation of the sedimentary system and indicate that drops in relative sea level must have occurred (Hardie et al., 1986; Strasser, 1991; D'Argenio et al., 1997). The hierarchical stacking of sequences further suggests that these accommodation changes were controlled at least partly by high-frequency sea-level changes reflecting Milankovitch cyclicity.

However, if sea-level amplitudes are not high enough to induce facies changes in subtidal environments, no sequence boundaries or maximum-flooding surfaces will be recognizable. In the studied sections, beds commonly are separated by marly joints. Clays that are washed into the system during a sea-level fall may create such a marly joint and define a sequence boundary, but clays can also be mobilized through climate change, which is not necessarily in tune with sea-level change. On the other hand, clays may be concentrated because of reduced carbonate production below the photic zone or low-energy conditions below wave base. Early diagenesis commonly enhances the marl–limestone contrast through migration of carbonate from marls to limestones (Ricken and Eder, 1991). Some bedding planes created by the marls can be traced from one section to the other, others for only a few meters. Localized marl seams may reflect depressions on the seafloor where clays accumulated preferentially. The number of beds counted in a section therefore does not necessarily correspond to the number of sea-level cycles recorded.

Small-scale sequences were defined in portions of the studied sections where facies changes permit their easy identification (Fig. 4). By piecing together such portions within a rough frame given by biostratigraphy and major lithologic changes, a best-fit

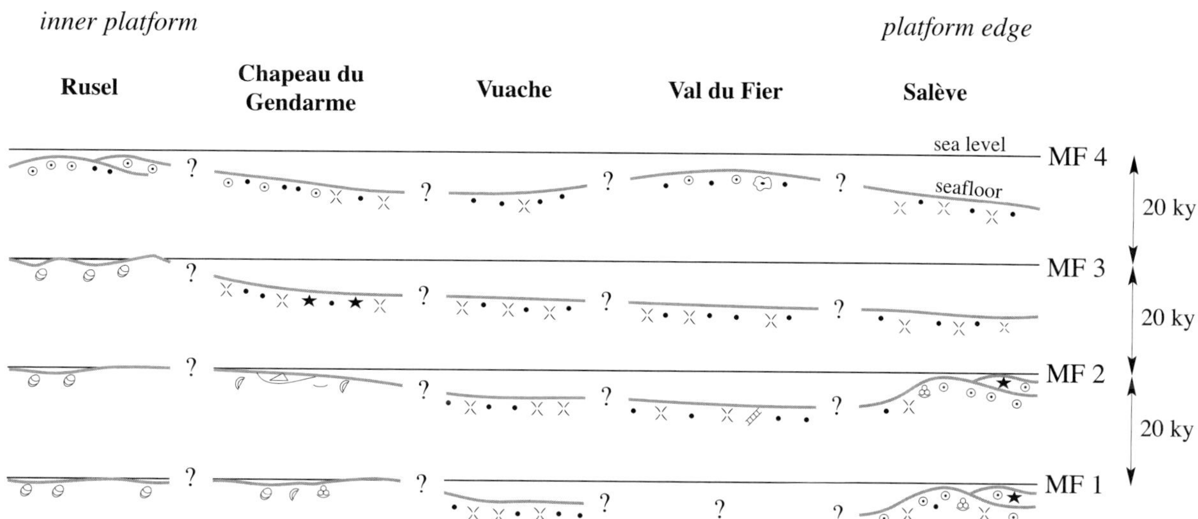

FIG. 11.—Hypothetical evolution of the Jura platform during the transgression that led to the deposition of the lower part of the Pierre-Châtel Formation. MF1 to MF4 correspond to the times of maximum flooding identified on the high-frequency sea-level curve in Figure 10. Symbols as in Figure 5. For discussion see text.

solution such as in Figure 5 can be proposed. Lateral correlation allows sequence boundaries to be inferred also in intervals where they are not clearly developed. However, especially in high-energy deposits, reactivation surfaces may simulate sequence boundaries. There, it may be easier to correlate maximum-flooding intervals (e.g., small-scale sequences 12 and 13 in the Salève section; Fig. 5).

Once the correlation is performed, and despite uncertainties in placing some small-scale sequence boundaries, it becomes clear that major lithological changes are not necessarily isochronous. For example, on the large scale (within one ammonite subzone), a major transgressive surface defines the base of the Pierre-Châtel Formation (Fig. 5). On the scale of elementary sequences, however, a stepwise flooding of the platform can be demonstrated (Figs. 8, 11).

Timing and Estimation of Sedimentation Rates

The timing of the depositional sequences defined in this study follows two lines of reasoning. First, dated ammonite zones and sequence boundaries give a rough estimate of the duration of the studied interval. Dividing this duration by the number of identified elementary and small-scale sequences places these within the Milankovitch frequency band. Second, the average of five elementary sequences building one small-scale sequence reflects the ratio between the precession cycle and the first eccentricity cycle. Medium-scale sequences, composed of four small-scale sequences and corresponding to the 400 ky eccentricity cycle, are poorly defined in the interval presented in Figure 5 but become apparent in the Upper Berriasian units (Strasser and Hillgärtner, 1998; Hillgärtner, 1999).

Uncertainties of course exist because the radiometric dating may include large error margins, and because bed thicknesses and stacking pattern do not always correspond to accommodation changes driven by sea-level fluctuations that are in tune with the Milankovitch cycles. In the platform sections, even if the sequences are differentially decompacted (e.g., Bond et al., 1993), their reconstructed thicknesses still do not reflect accommodation unless the top of the sequence shows intertidal facies. Tops of sequences may also be eroded, and erosion depth is difficult to evaluate. Estimation of water depths of subtidal facies is difficult, and error margins are large. Consequently, spectral analysis of bed thicknesses to test the Milankovitch hypothesis was not performed. In the Montclus section, slumps interrupt the limestone–marl alternations and preclude a spectral analysis based on a reasonably long time series.

It is evident from Figures 6 and 9 that the distribution of time within a sequence is not homogeneous. This is why it is difficult to estimate sedimentation rates. It is pointless to divide thickness by time because each facies may have a different production rate, because sediment may be deposited but then redistributed over the platform (e.g., Pratt and James, 1986) or exported to the basin (e.g., Schlager et al., 1994), and significant amounts of time may be spent without sediment accumulation or even with erosion, especially in the early-transgressive and late-highstand phases of a sea-level cycle (e.g., Goldhammer et al., 1990; Strasser, 1994). Consequently, one should distinguish clearly between sediment production rate, accumulation rate, and final preservation. Ancient accumulation rates are often calculated over long time intervals and indicated in meters per million years (e.g., Wilkinson et al., 1991; Bosscher and Schlager, 1993). Thus, changes in subsidence rate, hiatuses, and short-term variations in sediment production are not adequately taken into account. It is probably more realistic to estimate accumulation rates separately for each elementary or small-scale sequence.

Reconstruction of Eustatic Sea-Level Changes

On the scale of the platform studied, it can be assumed that eustatic sea-level fluctuations were synchronous. Subsidence rates, however, changed not only through time but also from one tectonic compartment to another. Assuming that the decompacted thicknesses and the correlation of the depositional sequences as presented in Figure 8 are correct, the thickness variations between Chapeau du Gendarme, Vuache, and Val du Fier appear to be minor and can be explained by changes in seafloor morphology. When the first three small-scale sequences were deposited, these sections were probably situated on the same tectonic compartment. However, the Rusel section experienced decreased subsidence or even uplift during the deposition of the third small-scale sequence, while the Salève section expresses increased subsidence at the platform margin for the first 250 ky. The sea-level reconstruction in Figure 10 is therefore based on the Chapeau-du-Gendarme section, where the sequences are clearly marked and at least two other sections show a similar thickness evolution. Many more sections in different locations would have to be analyzed to reconstruct a sea-level curve valid for the whole platform. Unfortunately, outcrop conditions in the Swiss and French Jura are such that good sections are rare.

The only tie points for the sea-level curve ("pinning points" of Goldstein and Franseen, 1995) are intertidal or marginal-marine facies, where mean sea level is at zero. If a vadose zone is developed, sea level must have fallen at least to the base of this zone. If only subtidal facies are present at the sequence boundary, water depth at that time has to be estimated from the facies and the paleoecology of the organisms present, which generally includes wide error margins (Steinhauff and Walker, 1995). The same holds true for water depth during maximum flooding. Fischer plots have not been employed in this study because they only trace changes in sequence thickness through time (Sadler et al., 1993) and do not account for accommodation space that is not filled (Osleger, 1991; Osleger and Read, 1991).

Sea-level changes during greenhouse conditions, as in the Early Cretaceous, were of lower amplitude and probably more regular than during glaciations (e.g., Koerschner and Read, 1989). However, orbitally driven changes in insolation translate into sea-level changes through several feedback mechanisms (Gornitz et al., 1982; Schulz and Schäfer-Neth, 1998; Jacobs and Sahagian, 1993). It can be assumed that irregularities caused by complex climate–ocean coupling occurred, even if their amplitudes were certainly less than in glacioeustatically dominated times such as the Pleistocene and Holocene (e.g., Kindler and Hearty, 1996; Peltier, 1988). Such irregularities could have led to facies changes that may simulate a Milankovitch cycle.

Potential of the Cyclostratigrapical Approach

Despite the uncertainties mentioned above, cyclostratigraphic analysis of carbonate platforms can greatly improve the understanding of their functioning. It is seen that the general trends of long-term (million-year scale) tectonic, climatic, and eustatic sea-level changes do not directly control the sedimentation but rather give the baseline for high-frequency environmental changes which are, directly or indirectly, linked to the orbital insolation changes. If Milankovitch cyclicity can be demonstrated, the duration of the observed depositional sequences is known. On the basis of detailed facies analysis, the rates of sedimentary, ecological, and diagenetic processes can then be estimated. Because the time intervals within which these processes are studied are relatively short (20 to 100 ky), a comparison with Pleistocene and Holocene rates can be made, where relevant data are much more abundant (e.g., Enos, 1991). Of course, it has to be considered that amplitudes of climate change and sea-level change differed greatly between greenhouse and icehouse worlds.

For example, during the first 300 ky of the transgression studied on the Jura platform, less than 2 m of sediment were preserved per 20 ky. In contrast, Holocene production rates of comparable lagoonal–peritidal facies are considerably higher (potentially 4 to 20 m in 20 ky; Enos, 1991). It is therefore suggested that sediment production is active during only part of a 20 ky cycle. A sea-level drop may cause erosion and subaerial exposure, and sediment production may need some time to start up when sea level rises again. Changes in climate periodically increase runoff of clays and nutrients from the hinterland, which hampers carbonate production (Hallock and Schlager, 1986). Also, part of the produced sediment is redistributed on the platform by currents and storms and exported towards the basin (e.g., Milliman et al., 1993).

The cyclostratigraphic approach further allows estimation of the time represented in stratigraphic gaps. In the Salève section, the 50 cm interval between sequence boundary Be4 and the transgressive surface at the base of the Pierre-Châtel Formation probably represents about 900 ky (Fig. 5). The karst surface underlining SB Be5 represents at least 300 ky in the section of Rusel, and at least 150 ky at Chapeau du Gendarme (Fig. 5).

The cyclostratigraphic analysis thus refines sequence-stratigraphic correlations. This has been demonstrated for sections on the platform (e.g., Goldhammer et al., 1994; D'Argenio et al., 1997; D'Argenio et al., 1999; Chen et al., 2001; Raspini, 2001) and for correlations from platform to ramp and basin (e.g., Elrick and Read, 1991; Pasquier and Strasser, 1997; Osleger, 1998; Pittet and Strasser, 1998; Betzler et al., 1999). The evolution of the sedimentary systems can be monitored in time steps of 20 or 100 ky, and this independently of the error margins of absolute dating.

CONCLUSIONS

Analysis of facies evolution and stacking pattern of depositional sequences in the Berriasian of the Swiss and French Jura Mountains suggests that high-frequency eustatic sea-level changes were an important factor controlling sedimentation on the shallow carbonate platform. Elementary and small-scale sequences are recognized. The biostratigraphic, sequence-stratigraphic, and cyclostratigraphic correlation with a well-dated deeper-water section implies that these sequences formed in tune with the 20 ky and 100 ky orbital cycles.

Three small-scale sequences constituting the early transgressive interval of the Pierre-Châtel Formation were analyzed in detail. Facies-dependent decompaction and correlation of the elementary sequences indicates that differential subsidence led to episodic accommodation changes. It also shows that the transgression flooded the platform in pulses: every 20 ky, marine conditions reached farther into the platform interior.

The reconstruction of a high-frequency sea-level curve cannot be more than a rough approximation but nevertheless opens the way to discuss rates of sediment production and accumulation on a scale that is comparable to that of the Pleistocene and the Holocene. This high time resolution also offers a potential to estimate rates of paleoecological and diagenetic processes.

However, considering the complexity of any sedimentary system, considering the multiple mechanisms that translate orbitally controlled insolation changes into sea-level fluctuations as well as into climatic and ecological changes at a given paleogeographical position, and considering the incompleteness of the sedimentary record, great care has to be taken in the interpretations. It is impossible to generalize from one study in one time interval and one paleogeographical and geodynamic setting. Cyclostratigraphy is a valuable tool that helps to establish a narrow time frame, but estimation of time involved in sedimentary processes also calls for detailed bed-by-bed analyses in order to understand the functioning of the corresponding depositional environments.

ACKNOWLEDGMENTS

Many of the ideas presented here have been elaborated together with Bernard Pittet, Christophe Dupraz, Wolfgang Hug, and Alex Waehry. We thank them for the animated discussions. We also thank Elias Samankassou for critically reading a first version of the manuscript. The constructive remarks of Hanspeter

Funk and an anonymous reviewer were greatly appreciated. The financial support of this research by the Swiss National Science Foundation (project No. 20-56491.99) is gratefully acknowledged.

REFERENCES

BATHURST, R.G.C., 1987, Diagenetically enhanced bedding in argillaceous platform limestones: stratified cementation and selective compaction: Sedimentology, v. 34, p. 749–778.

BERGER, A., LOUTRE, M.F., AND DEHANT, V., 1989, Astronomical frequencies for pre-Quaternary palaeoclimate studies: Terra Nova, v. 1, p. 474–479.

BERGGREN, W.A., KENT, D.V., AUBRY, M.P., AND HARDENBOL, J., EDS., 1995, Geochronology, Time Scales and Global Stratigraphic Correlation: SEPM, Special Publication 54, 386 p.

BETZLER, C., REIJMER, J.J.G., BERNET, K., EBERLI, G.P., AND ANSELMETTI, F.S., 1999, Sedimentary patterns and geometries of the Bahamian outer carbonate ramp (Miocene–Lower Pliocene, Great Bahama Bank): Sedimentology, v. 46, p. 1127–1143.

BLONDEL, T., 1984, Etude géologique de la partie septentrionale de la Montagne du Vuache (Haute Savoie—France): Unpublished Diploma Thesis, University of Geneva, 310 p.

BOND, G.C., DEVLIN, W.J., KOMINZ, M.A., BEAVAN, J., AND MCMANUS, J., 1993, Evidence of astronomical forcing of the Earth's climate in Cretaceous and Cambrian times: Tectonophysics, v. 222, p. 295–315.

BOSSCHER, H., AND SCHLAGER, W., 1993, Accumulation rates of carbonate platforms: Journal of Geology, v. 101, p. 345–355.

CHEN, D., TUCKER, M.E, JIANG, M., AND ZHU, J., 2001, Long-distance correlation between tectonic-controlled, isolated carbonate platforms by cyclostratigraphy and sequence stratigraphy in the Devonian of South China: Sedimentology, v. 48, p. 57–78.

CLAVEL, B., CHAROLLAIS, J., BUSNARDO, R., AND LE HEGARAT, G., 1986, Précisions stratigraphiques sur le Crétacé inférieur basal du Jura méridional: Eclogae Geologicae Helvetiae, v. 79, p. 319–341.

COTILLON, P., FERRY, S., GAILLARD, C., JAUTÉE, E., LATREILLE, G., AND RIO, M., 1980, Fluctuation des paramètres du milieu marin dans le domaine vocontien (France Sud-Est) au Crétacé inférieur: mise en évidence par l'étude des formations marno-calcaires alternantes: Société Géologique de France, Bulletin, v. 7/22, p. 735–744.

DARSAC, C., 1983, La plate-forme berriaso-valanginienne du Jura méridional aux massifs subalpins (Ain, Savoie): Ph.D. Dissertation, University of Grenoble, 319 p.

D'ARGENIO, B., FERRERI, V., AMODIO, S., AND PELOSI, N., 1997, Hierarchy of high-frequency orbital cycles in Cretaceous carbonate platform strata: Sedimentary Geology, v. 113, p. 169–193.

D'ARGENIO, B., FERRERI, V., RASPINI, A., AMODIO, S., AND BUONOCUNTO, F.P., 1999, Cyclostratigraphy of a carbonate platform as a tool for high-precision correlation: Tectonophysics, v. 315, p. 357–385.

DE GRACIANSKY, P.-C., AND LEMOINE, M., 1988, Early Cretaceous extensional tectonics in the southwestern French Alps: a consequence of North-Atlantic rifting during Tethyan spreading: Société Géologique de France, Bulletin, v. 8/4, p. 733–737.

DERCOURT, J., GAETANI, M., VRIELYNCK, B., BARRIER, E., BIJU-DUVAL, B., BRUNET, M.F., CADET, J.P., CRASQUIN, S., AND SANDULESCU, M., EDS., 2000, Atlas Peri-Tethys—Palaeogeographical Maps: Paris.

DÉTRAZ, H., AND MOJON, P.-O., 1989, Evolution paléogéographique de la marge jurassienne de la Téthys du Tithonique–Portlandien au Valanginien: corrélations biostratigraphique et séquentielle des faciès marins à continentaux: Eclogae Geologicae Helvetiae, v. 82, p. 37–112.

EINSELE, G., AND RICKEN, W., 1991, Limestone–marl alternation—an overview, *in* Einsele, G., Ricken, W., and Seilacher, A., eds., Cycles and Events in Stratigraphy: Berlin, Springer-Verlag, p. 23–47.

ELRICK, M., AND READ, J.F., 1991, Cyclic ramp-to-basin carbonate deposits, Lower Mississippian, Wyoming and Montana: a combined field and computer study: Journal of Sedimentary Petrology, v. 61, p. 1194–1224.

ENOS, PAUL, 1991, Sedimentary parameters for computer modeling, *in* Franseen, E.K., Watney, W.L., Kendall, C.G.St.C., and Ross, W., eds., Sedimentary Modeling: Computer Simulations and Methods for Improved Parameter Definition: Kansas Geological Survey, Memoir 233, p. 63–99.

EYLES, N., 1993, Earth's glacial record and its tectonic setting: Earth-Science Reviews, v. 35, p. 1–248.

FAIRBRIDGE, R.W., 1976, Convergence of evidence on climatic change and ice ages: New York Academy of Science, Annals, v. 91, p. 542–579.

FRAKES, L.A., FRANCIS, J.E., AND SYKTUS, J.I., 1992, Climate Modes of the Phanerozoic: Cambridge, U.K., Cambridge University Press, 274 p.

GOLDHAMMER, R.K., 1997, Compaction and decompaction algorithms for sedimentary carbonates: Journal of Sedimentary Research, v. 67, p. 26–35.

GOLDHAMMER, R.K., DUNN, P.A., AND HARDIE, L.A., 1990, Depositional cycles, composite sea-level changes, cycle stacking patterns, and the hierarchy of stratigraphic forcing: Examples from Alpine Triassic platform carbonates: Geological Society of America, Bulletin, v. 102, p. 535–562.

GOLDHAMMER, R.K., OSWALD, E.J., AND DUNN, P.A., 1994, High-frequency, glacio-eustatic cyclicity in the Middle Pennsylvanian of the Paradox Basin: an evaluation of Milankovitch forcing, *in* de Boer, P.L., and Smith, D.G., eds., Orbital Forcing and Cyclic Sequences: International Association of Sedimentologists, Special Publication 19, p. 243–283.

GOLDSTEIN, R.H., AND FRANSEEN, E.K., 1995, Pinning points: a method providing quantitative constraints on relative sea-level history: Sedimentary Geology, v. 95, p. 1–10.

GORNITZ, V., LEBEDEFF, S., AND HANSEN, J., 1982, Global sea-level trend in the past century: Science, v. 215, p. 1611–1614.

GRADSTEIN, F.M., AGTERBERG, F.P., OGG, J.G., HARDENBOL, J., VAN VEEN, P., THIERRY, J., AND HUANG, Z., 1995, A Triassic, Jurassic and Cretaceous time scale, *in* Berggren, W.A., Kent, D.V., Aubry, M.P., and Hardenbol, J., eds., Geochronology, Time Scales and Global Stratigraphic Correlation: SEPM, Special Publication 54, p. 95–126.

HÄFELI, C., 1966, Die Jura/Kreide-Grenzschichten im Bielerseegebiet (Kt. Bern): Eclogae Geologicae Helvetiae, v. 59, p. 565–696.

HALLEY, R.B., AND HARRIS, P.M., 1979, Fresh-water cementation of a 1,000-year-old oolite: Journal of Sedimentary Petrology, v. 49, p. 969–988.

HALLOCK, P., AND SCHLAGER, W., 1986, Nutrient excess and the demise of coral reefs and carbonate platforms: Palaios, v. 1, p. 389–398.

HARDENBOL, J., THIERRY, J., FARLEY, M.B., JACQUIN, T., DE GRACIANSKY, P.-C., AND VAIL, P.R., 1998, Charts, *in* de Graciansky, P.-C., Hardenbol, J., Jacquin, T., and Vail, P.R., eds., Mesozoic and Cenozoic Sequence Stratigraphy of European Basins: SEPM, Special Publication 60.

HARDIE, L.A., BOSELLINI, A., AND GOLDHAMMER, R.K., 1986, Repeated subaerial exposure of subtidal carbonate platforms, Triassic, Northern Italy: evidence for high frequency sea level oscillations on a 10^4 year scale: Paleoceanography, v. 1, p. 447–457.

HILLGÄRTNER, H., 1999, The evolution of the French Jura platform during the Late Berriasian to Early Valanginian: controlling factors and timing: GeoFocus, v. 1, 203 p.

JACOBS, D.K., AND SAHAGIAN, D.L., 1993, Climate-induced fluctuations in sea level during non-glacial times: Nature, v. 361, p. 710–712.

JACQUIN, T., AND DE GRACIANSKY, P.-C., 1998, Transgressive/regressive (second order) facies cycles: the effects of tectono-eustasy, *in* de Graciansky, P.-C., Hardenbol, J., Jacquin, T., and Vail, P.R., eds., Mesozoic and Cenozoic Sequence Stratigraphy of European Basins: SEPM, Special Publication 60, p. 31–42.

KINDLER, P., AND HEARTY, P.J., 1996, Carbonate petrography as an indicator of climate and sea-level changes: new data from Bahamian Quaternary units: Sedimentology, v. 43, p. 381–399.

KOERSCHNER, W.F., III, AND READ, J.F., 1989, Field and modelling studies of Cambrian carbonate cycles, Virginia Appalachians: Journal of Sedimentary Petrology, v. 59, p. 654–687.

LE HEGARAT, G., 1971, Le Berriasien du Sud-Est de la France: Université de Lyon, Faculté des Sciences, Laboratoire Géologique, Documents, v. 43, 576 p.

LOURENS, L.J., ANTONARAKOU, A., HILGEN, F.J., VAN HOOF, A.A.M., VERGNAUD-GRAZZINI, C., AND ZACHARIASSE, W.J., 1996, Evaluation of the Plio-Pleistocene astronomical timescale: Paleoceanography, v. 11, p. 391–413.

MEYER, M., 2000, Le complexe récifal kimméridgien–tithonien du Jura méridional interne (France), évolution multifactorielle, stratigraphie et tectonique: Terre & Environnement, v. 24, 179 p.

MILLIMAN, J.D., FREILE, D., STEINEN, R.P., AND WILBER, R.J., 1993, Great Bahama Bank aragonite muds: mostly inorganically precipitated, mostly exported: Journal of Sedimentary Petrology, v. 63, p. 589–595.

MONTAÑEZ, I.P., AND OSLEGER, D.A., 1993, Parasequence stacking patterns, third-order accommodation events, and sequence stratigraphy of Middle to Upper Cambrian platform carbonates, Bonanza King Formation, southern Great Basin, *in* Loucks R.G., and Sarg, J.F., eds., Carbonate Sequence Stratigraphy: American Association of Petroleum Geologists, Memoir 57, p. 305–326.

MOORE, C.H., 1989, Carbonate Diagenesis and Porosity: Amsterdam, Elsevier, Developments in Sedimentology, v. 46, 338 p.

OSLEGER, D.A., 1991, Subtidal carbonate cycles: Implications for allocyclic vs. autocyclic controls: Geology, v. 19, p. 917–920.

OSLEGER, D.A., 1998, Sequence architecture and sea-level dynamics of Upper Permian shelfal facies, Guadalupe Mountains, southern New Mexico: Journal of Sedimentary Research, v. 68, p. 327–346.

OSLEGER, D., AND READ, J.F., 1991, Relation of eustasy to stacking pattern of meter-scale carbonate cycles, Late Cambrian, U.S.A.; Journal of Sedimentary Petrology, v. 61, p. 1225–1252.

PASQUIER, J.-B., 1995, Sédimentologie, stratigraphie séquentielle et cyclostratigraphie de la marge nord-thétysienne au Berriasien en Suisse occidentale (Jura, Helvétique et Ultrahelvétique; comparaison avec les séries de bassin des domaines vocontien et subbriançonnais): Ph.D. Dissertation, University of Fribourg, 274 p.

PASQUIER, J.-B., AND STRASSER, A., 1997, Platform-to-basin correlation by high-resolution sequence stratigraphy and cyclostratigraphy (Berriasian, Switzerland and France): Sedimentology, v. 44, p. 1071–1092.

PELTIER, W.R., 1988, Lithospheric thickness, Antarctic deglaciation history, and ocean basin discretization effects in a global model of postglacial sea-level change: a summary of some sources of nonuniqueness: Quaternary Research, v. 29, p. 93–112.

PITTET, B., AND STRASSER, A., 1998, Long-distance correlations by sequence stratigraphy and cyclostratigraphy: examples and implications (Oxfordian from the Swiss Jura, Spain, and Normandy): Geologische Rundschau, v. 86, p. 852–874.

PITTET, B., STRASSER, A., AND MATTIOLI, E., 2000, Depositional sequences in deep-shelf environments: a response to sea-level changes and shallow-platform carbonate productivity (Oxfordian, Germany and Spain): Journal of Sedimentary Research, v. 70, p. 392–407.

PRATT, B.R., AND JAMES, N.P., 1986, The St George Group (Lower Ordovician) of western Newfoundland: tidal flat island model for carbonate sedimentation in shallow epeiric seas: Sedimentology, v. 33, p. 313–343.

PRICE, G.D., 1999, The evidence and implications of polar ice during the Mesozoic: Earth-Science Reviews, v. 48, p. 183–210.

RASPINI, A., 2001, Stacking pattern of cyclic carbonate platform strata: Lower Cretaceous of southern Apennines, Italy: Geological Society of London, Journal, v. 158, p. 353–366.

RICKEN, W., AND EDER, W., 1991, Diagenetic modification of calcareous beds—an overview, *in* Einsele, G., Ricken, W., and Seilacher, A., eds., Cycles and Events in Stratigraphy: Berlin, Springer-Verlag, p. 430–449.

SADLER, P.M., OSLEGER, D.A., AND MONTAÑEZ, I.P., 1993, On the labeling, length, and objective basis of Fischer plots: Journal of Sedimentary Petrology, v. 63, p. 360–368.

SCHLAGER, W., REIJMER, J.J.G., AND DROXLER, A., 1994, Highstand shedding of carbonate platforms: Journal of Sedimentary Research, v. B64, p. 270–281.

SCHULZ, M., AND SCHÄFER-NETH, C., 1998, Translating Milankovitch climate forcing into eustatic fluctuations via thermal deep water expansion: a conceptual link: Terra Nova, v. 9, p. 228–231.

SCHWARZACHER, W., 1993, Cyclostratigraphy and the Milankovitch Theory: Amsterdam, Elsevier, Developments in Sedimentology, v. 52, 225 p.

SHINN, E.A., AND ROBBIN, D.M., 1983, Mechanical and chemical compaction in fine-grained shallow-water limestones: Journal of Sedimentary Petrology, v. 53, p. 595–618.

SINCLAIR, I.K., SHANNON, P.M., WILLIAMS, B.P.J., HARKER, S.D., AND MOORE, J.G., 1994, Tectonic control on sedimentary evolution of three North Atlantic borderland Mesozoic basins: Basin Research, v. 6, p. 193–217.

STEINHAUFF, D.M., AND WALKER, K.R., 1995, Recognizing exposure, drowning, and "missed beats": platform-interior to platform-margin sequence stratigraphy of Middle Ordovician limestones, east Tennessee: Journal of Sedimentary Research, v. B65, p. 183–207.

STEINHAUSER, N., AND LOMBARD, A., 1969, Définition de nouvelles unités lithostratigraphiques dans le Crétacé inférieur du Jura méridional (France): Société Physique et Histoire Naturelle, Genève, Comptes Rendus, v. 4, p. 100–113.

STRASSER, A., 1988, Shallowing-upward sequences in Purbeckian peritidal carbonates (lowermost Cretaceous, Swiss and French Jura Mountains): Sedimentology, v. 35, p. 369–383.

STRASSER, A., 1991, Lagoonal–peritidal sequences in carbonate environments: autocyclic and allocyclic processes, *in* Einsele, G., Ricken W., and Seilacher, A., eds., Cycles and Events in Stratigraphy: Berlin, Springer-Verlag, p. 709–721.

STRASSER, A., 1994, Milankovitch cyclicity and high-resolution sequence stratigraphy in lagoonal–peritidal carbonates (Upper Tithonian–Lower Berriasian, French Jura Mountains), *in* de Boer, P.L., and Smith, D.G., eds., Orbital Forcing and Cyclic Sequences: International Association of Sedimentologists, Special Publication 19, p. 285–301.

STRASSER, A., AND DAVAUD, E., 1983, Black pebbles of the Purbeckian (Swiss and French Jura): lithology, geochemistry and origin: Eclogae Geologicae Helvetiae, v. 76, p. 551–580.

STRASSER, A., AND HILLGÄRTNER, H., 1998, High-frequency sea-level fluctuations recorded on a shallow carbonate platform (Berriasian and Lower Valanginian of Mount Salève, French Jura): Eclogae Geologicae Helvetiae, v. 91, p. 375–390.

STRASSER, A., PITTET, B., HILLGÄRTNER, H., AND PASQUIER, J.-B., 1999, Depositional sequences in shallow carbonate-dominated sedimentary systems: concepts for a high-resolution analysis: Sedimentary Geology, v. 128, p. 201–221.

STRASSER, A., HILLGÄRTNER, H., HUG, W., AND PITTET, B., 2000, Third-order depositional sequences reflecting Milankovitch cyclicity: Terra Nova, v. 12, p. 303–311.

STROHMENGER, C., AND STRASSER, A., 1993, Eustatic controls on the depositional evolution of Upper Tithonian and Berriasian deep-water carbonates (Vocontian Trough, SE France): Centres Recherche Exploration-Production Elf Aquitaine, Bulletin, v. 17, p. 183–203.

TRÜMPY, R., 1980, Geology of Switzerland, a Guide-Book. Part A: An Outline of the Geology of Switzerland: Basel, Wepf & Co. Publishers, 104 p.

VAIL, P.R., AUDEMARD, F., BOWMAN, S.A., EISNER, P.N., AND PEREZ-CRUZ, C., 1991, The stratigraphic signatures of tectonics, eustasy and sedimentology—an overview, *in* Einsele, G., Ricken W., and Seilacher, A., eds., Cycles and Events in Stratigraphy: Berlin, Springer-Verlag, p. 617–659.

WAEHRY, A., 1989, Faciès et séquences de dépôt dans la Formation de Pierre-Châtel (Berriasien moyen, Jura méridional / France): Unpublished Diploma Thesis, University of Geneva, 83 p.

WILDI, W., FUNK, H., LOUP, B., AMATO, E., AND HUGGENBERGER, P., 1989, Mesozoic subsidence history of the European marginal shelves of the Alpine Tethys (Helvetic realm, Swiss Plateau and Jura): Eclogae Geologicae Helvetiae, v. 82, p. 817–840.

WILKINSON, B.H., OPDYKE, B.N., AND ALGEO, T.J., 1991, Time partitioning in cratonic carbonate rocks: Geology, v. 19, p. 1093–1096.

ZIEGLER, P.A., 1988, Evolution of the Arctic–North Atlantic and the Western Tethys: American Association of Petroleum Geologists, Memoir 43, 198 p.

ORBITAL CHRONOSTRATIGRAPHY OF THE VALANGINIAN–HAUTERIVIAN BOUNDARY: A CYCLOSTRATIGRAPHIC APPROACH

VITTORIA FERRERI
Dipartimento di Scienze della Terra, Università "Federico II", Largo San Marcellino 10, 80138 Napoli, Italy
e-mail: ciclisti@gms01.geomare.na.cnr.it
SABRINA AMODIO
Istituto per l'Ambiente Marino Costiero, Geomare, National Research Council,
Calata Porta di Massa, Porto di Napoli, 80133 Napoli, Italy
e-mail: amodio@gms01.geomare.na.cnr.it
AND
Dipartimento di Scienze della Terra, Università "Federico II", Largo San Marcellino 10, 80138 Napoli, Italy
ROSARIA SANDULLI
Dipartimento di Scienze della Terra, Università "Federico II", Largo San Marcellino 10, 80138 Napoli, Italy
AND
BRUNO D'ARGENIO
Istituto per l'Ambiente Marino Costiero, Geomare, National Research Council,
Calata Porta di Massa, Porto di Napoli, 80133 Napoli, Italy
e-mail: dargenio@gms01.geomare.na.cnr.it
AND
Dipartimento di Scienze della Terra, Università "Federico II", Largo San Marcellino 10, 80138 Napoli, Italy

Abstract: On the basis of cyclostratigraphy and sequence stratigraphy criteria, an orbital chronostratigraphy is here proposed for the Upper Valanginian–Lower Hauterivian (Lower Cretaceous) stratigraphic interval. For such a purpose two biostratigraphically constrained carbonate platform sections were measured and sampled at centimeter scale in southern Italy: (1) San Lorenzello, Matese Mountains in the Campania Apennines, and (2) Sferracavallo, Palermo Mountains in northwestern Sicily. The former has a thickness of about 87 m, and the latter totals about 40 m. Analysis of depositional and early diagenetic features has shown that both the successions are characterized by internal cyclicity, consistent with orbital (Milankovitch) forcing. This high-frequency cyclicity, in which the elementary cycles are organized in bundles and in groups of bundles (superbundles), appears to be superimposed on three longer (from ~ 800 ky to ~ 1200 ky) transgressive–regressive facies trends (T/RFTs). Whereas the elementary cycles record the precession and/or the obliquity signal (or a combination of them), the bundles and the superbundles correspond to the short- and long-eccentricity signals (~ 100 ky and 400 ky cycles), respectively. Sequence stratigraphic criteria, used to interpret the superbundles and the T/RFTs in terms of depositional-sequence equivalents, allowed us to propose chronostratigraphic diagrams of the studied intervals and to attempt regional- to global-scale physical correlations.

At regional scale, the Sferracavallo section was correlated, bundle by bundle, with the time-correspondent segment of the San Lorenzello section. The chronostratigraphic correlation shows that only one main gap, calculated as about 200 ky, occurs in the studied intervals. This allows us to assume that the sedimentary record can be considered *quasi*-continuous, at least at the superbundle scale (~ 400 ky). The assembled orbital chronostratigraphy suggests that the time recorded in the San Lorenzello and in the Sferracavallo sections is 2.9 and 1.9 My, respectively. At global scale, a physical correlation is proposed with the time-equivalent third-order sequences of Haq et al. (1987) and Jacquin et al. (1998), taking into account the stratigraphic position of the Valanginian–Hauterivian (Early Cretaceous) boundary in the standard time scales, as well as in our orbital chronostratigraphy. This correlation shows that there is a good agreement between the 2.6 My time duration of the stratigraphic interval spanning from the Valanginian 4 to the Hauterivian 1 sequence boundaries (Jacquin et al., 1998) and the 2.8 My interval estimated on the basis of our orbital chronostratigraphy.

INTRODUCTION

Tropical carbonate platforms, whose predominantly organogenic sequences are widespread in the Mesozoic–Paleogene strata of the Periadriatic region, developed for more than 200 My along the southern side of the Tethys Ocean. They form large (up to 10^4 km^2) and thick (up to > 3–4 x 10^3 m) sedimentary bodies, which are well stratified and characterized by high rates of sedimentation (from < 30 up to > 100 m/My). These carbonate platforms contain minor or no terrigenous admixtures and were highly sensitive to both short-term and long-term climatic and oceanographic variations, as well as to high-frequency sea-level oscillation.

In the last few years we measured and sedimentologically studied, at centimeter scale, a total of about 1500 m of Cretaceous carbonate platform deposits cropping out in southern Italy (Apennines and Sicily) and Montenegro (Dinarides). These studies were carried out on particularly well exposed sections and on long commercial cores (Buonocunto et al., 1994; D'Argenio et al., 1997; D'Argenio et al., 1999; Raspini, 1998, 2001; Sandulli, 1999; Amodio, 2000). For each studied succession the microstratigraphic (centimeter-scale) analysis allowed us to recognize a number of lithofacies, organized in lithofacies associations and groups of lithofacies associations (clans), in turn suggesting carbonate-platform environments ranging from peritidal–supratidal to la-

Cyclostratigraphy: Approaches and Case Histories
SEPM Special Publication No. 81, Copyright © 2004
SEPM (Society for Sedimentary Geology), ISBN 1-56576-108-1, p. 153–166.

goonal and shallow ramp (for more detailed information see, e.g., D'Argenio et al., 1997; Buonocunto et al., 1999).

Stacking pattern of lithofacies and their grouping, including their early diagenetic features (from marine to meteoric), revealed a hierarchical organization composed of elementary cycles (average thickness about 1 m) grouped in bundles (average thickness 2–5 m) and in superbundles (average thickness 10–15 m). The upper boundaries of the superbundles are normally marked by deeply penetrative vadose caps suggesting discontinuity surfaces due to subaerial exposure (cf. Hillgärtner, 1998). It has been shown that the elementary cycles can be related to the precession or the obliquity signals (or a combination of them), whilst the bundles and superbundles appear to record the periodicities of the short- and long-eccentricity signals, respectively (D'Argenio et al., 1997; D'Argenio et al., 1999; Raspini, 1998, 2001; Buonocunto et al., 1999). The orbital nature of these cycles, which are superimposed on lower-frequency transgressive–regressive facies trends (T/RFTs, as defined in D'Argenio et al., 1999), was confirmed by mathematical treatment of the sedimentary data, including neural network and image analysis (detailed information on this kind of numerical approach can be found in Longo et al., 1994; Brescia et al., 1996; D'Argenio et al., 1997; Tagliaferri et al., 2001; see also Scargle, 1982, and Cottle, 1989).

On the basis of the hierarchical cyclic organization, carbonate-platform successions analyzed in the Cretaceous of the central and southern Apennines were interpreted in terms of sequence stratigraphy, in order to propose an orbital chronostratigraphy of the studied time intervals as well as to attempt high-resolution long-distance correlations (D'Argenio et al., 1999).

In this paper we present a case of such a high-resolution regional- to global-scale chronostratigraphic correlation carried out in the Lower Cretaceous (Valanginian–Hauterivian) of southern Italy. Here two carbonate-platform successions, well exposed at San Lorenzello (Matese Mountains, Campania Apennines) and at Sferracavallo (Palermo Mountains, northwestern Sicily), were analyzed (Fig. 1). Some of the results of this study are derived from the Ph.D. dissertations of Sandulli (1999) and Amodio (2000) and were presented briefly in Ferreri et al. (2001) and in Amodio et al. (2001). In addition, a cyclostratigraphic study of the upper half of San Lorenzello section was discussed in D'Argenio et al. (1997).

The San Lorenzello and the Sferracavallo successions belong to two different paleogeographic domains of the Periadriatic region: the Abruzzi–Campania Carbonate Platform and the Panormide Carbonate Platform, respectively (e.g., D'Argenio et al., 1975; Catalano and D'Argenio, 1978). During Early Cretaceous times these carbonate platforms were located in the tropical zone, isolated from the cratonic mainlands by basinal areas, and this precluded terrigenous influx. According to some paleogeographic restorations both the Panormide and the Abruzzi–Campania domains developed on the margins of Adria, a northern promontory of the African Plate (Channell et al., 1979). Other authors suggest that the Panormide Platform belongs to the true African margin and the Abruzzi–Campania Platform is part of a microplate that by the Cretaceous was almost totally separated from the mainland of Africa by an oceanic corridor (e.g., Catalano et al., 1996). In the latter restoration, the original distance between the two studied sections would have been much larger than the present one.

We analyzed, at a centimeter scale, the vertical evolution of depositional (lithofacies and their grouping) and early diagenetic features in order (1) to interpret the above successions in terms of cyclostratigraphy and sequence stratigraphy, (2) to assemble orbital chronostratigraphy of the studied stratigraphic intervals, and (3) to attempt a high-resolution regional-to global-scale correlation.

CYCLOSTRATIGRAPHY AND SEQUENCE STRATIGRAPHY

The main results of a centimeter-scale analysis carried out on the San Lorenzello (D'Argenio et al., 1997; Amodio, 2000) and Sferracavallo (Sandulli, 1999) sequences are summarized below.

- **Biostratigraphy:** According to Luperto-Sinni and Masse (1993), Chiocchini et al. (1994), Schindler and Conrad (1994), and Bucur et al. (1995) the age of the studied successions ranges from the Valanginian to the earliest Hauterivian; this is suggested by the recurrence, in both of the studied sections, of the benthonic foraminifers *Vercorsella camposaurii* Sartoni and Crescenti, *Vercorsella laurentii* Sartoni and Crescenti, *Vercorsella scarsellai* (De Castro), *Pseudotextulariella salevensis* Charollais, Zaninetti and Brönnimann, *Praecrysalidina infracretacea* Luperto-Sinni, as well as green algae, among which are *Actinoporella podolica* ALTH, *Salpingoporella annulata* Carozzi, and *Clypeina? solkani* Conrad and Radoičić (D'Argenio et al., 1997, Sandulli, 1999; Amodio, 2000). It is worth noting that in the Tethyan biochronostratigraphy (Hardenbol et al., 1998,

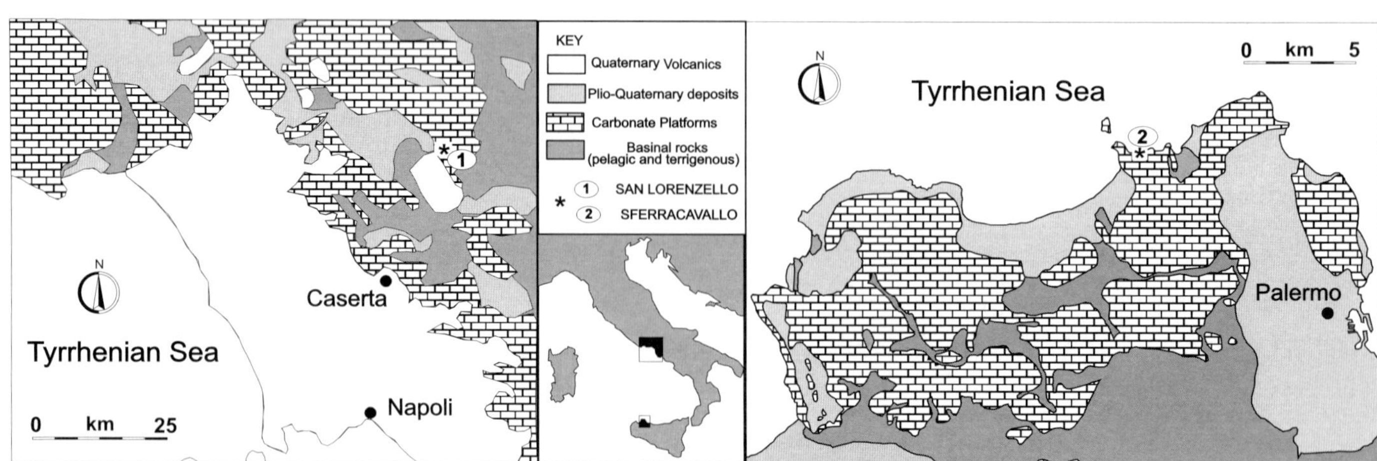

Fig. 1.—Simplified geologic maps of the studied areas showing the location of the San Lorenzello and Sferracavallo successions.

Chart 5) the last occurrence of the above mentioned *Pseudotextulariella salevensis* is reported at 132.03 Ma in the Gradstein et al. (1995) time scale and is located at the Valanginian–Hauterivian boundary. According to Hardenbol et al. (1998) and on the basis of the disappearance of *Pseudotextulariella salevensis* in the uppermost 20 m of the San Lorenzello and in the last 5 m of the Sferracavallo sections, we have tentatively located the Valanginian–Hauterivian boundary in both our sections, within a transitional interval 5–10 m thick, above the last occurrence of *P. salevensis* (Figs. 4A and 4B).

- **Sedimentology:** The San Lorenzello succession is 87 m thick and crops out in the Matese Mountains of the Campania region about 40 km northeast of Napoli; the Sferracavallo succession is 40 m thick and is well exposed in the Palermo Mountains, about 20 km west of Palermo, Sicily (Fig. 1). The San Lorenzello and Sferracavallo successions show, respectively, 7 and 8 lithofacies grouped in 4 lithofacies associations (FA to L in Table 1; see also Fig. 2).
Lithofacies and lithofacies associations suggest that, although belonging to two different paleogeographic domains (Abruzzi–Campania Carbonate Platform and Panormide Carbonate Platform), the analyzed strata formed under similar paleoenvironmental conditions (Fig. 3), characterized by more external and open lagoonal settings (lagoonal domain, Clan I deposits), passing towards more internal and protected areas (peritidal domain, Clan II deposits). Very shallow-water, sand-size sediments, locally showing an indistinct oblique to parallel lamination (lithofacies association BP, inner-sand-shoal deposits, Table 1) accumulated in more inland areas, at the transition between the more open lagoonal domain and the more restricted peritidal environments (Fig. 3). These deposits suggest depositional regimes characterized by higher hydrodynamic energy (probably the place where fair-weather waves start touching bottom and therefore carbonate production reaches a local maximum); they appear particularly widespread in the Sferracavallo section.

Cyclostratigraphy

Like other Cretaceous carbonate-platform deposits studied in southern Apennines, both of the successions exhibit paleoenvironmental oscillations expressed by a hierarchy of elementary

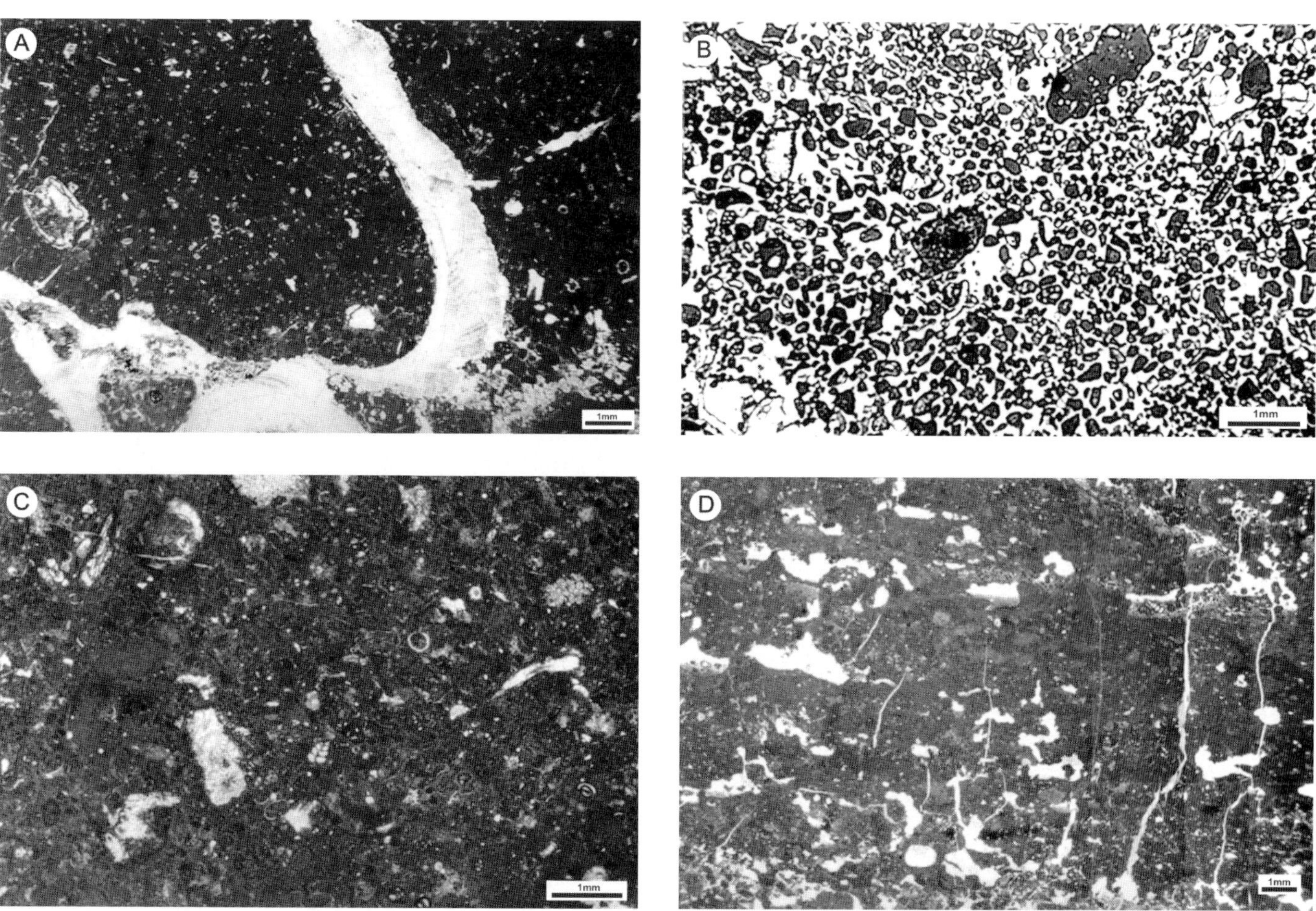

Fig. 2.—Some lithofacies recognized in the San Lorenzello and Sferracavallo sequences. **A)** Lithofacies FA1: benthic foraminifers, green algae wackestone with molluscan shells. **B)** Lithofacies BP1: bioclastic grainstone with intraclasts and peloids. **C)** Lithofacies MO1: wackestone with green algae, benthic foraminifers, and ostracods. **D)** Lithofacies L2: loferitic mudstone with small cavities, enlarged by karstic dissolution and occluded by calcite cement and/or geopetal infilling. Thin sections, positive prints.

TABLE 1.—Lithofacies, lithofacies associations, and their depositional environments of San Lorenzello (SL) and Sferracavallo (Sf) sections.

LITHOFACIES ASSOCIATIONS	LITHOFACIES	SECTION	DEPOSITIONAL ENVIRONMENT
FOR-ALGAL LIMESTONES (FA)	**FA1**: Benthic forams, green algae wackestone and bioturbated wackestone–packstone with molluscan shells (pelecypods and gastropods) and *Cayeuxia* sp.	Sf and SL sections	OPEN LAGOON
	FA2: Benthic forams wackestone and bioturbated wackestone–packstone with rare green algae, small molluscan shells and peloids	SL section	
BIO-PELOIDAL LIMESTONES (BP)	**BP1**: Benthic forams packstone, packstone–grainstone, and grainstone with bioclasts (molluscan shells, algae) and intraclasts	Sf and SL sections	INNER SHOALS
	BP2: Peloids, benthic forams packstone–wackestone, packstone, and packstone-grainstone.	Packstone and packstone–grainstone, locally with weack lamination and /or keystone vugs in the Sf section; wackestone–packestone and packstone with ostracods and small gastropods in the SL section.	
MILI-OSTRACOD LIMESTONES (MO)	**MO1**: Benthic forams wackestone–packstone and wackestone, locally bioturbated, with small gastropods, ostracods and rare green algae.	Sf and SL sections	RESTRICTED LAGOON
	MO2: Miliolids wackestone, wackestone–mudstone and mudstone locally bioturbated; ostracods and other rare benthic forams are also present	Sf and SL sections	
	MO3: Ostracods mudstone–wackestone and mudstone with rare small benthic forams.	Sf section	
LAMINATED LIMESTONES (L)	**L1**: Stromatolitic bindstone	Sf section	TIDAL FLAT
	L2: Loferitic mudstone and bindstone	Sf and SL sections	

cycles, bundles, and superbundles (Fig. 4). The San Lorenzello succession (Fig. 4A) has 86 elementary cycles (average thickness of 1.01 m), organized in 25 bundles (average thickness 3.24 m) and in 7 superbundles (SL1 to SL7, with an average thickness of 11.38 m). The Sferracavallo section (Fig. 4B) has 63 elementary cycles (average thickness of 0.64 m), grouped in 17 bundles (average thickness of 2.33 m) and in 5 superbundles (Sf1 to Sf5, with an average thickness of 7.47 m). Normally, thicker cycles suggest more open lagoonal environments whereas thinner cycles indicate inner (peritidal) settings. More than 95% of the elementary cycles identified in our sections exhibit exposure-related features directly superimposed on subtidal deposits (Fig. 5). This, combined with the well-ordered hierarchical organization of the cycles, precludes a prevailing autocyclic control and suggests that high-frequency eustatic changes, modulated by the Earth's orbital fluctuation, produced the boundaries of the elementary cycles and determined the cycle stacking patterns (Crevello, 1991; Fischer, 1991; de Boer and

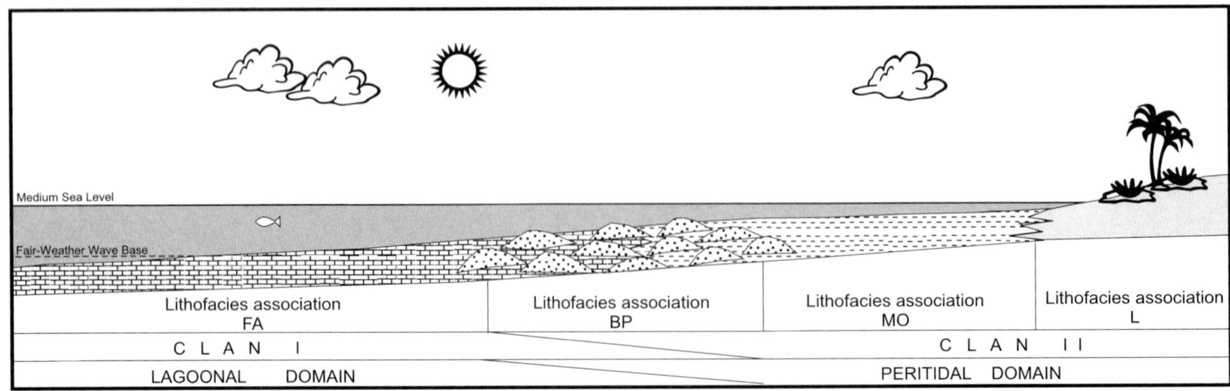

FIG. 3.—Schematic carbonate-platform model showing environmental distribution of the lithofacies associations (FA to L) and their grouping (Clan I and Clan II) in the San Lorenzello and Sferracavallo sections (see also Table 1). In this sedimentary model a discontinuous belt of sand shoals between peritidal and open lagoonal domains is suggested by the deposits of the lithofacies association BP.

Smith, 1994; Strasser, 1994; Grötsch, 1996; Lehmann et al., 1998). On this assumption the elementary cycles can be related to the precession and / or the inclination forcing (or to a combination of them), whereas bundles and superbundles are considered to record the short- and long-eccentricity periodicities, respectively. In the San Lorenzello succession this interpretation was supported by mathematical data treatment (spectral analysis, neural network) using data for thickness of the lithofacies and related early meteoric overprint, collected at a centimeter scale (D'Argenio et al., 1997; Pelosi and Amodio, 2001; Tagliaferri et al., 2001). The periodicity of the cycles was calculated by comparing the relative ratio sets of the thickness of the sedimentary features, (RRSs) expressed in centimeters, with the relative ratio sets of the orbital periodicities (RRSt) expressed in years (as calculated by Berger et al., 1992, for the corresponding Early Cretaceous time interval). The two ratio sets so obtained, being dimensionless numbers, can be then compared to each other. The correlation coefficient for all possible combinations of ratios (RRSs vs. RRSt), which determines the best agreement between the stratigraphic and the orbital periodicities, is in our case very high ($r \geq 0.99$) and confirms the reliability of our methodology. (Procedures for the mathematical treatment of the collected data are discussed extensively in Longo et al., 1994, Brescia et al., 1996, and D'Argenio et al., 1997.)

Sequence Stratigraphy

The paradigm of sequence stratigraphy in general does not apply directly to the flat carbonate-platform interiors, inasmuch as they are characterized by a prevailing aggradational organization of the strata (e.g., Goldhammer et al., 1990; Schlager, 1992, 1999; Embry, 1993). In this context, lateral variations in stratal geometry normally are not detectable at the outcrop scale, even if a few kilometers in extent.

Notwithstanding these constraints, a sequence stratigraphic approach to Mesozoic shallow-water carbonate rocks based on cyclostratigraphic analysis has been undertaken in the last decade in the Mesozoic of the southern Apennines (Italy), of the Jura Mountains (Switzerland and France) and of the southern Alps (e.g., Goldhammer et al., 1990; Strasser, 1994; D'Argenio et al., 1997; Pasquier and Strasser, 1997; D'Argenio et al., 1999; Strasser et al., 1999; Strasser et al., 2001).

Also the Cretaceous strata of San Lorenzello and Sferracavallo, formed in prevailingly aggrading depositional systems and showing a cyclic stratal pattern (Fig. 4), can be interpreted in terms of sequence stratigraphy, considering both superbundles and transgressive–regressive facies trends as depositional sequences (D'Argenio et al., 1997; D'Argenio et al., 1999; Buonocunto et al., 1999; Raspini, 2001).

For each superbundle, two basic systems-tracts equivalents (transgressive systems tract and highstand systems tract) can be recognized. The occurrence of the most open marine lithofacies suggests the position of the maximum flooding surface (MFS), whereas the superbundle limits are considered to be sequence boundaries (SB). The transgressive systems tracts are suggested by the sedimentary evolution of the lithofacies associations (becoming progressively more open marine upsection) and related early diagenetic features (increase in marine cementation, decrease in early meteoric overprint) accompanied by upward thickening of the elementary cycles and bundles. The highstand systems tracts of each superbundle are instead suggested by shoaling facies trends above the maximum flooding surface (increasing peritidal conditions, more deeply penetrating early meteoric diagenesis), as well as by upward-decreasing cycle thickness (Fig. 4).

In both the San Lorenzello and the Sferracavallo sections the superbundles (and related elementary cycles and bundles) are superimposed on longer transgressive–regressive facies trends (T / RFTs in Fig. 4). The latter, well evident in the field along natural and man-made profiles, were defined in D'Argenio et al. (1999) as discrete (a few tens of meters) shoaling-upward, unconformity-bounded stratigraphic intervals (Figs. 5A, C, E) and are considered to be an expression of lower-frequency relative sea-level changes (third-order cycles *sensu* Vail et al., 1991; see also Raspini, 2001).

Like the transgressive systems tracts of the superbundles, each transgressive facies trend (TFT) is characterized upsection by: (a) prevalence of lithofacies associations suggesting paleoenvironmental conditions controlled by increasingly open-marine circulation; (b) decreasing (or even lack of) early meteoric diagenesis; and (c) increasing thickness of beds (and cycles). In each TFT interpreted in terms of sequence stratigraphy, the transgressive peak is located at the maximum flooding surface of the most open superbundle of the related stratigraphic interval. In contrast, each regressive facies trend (RFT) shows upsection: (a) prevalence of lithofacies associations suggesting upward-increasing restricted conditions (restricted lagoon, tidal flat); (b) increasing enhancement of emersion-related features; and (c) decreasing thickness of beds (and cycles). The related regressive peak is located at the upper boundary of the most restricted (and normally thinnest) superbundle, locally topped by clayey horizons and showing downward-penetrating karstic and / or pedogenetic features (Fig. 5). On the basis of these characteristics we assume that at T / RFTs boundaries there is the highest probability that cycles, or groups of cycles, may be missing (e.g., D'Argenio et al., 1999).

Three transgressive–regressive facies trends (Fig. 4A) are recorded in the San Lorenzello section: T1 / R1 (superbundles SL1 and SL2), T2 / R2 (superbundles SL3 and SL4), and T3 / R3 (superbundles SL5 to SL7). The Sferracavallo section (Fig. 4B) shows only one regressive facies trend (superbundles Sf1 to Sf3), followed by one transgressive facies trend (superbundles Sf4 and Sf5).

ORBITAL CHRONOSTRATIGRAPHY

Chronostratigraphic diagrams, which quantify the minimum time required for each succession to accumulate, have been also assembled assuming that: (a) high-frequency cyclicity recognized in the studied successions is due to eustatic control, forced by the Earth's orbital fluctuations; (b) bundles (groups of 2–5 elementary cycles) and superbundles (groups of 2–4 bundles) have a duration of about 100 and 400 ky, respectively; and (c) the stratigraphic record can be considered to be continuous at the superbundle scale. According to D'Argenio et al. (1997); and D'Argenio et al. (1999), and taking into account the cyclic stacking pattern (upward-increasing frequency of peritidal deposits overprinted by early meteoric diagenesis and decrease in cycle thickness), the sedimentary gaps can be located in the chronostratigraphic schemes in terms of missing bundles. The orbital chronostratigraphy of the studied successions suggests two gaps (each representing a time duration of ~ 200 ky) in the San Lorenzello section (Fig. 6A), whilst only one gap (~ 200 ky) at Sferracavallo (Fig. 6B). We can then calculate that the minimum time for the successions to accumulate is 2.9 My for San Lorenzello and 1.9 My for Sferracavallo (Fig. 6).

On the basis of the thickness and time length of the superbundles, the accumulation rates for the above successions were also calculated (Table 2). Values range from 14.9 to 38.2 B (Bubnoff = 1 mm / ky), with an average of 28.5 B at San Lorenzello, whereas

A

AGE
METRIC SCALE
STRATA
GRAPHIC LOG
CLAN I
CLAN II
FA
BP
MO
L
FA1
FA2
BP1
BP2
MO1
MO2
L2
Em
1
2
ELEMENTARY CYCLES
BUNDLES
SUPER BUNDLES
SYSTEMS TRACTS
T / R Facies trends
HAUTERIVIAN
VALANGINIAN
0 m
10
20
30
40
50
60
70
80
25 262
24 272
23 355
22 411
21 474
20 131
19 377
18 291
17 336
16 350
15 401
14 343
13 403
12 211
11 396
10 304
9 292
8 450
7 315
6 247
5 147
4 229
3 511
2 238
1 342
7 1526
6 1135
5 751
4 1350
3 596
2 1284
1 1320
TR
SB
HS
mfs
SL7
SL6
SL5
SL4
SL3
SL2
SL1
R3
T3
R2
T2
R1
T1

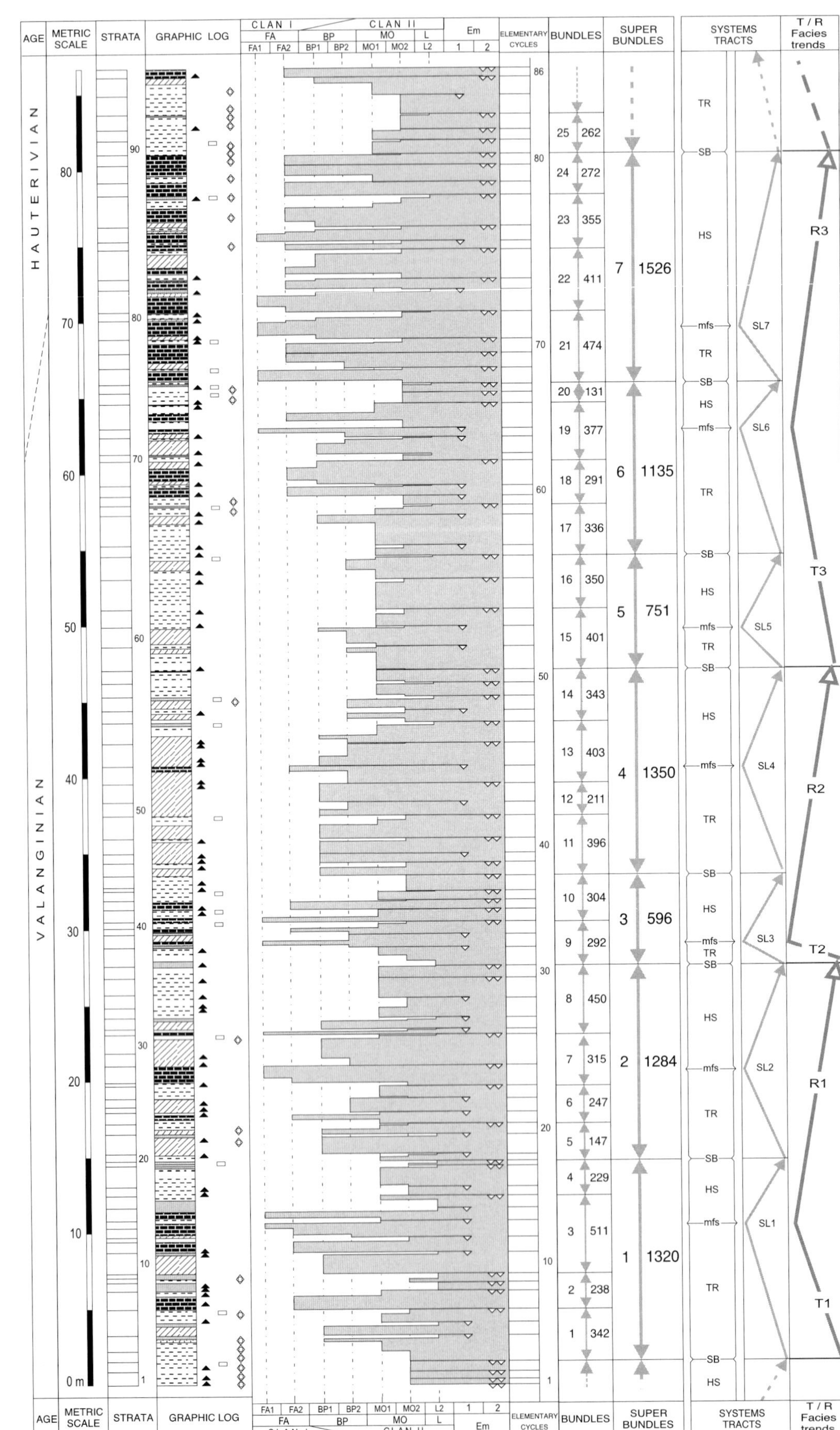

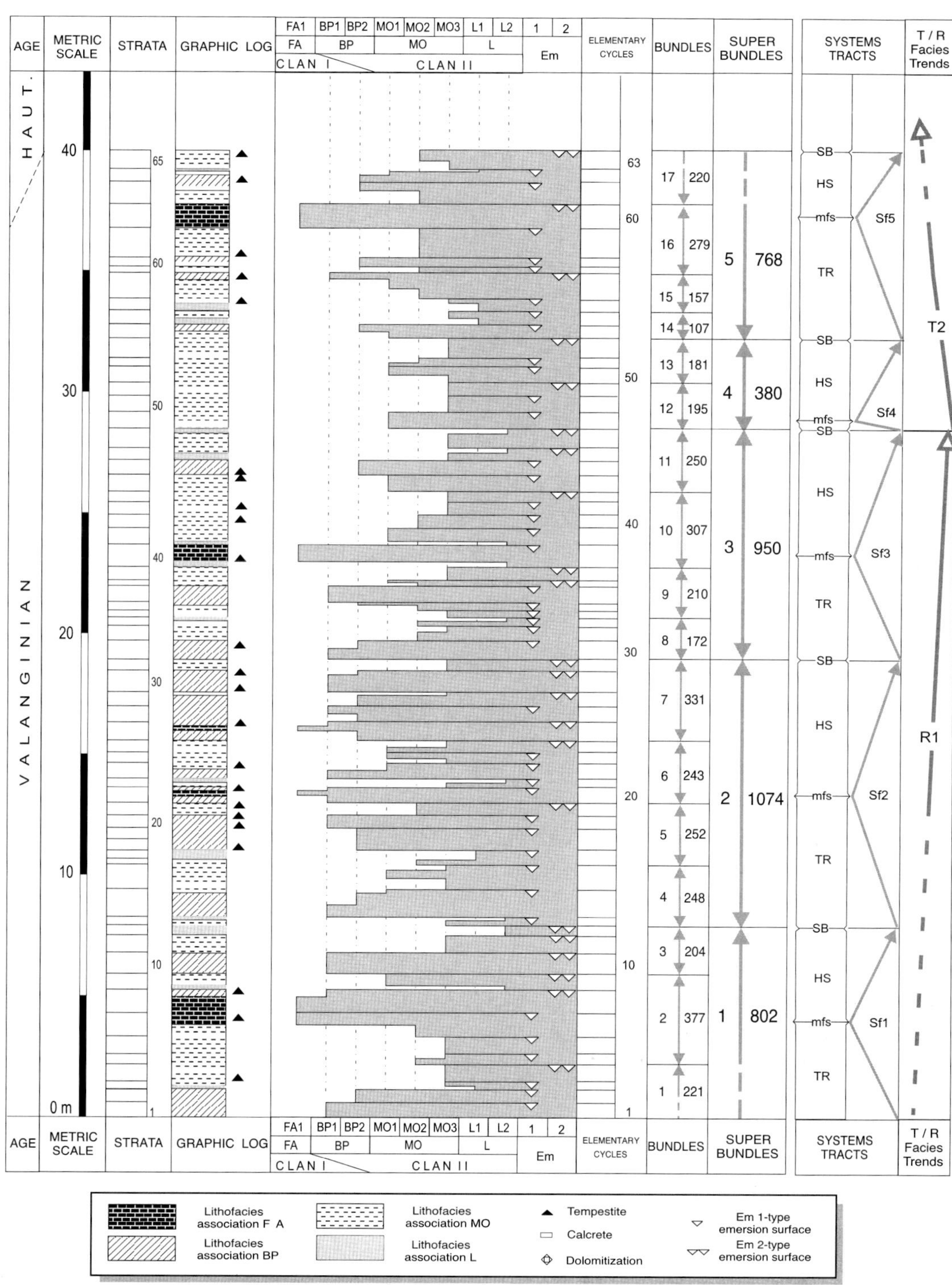

Fig. 4.—Lithostratigraphy, cyclostratigraphy, sequence stratigraphy, and transgressive–regressive facies trends (T/RFTs) of San Lorenzello and Sferracavallo sections. **A)** Composite log of San Lorenzello. The sharp-cornered curve (on the left) shows the elementary cycles as revealed by the vertical succession of lithofacies and lithofacies associations (FA to L) and related early diagenetic meteoric overprint. Em1 and Em2 indicate emersion-related surfaces whose features increasingly penetrate downwards. Note that the elementary cycles are organized in 25 bundles and 7 superbundles (thickness is indicated on the right side of the bundles and superbundles). **B)** Composite log of Sferracavallo. The sharp-cornered curve (on the left) shows the elementary cycles as revealed by the vertical succession of lithofacies and lithofacies associations (FA to L) and related early diagenetic meteoric overprint. Em1 and Em2 indicate emersion-related features increasingly penetrating downwards. Elementary cycles are organized in 17 bundles and 5 superbundles (thickness is indicated on the right side of the bundles and superbundles). SB, Sequence Boundary; MFS, maximum flooding surface; TR, transgressive systems tract; HS, highstand systems tract; T, transgressive facies trend; R, regressive facies trend.

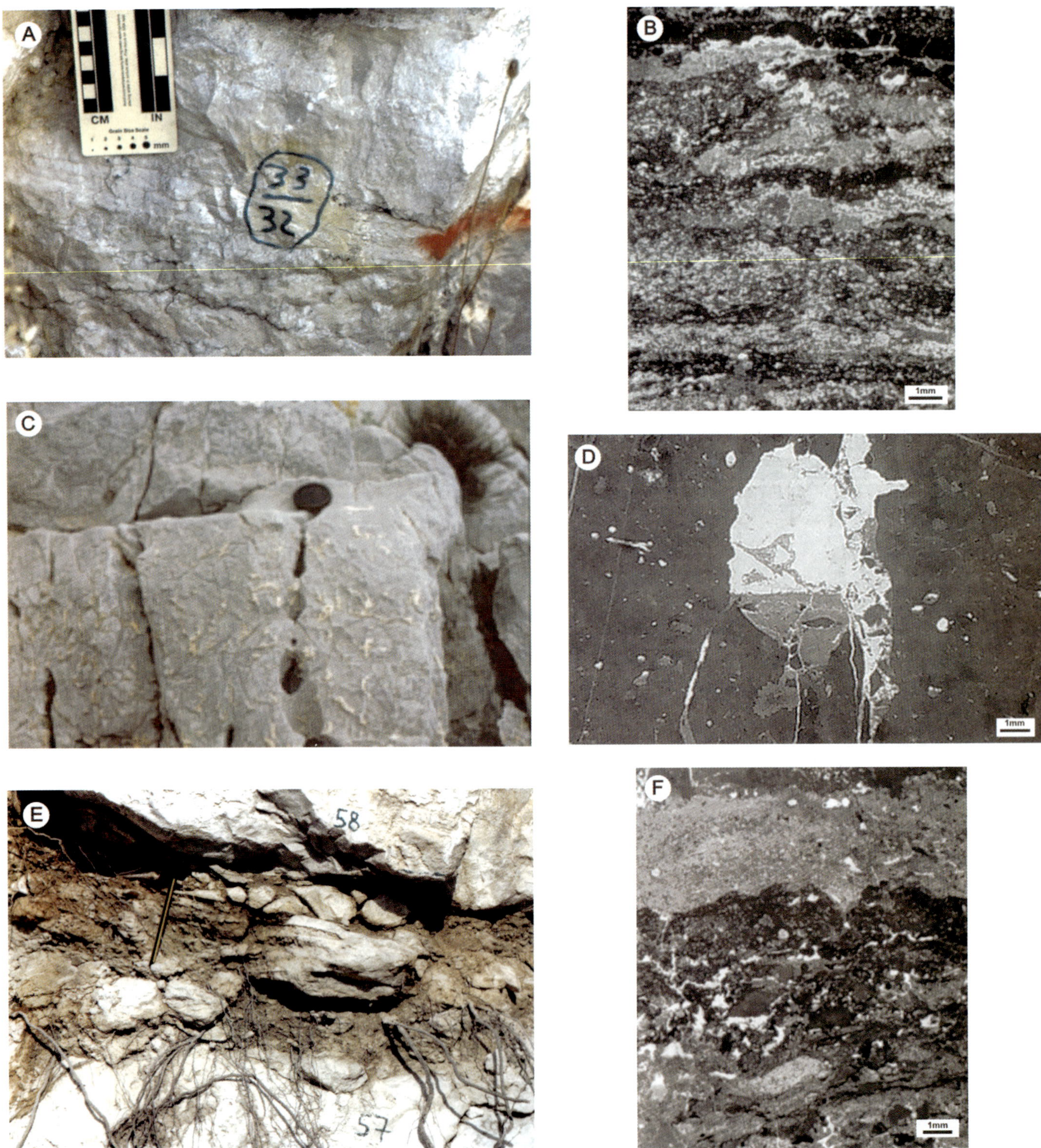

Fig. 5.—**A)** San Lorenzello. Regressive peak R1 at the SL2 sequence boundary (see also Fig. 4A) marked by a cm-thick calcrete horizon, irregularly laminated. **B)** The laminar calcrete shown in Part A formed of alternating micritic and microsparitic laminae. Thin section, positive print. **C)** Sferracavallo. Regressive peak R1 at the Sf3 sequence boundary (see also Fig. 4B) outlined by deeply penetrating paleokarst features. **D)** Detail of Part C showing a paleokarstic cavity occluded by geopetal infilling. Thin section, positive print. **E)** San Lorenzello. Regressive peak R2 at the SL4 sequence boundary (see also Fig. 4A) marked by a clayey paleosol superimposed on karstified subtidal deposits (lithofacies MO2). **F)** Detail of Part E showing the karstified deposits characterized by millimeter- to centimeter-size karst cavities, the latter occluded by geopetal crystal silt and calcite cements. Thin section, positive print.

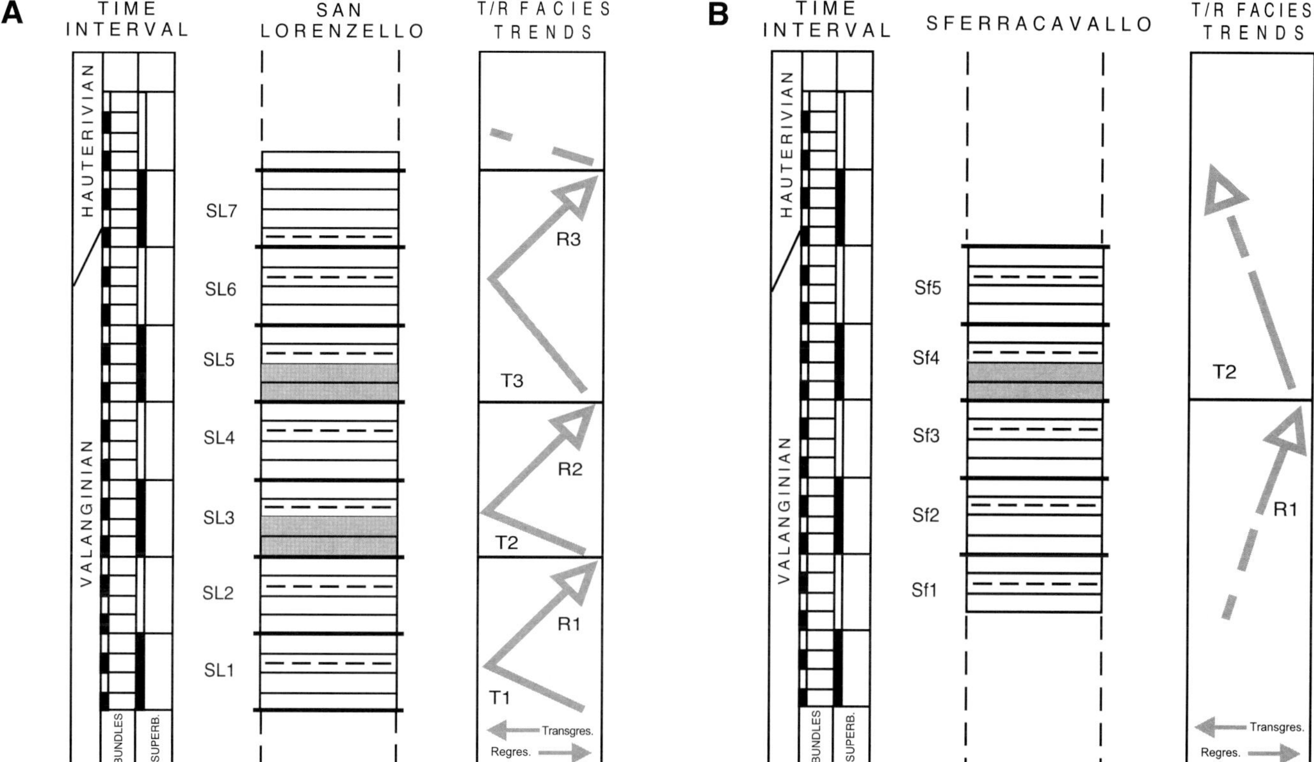

Fig. 6.—Chronostratigraphic scheme of **A)** San Lorenzello and **B)** Sferracavallo sequences. Missing bundles and duration of related stratigraphic gaps are shown in gray. See text for additional comments.

at Sferracavallo they vary from 9.5 to 26.9 B, with an average of 19.9 B. These values appear well comparable with those of other orbitally driven carbonate successions measured in the Cretaceous of southern Italy (Longo et al., 1994; Brescia et al., 1996; Raspini, 1998, 2001; Buonocunto et al., 1999; D'Argenio et al., 1997; D'Argenio et al., 1999).

REGIONAL AND GLOBAL CORRELATION

We used a centimeter-scale cyclostratigraphic analysis of two carbonate platform sections, both including the biostratigraphic Valanginian–Hauterivian boundary, to investigate: (a) the reliability of this approach in the high-precision (bundle level, less than 5 m thick in this case) regional correlation, and (b) the feasibility of an accurate linkage of the studied intervals to the available global sea-level oscillation curves at the scale of ~ 0.5–0.6 My (Haq et al., 1987; Jacquin et al., 1998).

Regional Correlation

Assuming the Valanginian–Hauterivian boundary as a biostratigraphic reference interval, taking into account the well-ordered hierarchical cyclic organization recognized in both the studied successions, and considering the superbundles to be physical correlation tools, the 40-m-thick Sferracavallo section (superbundles Sf1 to Sf5) was correlated, bundle by bundle (Fig. 7) with a 48-m-thick interval (superbundles SL2 to SL6) of the San Lorenzello section (see also Sandulli, 1999; Amodio, 2000). Moreover, the R2 peak of San Lorenzello (Fig. 4A) was correlated with the R1 peak of Sferracavallo (Fig. 4B). In particular, the correlation of superbundles SL3 and SL4 with Sf2 and Sf3 allowed us to locate a regressive peak in the lowermost part of the Sferracavallo succession, at 8 m from the base, corresponding to the upper boundary of Sf1. This regressive peak is correlative with the R1 peak of San Lorenzello (Fig. 7). In the correlated intervals, the San Lorenzello section shows, on the whole, accumulation rates higher than those of Sferracavallo (Table 2). The estimated average accumulation rates that emerge are: 23.9 B for San Lorenzello (superbundles SL2 to SL6) and 19.9 B for Sferracavallo (superbundles Sf1 to Sf5).

TABLE 2.—Accumulation rates of San Lorenzello and Sferracavallo sections estimated on the basis of the present-day thickness of the superbundles (~ 400 ky). The values are indicated in Bubnoffs, B. (B = 1 mm/ky).

SAN LORENZELLO SECTION		**SFERRACAVALLO SECTION**	
SUPERBUNDLES	Accumulation rates (1 B = 1 mm/ky)	SUPERBUNDLES	Accumulation rates (1 B = 1 mm/ky)
SL7	38.2 B	—	—
SL6	28.4 B	Sf5	19.2 B
SL5	18.8 B	Sf4	9.5 B
SL4	33.8 B	Sf3	23.8 B
SL3	14.9 B	Sf2	26.9 B
SL2	32.1 B	Sf1	20.1 B
SL1	33.0 B	—	—
Average Accumulation rate: 28.5 B		Average Accumulation rate: 19.9 B	

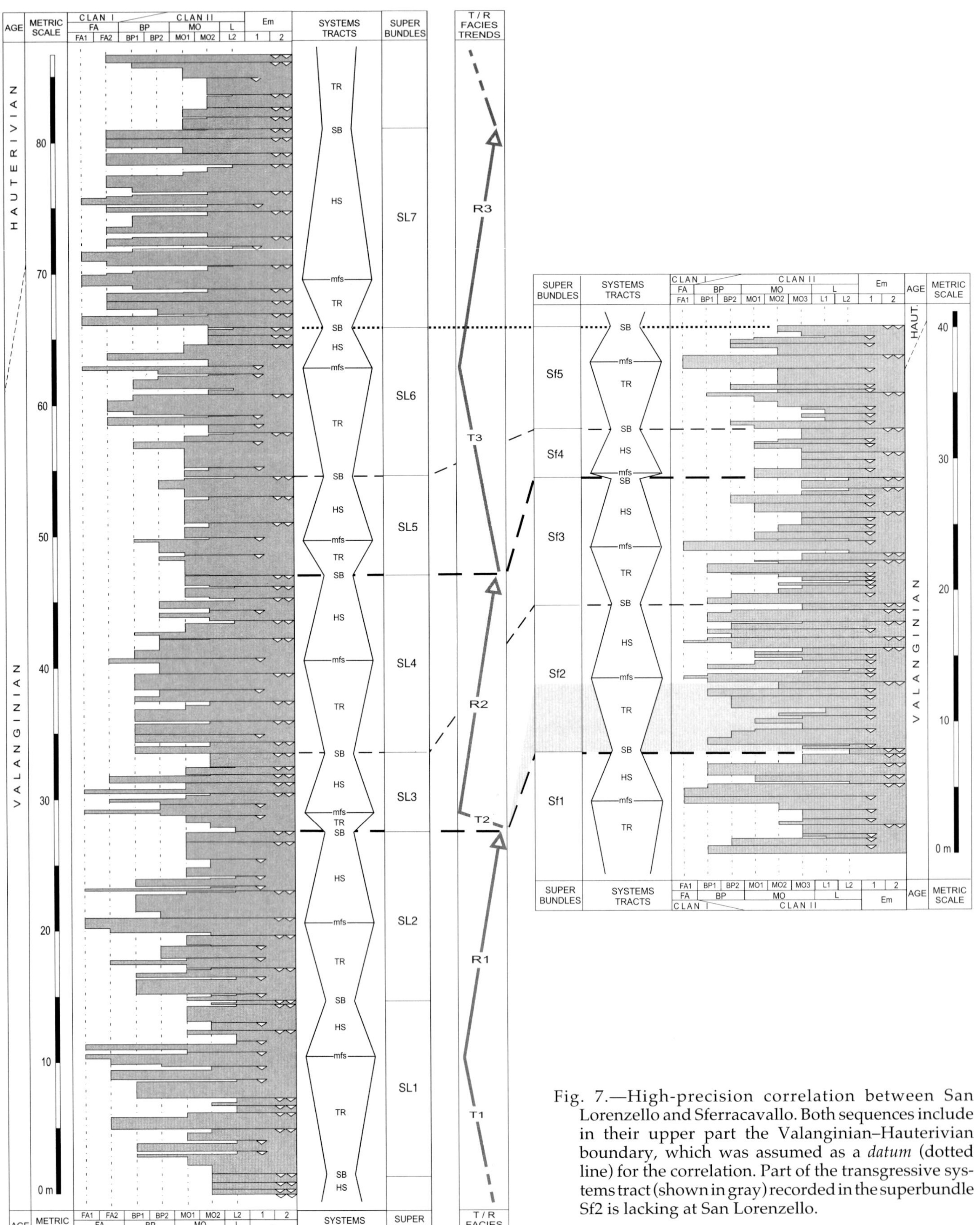

Fig. 7.—High-precision correlation between San Lorenzello and Sferracavallo. Both sequences include in their upper part the Valanginian–Hauterivian boundary, which was assumed as a *datum* (dotted line) for the correlation. Part of the transgressive systems tract (shown in gray) recorded in the superbundle Sf2 is lacking at San Lorenzello.

We now discuss briefly the control on the superbundle thickness and related accumulation rate. Superbundle thickness is supposed to be the actual expression of interaction among the production of accommodation space (subsidence + eustasy), carbonate productivity, and sediment compaction (e.g., Crevello, 1991; Schlager, 1991, 1999; Handford and Loucks, 1993; D'Argenio et al., 1999). Reduction in thickness due to compaction was not considered in this study and probably did not play a major role in these shallow-water deposits because of their early lithification by cementation, whereas stylolitization is apparently evenly distributed in the section (Iorio et al., 1996; D'Argenio et al., 1999).

On the other hand, it is reasonable to suppose that paleoenvironmental changes recorded in the San Lorenzello and Sferracavallo sections (two distant paleogeographic domains), if controlled only by the rate of eustatic change, may have produced comparable sequences. This allows us to infer that different thicknesses of the time-equivalent superbundles must basically reflect a different subsidence rate at the two localities. The latter, counterbalanced by carbonate productivity, was higher for the San Lorenzello succession.

Chronostratigraphic correlation between San Lorenzello and Sferracavallo is shown in Figure 8. Here the location of the gaps punctuating the stratigraphic record within time-equivalent superbundles is also indicated. We note that two main gaps occur, each corresponding to the omission of two bundles (~ 200 ky), associated with boundaries of transgressive–regressive facies trends. In particular, the older gap (lowermost part of T2 facies trend) is observed at San Lorenzello, whilst the younger gap (lowermost part of the T3 Facies Trend) is recorded in both the successions. We can suppose that local factors (different bottom topography and subsidence rate) could have controlled omission of elementary cycles and bundles in the sedimentary record (see also Strasser et al., 2001).

It is worth noting that the time resolution of shallow-water biozones based on algae and benthic foraminifers, unlike that of the pelagic realm, is in general > 1 My (e.g., Chiocchini et al., 1994), at least for the time interval and facies discussed here. Therefore, the regional correlation proposed here increases the accuracy of the stratigraphic correlation between these distant Valanginian–Hauterivian strata by at least one order of magnitude. Finally, the above correlation supports the view that (1) eustatic changes controlled the cyclic evolution of the age-equivalent strata now located more than 400 km from each other, and (2) the sedimentary record is almost fully preserved at the superbundle scale (~ 400 ky).

Correlation with Global Stratigraphic Charts

On the basis of sequence stratigraphic criteria and considering the location of the Valanginian–Hauterivian boundary in the standard time scales as well as in our orbital chronostratigraphy, a correlation with the "eustatic curve" of Haq et al. (1987) and with the chronostratigraphic schemes of Jacquin et al. (1998) is here proposed (Fig. 8). In this exercise, correlation of our transgressive–regressive facies trends with the third-order cycles of Jacquin et al. (1998), calibrated against the Gradstein et al. (1995) time scale, allowed us to test the validity of our orbital chronostratigraphy.

We suggest that the R2 regressive peak (at SL4 and Sf3 superbundle boundaries) corresponds to the boundary between depositional sequences LZB-2.2 and LZB-2.3 of Haq et al. (1987) and to sequence boundary Va7 of Jacquin et al. (1998).

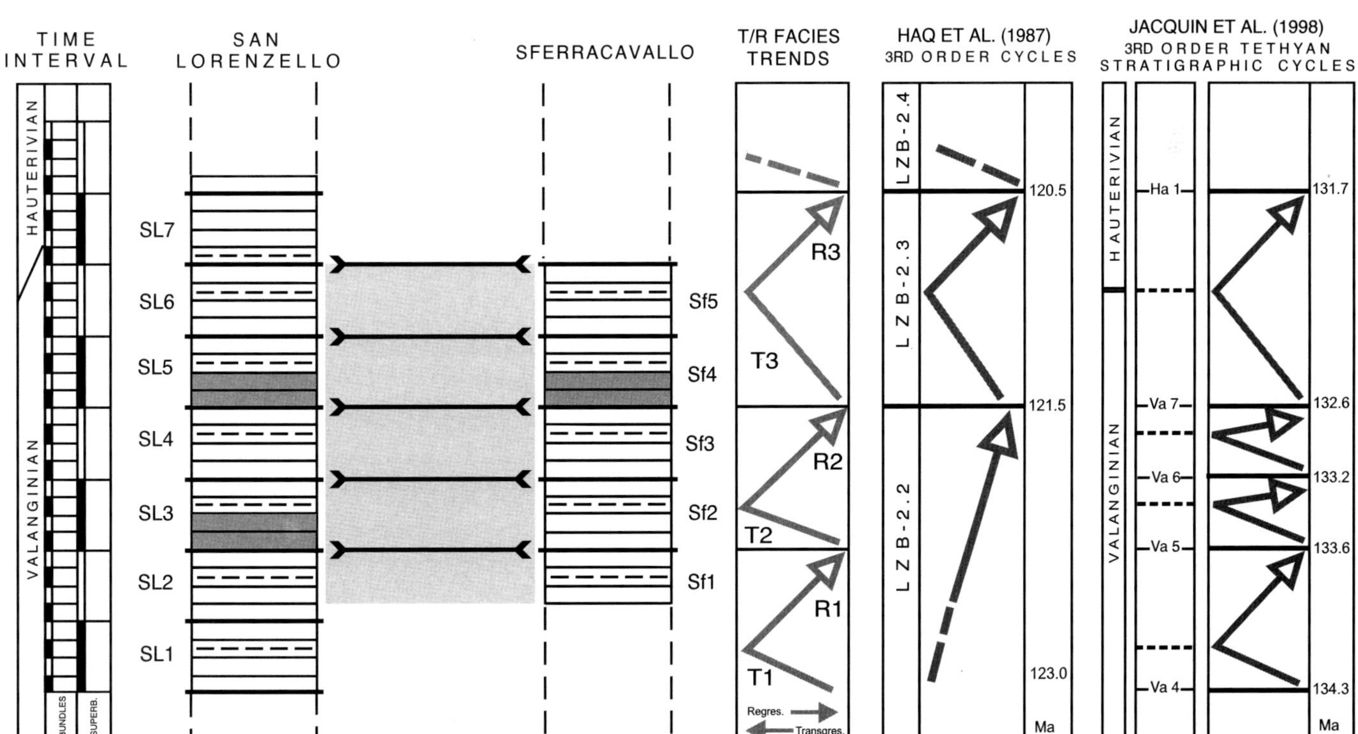

Fig. 8.—Chronostratigraphic correlation between San Lorenzello and Sferracavallo sections and their relationship to the third-order cycles of Haq et al. (1987) and Jacquin et al. (1998). Note that even if the absolute age figures in the above third-order cycles do not correspond to each other, their trends and durations are well comparable among them and with San Lorenzello and Sferracavallo T/R facies trends. See text for additional comments.

Moreover, the T3/R3 facies trend (SL5 to SL7 superbundles) matches both the third-order cycle LZB-2.3 of Haq et al. (1987) and depositional sequence Va7 of Jacquin et al. (1998), respectively. Figure 8 shows that the duration of 2.8 My, estimated on the basis of our Valanginian–Hauterivian orbital chronostratigraphy, can be compared with that of about 2.6 My and 2.5 My, respectively, referred by Jacquin et al. (1998) and Haq et al. (1987) for the time-equivalent stratigraphic intervals. The different number of lower-frequency cycles (Fig. 8), recognized in the same Late Valanginian–Early Hauterivian time interval (two third-order cycles in Haq et al., 1987, four depositional sequences in Jacquin et al., 1998, and three T/RFTs in our sections), may be a consequence of the different methodologic approaches and observation scales (meter versus centimeter) used in interpreting the stratigraphic evolution of coeval deposits.

ACHIEVEMENTS AND PROSPECTS

The centimeter-scale cyclostratigraphic analysis of two Lower Cretaceous carbonate-platform sections, cropping out in southern Italy and located at present more than 400 km apart, leads to the following conclusions and implications.

- **Hierarchical Cyclic Organization** – The total measured stratal thickness includes the San Lorenzello (Matese Mountains, Campania) and the Sferracavallo (Palermo Mountains, Sicily) successions (Fig. 1). Both successions exhibit a hierarchical cyclic organization of textures and sedimentary features, physically expressed by elementary cycles, bundles, and superbundles (Fig. 4). High-frequency external control is suggested by early meteoric diagenesis directly superimposed on subtidal deposits, as well as by the well-ordered cycle hierarchy. On the basis of this evidence, elementary cycles can be linked to the precession or inclination signals, or to a combination of them (D'Argenio et al., 1997), whereas bundles and superbundles are considered to record the short- and long-eccentricity signals, respectively.

- **Lower-Frequency Cycles** – High-frequency cycles are superimposed on longer (from ~ 800 ky to ~ 1200 ky) transgressive–regressive facies trends (T/RFTs) that can be considered as third-order sequences of the standard model (Vail et al., 1991). It can be inferred that also these cycles have an orbital control, because of the considerable power of the eccentricity in the range of 0.8–1.2 My (e.g., Fischer et al., 1990; Fischer, 1991). Three T/RFTs were recognized in the San Lorenzello section (Fig. 4A), whilst only one regressive trend, followed by one transgressive trend, is recorded at Sferracavallo (Fig. 4B). The uppermost segment at Sferracavallo includes the ~ 3-m-thick Valanginian–Hauterivian biostratigraphic boundary zone.

- **Sequence Stratigraphy** – On the basis of cyclic stacking patterns, both sections were interpreted in terms of sequence stratigraphy, to assemble chronostratigraphic diagrams and to investigate regional- to global-scale correlation, the latter using the eustatic curves of Haq et al. (1987) and the third-order stratigraphic cycles of Jacquin et al. (1998).

- **Chronostratigraphic Diagrams** – Chronostratigraphic diagrams of the studied sections with the location of the recognized gaps (missing bundles) are presented in Figure 6. On the basis of our orbital time scale, we suggest that the minimum time duration required for the sedimentation of San Lorenzello and Sferracavallo successions is 2.9 and 1.9 My, respectively.

- **Regional Correlation** – Considering the Valanginian–Hauterivian boundary as biostratigraphically well constrained and using the superbundles as tools for high-precision regional correlation, the Sferracavallo section has been correlated, bundle by bundle, with the corresponding segment (superbundles SL2 to SL6) of San Lorenzello (Fig. 7). In particular, the correlation of superbundles SL3 and SL4 with Sf2 and Sf3 also suggests that a regressive peak, correlative with R1 peak of San Lorenzello, may be located in the lowermost part of the Sferracavallo succession. Moreover, on the basis of the accumulation rates, as deduced by the present-day thickness of the superbundles, we note that their values at San Lorenzello are higher than at Sferracavallo (Table 2). Because we can assume that a comparable eustatic variation rate controlled the time-equivalent paleoenvironmental changes recorded in these two distant paleogeographic domains, we conclude that the different thickness of the correlated superbundles is linked to different subsidence rates, higher at San Lorenzello than at Sferracavallo.

- **Global-Scale Correlation** – On the basis of sequence stratigraphy criteria and taking into account the stratigraphic position of the Valanginian–Hauterivian boundary in the standard time scales as well as in our orbital chronostratigraphy, a global-scale correlation is proposed (Fig. 8). It is suggested that the R2 regressive peak (at SL4 and Sf3 superbundle boundaries) corresponds to the boundary between the depositional sequences LZB-2.2 and LZB-2.3 of Haq et al. (1987) and to the sequence boundary Va7 of Jacquin et al. (1998). Moreover, the T3/R3 facies trend (SL5 to SL7 superbundles) matches both the depositional sequences LZB-2.3 of Haq et al. (1987) and Va7 of Jacquin et al. (1998).

- **Orbital Time Scale** – Finally, the high-resolution physical correlation shown in Figure 8 suggests that the time duration of 2.8 My estimated in our orbital chronostratigraphy fits well with the 2.6 My estimated by Jacquin et al. (1998) between their Va4 and Va7 sequence boundaries on the basis of the Gradstein et al. (1995) time scale.

- **Prospects** – We have seen that the investigated carbonate platforms, in spite of being prone to frequent gaps in their sedimentary record, offer a good opportunity to document, for long time intervals, the variation of climate, eustasy, and other paleoceanographic characteristics. In addition to demonstrating the opportunity given by the analytical methodology used here in chronostratigraphy and in regional-scale and global-scale correlation, we have also shown that the lack of information at the stratigraphic breaks is not random, because the gaps are distributed along the sections following predictable rules imposed by the orbital control. So, we can assume that size of stratigraphic gaps and their location along the sedimentary record are *per se* meaningful.

ACKNOWLEDGMENTS

We thank R. Radoičić and R. Schroeder for their helpful assistance and suggestions in the biostratigraphic analysis. We also thank our reviewers W. Schlager and E. Anderson for their much appreciated suggestions and comments. This research was financially supported by the Instituto per l'Ambiente Marino Costiero (Geomare), National Research Council, Naples.

REFERENCES

AMODIO, S., 2000, Stratigrafia sequenziale dei sistemi carbonatici neritici aggradanti. Micro- e biostratigrafia nel Cretacico inferiore dell'Appennino Meridionale: Ph.D. Dissertation, Università di Napoli "Federico II", 136 p.

AMODIO, S., D'ARGENIO, B., AND FERRERI, V., 2001, San Lorenzello: High-frequency cyclicity in a Valanginian–Hauterivian sequence, *in* D'Argenio, B., and Ferreri, V., eds., Cyclostratigraphy of Carbonate Platform Strata. Cretaceous of Matese Mountains of Southern Apennines: SEPM, International Workshop on: "Multidisciplinary Approach to Cyclostratigraphy", Sorrento, Italy, p. 22–29, Field Trip Guidebook, Stop 2.

BERGER, A., LOUTRE, M.F., AND LASKAR, J., 1992, Stability of the astronomical frequencies over the Earth's history for palaeoclimate studies: Science, v. 255, p. 560–565.

BRESCIA, M., D'ARGENIO, B., FERRERI, V., PELOSI, N., RAMPONE, S., AND TAGLIAFERRI, R., 1996, Neural net aided detection of astronomical periodicities in geologic records: Earth and Planetary Science Letters, v. 139, p. 33–45.

BUCUR, U., CONRAD, M.A., AND RADOIČIĆ, R., 1995, Foraminifers and calcareous algae from Valanginian limestones in the Jerma River Canyon, Eastern Serbia: Revue de Paléobiologie, v. 11, p. 349–377.

BUONOCUNTO, F.P., D'ARGENIO, B., FERRERI, V., AND RASPINI, A., 1994, Microstratigraphy of highly organized carbonate platform deposits of Cretaceous age. The case of Serra Sbregavitelli, Matese (Central Apennines): Giornale di Geologia, v. 56, p. 179–192.

BUONOCUNTO, F.P., D'ARGENIO, B., FERRERI, V., AND SANDULLI, R., 1999, Orbital cyclostratigraphy and sequence stratigraphy of Upper Cretaceous platform carbonates at Monte Sant'Erasmo, southern Apennines, Italy: Cretaceous Research, v. 20, p. 81–95.

CATALANO, R., AND D'ARGENIO, B., 1978, An essay of palinspastic restoration across Western Sicily: Geologica Romana, v. 17, p. 145–159.

CATALANO, R., DI STEFANO, P., SULLI, A., AND VITALE, F.P., 1996, Palaeogeography and structure of the central Mediterranean: Sicily and its offshore area: Tectonophysics, v. 260, p. 291–323.

CHANNELL, J.E.T., D'ARGENIO, B., AND HORVATH, F., 1979, Adria, the African promontory in Mesozoic Mediterranean palaeogeography: Earth-Science Reviews, v. 15, p. 213–292.

CHIOCCHINI, M., FARINACCI, A., MANCINELLI, A., MOLINARI, V., AND POTETTI, M., 1994, Biostratigrafia a foraminiferi, dasicladali e calpionelle delle successioni carbonatiche Mesozoiche dell'Appennino Centrale, *in* Mancinelli, A., ed., Biostratigrafia dell'Italia Centrale: Università di Camerino, Studi Geologici Camerti, Volume Speciale 9, p. 9–128.

COTTLE, R.A., 1989, Orbitally mediated cycles from the Turonian of Southern England: their potential for high-resolution stratigraphic correlation: Terra Nova, v. 1, p. 426–431.

CREVELLO, P.D., 1991, High-frequency carbonate cycles and stacking patterns: interplay of orbital forcing and subsidence on Lower Jurassic rift platforms, High Atlas Marocco, *in* Franseen, E.K., Watney, W.L., Kendall, C.G.St.C., and Ross, W., eds., Sedimentary Modeling: Computer Simulations and Methods for Improved Parameter Definition: Kansas Geological Survey, Bulletin, v. 233, p. 207–230.

D'ARGENIO, B., FERRERI, V., AMODIO, S., AND PELOSI, N., 1997, Hierarchy of high-frequency cycles and time calibration in Cretaceous carbonate platform strata: Sedimentary Geology, v. 113, p. 169–193.

D'ARGENIO, B., FERRERI, V., RASPINI, A., AMODIO, S., AND BUONOCUNTO, F.P., 1999, Cyclostratigraphy of a carbonate platform as a tool for high-precision correlation: Tectonophysics, v. 315, p. 357–385.

D'ARGENIO, B., PESCATORE, T., AND SCANDONE, P., 1975, Structural pattern of the Campania–Lucania Apennines: National Research Council (CNR) Quaderni de "La Ricerca Scientifica", no. 90, 17 p.

DE BOER, P.L., AND SMITH, D.G., EDS., 1994, Orbital Forcing and Cyclic Sequences: International Association of Sedimentologists, Special Publication 19, 559 p.

EMBRY, A.F., 1993, Transgressive–regressive (T–R) sequence analysis of the Jurassic succession of the Sverdrup Basin, Canadian Arctic Archipelago: Canadian Journal of Earth Sciences, v. 30, p. 301–320.

FERRERI, V., AMODIO, S., D'ARGENIO, B., AND SANDULLI, R, 2001, High-resolution correlation and orbital chronostratigraphy at the Valanginian–Hauterivian boundary (abstract): SEPM International Workshop on: "Multidisciplinary Approach to Cyclostratigraphy", Sorrento, Italy, Abstract Volume, p. 27–28.

FISCHER, A.G., 1991, Orbital cyclicity in Mesozoic strata, *in* Einsele, G., Ricken, W., and Seilacher, A., eds., Cycles and Events in Stratigraphy: Berlin, Springer-Verlag, p. 48–62.

FISCHER, A.G., DE BOER, P.L., AND PREMOLI SILVA, I., 1990, Cyclostratigraphy, *in* Ginsburg, R.N., and Beaudoin, B., eds., Cretaceous Resources, Events and Rhythms: Dordrecht, The Netherlands, Kluver Academic Publishers, p. 139–172.

GOLDHAMMER, R.K., DUNN, P.A., AND HARDIE, L.A., 1990, Depositional cycles, composite sea-level changes, cycle stacking pattern, and the hierarchy of stratigraphic forcing: Example from Alpine Triassic platform carbonates: Geological Society of America, Bulletin, v. 102, p. 535–562.

GRADSTEIN, F.M., AGTERBERG, F.P., OGG, J.G., HARDENBOL, J., VAN VEEN, P., THIERRY, J., AND HUANG, Z., 1995, A Triassic, Jurassic and Cretaceous time scale, *in* Berggren, W.A., Kent, D.V., Aubry, M.P., and Hardenbol, J., eds., Geochronology, Time Scales and Global Stratigraphic Correlation: SEPM, Special Publication 54, p. 95–126.

GRÖTSCH, J., 1996, Cycle stacking and long-term sea-level history in the Lower Cretaceous (Gavrovo Platform, NW Greece): Journal of Sedimentary Research, v. 66, p. 723–736.

HANDFORD, C.R., AND LOUCKS, R.G., 1993, Carbonate depositional sequences and systems tracts—Responses of carbonate platforms to relative sea-level changes, *in* Loucks, R.G., and Sarg, J.F., eds., Carbonate Sequence Stratigraphy; Recent Developments and Applications: American Association of Petroleum Geologists, Memoir 57, p. 3–41.

HAQ, B.U., HARDENBOL, J., AND VAIL, P.R., 1987, Chronology of fluctuating sea levels since the Triassic: Science, v. 237, p. 1156–1167.

HARDENBOL, J., THIERRY, J., FARLEY, M.B., JACQUIN, T., DE GRACIANSKY, P.C., AND VAIL, P.R., 1998, Cretaceous sequence chronostratigraphy, *in* de Graciansky, P.C., Hardenbol, J., Jacquin, T., Vail, P.R., and Farley, M.B., eds., Mesozoic and Cenozoic Sequence Stratigraphy of European Basins: SEPM, Special Publication 60, Charts.

HILLGÄRTNER, H., 1998, Discontinuity surfaces on shallow-marine carbonate platform (Berriasian, Valanginian, France and Switzerland): Journal of Sedimentary Research, v. 68, p. 1093–1108.

IORIO, M., TARLING, D.H., D'ARGENIO, B., AND NARDI, G., 1996, Ultra-fine magnetostratigraphy of Cretaceous shallow-water carbonates, Monte Raggeto, Southern Italy, *in* Morris, A., and Tarling, D.H., eds., Palaeomagnetism and Tectonics of the Mediterranean Region: Geological Society of London, Special Publication 105, p. 195–203.

JACQUIN, T., RUSCIADELLI, G., AMEDRO, F., DE GRACIANSKY, P.C., AND MAGNIEZ-JANNIN, F., 1998, The North Atlantic cycle: an overview of 2nd order transgressive / regressive facies cycles in the Lower Cretaceous Western Europe, *in* de Graciansky, P.C., Hardenbol, J., Jacquin, T., Vail, P.R., and Farley, M.B., eds., Mesozoic and Cenozoic Sequence Stratigraphy of European Basins: SEPM, Special Publication 60, p. 397–409.

LEHMANN, C., OSLEGER, D.A., AND MONTAÑEZ, I., 1998, Controls on cyclostratigraphy of Lower Cretaceous carbonates and evaporites, Cupido and Coahuila platforms, northeastern Mexico: Journal of Sedimentary Research, v. 68, p. 1109–1130.

LONGO, G., D'ARGENIO, B., FERRERI, V., AND IORIO, M., 1994, Fourier evidence for high-frequency astronomical cycles recorded in Lower Cretaceous carbonate platform strata. Monte Maggiore, Southern Apennines, Italy, *in* de Boer, P.L., and Smith, D.G., eds., Orbital Forcing and Cyclic Sequences: International Association of Sedimentologists, Special Publication 19, p. 77–85.

LUPERTO-SINNI, E., AND MASSE, J.P., 1993, The Early Cretaceous Dasycladales from the Apulia Region (Southern Italy): biostratigraphic distribution and palaeogeographic significance, *in* Barattolo, F., De Castro, P., and Parente, M., eds., Studies on Benthic Fossil Algae: Società Palaeontologica Italiana, Bollettino, Special Volume 1, Mucchi Modena, p. 295–309.

PASQUIER, J.B., AND STRASSER, A., 1997, Platform-to-basin correlation by high resolution sequence stratigraphy and cyclostratigraphy (Berriasian, Switzerland and France): Sedimentology, v. 44, p. 1071–1092.

PELOSI, N., AND AMODIO, S., 2001, Wavelet application to cyclostratigraphy signals. The San Lorenzello section, Lower Cretaceous, Italy (abstract): SEPM, International Workshop on: "Multidisciplinary Approach to Cyclostratigraphy", Sorrento, Italy, Abstract Volume, p. 39–40.

RASPINI, A., 1998, Microfacies analysis of shallow-water carbonates and evidence of hierarchically organized cycles: Aptian of Monte Tobenna, southern Apennines, Italy: Cretaceous Research, v. 19, p. 197–223.

RASPINI, A., 2001, Stacking pattern of cyclic carbonate platform strata: Lower Cretaceous of southern Apennines, Italy: Geological Society of London, Journal, v. 158, p. 353–366.

SANDULLI, R., 1999, Studi microstratigrafici e correlazioni di elevata risoluzione nel Cretacico inferiore di piattaforma carbonatica. Montenegro e Sicilia: Ph.D. Dissertation, Università di Napoli "Federico II", 135 p.

SCARGLE, J.D., 1982, Studies in astronomical time series analysis II, Statistical aspects of spectral analysis of unevenly spaces data: Astrophysical Journal, v. 263, p. 835–853.

SCHINDLER, U., AND CONRAD, M.A., 1994, The Lower Cretaceous dasycladales from the northwestern Friuli platform and their distribution in chronostratigraphic and cyclostratigraphic units: Revue de Paléobiologie, v. 13/1, p. 59–96.

SCHLAGER, W., 1991, Depositional bias and environmental change—important factors in sequence stratigraphy: Sedimentary Geology, v. 70, p. 109–130.

SCHLAGER, W., 1992, Sedimentology and sequence stratigraphy of reefs and carbonate platforms: American Association of Petroleum Geologists, Continuing Educational Course Notes, Series, no. 34, 71 p.

SCHLAGER, W., 1999, Type 3 Sequence Boundaries, *in* Harris, P.M., Saller, A.H., and Simo, J.A., eds., Advances in Carbonate Sequence Stratigraphy: Application to Reservoirs, Outcrops, and Models: SEPM, Special Publication 63, p. 35–45.

STRASSER, A., 1994, Milankovitch cyclicity and high-resolution sequence stratigraphy in lagoonal–peritidal carbonates (Upper Tithonian–Lower Berriasian, French Jura Mountains), *in* de Boer, P.L., and Smith, D.G., eds., Orbital Forcing and Cyclic Sequences: International Association of Sedimentologists, Special Publication 19, p. 285–301.

STRASSER, A., CARON, M., AND GJERMENI, M., 2001, The Aptian, Albian and Cenomanian of Roter Sattel, Romandes Prealps, Switzerland: a high-resolution record of oceanographic changes: Cretaceous Research, v. 22, p. 173–199.

STRASSER, A., PITTET, B., HILLGÄRTNER, H., AND PASQUIER, J-B., 1999, Depositional sequences in shallow carbonate-dominated sedimentary systems: concepts for a high-resolution analysis: Sedimentary Geology, v. 128, p. 201–221.

TAGLIAFERRI, R., PELOSI, N., CIARAMELLA, A., LONGO, N., MILANO, M., AND BARONE, F., 2001, Soft computing methodologies for spectral analysis in cyclostratigraphy: Computers & Geoscience, v. 27, p. 535–548.

VAIL, P.R., AUDEMARD, F., BOWMAN, S.A., EISNER, P.N., AND PEREZ-CRUZ, C., 1991, The stratigraphic signatures of tectonics, eustasy and sedimentology—an overview, *in* Einsele, G., Ricken, W., and Seilacher, A., eds., Cycles and Events in Stratigraphy: Berlin, Springer-Verlag, p. 617–659.

THE MILANKOVITCH INTERPRETATION OF THE LATEMAR PLATFORM CYCLES (DOLOMITES, ITALY): IMPLICATIONS FOR GEOCHRONOLOGY, BIOSTRATIGRAPHY, AND MIDDLE TRIASSIC CARBONATE ACCUMULATION

NEREO PRETO

Dipartimento di Geologia, Paleontologia e Geofisica, Università di Padova, Via Giotto, 1, and Centro di Studio per i Problemi della Geodinamica Alpina - CNR, Corso Garibaldi, 37, 35137 Padova, Italy

e-mail: nereo@geol.unipd.it

LINDA A. HINNOV

Morton K. Blaustein Department of Earth and Planetary Sciences, Johns Hopkins University, 3400 North Charles Street, Baltimore, Maryland 21218, U.S.A.

e-mail: hinnov@jhu.edu

VITTORIO DE ZANCHE AND PAOLO MIETTO

Dipartimento di Geologia, Paleontologia e Geofisica, Università di Padova, Via Giotto, 1, and Centro di Studio per i Problemi della Geodinamica Alpina - CNR, Corso Garibaldi, 37, 35137 Padova, Italy

e-mail: vittorio@geol.unipd.it; mietto@geol.unipd.it

AND

LAWRENCE A. HARDIE

Morton K. Blaustein Department of Earth and Planetary Sciences, Johns Hopkins University, 3400 North Charles Street, Baltimore, Maryland 21218, U.S.A.

e-mail: hardie@ekman.eps.jhu.edu

ABSTRACT: A 160-m-long section measured in the lagoonal facies of the Middle Triassic Latemar platform (Dolomites, Italy) reveals a set of frequency components that we interpret as a strong Milankovitch signal. In this interpretation, all principal frequencies associated with the theoretical Middle Triassic precession index, P1 = 1/(21.7 ky), P2 = 1/(17.6 ky), and its modulations, E1 = 1/(400 ky), E2 = 1/(95 ky), and E3 = 1/(125 ky), were detected in a time–frequency evaluation of the cycles. A weak obliquity signal is also present in part of the section. Thus, the Latemar cycles appear to have recorded the clearest orbital forcing signal yet found in a carbonate platform. This astronomical calibration indicates that the section was deposited in ca. 3.1 My and therefore that the entire Latemar cyclic succession (~ 470 m) took at least 9 My to form. However, the calibration also leads to serious conflicts with other interpreted geological data: U/Pb radiometric ages of zircons collected from tuffites within the Latemar lagoon and in coeval basinal sediments point to a timescale that is five times shorter than this astronomically calibrated estimate; similar discrepancies arise when the average duration of Triassic ammonoid biozones or the sedimentation rates of coeval basinal series are considered. Nonetheless, all of the methods that have been used to estimate the time of formation of the Latemar platform continue to have shortcomings, and the contradictions among these different geological calibrations remain unresolved.

INTRODUCTION

The Latemar platform of the western Dolomites, northern Italy (Fig. 1), is the relict of a small (~ 2 km diameter) Middle Triassic carbonate atoll with a flat-lying internal lagoon, reef, and slope facies (e.g., Gaetani et al., 1981; Goldhammer et al., 1987; Goldhammer and Harris, 1989; De Zanche et al., 1995; Egenhoff et al., 1999). The platform interior is composed of nearly 500 m of stacked meter-scale, shallow-water carbonate cycles, each cycle usually capped by subaerial exposure surfaces. Early studies (Hardie et al., 1986; Goldhammer et al., 1987, 1990; Hinnov and Goldhammer, 1991) suggested that the Latemar cycles were Milankovitch-driven, each cycle corresponding to a ca. 20-ky-long precession cycle. Accordingly, it was hypothesized that the deposition of the Latemar took place over ca. 10–12 My (Goldhammer et al., 1993).

This interpretation was soon disputed. Ammonoid findings within the Latemar platform and the coeval basinal Livinallongo–Buchenstein beds suggested that the bulk of the Latemar platform accounts for just a small portion of the lowermost Ladinian (Brack and Rieber, 1993; De Zanche et al., 1993; De Zanche et al., 1995). Assuming equal biozone durations, if the Latemar accounted for 12 My, then the Ladinian stage alone would represent a ca. 40 My (i.e., almost the duration of the entire Triassic). Shortly thereafter, high-precision radiometric dates of volcaniclastics in the coeval Livinallongo–Buchenstein beds became available (Mundil et al., 1996; Brack et al., 1996), suggesting that the deposition of the Latemar platform took place in less than 2 My. This result was apparently in agreement with the crude indications from biostratigraphy but clearly in conflict with the proposed astronomical calibration. These radiometric ages, which were based upon U/Pb dating of zircons, were later questioned by Hardie and Hinnov (1997). Subsequent U/Pb dating of zircons collected from "tuffite" beds within the Latemar platform interior appear to corroborate the Buchenstein results (Zühlke et al., 2002). Finally, Egenhoff et al. (1999) found that individual Latemar cycles contain lateral facies changes, implying that their formation may not have been controlled solely by a global phenomenon such as orbital forcing but also by autocyclic processes such as local progradation. However, although some Latemar cycles

Cyclostratigraphy: Approaches and Case Histories
SEPM Special Publication No. 81, Copyright © 2004
SEPM (Society for Sedimentary Geology), ISBN 1-56576-108-1, p. 167–182.

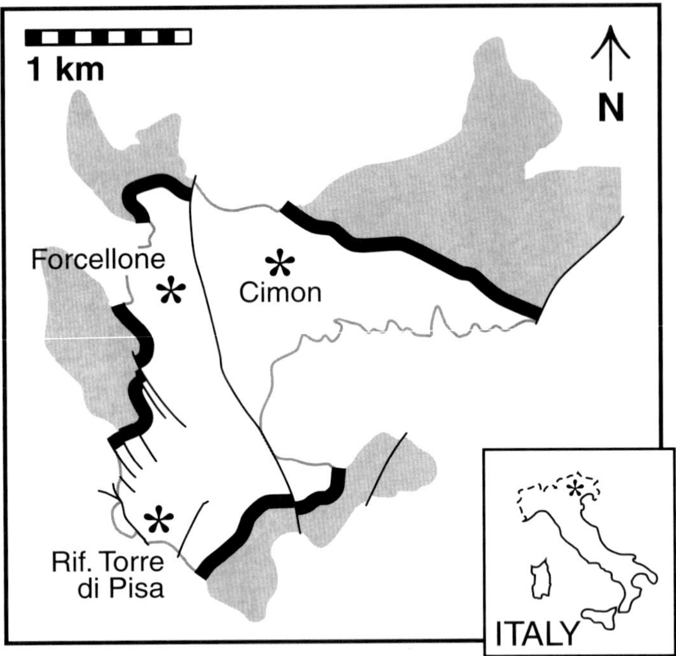

FIG. 1.—Map of the study area. Positions of the Goldhammer–Dunn section at Forcellone (and, for the lower part, at Rif. Torre di Pisa) and CDL section at Cimon del Latemar are also indicated. Light gray, slope facies; black belt, reef margin; lagoonal facies outcrop in the area enclosed by the reef margin. Black thin lines: faults. Redrawn from Egenhoff et al. (1999).

appeared to be discontinuous, most of them were traceable across the platform, in accordance with the earlier studies (e.g., Goldhammer et al., 1990). This is a strong indication that the platform had indeed responded mainly to an external cyclic forcing process.

Here, we present the cyclostratigraphic analysis of the recently measured Cimon del Latemar (CDL) section (Preto et al., 2001), which encompasses ca. one-third of the total Latemar platform-interior thickness at Cimon del Latemar (Figs. 1, 2). The section was measured with the specific aim of critically testing the periodic and Milankovitch nature of the Latemar cycles. The new high-definition stratigraphic measurements were subjected to time–frequency analysis, which revealed that the Latemar cycles developed with two concurrent, narrowly spaced frequency modes consistent with the modes of long and short precession. Through time, these modes underwent amplitude modulations that closely match those of the precession index, i.e., the orbital eccentricity. All three principal eccentricity components (near 1/400, 1/125, and 1/98 cycles/ky) are present at their predicted relative power levels. This new result strongly supports the original Milankovitch hypothesis proposed by Goldhammer et al. (1987) but at the same time heightens the discrepancy of the Milankovitch calibration with the other chronostratigraphic evidence.

The Latemar represents a rare opportunity to compare within a single stratigraphic section the independent results of biostratigraphy, cyclostratigraphy, and radiometric ages during a very complicated geologic age. Indeed, the Latemar has become a veritable "hot spot" for fundamental stratigraphy. But instead of furnishing answers, it has thus far brought researchers to an impasse. We critically discuss all of the chronostratigraphic data related to this still developing debate and conclude with some recommendations for future studies that might help lead to a resolution of the "Latemar controversy."

SEDIMENTOLOGY

Lithofacies and Comparison with Previous Studies

The Latemar cyclic succession was originally examined as a *cycle-thickness series*, and analyzed for nonrandom stacking patterns (Goldhammer et al., 1987; Hinnov and Goldhammer, 1991). However, the cycle thickness approach has limitations: (1) cycle definition requires two levels of interpretation (facies recognition, cycle delineation), (2) spectral analysis of a cycle thickness series cannot test for the presence of orbital signals directly, but only indirectly through the detection of signal modulation frequencies (cycle "bundling"); and (3) mixed orbital signals, i.e., those with comparable forcing contributions from both the obliquity and precession index, cannot be evaluated using a cycle-thickness series. An alternative way to represent cyclic carbonates is to identify the different subfacies constituting the cycles and rank them in order of inferred paleodepth, proximality, or other paleoenvironmental parameter (e.g., Olsen, 1986; Bond et al., 1991; D'Argenio et al., 1997). This approach produces a *lithofacies series*, in which the stratigraphic succession of lithofacies is mapped in terms of rank number. The advantages of the lithofacies approach are: (1) the degree of interpretation is limited to the correct identification and ranking of lithofacies; and (2) the presence of mixed Milankovitch signals can be evaluated directly.

The development of a lithofacies series for the Latemar was undertaken in order to obtain the highest-resolution cyclostratigraphy possible. The lithofacies that we have defined here are not purely descriptive categories but invoke interpretations of sedimentary environment. In this regard, they closely follow the concept of "depth rank" as given in Olsen (1986). At Cimon del Latemar, lithofacies were identified from field observations, polished hand samples, and thin sections (an average of 0.5–1 per meter). Comparison of field observations and thin sections for the same samples showed that the field observations were almost always reliable (discrepancies between field and thin-section observations occurred less than 5% of the time).

Lithofacies 1.—

Lithofacies 1 consists of yellowish, yellow, or sometimes pale orange dolostones with vadose pisoids, commonly coalescent or with multiple-phase growth features (e.g., broken pisoids coated with younger growth layers), and millimeter- to centimeter-scale crusts composed of micrite or cement, in a micritic matrix (Fig. 3D). Pendant and meniscus cements, and dissolution cavities, are abundant. Marine allochems (e.g., gastropods, dasyclads) are also present but usually are affected by dissolution or constitute the pisoid nuclei. Lithofacies 1 is found in association with tepee structures, and may show, at the thin-section scale, millimeter- to centimeter-scale microtepees.

Interpretation.—Caliche soil. Well developed caliches present all the characteristics of this lithofacies and may reach a thickness of more than 10 cm; less developed caliches are usually recognized only from the abundant vadose early cements and dissolution vugs. Beds of lithofacies 1 are commonly associated with

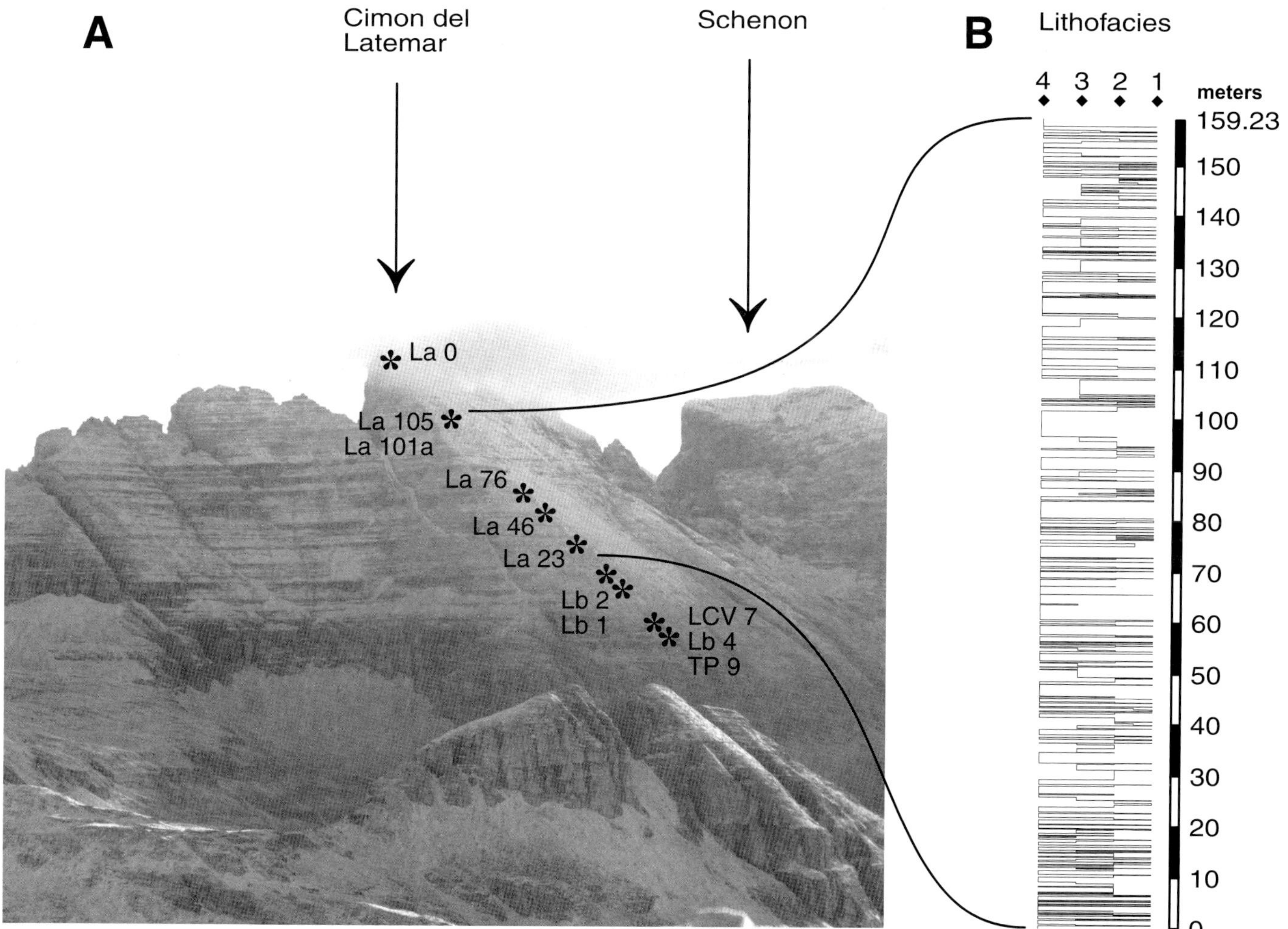

FIG. 2.—The Cimon del Latemar and the CDL section. **A)** Cimon del Latemar; * indicates the approximate position of ammonoid-bearing horizons (projected) which constrain the CDL series to the two lowermost subzones of the Ladinian (*sensu* Mietto and Manfrin, 1995); **B)** CDL series (see the main text for the lithofacies description) and its approximate position at Cimon del Latemar.

storm layers, suggesting that the exposed platform top was flooded occasionally under extreme weather conditions (major storms). The material constituting the caliches derives from these storm layers deposited on the exposed platform top, and only partially from underlying limestones. This latter interpretation coincides only partially with the previous one by Goldhammer et al. (1987), who proposed that caliches formed completely at the expense of underlying subtidal sediments.

Lithofacies 2.—

Lithofacies 2 consists of weakly laminated, mostly peloidal packstones–grainstones, with tufts of cyanobacteria in growth position, partly developed stromatolites, and planar fenestrae (Fig. 3C). Pendant and meniscus cements may be present but are scarce and weakly developed. Layers of Lithofacies 2 rarely exceed a few tens of centimeters in thickness.

Interpretation.—Supratidal-flat sediment. This lithofacies is clearly distinguished from lithofacies 1 because of the virtual absence of pedogenetic features and the rather continuous supply of marine-derived sediments.

Lithofacies 3.—

Lithofacies 3 consists of fine-grained peloidal packstones–wackestones (Fig. 3B). Faunal content is low: this facies typically lacks biota but may carry an oligotypic biota consisting mostly of ostracods and / or agglutinated foraminifers. Gastropods may be present; dasycladacean algae are typically absent. Oncoids can also be present. Bioturbation is detectable as an overall mottled appearance or as open burrows, and it appears to have destroyed any primary sedimentary structures. Lithofacies 3 layers alone can reach meter scale.

Interpretation.—Restricted lagoon sediment, i.e., shallow subtidal lagoon with perhaps unstable salinity, temperature and / or oxygenation, resulting in strong limitations for the survival of a diverse biota. Deposition also in the lower intertidal zone cannot be ruled out.

Lithofacies 4.—

Lithofacies 4 consists of peloidal–bioclastic packstones–wackestones (Fig. 3A). Diverse biota given by abundant dasycla-

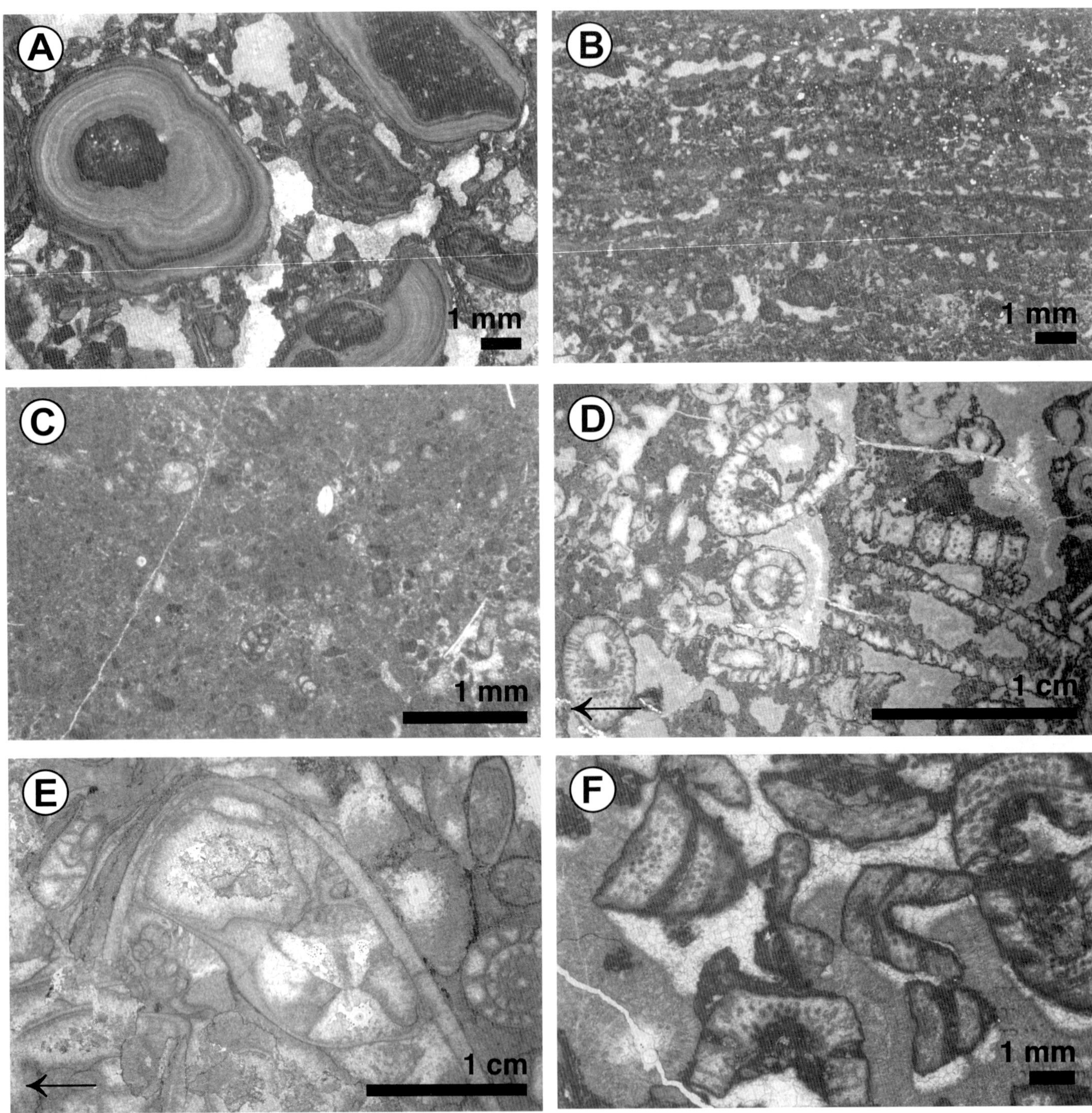

FIG. 3.—Lithofacies of the Latemar platform interior; stratigraphic direction is up where not indicated. **A)** Lithofacies 1: vadose pisoids in dismicrite with abundant dissolution cavities (caliche). **B)** Lithofacies 2: laminated peloidal packstone with elongate cavities (planar fenestrae). This sample is unusually well laminated with respect to the typical Lithofacies 2. **C)** Lithofacies 3: peloidal packstone–wackestone with agglutinate and sparse hyaline foraminifers. **D)** Lithofacies 4: peloidal packstone with abundant dasyclads and foraminifers, and large cavities filled by cement, at least partially due to bioturbation. **E)** Ammonoid coquina (storm deposit): note the random orientation of the ammonoid shells. **F)** Pendant cement in a dasyclad coquina (storm deposit within Lithofacies 1). Such a storm deposit is ranked 1 because of its stratigraphic position within Lithofacies 1.

dacean algae, agglutinated and hyaline foraminifers, molluscs (mainly gastropods), ostracods, and various problematica. Oncoids can be present. Bioturbation is common, and is recognized as open burrows or irregular vugs; primary sedimentary structures are completely absent, probably because of intense bioturbation. Dasycladacean algal remains are commonly accumulated in "dasyclad coquinas" containing abundant peloids. Layers of Lithofacies 4 usually range from some tens of centimeters to a few meters in thickness.

Interpretation.—Open-lagoon sediment, i.e., well oxygenated shallow-water (photic zone) lagoon with normal marine waters. The "dasyclad coquinas" imply a high-energy environment.

Storm Layers.—

Storm layers and storm-derived sediments are most commonly found associated with Lithofacies 1 and in tepee cavities, although they can also be associated with other lithofacies. They are recognized mainly by the presence of pelagic faunas (various cephalopods, with a diverse ammonoid fauna; Figs. 3E and 10) mixed with a shallow-water fauna composed of gastropods, dasycladacean algae, and many other, less common elements; by the upended positions of some fossils, especially ammonoids (Fig. 3E); by the normal grading with jumps in grain size observed in some layers associated with Lithofacies 1 (Seilacher and Aigner, 1991). Storm layers are not associated with any particular lithofacies, and so have been ranked together with the lithofacies of the underlying and overlying layers (e.g., Fig. 3F). Their scarcity in Lithofacies 3 and 4 is most probably due to bioturbation and continuous wave action, which obliterated any recognizable features indicating storm influence and removed storm-derived allochthonous grains from the lagoon.

Comparison with Previous Work.—

Two previous studies (Goldhammer et al., 1987; Egenhoff et al., 1999) have proposed different lithofacies systems. A comparison of the lithofacies defined here with the equivalent lithofacies of these earlier studies is given in Table 1. Probably because those authors had different objectives and measured different sections, they proposed lithofacies that do not always correspond to what we have recognized. Goldhammer et al. (1987) focused on the superposition of caliche soils directly above subtidal sediments, which indeed occurs very often (but not always) at Cimon del Latemar. Thus, in their work only subtidal sediments and caliches were distinguished. In contrast, Egenhoff et al. (1999) endeavored to provide a complete description of facies without regard to an interpretation in terms of inferred depth of deposition or proximality. Thus, they defined lithofacies that are mainly descriptive and may represent different environments from case to case. In this study, we were able to define lithofacies for the purpose of constructing a series that approximates depth of deposition, in order to obtain a record of relative sea level.

Cycle Definition

The four lithofacies alternate in various orders, resulting in the stacking of meter-scale sedimentary cycles. Although the cycles are usually shallowing-upward, there is no fixed rule for the order of the lithofacies within a cycle. The only common feature is that all recognizable sedimentary cycles terminate with a subaerial exposure surface (Lithofacies 1 or 2, or in some cases a seam laterally corresponding to discontinuous patches of Lithofacies 1). Rather than define a "type sedimentary cycle," we have placed cycle boundaries at every first change to a deeper (i.e., higher-ranked) lithofacies. With this procedure a cycle-thickness series was produced (Fig. 4). This cycle-thickness series is useful for broad correlations (see next section) and preliminary time-series analysis (e.g., Hinnov and Goldhammer, 1991). On the other hand, the lithofacies series can be sampled at a very fine spacing and thus permits direct analysis of high-frequency components, which the cycle-thickness series cannot represent. Accordingly, detailed time-series analysis was performed on the lithofacies series only.

Continuity of Cycles

If the Latemar cycles were Milankovitch-driven, i.e., by global sea-level oscillations, then they should be continuous across the platform (Goldhammer et al., 1987). Egenhoff et al. (1999) found that many cycles are laterally continuous but that the lithofacies within single cycles vary laterally; some cycles were physically traced several kilometers around the platform. This is a remarkable outcome, considering that at any given time the platform top would not have been totally flat and more likely to have accumulated sediment unevenly at different locations.

The question arises: are the patterns of cyclic deposition also correlative at different locations? Comparison of the Cimon del Latemar cycle-thickness series with cycle-thickness data mainly from Cima del Forcellone and, for the lowermost part, from Rif. Torre di Pisa (Fig. 1), provides additional evidence not only for lateral continuity of the cycles but also that the cycles experienced similar forcing during deposition. For this analysis, we used a modified version of the cycle-thickness series in Goldhammer et al. (1987) where the major tepee belt has been replaced with correlative cycle thicknesses measured at Cima del Forcellone (Dunn, 1991).

This comparison was performed using a semi-automated procedure: "running" correlation coefficients (r) (e.g., Holland et al., 2000) were calculated between the shorter Cimon del Latemar

TABLE 1.—Equivalence of lithofacies ranks with previous work (Goldhammer et al., 1987; Egenhoff et al., 1999). Goldhammer et al. (1987) is less detailed: only caliche crusts and subtidal limestones are distinguished; equivalence with Egenhoff et al. (1999) is based on their lithofacies descriptions and a densely sampled portion of the CDL series overlapping with their Figure 14.

Lithofacies rank (this work)	Egenhoff et al. (1999)	Goldhammer et al. (1987)
"storm layers" (not ranked)	lithofacies 5 p.p.	- not included -
1 (subaerial exposure, soils)	lithofacies 4 and 5 p.p. (red subaerial exposure horizons)	"caliche crusts"
2 (supratidal)	lithofacies 4, sometimes lithofacies 3.	- not included -
3 (restricted subtidal)	lithofacies 3 p.p.	"subtidal limestones"
4 (open subtidal)	lithofacies 1 and 2	"subtidal limestones"

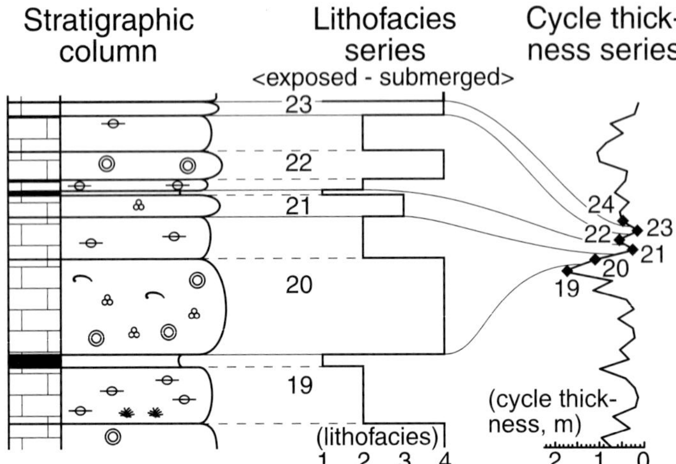

FIG. 4.—Definition of sedimentary cycles with an example from the lower Cimon del Latemar series. A cycle boundary is positioned at the top of each shallowing sequence, allowing reconstruction of a cycle-thickness series. Cycle numbers are indicated. Note symmetric cycle #22 and cycles #21 and #23, with caliche caps directly overlying subtidal lithofacies.

series and successive portions of the much longer Cima del Forcellone series; the highest value of *r* is expected when the Cimon del Latemar series matches the correlative portion of the Cima del Forcellone series.

The procedure identifies the most probable correlation between the two series (Fig. 5) at a level that is confirmed further by marker beds in the field (e.g., layer M of Egenhoff et al., 1999). In addition, the patterns of the variation of cycle thickness show time–frequency similarities between the two series. Finally, there is a close correspondence between the thickest cycles and tepee-rich intervals in the two series, which, together with the correlation test results, indicates an almost cycle-to-cycle correlation between these two long sections measured more than 1 km apart.

Origin and Significance of the Latemar Cycles

Although there are several views about what the Latemar sedimentary cycles exactly represent, there is a general agreement about their allocyclic origin (e.g., Goldhammer et al., 1987, 1990; Hinnov and Goldhammer, 1991; Egenhoff et al., 1999). This is supported mainly by two observations:

(1) Latemar sedimentary cycles are laterally continuous (see previous section). Furthermore, the cycles are continuous despite lateral facies changes, so the formation of cycles does not depend on the local sub-environment.

(2) Walther's law of the superposition of facies does not apply to most of the cycles, i.e., in most cycles subaerial exposure features are directly superimposed on subtidal sediments (cf. diagenetic cycles of D'Argenio, 1974; Hardie et al., 1986; Goldhammer et al., 1987). This violation of Walther's law requires an allocyclic forcing mechanism—in particular, a relative sea-level drop.

In sum, the repeated subaerial exposure of subtidal carbonates indicates that the Latemar sedimentary cycles are related to externally driven sea-level changes (cf. Hardie et al., 1986).

TIME-SERIES ANALYSIS

Time–Frequency Analysis

As for any other carbonate platform, sedimentation rates at the Latemar are expected to have varied over the long term (> 10^6 years) as a result of changes in subsidence rate and sea level (e.g., Goldhammer and Harris, 1989; Goldhammer et al., 1993; Egenhoff et al., 1999). Under such conditions, constant-frequency sea-level oscillations, for example, would have been transformed into lithofacies cycles of varying thickness (wavelengths). Consequently, an average power spectrum of a stratigraphic series of the lithofacies cycles will contain multiple frequencies resulting from timewise migration of actual stable frequency components. These effects can be evaluated through time–frequency analysis using spectrograms (see recent examples for cyclic basinal series in Hinnov et al., 2000; Meyers et al., 2001).

The spectrogram of the CDL lithofacies series (Fig. 6) shows at least one continuous spectral line (labeled P1) slowly oscillating upsection. This particular line corresponds to a wavelength close to the average sedimentary cycle thickness of 0.87 m. Other lines are less easily recognizable but show a similar oscillation. These time–frequency attributes suggest that stable frequency components made up a substantial part of the forcing that generated the Latemar cycles, but that long-term changes in accumulation rate have modified their appearance in the stratigraphic depth representation of the CDL series. Estimation of these accumulation rates is needed to realign the frequencies to their true positions prior to hypothesis testing against theoretical Milankovitch signals, as described below.

Testing the Milankovitch Hypothesis

Background.—

The original Milankovitch hypothesis contemplated by Goldhammer et al. (1987) was based upon observations that the Latemar cycle thicknesses occurred stratigraphically in a basic ca. 5:1 bundling pattern. These authors hypothesized that the cycles were the combined products of long-term subsidence superimposed by rapidly changing sea levels that were under Milankovitch control. The 5:1 bundling, they argued, resulted from a combination of weak eccentricity and dominant precession forcing. Such a signature is actually the hallmark of the precession index, which dominates the orbitally forced seasonal solar radiation over most of the Earth (Berger et al., 1993). The precession index is a quasi-periodic signal with principal frequencies estimated for the Earth's current orbit-rotational configuration at 1/(24 ky) and 1/(22 ky), commonly reported as one frequency at P1 = 1/(23 ky), and P2 = 1/(19 ky). Over the long term these frequencies modulate, which are caused by variations in orbital eccentricity, with frequencies E1 = 1/(400 ky) ("long eccentricity"), and E2 = 1/(95 ky) and 1/(125 ky) ("short eccentricity").

Middle Triassic Orbital Parameters.—

The Earth's orbit–rotational parameters are predicted to have been somewhat different in the geological past. The presence of tidal dissipation suggests that faster spin rates, closer Earth–Moon distances, and a flatter Earth may have characterized more remote times. This would have translated into faster precession rates, and accordingly, shorter periodicities for the precession index and obliquity (e.g., Berger et al., 1992). To model the Middle Triassic orbital parameters, we interpolated the precession and obliquity frequencies predicted by Berger and Loutre (1994, Table 3, Model 2) to 240 Ma. This gives 1 / (21.7 ky) and 1 / (17.6 ky) for the major precession frequencies (P1 and P2), and 1 / (35.8 ky) for the principal obliquity mode. The Earth's orbital eccentricity, on the other hand, is thought to have been relatively invariant over geological timescales, although chaotic motions predicted for some of the inner planets (Laskar, 1990) may have altered some of the short-eccentricity modes (Laskar, 1999). Noteworthy is Late Triassic evidence in the continental Newark basin series that E1, E2, and E3 have apparently the same values then as they do under present the orbit–rotational configuration (Olsen and Kent, 1999).

Hypothesis Testing.—

As a working hypothesis, the continuous spectral line (P1 in Fig. 6B) was interpreted as representing the long precession. This interpretation is corroborated by the presence of a less strong and somewhat less continuous spectral line to the right of P1, namely P2, which occupies a position that is consistent with the short precession. Following this hypothesis, a conservative realignment procedure was carried out by imposing the constant frequency of the Middle Triassic long precession, 1 / (21.7 ky), on P1. More precisely, by imposing a fixed frequency on P1 it is possible to calculate a point-by-point accumulation rate using the formula $s = 1/(21.7fa)$, and where s = accumulation rate, 21.7 ky = Middle Triassic long precession period, and fa = apparent frequency, i.e., the trace of P1 in the spectrogram, in cycles / m. The effect of the variable s is subtracted by using s to transform the CDL series into a tuned time series. If a Milankovitch signal is present and the P1 is correctly identified as the long precession, then other frequencies related to orbital forcing should come into focus in the spectrum of this "minimally tuned" (Muller and MacDonald, 2000) time series.

The tuned spectrum is shown in Fig. 7A. The highest peak at 1 / (21.7 ky) is the tuned long precession and, because it is the product of the tuning, does not provide evidence for Milankovitch forcing. The narrow peak at 1 / (17.6 ky), corresponding precisely to the Middle Triassic short precession, and the peak at 1 / (98 ky), corresponding to short eccentricity, are indications that a Milankovitch signal is present and that the tuning hypothesis is viable. A further peak which could be related to orbital forcing is present at 1 / (400 ky) (i.e., E1), although this has very low power. Other large peaks, at ca. 1 / (200 ky) and ca. 1 / (63 ky), may be related to transient components of the eccentricity (e.g., Hinnov, 2000), although the former has anomalously amplified power. Further examination (Fig. 7B) reveals that the 1 / (200 ky) component appears only in the upper portion of the tuned series, where the trace of P1 was less clear (cf. Fig. 6C); there may be errors in our decisions for tracking P1 at all stratigraphic

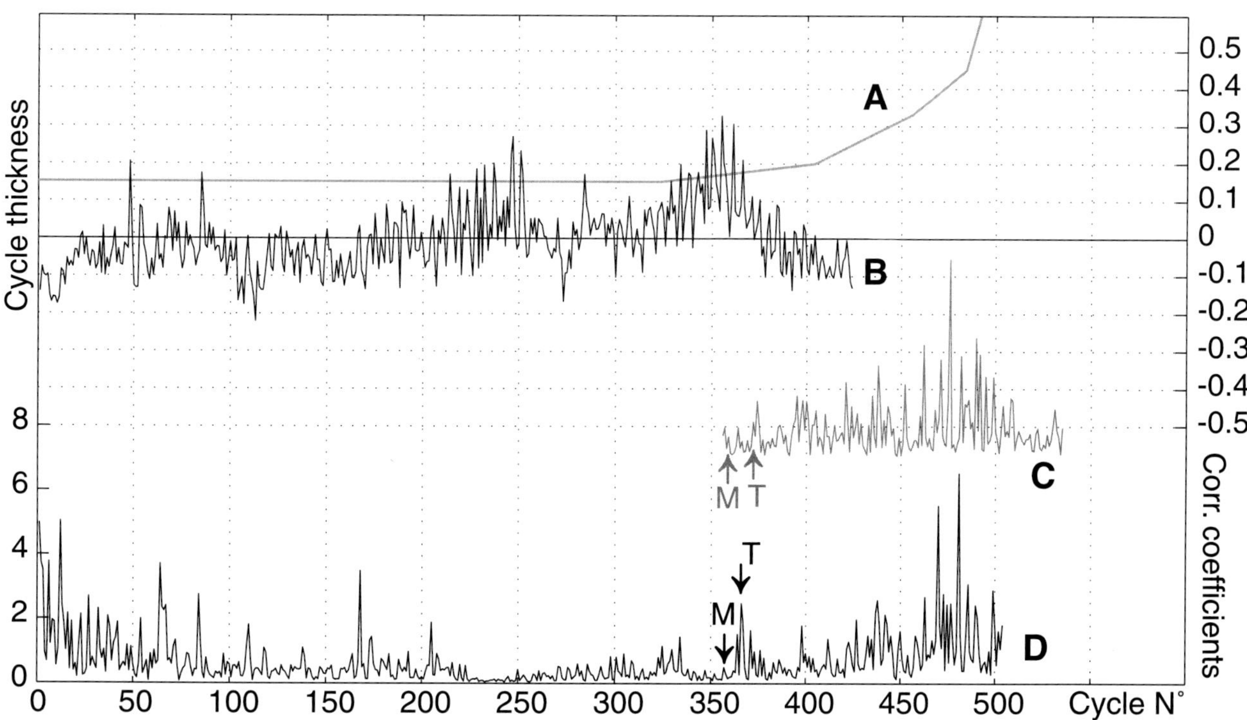

FIG. 5.—Semi-automated correlation between the Cimon del Latemar and Forcellone cycle thickness series (see Fig. 1 for location). Correlation coefficients between the shorter Cimon del Latemar series and successive portions of the Forcellone series are computed for different positions of the Cimon del Latemar series; the highest r is expected between the Cimon del Latemar series and the correlative portion of the Forcellone series. **A, B)** Correlation coefficients r (*b*) and +95% confidence limit (*a*) in the hypothesis of two random series (Taylor, 1990, Table 8.6). Where r exceeds the limit, the two series may be said to be significantly correlated; this occurs most significantly at lag no. 356. **C)** Cimon del Latemar series, at the position suggested by the semi-automated correlation procedure. **D)** Forcellone series (Goldhammer, 1987; Dunn, 1991); the uppermost 140 cycle thicknesses were subjected to spectral analysis in Hinnov and Goldhammer (1991). Arrows indicate the position of marker bed M (Egenhoff et al., 1999) and of a correlatable tepee belt (T).

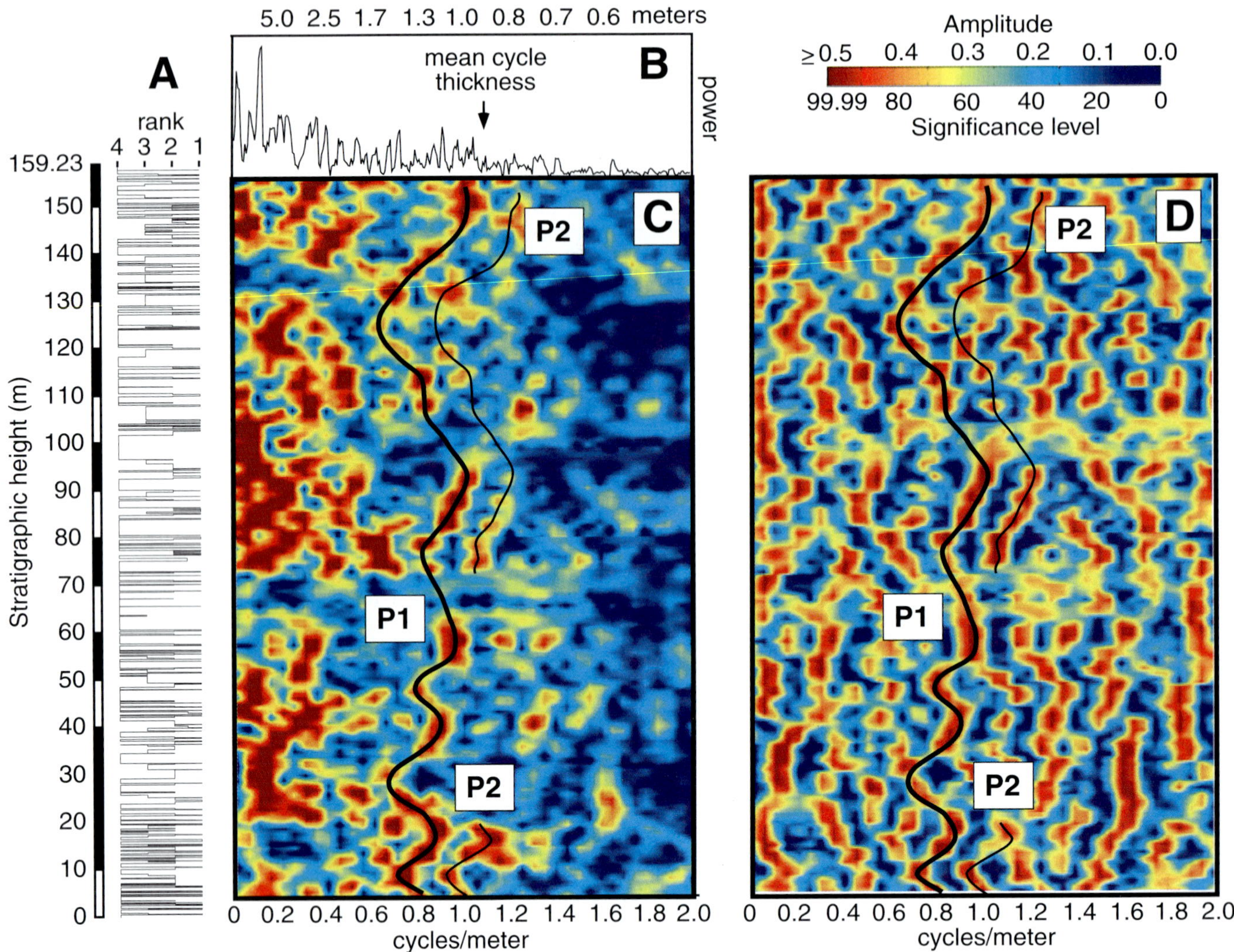

FIG. 6.—Spectral analysis of the Cimon del Latemar lithofacies series. **A)** The 160-meter long CDL lithofacies series, sampled at Δd = 0.2 cm intervals. **B)** 2π multitapered power spectrum of the CDL series. **C)** 2π multitapered line spectrogram of the CDL series, composed of harmonic amplitudes estimated over 10-meter windows in 1-meter steps. **D)** 2π multitapered F-test spectrogram, which depicts results of Thomson harmonic F-testing of the line spectrogram in Part C. The heavy black curve labeled P1 tracks a significant harmonic that modulates slowly up the section, and is interpreted as evidence for the presence of long precession. The line designated P2 is interpreted as evidence for short precession. The decline of P2 over the first part of the section is consistent with the time–frequency behavior of theoretical P2, which experiences minimum power over 400 ky intervals once every ca. 2.3 My.

heights, particularly above 100 m. Thus, the 1/(200 ky) component may be "misplaced," related instead to one of the principal eccentricity modes.

The significance of the spectral peaks in Fig. 7A was evaluated using harmonic F-tests (Thomson, 1982). Because hundreds of F-tests were performed for the analysis, a certain number of false alarms, i.e., positive F-tests not related to true periodicities, are to be expected (Thomson, 1990). False positives can also occur in the absence of true lines when average power is very close to zero (Preto et al., 2001). The tuned long-precession frequency at 1/(21.7 ky) does not pass the F-test at the 95% level, possibly because the realignment procedure has forced two close frequencies that constitute true P1 into a single component (see Preto et al., 2001 for details). A high level of significance of ca. 99% characterizes instead the short-precession peak at 1/(17.6 ky), and a peak at 1/(98 ky) passes the F-test at the 95% level. Among other high F-tests, the one at 1/(35.6 ky) closely corresponds to theoretical Middle Triassic obliquity.

Filtering and Comparison with a Model Precession Index

A Middle Triassic precession index $e\sin\varpi$ (where e is the orbital eccentricity and ϖ is the longitude of the vernal equinox with respect to the perihelion), can be recovered by filtering the precession band (see Fig. 7A). Comparison of the filtered CDL series (Fig. 8A) with the Middle Triassic precession index (Fig. 8B) reveals numerous similarities. The first (i.e., lower) 1200 ky of the CDL series are similar to a portion of the precession index with low-power short precession (indicated by shading in Figs. 8A and

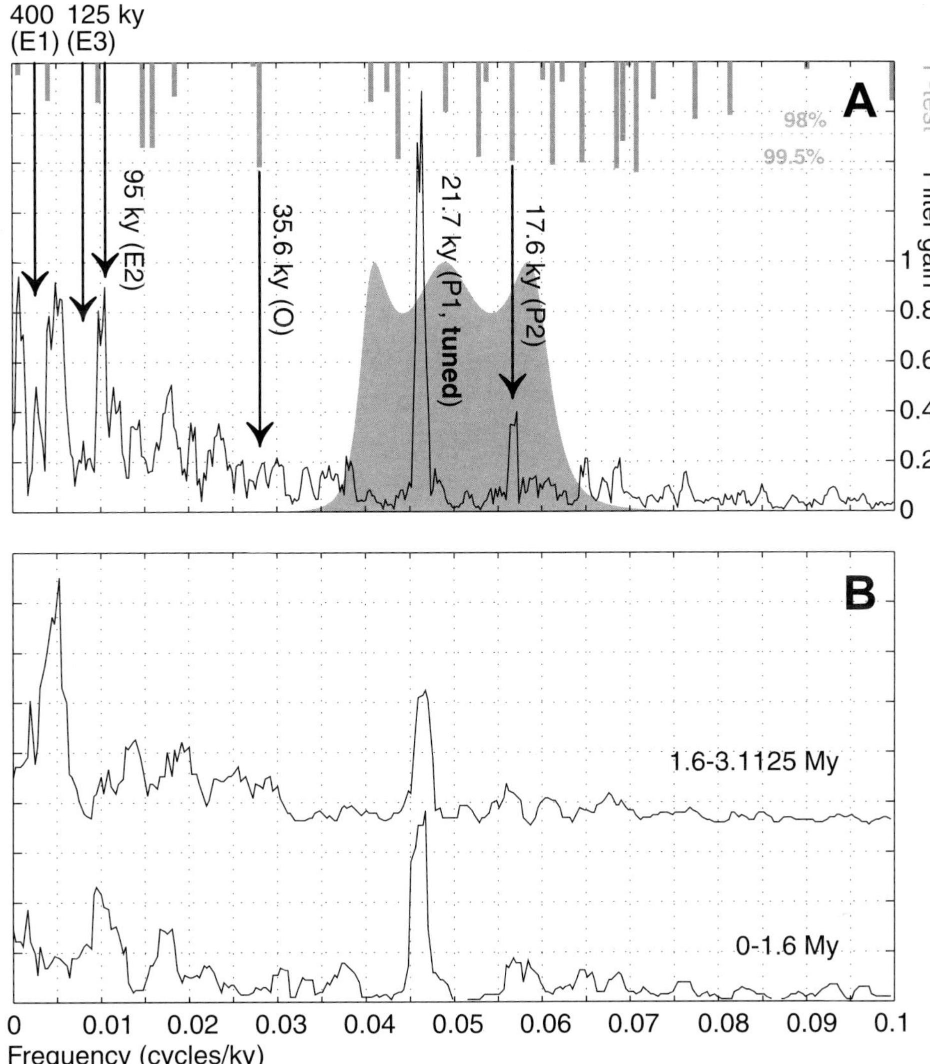

FIG. 7.—Multitaper power spectra (Thomson, 1982) of the P1-tuned Cimon del Latemar time series. **A)** 2π multitaper spectrum for the full time series. Arrows indicate expected positions of Milankovitch frequencies (E1, E2, E3: eccentricity components; O: obliquity; P1, P2: precession components). On the upper side of the graphic, results of the harmonic analysis are provided as F-test results (Thomson, 1990). Frequencies that passed the F-test at the 95% level of significance are: 1/(200 ky), 1/(98.8 ky), 1/(53.5 ky), 1/(35.4 ky), 1/(21.7 ky) ("tuned"), 1/(17.6 ky), 1/(15.4 ky), and 1/(14.6 ky). Filter transfer function is plotted as shaded area. **B)** 2π multitaper spectra of the lower (0–1.6 My) and upper (1.6–3.1125 My) halves of the time series. The 200 ky component is restricted to the upper part, whereas the components at 17.6, 98.8, and ca. 400 ky are present in both parts of the series.

B); the rest of the CDL series shows mainly a qualitative similarity to the precession index.

The amplitude modulations of the recovered precession index are effectively a direct measure of the eccentricity, i.e., in $e\sin\varpi$, $e = e(t)$, as shown in Figure 8C, together with the theoretical eccentricity in Figure 8D. The spectrum of the filtered CDL amplitude modulation series (Fig. 9) reveals spectral peaks at ca. 1/(400 ky), 1/(125 ky), and 1/(95 ky), i.e., the three major components of the eccentricity. The peak at ca. 1/(200 ky) may correspond to a transient component of the eccentricity (Hinnov, 2000), although with a much higher than predicted power level; we suspect that a possible time misalignment in the upper part of the section, as explained previously, may instead be the cause.

An Astronomical Timescale for the Latemar Platform

These results support the hypothesis that the Latemar platform formed under strong orbital control. The recovery of a precession index and the overall clarity of the Milankovitch signal in the Latemar cycles are unprecedented in the cyclostratigraphy of carbonate platforms. We conclude that a true Milankovitch signal has been identified at Cimon del Latemar, allowing the development of an astronomical timescale for the Latemar cyclic succession.

The tuning procedure produced a time series that is 3.1125 My long over the 160 m section. Assuming an average accumulation rate of 160 m/3.1125 My = 51.4 m/My, then the entire ca. 470 meters of the cyclic lagoonal succession (at Cimon del

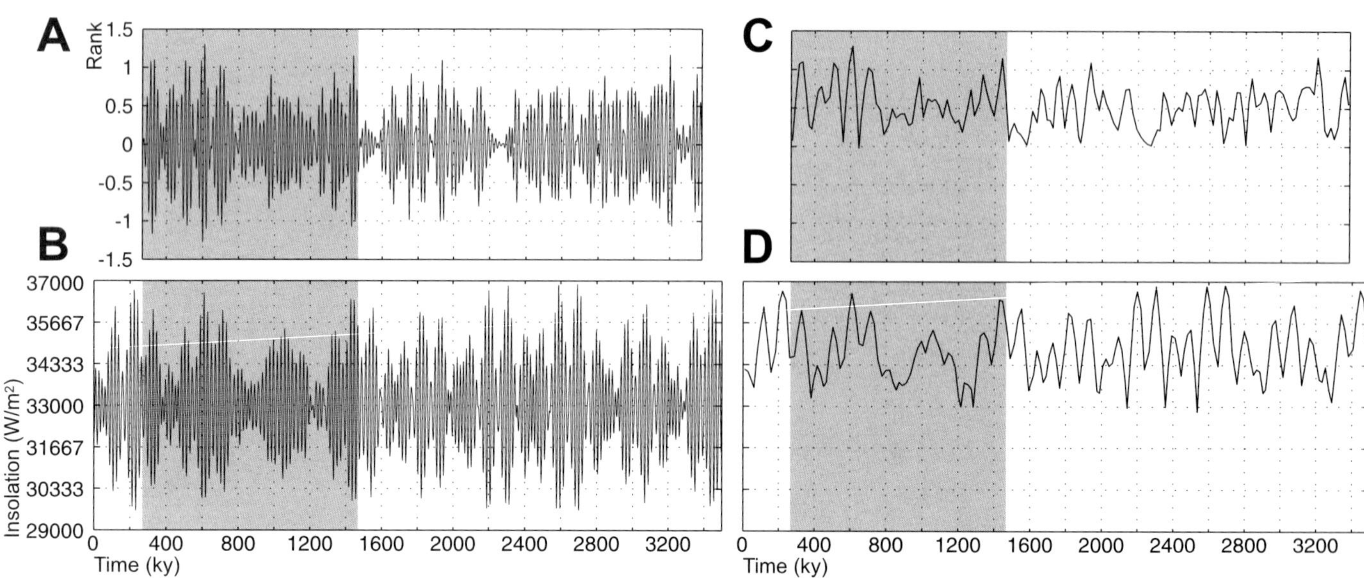

FIG. 8.—The P1-tuned Cimon del Latemar series, bandpassed by the filter depicted in Figure 7A, compared with a model Middle Triassic precession index. Scales in ky. **A)** The tuned, filtered Cimon del Latemar series, interpreted as a record of the precession index. In the lower 1200 ky, highlighted in gray, the precession index is best preserved (compare with the gray box in Part B). **B)** Middle Triassic precession index, from Preto et al. (2001). (Note in that figure caption: k = 56.070946 in/yr should have the dimension arcsec/yr). **C)** Amplitude modulations (AM) of Part A. **D)** Amplitude modulations (AM) of Part B. Because the precession index is equal to $e\sin\varpi$, its amplitude modulations are equal to the eccentricity $e(t)$.

Latemar) should have taken ca. 9.14 My to form. This timescale is the consequence of the foregoing Milankovitch interpretation of the Latemar cycles, which is well supported by the spectral analysis.

Of additional significance is that we have interpreted the CDL series to have 180 *sedimentary* cycles (cf. Figs. 4 and 5). Assuming that each bed represents 20 ky, then the CDL section would calibrate to a duration of 3.6 My, a value that is nearly 0.5 My (16%) longer than the minimally tuned 3.1125 My. Using a simple cycle count for a Milankovitch calibration in the case of the Latemar would yield a biased result, i.e., an overestimation of time or underestimation of the average accumulation rate. This discrepancy is even more pronounced in the lower part of the CDL section.

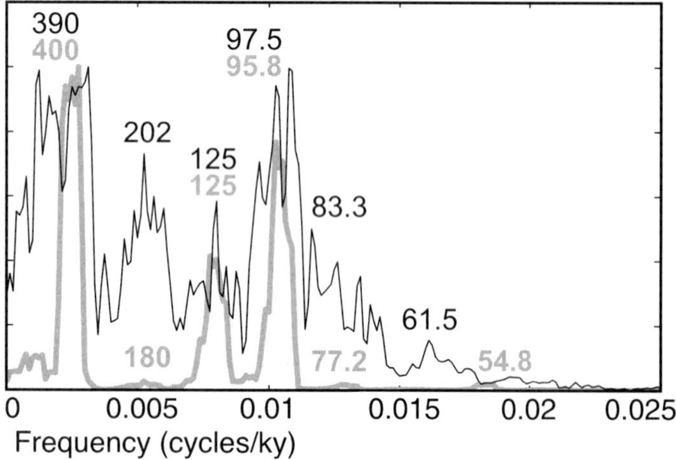

FIG. 9.—AM spectra compared (using 2π multitapers; see Thomson, 1982). Thick gray line: spectrum of the model precession index amplitude modulations of Figure 8D, showing the spectral components of eccentricity. Thin black line: spectrum of the CDL amplitude modulation series of Figure 8C. All components of the eccentricity are recorded: 1/(400 ky), 1/(125 ky), and 1/(95 ky), plus an expected transient component (1/200 ky) with anomalously high power (see text). Labels indicate periodicity in ky.

DISCUSSION

Comparison of this Milankovitch-calibrated timescale with radiometric dating, as well as other interpreted geological data, reveals major discrepancies, as follows.

Average Duration of Ammonoid Biozones

Ammonoids are abundant in several locations in the Latemar platform interior (Brack and Rieber, 1993; De Zanche et al., 1995; Egenhoff et al., 1999; see also Figs. 2 and 10). Sampling of several *in situ* horizons reveals that three ammonoid subzones (namely the Avisianum, Crassus, and Serpianensis Subzones) are documented in the Latemar platform. In the biostratigraphic scale adopted here (Mietto and Manfrin, 1995), the Anisian–Ladinian boundary is placed at the base of the Crassus Subzone, so approximately the lower half of the Latemar platform is Anisian and the upper half is Ladinian (De Zanche et al., 1995).

The published chronostratigraphic scale of Gradstein et al. (1994) indicates that the average duration for the ensemble of the 32 Anisian through Carnian subzones is 0.66 ± 0.20 Ma (2σ level). Accordingly, deposition of the Latemar platform would have occurred in 2.0 ± 0.35 Ma (2σ level), in good agreement with radiometric ages but clearly in conflict with the astronomical calibration.

However, the average duration of bio(chrono)zones provides only a crude estimate of geologic time. Comparison of biostrati-

rates of caliche formation and tepee cement growth estimated from Quaternary analogs.

James (1972) estimated rates of 0.6–3.0 cm / 1000 yr for Pleistocene caliche in Barbados. In the same range are the growth rates of ca. 0.2 cm / 1000 yr for small Holocene caliche nodules (~ 1 cm) that have grown in the buried mud brick walls of Late Bronze age dwellings in the arid Nile delta (Demicco and Hardie, 1994, p. 163). At such rates it would not be possible to accrete more than 6–12 cm of crust in a cycle of 4 to 8 ky (as called for by Brack et al., 1996), assuming that half the cycle time was spent subaerially exposed. The Latemar crusts range in thickness from 5 to 35 cm, so that faster caliche growth rates would be needed to account for the thicker crusts. This, of course, may actually have been the case, so that on its own this argument is not compelling.

However, when we look at the massive tepee cements, then the need for long exposure times is compelling. Typically the ridges of Latemar tepees are arched tens of centimeters to several meters upward with successive caliche-capped cycles onlapping the ridges, demonstrating that much of the ridge growth had taken place before the next cycle was deposited. These ridges are fractured and uparched by expansive cement growth, which isotopic data indicate to have precipitated in the marine vadose and phreatic zones from Triassic seawater (Dunn, 1991). More than 70% of the tepees are composed of cement that was originally aragonite, now pseudomorphed by calcite. Bands of "coconut-meat" cements in sheet cracks in the tepee ridges are typically 2–10 cm wide (examples in Plates 7A–B of Hardie et al., 1991). Handford et al. (1984), using carbon-14 dating, measured rates of aragonite cement growth of 0.2 cm / 1000 yr from seawater brines within modern tepees in the arid McLeod Basin in northern Shark Bay, Australia. If these rates are anywhere near typical, then the massive cements of each Latemar tepee "horizon" would have required tens of thousands of years to form.

Nearly half of the 120 m thickness of the Tepee Facies is taken up by these tepee horizons (a 15 m detail appears in Fig. 21 of Hardie et al., 1991). Using a conservative estimate of 50 m for the aggregate thickness of the tepee horizons in the Tepee Facies alone, ~ 1 million years needs to be added to the time it took to form the primary cyclic bedding (cf. the "missed-beats concept" of Goldhammer et al., 1990).

Alternative Hypotheses

A number of alternative hypotheses have been proposed that might resolve the "Latemar controversy." Brack et al. (1996) suggested that only limited portions of the Latemar cyclic succession may have been controlled by the orbital forcing. However, the basic cyclic theme pervades the Latemar succession, implying that the forcing persisted throughout the formation of the cycles. Schwarzacher (1998) proposed that the Latemar cycles might instead represent products of a millennial-scale forcing, and were modulated by the precession. Such a calibration would neatly resolve all of the geological discrepancies discussed above. This "sub-Milankovitch" hypothesis has been explored in detail by Zühlke et al. (2002), who analyzed cycle-thickness spectra of sections from various locations around the platform, calibrating each cycle to the U / Pb zircon timescale, i.e., a duration of 4 ky.

The sub-Milankovitch hypothesis can be tested against the CDL series by reducing the timescale developed here by a factor of 5. Thus, in Figure 7A, the major components become, from low to high frequency: 1 / (40 ky), 1 / (19.8 ky), 1 / (13.3 ky), 1 / (10.7 ky), 1 / (7.1 ky), 1 / (4.3 ky) ("tuned"), and 1 / (3.5 ky). Although the 1 / (40 ky) component is suspect because of the P1 tracking problems discussed above, the 1 / (19.8 ky) component falls close to Triassic long precession. In this time calibration, the total length of the series is only 622 ky, which may not be long enough to resolve a possible short precession.

Similarly, the AM spectrum (Fig. 8) should contain all of the orbital frequencies involved in the modulation of the hypothesized sub-Milankovitch cycles. Dividing the periodicity labels in Figure 8 by 5 gives, from right to left: 78 ky, 40.4 ky, 25 ky, 19.5 ky, 16.7 ky, and 12.3 ky. It could be argued that the 19.5 ky and 16.7 ky components represent the long and short precession, but then the 40.4 ky component would be too long to represent the obliquity variation, and the 78 ky component too short to be eccentricity.

In sum, although the sub-Milankovitch calibration does not produce periodicities that coincide as closely with the orbital parameters as does the Milankovitch calibration, the uncertainties in our tuning approach and the brevity of the CDL series cannot rule out this calibration on spectral "mismatching" alone. Given the weight of the geological evidence as it presently stands, the sub-Milankovitch hypothesis deserves full scrutiny.

CONCLUSIONS

The objective of this study was to measure and analyze a new section from the Middle Triassic Latemar platform to test the validity of earlier interpretations that the meter-scale cycles constituting the platform were forced by Milankovitch climate variations. The more than 600 cycles of the platform buildup had been proposed to represent ~ 12 million years, but this now stands in serious conflict with U / Pb dating of single zircons that points to a much shorter ~ 3 million year duration.

The principal results are:

1. The sedimentary cycles of the Middle Triassic Latemar platform interior are allocyclic in origin and were caused by relative sealevel changes.

2. Semi-automated correlation analysis shows that the cycles are mostly continuous over significant (~1 km) distances across the platform, pointing to external control on these sealevel changes.

3. Time–frequency signal analysis and a conservative tuning procedure of a 160 m cyclic section at Cimon del Latemar supports an interpretation of strong Milankovitch control on the formation of the Latemar cycles, with dominant forcing from the climatic precession.

4. Astronomical calibration of the 470-m-thick lagoonal succession yields a timescale of 9.14 My, and so remains in serious disagreement with U / Pb-determined ages of zircons found in the Latemar platform and coeval basinal deposits.

5. The astronomical calibration indicates a "normal" 50 m / My accumulation rate for the platform cycles but an unusually low ca. 1 m / My rate for the coeval basinal sediments.

6. The very slow accretion rates of Quaternary caliches and tepees point to many millions of years for the formation of the hundreds of 5 to 35 cm scale Latemar caliches and the ca. 50 meters of massive tepee cement in the Latemar's Tepee Facies.

At present, none of the methods that have been used to estimate the duration of the Latemar buildup have given consistent results, and a Milankovitch interpretation of the Latemar cycles cannot be rejected yet. Therefore, we recommend the following studies that might help lead to a resolution between the radiometric and astronomical calibrations of the Latemar timescale:

1. A reevaluation of the estimated zircon ages of the Buchenstein and Latemar tuffs needs to be undertaken. It is possible that the limited number of zircons reported by Mundil et al. (1996) do not reflect the true range of zircon ages that are present in the sampled tuffs; the potential for having missed those zircons that represent the actual date of deposition of the tuff remains high. An assumption of Pb loss from the outer rims of a significant number of the analyzed zircons prompted those authors to reject younger—yet concordant—zircons that did not cluster. This assumption needs to be tested, because incorporating these younger, concordant zircons would reduce the estimated ages slightly and enlarge the estimated uncertainties significantly. In particular, if the true uncertainty of the U/Pb age estimates is actually on the order of several million years, then the U/Pb vs. the Milankovitch timescale presented in this study would no longer be inconsistent. Therefore, it must be determined whether the zircon rims experienced Pb loss or are simply younger, concordant crystal overgrowths.

2. By the same token, an extension of the CDL series, both upward and downward, is desirable to seek for evidence for a breakdown in the signal and to focus on the origin of the some 15–20% of beds that do not appear to be aligned with the main signal. If "misaligned" beds can be shown to be the products of random events, e.g., storm layers (small beds) or tectonic events (thick beds), as opposed to continuously shallowing-upward sedimentary cycles, then they, and any intervals exhibiting signal loss, can be subtracted from the total time estimate.

3. Finally, improvements to the definition of the ammonite zones in the Latemar are greatly needed. Although critical new information has now been brought to light for defining the upper limit of the cyclic succession, biozone boundaries within the succession have not yet been agreed upon. This information would greatly improve the correlation of U/Pb dated levels from the nearby basinal sections to specific horizons within the platform interior and permit duration estimates for exact stratigraphic intervals.

ACKNOWLEDGMENTS

Research supported by the Università di Padova, fondi di Ateneo 2000 (to N. Preto, V. De Zanche and P. Mietto) and National Science Foundation grant EAR-9909528 (to L.A. Hinnov and L.A. Hardie). Elio Dell'Antonio (Museo Geologico di Predazzo, Italy) kindly provided samples of Anisian ammonoids from the Latemar platform. W. Schlager provided useful suggestions by giving a critical and impartial reading of an early draft of this work. We thank also B. D'Argenio, S. Manfrin, P. Gianolla, G. Roghi, and F. Maurer for assistance in the field and/or for the useful discussions.

REFERENCES

BERGER, A., LOUTRE, M.-F., AND TRICOT, C., 1993, Insolation and Earth's orbital periods: Journal of Geophysical Research, v. 98, p. 10,341–10,362.

BERGER, A., LOUTRE, M.-F., AND LASKAR, J., 1992, Stability of the astronomical frequencies over the Earth's history for paleoclimate studies: Science, v. 255, p. 560–566.

BOWRING, S.A., ERWIN, D.H., JIN, Y.G., MARTIN, M.W., DAVIDEK, K., AND WANG, W., 1998, U/Pb zircon geochronology and tempo of the end-Permian mass extinction: Science, v. 280, p. 1039–1045.

BRACK, P., AND RIEBER, H., 1993, Towards a better definition of the Anisian/Ladinian boundary: New biostratigraphic data and correlations of boundary sections from the Southern Alps: Eclogae Geologicae Helvetiae, v. 86, p. 415–527.

BRACK, P., MUNDIL, R., OBERLI, F., MEIER, M., AND RIEBER, H., 1996, Biostratigraphic and radiometric age data question the Milankovitch characteristics of the Latemar cycles (Southern Alps, Italy): Geology, v. 24, p. 371–375.

BRACK, P., AND MUTTONI, G., 2000, High-resolution magnetostratigraphic and lithostratigraphic correlations in Middle Triassic pelagic carbonates from the Dolomites (northern Italy), Palaeogeography, Palaeoclimatology, Palaeoecology, v. 161, p. 361–380.

BUKRY, D., 1991, Paleoecological transect of western Pacific Ocean late Pliocene coccolith flora, Part I. Tropical Ontong–Java Plateau at ODP 806b: U.S. Geological Survey, Open-File Report 91-552, p. 1–35.

CHERNIAK, D.J., AND WATSON, E.B., 2000, Pb diffusion in zircon: Chemical Geology, v. 172, p. 5–24.

DEMICCO, R.V., AND HARDIE, L.A., 1994, Sedimentary Structures and Early Diagenetic Features in Shallow Marine Platform Carbonate Deposits: SEPM, Atlas Series no. 1, 255 p.

DE ZANCHE, V., GIANOLLA, P., MIETTO, P., SIORPAES, C., AND VAIL, P.R., 1993, Triassic sequence stratigraphy in the Dolomites (Italy): Memorie di Scienze Geologiche, v. 45, p. 1–27.

DE ZANCHE, V., GIANOLLA, P., MANFRIN, S., MIETTO, P., AND ROGHI, G., 1995, A Middle Triassic back-stepping carbonate platform in the Dolomites (Italy): sequence stratigraphy and biochronostratigraphy: Memorie di Scienze Geologiche, v. 47, p. 135–155.

DUNN, P.A., 1991, Cyclic stratigraphy and early diagenesis: an example from the Triassic Latemar platform, northern Italy: unpublished Ph.D. dissertation, The Johns Hopkins University, Baltimore, 836 p.

EGENHOFF, W., PETERHÄNSEL, A., BECHSTÄDT, T., ZÜHLKE, R., AND GRÖTSCH, J., 1999, Facies architecture of an isolated carbonate platform: Tracing the cycles of the Latemàr (Middle Triassic, northern Italy): Sedimentology, v. 46, p. 893–912.

ENOS, P., 1991, Sedimentary parameters for computer modeling, *in* Franseen, E.K., Watney, W.L., and Ross, W., eds., Sedimentary Modeling: Computer Simulations and Methods for Improved Parameter Definition: State Geological Survey of Kansas, Bulletin 233, p. 63–99.

GAETANI, M., FOIS, E., JADOUL, F., AND NICORA, A., 1981, Nature and evolution of Middle Triassic carbonate buildups in the Dolomites (Italy): Marine Geology, v. 44, p. 25–57.

GOLDHAMMER, R.K., 1987, Platform carbonate cycles, middle Triassic of northern Italy: the interplay of local tectonics and global eustasy: unpublished Ph.D. dissertation, The Johns Hopkins University, Baltimore, 468 p.

GOLDHAMMER, R.K., AND HARRIS, M.T., 1989, Eustatic control on the stratigraphy and geometry of the Latemar buildup (Middle Triassic), the Dolomites of Northern Italy, *in* Crevello, P.D., Wilson, J.L., Sarg, J.F., and Read, J.F., eds., Controls on Carbonate Platform and Basin Development: SEPM, Special Publication 44, p. 323–338.

GOLDHAMMER, R.K., DUNN, P.A., AND HARDIE, L.A., 1987, High frequency glacio-eustatic sealevel oscillations with Milankovitch characteristics recorded in Middle Triassic platform carbonates in Northern Italy: American Journal of Science, v. 287, p. 853–892.

GOLDHAMMER, R.K., DUNN, P.A., AND HARDIE, L.A., 1990, Depositional cycles, composite sea-level changes, cycle stacking patterns, and the hierarchy of stratigraphic forcing: Examples from Alpine Triassic platform carbonates: Geological Society of America, Bulletin, v. 102, p. 535–562.

GOLDHAMMER, R.K., HARRIS, M.T., DUNN, P.A., AND HARDIE, L.A., 1993, Sequence stratigraphy and systems tract development of the Latemar platform, Middle Triassic of the Dolomites (northern Italy): outcrop calibration keyed by cycle stacking patterns, *in* Loucks, R.G., and Sarg, J.F., eds., Carbonate Sequence Stratigraphy, Recent Developments and Applications: American Association of Petroleum Geologists, Memoir 57, p. 353–387.

GRADSTEIN, F.M., AGTERBERG, F.P., OGG, J.C., HARDENBOL, J., VAN VEEN, P., THIERRY, J., AND HUANG, Z., 1994, A Triassic, Jurassic and Cretaceous time scale, *in* Berggren, W.A., Kent, D.A., Aubry, M.-P., and Hardenbol, J., eds., Geochronology, Time Scales and Global Stratigraphic Correlation: SEPM, Special Publication 54, p. 95–128.

HANCHAR, J.M., AND MILLER, C.F., 1993, Zircon zonation patterns as revealed by cathodoluminescence and backscattered electron images: implications for interpretation of complex crustal histories: Chemical Geology, v. 110, p. 1–13.

HANCHAR, J.M., AND RUDNICK, R.L., 1995, Revealing hidden structures: the application of cathodoluminescence and back-scattered electron imaging to dating zircons from lower crustal xenoliths: Lithos, v. 36, p. 289–303.

HANDFORD, C.R., KENDALL, A.C., PREZBINDOWSKI, D.R., DUNHAM, J.B., AND LOGAN, B.W., 1984, Salina-margin tepees, pisolites, and aragonite cements, Lake MacLeod, Western Australia: their significance in interpreting ancient analogs: Geology, v. 12, p. 523–527.

HARDIE, L.A., AND HINNOV, L.A, 1997, Biostratigraphic and radiometric age data question the Milankovitch characteristics of the Latemar cycles (Southern Alps, Italy), Discussion: Geology, v. 25, p. 470–472.

Hardie, L.A., Wilson, E.N., and Goldhammer, R.K., 1991, Cyclostratigraphy and dolomitization of the Middle Triassic Latemar buildup, the Dolomites, northern Italy: Dolomieu Conference on Carbonate Platforms and Dolomitization, Ortisei, Italy, Guidebook Excursion F, 56 p.

HARDIE, L.A., BOSELLINI, A., AND GOLDHAMMER, R.K., 1986, Repeated subaerial exposure of subtidal carbonate platforms, Triassic, northern Italy: evidence for high frequency sea level oscillations on a 10^4 year scale: Paleoceanography, v. 1, p. 447–457.

HINNOV, L.A., 2000, New perspectives on orbitally forced stratigraphy: Annual Review of Earth and Planetary Sciences, v. 28, p. 419–475.

HINNOV, L.A., PARK, J., AND ERBA, E., 2000, Lower–Middle Jurassic rhythmites from the Lombard Basin, Italy: a record of orbitally forced carbonate cycles modulated by secular environmental changes in West Tethys, *in* Hall, R.L., and Smith, P.L., eds., Advances in Jurassic Research 2000: Switzerland, Trans Tech Publications Ltd., p. 437–454.

HINNOV, L.A., AND GOLDHAMMER, R.K., 1991, Spectral analysis of the Middle Triassic Latemar Limestone: Journal of Sedimentary Petrology, v. 61, p. 1173–1193.

HOLLAND, S.M., MEYER, D.L., AND MILLER, A. I., 2000, High-resolution correlation in apparently monotonous rocks: Upper Ordovician Kope Formation, Cincinnati Arch: Palaios, v. 15, p. 73–80.

JAMES, N., 1972, Holocene and Pleistocene calcareous crust (caliche) profiles: criteria for subaerial exposure: Journal of Sedimentary Petrology, v. 42, p. 817–836.

LANDING, E., BOWRING, S.A., DAVIDEK, K.L., WESTROP, S.R., GEYER, G., AND HELDMAYER, W, 1998, Duration of the Early Cambrian: U–Pb ages of volcanic ashes from Avalon and Gondwana: Canadian Journal of Earth Sciences, v. 35, p. 329–338.

LASKAR, J., 1999, The limits of Earth orbital calculations for geological time-scale use: Royal Society (London), Philosophical Transactions, Series A, v. 357, p. 1785–1759.

LASKAR, J., 1990, The chaotic motion of the solar system: a numerical estimate of the size of the chaotic zones: Icarus, v. 88, p. 266–291.

LOURENS, L.J., ANTONARAKOU, A., HILGEN, F.J., VAN HOOF, A.A.M., VERGNAUD-GRAZZINI, C., AND ZACHARIASSE, W.J., 1996, Evaluation of the Plio-Pleistocene astronomical timescale: Paleoceanography, v. 11, p. 391–413.

MARTIRE, L., AND CLARI, P., 1995, Evaluation of sedimentation rates in Jurassic–Cretaceous pelagic facies of the Trento Plateau: relevance of discontinuities and compaction: Giornale di Geologia, s. 3, v. 56 (1994), p. 193–209.

MIETTO, P., AND MANFRIN, S., 1995, A high resolution Middle Triassic ammonoid standard scale in the Tethys Realm. A preliminary report: Société Gèologique de France, Bulletin, v. 166, p. 539–563.

MILLER, C.F., HATCHER, R.D., JR., HARRISON, T.M., COATH, C.D., AND GORISCH, E.B., 1998, Cryptic crustal events elucidated through zone imaging and ion microprobe studies of zircon, southern Appalachian Blue Ridge, North Carolina–Georgia: Geology, v. 26, p. 419–422.

MULLER, R.A., AND MACDONALD, G.J., 2000, Ice Ages and Astronomical Causes: Chichester, U.K., Springer-Praxis Publishing, 318 p.

MUNDIL, R., BRACK, P., MEIER, M., RIEBER, H., AND OBERLI, F., 1996, High resolution U–Pb dating of Middle Triassic volcaniclastics: Time scale calibration and verification of tuning parameters for carbonate sedimentation: Earth and Planetary Science Letters, v. 141, p. 137–151.

MUNDIL, R., METCALFE, I., LUDWIG, K.R., RENNE, P.R., OBERLI, F., AND NICOLL, R.S. , 2001, Timing of the Permian–Triassic biotic crisis: implications from new zircon U/Pb age data (and their limitations): Earth and Planetary Science Letters, v. 187, p. 131–145.

MURAKAMI, T., CHAKOUMAKOS, B.C., EWING, R.C., LUMPKIN, G.R., AND WEBER, W.J., 1991, Alpha-decay event damage in zircon: American Mineralogist, v. 76, p. 1510–1532.

OKADA, H., AND BUKRY, D., 1980, Supplementary modification and introduction of code numbers to the low-latitude coccolith biostratigraphic zonation (Bukry, 1973; 1975): Marine Micropaleontology, v. 5, p. 321–325.

OLSEN, P.E., 1986, A 40-million-year lake record of Early Mesozoic orbital climatic forcing: Science, v. 234, p. 842–848.

OLSEN, P.E., AND KENT, D.V., 1999, Long-period Milankovitch cycles from the Late Triassic and Early Jurassic of eastern North America and their implications for the calibration of the Early Mesozoic time-scale and the long-term behaviour of the planets: Royal Society (London), Philosophical Transactions, Series A, v. 357, p. 1761–1786

PRETO, N., HINNOV, L.A., HARDIE, L.A., AND DE ZANCHE, V., 2001, Middle Triassic orbital signature recorded in the shallow-marine Latemar carbonate buildup (Dolomites, Italy): Geology, v. 29, p. 1123–1126.

SCHALTEGGER, U., FANNING, C.M., GÜNTHER, D., MAURIN, J.C., SCHULMANN, K., AND GEBAUER, D., 1999, Growth, annealing and recrystallization of zircon and preservation of monazite in high-grade metamorphism: conventional and in-situ U–Pb isotope, cathodoluminescence, and microchemical evidence: Contributions to Mineralogy and Petrology, v. 134, p. 186–201.

SCHLAGER, W., 1981, The paradox of drowned reefs and carbonate platforms: Geological Society of America, Bulletin, v. 92, p. 197–211.

SCHWARZACHER, W., 1998, Bed thickness measurements and the cyclostratigraphy of the Latemar Limestone: 15th International Sedimentological Congress of the International Association of Sedimentologists, April 12–17, Alicante, Spain, p. 707.

SEILACHER, A., AND AIGNER, T., 1991, Storm deposition at the bed, facies, and basin scale: the geologic perspective, *in* Einsele, G. Ricken, W., and Seilacher, A., eds., Cycles and Events in Stratigraphy: Amsterdam, Elsevier, p. 249–267.

SPAAK, P., 1983, Accuracy in correlation and ecological aspects of the planktonic foraminiferal zonation of the Mediterranean Pliocene: Utrecht Micropaleontology Bulletin, v. 28, p. 1–159.

TAYLOR, J.K., 1990, Statistical Techniques for Data Analysis: Chelsea, Michigan, Lewis Publishers, Inc., 200 p.

THOMSON, D.J., 1982, Spectrum estimation and harmonic analysis: Institute of Electrical and Electronics Engineers, Proceedings, v. 70, p. 1055–1096.

THOMSON, D.J., 1990, Time series analysis of Holocene climate data: Royal Society (London), Philosophical Transactions, Series A, v. 330, p. 601–616.

Weedon, G.P., and Jenkins, H.C., 1999, Cyclostratigraphy and the Early Jurassic timescale: data from the Belemnite marls, Dorset, southern England: Geological Society of America, Bulletin, v. 111, p. 1823–1840.

Zühlke, R., Bechstädt, T., and Mundil, R., 2002, Integrated cyclostratigraphy of a model Mesozoic carbonate platform—the Latemar (Middle Triassic, Italy): 16th International Sedimentological Congress, July 8–12, Johannesburg, South Africa, p. 427–428.

INTEGRATED CYCLOSTRATIGRAPHY OF A MODEL MESOZOIC CARBONATE PLATFORM—THE LATEMAR (MIDDLE TRIASSIC, ITALY)

RAINER ZÜHLKE

Institute of Geology and Palaeontology, Ruprecht Karl University, Im Neuenheimer Feld, 234, D-69120 Heidelberg, Germany
e-mail: zuehlke@uni-hd.de, www.rainer-zuehlke.de

ABSTRACT: An integrated cyclostratigraphic approach was applied to the 460-m-thick succession in the Latemar platform interior. The approach uses new high-resolution cyclostratigraphic data from vertical sections, lateral tracing of physical surfaces over the platform top, new and existing biostratigraphic data, existing isotopic ages from volcanic ash layers, and new spectral analyses in order to develop a genetic cyclostratigraphic model. Hierarchical cycles include meter-scale shallowing-upward microcycles and 2–6 bundled thinning-upward macrocycles. Lateral tracing and correlation of microcycles and macrocycles provides a high-resolution 2D architectural model of the platform interior. The large majority of microcycles and macrocycles is physically persistent over the platform top with only moderate changes in thickness and internal facies. The platform top showed simultaneous vertical aggradation controlled by low-amplitude, high-frequency sea-level changes. Tied-in cyclostratigraphic and biochronostratigraphic data indicate that the 619–701 microcycles in the platform interior include little more than a single ammonoid biozone (*Secedensis* Zone), that the total time interval is shorter than 4.10 Ma (average total time interval = 1.88 My), and that the interpolated microcycle period is shorter than 5.85 ky (average interpolated microcycle period = 2.68 ky). Microcycles cannot be reconciled with precession forcing but reflect sub-Milankovitch forcing. Spectral analysis is based exclusively on accommodation cycles, which represent the only direct indication of external control on cyclic deposition. Blackman–Tukey spectral, multi-taper spectral, and harmonic analyses indicate highly similar and significant frequencies and amplitudes which are largely stationary over all subsets applied to the cyclic series. Ratios and periods indicative of orbital forcing in the Milankovitch band potentially exist at (very) high significance points with Δt = 4.2 ky. In the Latemar cyclic succession, basic microcycles represent sub-Milankovitch forcing (4.2 ky), thinning-upward, 2–6 bundled macrocycles short- and long-precession forcing (18, 21 ky), and higher-order cycle bundles short and long obliquity (35, 45 ky) as well as short-eccentricity forcing (95–105 ky). Because of significant latitudinal temperature gradients and seasonal climate differences, the Triassic period held a significant potential for sub-Milankovitch fluctuations in coupled ocean–atmosphere circulation. They probably triggered low-amplitude, high-frequency changes in sea level and controlled the deposition of sub-Milankovitch microcycles. Previous studies of the Latemar carbonate platform favored a model-dependent approach based on smaller cyclostratigraphic datasets from single sections and spectral analyses. The resulting orbital-forcing models could not be reconciled with the existing biochronostratigraphic framework for the Triassic and the Anisian to Ladinian stages. They left a widely noted deep disagreement between biochronostratigraphic and cyclostratigraphic time scales. In contrast, the new forcing model of this study is based on a complete 2D cyclostratigraphic dataset, considers all biochronostratigraphic constraints, and includes time-calibrated spectral analyses. The model reconciles biochronostratigraphic and cyclostratigraphic time scales. The Latemar cyclic series includes the oldest explicit sub-Milankovitch signal and the oldest set of both sub-Milankovitch and Milankovitch signals yet observed in the geologic record.

INTRODUCTION

Concepts in Cyclostratigraphy

Studies in ancient cyclic carbonate platforms started with the pioneer works of Sander (1936), Schwarzacher (1947), and Fischer (1964). Especially since the late 1980s to early 1990s a large number of detailed studies (e.g., Grotzinger, 1986; Hardie, 1986; Schwarzacher and Haas, 1986; Strasser, 1988; Read, 1989; Goldhammer and Harris, 1989; Bond et al., 1991; Osleger and Read, 1991; Montañez and Read, 1992; Goldhammer et al., 1993) focused on specific aspects of cyclostratigraphy in shallow marine carbonate-platform settings. Of prime interest has been the recognition of cycle stacking patterns, their relation to orbital forcing in the Milankovitch band (Milankovitch, 1941; Berger and Loutre, 1994), and the resulting implications for geological time. Up to now, there have been very few integrated cyclostratigraphic studies incorporating high-resolution (1) cyclostratigraphic data including the spatial architecture of cycles; (2) model-independent time constraints from biostratigraphy, chronostratigraphy, and / or magnetostratigraphy; (3) geochemical proxies like mineralogy and stable isotopes; and (4) spectral analysis.

Cyclostratigraphic models are often based on a single vertical section or well. They include little or no spatial information on (1) the lateral physical continuity of cycles over the carbonate platform or ramp; (2) the lateral facies variability within individual cycles; (3) the lateral variations in the degree of amalgamation or condensation of cycles; and (4) stratal terminations of cycle packages. Few cyclostratigraphic models were based on platform- or ramp-to-basin transects and included several vertical sections (e.g., Montañez and Osleger, 1993; Harris et al., 1993; Goldhammer et al., 1994; Read et al., 1995; Yang and Kominz, 1999). In these studies, the correlation of cycle packages depends largely to entirely on lithofacies correlation, the recognition of large-scale cycle stacking patterns, and few disconformities or marker beds. High-resolution tracing of physical surfaces (cycle tops) and correlations based on detailed small-scale cycle stacking patterns and numerous marker beds has never been performed across a complete carbonate ramp or platform. Spatial information on physical cycle continuity and internal facies variability is decisive for genetic cyclostratigraphic models. Predominantly external (allocyclic) control of rhythmic deposition requires that the large majority of sedimentary cycles physically extend over the platform top or upper ramp. Predominantly internal (autocyclic) control of rhythmic deposition is characterized by progradation from topographic and / or bathymetric highs and by stratal terminations (downlap, onlap) against preexisting surfaces (Ginsburg, 1971; Pratt et al., 1992; Read et al., 1995).

Cyclostratigraphic models for pre-Quaternary platform or ramp settings routinely use cycle stacking patterns and / or spec-

Cyclostratigraphy: Approaches and Case Histories
SEPM Special Publication No. 81, Copyright © 2004
SEPM (Society for Sedimentary Geology), ISBN 1-56576-108-1, p. 183–211.

tral analyses for time calibration. With few exceptions (Drummond and Wilkinson, 1993; Peper and Cloetingh, 1995; Hüssner et al., 2000), they regard specific cycle stacking patterns as proof for orbital forcing in the Milankovitch bandwidth, which triggered rhythmic deposition via climatic and associated sea-level changes. It is usually assumed that the basic meter-scale shallowing-upward cycles represent the shortest orbital-forcing signal, i.e., the precession in the Milankovitch bandwidth. For instance, cycle stacking patterns of five shallowing-upward meter-scale cycles arranged in thinning-upward trends (macrocycles) indicate eccentricity superimposed on precession forcing (Goldhammer et al., 1993; Goldhammer et al., 1994). For spectral analyses of cyclic series, this approach requires calibration of the time series to the signal, which is assumed to be the 20 ky precession signal. A number of studies have compared ancient cyclic series to a (sub-) recent model time series for calibration purposes (e.g., Fischer, 1986; Hinnov and Goldhammer, 1991; Yang and Kominz, 1999; Preto et al., 2001). This approach basically represents a Milankovitch-model-dependent calibration. It requires that (1) orbital forcing in the Milankovitch band is present and is the dominating trigger for rhythmic deposition; (2) each basic cycle reflects a single beat of orbital forcing; (2) other, non-Milankovitch controls on rhythmic deposition were not active, although they are known to exist in the Quaternary; and (3) the shortest period documented in the cyclic platform or ramp succession is the precessional period of approximately 20 ky.

Model-independent time calibrations of cyclic ramp/basin successions or basic shallowing-upward cycles have rarely been performed for pre-Quaternary settings. This was due to relatively large uncertainties of isotopic ages and insufficient biostratigraphic data. However, recent high-resolution age-dating techniques based on single grains (e.g., Steiger et al., 1993; Mundil et al., 1996b; Hilgen et al., 1997; Bowring et al., 1998) have reduced the uncertainties in isotopic ages. The biochronostratigraphic framework of thick cyclic successions is important, because cycles are used as a tool for improving the internal resolution and accuracy of the geological time in the succession and for understanding the variability of other geological factors (e.g., climate) in time.

Miall (1997, p. 208) has summarized the requirements for the establishment of orbital forcing: (1) study of numerous vertical sections; (2) demonstration of the long-distance lateral persistence of cycles, e.g., by the tracing of physical sedimentary surfaces; (2) demonstration of the persistent regularity of cycle periods and bundling; (3) spectral mapping; (4) radiometric dating of bentonites; (5) relation of absolute ages to orbital frequencies.

Objectives

This study follows an integrated approach in order to develop a well constrained genetic cyclostratigraphic model for an isolated carbonate platform in the early Mesozoic. It includes:

- a sub-meter-scale architectural and facies model of a carbonate platform based on laterally persistent physical sedimentary surfaces;
- a high-resolution biostratigraphic and chronostratigraphic framework utilizing existing and new age constraints;
- minimum or maximum datasets for key parameters, e.g., cycle number and period, in order to quantify the cumulative error intervals in the resulting cyclostratigraphic model;
- power and amplitude spectra calibrated to both thickness and time.

This integrated approach was applied to the Latemar carbonate platform in the Southern Alps (Fig. 1). Within the last 15 years it has become a model example of Mesozoic platforms because (1) the platform geometries are completely exposed, and (2) the platform interior shows an extraordinarily well developed cyclicity. A large number of cyclostratigraphic, biostratigraphic, and chronostratigraphic data exist for the Latemar platform and have led to the development of contradicting forcing models. This has triggered an intense debate, informally called the "Latemar controversy".

Geological Setting

The middle Triassic Latemar platform is situated in the central part of the Southern Alps of Northern Italy, commonly referred to as the Dolomites. In the middle Triassic, the Dolomites formed part of a wide continental shelf in the westernmost Tethys. Carbonate ramps and deeper marine straits (100–200 m) prevailed in the Anisian (Zühlke, 2000; Masetti and Trombetta, 1998). In the latest Anisian to early Ladinian, isolated carbonate platforms developed (Rüffer and Zühlke, 1995; Gianolla et al., 1998). They were separated by irregularly shaped basins with water depths of several hundred meters. Upper Anisian to Lower Ladinian platform successions belong to the Schlern Fm. (Sciliar Fm.), time-equivalent basinal successions to the Buchenstein Fm. (Livinallongo Fm.; Bosellini and Rossi, 1974; Gaetani et al., 1981).

The Latemar represents a small isolated carbonate platform in the southwestern part of the Dolomites. The diameter of the platform top ranges between 2.5 and 3.0 km. Dunn (1992), Goldhammer et al. (1993), Harris (1994), and Egenhoff et al. (1999) described the large-scale platform architecture. The major part of the platform interior is built by shallow subtidal to peritidal and supratidal cycles. The platform margin, a tepee belt, was little elevated in terms of topography. A narrow transitional zone separates the platform margin from a discontinuous reef belt, which is situated on the upper foreslope.

EXISTING DATASETS AND MODELS

Figure 2 presents a concise overview on the existing and new cyclostratigraphic and biochronostratigraphic constraints for the Latemar cyclic series. A first set of cyclostratigraphic studies was presented by Hardie et al. (1986), Goldhammer (1987), Goldhammer et al. (1987), Goldhammer and Harris (1989), Goldhammer et al. (1990), Hardie et al. (1991), Hinnov and Goldhammer (1991), and Goldhammer et al. (1993). They described the lithofacies and cycle stacking patterns in the platform-interior succession. Basic fifth-order cycles document an internal shallowing-upward trend from subtidal to peritidal or supratidal conditions. The platform-interior succession was subdivided into four units based on the occurrence or absence of tepees and cycle stacking patterns (Fig. 2, column A; Chart 1 (in pocket in back of book), column A1; Goldhammer et al., 1993, p. 362–365): (1) Lower Platform Facies (LPF); (2) Lower Cyclic Facies (LCF); (3) Tepee Facies (TF); and (4) Upper Cyclic Facies (UCF).

The cyclic succession of the Latemar platform interior comprised 488 measured shallowing-upward cycles in the LCF to lower UCF plus an estimated 110 cycles in the upper UCF. Five basic, fifth-order shallowing-upward cycles constitute a thinning-upward trend and form macrocycles, which was interpreted as 100 ky eccentricity forcing superimposed on 20 ky precessional forcing. Spectral analyses based on a Milankovitch model-dependent time calibration supported this model. The authors assumed an Anisian age for the LPF and a Ladinian age for the LCF to UCF. Existing time scales (e.g., Harland et al., 1982; Palmer, 1983; Harland et al., 1990) proposed a Ladinian stage duration of up to 7 ± 14 My (sic, 238 ± 5 My to 231 ± 13 My; Hinnov and Goldhammer, 1991;

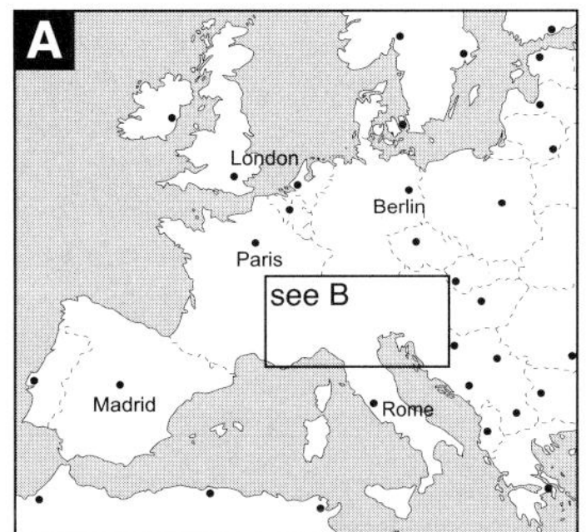

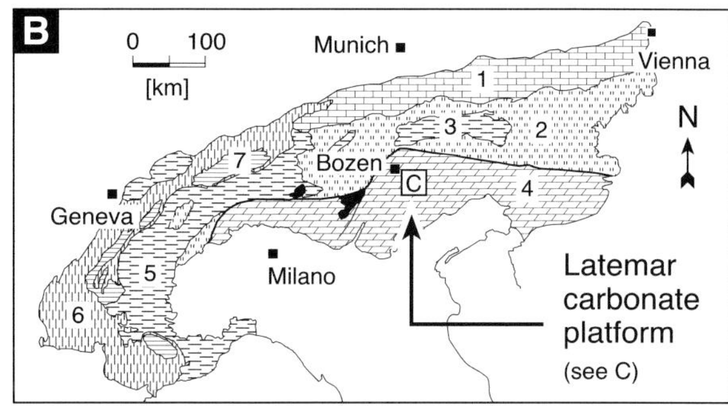

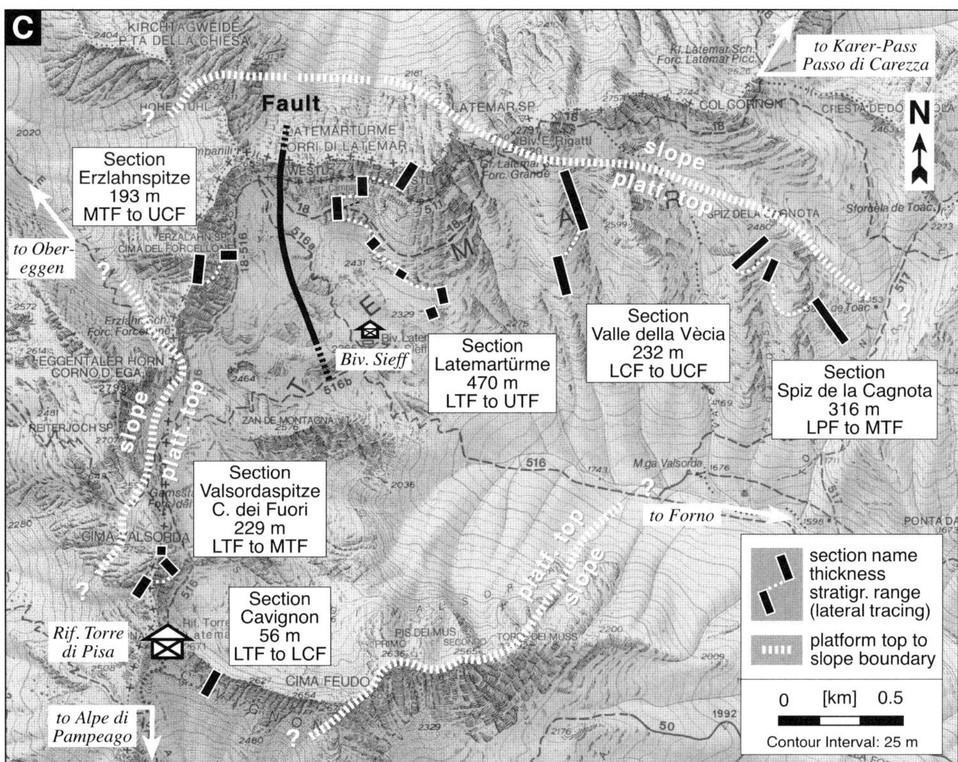

FIG. 1.—Location of the Latemar carbonate platform in the Southern Alps of Northern Italy. **A)** Overview. **B)** Major structural units in the Alps including the position of the Southern Alps and the Latemar carbonate platform. **C)** Topographic map of the Latemar Mountain Range, with the locations of vertical sections (see Chart 1, A3-A7). The boundary between platform top and slope refers to LCF to MTF times. Legend for Part B: 1, eastern Austroalpine unit, Mesozoic sediments; 2, eastern Austroalpine unit, Paleozoic sediments and basement; 3, Penninic unit (eastern Alps); 4, Southern Alpine unit (Southern Alps); 5, Penninic unit (western Alps); 6, Helvetic unit; 7, External massifs; black, Alpine intrusives.

their table 1). This was sufficient to accommodate the necessary < 14.0 My, but at least 11–12 My, for the Latemar cyclic series.

Brack and Rieber (1993) presented the first biostratigraphic ages for the Latemar from the uppermost LPF. According to Brack et al. (1996) the cyclic series covers the upper *Reitzi* to upper *Gredleri* or basal *Archelaus* Zone. Mundil et al. (1996b) and Brack et al. (1996) presented the first radiometric ages for the Latemar series (Fig. 2, column C). On the basis of biostratigraphic data, they correlated U–Pb ages measured for volcaniclastic layers in the basinal Buchenstein Fm. at the Seceda (Northern Dolomites) to the Latemar. These ages constrained the maximum duration of the LCF to UCF interval to < 4.7 My, probably between 2 to 4 My. However, the correlation of the younger radiometric ages to the Latemar platform interior was loosely constrained. It was based on ammonoid faunas in two platform-slope localities (L3 and L4 of Brack et al., 1996), which were difficult to project to the platform interior.

The studies of Brack et al. (1996) and Mundil et al. (1996b) included the first biostratigraphic and chronostratigraphic framework for the Latemar cyclic series. It raised serious doubts about the Hinnov and Goldhammer (1991) and Goldhammer et al. (1993) model. Given a maximum duration of < 4.7 My for the LCF to UCF, basic shallowing-upward microcycles could not represent 20 ky precessional cycles.

De Zanche et al. (1995) presented biostratigraphic data for four levels (TP9 to L3) within the LCF and TF of Goldhammer et al. (1993). They assigned the two cyclostratigraphic units (Fig. 2, column B) to the *Avisianum* to *Crassus* subzones, uppermost *Hungarites* to lowermost *Nevadites* Zone, in the biostratigraphic scheme of Mietto and Manfrin (1995). According to Preto et al.

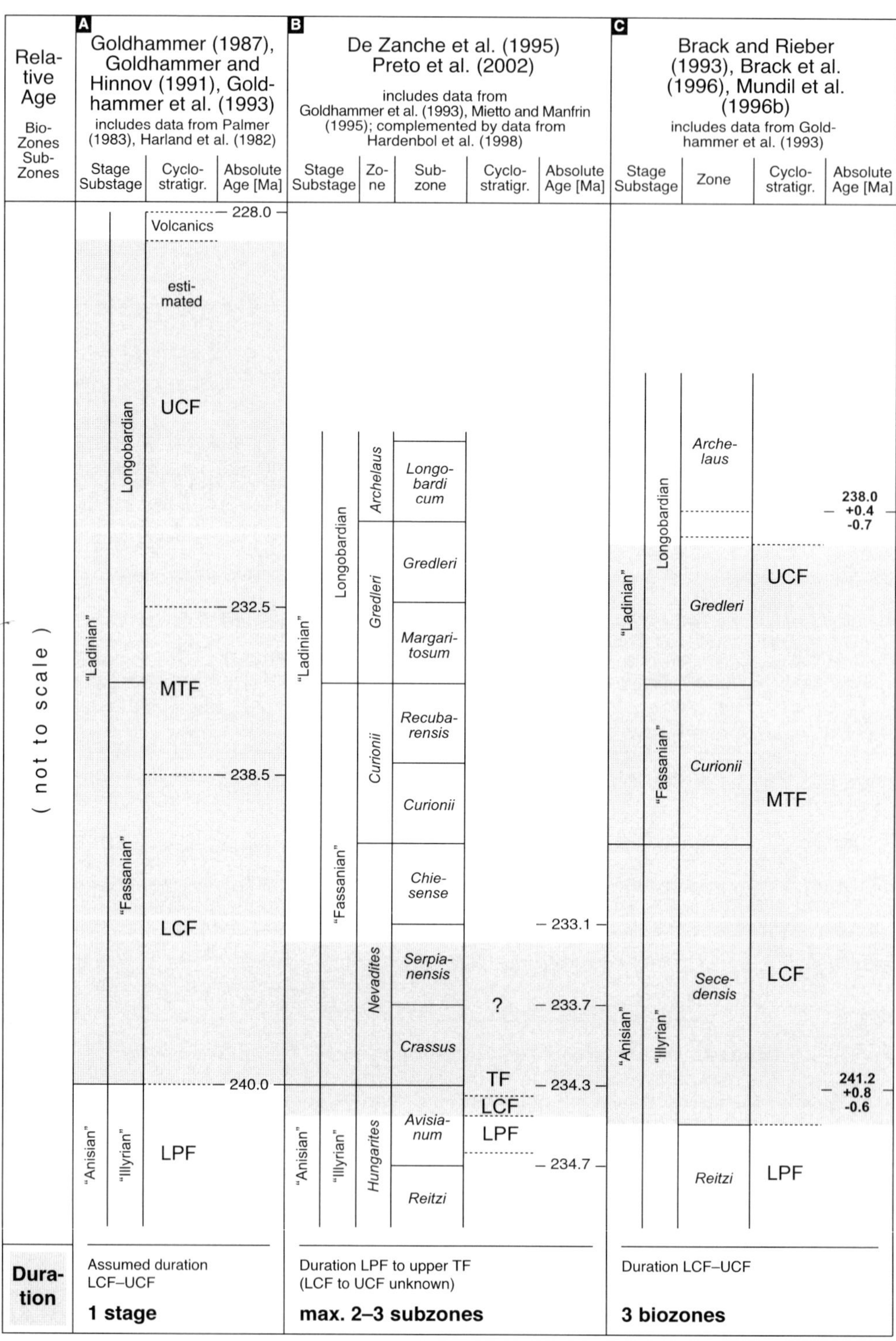

FIG. 2.—Cyclostratigraphic, biostratigraphic, and chronostratigraphic constraints on the Latemar platform interior scaled to relative time. Columns A–D list datasets of previous studies, column E datasets of this study. Gray-shaded areas indicate the biostratigraphic range of the cyclic succession (LCF–UTF, this study) or parts of its (previous studies). Ages in boldface type represent isotopic ages measured from single zircons in volcanic ash layers intercalated in the cyclic series (Mundil et al., 1996b; Mundil et al., 2003). Ages in normal typeface represent interpolated ages from biochronostratigraphic charts (Palmer, 1983; Harland et al., 1982; Hardenbol et al., 1998; Columns A–B). The basal row indicates the biostratigraphic duration of (parts of) the cyclic series. Column E ties all biochronostratigraphic information available to this study to the biostratigraphic schemes of (1) Mietto and Manfrin (1995), de Zanche et al. (1995) (Column E1); (2) Brack and Rieber (1993), Brack et al. (1996) (Column E2). The biostratigraphic duration of the Latemar cyclic series (LCF–UTF), approximately a single ammonoid biozone, is identical in the two schemes.

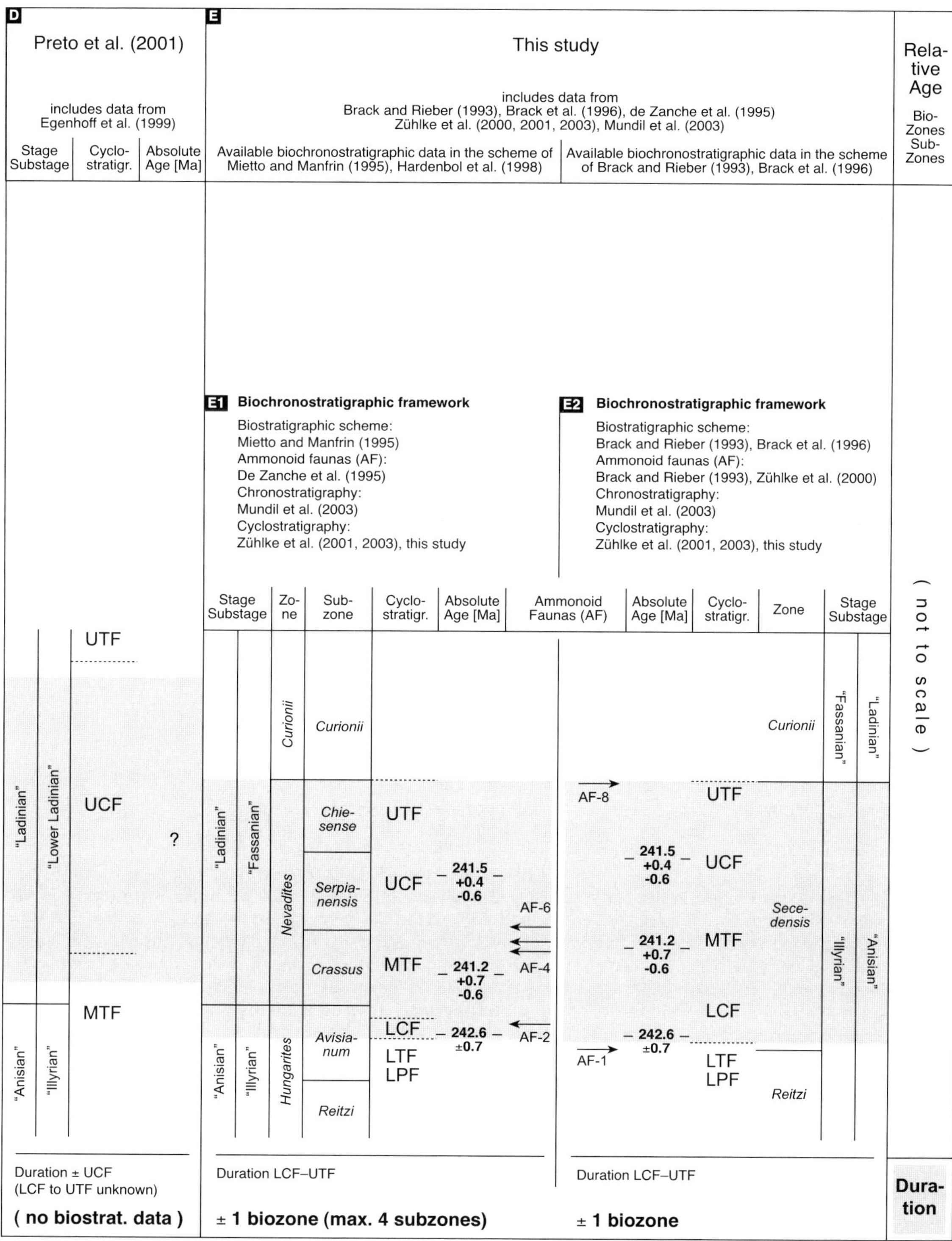

FIG. 2 (continued).—

(2002), the Latemar cyclic series additionally covers the *Serpianensis* subzone.

Egenhoff et al. (1999) proposed a modified subdivision of the Latemar cyclic series, which was based on the observation that tepees also existed at the top of the LPF and in the UCF (*sensu* Goldhammer et al., 1993). They introduced two new units, termed the Lower Tepee Facies (LTF) and the Upper Tepee Facies (UTF).

Zühlke et al. (2000), Zühlke and Bechstädt (2000), and Zühlke et al. (2001, 2003) have presented new cyclostratigraphic data, spectral analyses, and new and existing biostratigraphic data in combination with chronostratigraphic data by Mundil et al. (2003) for the Latemar platform interior. Mundil et al. (2003) have measured high-resolution U–Pb ages from single zircons from three volcaniclastic ash layers intercalated in the cyclic succession. According to the new cyclostratigraphic, biostratigraphic, and chronostratigraphic data the basic shallowing-upward microcycle in the Latemar represents a sub-Milankovitch signal. Milankovitch forcing, including precession, obliquity, and short eccentricity, potentially exists in the larger-scale cycle bundlings.

Preto and Hinnov (2000) compared an 84-m-thick section at the Latemartürme with the section measured by Goldhammer et al. (1987) 1 km to the west at the Erzlahnspitze (in Italian, Cima Forcellone). Preto et al. (2001) measured a single 160-m-thick cyclostratigraphic section in the uppermost MTF to basal UTF (sic) segment of the cyclic series at the Latemartürme (in Italian, Torri di Latemar) (see Chart 1, column A2). They rejected existing chronostratigraphic and biostratigraphic constraints in favor of a Milankovitch-model-dependent approach. The resulting cyclostratigraphic model for the UCF includes short and long precessional, obliquity, and eccentricity forcing.

The Latemar Controversy

In summary, six forcing models exist for the Latemar cyclic series: (1) combined eccentricity and precession forcing for the 1:5 stacking pattern between megacycles and basic shallowing-upward cycles (e.g., Goldhammer et al., 1993); (2) cycles of identical periods reflecting sea-level oscillations of less than 8 ky (Brack et al., 1996); (3) aperiodic "cycles" (Brack et al., 1996); (4) cycles are not allocyclic and of Milankovitch frequency (in Brack et al., 1996; Peterhänsel and Egenhoff, 2000); (5) sub-Milankovitch control of the basic shallowing-upward microcycles, precession forcing in the thinning upward 2–6 bundled macrocycles and obliquity and short-eccentricity forcing in the larger-scale cycle bundlings (e.g., Zühlke and Bechstädt, 2000; Zühlke et al., 2001, 2003); (6) precession forcing of the basic shallowing-upward cycles and obliquity and eccentricity forcing in the large-scale cycle bundlings (e.g., Preto et al., 2001). Bechstädt et al. (2003) have given an overview on the current status of the Latemar controversy.

Open Questions.—

The Latemar controversy has serious implications for carbonate-platform models in the Mesozoic, including the following: (1) Were the basic shallowing-upward microcycles in the Latemar actually triggered by external controls? Similar well developed shallowing-upward cycles exist in many Triassic carbonate platform interiors, e.g., the Dachstein Fm. (Northern Calcareous Austrian Alps; Satterly, 1996; Enos and Samankassou, 1998). (2) Do basic shallowing-upward cycles in the Latemar reflect Milankovitch forcing? (3) Do 1:4–1:5 cycle bundlings between macrocycles and basic shallowing-upward microcycles necessarily indicate eccentricity superimposed on precessional forcing, as usually proposed? (4) Do indications of Milankovitch forcing in the Latemar on a larger scale actually exist if the basic shallowing-upward cycles do not represent precessional forcing?

Current Restrictions.—

The published datasets and cyclostratigraphic models for the Latemar cyclic series include several restrictions, as follows: (1) The total number of cycles is unknown. Cyclostratigraphic information covering a major part of the cyclic series is derived from a single section (Goldhammer, 1987). The section presented by Preto et al. (2001) covers only one-third of the cyclic succession. (2) The lateral continuity of cycles in the platform interior is uncertain. Lateral facies and thickness variations have been demonstrated in four 9-m-thick intervals (Egenhoff et al., 1999). Additional information on cycle continuity is based on an entirely statistical approach (Preto and Hinnov, 2000). (3) Spectral analyses have been applied to only a small portion of the cyclic succession (UCF; Hinnov and Goldhammer, 1991; Preto et al., 2001). (4) biostratigraphic and chronostratigraphic data have not been integrated with high-resolution cyclostratigraphic data.

NEW DATASETS AND MODELS

Cyclostratigraphy

Lithofacies.—

Five vertical sections (Fig. 1) with thicknesses of 194–469 m cover the complete preserved cyclic succession of the Latemar platform interior. Sections include lithofacies, basic shallowing-upward cycles (microcycles), and large stacking patterns (macrocycles). Chart 1 includes all five sections (columns A3–A7) and presents a cyclostratigraphic transect of 4.3 km length over the whole Latemar platform interior. Existing and new biochronostratigraphic data have been tied into this new high-resolution cyclostratigraphic framework on the microcycle level. Seven lithofacies types (LFT) occur in the Latemar platform-interior succession:

- subtidal to shallow subtidal wackestones to grainstones (LFT 1);
- peritidal to intertidal dolomitic caps (LFT 2);
- intertidal to supratidal tepees (LFT 3);
- intertidal to supratidal floatstones, rudstones, or bindstones (LFT 4), often associated with residual sediment;
- supratidal red residual sediment (LFT 5), usually with thicknesses below 5 cm;
- sheet cracks (LFT 6);
- volcaniclastics (LFT 7).

Lithofacies types LFT 1–6 build meter-scale shallowing upward successions that constitute the basic microcycle in the Latemar succession. Two to six microcycles constitute a thinning-upward trend and form macrocycles. Larger-scale stacking patterns of microcycles are well visible in some cyclostratigraphic units and parts of the platform interior. However, a systematic approach to cycle stacking patterns requires spectral analyses.

Wackestones to grainstones (LFT 1) constitute the major part of the basic microcycles in the Latemar. Dolomitic caps (LFT 2), tepees (LTF 3), floatstones–rudstones–bindstones (LFT 4), and residual sediment (LFT 5) characterize cycle tops. Sheet cracks (LFT 6) are concentrated at or directly below cycle tops. They occur predominantly in association with tepees (LFT 3) and floatstones–rudstones–bindstones (LFT 4). Subordinately, sheet cracks occur with dolomitic caps (LFT 2) and residual sediment (LFT 5). Volcaniclastics (LFT 7) represent primary deposits from ash clouds.

Cyclostratigraphic analysis in this study relies exclusively on changes in accommodation. In carbonate settings supratidal lithofacies represents a clear indicator of a decrease in accommodation, because in-situ carbonate production cannot build up above sea level. In the Latemar platform interior, the vertical succession of (shallow) subtidal wackestones and grainstones (LFT 1) overlain by intertidal to supratidal tepees (LFT 3), intertidal to supratidal floatstones–rudstones–bindstones (LFT 4), or supratidal residual sediment (LFT 5) indicates a clear decrease in accommodation. The vertical succession of (shallow) subtidal wackestones and grainstones (LFT 1) overlain by peritidal to intertidal dolomitic caps (LFT 2) represents a further sufficiently reliable indicator of a decrease in accommodation.

Lithofacies type LFT 1 includes several subtypes:

- fine-grained wackestones with bioclasts and peloids (LFT 1a);
- packstones to grainstones with peloids and abundant bioclasts, especially dasycladaceans (LFT 1b);
- intensely bioturbated wackestones to packstones with oncoids, pisoids, and aggregate grains (LFT 1c).

Subtypes of lithofacies type 1 occur in:

- vertical successions of LFT 1a to 1b to 1c within the lower to upper portion of a microcycle, representing a continuous shallowing-upward trend in the subtidal environment;
- vertical successions of LFT 1b to 1a at the base of a shallowing-upward microcycle, representing an initial deepening-upward trend in the subtidal environment (see Chart 1, column A2; Preto et al. 2001, ranks 3 and 4)
- several vertical successions of LFT 1a to 1b to 1c within the lower to upper portion of a microcycle, representing repeated shallowing-upward trends in the subtidal environment.

Within the subtidal part of shallowing-upward microcycles, vertical successions of LFTs 1a–c change laterally in an unsystematic manner over tens to hundreds of meters. Continuous subtidal shallowing-upward trends change laterally to subtidal deepening-upward or to repeated shallowing-upward trends. Lateral and vertical variations of fine-grained wackestones (LFT 1a), packstones and grainstones (LFT 1b), and bioturbated wackestones and packstones (LFT 1c) in the subtidal parts of microcycles reflect local, unsystematic changes in paleobathymetry, water energy, or paleoecologic conditions.

Cyclostratigraphic Units.—

Cyclostratigraphic units in the Latemar platform interior have been redefined on the basis of (1) statistical parameters (average cycle thickness); (2) cycle stacking patterns and the degree of amalgamation or condensation of cycles; (3) relative amounts of lithofacies types LFTs 1–6; and (4) changes in accommodation. Figure 3 shows four 30-m-thick intervals from the Latemartürme section which are representative of the cyclostratigraphic units LCF to UTF. Table 1 lists key parameters of the LCF to UTF in each measured section and compares them to the values given by Goldhammer (1987) and Goldhammer et al. (1993). Cyclostratigraphic units in the Latemar comprise (Figs. 3–6):

- Lower Platform Facies (LPF), total thickness 235–255 m. The following values refer to the upper 90–100 m of the LPF, which have been measured in detail: 58–70 cycles, average microcycle thickness 1.30–1.50 m. In contrast to the cyclic succession (LCF–UTF), cycles in the LPF are usually subtidal to shallow subtidal. Upward thinning of microcycles occurs over thicknesses of 2–12 m (average 5.50 m). Microcycles are often amalgamated beyond recognition; macrocycles are usually well developed. Small-scale tepees, peritidal facies, or subaerial exposure are rare and restricted to the tops of macrocycles.
- Lower Tepee Facies (LTF), thickness 28–30 m, 9–10 microcycles (minimum/maximum numbers); average cycle thickness, 2.84–3.15 m. Thinning-upward microcycles are moderately amalgamated. Subtidal to peritidal tepees with heights of up to 4 m occur at the tops of macrocycles. Clear evidence for subaerial exposure is rare or is lacking completely. Cycle stacking patterns change laterally.
- Lower Cyclic Facies (LCF), thickness 83–100 m, 75–115 cycles (minimum to maximum numbers for all sections); average cycle thickness 0.87–1.16 m. Microcycles mainly include shallowing-upward subtidal to intertidal (lithofacies type LFT 2) or subtidal to supratidal (lithofacies types LFTs 4, 5) successions. Thinning-upward macrocycles are more complete in the southwestern part of the Latemar platform and are moderately amalgamated in the northeast. Few small-scale tepees occur and are usually restricted to the tops of macrocycles.
- Middle Tepee Facies (MTF), thickness 119–134 m, 201–268 cycles (minimum to maximum numbers for all sections except Valle della Veccia); average cycle thickness 0.49–0.52 m. Microcycles mainly include shallowing-upward subtidal to supratidal (lithofacies types LFTs 3, 5) successions. Thinning-upward macrocycles are condensed and grade laterally to large tepee antiforms. The abundance of single and stacked tepees is generally high but changes laterally over the platform interior. In general, the amount of tepees increases from West to East.
- Upper Cyclic Facies (UCF), thickness 169 m, 224–284 cycles (minimum to maximum numbers for the Latemartürme and Erzlahnspitze vertical sections, including projected data), average cycle thickness 0.60–0.75 m. Microcycles mainly include shallowing-upward subtidal to intertidal/supratidal (lithofacies types LFTs 2, 4–6) successions. Thinning-upward macrocycles are usually complete. In many macrocycles, the uppermost 1–2 microcycle(s) grade laterally into tepee antiforms. Stacked tepees do not exist.
- Upper Tepee Facies (UTF), thickness 75 m (cut by the recent erosional surface at the summit of the Latemartürme), 102–121 cycles (minimum to maximum numbers for the Latemartürme section); average cycle thickness 0.62–0.73 m. Microcycles mainly include shallowing-upward subtidal and intertidal to supratidal (lithofacies types LFTs 3–6) successions. Thinning-upward macrocycles are complete to condensed. In almost every macrocycle, the uppermost 2–3 microcycle(s) grade laterally into tepee antiforms. Stacked tepees, equivalent to at least two macrocycles, exist. The UTF develops gradually from the UCF.

Tepees occur over the complete cyclic Latemar succession (LCF–UTF), although in differing abundance. The presence or absence of tepees does not constitute an adequate feature by which to subdivide the succession. This is in clear contrast to the previous studies by Goldhammer (1987), Goldhammer et al. (1993), and Preto et al. (2001). They were based on single sections and did not take into account lateral variations in cycle stacking patterns or lithofacies over tens to hundreds of meters or over the whole platform top. According to Goldhammer (1987) (see Chart 1, Column A1) and Goldhammer et al. (1993), tepees occurred exclusively in the TF (MTF of this study). Preto et al. (2001) did not include any tepees in their rank series of the UCF (Chart 1, Column A2).

Cycle Numbers.—

The Latemartürme section covers the complete cyclic series and serves as the new reference section for the Latemar platform interior (Chart 1, Column A5; Fig. 6). Thickness of the LCF to UTF reaches 456 m. The minimum to maximum number of cycles is 619 to 701. Numbers of cycles in the Valsordaspitze,

LCF Lower Cyclic Facies

- shallow subtidal to peritidal microcycles
- complete to amalgamated thinning-upward macrocycles
- few tepees
- number of microcycles = 90–91
- average thickness of microcycles = 1.02–1.03 m

MTF Middle Tepee Facies

- shallow subtidal to supratidal microcycles
- condensed thinning-upward macrocycles
- stacked tepees
- number of microcycles = 201–221
- average thickness of microcycles = 0.54–0.59 m

UCF Upper Cyclic Facies

- shallow subtidal to peritidal microcycles
- largely complete thinning-upward macrocycles
- tepees
- number of microcycles = 226–286
- average thickness of microcycles = 0.63–0.75 m

UTF Uppper Tepee Facies

- shallow subtidal to supratidal microcycles
- complete to condensed thinning-upward macrocycles
- abundant tepees
- number of microcycles = 102–121
- average thickness of microcycles = 0.62–0.73 m

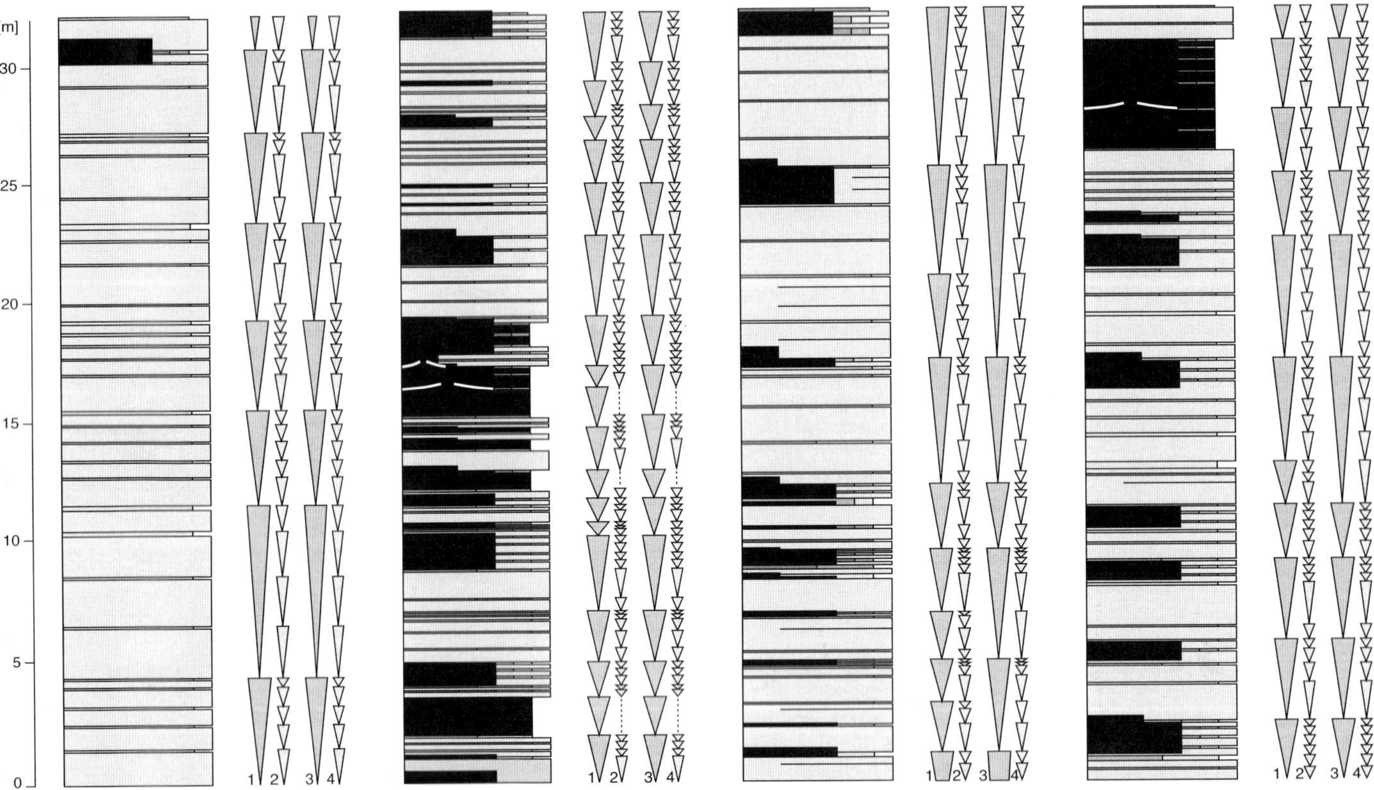

FIG. 3.—Features of cyclostratigraphic units in the Latemar cyclic succession. The figure depicts four 30-m-thick intervals that are representative of the LCF to UTF. Sections have not been idealized or schematized. Cycle numbers and average microcycle thicknesses refer exclusively to the Latemartürme section. Legend: 1, maximum number of macrocycles; 2, maximum number of microcycles; 3, minimum number or macrocycles; 4, minimum number of microcycles; see Chart 1 for color legend.

Erzlahnspitze, and Spiz de la Cagnota sections, where the recent erosional surface cuts deeper into the cyclic succession, ranges between 621 and 745 (including projections from the Latemartürme section). The error interval for the total minimum number of cycles is < ± 1% (619–629) and < ± 3% (699–745) for the total maximum number of cycles. The error interval for minimum and maximum number of cycles is < ± 9% (619–745). Minimum and maximum numbers of cycles account for uncertain cycle tops. As in every cyclostratigraphic study, vertical cycle definition is not always certain where individual shallowing-upward microcycles are condensed or amalgamated. Thin cycle tops built by dolomitic caps (LFT 2) and by floatstones–rudstones–bindstones (LFT 4) have been obliterated in places by postdepositional pressure solution along physical surfaces or by lateral variations in diagenetic overprint. Cycle tops characterized by small-scale tepees (LFT 3) are difficult to define in the inter-tepee areas.

Physical Continuity of Cycles.—

Chart 1 shows the internal architecture of the Latemar platform interior at microcycle and macrocycle resolution. Lagoonal cyclic successions in the Valsordaspitze, Erzlahnspitze, Latemartürme, and Valle della Veccia areas are partly separated by platform slope areas (Reiterjochspitze), topographic passes, and minor faults associated to postdepositional volcanic dikes (early late Ladinian). Several techniques were used to analyze the internal architecture of the Latemar platform:

- physical tracing of 30–50 macrocycle tops in each area. Traced horizons were tied both to vertical sections and high-resolution tele lens images and panorama photographs;
- within the framework of physically traced horizons, the remaining macrocycle tops were visually traced on high-resolution images;

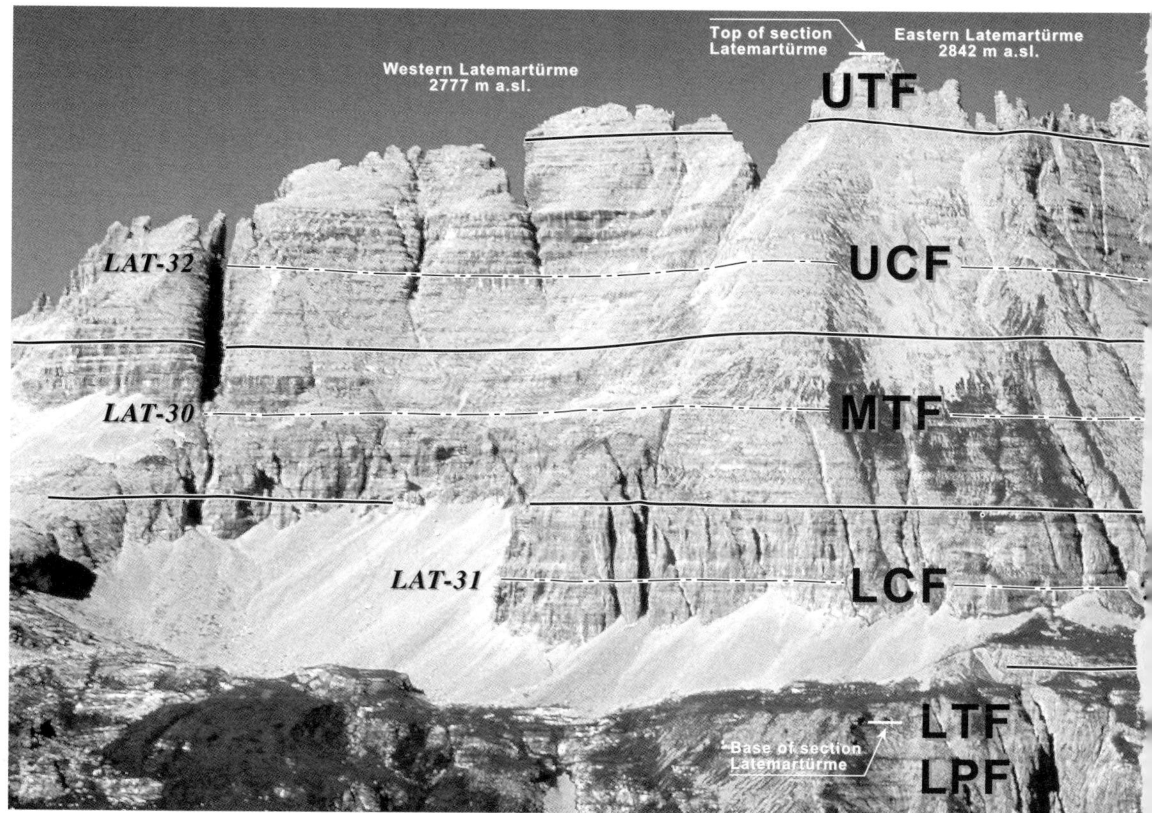

FIG. 4.—Photographic panorama showing the cyclostratigraphic units (LPF–UTF) in the northern part of
(Chart 1, Column A6) sections. Boundaries between cyclostratigraphic units are indicated by thick u

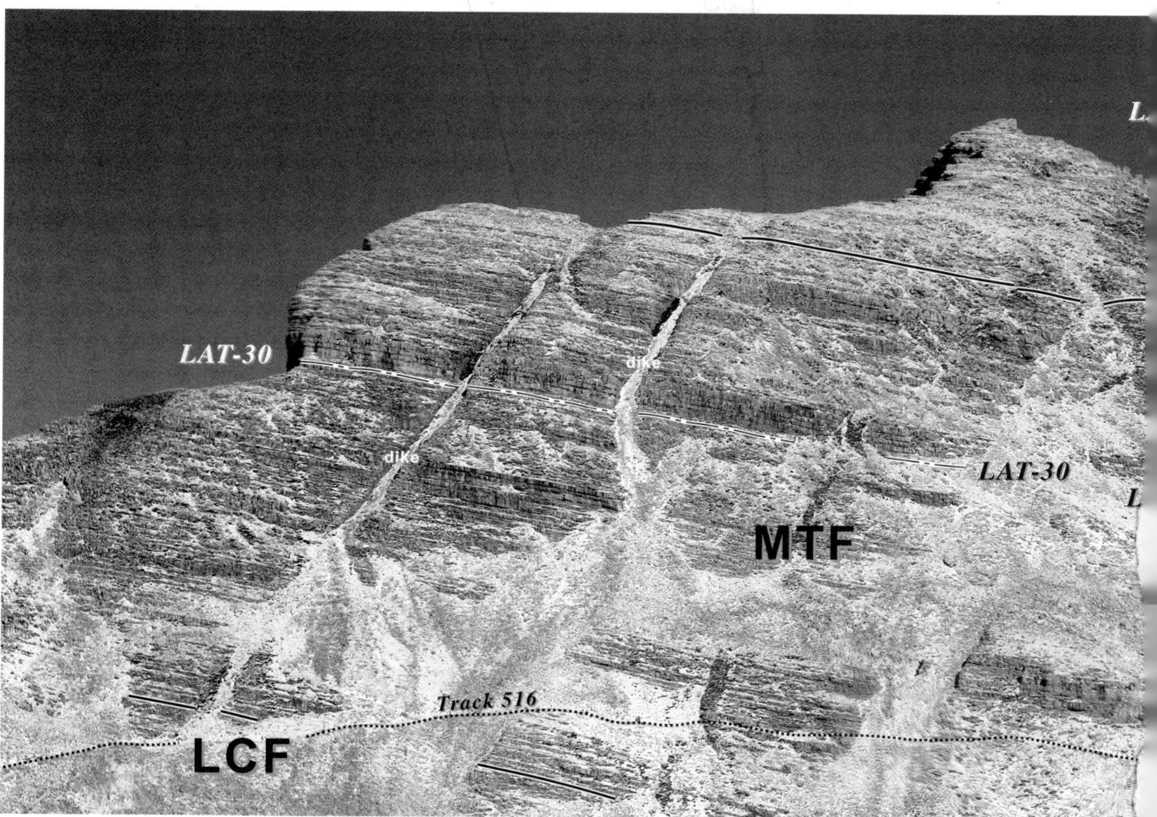

FIG. 5.—Photographic panorama showing the cyclostratigraphic units (LPF–UTF) in the northeastern part
between cyclostratigraphic units are indicated by thick unbroken lines. Age-dated volcanic ash-layers

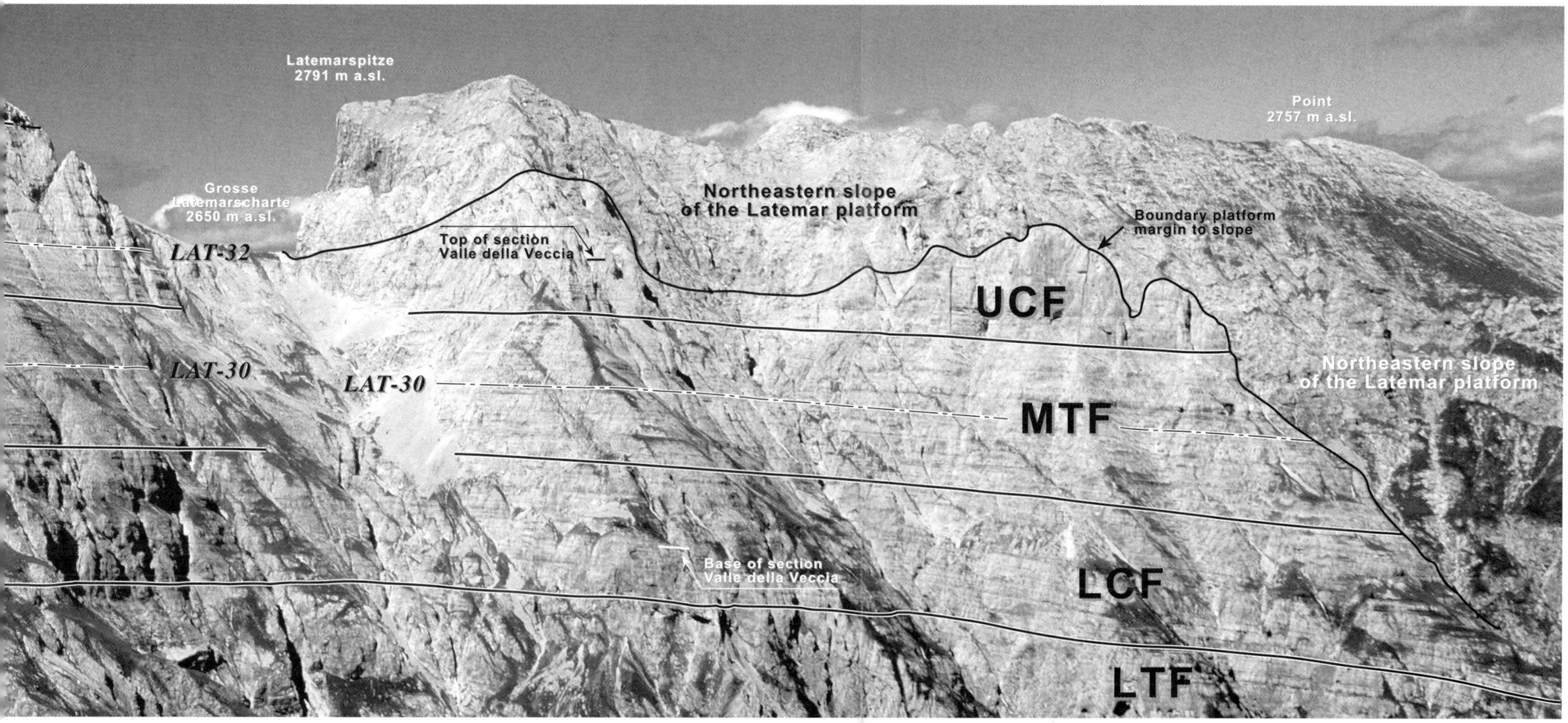

he Latemar (Fig. 1). It includes the Latemartürme (Fig. 6; Chart 1, Column A5) and Valle della Veccia
broken lines. Age-dated volcanic ash layers (Mundil et al., 2003) are indicated by thin broken lines.

of the Latemar (Fig. 1). It includes the Erzlahnspitze section (Chart 1, Column A4). Boundaries
(Mundil et al., 2003) are indicated by thin broken lines.

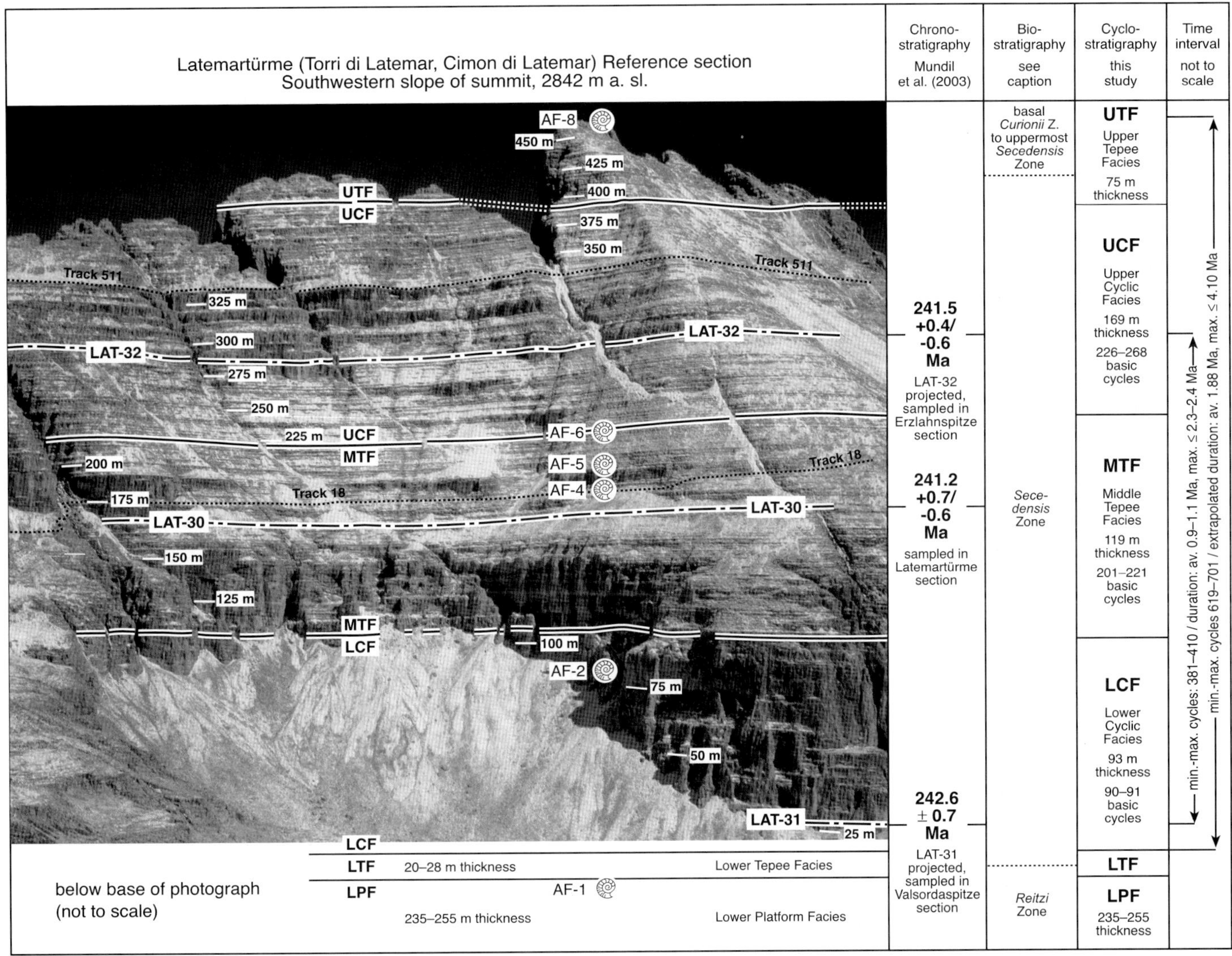

FIG. 6.—Reference section at the southwestern slope of the Latemartürme (in Italian, Torri di Latemar) with the cyclic series between the LCF and UTF. The lowermost 25 m of the LCF, the LTF (Lower Tepee Facies) and LPF (Lower Platform Facies) are below the base of the photograph. The trace of the section is indicated by the 25-m-spaced thickness ticks (see Chart 1, Column A5). Different parts of the section have been assembled along physically traced cycle tops. Cyclostratigraphic levels of biostratigraphic and chronostratigraphic data have been measured and projected into the Latemartürme reference section. Only isotopically dated ash layers and the stratigraphic position of age-indicative ammonoid faunas are shown. Biostratigraphic ages are according to Brack and Rieber (1993), Brack et al. (1996), Rieber (personal communication), and Zühlke et al. (2000). Isotopic ages of ash layers LAT-30 to -32 are according to Mundil et al. (2003).

- where the physical tracing between the four areas was impossible, the correlation of identical macrocycles was based on clearly identifiable marker horizons and recognition of detailed cycle stacking patterns;
- within selected macrocycles, individual shallowing-upward microcycles were traced laterally.

Physical tracing of top surfaces of accommodation cycles is decisive, because these surfaces represent high-resolution time lines. Boundaries between subtidal lithofacies types (LFTs 1a–c) neither necessarily indicate changes in accommodation nor represent time lines. This study exclusively considers accommodation cycles that built up to sea level at the end of each cycle and experienced supratidal conditions during peak relative sea-level fall. No accommodation space was left over from the previous microcycle at the start of the new microcycle.

More than 90% of all macrocycle tops in the Latemar platform interior show physical continuity without stratal terminations across the whole preserved platform interior. The number of thinning-upward microcycles within macrocycles includes subordinate variations caused by varying degrees of internal amalgamation or condensation. In the framework of minimum and maximum number of cycles, 80–95% of all microcycles within laterally continuous macrocycles show physical continuity. Both microcycles and macrocycles include moderate lateral variations in thickness, internal facies, diagenetic overprint, and the nature of cycle tops (lithofacies types LFTs 2–5). Lateral tracing does not provide indications of internal progradation within macrocycles.

TABLE 1.—Cyclostratigraphic data and calculated cycle periods for the Latemar platform interior. Minimum and maximum intervals (Δt = 2.68-5.85 ky; cf. Figure 7, Columns E1–E2) in the Latemartürme section represent values used for the time calibration in spectral analyses (Figs. 11–12). The trace of the section Valle delle Veccia approaches the vertical facies boundary between the cyclic platform-interior succession and the massive platform margin at a very oblique angle. Cycle tops are partly indistinct. Therefore, cycle numbers in the upper part of this section appear to be relatively low and are not representative of the platform interior.

This study	Section					Section	Goldhammer 1987
	Valsordaspitze Cima Valsorda	Erzlahnspitze Cima Forcellone	Latemartürme Torri di Latemar	Valle della Vecia	Spiz de la Cagnota	Cima Forcellone Erzlahnspitze	Goldhammer et al. 1993
complete cyclic series - LCF to UTF							cyclic series - LCF to UCF [10, 11, 12]
thickness [m]	474.20 [1]	456.60 [1]	455.80	463.60 [1]	460.70 [1]	321.78 / 420.00 [12]	thickness meas. / estim. [m]
min.-max. number of macrocycles	148 - 176 [1]	144 - 176 [1]	134 - 167	136 - 160 [1]	135 - 156 [1]	97 / 119 [12]	number of macrocycles meas. / estim.
min.-max. number of microcycles	621 - 745 [1]	629 - 738 [1]	619 - 701	579 - 660 [1]	621 - 699 [1]	488 / 598 [12]	number of microcycles meas. / estim.
min.-max. ø thickness/microcycle [m]	0.64 - 0.76 [1]	0.62 - 0.73 [1]	0.65 - 0.74	0.70 - 0.80 [1]	0.66 - 0.74 [1]	0.66 / 0.70 [12]	ø thickness/microcycle [m]
aver.-max. time interval (Δt=2.68-5.85 ky) [Ma]	1.66 - 4.36 [1]	1.69 - 4.32 [1]	1.66 - 4.10	1.55 - 3.86 [1]	1.66 - 4.09 [1]	-	-
probable time interval (Δt=4.2 ky, Fig. 12B) [Ma]	2.61 - 3.13 [1]	2.64 - 3.10 [1]	2.60 - 2.95	2.43 - 2.77 [1]	2.61 - 2.94 [1]	9.76 / 11.96 [12]	time interval (Δt=20 ky) [My]
Upper Tepee Facies - UTF							
thickness [m]	– [2]	– [2]	74.60	– [2]	– [2]	– [11]	– [11]
min.-max. number of macrocycles	– [2]	– [2]	21 - 26	– [2]	– [2]	– [11]	– [11]
min.-max. number of microcycles	– [2]	– [2]	102 - 121	– [2]	– [2]	– [11]	– [11]
min.-max. ø thickness/microcycle [m]	– [2]	– [2]	0.62 - 0.73	– [2]	– [2]	– [11]	– [11]
aver.-max. time interval (Δt=2.68-5.85 ky) [Ma]	– [2]	– [2]	0.27 - 0.71	– [2]	– [2]	-	-
probable time interval (Δt=4.2 ky, Fig. 12B) [Ma]	– [2]	– [2]	0.43 - 0.51	– [2]	– [2]	– [11]	– [11]
Upper Cyclic Facies - UCF							Upper Cyclic Facies - UCF [10]
thickness [m]	– [2]	169.60 [4]	169.40	169.20 [7]	– [2]	114.18 / 210.00 [10]	thickness meas. / estim. [m]
min.-max. number of macrocycles	– [2]	50 - 63 [4]	45 - 58	40 - 54 [7]	– [2]	22 / 46 [10]	number of macrocycles meas. / estim.
min.-max. number of microcycles	– [2]	224 - 284 [4]	226 - 268	191 - 235 [7]	– [2]	118 / 230 [10]	number of microcycles meas. / estim.
min.-max. ø thickness/microcycle [m]	– [2]	0.60 - 0.76 [4]	0.63 - 0.75	0.72 -0.89 [7]	– [2]	0.96 / 0.96 [10]	ø thickness/microcycle [m]
aver.-max. time interval (Δt=2.68-5.85 ky) [Ma]	– [2]	0.60 - 1.66 [4]	0.61 - 1.57	0.51 - 1.37 [7]	– [2]	-	-
probable time interval (Δt=4.2 ky, Fig. 12B) [Ma]	– [2]	0.94 - 1.19 [4]	0.95 - 1.13	0.80 - 0.99 [7]	– [2]	2.40 / 4.60 [10]	time interval (Δt=20 ky) [My]
Middle Tepee Facies - MTF							Tepee Facies - TF
thickness [m]	123.90 [3]	119.50 [5]	118.90	101.50	133.70 [9]	116.72	thickness [m]
min.-max. number of macrocycles	60 - 65 [3]	55 - 67 [5]	50 - 63	44 - 46	60 - 61 [9]	59	number of macrocycles
min.-max. number of microcycles	220 - 232 [3]	213 - 242 [5]	201 - 221	156 - 160	256 - 268 [9]	295	number of microcycles
min.-max. ø thickness/microcycle [m]	0.53 - 0.56 [3]	0.49 - 0.56 [5]	0.54 - 0.59	0.63 - 0.65	0.50 - 0.52 [9]	0.40	ø thickness/microcycle [m]
aver.-max. time interval (Δt=2.68-5.85 ky) [Ma]	0.59 - 1.36 [3]	0.57 - 1.42 [5]	0.54 - 1.29	0.42 - 0.94	0.69 - 1.57 [9]	-	-
probable time interval (Δt=4.2 ky, Fig. 12B) [Ma]	0.92 - 0.97 [3]	0.89 - 1.02 [5]	0.84 - 0.93	0.66 - 0.67	1.08 - 1.13 [9]	5.90	time interval (Δt=20 ky) [My]
Lower Cyclic Facies - LCF							Lower Cyclic Facies - LCF
thickness [m]	99.80	– [6]	92.90	92.20 [8]	83.20	89.73	thickness [m]
min.-max. number of macrocycles	19 - 24	– [6]	18 - 20	23 - 25 [8]	15 - 16	14	number of macrocycles
min.-max. number of microcycles	97 - 115	– [6]	90 - 91	100 - 111 [8]	72 - 75	73	number of microcycles
min.-max. ø thickness/microcycle [m]	0.87 - 1.03	– [6]	1.02 - 1.03	0.83 - 0.92 [8]	1.11 - 1.16	1.24	ø thickness/microcycle [m]
aver.-max. time interval (Δt=2.68-5.85 ky) [Ma]	0.26 - 0.67	– [6]	0.24 - 0.53	0.27 - 0.65 [8]	0.19 - 0.44	-	-
probable time interval (Δt=4.2 ky, Fig. 12B) [Ma]	0.41 - 0.48	– [6]	0.38 - 0.38	0.42 - 0.47 [8]	0.30 - 0.32	1.46	time interval (Δt=20 ky) [My]

Legend. 1, including projections from Latemartürme reference section; 2, cyclostratigraphic unit has been eroded; 3, present erosional surface at the summit of Valsordaspitze is at 121.5 m above base MTF; uppermost MTF = 1.1 m projected from Latemartürme section; 4, present erosional surface at the summit of Erzlahnspitze is at 93.5 m above base UCF; upper UCF = 77.6 m projected from Latemartürme section; 5, base of section is at 100.5 m below top MTF; lower MTF = 19.6 m projected from Latemartürme section; 6, outcrops not suited for safe recognition of cycles; previous section (Goldhammer, 1987) was measured here (LCF = 89.73 m); 7, limit of the cyclic lagoonal succession to the massive platform margin is at 74.70 m above base UCF, upper UCF = 94.50 m projected from Latemartürme section; 8, base of section is at 42.70 m below top LCF; lower LCF = 45.50 m projected from Latemartürme section; 9, sparse outcrops (approx. 30 m); the limit of the cyclic lagoonal succession to the massive platform margin is at 112.20 m above base MTF; uppermost MTF = 18.40 m projected from Latemartürme section; 10, measured (meas.) a thickness of 114.18 m and 118 cycles (Goldhammer, 1987); estimated a total number of 230 cycles and a total thickness of 210 m (Goldhammer et al., 1993); 11, UTF of this study was not defined and measured (Goldhammer, 1987; Goldhammer et al., 1993); for estimates, see footnote 10; 12, including estimates on cycle number and thickness of the UCF detailed in footnote 10.

In the MTF, packages of 2–5, but usually 3–4, thinning-upward microcycles build macrocycles. In the LCF and UCF–UTF, packages of 3–6, but usually 4–5, thinning-upward microcycles build macrocycles. In general a well developed layer-cake geometry characterizes the Latemar platform interior. This depositional architecture, the lack of stratal terminations (onlap, downlap), and progradational features indicate an external (allocyclic) control on platform development.

Biostratigraphy

Ammonoid faunas in the Latemar platform interior constrain the biostratigraphic correlation of the cyclic succession. They occur on at least eight levels within the cyclic succession. As part of this study, ammonoid faunas described from various platform localities (Brack and Rieber, 1993; De Zanche et al., 1995; Brack et al., 1996) were projected along traced and correlated macrocycle tops into the Latemartürme reference section (Fig. 6). Two ammonoid faunas bracket the complete cyclic succession. The ammonoid fauna AF-1 of Brack and Rieber (1993) indicates an upper *Reitzi* Zone age for the base of the cyclic succession. The rich ammonoid fauna AF-8 has recently been found approximately 1 m below the preserved top of the UTF (Rieber, personal communication; Zühlke et al., 2000). It includes *Chieseiceras* sp., which indicates uppermost *Secedensis* to (?) lowermost *Curionii* Zone (cf. Vörös, 1998).

Thus biostratigraphic constraints indicate that the Latemar cyclic series (LCF to UTF) is of considerably shorter duration than previously assumed. It covers little more than a single ammonoid biozone of the late "Anisian" to earliest "Ladinian" instead of the largest part of the "Ladinian" (Goldhammer et al., 1993, equal to at least five ammonoid biozones) or three to four biozones (Brack et al., 1996).

The new biostratigraphic data for the Latemar cyclic series imply that a considerable part of the northeastern slope of the Latemar platform postdates the preserved cyclic platform interior (LCF–UTF), which terminates in the uppermost *Secedensis* to (?) lowermost *Curionii* Zone. The ammonoid and Daonella faunas described by Brack et al. (1996; L3, L4) indicated upper *Gredleri* to *Archelaus* Zone for the upper part of the northeastern slope succession east of Col Cornon. No platform-interior successions equivalent to this part of the Latemar slope have been preserved because of Miocene to recent uplift and erosion of the Southern Alps.

Radiometric Stratigraphy

Volcaniclastic ash layers intercalated in the cyclic succession are present at at least eight stratigraphic levels. Lateral tracing of ash layers shows that three of them are platform-wide marker horizons. Four layers pinch out laterally, owing to strong dilution or loss by storms. Mundil et al. (2003) have sampled volcanic ash layers at three cyclostratigraphic levels and localities (Chart 1, Columns A3, A5): (1) Sample LAT-31 in the lower LCF in the Valsordaspitze section; (2) Sample LAT-30 in the middle MTF in the Latemartürme section; (3) Sample LAT-32 in the middle to upper UCF in the Erzlahnspitze section. On the basis of the dataset of this study, the sampling levels were tied to the cyclostratigraphic framework of the platform. The two sampling levels LAT-30 and LAT-31 bracket a minimum of 381 and a maximum of 410 shallowing-upward microcycles.

Mundil et al. (2003) present $^{206}Pb/^{238}U$ ages based on single-zircon techniques with precisions at the 2-3 permil level: (1) LAT-31; 242.6 ± 0.7 Ma; (2) LAT-30; 241.2 +0.7/-0.6 Ma; (3) LAT-32; 241.5 +0.4/-0.6 Ma (median age based on a normal distribution of U concentrations and/or Th/U ratios), 241.7 + 1.5/-0.7 Ma (including additional ages deviating from a normal distribution). Individual U/Pb ages, analytical procedures, and statistical methods are discussed by Mundil et al. (2003). Brack et al. (1996) and Mundil et al. (1996b) described a specific succession of volcaniclastic layers (Tc–Te) in the basinal Seceda section, situated 28 km to the northeast in the northwestern Dolomites (Gröden Valley, Val Gardena). Horizon Tc, a crystal tuff, represents a regional marker horizon in the lower *Secedensis* Zone and can be traced as far as 130 km to the southwest into the Lombardian Alps. In the Latemar, the Valsordaspitze section includes a crystal tuff layer in the upper LCF, the biostratigraphic age of which is lower *Secedensis* Zone. Single zircons from the Tc horizon in the Seceda section yielded an age of 241.2 + 0.8/-0.6 Ma. However, it is still uncertain whether the crystal tuff in the Latemar platform is laterally correlative to the Tc crystal tuff at the Seceda section.

Constraints on the Basic Microcycle and the Latemar Cyclic Succession

High-resolution cyclostratigraphic, biostratigraphic, and chronostratigraphic data constrain the average microcycle duration in the Latemar cyclic succession, provided that (1) isotopic ages reflect true ages of deposition; (2) the individual microcycle period remained constant during the LCF to UCF; (3) each external pulse triggering cyclic deposition is actually preserved in the cyclic succession; and (4) each shallowing-upward microcycle represents a single cycle of accommodation change. The time interval enclosed by the two ash layers LAT-31 and LAT-32 ranges between 0.9 and 1.1 Ma on average and between 2.3 and 2.4 My at the most. Given the cycle number of 381–410 between the two layers, the average and maximum periods of each microcycle are Δt_{av} = 2.20–2.89 ky and $\Delta t_{max} \leq$ 5.61–6.30 ky. Further cyclostratigraphic and spectral analyses in this study are based on cycle periods of Δt_{av} = 2.68 ky to $\Delta t_{max} \leq$ 5.85 ky. Even when maximum uncertainties on the ages are applied, microcycle periods cannot be reconciled with precession forcing in the Triassic, which requires a cycle period of at least 17.8 ky. Therefore basic-shallowing upward microcycles (accommodation cycles) represent sub-Milankovitch forcing.

Two approaches can be applied to extrapolate the total time interval for the complete cyclic series (base LCF to top UTF). The first approach applies a linear regression of age as a function of stratigraphic height extrapolated to the base and the top of the succession (Mundil et al., 2003). It indicates a time interval of ≤ 6.3 My if maximum error bounds are applied, and a mean of 2.2 My. The regression is based on a depositional rate that has not been corrected for compaction and is averaged over the whole cyclic succession (cf. Zühlke et al., 2003, their fig. 4). The second approach (this study) applies an extrapolation that is based on the cyclostratigraphic and chronostratigraphic framework. The average and maximum cycle periods calculated for the succession between LAT-31 and LAT-32 are attributed to all cycles in the cyclic succession. As a result, the total time interval amounts to ≤ 4.1 My at the most (Fig. 7, Column E2), and 1.88 My (Fig. 7, Column E1) on average. The variant total durations of the cyclic series from the two extrapolations do not affect the genetic interpretation of the basic accommodation cycle and superimposed cycle bundles in the Latemar carbonate platform.

Spectral Analysis

Five methods of spectral analysis were applied to the complete 458-m-thick cyclic series (LCF–UTF) in the Latemartürme

Column A

Constraints.—Cyclostratigraphy: single, 322-m-thick section. Biostratigraphy: none. Chronostratigraphy: none.

Remarks.—Chronostratigraphic framework was inferred from Milankovitch model. Spectral analysis was restricted to UCF.

Column B

Constraints.—Cyclostratigraphy: approximate boundaries of units according to Goldhammer (1987) and Goldhammer et al. (1993). Biostratigraphy: ammonoid faunas in the LCF–TF units, biostratigraphic scheme of Mietto and Manfrin (1995). Chronostratigraphy: none.

Remarks.—Absolute ages were not included in De Zanche et al. (1995). Ages have been added for comparison because the Mietto and Manfrin (1995) biostratigraphic scheme has been incorporated in the time scales of Gradstein et al. (1994, 1995) and Hardenbol et al. (1998).

Column C

Constraints.—Cyclostratigraphy: boundaries of units according to Goldhammer (1987) and Goldhammer et al. (1993). Biostratigraphy: ammonoid and Daonella faunas from the upper LPF and the platform slope; biostratigraphic scheme of Brack and Rieber (1993). Chronostratigraphy: U–Pb ages from the northern Dolomites (Seceda Section, Buchenstein Fm.) were correlated to the Latemar on the basis of biostratigraphic data.

Remarks.—The study includes the first biostratigraphic and chronostratigraphic framework for the Latemar cyclic series.

Column D

Constraints.—Cyclostratigraphy: single 160 m section in ± UCF; cyclostratigraphic scheme according to Egenhoff et al. (1999). Biostratigraphy: none. Chronostratigraphy: none.

Remarks.—Existing chronostratigraphic and biostratigraphic constraints were rejected in favor of a Milankovitch-model-dependent calibration of the

(continued at the top of page 197)

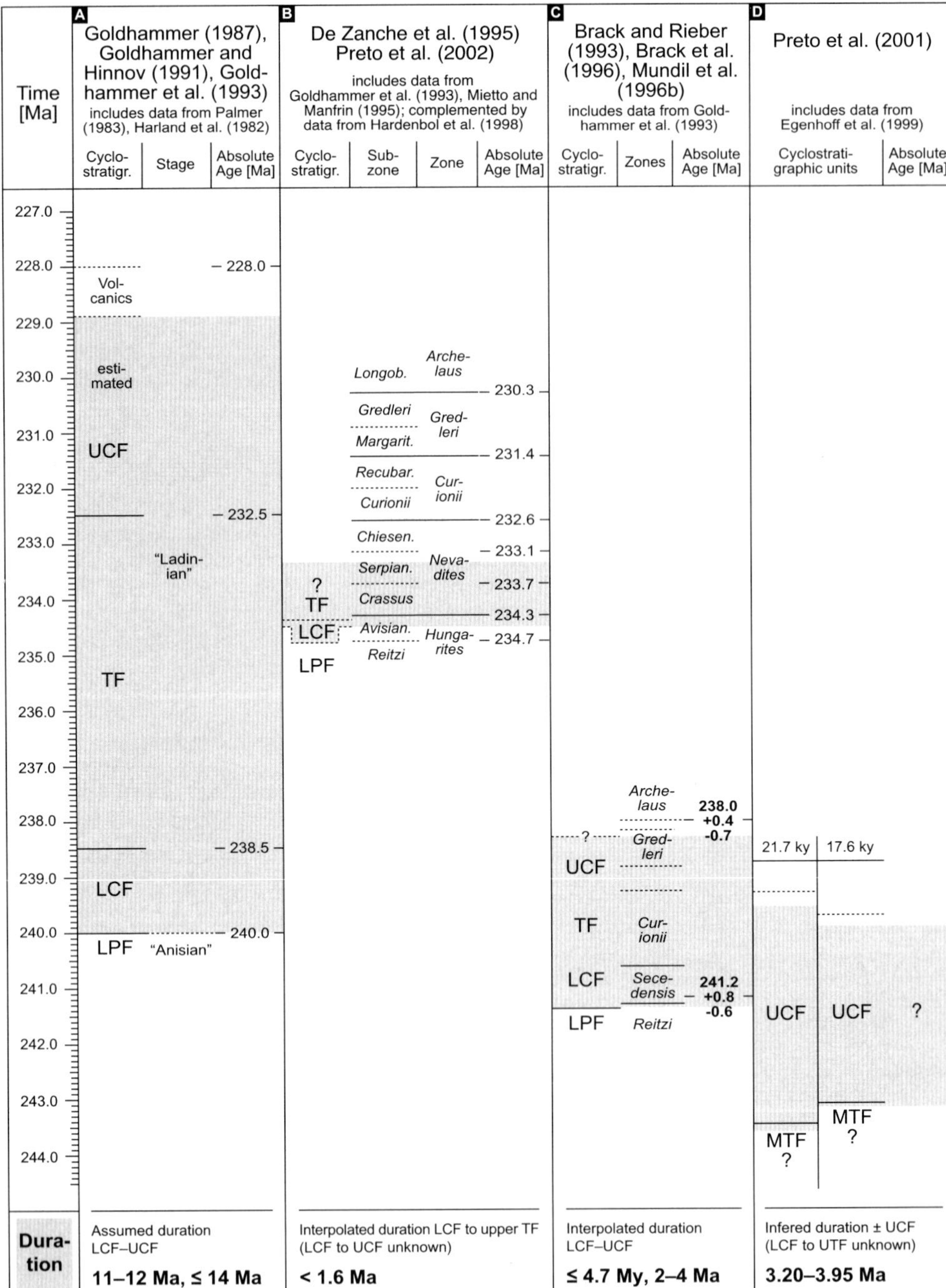

FIG. 7.—Cyclostratigraphic, biostratigraphic, and chronostratigraphic constraints on the Latemar platform interior scaled to absolute time (in full scale). Columns A–D list constraints of previous studies, and column E, datasets of this study. Gray-shaded areas indicate the chronostratigraphic range of the cyclic succession or parts of its. Ages in boldface type represent isotopic ages measured from single zircons in volcanic ash layers intercalated in the cyclic series (Mundil et al., 1996b, 2003). Ages in normal typeface represent interpolated ages from biochronostratigraphic charts (Palmer, 1983; Harland et al., 1982, 1990; Hardenbol et al., 1998, Columns A–B) or extrapolated ages based on cyclostratigraphic and biochronostratigraphic constraints (this study, Column E). The basal row indicates the extrapolated absolute duration of (parts of) the cyclic series (LCF–UTF).

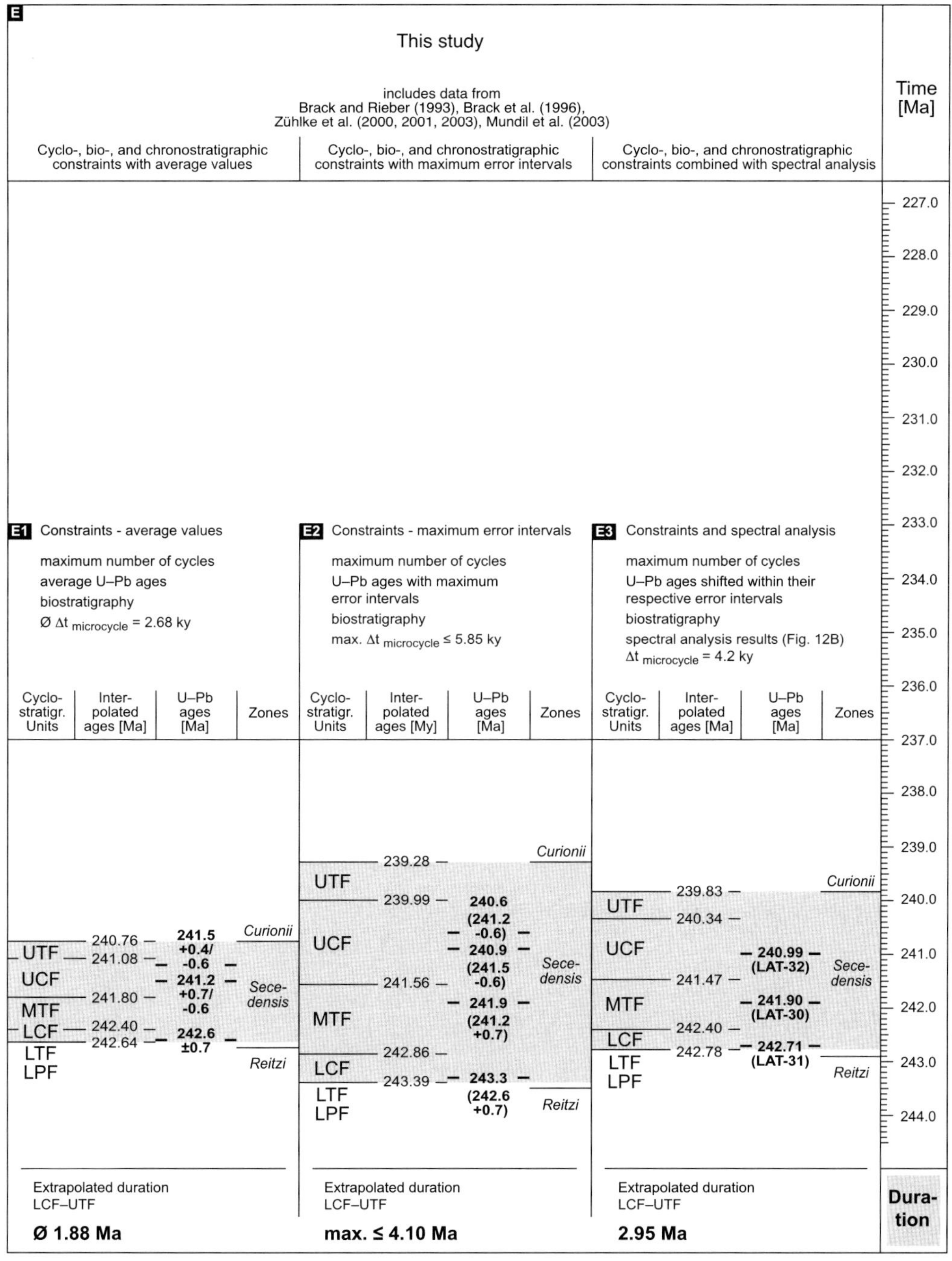

time series. Spectral analysis was applied to the uppermost MTF to the upper UCF, based on a rank series of four lithofacies types reflecting paleobathymetric changes. Preto et al. (2001) did not present any microcycle numbers. Their rank series includes at least 182 cycles (cf. Chart 1). On the basis of their model of precession forcing (Δt = 17.7–21.6 ky) the section has been centered to the U–Pb age of 241.5 ± 0.4 My (see Preto et al., 2001, fig. 1) cited from Zühlke (personal communication, 2001).

Column E

Constraints.—Cyclostratigraphy: complete cyclic Latemar series (LCF–UTF) measured in five sections, 190–470 m thick, across the platform interior. Biostratigraphy: ammonoid faunas from the top, internal succession and base of the cyclic series described by Brack and Rieber (1983), De Zanche et al. (1995), Brack et al. (1996), and Zühlke et al. (2003). Chronostratigraphy: U–Pb ages of Mundil et al. (2003). Absolute ages (sampling horizons) were tied to the cyclostratigraphic data of this study.

Remarks.—Extrapolated ages shown in columns E1 to E3 are based on the assumption that individual small-scale shallowing-upward cycles in the Latemar represent largely constant periods (Δt_{av} = 2.68, Δt_{max} ≤ 5.85) and that all beats are documented in the cyclic succession. Spectral analysis were applied to the complete cyclic Latemar succession (LCF–UTF), on the basis of thickness of accommodation cycles. The duration of the Latemar cyclic series has been constrained by cyclostratigraphic, biostratigraphic, and chronostratigraphic data as well as spectral-analysis data.

Columns E1–E3 compare different sets of constraints used in this study in order to account for maximum cumulative error intervals in the resulting forcing model.

E1: maximum number of cycles in the Latemar cyclic series (701); average ages (Mundil et al. 2003); biostratigraphic data (Fig. 2, Columns E1–E2); average period of the basic shallowing-upward microcycle (Δt = 2.68 ky).

E2: maximum number of cycles in the Latemar cyclic series (701); maximum uncertainties in ages (Mundil et al., 2003); biostratigraphic data (Fig. 2, columns E1–E2), maximum period of the basic shallowing-upward microcycle (Δt = 5.85 ky).

E3: maximum number of cycles in the Latemar cyclic series (701), spectral analysis (Fig. 12B); ages for LAT-30 to LAT-32 (Mundil et al., 2003) were shifted within their respective error intervals; biostratigraphic data (Fig. 2, Columns E1–E2); most probable duration of the basic shallowing-upward microcycle (Δt = 4.2 ky; Figure 12B).

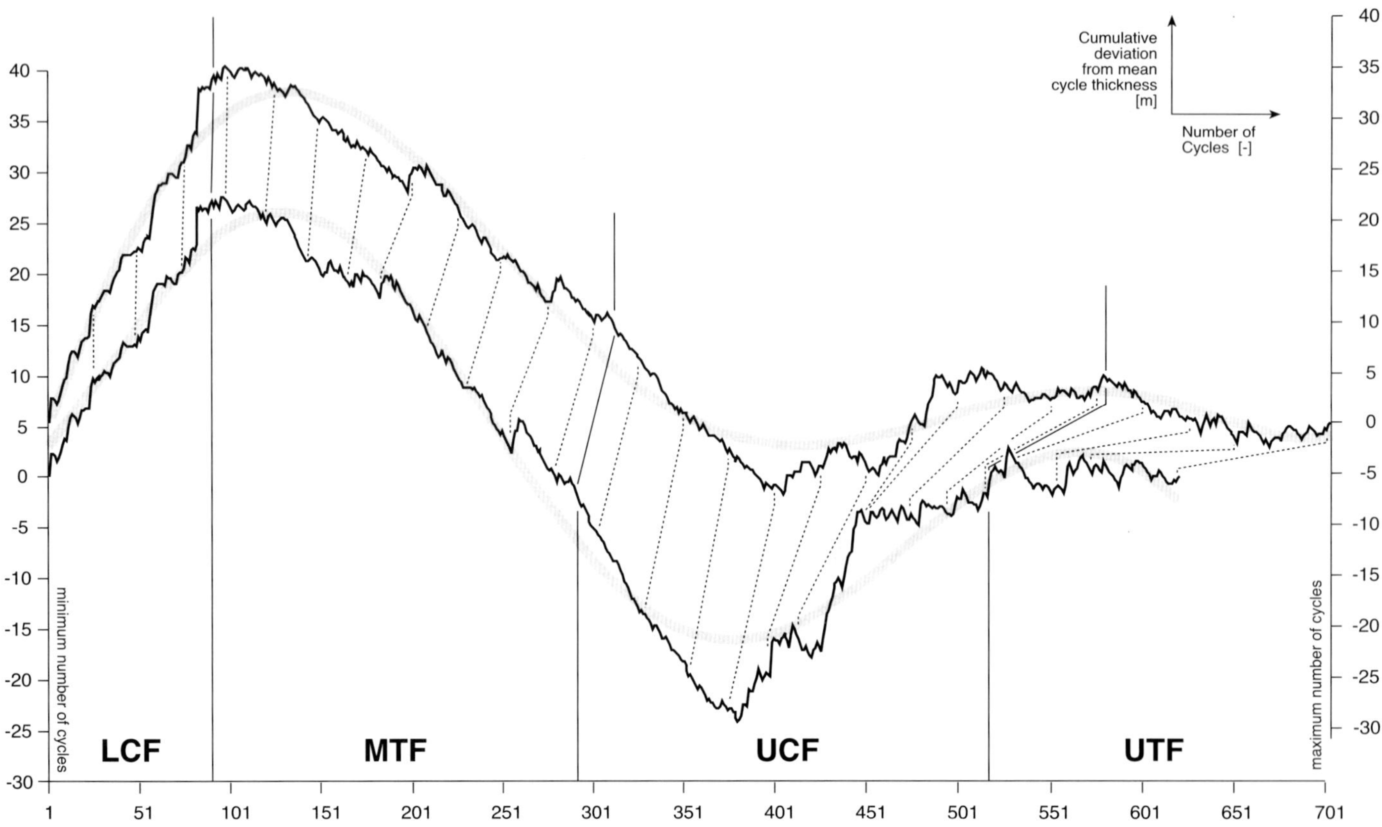

FIG. 8.—Raw thicknesses of accommodation cycles in the Latemartürme reference section (Chart 1, Column A3). The Fischer plot shows cumulative departures from mean cycle thickness versus cycle number (Fischer, 1964; Read and Goldhammer, 1988; Sadler et al. 1993). The upper curve indicates maximum number of cycles (701), and the lower curve, the minimum numbers of cycles (619). Dotted lines show identical cyclostratigraphic levels. Thick gray lines represent fourth-order polynomial trendlines. Fischer plots provide a first indication for higher-order cycle stacking patterns.

reference section. Previous spectral analyses include only the uppermost MTF and UCF (Hinnov and Goldhammer, 1991; Preto et al., 2001). In this study, spectral analysis is based exclusively on accommodation cycles, which built up to sea level at the end of each cycle. Shallowing-upward successions within subtidal depths have not been included, because unknown unfilled accommodation space from the previous microcycle may have existed at the beginning of a new microcycle starting in subtidal depths. Spectral-analysis methods include (1) periodogram, (2) maximum entropy, (3) Blackman–Tukey, (4) multi-taper spectral analysis, and (5) multi-taper harmonic analysis. Special focus is placed on Blackman–Tukey, multi-taper spectral, and harmonic analyses. The raw thicknesses of accommodation cycles (Figs. 8, 9) were subject to three transformations (Fig. 10): (1) linear trend removed, (2) unit variance normalized (even sampling), and (3) Box–Cox transformation. For spectral analysis the cyclic series was subdivided into seven spectral windows with 50% overlap each to the previous and next spectral window. Spectral windows include between 125 and 220 accommodation cycles each and therefore represent sufficiently large, reliable data windows for spectral analyses. All power and amplitude spectra were thickness and time calibrated, constrained by the maximum numbers of cycles in the Latemartürme section and the biochronostratigraphic data presented above. The time calibrations indicated in Figures 11A–B and 12A consider both the average and the maximum period of individual microcycles ($\Delta t_{av} = 2.68$ ky, $\Delta t_{max} \leq 5.85$ ky). Spectral analyses were performed with Mathematica® (Wolfram, 1996) including the Time Series Pack (Yu, 1995), and AnalySeries© (Paillard et al., 1996).

Blackman–Tukey Spectral Analysis.—The Blackman–Tukey method (Blackman and Tukey, 1958) is the classical method of spectral analysis. The algorithm computes first the autocovariance of the data, then applies a window, and finally Fourier-transforms it to compute the spectrum. According to Paillard et al. (1996) it is a very robust method, unlikely to present spurious spectral features. The main drawback is its poor resolution in the spectral domain; sharp features are considerably smoothed. Blackman–Tukey analysis of the complete Latemar cyclic series (LCF–UTF) reveals six clusters of frequencies with moderate to strong power (Fig. 11A). The distribution of clusters varies over the cyclic succession:

- Average ratio of 1:3.7—prominent in the LCF to upper MTF, gradually disappears in the UCF to UTF
- Average ratio of 1:5.9—significant in the UCF to lower UTF. Both ratios are largely lacking in the LCF. They partly appear as a double peak with an average ratio of 1:5.2 (upper MTF to lower UCF, upper UCF)
- Average ratios of 1:9.1 and 1:19.4—strong spectral power in the MTF to UCF. Both ratios are completely lacking in the LCF and middle to upper UTF.

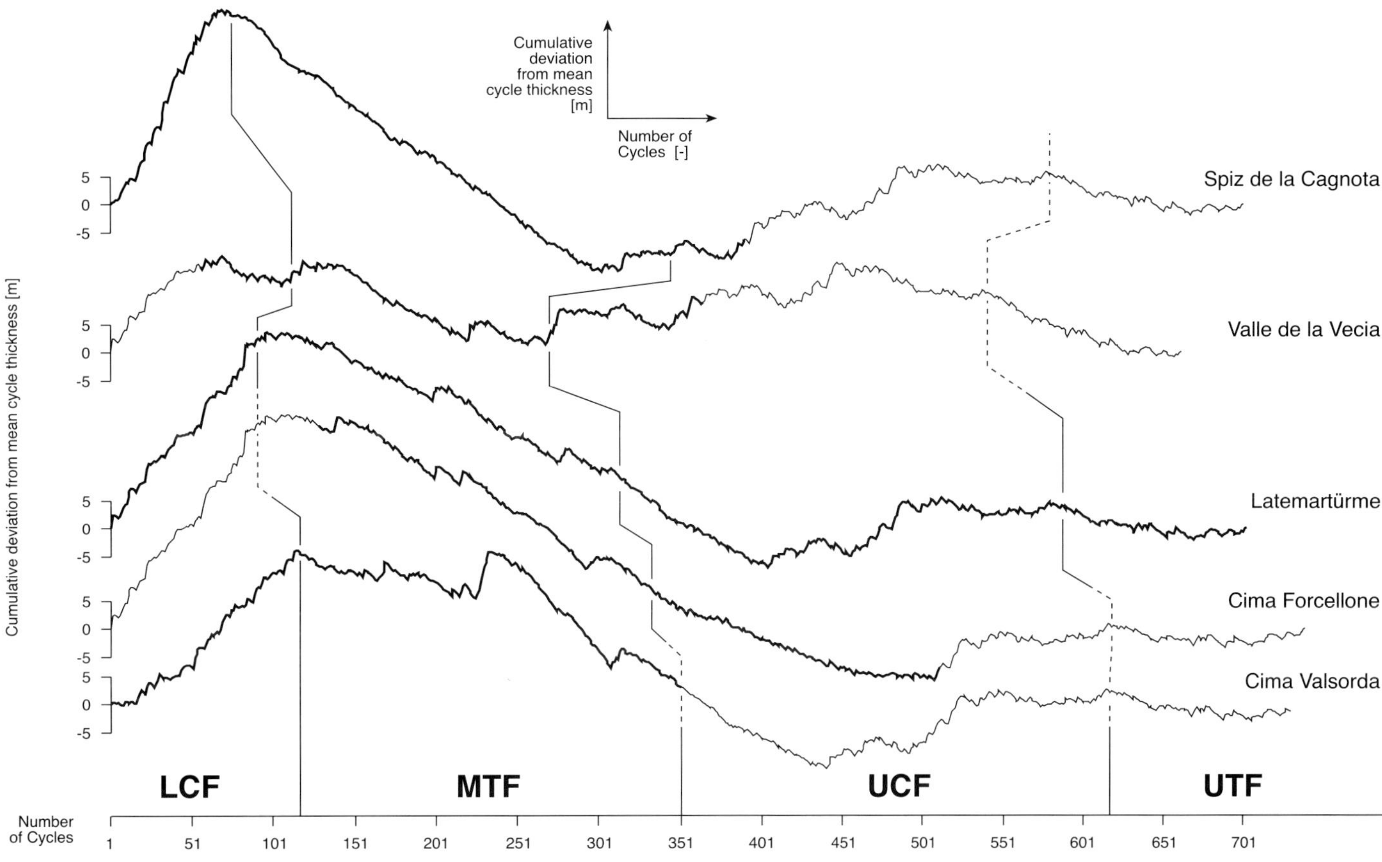

FIG. 9.—Raw thickness of accommodation cycles in the five vertical sections of this study (Chart 1, Columns A3–A7) as indicated by Fischer plots (cf. Figure 8). Thick black lines indicate the cyclic succession preserved in each section. Thin black lines indicate cycles projected from the Latemartürme reference section.

- Average ratio of 1:99.2—low resolution; the cluster shows considerable variations over the spectral windows.

Multi-Taper Spectral Analysis.—

The multi-taper method (Thomson, 1982) offers high resolution and statistical estimates that are independent of the spectral power (small-amplitude oscillations may have a high significance level; Yiou et al., 1994). Both multi-taper spectral analysis and harmonic analysis were applied to the complete Latemar cyclic series (LCF–UTF). Multi-taper spectral analysis reveals six clusters of frequencies with moderate to strong power (Fig. 11B). The variation in the distribution of clusters through the cyclic succession is similar to that revealed by the Blackman–Tukey analysis:

- Average ratio of 1:3.8—prominent in the LCF to middle MTF. The ratio disappears completely in the middle UCF to middle UTF and reappears with moderate power in the middle to upper UTF.
- Average ratio of 1:5.0—especially significant in the UCF to lower UTF. The ratio is subdued or absent in the LCF to middle MTF and the middle to upper UTF.
- Average ratio of 1:6.2—only moderate power and without any clear distribution over the cyclic series.
- Average ratio of 1:10.1—shows strong power in the upper LCF to UCF. The ratio is lacking in the lower LCF and the middle to upper UTF.
- Average ratio of 1:24.7—elevated to high power in the MTF to UCF. The distribution of this ratio is similar to that of the average ratio of 10.1.
- Average ratio of 1:115.2—low resolution as in Blackman–Tukey analysis. The variation of frequencies is comparatively low.

Multi-Taper Harmonic Analysis.—

In contrast to Blackman–Tukey analysis and multi-taper spectral analysis, multi-taper harmonic analysis considers exclusively statistical confidence levels (F-Test), which are independent of spectral power (Yiou et al., 1994). It allows identifying low-amplitude peaks with high significance or high-amplitude peaks with low significance. F-test results for individual amplitude peaks may vary between high and highest significance, depending on the taper type and the moving window applied. Multi-taper harmonic analysis has the advantage of high resolution even in series with high background noise. Background noise is potentially caused by changes in sedimentation rates over the interval to be analyzed or by (moderately) nonstationary cycle periods, especially in the high-frequency range. Multi-taper harmonic analysis of the complete Latemar cyclic series (LCF–UTF, Figure 12A) reveals seven clusters of frequencies with high significance (> 0.9) to highest significance ($1/(1-n)$, > 0.994). The variation in the distribution of clusters over the cyclic succession is similar to that revealed by Blackman–Tukey and multi-taper harmonic analysis:

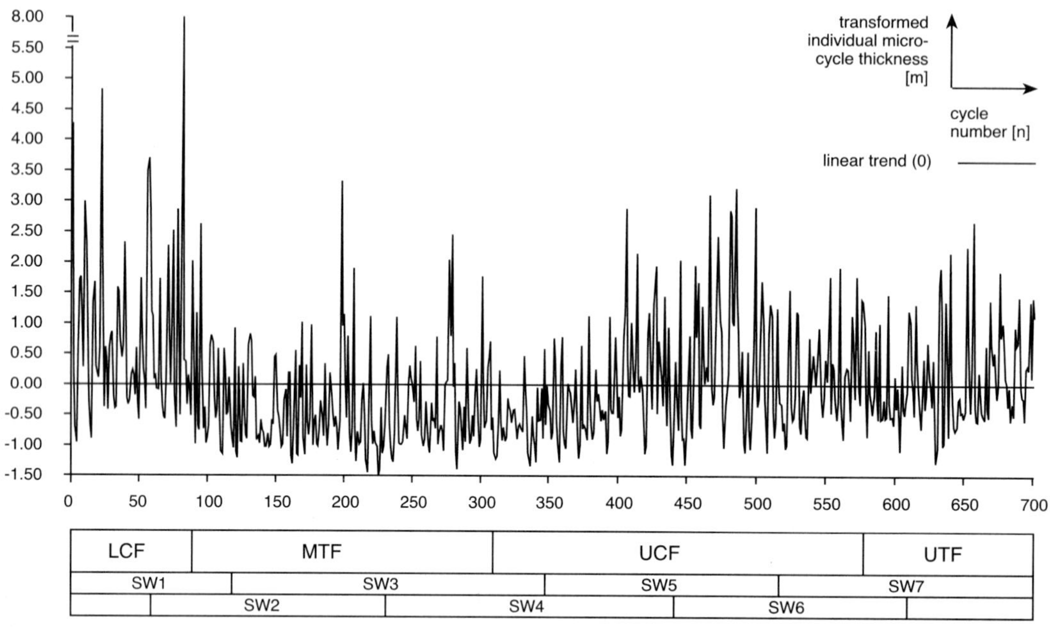

FIG. 10.—Basic dataset for spectral analyses in the Latemartürme reference section. The graph plots transformed individual accommodation cycle thickness (Chart 1, Column A5) versus cycle number. Raw data were subject to the following transformations (1) linear trend removed; (2) unit variance normalized (even sampling); (3) Box-Cox transformation (not included in this graph). SW indicates sliding windows used in spectral analyses (see Figures 11 and 12).

- Average ratios of 1:2.2 and 1:2.7—occur only in multi-taper harmonic analysis but not in the other four analysis methods. The ratios may be produced by the different statistical null hypothesis of the method.
- Average ratio of 1:3.7—highly significant between the upper LCF to UTF. The ratio is especially strong in the MTF.
- Average ratio of 1:5.2—moderate significance; occurs throughout the cyclic series.
- Average ratio of 1:6.8—occurs only in multi-taper harmonic analysis but not in the other four analysis methods. The significance is moderate in the LCF to MTF and upper UCF to lower UTF.
- Average ratio of 1:9.9—significant between the middle LCF to UTF. Highest significance occurs in the lower to middle MTF and the UCF.
- Average ratio of 1:19.5—significant between the middle LCF and lower UTF. It includes a double peak between the upper MTF and the UCF.
- Average ratio of 1:105—significant in the MTF to UTF. High-significance points occur in the UCF to lower UTF.

Results of Spectral Analysis.—

All five analysis methods indicate five ratios with strong power and/or high significance over the largest part of the Latemar cyclic series with average ratios of: (1) 1:3.7 to 1:3.8; (2) 1:4.7 to 1:5.2; (3) 1:9.1 to 1:10.1; (4) 1:19.4 to 1:24.7; (5) 1:99.2 to 1:105. Basic shallowing-upward microcycles in the Latemar cyclic succession cannot be reconciled with precession forcing but represent sub-Milankovitch control. In order to check whether Milankovitch forcing is additionally present in the larger-scale cycle bundlings of the Latemar cyclic series, the results of multi-taper harmonic analysis were tested for a mathematical solution that meets all of the following four requirements:

- ratios indicative of orbital forcing in the Triassic (Berger and Loutre 1994). With the short precession period set to 1, ratios of 1:1.9 indicate long precession, 1:2.01 short obliquity, 1:2.54 long obliquity, 1:5.63 short eccentricity, and 1:22.5 long eccentricity (Table 2, Column G).
- time intervals indicative of orbital forcing in the Triassic (Berger and Loutre, 1994). Intervals include 17.78 ky and 21.23 ky for short and long precession, 35.73 ky and 45.24 ky for short and long obliquity, and 100 and 400 ky for long eccentricity (Table 2, Column J)
- indicative ratios and time intervals must be compatible with the chronobiostratigraphic data, i.e., must be applicable with $\Delta t_{max} \leq 5.85$ ky.
- the first three requirements must be met by significant to highly significant amplitude peaks in multi-taper harmonic analysis.

In contrast to the approach of Preto et al. (2001), which was based exclusively on spectral analyses with a Milankovitch-model-dependant calibration and which considered only a minor part of the Latemar cyclic succession, this approach considers all available constraints: (1) the complete 458-m-thick cyclic succession in the Latemar with 701 accommodation cycles; (2) relative ages and durations indicated by biostratigraphy; (3) absolute ages derived from chronostratigraphy, which were extrapolated to the complete LCF–UTF succession; (4) spectral analyses with a model-independent time calibration of microcycles.

The approach uses a goal seek script that includes the four requirements stated above as nested conditions. The script runs recursively on all seven sliding windows with steps of 0.1 ky from $\Delta t > 0$ ky to $\Delta t_{max} \leq 5.85$ ky. Bandwidths of Milankovitch signals include an error interval of $\leq \pm 6\%$ to account for geological distortions (e.g., hiatuses, varying sedimentation rates; cf. Stage,

1999). In fact there is a single mathematical solution that meets all four requirements stated above (Fig. 12B). Ratios and time intervals indicative of orbital forcing in the Milankovitch band exist in the Latemar cyclic series with high and very high significance points when $\Delta t = 4.2$ ky. The period of $\Delta t = 4.2$ ky for each sub-Milankovitch shallowing-upward microcycle is well within the range of the average to maximum period indicated by the chronostratigraphy and biostratigraphy ($\Delta t_{av} = 2.68$ ky, $\Delta t_{max} \leq 5.85$ ky).

Provided that isotopic ages reflect true ages of deposition, and on the basis of solution above, at least five different hierarchic cyclicities exist in the Latemar cyclic series (Table 2). Ratios are expressed both with the sub-Milankovitch basic microcycle of 4.2 ky set to 1 (Table 2, Columns C–D) and the higher-order short-precession cycle of 17.8 ky set to 1 (Table 2, Columns E–F):

- sub-Milankovitch cyclicity: basic shallowing-upward microcycle, average microcycle thickness 0.63 m (e.g. in UCF), period $\Delta t = 4.2$ ky. The basic shallowing-upward microcycle was previously interpreted as a precessional cycle of $\Delta t \sim 20$ ky (Goldhammer et al., 1993). Preto et al. (2001) attributed a periodicity of $\Delta t = 17.6$–21.7 ky (precession forcing) to the shortest cyclic signal in their rank series.
- condensed precession cycle or higher-order sub-Milankovitch cycle: stacking pattern of 1:3.2 to 1:4.0 microcycles, average cycle thickness 2.27 m (e.g., in UCF), $\Delta t_{max} = 13.6$–16.7 ky.
- short and long precessional cycle: stacking patterns of 1:4.3 to 1:5.1 microcycles, average cycle thickness 2.96 m (e.g., in UCF), $\Delta t_{max} = 17.1$–21.5 ky. This stacking pattern was previously interpreted as an eccentricity cycle of $\Delta t \sim 100$ ky (Goldhammer et al., 1993) or $\Delta t = 98$ ky (Preto et al., 2001).
- short and long obliquity cycle: stacking patterns of 1:8.5 to 1:11.4 microcycles, stacking patterns of 1:2.0 to 1:2.7 short-precession cycles, average thickness 6.24 m (e.g., in UCF), period $\Delta t_{max} = 35.5$–48.0 ky.
- short-eccentricity cycle: stacking patterns of 1:22.7 to 1:25.2 microcycles, stacking patterns of 1:5.4 to 1:6.0 precession cycles, average cycle thickness 15.10 m (e.g., in UCF), period $\Delta t_{max} = 95.4$–105.9 ky.
- long-eccentricity cycle?: stacking pattern of 1:64.3 to 1:147.0 microcycles (average 1:105.7), stacking pattern of 1:15.2 to 1:34.7 precession cycles, average cycle thickness 66.6 m (e.g., in UCF), period $\Delta t_{max} = 270.1$–617.3 ky ($\Delta t_{av} = 444$ ky). The existence of the long-eccentricity signal remains speculative because ratios show considerable variations.

High-resolution cyclostratigraphic, biostratigraphic, and chronostratigraphic data in combination with the results of spectral analyses provide an integrated set of constraints on the duration of the Latemar cyclic succession (Fig. 7, Column E3). The extrapolated duration for the LCF to UTF amounts to 2.95 My. This equals the *Secedensis* Zone in the biostratigraphic scheme of Brack and Rieber (1993) and Brack et al. (1996). For comparison, Figure 7, Columns A–D, list the constraints and resulting absolute durations of the Latemar cyclic series as assumed in previous studies.

DISCUSSION

Platform Architecture, Allocyclicity, and Autocyclicity Models

The internal architecture of the Latemar platform interior is characterized by laterally continuous shallowing-upward microcycles and thinning-upward macrocycles, the large majority of which are platform-wide and physically traceable or correlative. This indicates a simultaneously aggrading sheet model *sensu* Pratt et al. (1992), which was described for ancient platform settings by, e.g., Koerschner and Read (1989). Other peritidal platform models like tidal wedges or multiple tidal-flat islands include meter-scale successions that are laterally discontinuous and cannot be traced or correlated over kilometer distances (except for their basal disconformity). The tidal-wedge model necessitates pronounced lateral progradation from some sort of nucleus. Progradation typically shows simple or staggered offlap with progradation jumps between barriers (Hardie and Shinn, 1986). These features are absent from the Latemar platform interior. The tidal-flat-island model requires complex lateral and vertical shifting of depositional facies in response to local variations in topography, water energy, and currents. Because individual microcycles in the Latemar platform interior show moderate lateral variations in lithofacies, it cannot be completely ruled out that minor, low-angle progradation took place within single microcycles. However, no progradational features including several microcycles, macrocycles, or larger-scale cycle bundles exist.

Models of internal control (autocyclicity models) assume localized changes in carbonate accumulation rates as the driving force for shallowing-upward cycles. Autocyclic processes may generate successions that are similar to those that are inferred to have formed by eustatic, orbitally driven sea-level changes. The original model of Ginsburg (1971) was based essentially on a prograding-wedge model. On platform tops with simultaneous vertical aggradation, an autocyclic model implies that sediment production rates in subtidal and peritidal to intertidal depths are sufficiently high to aggrade to sea level. This is difficult to achieve with an exclusively autocyclic model because the carbonate factory is shut down progressively with increasing extension of intertidal areas. On small platforms with an only slightly elevated topographic margin like the Latemar, wave activity occurs over the whole platform top and inhibits cyclic aggradation to sea level exclusively driven by autocyclic processes.

Models of external control (allocyclic models) include structural processes or low-amplitude, high-frequency sea-level changes as the dominant driving force for rhythmic deposition.

Structural processes that operate on a 10–100 ky scale include the inflation of magma chambers, accompanied by doming of a few tens of meters of kilometers across and eruption followed by deflation (Sloan and Williams, 1991). The Latemar is situated in the vicinity of the magmatic center of Predazzo and Monzoni (Bosellini et al., 1982). However, significant magmatic and volcanic activity occurred only in the early late Ladinian (Longobardian, upper *Gredleri* to *Archelaus* Zone) at an age of 237.2 +0.4/-1.0 Ma (Mundil et al. 1996a). It postdated the deposition of the preserved Latemar carbonate platform interior for at least 2.3 My and 4.6 My at maximum. High-frequency structural processes, e.g., stick-slip faulting at the platform margin (Cisne, 1986), can be ruled out, because they depend on lithospheric flexure on a lateral scale of > 100 km.

The internal architecture of the Latemar platform interior favors a predominantly allocyclic control of individual microcycles, by high-frequency sea-level changes in the sub-Milankovitch bandwidth. Each microcycle reflects a single excursion of sea level. Low-amplitude, high-frequency sea-level changes triggered by orbital forcing via climatic changes predominantly controlled the large stacking patterns of 1:4.3 to 1:5.1 and higher.

Biostratigraphy and Chronostratigraphy

Biostratigraphic information on the ammonoid biozonal level does provide relative estimates of time covered by the Latemar

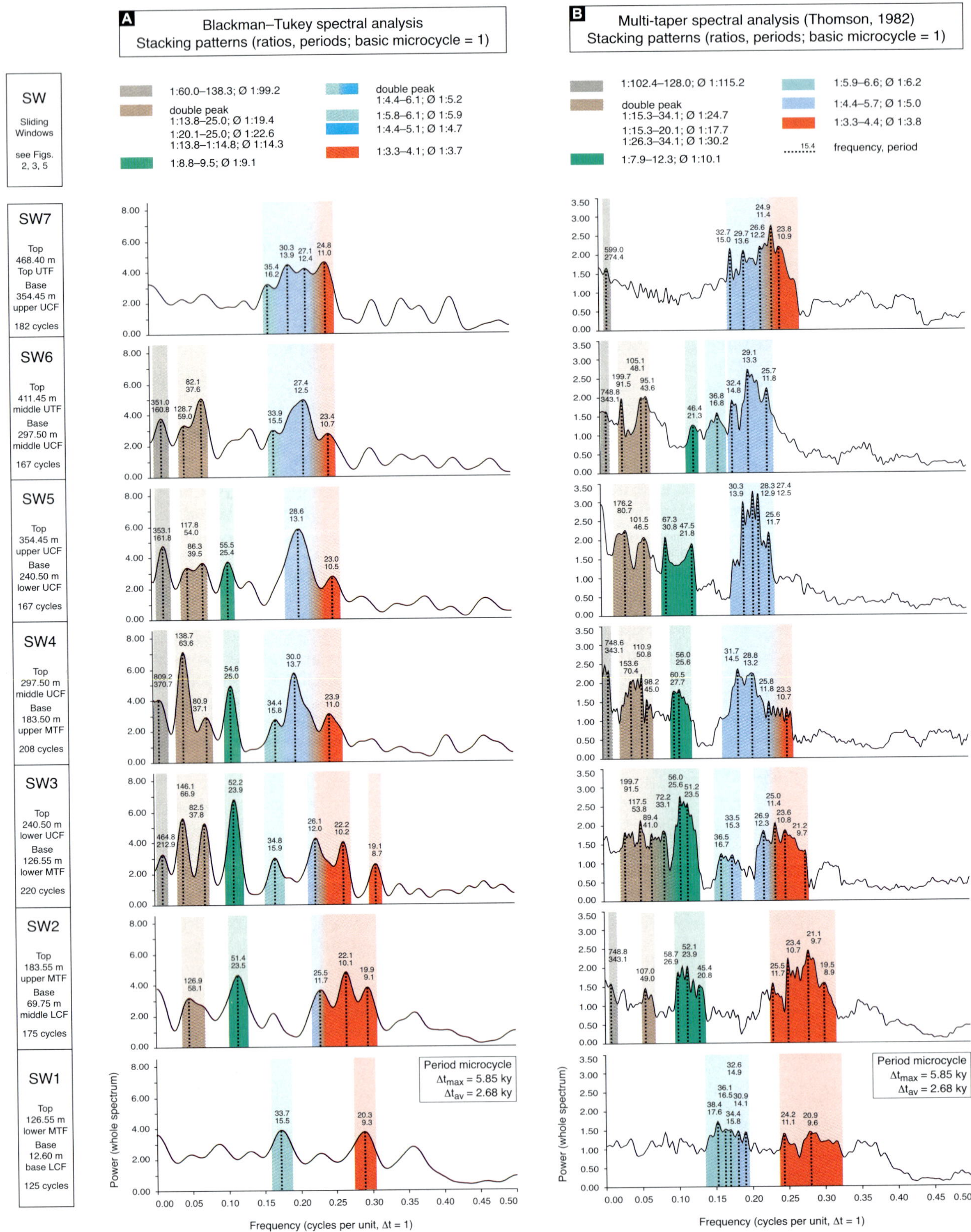

FIG. 11.—**A)** Blackman–Tukey spectral analysis and **B)** multi-taper harmonic analysis of the complete cyclic series in the Latemartürme reference section (LCF–UTF). Color bars indicate ratios of prominent frequencies to the basic microcycle. Periods are based on the average and maximum cycle periods of $\Delta t_{av} = 2.68$ ky and $\Delta t_{max} \leq 5.85$ ky calculated from biochronostratigraphic constraints and the maximum number of cycles bracketed by ash layers LAT-31 and LAT-32 (Fig. 7, Columns E1, E2; Chart 1, Column A5). Legend: SW1-7 = sliding windows 1–7 (Fig. 10).

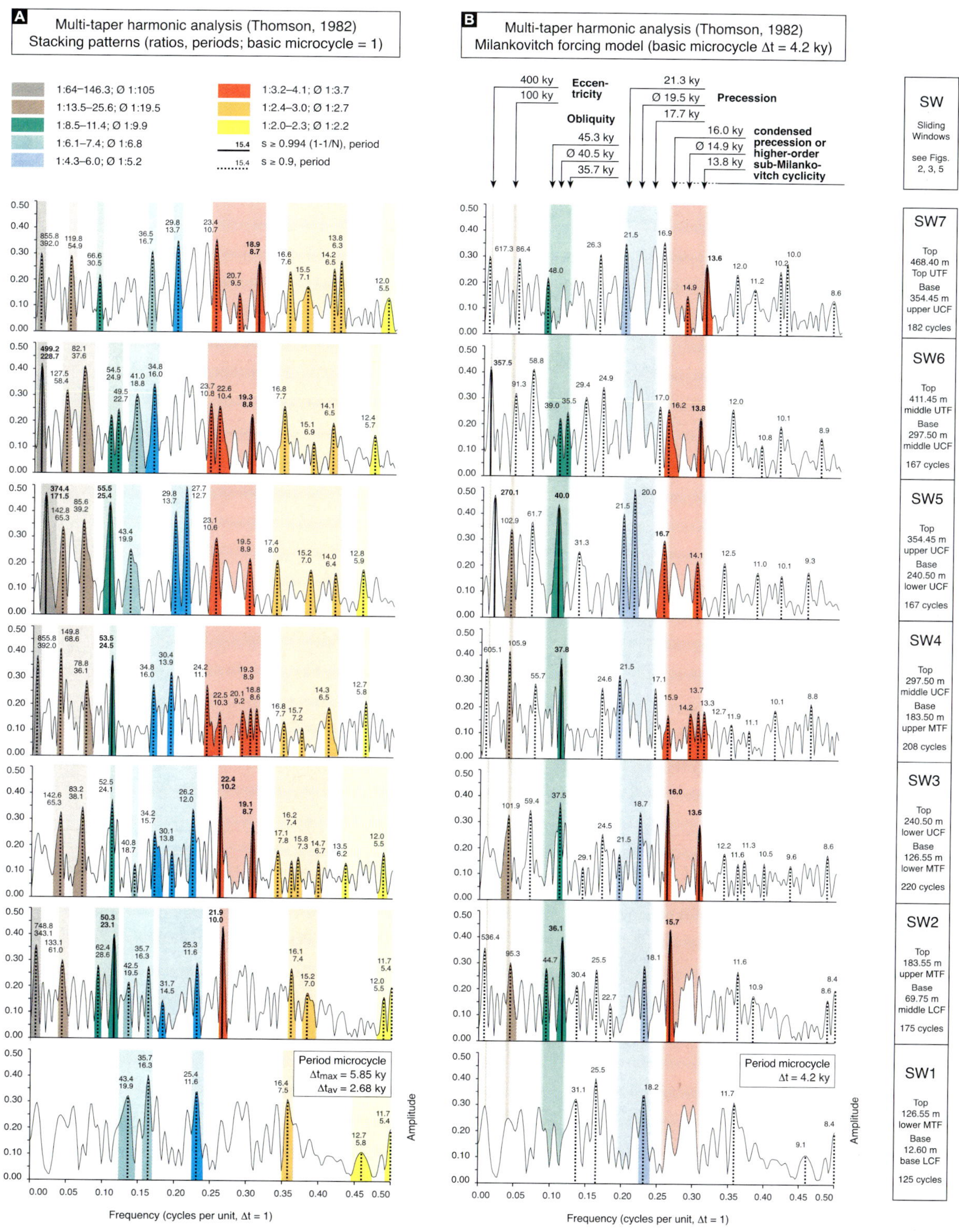

FIG. 12.—Multi-tapered harmonic analyses of the complete cyclic series in the Latemartürme reference section (LCF–UTF). **A)** Color bars indicate ratios of (highly) significant amplitudes compared to the basic microcycle. Periods are based on the average and maximum cycle periods of $\Delta t_{av} = 2.68$ ky and $\Delta t_{max} \leq 5.85$ ky calculated from biochronostratigraphic constraints and the maximum number of cycles bracketed by ash layers LAT-31 and LAT-32 (Fig. 5, Columns E1 and E2; Figure 4; Chart 1, Column A5). **B)** Milankovitch forcing model of the cyclic series based on $\Delta t = 4.2$ ky (Fig. 5, Column E3; Table 2, Columns A–F). Color bars indicate condensed precession or higher-order sub-Milankovitch cyclicity (red, 13.8–16.0 ky), short and long precession (blue, 17.7 ky, 21.3 ky); short and long obliquity (green, 35.7 ky, 45.3 ky), short eccentricity (brown, 100 ky) and questionable long eccentricity (gray, 400 ky). Legend: s = significance point, N = sampling number, SW1–7 = sliding windows 1–7 (Fig. 8).

TABLE 2.—Sub-Milankovitch and Milankovitch cycles in the Latemar platform interior compared to Triassic orbital parameters. The table lists statistically significant periods (column A), cyclicities (B), ratios (C), and average ratios (D) of cycle bundles in the Latemar when the basic shallowing upward microcycle is set to 1 and $\Delta t = 4.2$ ky based on biochronostratigraphic data and spectral analyses (see Figure 5, Column E3). Columns G–J list average ratios, ratios, cyclicities, and periods of Triassic orbital forcing (Berger and Loutre, 1994; interpolation between 270 and 72 Ma). Average ratios (G) and ratios (H) are based on the short precession set to 1 and $\Delta t = 17.78$ ky. Given the solution presented above, the duration of the 1:4.7 stacking pattern (see Column D) in the Latemar closely matches the precessional period in the Triassic. Setting this precessional signal to 1 and $\Delta t = 17.78$ ky, periods (A), ratios (E), and average ratios (F) of the large-scale cycle bundles in the Latemar closely match periods (J), average ratios (G), and ratios (H) of Triassic orbital parameters. Columns F and G indicate the consistent average ratios of orbitally forced cycle bundles in the Latemar and of Triassic orbital parameters (in boldface type).

This study - Latemar carbonate platform interior						Orbital parameters - Triassic			
(A)	(B)	(C)	(D)	(E)	(F)	(G)	(H)	(I)	(J)
Δt min. Δt max. [ky]	Cyclicity	Ratio	Ø Ratio	Ratio	Ø Ratio	Ø Ratio	Ratio	Cyclicity	Δt [ky]
		microcycle = 1 Δt = 4.2 ky		precession cycle = 1 Δt = 17.78 ky		precession cycle = 1 Δt = 17.78 ky			
4.2	basic microcycle sub-Milankovitch	1 : 1.00	1 : 1.00	-	-	-	-	-	-
13.63 16.68	cond. precession or higher order sub-Milankovitch	1 : 3.24 1 : 3.98	1 : 3.61	-	-	-	-	-	-
18.10 21.50	precession	1 : 4.31 1 : 5.12	1 : 4.72	1 : 1.02 1 : 1.21	**1 : 1.11**	**1 : 1.10**	1 : 1.00 1 : 1.19	short precession long precession	17.78 21.23
35.46 48.01	obliquity	1 : 8.45 1 : 11.43	1 : 9.94	1 : 2.00 1 : 2.70	**1 : 2.35**	**1 : 2.28**	1 : 2.01 1 : 2.54	short obliquity long obliquity	35.73 45.24
95.35 105.89	short eccentricity	1 : 22.69 1 : 25.21	1 : 23.95	1 : 5.36 1 : 5.96	**1 : 5.66**	**1 : 5.63**	1 : 5.63	short eccentricity	100.00
270.08 617.32	long eccentricity ?	1 : 64.31 1 : 146.98	1 : 105.65	1 : 15.19 1 : 34.71	**1 : 24.95**	**1 : 22.50**	1 : 22.50	long eccentricity	400.00

cyclic series. It has been argued that this estimate depends on whether it is based on the biostratigraphic dataset of Brack and Rieber (1993), Brack et al. (1996), and Zühlke et al. (2000), or on the dataset of de Zanche et al. (1995). Because the boundary between the Middle Triassic stages of the Anisian and the Ladinian (Illyrian to Fassanian substage boundaries) has not yet been formally defined (see Brack and Rieber, 1994), all biostratigraphic ages have to refer to the ammonoid biozone level. It is necessary to correlate the biozones in the two schemes on the basis of indicative faunas (Fig. 2, Columns E1, E2).

The ammonoid fauna AF-1, including *Aploceras avisianum*, *Parakellnerites rothpletzi*, *Flexoptychites noricus*, and *Latemarites latemarensis*, occurs in the uppermost part of the LPF. In the scheme of Brack and Rieber (1993) and Brack et al. (1996) this fauna indicates upper *Reitzi* Zone. In the scheme of Mietto and Manfrin (1995) and De Zanche (1995) it indicates the middle to upper *Avisianum* subzone/upper *Hungarites* Zone. *Chieseiceras* sp., as included in the ammonoid fauna AF-8, at the top of the UTF indicates the uppermost *Secedensis* to (?) lowermost *Curionii* Zone in the scheme of Brack and Rieber (1993) and Brack et al. (1996). In the scheme of Mietto and Manfrin (1995) and De Zanche et al. (1995) it indicates the *Chiesense* subzone to upper *Nevadites* Zone. Both schemes define the base of their *Curionii* Zone, *Curionii* subzone, respectively, by the first occurrence of *Eoprotrachyceras curionii*. The top of the UTF (AF-8) is slightly below or above this well defined boundary.

Therefore the relative biostratigraphic ranges of the Latemar cyclic succession are almost identical in the two biostratigraphic schemes—little more than a single biozone: upper *Reitzi Zone* to uppermost *Secedensis Zone* or (?) lowermost *Curionii* Zone (Brack and Rieber, 1993; Brack et al., 1996), upper *Hungarites Zone* to middle *Nevadites* Zone (Mietto and Manfrin, 1995, De Zanche et al., 1995). This consistent estimate is independent of biozonal or subzonal definitions or different levels proposed for the Anisian—Ladinian boundary.

Internal plausibility checks on the genetic cyclostratigraphic models for the Latemar relate (1) the durations of the Latemar cyclic series as proposed or implied by the different cyclostratigraphic models; (2) biostratigraphic data for the Triassic system; and (3) chronostratigraphic data for the Triassic System. According to the time scale of Gradstein et al. (1994, 1995) and the biochronostratigraphic chart of Hardenbol et al. (1998), the Permian–Triassic boundary has an age of 248.2 ± 4.8 My, and the Triassic–Jurassic boundary an age of 205.7 ± 4.0 My. The duration of the Triassic is 42.5 My (based on average ages) and ≥ 33.7 My or ≤ 51.3 My (based on maximum uncertainties). The Triassic series in the Alpine–Tethyan domain includes 31–34 ammonoid biozones. The Latemar cyclic series (LCF–UTF) covers about a single ammonoid biozone. This Middle Triassic biozone is calibrated by the absolute durations that the different cyclostratigraphic models require for the cyclic series (Fig. 7): (1) ≤ 14 My, but at least 12 My according to Hinnov and Goldhammer (1991) and Goldhammer et al. (1993); (2) 3.20–3.95 My for the uppermost MTF to upper UCF according to Preto et al. (2001), which implies at least 10 My for the LCF–UTF; (3) 1.88–4.10 My (cyclostratigraphic, biostratigraphic, and chronostratigraphic constraints) and 2.95 My (in combination with spectral analysis data) according to this study.

On the basis of the forcing models of Goldhammer et al. (1993) and Preto et al. (2001) the complete Latemar cyclic series or the single Middle Triassic Latemar biozone represents min.-max. 19–36% (including all error intervals) or an average of 24–28% of the complete Triassic duration. The *Secedensis* Zone or the *Nevadites* Zone in the schemes of Brack and Rieber (1993) and Mietto and Manfrin (1995) was min.-max. 7.0–8.3 times as long as the average duration of the other 30–33 Triassic biozones.

On the basis of the forcing model of this study, the Latemar cyclic series or the single Middle Triassic Latemar biozone represents min.-max. 4–12%, probably 7%, of the complete Triassic duration (based on a total duration of 2.95 My of the Latemar platform interior; see Fig. 7, column E3). The *Secedensis* Zone or the *Nevadites* Zone in the schemes of Brack and Rieber (1993) and Mietto and Manfrin (1995) was min.-max. 1.3–2.9 times, probably 2.1 times, as long as the average duration of the other 30–33 Triassic biozones.

Mundil et al. (2003, their table 1) document isotopic data analyzed by single-crystal techniques for the three ash-fall horizons LAT-30 to LAT-32 in detail and discuss potential and existing uncertainties in $^{206}Pb/^{238}U$ dating. Uncertainties are caused by secondary Pb loss, the statistical approaches applied to calculate mean ages and maximum errors, and the presence of inherited cores in zircons. Mundil et al. (2003) present clearly identifiable age clusters for LAT-30 to LAT-32 and conservative criteria for the data selection in isochore plots.

Radiometric ages for the Latemar, measured by Mundil et al. (2003), are supported by integrated U–Pb geochronology and ammonoid biochronology in the Balaton Highlands, Hungary (Palfy et al., 2002). Multiple concordant analyses of multi-grain zircon fractions in volcanic ash layers in the Felsöors section yield ages of 241.1 ± 0.5 Ma to 240.4 ± 0.4 Ma in the *Reitzi* Zone, and 238.7 ± 0.6 Ma in the *Gredleri* Zone. Palfy et al. (2002) estimate a duration of less than 5 My for the stratigraphic interval *Reitzi* Zone to Top Ladinian (cf. Fig. 7, Columns C, E1–E3). The Latemar cyclic series, with 701 basic shallowing-upward microcycles, covers only a single biozone within this stratigraphic interval of 5 My. The age data for the Middle Triassic from Hungary provide independent evidence that microcycles in the Latemar do not represent precession forcing, but instead sub-Milankovitch forcing.

In summary, the orbital forcing model of Goldhammer et al. (1993) and Preto et al. (2001) cannot be reconciled with the existing high-resolution biochronostratigraphic framework for the Triassic and the Anisian–Ladinian. This framework includes: (1) the biostratigraphic dataset existing for the Latemar platform; (2) biostratigraphic schemes for the (Middle) Triassic; (3) the isotopic ages according to Mundil et al. (2003); and (4) the biochronostratigraphic chart of Hardenbol et al. (1998). In contrast, the combined sub-Milankovitch and Milankovitch forcing model of this study for the Latemar cyclic series is well compatible with the high-resolution biochronostratigraphic framework.

Sub-Milankovitch Cyclicity

The new model for the Latemar cyclic series bears important implications both for early Mesozoic platforms and for carbonate platforms during greenhouse climate conditions in general. The strongest control on platform-interior deposition was exerted by sub-Milankovitch processes with periods of 33% (based on cumulative errors, $\Delta t_{max} \leq 5.85$ ky), most probably only 24% (based on cumulative errors and spectral-analysis data, $\Delta t \leq 4.2$ ky) of the short-precession period in the Triassic. Paleozoic and Mesozoic sub-Milankovitch periodicities have been reported, e.g., by Bond et al. (1993) for the Cambrian and Yang and Kominz (1999) for the Pennsylvanian. However, these periodicities were inferred exclusively from spectra that were calibrated to match the Milankovitch spectrum. No biostratigraphic or chronostratigraphic ages existed, which could have been used to perform a model-independent calibration. Yang and Kominz (1999) therefore assumed that the sub-Milankovitch peaks in their spectra represented random components. In contrast, the existence of sub-Milankovitch periodicities in the Latemar cyclic succession is substantiated both by high statistical significance in spectral analyses and by high-resolution biochronostratigraphic ages.

Quaternary sub-Milankovitch cyclicities with periods of 1–10 ky have been studied in detail since the late 1980s. They include the Dansgaard–Oeschger (D–O) cycles (Broecker and Denton, 1989), Heinrich events (Heinrich, 1988; Broecker et al., 1992) and Bond cycles (Bond et al., 1992). D–O cycles were initially detected in cores from the Greenland ice shield (Broecker and Denton, 1989). Subsequent research has proved that D–O cycles are widespread in marine sediments on northern Atlantic continental shelves (Rasmussen et al. 1998). Furthermore, D–O cycles have also been found in deep-sea sediments off Namibia and in cores from the Antarctic ice sheet. D–O cycles, like other sub-Milankovitch cycles, are strongly asymmetric, with periods between 1.5 to 5.0 ky over the last 50 ky. Nonstationary periodicity is probably caused by stochastic resonance, i.e., the combination of a dominating sinusoidal fluctuation in ocean–atmosphere circulation with another subordinate arbitrary fluctuation. According to existing models, D–O cyclicity occurs on an interhemispheric, global scale. It is controlled by coupled changes in the ocean–atmosphere circulation, which in turn result in high-frequency climate changes.

Heinrich events (Heinrich, 1988) are documented in the deep-sea record by the cyclic occurrence of levels rich in ice-rafted debris. They are controlled by rhythmic instabilities in the Arctic ice shields leading to melting–glaciation pulses. Bond cycles (Bond et al., 1992) are saw-toothed cycles of Heinrich events with a 7.2 ky oscillation. Both Heinrich events and Bond cycles are inevitably linked to the (sub-)recent wide polar ice sheets. In the Triassic, significant polar ice sheets apparently did not exist (Wilson et al., 1994).

In general, Quaternary sub-Milankovitch cycles are linked to modern icehouse climate conditions, the current plate-tectonic configuration, and associated oceanic circulation patterns. Therefore, modern sub-Milankovitch cycles like the Dansgaard–Oeschger cycles should not be taken as direct analogies for sub-Milankovitch cycles in the early Mesozoic. The latter occurred under relative greenhouse climate conditions and a supercontinent plate-tectonic configuration with oceanic circulation patterns different than today. However, similarly to the recent climate conditions, Triassic climate was characterized by large seasonal and latitudinal temperature gradients (Wilson et al., 1994). Atmospheric general circulation models (AGCMs) of Scythian and Carnian climate indicate latitudinal surface–air temperature gradients of up to 60°C between 0° to 60°N. Widespread and dominant seasonal aridity characterized Triassic continental interior lowlands. Ranges in seasonal temperature are above 45°C and reach up to 70°, which in the recent icehouse climate is approached only in Central Siberia, Canada, and Mongolia. Low–latitude areas and marine shelves bordering the western and northwestern Tethys showed ranges in seasonal temperature between 10 and 30°C. Strong monsoonal circulation with very high precipitation reached the western Tethys and adjacent northern highlands in the summer. Latitudinal temperature gradients and strong seasonal climate differences represent an important control on large-scale and strong coupled ocean–atmosphere circulation. The Triassic Period held a very significant potential for fluctuations in coupled ocean–atmosphere circulation in the sub-Milankovitch bandwidth. Similarly to the recent climate and D–O cycles, the exact link in the Triassic between fluctuations in coupled ocean–atmosphere circulation and sub-Milankovitch sea-level changes is unclear. However, it is well possible that rapid oceanographic and/or atmospheric changes triggered a strong sub-Milankovitch signal in the Middle Triassic.

Milankovitch Cyclicity

Orbital-forcing models for icehouse climate periods (late Cenozoic, late Paleozoic) are based on the direct coupling between changes in insolation, climate change, waning and waxing of polar ice sheets, and sea-level change. The coupling of changes in insolation and sea level via polar ice sheets cannot be applied to (relative) greenhouse climate periods like the Triassic. Other causal links between Milankovitch forcing and cyclic deposition on carbonate platforms must have existed. The global sensitivity to climate changes in the Milankovitch bandwidth during the Triassic can be assessed on the grounds of seasonal variations in temperature, precipitation, atmospheric circulation, and oceanic heat flux. Seasonal variations in these parameters serve as a proxy for climate changes during Milankovitch cycles in one hemisphere.

Given the absence of polar ice sheets in the Triassic, Hay and Leslie (1990) propose that changes in the volume of water stored in groundwater reservoirs may have triggered eustatic sea-level changes. Volume of stored water may have been controlled by plate-tectonically driven changes in continental elevation (Triassic: approx. 1.2–1.4 km, Hay et al. 1981). However, this process refers rather to sea-level changes of ≥ 1 My (third-order to second-order sea-level changes). Donovan and Jones (1979) and Elder et al. (1994) suggested that, during high insolation, surface waters thermally expanded and contributed to sea-level changes. Schulz and Schäfer-Neth (1997) further developed this thermal-expansion model for Mesozoic greenhouse climates. On the basis of atmospheric general circulation models for the Triassic, they proposed that warm, saline deep waters formed in summer when net evaporation rates were high. In winter, saline deep waters did not form in the same hemisphere. However, warm, saline deep water was formed throughout the year alternating between the two hemispheres. Schulz and Schäfer-Neth (1997) transferred this summer–winter situation to the hot–cold orbit situation during a Milankovitch cycle. Temperature variations of 2–4°C between hot and cold orbits produced sea-level changes with an amplitude of 1.7 m. Sea-level changes of this amplitude may well have controlled or contributed to the thinning-upward microcycles and higher-order cycles in the Latemar platform interior. Thermal expansion of the water column because of non-orbitally forced, high-frequency climatic changes also represents a potential driving force for sub-Milankovitch sea-level fluctuations in the Triassic, resulting in the development of shallowing-upward microcycles on carbonate-platform interiors.

Spectral Analysis

Data Types and Input.—

Spectral analysis studies in the Latemar platform interior have used three different data types. Hinnov and Goldhammer (1991) used the cycle-thickness series included in Goldhammer (1987, their Appendix 2, 4, L-13; see Chart 1, Column A1). The definition of microcycles in Goldhammer (1987) was based partly on a model-dependent approach. Where individual shallowing-upward cycles could not be recognized because thinning-upward megacycles (macrocycles of this study) were condensed, five microcycles were assumed to exist. For instance, Goldhammer (1987, App. 4) recognized 6–7 megacycles in the stacked tepee zone (middle TF) and schematically attributed 5 microcycles to each of them, although he could not recognize or measure them in the section. In fact, every single megacycle of the 97 megacycles described by Goldhammer (1987) includes five shallowing-upward microcycles.

Preto et al. (2001) used a four-rank series based on lithofacies types (Chart 1A, Column A2). Lithofacies types included (1) supratidal caliches and soils; (2) supratidal-flat laminated dolomites; (3) restricted subtidal wackestones; (4) open subtidal packstones to grainstones. The authors stressed that their continuously depth-ranked series permitted a detailed, better comparison of the series with a full theoretical Milankovitch spectrum.

Analyzing external controls on cyclic deposition, e.g., orbital forcing, requires an input data type that is able to directly reflect the external signal. Orbital forcing in the Milankovitch band triggers climate changes, which result in high-frequency, low-amplitude sea-level changes. In combination with total subsidence, fluctuating sea level produces changes in accommodation space. Temporal variations in accommodation space cannot be derived from lithofacies analysis, because lithofacies essentially reflects paleobathymetry and not accommodation. In contrast to accommodation, paleobathymetry is additionally influenced by localized, short-term changes in sediment accumulation, triggered for instance by biological, hydrodynamic, or paleoecologic variations. Subtidal changes in paleobathymetry may occur, although the rate of accommodation has remained constant. On the large scale, Posamentier et al. (1988) and Schlager (1993) have demonstrated the fundamental difference between changes in paleobathymetry and accommodation. Bond and Kominz (1991) have described and analyzed the problem of using vertical facies changes to infer accommodation and eustatic sea-level histories. They stressed that because of the effects of localized sediment accumulation rates on facies patterns, the interpretation of accommodation, i.e., external controls on deposition, from vertical facies is far from straightforward. In order to correctly identify an orbitally forced sea-level signal, accommodation change and not water-depth changes have to be analyzed.

Data on lithofacies rank or water-depth introduce a high-frequency signal into the series, which is internally controlled by sediment accumulation. The rank series of Preto et al. (2001) mixes external controls and internal controls on cyclic deposition. Internal control includes vertical successions of open and restricted subtidal facies (ranks 3, 4) and of supratidal facies (ranks 1, 2). Of the 471 boundary conditions (rank boundaries) included in the series of Preto et al. (2001), only 39% (182 shallow subtidal to supratidal basic cycles) unequivocally indicate a change in accommodation space. The remaining 61% (289) introduce a wide and laterally variable range of high-frequency signals, which are probably due to internal control, to the series.

The individual thicknesses of accommodation cycles constitute the exclusive data input to the spectral analyses presented in this study. In contrast to Goldhammer et al. (1993), measured sections do not include a model-dependent definition of microcycles. Where microcycles have been condensed beyond recognition, no microcycles were defined. In such cases, it is preferable to miss beats rather than schematically introduce a model-specific stacking pattern like 1:5 into the series. Intervals in the Latemartürme section where microcycles have been condensed beyond recognition may include another 25–35 microcycles in addition to the 619 (min.) to 701 (max.) cycles actually measured. Apart from strongly condensed intervals, thinning-upward macrocycles as defined in this study include between two to six microcycles.

Power and Amplitude Spectra.—

The identification of prominent frequencies in Blackman–Tukey power spectra represents a straightforward procedure that is based exclusively on the (relative) spectral power in each sliding window. Frequencies above a power level of 2.5–3.0 were selected.

Potential perturbations of the primary Milankovitch spectrum in the Latemar cyclic series include (1) nonstationary periods of sub-Milankovitch cyclicity; (2) hierarchical sub-Milankovitch cyclicities; (3) variations in the rate of deposition; (4) autocyclic input in the series. These potential perturbation are not specific to the Latemar series but can be expected to exist in most cyclic carbonate-platform series.

Similarly to recent sub-Milankovitch cyclicities (D–O cycles), Triassic sub-Milankovitch cycle periods were probably less stationary than Triassic Milankovitch cycle periods. Moderately shifting sub-Milankovitch cyclicity could raise the level of noise in the Latemar series. Similarly to the Heinrich events and superimposed Bond cycles in recent icehouse climate conditions, Triassic sub-Milankovitch cyclicities of different orders may have been superimposed on each other. The significant to highly significant frequencies and amplitudes with an average ratio of 1:3.7 to 1:3.8 microcycles in Blackman–Tukey and multi-taper spectral and harmonic analysis (Figs. 11A–B, 12A) possibly represent higher-order sub-Milankovitch cyclicity (Fig. 12B, 13.8–16.0 ky). This ratio occurs in the same sliding windows irrespective of the method of spectral analysis applied.

Short-term and long-term variations in the sedimentation rate may change the spectrum by increasing the signal-to-noise ratio. Reverse modeling for the Latemar platform (Zühlke et al., 2003) indicates that the decompacted maximum sedimentation rates varied between 230 mm / ky (MTF) and 470 mm / ky (LCF), i.e., for a factor of 2 during a time interval of 2.95 My (LCF–UTF; cf. Fig. 7, column E3). Because the cyclic series has been subdivided into sliding windows, variations in the sedimentation rate are small within single windows (0.5–0.9 My) and do not increase the signal-to-noise ratio significantly. It is probable that short-term sedimentation rates varied over 4.2 ky (single microcycle) and 18.1–21.5 ky (1:4.3 to 1:5.1 bundled thinning-upward macrocycles). Sedimentation rates were high in the subtidal part of a microcycle (LFT 1, after the initial lag time) compared to the supratidal top (LFTs 3–5). High average sedimentation rates prevailed in the basal microcycles and reduced sedimentation rates in the upper microcycles of thinning-upward macrocycles. Schwarzacher (1989) and Stage (1999) have demonstrated that moderate random variations from the mean sedimentation rate alter the spectra slightly but original single frequency peaks are still highly significant. Random small variations in the sedimentation rate tend to attenuate high frequencies especially.

In summary, the spectral-analysis results of this study reflect external control. The following criteria indicate that the spectra hold high confidence levels and that the selected frequencies and amplitudes are real:

- the same prominent frequencies and amplitudes occur with all spectral-analysis methods applied, especially Blackman–Tukey, multi-taper harmonic, and spectral analysis. Significant average ratios show little variation.
- the results of spectral analysis do not change if spectral power (Blackman–Tukey, multi-taper spectral analysis), amplitude or significance levels from F-tests (multi-taper harmonic analysis) are considered.
- the same peaks are present in the power and amplitude spectra for all sliding windows (subsets). According to Stage (1999), this indicates that the cycles are real.
- the model resulting from spectral analyses is well compatible with the high-resolution biochronostratigraphic framework constrained by isotopic ages from the Latemar cyclic series, interpolated ages in current chronostratigraphic charts, the complete set of biostratigraphic data for the Latemar, and various biostratigraphic schemes.

CONCLUSIONS

Individual shallowing-upward, meter-scale microcycles in the Latemar clearly reflect allocyclic control by changes in sea level. Individual cycles show moderate lateral variations in thickness, internal facies, and diagenetic overprint. More than 90% of macrocycles can be physically traced all over the platform. Within laterally persistent macrocycles, 80–95% of all microcycles equally show lateral physical persistence. Except for the basal LCF, no stratal terminations exist on the microcycle to macrocycle scale. The small diameter of the platform and the high average accommodation rates favored a simultaneous vertical aggradation of the platform top. This allocyclic model for the Latemar platform should not automatically be transferred to other Triassic or Mesozoic platforms with similar subtidal to peritidal cycles and hierarchical cycle stacking patterns (e.g., Upper Triassic Dachstein Fm.). Each platform requires detailed cyclostratigraphic logging, lateral tracing of cycles, and biochronostratigraphic and spectral analyses in order to develop a reliable forcing model. The Latemar represents a model platform because it offers a unique combination of features: well developed cyclicity, excellent outcrops for the lateral physical tracing of cycles, abundant ammonoid faunas, and volcanic ash layers intercalated in the platform interior.

Basic meter-scale shallowing-upward microcycles in the Latemar platform reflect sub-Milankovitch control. High-resolution biochronostratigraphic data indicate clearly that basic microcycles cannot be reconciled with precession forcing. The biochronostratigraphic framework includes a large set of biostratigraphic data from different studies, isotopic ages from the Latemar, and general data from current biochronostratigraphic charts (Hardenbol et al., 1998). The period of each microcycle ranges between $\Delta t_{av} = 2.68$ ky and $\Delta t_{max} \leq 5.85$ ky, compared to the Triassic short-precession period of $\Delta t = 17.78$ ky. Spectral analyses show that the most probable microcycle period in the Latemar platform interior is $\Delta t = 4.2$ ky, which is less than a quarter of the precessional period. These results are in clear contrast to studies of, e.g., Goldhammer et al. (1993) and Preto et al. (2001), which assumed that individual microcycles or the shortest periodicity in rank series of the Latemar platform interior presented precession forcing of $\Delta t = 17.78$ ky. These studies relied exclusively on the spectral analysis of series that covered a small part (one-third) of the cyclic succession preserved in the Latemar.

In the Latemar, cycle bundles of 1:4.3 to 1:5.1 shallowing-upward microcycles do not reflect eccentricity forcing superimposed on precession forcing. Instead, the ratio reflects short and long precession forcing of 18 ky to 21 ky superimposed on sub-Milankovitch forcing of 4.2 ky. At least for Mesozoic platform-interior successions, the interpretation of Milankovitch forcing and equivalent time intervals from specific cycle stacking patterns is unwarranted. Establishing orbital forcing in Mesozoic series requires a model-independent calibration of time by biochronostratigraphic data. If sufficiently high-resolution data are not available, the time intervals implied by the orbital-forcing model can be compared to current biochronostratigraphic charts in order to check whether these time intervals are internally plausible.

Spectral analyses in this study have been both thickness-calibrated and time-calibrated, on the basis of high-resolution biochronostratigraphic framework available for the Latemar. Several spectral-analysis methods (Blackman–Tukey, multi-taper spectral, and harmonic analysis) show highly similar significant ratios that are largely stationary over all seven subsets. The platform-interior cyclic series reveals higher-order significant amplitudes at both ratios and time intervals indicative of orbital forcing in the Milankovitch band, when the microcycle period is set to $\Delta t = 4.2$ ky. This cyclostratigraphic, biochronostratigraphic, and spectral-analysis model indicates that in addition to sub-Milankovitch forcing, short and long precession, short and long obliquity, and short eccentricity forcing are present in the large-scale cycle bundlings. The Latemar cyclic series includes the oldest explicit sub-Milankovitch signal and the oldest set of both sub-Milankovitch and Milankovitch signals yet observed in the geologic record.

The forcing of sub-Milankovitch cyclicity in the early Mesozoic is yet unclear. Quaternary sub-Milankovitch cycles (e.g., Dansgaard–Oeschger cycles) are linked to icehouse climate conditions with polar ice caps and oceanic circulation patterns influenced by the current plate-tectonic configuration. In contrast, the early Mesozoic was characterized by (relative) greenhouse climate conditions and the Pangea supercontinent plate-tectonic configuration. Therefore, Quaternary sub-Milankovitch cycles cannot serve as direct analogies for (early) Mesozoic sub-Milankovitch cycles. In general, modern D–O cycles are caused by rapid changes in coupled ocean-atmosphere circulation patterns. Strong latitudinal temperature gradients and seasonal climate differences represent the major driver for modern oceanic–atmospheric circulation patterns and sub-Milankovitch cycles. Triassic climate was equally characterized by large latitudinal temperature gradients and seasonal climate differences. Therefore, the Triassic period held a very significant potential for fluctuations in ocean–atmosphere circulation.

The most probable link between high-frequency, low-amplitude sea-level changes and both sub-Milankovitch and Milankovitch forcing in the Triassic is the thermal contraction and expansion of the oceanic water column. Global seizure or release of water stored in groundwater reservoirs and associated changes in the runoff–evaporation–precipitation cycle may have triggered lower-frequency sea-level changes and contributed to higher-order cycles on carbonate platforms.

ACKNOWLEDGMENTS

I am greatly indebted to Thilo Bechstädt, who initiated this project and supported it in many ways. Peter Brack, Hans Rieber, John Reijmer, Wolfgang Schwarzacher, and Alton Brown are thanked for discussions and comments. Alfred Fischer and two anonymous reviewers provided many valuable comments. Claudia Seibold detected the ammonoid fauna below the summit of the Latemartürme. The author is very grateful to the Gabrielli family from the Rifugio Torre di Pisa on the Latemar for their hospitality and logistical support. Part of this study was funded by the German Research Fund (DFG, Be 641/28).

REFERENCES

Bechstädt, T., Brack, P., Preto, N., Rieber, H., and Zühlke, R., 2003, Fieldtrip to Latemar: Guidebook, Triassic Geochronology and Cyclostratigraphy—A Field Symposium, St. Christina / Val Gardena, Italy, September 11–13, 72 p. (see also www.Latemar.sedimentology.info)

Berger, A., and Loutre, M.F., 1994, Astronomical forcing through geological time, *in* de Boer, P.L., and Smith, D.G., eds., Orbital Forcing and Cyclic Sequences: International Association of Sedimentologists, Special Publication 19, p. 15–24.

Blackman, R.B., and Tukey, J.W., 1958, The Measurement of Power Spectra from the Point of View of Communication Engineering: New York, Dover Publications, 190 p.

Bond, G.C., Heinrich, H., Broecker, W., Labeyrie, L., McManus, J., Andrews, J., Huon, S., Jantschik, R., Clasen, S., Simet, C., Tedesco, K., Klas, M., Bonani, G., and Ivy, S., 1992, Evidence for massive discharges of icebergs into the North Atlantic ocean during the last glacial period: Nature, v. 360, p. 245–249.

Bond, G.C., and Kominz, M.A., 1991, Some comments on the problem of using vertical facies changes to infer accommodation and eustatic sea-level histories with examples from Utah and the southern Canadian Rockies, *in* Franseen, E.K., Watney, W.L., Kendall, C.G.St.C., and Ross, W., eds., Sedimentary Modeling: Computer Simulations and Methods for Improved Parameter Definition: Kansas Geological Survey, Bulletin 233, p. 273–291.

Bond, G.C., Kominz, M.A., and Bevan, J., 1991, Evidence for orbital forcing of Middle Cambrian peritidal cycles, Wah Wah Range, south-central Utah, *in* Franseen, E.K., Watney, W.L., Kendall, C.G.St.C., and Ross, C.A., eds., Sedimentary Modeling: Computer Simulations and Methods for Improved Parameter Definition: Kansas Geological Survey, Bulletin 233, p. 293–318.

Bond, G.C., Kominz, M.A., Devlin, W.J., and Beavan, J., 1993, Evidence of astronomical forcing of the Earth´s climate in Cretaceous and Cambrian times, *in* Anderson, D.M., and Eaton, G.P., eds., Geological Perspectives on Global Change: Tectonophysics, v. 222, p. 295–315.

Bosellini, A., and Rossi, D., 1974, Triassic carbonate buildups of the Dolomites, northern Italy, *in* Laporte, L.F., ed., Reefs in Time and Space: Society of Economic Paleontologists and Mineralogists, Special Publication 18, p. 209–233.

Bosellini, A., Castellarin, A., Doglioni, C., Guy, F., Lucchini, F., Perri, M.C., Rossi, P.L., Simboli, G., and Sommavilla, E., 1982, Magmatismo e tettonica nel Trias delle Dolomiti, *in* Castellarin, A., and Vai, G.B., eds., Guida alla Geologia del Sudalpino centro-orientale: Società Geologica Italiana, Guide Geologiche Regionali, p. 189–210.

Bowring, S.A., Erwin, D.H., Jin, Y.G., Martin, M.W., Davidek, K., and Wang, W., 1998, Geochronology of the end-Permian mass extinction: Science, v. 280, p. 1039–1045.

Brack, P., Mundil, R., Oberli, F., Meier, M., and Rieber, H., 1996, Biostratigraphic and radiometric age data question the Milankovitch characteristics of the Latemar cycles (Southern Alps, Italy): Geology, v. 24, p. 371–375.

Brack, P., and Rieber, H., 1993, Towards a better definition of the Anisian/Ladinian boundary: new biostratigraphic data and correlations of boundary sections from the Southern Alps: Eclogae Geologicae Helvetiae, v. 86, p. 415–527.

Brack, P., and Rieber, H., 1994, The Anisian/Ladinian boundary: retrospective and new constraints: Albertiana, v. 13, p. 25–36.

Broecker, W., and Denton, G.H., 1989, The role of ocean–atmosphere reorganizations in glacial cycles: Geochimica et Cosmochimica Acta, v. 53, p. 2465–2501.

Broecker, W.S., Bond, G.C., Klas, M., Clark, E.A., and McManus, J., 1992, Origin of the northern Atlantic's Heinrich events, *in* Kelts, K.R., ed., Past and Present Climate Dynamics; Reconstruction of Rates of Change: Climate Dynamics, v. 6, p. 265–273.

CISNE, J.L., 1986, Earthquakes recorded stratigraphically on carbonate platforms: Nature, v. 323, p. 320–322.

DE ZANCHE, V., GIANOLLA, P., MANFRIN, S., MIETTO, P., AND ROGHI, G., 1995, A Middle Triassic back-stepping carbonate platform in the Dolomites (Italy): sequence stratigraphy and biochronostratigraphy: Memorie di Scienze Geologiche, v. 47, p. 135–155.

DONOVAN, D.T., AND JONES, E.J.W., 1979, Causes of world-wide changes in sea-level: Geological Society of London, Journal, v. 136, p. 187–192.

DRUMMOND, C.N., AND WILKINSON, B.H., 1993, Carbonate cycle stacking patterns and hierarchies of orbitally forced eustatic sea level change: American Association of Petroleum Geologists, Bulletin, v. 63, p. 369–377.

DUNN, P.A., 1992, Cyclic stratigraphy and early diagenesis; an example from the Triassic Latemar Platform, northern Italy: Unpublished Ph.D. Thesis, John Hopkins University, Baltimore, 865 p.

EGENHOFF, S., PETERHÄNSEL, A., BECHSTÄDT, T., ZÜHLKE, R., AND GRÖTSCH, J., 1999, Facies architecture of an isolated carbonate platform: tracing the cycles of the Latemar (Middle Triassic, northern Italy): Sedimentology, v. 46, p. 893–912.

ELDER, W.P., GUSTASON, E.R., AND SAGEMANN, B.B., 1994, Correlation of basinal carbonate cycles to nearshore parasequences in the late Cretaceous Greenhorn seaway, Western interior, USA: Geological Society of America, Bulletin, v. 106, p. 892–902.

ENOS, P., AND SAMANKASSOU, E., 1998, Lofer cyclothems revisited (Late Triassic, Northern Alps, Austria): Facies, v. 38, p. 207–228.

FISCHER, A.G., 1964, The Lofer cyclothems of the Alpine Triassic: Kansas Geological Survey, Bulletin 169, p. 107–149.

FISCHER, A.G., 1986, Climatic rhythms recorded in strata: Annual Review of Earth and Planetary Sciences, v. 14, p. 351–376.

GAETANI, M., FOIS, E., JADOUL, F., AND NICORA, A., 1981, Nature and evolution of the middle Triassic carbonate buildups in the Dolomites (Italy): Marine Geology, v. 44, p. 25–57.

GIANOLLA, P., DE ZANCHE, V., AND MIETTO, P., 1998, Triassic sequence stratigraphy in the Southern Alps (northern Italy): definition of sequences and basin evolution, *in* de Graciansky, P.C., Hardenbol, J., Jacquin, T., and Vail, P.R., eds., Mesozoic and Cenozoic Sequence Stratigraphy of European Basins: SEPM, Special Publication 60, p. 719–747.

GINSBURG, R.N., 1971, Landward movement of carbonate mud: a new model for regressive cycles in carbonates (abstract): American Association of Petroleum Geologists, Bulletin, v. 55, p. 340.

GOLDHAMMER, R.K., 1987, Platform carbonate cycles, middle Triassic of northern Italy: the interplay of local tectonics and global eustasy: Unpublished Ph.D. Thesis, John Hopkins University, Baltimore, 468 p.

GOLDHAMMER, R.K., DUNN, P.A., AND HARDIE, L.A., 1987, High-frequency glacio-eustatic oscillations with Milankovitch characteristics recorded in northern Italy: American Journal of Science, v. 287, p. 853–892.

GOLDHAMMER, R.K., DUNN, P.A., AND HARDIE, L.A., 1990, Depositional cycles, composite sea-level changes, cycle stacking patterns, and the hierarchy of stratigraphic forcing: Geological Society of America, Bulletin, v. 102, p. 535–562.

GOLDHAMMER, R.K., AND HARRIS, M.T., 1989, Eustatic controls on the stratigraphy and geometry of the Latemar buildup (Middle Triassic), the Dolomites of northern Italy, *in* Crevello, P.D., Wilson, J.L., Sarg, J.F., and Read, J.F., eds., Controls on Carbonate Platform and Basin Development: SEPM, Special Publication 44, p. 323–338.

GOLDHAMMER, R.K., HARRIS, M.T., DUNN, P.A., AND HARDIE, L.A., 1993, Sequence stratigraphy and system tract development of the Latemar Platform, middle Triassic of the Dolomites (Northern Italy): outcrop calibration keyed by cycle stacking patterns, *in* Loucks, R.G., and Sarg, J.F., eds., Carbonate Sequence Stratigraphy: Recent Developments and Applications: American Association of Petroleum Geologists, Memoir 57, p. 353–388.

GOLDHAMMER, R.K., OSWALD, E.J., AND DUNN, P.A., 1994, High-frequency, glacio-eustatic cyclicity in the Middle Pennsylvanian of the Paradox Basin: an evaluation of Milankovitch forcing, *in* de Boer, P.L., and Smith, D.G., eds., Orbital Forcing and Cyclic Sequences: International Association of Sedimentologists, Special Publication 19, p. 243–283.

GRADSTEIN, F.M., AGTERBERG, F.P., OGG, J.G., HARDENBOL, J., VAN VEEN, P., THIERRY, J., AND HUANG, Z., 1994, A Mesozoic time scale: Journal of Geophysical Research, v. 99, p. 24,051–24,074.

GRADSTEIN, F.M., AGTERBERG, F.P., OGG, J.G., HARDENBOL, J., VAN VEEN, P., THIERRY, J., AND HUANG, Z., 1995, A Triassic, Jurassic and Cretaceous time scale, *in* Berggren, W.A., Kent, D.V., and Aubry, M.-P., eds., Geochronology, Time Scales and Global Stratigraphic Correlation: SEPM, Special Publication 54, p. 95–126.

GROTZINGER, J.P., 1986, Upward shallowing platform cycles: a response to 2.2 billion years of low-amplitude, high-frequency (Milankovitch band) sea level oscillations: Paleoceanography, v. 1, p. 403–416.

HARDENBOL, J., THIERRY, J., FARLEY, M.B., JACQUIN, T., DE GRACIANSKY, P.-C., AND VAIL, P.R., 1998, Mesozoic and Cenozoic sequence chronostratigraphic framework of European basins, *in* de Graciansky, P.-C., Hardenbol, J., Jaquin, T., and Vail, P.R., eds., Mesozoic and Cenozoic Sequence Stratigraphy of European Basins: SEPM, Special Publication 60, p. 3/15.

HARDIE, L.A., 1986, Stratigraphic Models for carbonate tidal-flat deposition, *in* Hardie, L.A., and Shinn, E.A., eds., Carbonate Depositional Environments, Part 3: Tidal Flats: Colorado School of Mines, Quarterly, v. 81, p. 59–73.

HARDIE, L.A., AND SHINN, E.A., 1986, Carbonate depositional environments: Colorado School of Mines, Quarterly, v. 81, 74 p.

HARDIE, L.A., BOSELLINI, A., AND GOLDHAMMER, R.K., 1986, Repeated subaerial exposure of subtidal carbonate platforms, Triassic, northern Italy: evidence for high-frequency sea-level oscillations on a 10^4 year scale: Paleoceanography, v. 1, p. 447–457.

HARDIE, L.A., WILSON, E.M., AND GOLDHAMMER, R.K., 1991, Cyclostratigraphy and dolomitization of the Middle Triassic Latemar buildup, the Dolomites, northern Italy: Dolomieu Conference on Carbonate Platforms and Dolomitization, September, St. Ullrich/Gröden, Guidebook Excursion F, p. 1–56.

HARLAND, W.B., ARMSTRONG, R.L., COX, A.V., CRAIG, L.E., SMITH, A.G., AND SMITH, D.G., 1990, A Geologic Time Scale 1989: Cambridge, U.K., Cambridge University Press, 263 p.

HARLAND, W.B., COX, A.V., LLEWELLYN, P., THICKTON, C.A., SMITH, A.G., AND WALTERS, R., 1982, A Geological Time Scale: Cambridge, U.K., Cambridge University Press, 131 p.

HARRIS, M.T., 1994, The foreslope and toe-of-slope facies of the middle Triassic Latemar buildup (Dolomites, northern Italy): Journal of Sedimentary Research, v. B64, p. 132–145.

HARRIS, M.T., KERANS, C., AND BEBOUT, D.G., 1993, Ancient outcrop and modern examples of platform carbonate cycles—implications for understanding subsurface correlation and understanding reservoir heterogeneity, *in* Loucks, R.G., and Sarg, J.F., eds., Carbonate Sequence Stratigraphy: Recent Developments and Applications: American Association of Petroleum Geologists, Memoir 57, p. 475–492.

HAY, W.W., BARRON, E.J., SLOAN, J.L., AND SOUTHAM, J., 1981, Continental drift and the global pattern of sedimentation: Geologische Rundschau, v. 70, p. 302–315.

HAY, W.W., AND LESLIE, M.A., 1990, Could possible changes in global groundwater reservoir cause eustatic sea-level fluctuations?, *in* Revelle, R., ed., Sea-Level Change: National Research Council, Studies in Geophysics, p. 161–170.

HEINRICH, H., 1988, Origin and consequences of cyclic ice rafting in the Northeast Atlantic Ocean during the past 130,000 years: Quaternary Research, v. 29, p. 142–152.

HILGEN, F.J., KRIJGSMAN, W., LANGEREIS, C.G., AND LOURENS, L.J., 1997, Breakthrough made in dating of the geological record: Eos, Transactions, American Geophysical Union, v. 78, p. 285–289.

HINNOV, L.A., AND GOLDHAMMER, R.K., 1991, Spectral analysis of the Middle Triassic Latemar Limestone: Journal of Sedimentary Petrology, v. 61, p. 1173–1193.

HÜSSNER, H.M., ROESSLER, J., BETZLER, R., AND PETSCHICK, R., 2000, Milankovitch-Steuerung ohne orbitale Steuerung?: Gesellschaft der Geologie- und Bergbaustudenten Österreichs, Mitteilungen, v. 43, p. 66.

KOERSCHNER, W.F., AND READ, J.F., 1989, Field and modelling studies of Cambrian carbonate cycles, Virginia Appalachians: Journal of Sedimentary Petrology, v. 59, p. 654–687.

MASETTI, D., AND TROMBETTA, G.L., 1998, L´eredita anisica nella nascita ed evoluzione delle piattaforme medio-triassiche delle dolomiti occidentali: Memorie di Scienze Geologiche, v. 50, p. 213–237.

MIALL, A.D., 1997, The Geology of Stratigraphic Sequences: Berlin, Springer, 433 p.

MIETTO, P., AND MANFRIN, S., 1995, A high-resolution ammonoid standard scale in the Tethys realm. A preliminary report: Société Géologique de France, Bulletin, v. 166, p. 539–563.

MILANKOVITCH, M., 1941, Kanon der Erdbestrahlung und seine Anwendung auf das Eiszeitenproblem: Royal Serbian Academy, Section of Mathematics and Natural Sciences, Special Publication, v. 132.

MONTAÑEZ, I.P., AND OSLEGER, D.A., 1993, Parasequence stacking patterns, third-order accommodation events, and sequence stratigraphy of Middle to Upper Cambrian platform carbonates, Bonanza King Formation, Southern Great Basin, *in* Loucks, R.G., and Sarg, J.F., eds., Carbonate Sequence Stratigraphy: Recent Developments and Applications: American Association of Petroleum Geologists, Memoir 57, p. 305–326.

MONTAÑEZ, I.P., AND READ, J.F., 1992, Eustatic control on early dolomitization of cyclic peritidal carbonates: evidence from the early Ordovician Upper Knox Group, Appalachians: Geological Society of America, Bulletin, v. 104, p. 872–886.

MUNDIL, R., BRACK, P., AND LAURENZI, M.A., 1996a, High-resolution U/Pb single-zircon age determinations: new constraints on the timing of Middle Triassic magamatism in the Southern Alps (abstract): Società Geologica Italiana, 78a Riunione Estiva, September 16–18, San Cassiano, Italy, Abstract Volume.

MUNDIL, R., BRACK, P., MEIER, M., RIEBER, H., AND OBERLI, F., 1996b, High resolution U–Pb dating of middle Triassic volcaniclastics: time-scale calibration and verification of tuning parameters for carbonate sedimentation: Earth and Planetary Science Letters, v. 141, p. 137–151.

MUNDIL, R., ZÜHLKE, R., BECHSTADT, T., BRACK, P., EGENHOFF, S., MEIER, M., OBERLI, F., PETERHÄNSEL, A., AND RIEBER, H., 2003, Cyclicities in Triassic platform carbonates: synchronizing radio-isotopic and orbital clock: Terra Nova, in press.

OSLEGER, D.A., AND READ, J.F., 1991, Relation of eustasy to stacking patterns of meter-scale carbonate cycles, late Cambrian, U.S.A: Journal of Sedimentary Petrology, v. 61, p. 1225–1252.

PAILLARD, D., LABEYRIE, L., AND YIOU, P., 1996, Macintosh program performs time-series analysis: Eos, Transactions, American Geophysical Union, v. 77, p. 379.

PALFY, J., PARRISH, R.R., AND VÖRÖS, A., 2002, Integrated U–Pb geochronology and ammonoid biochronology from the Anisian/Ladinian GSSP candidate section at Felsöörs (abstract), *in* Piros, O., ed., I.U.G.S Subcommission on Triassic stratigraphy, STS/IGCP 467 field meeting: Hungarian Geological Society, 5–8 September, Veszprem, Hungary, Abstract Volume, p. 28.

PALMER, A.R., 1983, The Decade of North American Geology, 1983 geologic time scale: Geology, v. 11, p. 503–504.

PEPER, T., AND CLOETINGH, S., 1995, Autocyclic perturbations of orbitally forced signals in the sedimentary record: Geology, v. 23, p. 937–940.

PETERHÄNSEL, A., AND EGENHOFF, S., 2000, Cyclomania—Milankovitch cycles in the Triassic of northern Italy refuted (abstract): International Association of Sedimentologists, 20th Regional Meeting, September 13–15, Dublin, Ireland, Abstract Volume, p. 46.

POSAMENTIER, W., JERVEY, M.T., AND VAIL, P.R., 1988, Eustatic controls on clastic deposition I—conceptual framework, *in* Wilgus, C.K., Hastings, B.S., Kendall, C.G.St.C., Posamentier, H.W., Ross, C.A., and Van Wagoner, J.C., eds., Sea-Level Changes: An Integrated Approach: SEPM, Special Publication 42, p. 109–124.

PRATT, B.R., JAMES, N.P., AND COWAN, C.A., 1992, Peritidal carbonates, *in* Walker, R.G., and James, N.P., eds., Facies Models: Response to Sea Level Change, Geological Association of Canada, p. 303–322.

PRETO, N., AND HINNOV, L., 2000, Semi-automated correlation and the cyclicity of the Latemar platform (Middle Triassic, Southern Alps) (abstract): International Association of Sedimentologists, 20th Regional Meeting, 13–15 September, Dublin, Ireland, Abstract Volume, p. 127.

PRETO, N., MIETTO, P., AND MANFRIN, S., 2002, Ammonoid biostratigraphy of the Latemar platform and its significance for the A/L boundary (abstract), *in* Piros, O., ed., I.U.G.S. Subcommission on Triassic stratigraphy, STS/IGCP 467 field meeting: Hungarian Geological Society, 5–8 September, Veszprem, Hungary, Abstract Volume, p. 21.

PRETO, N., HINNOV, L.A., HARDIE, L.A., AND DE ZANCHE, V., 2001, Middle Triassic orbital signature recorded in the shallow marine Latemar carbonate buildup (Dolomites, Italy): Geology, v. 29, p. 1123–1126.

RASMUSSEN, T.L., THOMSEN, E., AND VAN WEERING, T.C.E., 1998, Cyclic sedimentation on the Faeroe rift 53-10 ka related to climatic variations, *in* Stroker, M.S., Eavens, D., and Cramp, A., eds., Geological Processes on Continental Margins: Sedimentation, Mass-Wasting and Stability: Geological Society of London, Special Publication 129, p. 225–267.

READ, J.F., 1989, Controls on evolution of Cambrian–Ordovician passive margin, U.S. Appalachians, *in* Crevello, P.D., Sarg, J.F., Read, J.F., and Wilson, J.L., eds., Controls on Carbonate Platform and Basin Development: SEPM, Special Publication 44, p. 147–166.

READ, J.F., AND GOLDHAMMER, R.K., 1988, Use of Fischer plots to define third-order sea-level curves in Ordovician peritidal cyclic carbonates, Appalachians: Geology, v. 16, p. 895–899.

READ, J.F., KERANS, C., WEBER, L.J., SARG, J.F., AND WRIGHT, F.M., 1995, Milankovitch sea-level changes, cycles, and reservoirs on carbonate platforms in greenhouse and icehouse worlds: SEPM, Short Course Notes 35, 203 p.

RÜFFER, T., AND ZÜHLKE, R., 1995, Sequence stratigraphy and sea-level changes in the Early to Middle Triassic of the Alps: a global comparison, *in* Haq, B.U., ed., Sequence Stratigraphy and Depositional Response to Eustatic, Tectonic and Climatic Forcing: Dordrecht, The Netherlands, Kluwer Academic Publishers, p. 161–207.

SADLER, P.M., OSLEGER, D.A., AND MONTAÑEZ, I.P., 1993, On the labeling, length and objective basis of Fisher plots: Journal of Sedimentary Petrology, v. 63, p. 360–368.

SANDER, B., 1936, Beiträge zur Kenntnis der Anlagerungsgefüge (Rhythmische Kalke und Dolomite aus der Trias); II, Südalpine Beispiele, Hauptdolomit, Allgemeines: Mineralogische und Petrographische Mitteilungen, v. 48, p. 141–209.

SATTERLY, A.K., 1996, Cyclic sedimentation in the upper Triassic Dachstein Limestone, Austria: the role of patterns of sediment supply and tectonics in a platform–reef–basin system: Journal of Sedimentary Research, v. 66, p. 307–323.

SCHLAGER, W., 1993, Accommodation and supply—a dual control on stratigraphic sequences: Sedimentary Geology, v. 86, p. 111–136.

SCHULZ, M., AND SCHÄFER-NETH, C., 1997, Translating Milankovitch climate forcing into eustatic fluctuations via deep water expansion: a conceptual link: Terra Nova, v. 9, p. 228–231.

SCHWARZACHER, W., 1947, Über die sedimentäre Rhythmik des Dachsteinkalks von Lofer: Vienna, Geologische Bundes-Anstalt, Verhandlungen, v. H10-12, p. 175–188.

SCHWARZACHER, W., 1989, Milankovitch cycles and the measurement of time: Terra Nova, v. 15, p. 405–408.

SCHWARZACHER, W., AND HAAS, J., 1986, Comparative statistical analysis of some Hungarian and Austrian Upper Triassic peritidal carbonate sequences: Acta Geologica Hungarica, v. 29, p. 175–196.

SLOAN, R.J., AND WILLIAMS, B.P.J., 1991, Volcano-tectonic control of offshore to tidal-flat regressive cycles from the Dunquin Group (Silurian) of southwest Ireland, *in* Macdonald, D.I.M., ed., Sedimentation, Tectonics and Eustasy: Sea-Level Changes at Active Margins: International Association of Sedimentologists, Special Publication 12, p. 105–119.

STAGE, M., 1999, Signal analysis of cyclicity in Maastrichtian pelagic chalks from the Danish North Sea: Earth and Planetary Science Letters, v. 173, p. 75–90.

STEIGER, R., BICKEL, R.A., AND MEIER, M., 1993, Conventional U–Pb dating of single fragments of zircon for petrogenetic studies of Phanerozoic granitoids: Earth and Planetary Science Letters, v. 115, p. 197–209.

STRASSER, A., 1988, Shallowing upward sequences in Purbeckian peritidal carbonates (lower Cretaceous, Swiss and French Jura Mountains): Sedimentology, v. 35, p. 369-383.

THOMSON, D.J., 1982, Spectrum estimation and harmonic analysis: IEEE Proceedings, v. 70, p. 1055-1096.

VÖRÖS, A., 1998, Triassic ammonoids and biostratigraphy of the Balaton Highland: Studia Naturalis Magyar Természettudományi Múzeum, v. 12.

WILSON, K.M., POLLARD, D., HAY, W.W., THOMPSON, S.L., AND WOLD, C.N., 1994, General circulation model simulations of Triassic climates; preliminary results, *in* Klein, G.D., ed., Pangea—Paleoclimate, Tectonics, and Sedimentation during Accretion, Zenith, and Breakup of a Supercontinent: Geological Society of America, Special Paper 288, p. 91–116.

WOLFRAM, S., 1996, The Mathematica Book: Wolfram Media / Cambridge University Press, 1403 p.

YANG, H., AND KOMINZ, M.A., 1999, Testing periodicity of depositional cyclicity, Cisco Group (Virgilian and Wolfcampian), Texas: Journal of Sedimentary Research, v. 69, p. 1209–1231.

YIOU, P., GHIL, M., JOUZEL, J., PAILLARD, D., AND VAUTARD, R., 1994, Nonlinear variability of the climatic system, from singular and power spectra of Late Quaternary records: Climate Dynamics, v. 9, p. 371–389.

YU, H., 1995, Mathematica Time Series Pack—Reference and User´s Guide: Champaign, Illinois, Wolfram Research Inc., 168 p.

ZÜHLKE, R., 2000, Fazies, hochauflösende Sequenzstratigraphie und Beckenentwicklung im Anis (mittlere Trias) der Dolomiten (Südalpin, N-Italien): Gaea Heidelbergensis, v. 6, p. 1–368.

ZÜHLKE, R., AND BECHSTÄDT, T., 2000, High-resolution cyclostratigraphy and architecture of a carbonate platform interior—the Latemar, Middle Triassic, Southern Alps (abstract): International Association of Sedimentologists, 20th Regional Meeting, 13–15 September, Dublin, Abstract Volume, p. 68.

ZÜHLKE, R., BECHSTÄDT, T., AND MUNDIL, R., 2001, Sub-Milankovitch and Milankovitch cyclicity in an model carbonate platform—Latemar, Triassic, Italy (abstract): American Association of Petroleum Geologists, Annual Meeting, June 3–6, Denver, Abstract Volume, p. A226.

ZÜHLKE, R., BECHSTADT, T., AND MUNDIL R., 2003, Sub-Milankovitch and Milankovitch forcing on a model Mesozoic carbonate platform—the Latemar (Middle Triassic, Italy): Terra Nova, in press.

ZÜHLKE, R., BECHSTÄDT, T., BRACK, P., MUNDIL, R., AND RIEBER, H., 2000, Die Latemar-Kontroverse: neue Daten zur Geometrie, zeitlichen Entwicklung und Interpretation lagunärer Zyklen: Gesellschaft der Geologie- und Bergbaustudenten in Österreich, Mitteilungen, v. 43, p. 154.

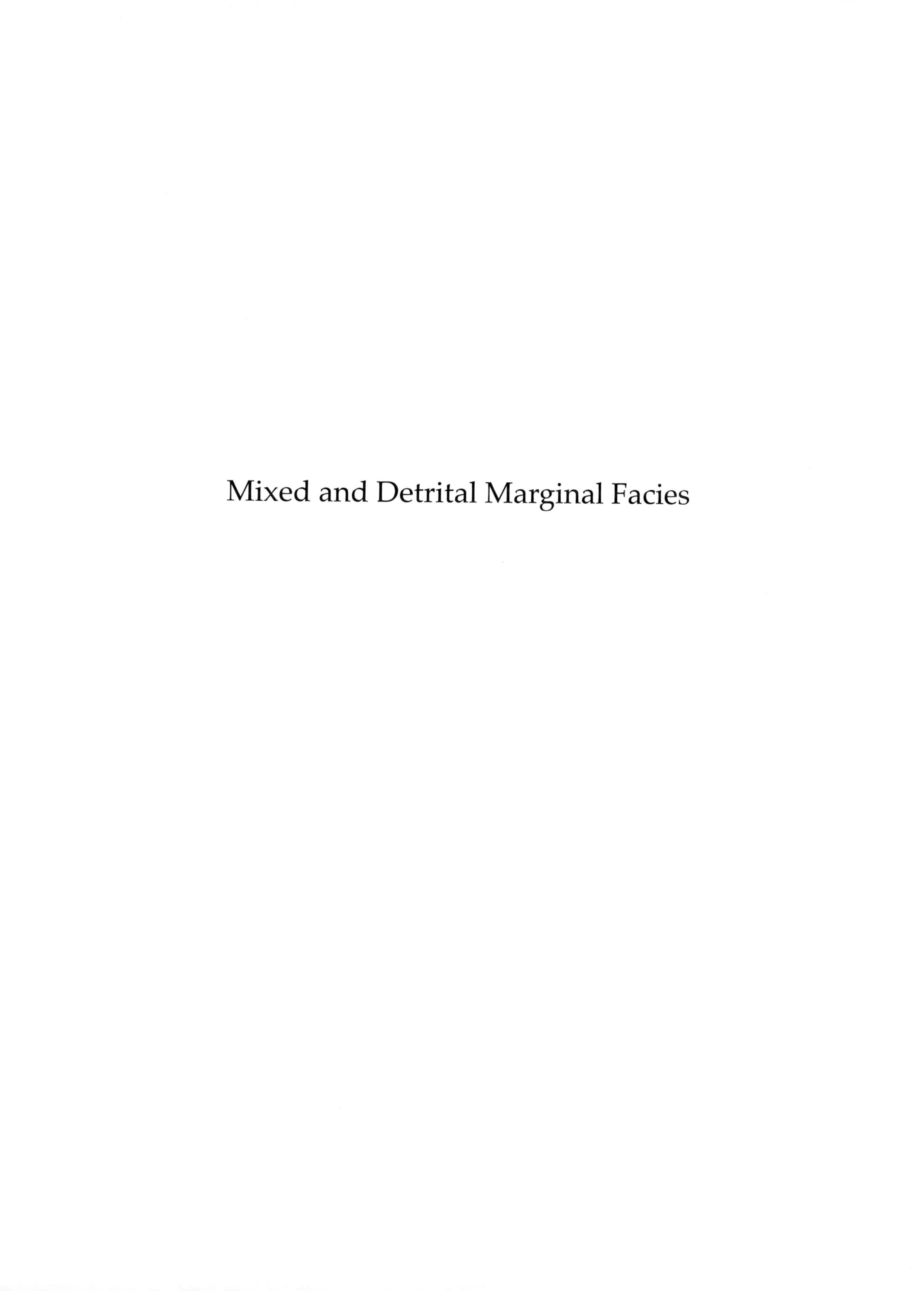

Mixed and Detrital Marginal Facies

LOWSTAND PROGRADING WEDGES AS FOURTH-ORDER GLACIO-EUSTATIC CYCLES IN THE PLEISTOCENE CONTINENTAL SHELF OF APULIA (SOUTHERN ITALY)

GEMMA AIELLO AND FRANCESCA BUDILLON
Istituto per l'Ambiente Marino Costiero, Geomare, National Research Council,
Calata Porta di Massa, Porto di Napoli, 80133 Napoli, Italy
e-mail: aiello@gms01.geomare.na.cnr.it

ABSTRACT: The Salento continental shelf of Apulia (southern Adriatic) shows a complex stratigraphic architecture of Pleistocene prograding wedges that records the effects of the interplay of sediment accumulation, glacio-eustatic sea level changes, and regional tectonics. The interpretation of high-resolution seismic reflection profiles reveals three ravinement surfaces, extending landwards from the shelf break and showing low gradients. They have been considered to be stratigraphic markers and tentatively correlated to the oxygen curves of the isotopic stratigraphy. Even if direct dating of seismic sequences and corresponding unconformities is lacking, this correlation is well supported by the lateral extent and continuity of the ravinement surfaces, recognized in the whole investigated area. Similar qualitative correlation of unconformities with the curves of oxygen-isotope stratigraphy has been already used by other authors in different case histories of continental shelves of both active and passive margins. Moreover, in the study area the regional geological framework suggests that during the Middle Pleistocene the rate of tectonic uplift interacted with the rate of glacio-eustatic fluctuations, producing the deposition of forced-regression systems tracts. The forced-regression prograding wedge that enlarged the shelf by about fifteen kilometers is here dated as Middle Pleistocene. Starting from the upper part of the Middle Pleistocene, the stratigraphic architecture of the prograding wedges was controlled mainly by glacio-eustatic sea-level changes forced by short-eccentricity cycles. This is suggested by the stratigraphic architecture of the last two prograding wedges, interpreted as incomplete fourth-order depositional sequences and consisting of forced-regression, lowstand, and transgressive systems tracts. The eustatic signal as an expression of the Earth's orbital cyclicity (short eccentricity) appears overwhelming with respect to the tectonic one, and its prominence suggests a decrease in the rate of uplift of the Apulian foreland during the last 250 ky.

INTRODUCTION

Recent studies carried out on active and passive margins show that the stratigraphic architecture of the continental shelves consists mostly of lowstand, shelf-margin, and transgressive system tracts deposits (Suter and Berryhill, 1985; Suter et al., 1987; Saito, 1991; Tesson et al., 1990; Tesson et al., 1993; Okamura and Blum, 1993; Trincardi and Correggiari, 2000). Highstand deposits form mainly on the inner shelf, where they appear as thick wedges thinning significantly basinwards. Moreover, in the outer shelf, the preservation of highstand deposits is often difficult, due to the erosive action induced by emersions and wave currents. On the inner shelf, relative sea-level falls, subsequent to phases of rising, often cause partial erosion of highstand deposits, along with sediment reworking in the coastal areas (Field and Trincardi, 1992; Gensous et al., 1993).

Isotopic analysis of deep-sea sediments and corals (Bard et al., 1990a; Bard et al., 1990b) shows that, during the Pleistocene, glacio-eustatic changes clearly exhibit high-frequency (about 100 ky) oscillations; cycles are highly asymmetrical, showing rapid rises and slow falls of the sea level (Shackleton and Opdyke, 1973; Martinson et al., 1987). The stratigraphic architecture of Pleistocene continental margins is characterized by a relative abundance of lowstand prograding wedges with respect to transgressive and highstand deposits, as is well known in the Mediterranean region (Trincardi and Field, 1991; Tesson et al., 1993; Chiocci et al., 1997; Catalano et al., 1998). Shelf sediments, deposited during Pleistocene glacio-eustatic cycles, accumulated on the outer shelf and upper slope during subaerial exposures and formed thick sedimentary wedges (shelf-perched lowstand wedges; Posamentier and Vail, 1988).

The aim of this paper is to discuss the role of fourth-order glacio-eustasy *sensu* Vail et al. (1991) vs. regional tectonics in controlling the stacking patterns of lowstand prograding wedges and their cyclicity. In the study area the Apulia (Puglia) shelf shows a Middle–Late Pleistocene stratigraphic record with seismic reflectors of high lateral continuity, being part of a relatively undeformed area ("Apulian foreland"; Fig. 1; D'Argenio et al., 1973; Ricchetti et al., 1992). In this area the methods of sequence stratigraphy (Mitchum et al., 1977; Posamentier and Vail, 1988; Posamentier et al., 1992) can easily be applied in the interpretation of seismic sections.

High-resolution reflection seismic profiles (Sparker 1 kJ seismic source) analyzed in this paper were recorded in 1994, during the GMS94-01 cruise, carried out by the Geomare Sud Research Institute, National Research Council, Naples, Italy, on the R/V Urania (Aiello et al., 1994, 1995; Budillon and Aiello, 1999). The study area (Fig. 2) is located in the southernmost sector of the continental shelf of the southern Adriatic sea (Fig. 1), offshore of the Apulia region. The investigated area is located offshore of the Salento Peninsula between Capo d'Otranto and Capo S. Maria di Leuca (Fig. 1). The seismic grid, covering an area of about 1200 km^2 , consists of NW–SE and NE–SW oriented seismic profiles (line LC8; line LC9; line LC10; line LC11; Fig. 2) and of two regional seismic profiles (line LC4-5 and line LC1-2; Fig. 2) recorded in order to provide crossing points with the other lines. The seismic grid is located mostly in water depths ranging between the –50 m and –200 m isobaths (Fig. 2). Two E–W oriented seismic profiles (line LC7 and line LC12; Fig. 2) were acquired off Capo d'Otranto and Capo S. Maria di Leuca, respectively, (Figs. 1, 2). The resolution of Sparker 1 kJ profiles is about 3 m; the seismic sections have a penetration corresponding to a vertical scale of 0.5 s.

GEOLOGIC SETTING

The Salento continental shelf is part of the Apulia foreland (Fig. 1), which corresponds to a wide antiformal structure, WNW–

Cyclostratigraphy: Approaches and Case Histories
SEPM Special Publication No. 81, Copyright © 2004
SEPM (Society for Sedimentary Geology), ISBN 1-56576-108-1, p. 215–230.

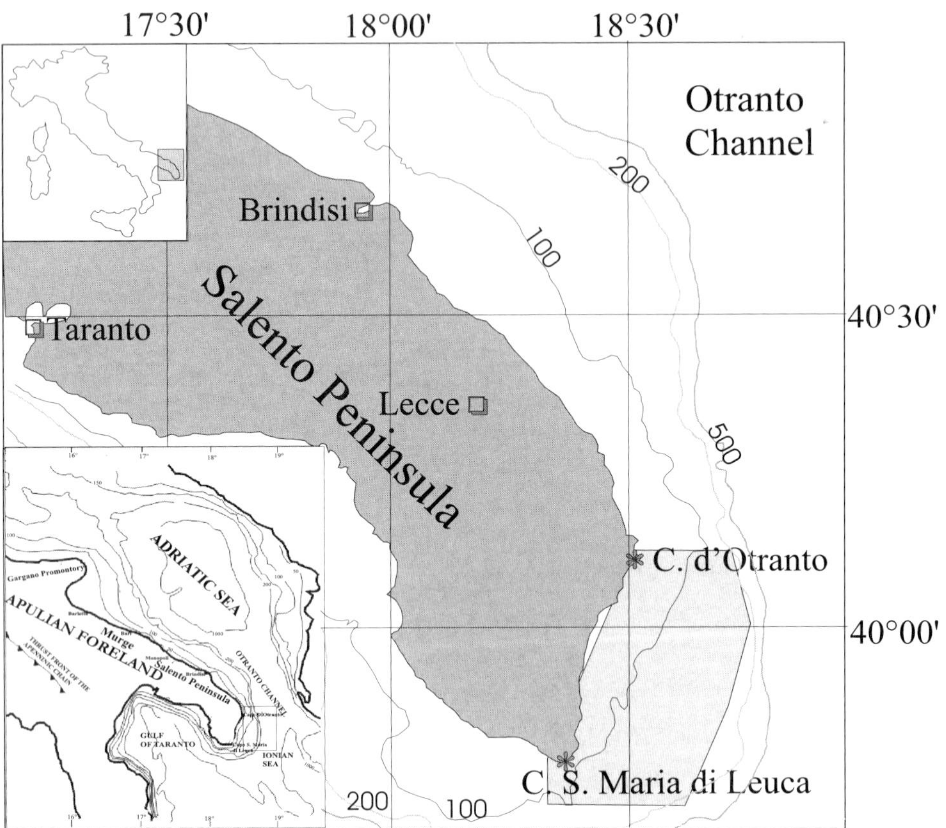

FIG. 1.—Study area in its regional frame: the southern Adriatic Sea and the foreland of the Southern Apennines chain (Apulia). The locations of the Gargano Promontory, the Murge region, and the Salento Peninsula are also shown. The triangles separate the front of the Southern Apennines fold-and-thrust belt from the foredeep system.

ESE trending, block-faulted, and variably uplifted during late Neogenic times (Ricchetti et al., 1992). Transfer faults, striking oblique or perpendicular to the main antiformal structure, segmented the Apulia into three main blocks (Gargano, Murge, and Salento blocks; Fig. 1), which had different rates of uplift during the Plio-Pleistocene, when the central Adriatic Sea underwent high subsidence rates, interpreted as the effect of the eastward rollback of the hinge of the west-dipping Apenninic subduction zone (Doglioni et al., 1994).

The occurrence of regressive sedimentary sequences, overlying erosional surfaces, suggests a uniform uplift of the Murge and Salento blocks after the Middle Pleistocene (Fig. 1), also supported by the regional distribution of the fossil shorelines (Cosentino and Gliozzi, 1992). The eustatic sea-level oscillations and tectonic uplift have produced various terraces, reaching even 200 m above sea level (Ciaranfi et al., 1992; Cosentino and Gliozzi, 1992; Ricchetti et al., 1992). Average values of about 0.23 m/ky have been suggested for the rate of uplift of the Salento Peninsula during the Late Pleistocene, on the basis of the age of the Tyrrhenian deposits (Hearty and Dai Pra, 1992; Cosentino and Gliozzi, 1992).

Previous studies of the geology of the southern Salento Peninsula (Bossio et al., 1988; Ricchetti et al., 1992) recognize carbonate stratal packages in the onshore sector of the investigated area (Figs. 1, 2). These rocks, shown in the geological sketch map in Figure 3, range in age from the Middle–Late Miocene (Calcareniti di Andrano Formation; Langhian to Tortonian; no. 3 in Fig. 3) to the Pleistocene (Calcareniti del Salento Formation; no. 1 in Fig. 3). Their substrate, cropping out widely in the mainland, is represented by Upper Cretaceous limestones and dolomitic limestones (Calcare di Altamura Formation; no. 5 in Fig. 3), overlain unconformably by bioclastic limestones, Paleocene–Oligocene in age (Calcari di Castro and Calcareniti di Porto Badisco formations; no. 4 in Fig. 3). The Mesozoic–Cenozoic deposits cropping out in the Salento Peninsula were eroded during phases of tectonic uplift of the foreland, supplying sediments for the deposition of the lowstand prograding wedges discussed in this paper. The geologic map in Figure 3 also gives some information on the offshore area based on the interpretation of the seismic profiles (see the subsequent description of the seismic units and the interpretation of data). This map shows the main outcrops of the basal unit, which forms the substrate of the Pleistocene prograding wedges on land and at the sea bottom, bounded seawards by normal faults. In Figure 3 the main axis of antiformal and synformal structures, involving the Middle–Late Pleistocene prograding wedges, and the trends of the ravinement surfaces RS1, RS2, and RS3 are also represented.

DESCRIPTION OF SEISMIC UNITS

The stratigraphic architecture of the Salento continental shelf is characterized by thick prograding wedges unconformably overlying a basal unit, forming a substrate over which progradation of the platform occurred during the Pleistocene (Fig. 4). This basal unit crops out close to the coastline, where it is bounded

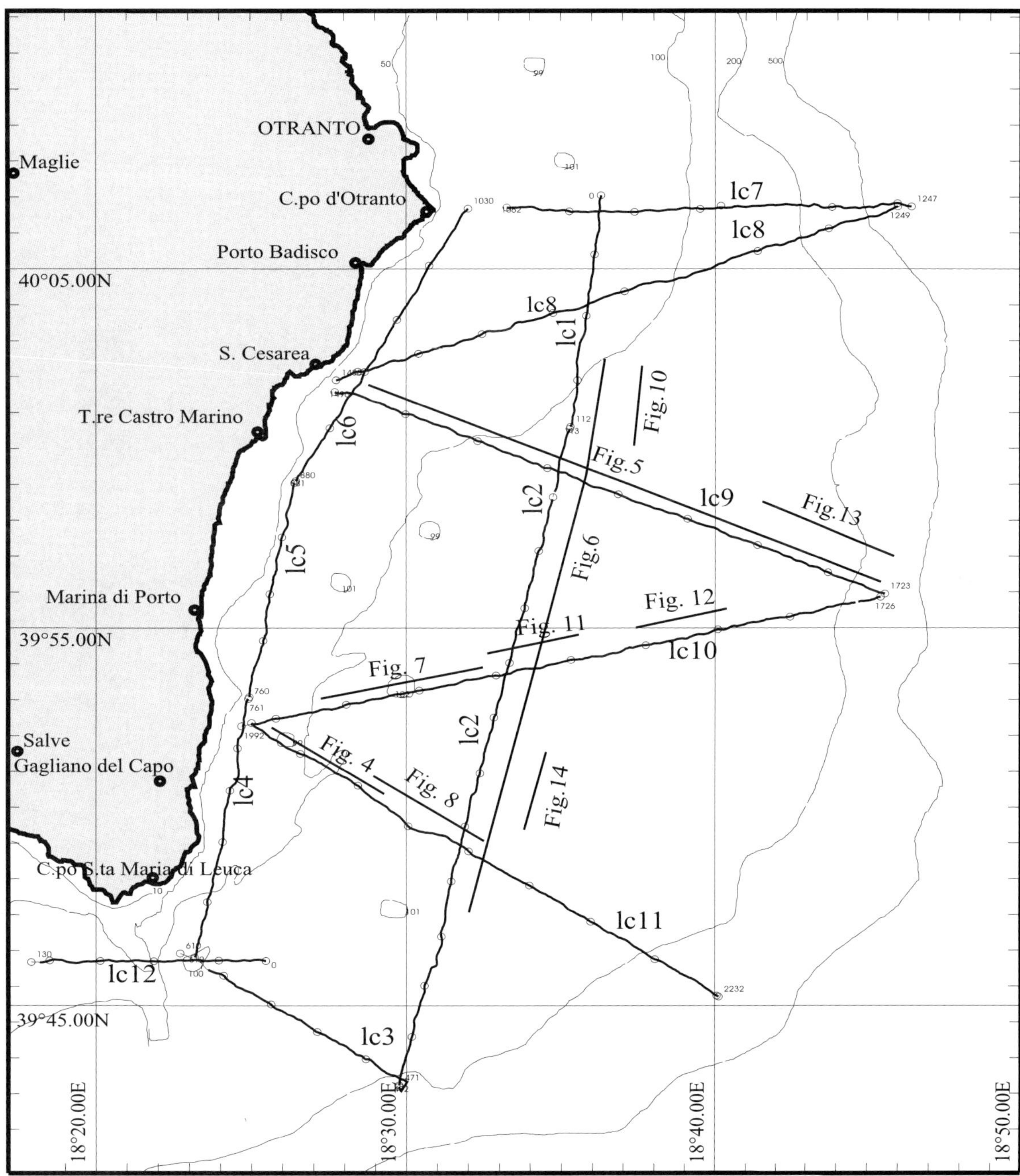

FIG. 2.—Seismic grid of Sparker 1 kJ profiles analyzed in this paper (see the text for further details).

seawards by normal faults (Fig. 3) and correlates to the units forming the present-day coastal cliffs (Fig. 3; Aiello et al., 1995; Budillon and Aiello, 1999).

The coastal cliffs are formed by the Calcari di Castro and Calcareniti di Porto Badisco formations (Paleocene to Oligocene in age; Fig. 3) and by the "Calcareniti di Andrano" Formation (Langhian to Tortonian; Fig. 3). The seismic expression of the basal unit shows a complex morphostructural evolution. This unit, characterized by a chaotic acoustic facies (Fig. 4), is overlain by several erosional surfaces, probably formed above sea level (Fig. 4; Aiello et al., 1995; Budillon and Aiello, 1999).

Seismic Units

Depositional geometries and stratal patterns of seismic units are described below, as shown by the interpretation of two regional seismic profiles, located parallel and perpendicular to the coastline between Otranto and S. Maria di Leuca, respectively (line LC9 and line LC1-2, in Figs. 5 and 6, respectively). Good control of seismic reflectors and unconformities at the crossing points between the two profiles (Figs. 5 and 6) provided a reliable stratigraphic interpretation of Pleistocene prograding wedges based on seismic stratigraphy criteria (Mitchum et al., 1977; Vail et al., 1991). Another regional seismic profile (line LC10) is shown

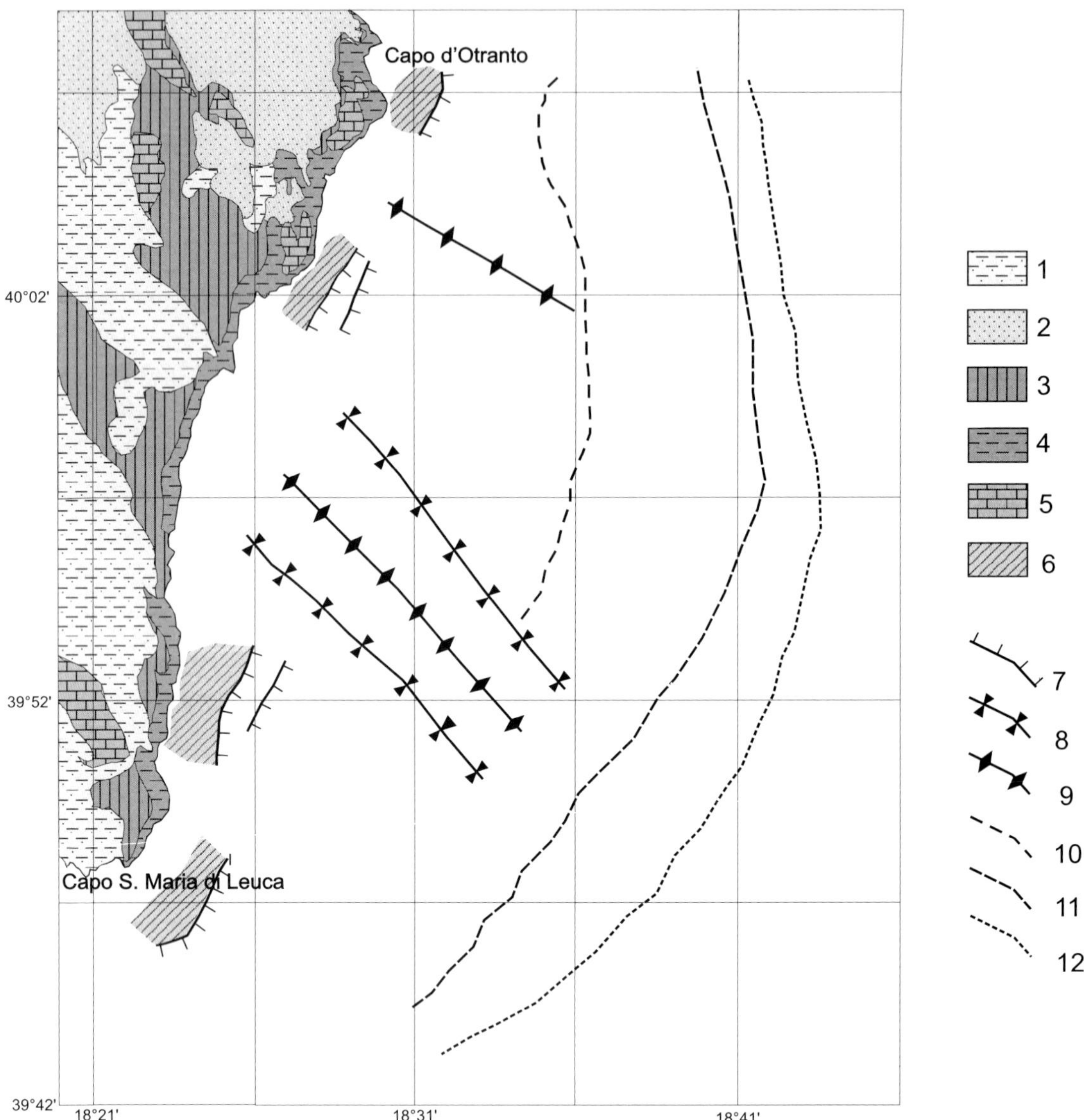

FIG. 3.—Simplified geologic map of the Salento Peninsula (see Fig. 1 for location). Onshore geology is from National Geological Survey of Italy, Geological Map 1:100.000, no. 223 "Capo S. Maria di Leuca". Axes of antiformal and synformal structures involving Pleistocene prograding wedges, and normal faults bounding seaward outcrops of the basal unit located in the inner shelf, have also been reported. Trends of the ravinement surfaces RS1, RS2, and RS3, related to the Pleistocene prograding wedges, show that the shelf progradation in the Salento offshore is on the order of several kilometers. Legend: 1, calcarenites, bioclastic limestones, and calcareous sands (Calcareniti del Salento Formation; Pleistocene); 2, calcareous sands and marly calcarenites (Sabbie di Uggiano La Chiesa Formation; Early Pliocene); 3, organogenic calcarenites (Calcareniti di Andrano Formation; Langhian to Tortonian); 4, bioclastic limestones (Calcari di Castro and Calcareniti di Porto Badisco Formations; Paleocene to Oligocene); 5, limestones and dolomitic limestones (Calcare di Altamura Formation; Upper Cretaceous); 6, outcrops of the basal unit AB at the sea bottom (Upper Cretaceous to Paleocene; Upper Cretaceous to Oligocene); 7, normal faults; 8, synform axes; 9, antiform axes; 10, paleo–shelf break of ravinement surface RS1; 11, paleoshelf break of ravinement surface RS2; 12, paleo–shelf break of ravinement surface RS3.

in Figure 7 in order to give further information on the stratigraphic framework of the area.

Details of critical segments of the seismic lines are shown in Figures 4 and 8 and in Figures 10 to 14, as follows:

AB: basal unit. It is characterized by chaotic seismic facies (Fig. 4). Landwards it is bounded by subaerial unconformities (Fig. 4), and basinwards it shows relationships of paraconformity with the overlying unit U1 (Fig. 8). The main outcrops of the basal unit are located close to the coastline and are bounded seawards by normal faults (Fig. 3). The basal unit correlates to the rocks forming the present-day coastal cliffs, i.e., the Calcari di Castro and Calcareniti di Porto Badisco formations (Paleocene to Oligocene) and the Calcareniti di Andrano Formation (Langhian to Tortonian).

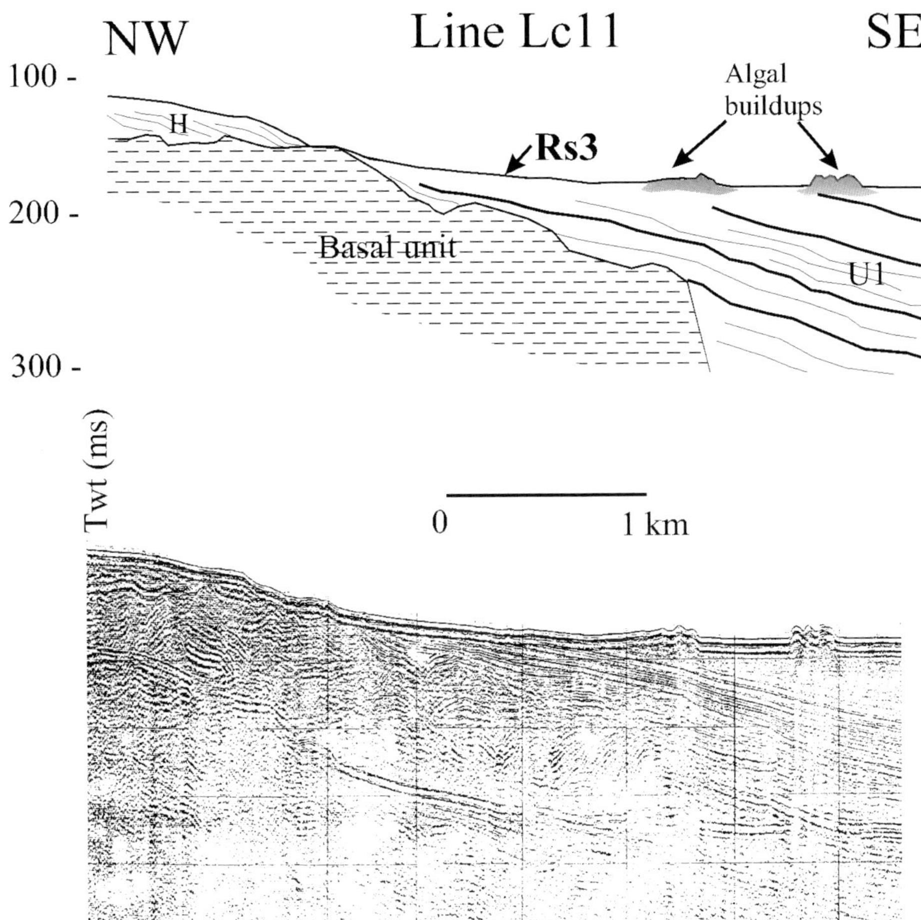

FIG. 4.—Seismic profile LC11 (see Fig. 2 for location), showing the basal unit (AB on the left in the profile), characterized by a chaotic seismic facies and slightly deformed by normal faulting. The unit crops out close to the coastline (see also Fig. 3) and correlates with Paleocene to Oligocene limestones and Middle–Late Miocene calcarenites cropping out in the coastal cliffs of the Salento Peninsula (Fig. 3). Unit AB is unconformably overlain by a Pleistocene prograding wedge (seismic unit U1, on the right of the profile; see the text and Fig. 8 for further details).

U1: This seismic unit is characterized by alternations of reflector packages with high amplitude and coherence and of acoustically transparent intervals (Fig. 8). Such a seismic facies suggests the alternation of coarse-grained deposits (corresponding to the transparent intervals) and fine-grained sediments (reflectors with high amplitude; Fig. 8). In the lower part of the unit the prograding clinoforms show a more pronounced aggradational component, whereas progradational geometries prevail in the upper part (Fig. 8). Seismic profiles show a paraconformity between the unit U1 and the underlying unit AB (Fig. 6). The prograding clinoforms of the unit U1 are truncated near the sea bottom by the ravinement surface RS3 (Figs. 7, 8).

As shown by the regional profile LC1-2 (Figs. 6, 9), Unit U1 is affected by compressional deformation, forming large antiformal and synformal structures, whose axes are represented in plan view in Figure 3. These structures also affect Units U2, U3, and U4 and represent the seaward extension of compressional trends documented in the Salento Peninsula (Bossio et al., 1988). The antiforms are cut by normal faults (Figs. 6, 7). The ravinement surface RS1 and the unit U5 cut the antiforms and the synforms probably at the end of the Middle Pleistocene (Fig. 6).

U2: This seismic unit shows continuous reflectors overlying the structural highs with onlap terminations, as shown by the regional profile LC1-2 (Figs. 6, 9) and by its details (Fig. 10), as well as by the downlap geometries at the top of the underlying unit U1 (Fig. 11).

U3: This seismic unit shows acoustic facies characterized by high amplitude and low lateral continuity, with an aggradational component particularly evident in the N–S direction (Fig. 6).

U4: seismic unit thinning rapidly onto the structural highs (Fig. 6) with onlap geometries on the ravinement surface RS1 (Fig. 12).

U5: seismic unit onlapping the erosional surface RS1 (Fig. 12) and thinning laterally (along strike) with downlap geometries (Fig. 6).

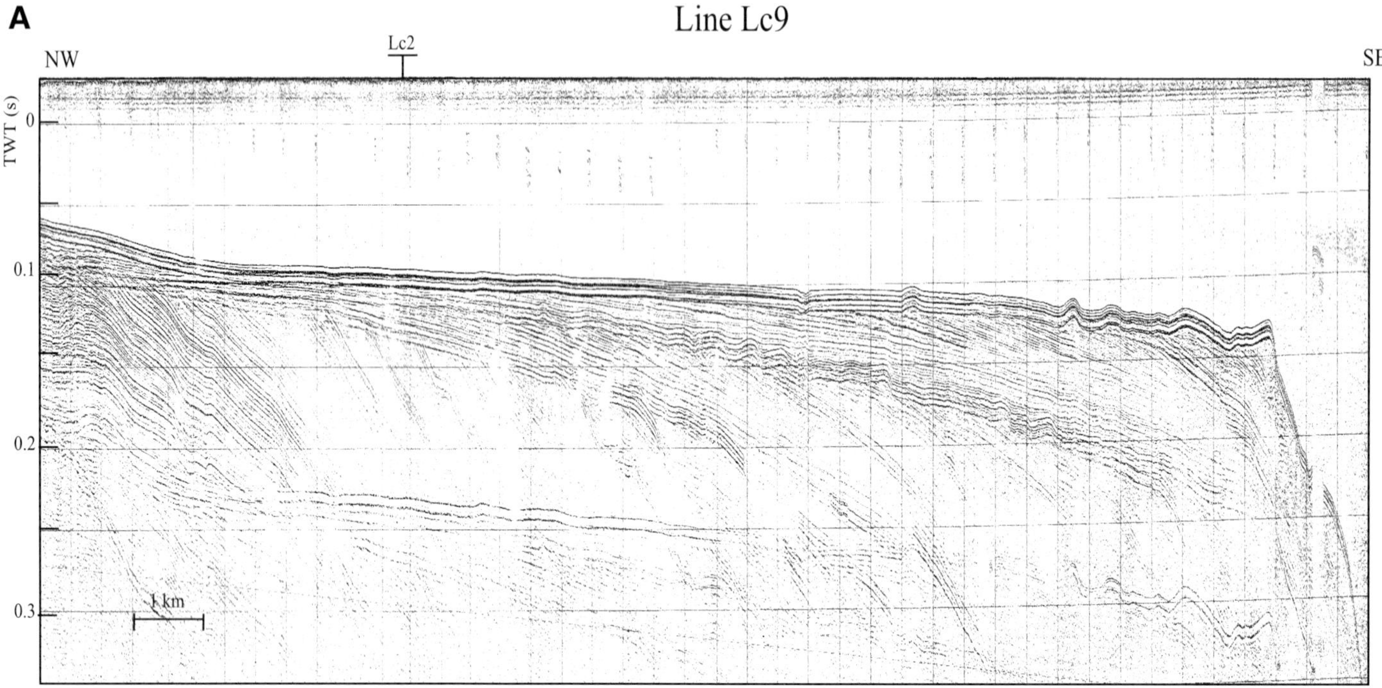

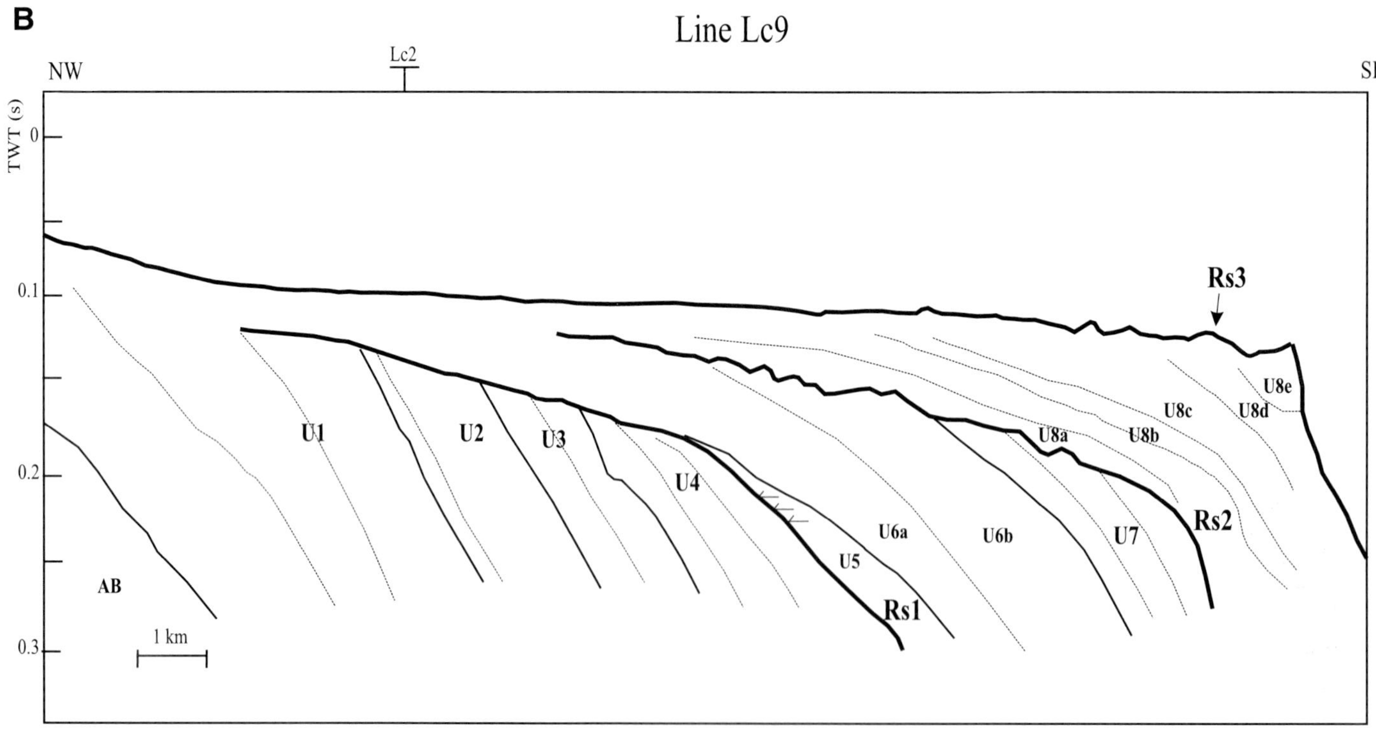

FIG. 5.—**A)** Seismic profile LC9 (see Fig. 2 for its location) and **B)** corresponding interpretation. The seismic profile and its interpretation show the Pleistocene prograding wedges of the Salento continental shelf, overlying an acoustic basement (unit AB; see also Figs. 4 and 7) and the stratigraphic relationships between the ravinement surfaces and the seismic units. The Middle Pleistocene prograding wedge is characterized by two episodes of forced regression, separated by a phase of marine ingression, evidenced by ravinement surface RS1 and by seismic unit U5. The units of forced regression are interpreted on the basis of the erosional truncations of internal reflections, as well as of the seaward shifting of the offlap breaks and of the occurrence of chaotic seismic facies suggesting mass transport at the shelf margin.

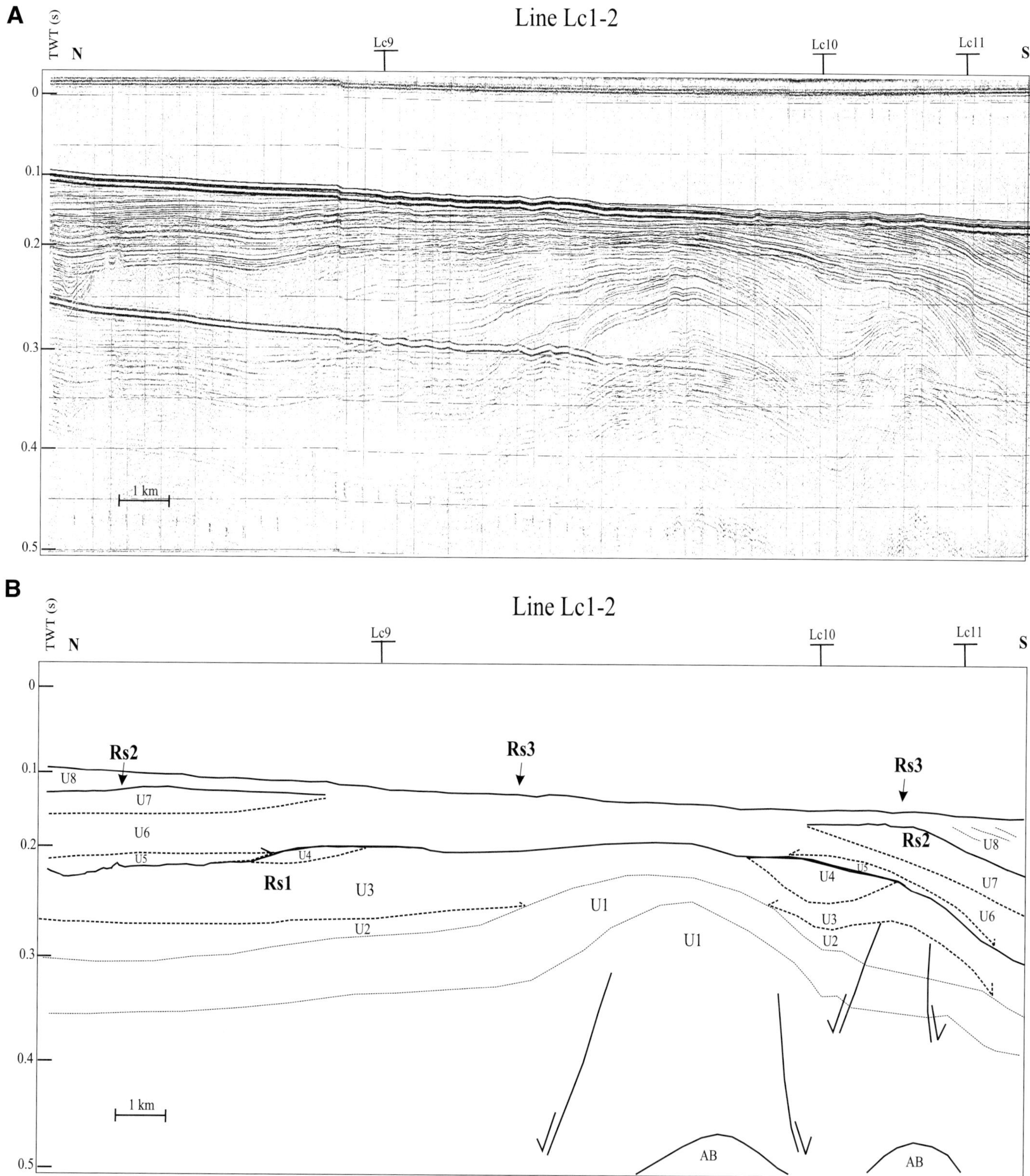

FIG. 6.—**A)** Seismic profile LC1–LC2 (see Fig. 2 for its location) and **B)** corresponding interpretation. Seismic units and ravinement surfaces were also interpreted on the basis of the crossing points with the line LC9 (Fig. 5) and with the perpendicular lines. The seismic profile and its interpretation show antiformal and synformal structures affecting units U1, U2, U3, and U4 and representing the seaward extent of structural compressional trends documented in the adjacent mainland. Axes of the antiformal and synformal structures are shown in Figure 3. Ravinement surface RS1 and Unit U5 fossilize these structures at the end of the Middle Pleistocene. Antiforms are slightly cut by normal faulting.

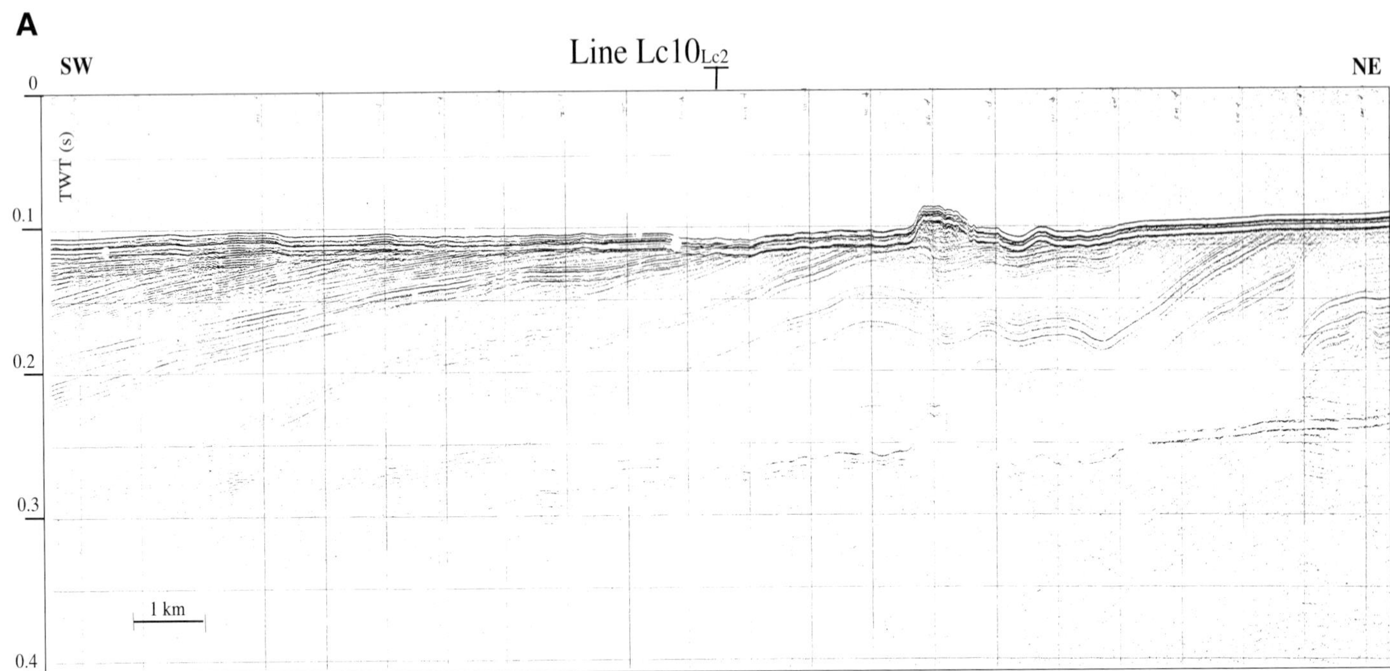

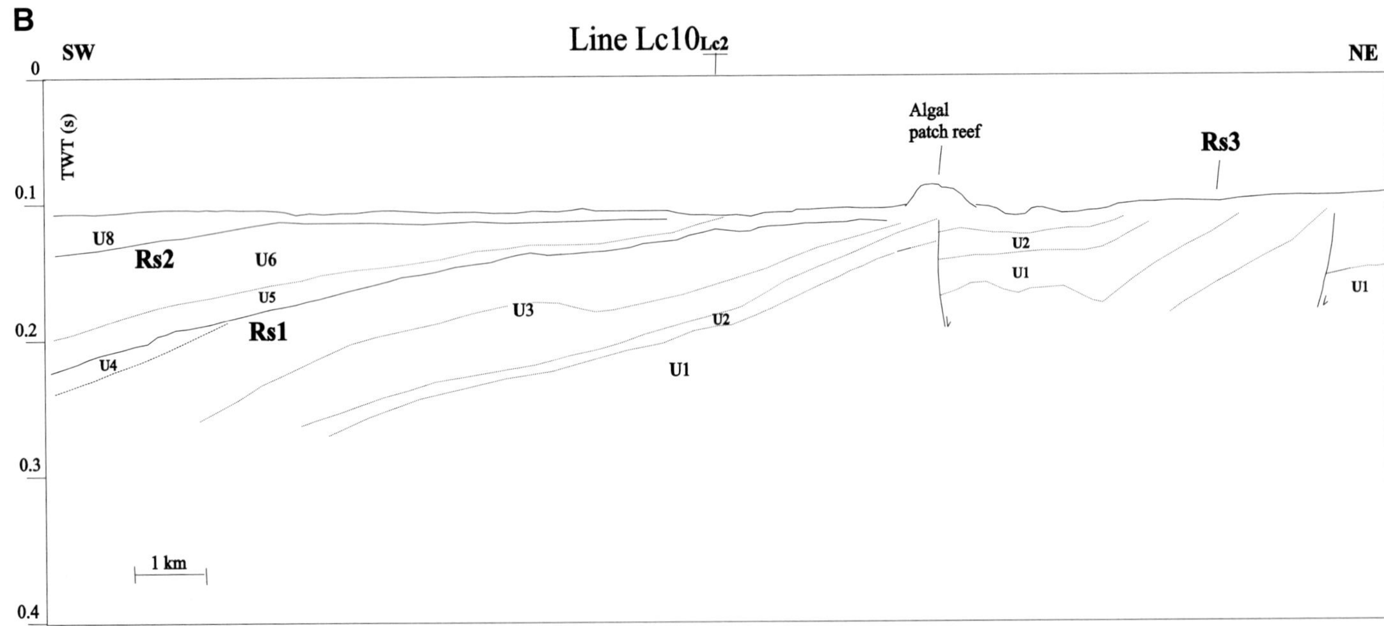

FIG. 7.—**A)** Seismic profile LC10 and **B)** corresponding interpretation (see Fig. 2 for location). The seismic profile and its interpretation show the Middle Pleistocene prograding wedge (seismic units U1–U4) truncated by ravinement surface RS1 and slightly deformed by normal faulting. Seismic units U5 and U6 overlie RS1 and are bounded in the uppermost part by RS2. Algal mounds have been recognized on ravinement surface RS3, cropping out at the sea bottom.

U6: progradational seismic unit conformably overlying the top of the unit U5 (Figs. 5, 6, and 7); two minor units (U6a and U6b) were also distinguished (Fig. 5). In some sectors of the inner shelf, where unit U7 is not yet developed, unit U6 is directly truncated by ravinement surface RS2 (Fig. 7). There is paraconformity between the two units at the shelf margin (Fig. 13).

U7: progradational unit, conformably overlying the U6 seismic unit; in its upper part it is bounded by ravinement surface RS2 (Figs. 5, 6, 13). The unit is well developed at the shelf margin; on the contrary, in the inner part of the shelf unit U7 is missing and the underlying unit U6 is directly truncated by ravinement surface RS2 (Fig. 14).

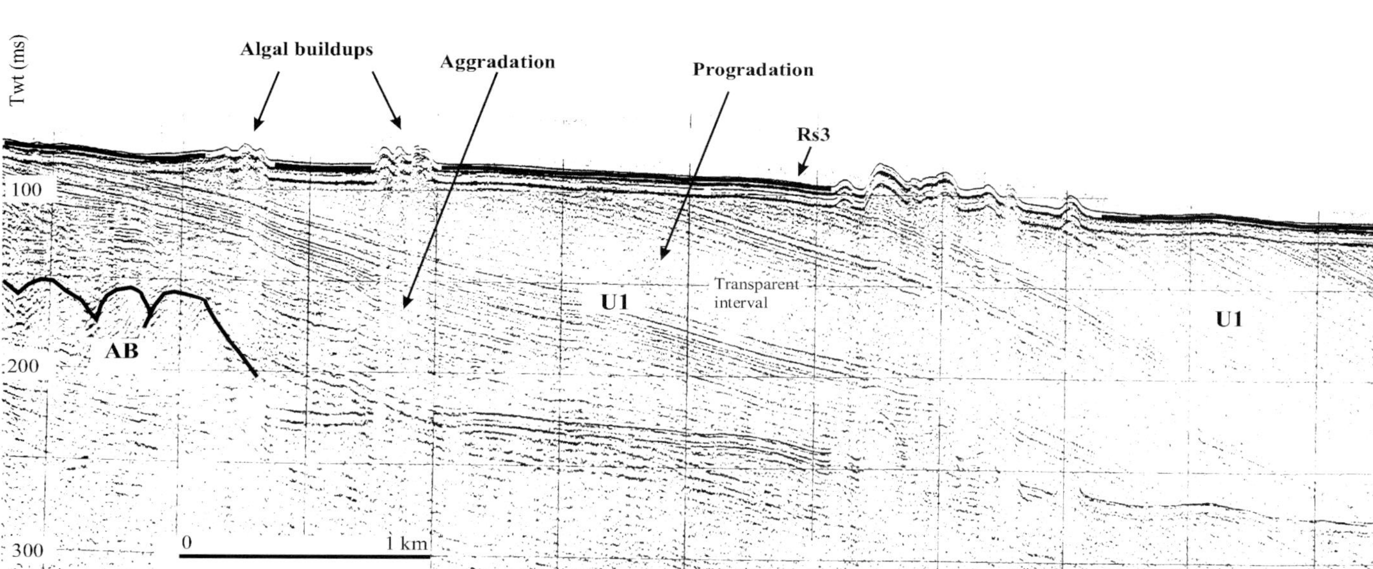

FIG. 8.—Seismic profile LC11 (see Fig. 2 for location), showing the geometry and the seismic facies of seismic unit U1. The seismic facies is characterized by the alternation of reflector packages with high amplitude and coherence and of acoustically transparent intervals. Note that the basal part of the unit shows mainly an aggradational component (on the left of the profile), whereas progradational geometries have been identified in the upper part of the unit (on the right of the profile).

U8: seismic unit located at the present-day shelf margin and characterized by reflectors onlapping RS2 and truncated by ravinement surface RS3 (Figs. 13, 14). The stratigraphic architecture of the unit is well shown in Figures 13 and 14. Here, five minor seismic units (U8a, U8b, U8c, U8d, and U8e) were identified by analysis of seismic facies. In the external part of the shelf margin the offlap breaks are well preserved and migrate progressively towards the shelf margin at increasing depths (Fig. 13). Chaotic intervals, probably corresponding to continental deposits filling incised valleys, formed during subaerial exposures of the shelf. They occur especially where the steepness of the reflectors increases and near the present-day shelf break (units U8d and U8e; Fig. 13). The top of Unit U8 is an erosional surface (RS3 in Figs. 5–9, 13, 14), developing with low gradients and eroding reflectors belonging to the underlying units up to the inner shelf. Ravinement surface RS3 coincides with the present-day sea bottom (Figs. 5–9, 13, 14).

INTERPRETATION

A forced-regression systems tract, formed during phases of relative sea-level fall, was proposed by Hunt and Tucker (1992) as

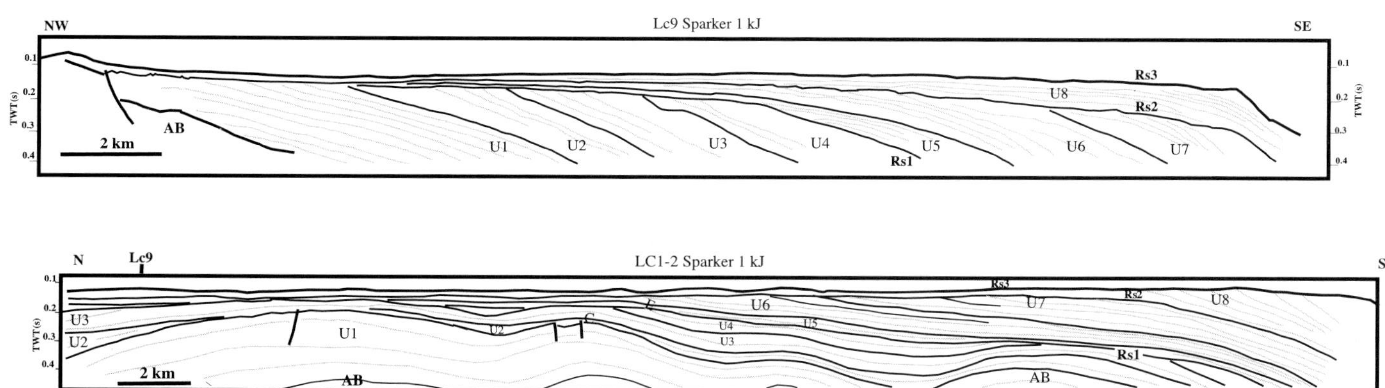

FIG. 9.—Regional seismic stratigraphy of the Salento continental shelf, as interpreted on the line drawings of seismic profiles LC9 (Fig. 5) and LC1-2 (Fig. 6). Both of the profiles show the stratigraphic relationships between the ravinement surfaces and the seismic units and are parallel (LC9 profile) and perpendicular (LC1-2 profile) to the progradation of wedges of the Salento shelf. The basal unit (AB) is reported on both the regional line drawings.

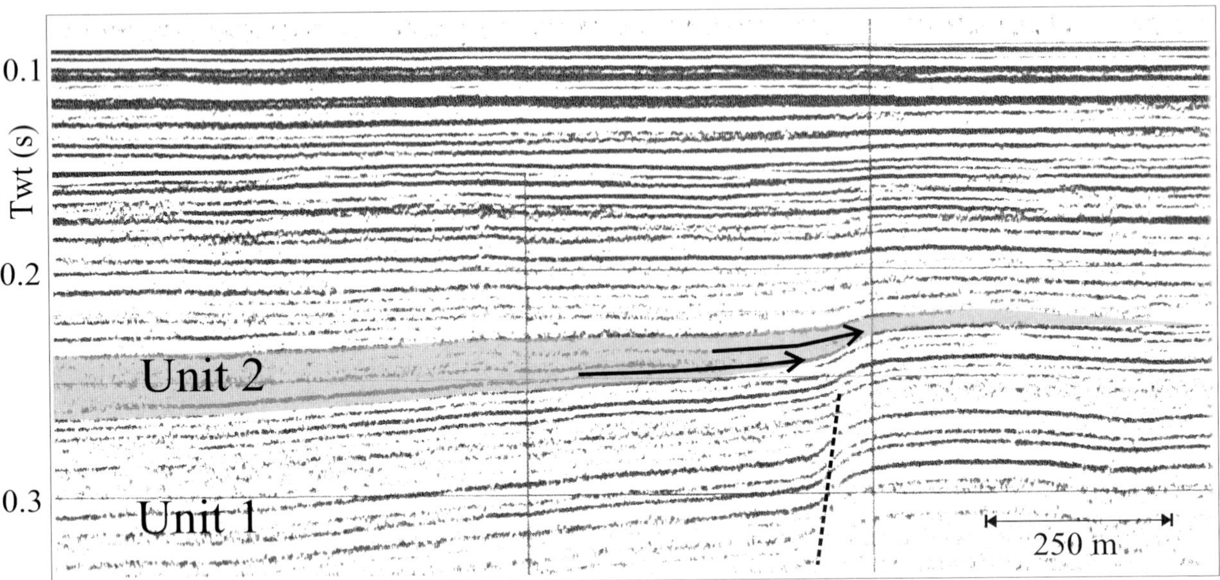

FIG. 10.—Detail of seismic line LC1-2 (see Fig. 2 for location and Figs. 6 and 9 for regional seismic stratigraphy of the units), showing onlap terminations of seismic unit U2 on the underlying unit U1.

a variation to the classical sequence-stratigraphic model, considering the relative abundance of lowstand prograding wedges in the stratigraphic architecture of the continental shelves. The parasequences, related to forced regressions, develop as a response to the decrease of the rate of sea-level fall; subsequently, they are "abandoned" and "isolated" when the rate of eustatic fall increases. The scientific debate mainly concerns the location of the sequence boundary. In the model of Hunt and Tucker (1992) the boundary is located above the forced-regression deposits, up to the point of maximum forced regression. In the classical model of sequence stratigraphy (Posamentier and Vail, 1988), it is located at the base of the lowstand "shelf-perched" prograding wedges (Tesson et al., 1993).

Polycyclic erosional surfaces *sensu* Posamentier and James (1993) often characterize the stratigraphic framework of continental shelf areas: they form during phases of relative sea-level

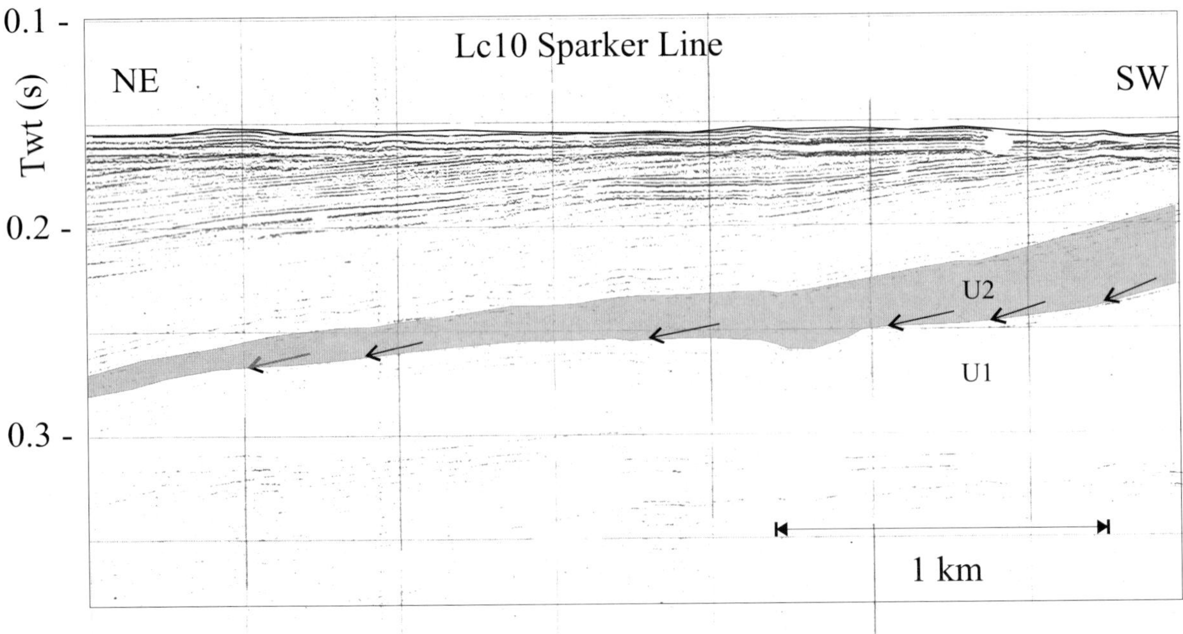

FIG. 11.—Detail of seismic line LC10 (see Fig. 2 for location and Fig. 7 for regional seismic stratigraphy of the units), showing downlap terminations of unit U2 on the underlying unit U1.

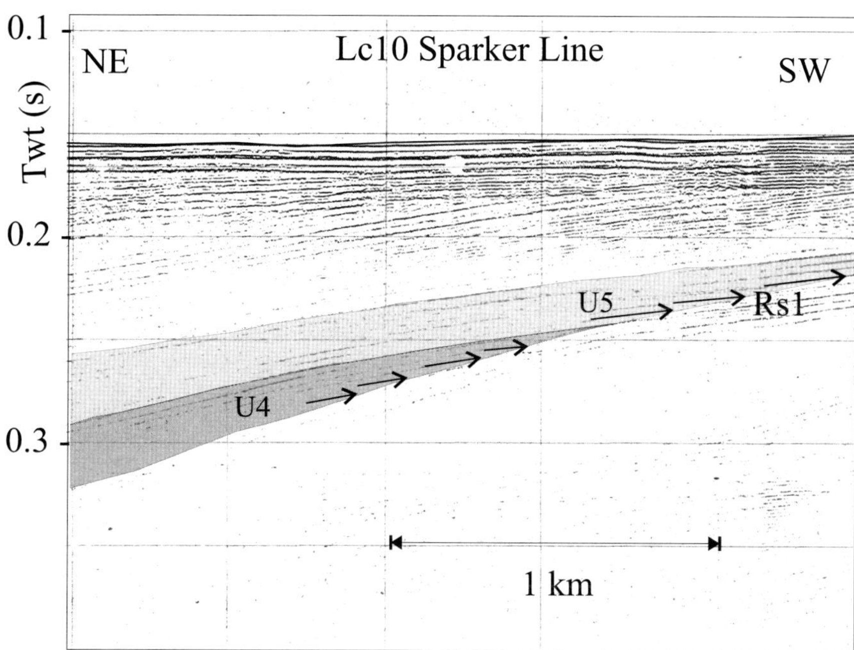

FIG. 12.—Detail of seismic line LC10 (see Fig. 2 for location and Fig. 7 for regional seismic stratigraphy of the units), showing onlap terminations of units U4 and U5 on ravinement surface RS1.

fall (lowstand surface of erosion) and are subsequently reworked during relative sea-level rise (ravinement surfaces). The frequent lack of transgressive paralic deposits, owing to their low thickness and poor potential for preservation, often forces the sequence boundaries to coincide with transgressive surfaces of erosion. These unconformities, namely the ravinement surfaces (Posamentier and James, 1993) develop because the erosional landward shift of the shoreline and their formation is due to reworking by wave action. The lowstand systems tract (LST) occurs between the sequence boundary and the ravinement surface only at the shelf margin (see also Fig. 5). The ravinement surface, marking the rise of sea level, represents the base of a marine succession characterized by parallel reflectors, diachronous in time, onlapping the underlying paralic sequence. When the paralic sequence is lacking, the ravinement surface coincides with the base of the transgressive systems tract. Then the ravinement surface (RS) and the transgressive surface (TS) are represented by the same unconformity, as in the study area (Correggiari et al., 1992; Trincardi et al., 1994; Budillon and Aiello, 1999; Trincardi and Correggiari, 2000; Ridente and Trincardi, 2002).

In the sedimentary sequence of the Salento continental shelf, several erosional unconformities are present, which outline the stratigraphic evolution of the area (Figs. 5–14). These surfaces are linked to large-scale erosional events and, on the basis of their geometry and depth, can be correlated at a regional scale along the Salento offshore.

The youngest ravinement surface (RS3; Figs. 5–9, 13, 14), which appears almost horizontal, has been observed at the sea bottom down to a maximum depth of –120 to –130 m, and only in the inner shelf has it been buried by the recent sediments of the Holocene wedge and by patch-reef structures, interpreted as relicts of algal buildups (Aiello et al., 1994; 1995). RS3 represents a main erosional surface, close to the physiographic shelf break, truncating the seismic reflectors of the prograding unit U8 (Figs. 5–9, 13, 14).

The second ravinement surface (RS2; Figs. 5–9, 13, 14) is observed down to a maximum depth of –175 to 180 m and develops landwards with higher gradients than RS3. The last surface truncates RS2 in the inner shelf and/or on structural highs (Fig. 7), eroding an onlapping sequence (unit U5 in Figs. 5–9) and the overlying prograding wedge, composed of the seismic units U6 and U7 (Figs. 5–9, 13).

Finally, the oldest ravinement surface (RS1; Figs. 5–9, 12) is observed down to a maximum of –180 to –200 m of water depth and presents a well preserved paleo–shelf break (Fig. 5). On the inner shelf RS1 was eroded by RS2. Compressional deformation, which produced the antiformal and synformal structures shown in the Figures 6 and 9, affected the ravinement surface RS1. The wide prograding wedge, composed by seismic units U1, U2, U3, and U4 and bounded above by RS1 (Figs. 5, 9), has widened the Salento shelf by approximately fifteen kilometers (Fig. 3).

The steepness of the ravinement surfaces increases slightly from the youngest (RS3) to the oldest (RS1). This can be interpreted as an effect of the subsidence of the South Adriatic continental margin, due both to sediment loading and to compaction (Allen and Allen, 1990). However, part of this steepness enhancement may also be explained as an effect of tectonic subsidence, controlled by the regional evolution of the South Adriatic–Ionian bathyal plain (Moretti and Royden, 1987).

DISCUSSION

The concept of a ravinement surface in sequence stratigraphy is really powerful, because such a surface can be considered to be a stratigraphic marker, notwithstanding its diachronous character (Posamentier and James, 1993; Trincardi et al., 1994).

Correlation to Oxygen Isotope Stratigraphy

The ravinement surfaces and the seismic units of the Salento continental shelf can be correlated with isotope stratigraphy

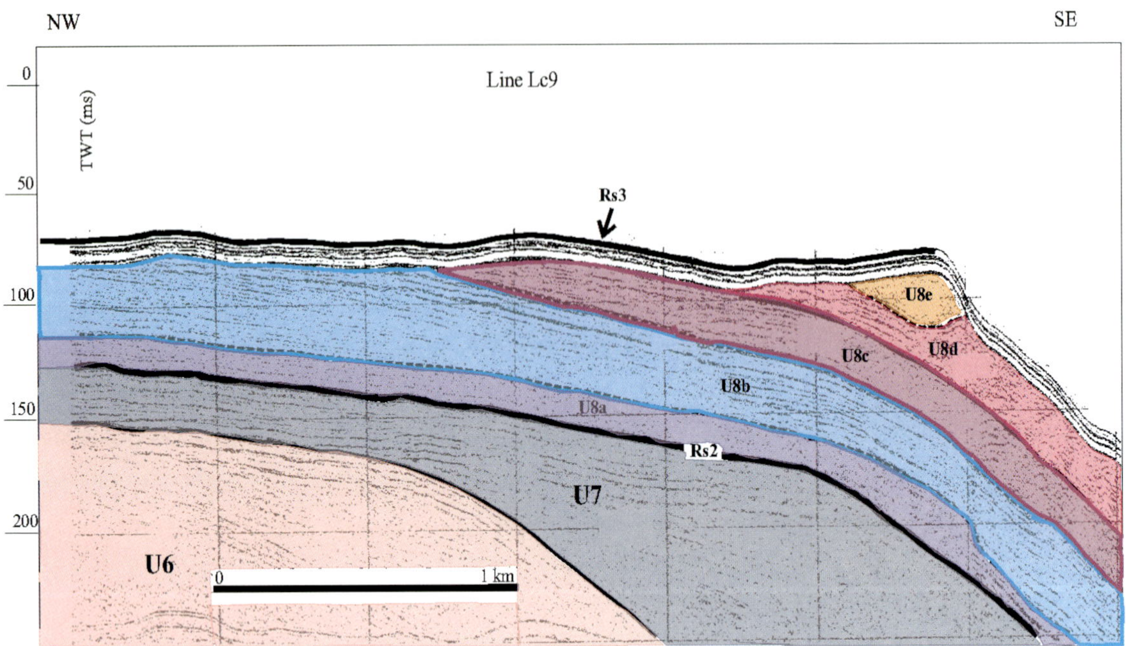

FIG. 13.—Detail of seismic line LC9 (see Fig. 2 for location and Figs 5 and 9 for regional seismic stratigraphy of the units), showing stratigraphic relationships between seismic units U6 and U7 at the shelf margin. The analytical seismic stratigraphic interpretation of the Late Pleistocene prograding wedge (Unit U8), bounded by ravinement surfaces RS2 and RS3, shows the presence of five seismic units (U8a, U8b, U8c, U8d, and U8e), interpreted as transgressive, lowstand, and forced-regression systems tracts.

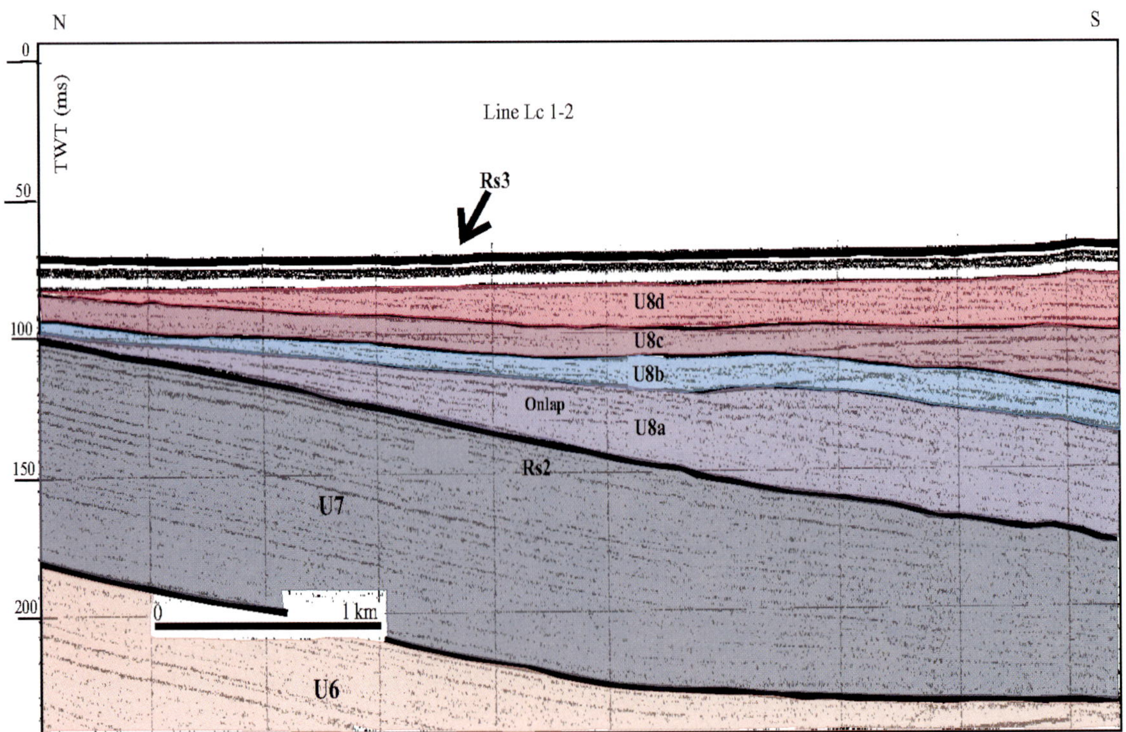

FIG. 14.—Detail of seismic line LC1-2, showing the onlap of Unit U8 on ravinement surface RS2 and the stratigraphic architecture of Unit U8 in the inner shelf. Note the lack of the minor seismic unit U8e, deposited only at the shelf margin, whereas Units U8a–U8d appear well developed.

because, starting from the Middle Pleistocene, they formed during relative sea-level rises, very rapid compared with the subsequent sea-level falls (Shackleton and Opdyke, 1973; Chappell and Shackleton, 1986; Martinson et al., 1987; Bard et al., 1990a; Bard et al., 1990b; Pirazzoli, 1993). In fact, the transgressive erosional unconformities formed during time intervals corresponding to the transition from even to odd stages on the isotopic curve. Accordingly, the isotope curve shows that, for the glacial Pleistocene, sea-level rise was very rapid and more or less comparable in amplitude to the most recent sea-level rise (about 120 m; Bonifay, 1975), with periodicities of about 100 ky.

On the Salento shelf, we tentatively correlate the ravinement surfaces with the oxygen isotope stratigraphy of the last glacio-eustatic cycle, as proposed by Martinson et al. (1987) and shown in Figure 15.

Ravinement surface RS3 (Figs. 5–9, 13, 14) was produced by the last significant sea-level rise, which developed from 18 ka and is well known and documented along all of the Italian offshore (Borsetti et al., 1984; Marani et al., 1986; Chiocci et al., 1989; Colantoni et al., 1989). On the isotope curve it corresponds to the transition from stage 2 to stage 1 (Fig. 15); therefore we consider seismic unit U8b–U8e (Figs. 5, 13, 14) to have been formed during the last phases of the recent pleniglacial (isotope stages 3 and 2), and seismic unit U8a to represent a transgressive sequence (Fig. 5).

Ravinement surface RS2 (Figs. 5–9, 13, 14) formed during previous sea-level rise. For this rise the best choice considered is the transition from isotope stage 6 to 5 (Fig. 15). We support this interpretation for the clear continuity and conformity among the upper-slope units. Accordingly, the prograding wedge, limited in its uppermost part by ravinement surface RS3 and in its lowermost part by ravinement surface RS2 (corresponding to seismic unit U8 in Figs. 5–9, 13, 14), can be attributed to the Late Pleistocene. The minor seismic units distinguished in the wedge (seismic units U8b–U8e; Figs. 5, 13, 14) are here referred to forced-regression and lowstand systems tracts, whereas unit U8a is considered to represent a transgressive systems tract.

Finally, RS1 erodes a prograding wedge composed of units U1–U4 and is overlain by seismic unit U5, showing stacking patterns typical of a transgressive unit (Figs. 5–9). We consider RS1 to be due to a sea-level rise, antecedent to the rise suggested by RS2 and developing between isotope stages 8 and 7. This interval corresponds to a relative sea-level rise at about 250 ka (Martinson et al., 1987). Therefore the seismic units from U1 to U4 are referred to forced-regression systems tracts.

On the basis of the above considerations we assume that during the Middle Pleistocene a fourth-order glacio-eustatic oscillation related to the short-eccentricity cycles controlled the deposition of the forced-regression systems tracts on the Salento continental shelf. Furthermore, geological and geomorphological evidence suggests appreciable rates of tectonic uplift of the Apulian foreland during the Early–Middle Pleistocene (Cosentino and Gliozzi, 1992; Doglioni et al., 1994). As a consequence, in the Middle Pleistocene the rate of tectonic uplift of the Apulian foreland prevailed over the rate of glacio-eustatic variation and produced a relative sea-level fall controlling the formation of a

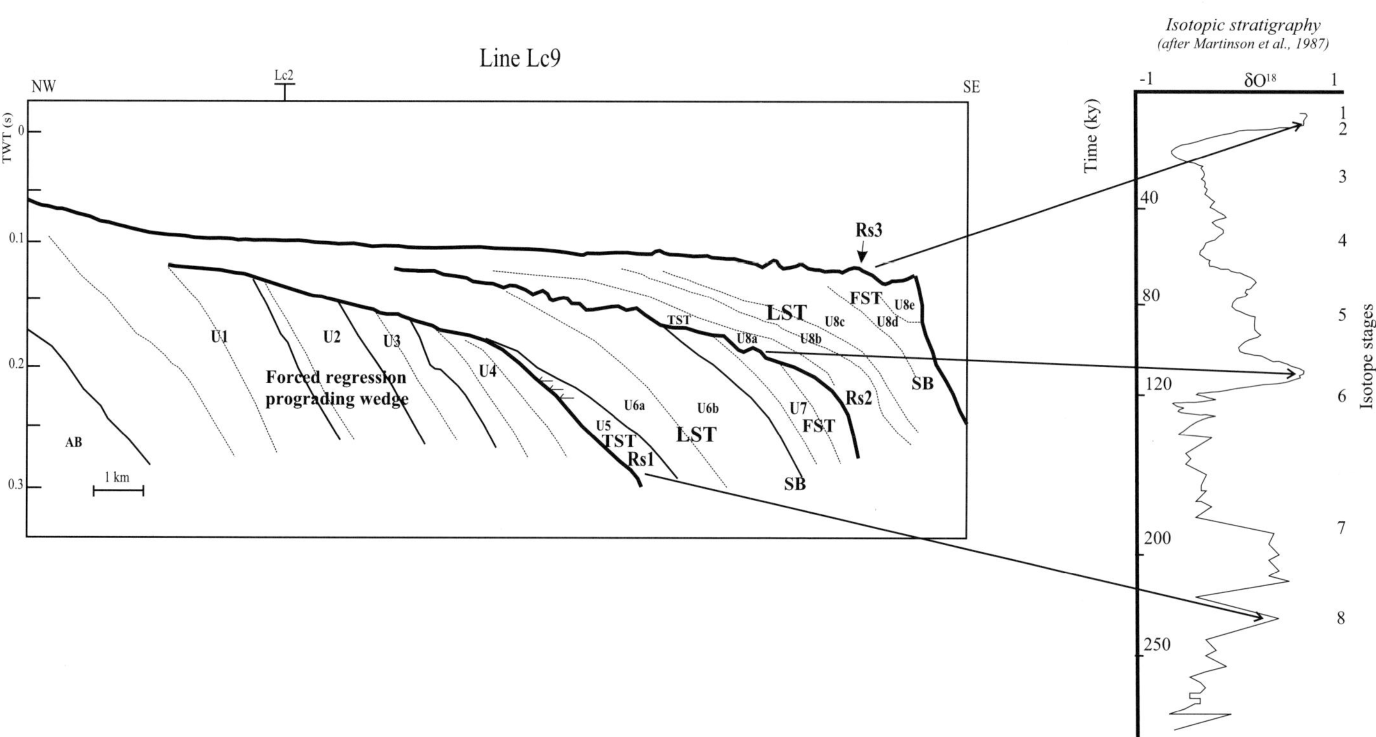

FIG. 15.—High-resolution stratigraphic correlation between the ravinement surfaces recognized in the Pleistocene succession of the Salento continental shelf and the curve of oxygen-isotope stratigraphy of the last glacio-eustatic cycle (Martinson et al., 1987). Unconformity RS1 is correlated with a sea-level rise at about 250 ka, corresponding to the transition from isotope stage 8 to 7. Unconformity RS2, overlying the Middle Pleistocene prograding wedge, is considered to correspond to the transition between isotope stages 6 and 5 and unconformity RS3, overlying the Late Pleistocene prograding wedge and corresponding to the transition from isotope stage 2 to 1.

forced-regression prograding wedge (seismic units U1, U2, U3, U4; Figs. 5–9, 10, 11, 12).

We now consider other implications of the seismostratigraphical architecture of the studied sections. Starting from RS1 (formed at about 250 ka) the depositional sequences consist not only of forced-regression deposits but also show a transgressive and lowstand systems tract. Transgressive systems tracts are very thin and composed of a single retrogradational parasequence (units U5 and U8a in Figs. 5–14), or they coincide with a transgressive surface of erosion, because of the starvation of the platform during the previous sea-level rise. Therefore, the transgressive systems tracts scarcely contributed in terms of thickness to the growth of the platform; instead they controlled the formation of important unconformities (ravinement surfaces; Posamentier and James, 1993), which are particularly evident when associated with asymmetric sea-level fluctuations, like those due to the short-eccentricity (fourth-order) cycles.

Starting at about 250 ka, geometries and stacking patterns suggest phases of rises and falls of relative sea level. Hence if the correlation of the RS surfaces with the curves of oxygen-isotope stratigraphy is consistent, the periodicity of the depositional sequences is about 100 ky and is linked to short-term glacio-eustatic cycles. Even if based on indirect dating of seismic units and corresponding unconformities, this correlation is well supported by the lateral extent and continuity of the ravinement surfaces, recognized in the whole investigated area. We note also that similar qualitative correlations of unconformities with tracts of the curves of oxygen-isotope stratigraphy have already been carried out by other authors, in different case histories of continental shelves of both active and passive margins in the same interval (Ashley et al., 1991; Okamura and Blum, 1993).

Control Mechanisms

The sedimentary supply feeding the forced-regression prograding wedge of the seismic units U1, U2, U3, and U4 is significantly rich, because it allowed a platform progradation of about fifteen kilometers (Fig. 3). Because a well developed river system was lacking in the Salento Peninsula, we suppose that the high sediment supply was due to the tectonic uplift affecting this area, lasted up to the Middle Pleistocene (Cosentino and Gliozzi, 1992; Doglioni et al., 1994). Moreover, antiformal and synformal structures shown by the LC1-2 profile (Figs. 6, 9), which deform units U1–U4, appear to be synsedimentary, as evidenced by the presence of growth structures in unit U2 (Figs. 6, 9). On the basis of the correlation of these compressional trends with similar structures recognized onland in the Salento Peninsula by Bossio et al. (1988), we are led to think that the prograding wedge formed during the late Middle Pleistocene. All these observations also suggest that the mechanisms of onshore sequences and of contemporaneous development of the prograding wedge were very fast.

The upper-slope progradation could also have been fed by the biodetrital production in the inner shelf, whose resedimented materials may have contributed to form the prograding wedges.

Starting from seismic unit U5, onlapping RS1 and interpreted as a transgressive systems tract (Figs. 5–9), the eustatic control overwhelms the tectonic one, creating fourth-order incomplete depositional sequences. The sequences are incomplete because the stratigraphic record is fully preserved only at the shelf margin; behind this margin we observe only transgressive, forced-regression, and lowstand systems tracts. In the wedge between RS1 and RS2 (Figs. 5–9), Unit U5 represents a transgressive systems tract, bounded above by a sequence boundary, whereas Unit U6 forms a forced-regression systems tract and Unit U7 a lowstand systems tract. Moreover, in the wedge between RS2 and RS3 (seismic Unit U8; Figs. 5–9, 13, 14) Unit U8a is interpreted as a transgressive systems tract (Fig. 5), Units U8b and U8c are read as a forced regression systems tract, bounded above by the sequence boundary, and Units U8d and U8e are interpreted as a lowstand systems tract, where chaotic deposits have also been identified (Fig. 5).

In conclusion, eustatic control on the stratigraphic architecture of the Late Pleistocene wedge is supported by the correlation of RS1, RS2, and RS3 with the curve of isotope stratigraphy (Fig. 15). Accordingly a fourth-order cyclicity with a periodicity of about 100 ky (short-eccentricity cycles), which may have controlled the formation of fourth-order incomplete depositional sequences (seismic Units U5–U8 in Figs. 5–9, 13, 14), is inferred and related to the oxygen isotope stages from 7/8 to 2 (Martinson et al., 1987).

Quaternary forced-regression deposits formed in response to fourth-order (100 ky) cyclicity have been also investigated in the Central Adriatic basin (Trincardi et al., 1994; Trincardi and Correggiari, 2000; Ridente and Trincardi, 2002) and a composite nature of oscillation cycles of relative sea level, where a longer-term relative sea-level rise interacted with the shorter-term (fourth- to fifth-order) sea-level cycles, has been suggested in order to explain the formation of forced-regression deposits and their preservation (Trincardi and Correggiari, 2000). The progradational units recorded a fourth-order Quaternary cyclicity in which transgressive erosional surfaces punctuated the stratigraphic record with a periodicity of about 100 ky during isotope stage boundaries 2/1, 6/5, 8/7, 10/9, and 12/11, as suggested also by core analysis (Trincardi and Correggiari, 2000).

CONCLUSIONS

1. The Salento continental shelf is an example of complex stratigraphic architecture of Pleistocene prograding wedges, recording the effects of eustatic oscillation of sea level and of changes in rate of sedimentation, driven by tectonic activity in the source zone. Moderate deformation in the depositional areas does not obscure the eustatic signal.

2. In the stratigraphic sequence analyzed, three ravinement surfaces were recognized, extending landwards from the shelf break and showing low gradients. The three ravinement surfaces (RS1, RS2, and RS3) have been considered to be stratigraphic markers and correlated to the curves of oxygen-isotope stratigraphy (Martinson et al., 1987).

3. The regional geological framework suggests that, during the Middle Pleistocene, tectonic uplift of the Apulian foreland interacted with high-frequency fourth-order glacio-eustatic fluctuations, producing the deposition of forced-regression systems tracts. A wide forced-regression prograding wedge (Units U1–U4), deposited during the Middle Pleistocene, enlarged the Salento shelf by about 15 kilometers (Fig. 3).

4. During the Late Pleistocene–Early Holocene, glacio-eustatic sea-level changes with a periodicity of about 100 ky controlled the stratigraphic architecture of the Salento continental shelf, creating fourth-order incomplete depositional sequences in the two upper prograding wedges (Units U5–U7, bounded in their uppermost part by ravinement surface RS2 and in its lowermost part by ravinement surface RS1; Unit U8, bounded in its uppermost part by ravinement surface RS3 and in its lowermost part by ravinement surface RS2). Deposition of forced-regression (seismic units U6, U8b–U8c), lowstand (seis-

mic units U7 and U8d-U8e) and transgressive (seismic units U5 and U8a) systems tracts, which were distinguished in the two incomplete depositional sequences, is controlled by fourth-order glacio-eustasy modulated by short-eccentricity cycles of the Earth's orbit.

5. The eustatic signal, overwhelming with respect to the tectonic one, suggests a decrease in the rate of uplift of the Apulian foreland during the last 250 ky.

ACKNOWLEDGMENTS

This work was carried out with the financial contribution of the Istituto per l'Ambiente Marino Costiero, IAMC, National Research Council, Naples, Italy (formerly Geomare sud). We wish to thank the colleagues participating in the GMS94-01 oceanographic cruise (R/V Urania, CNR, Italy), during which seismic profiles interpreted in this paper were recorded. We thank Prof. Bruno D'Argenio and Prof. Vittoria Ferreri (Dipartimento di Scienze della Terra, Università degli Studi di Napoli "Federico II", Naples, Italy) for suggestions at different steps of the manuscript preparation. We also thank the two referees, Dr. Bernard Gensous (Université de Perpignan, CEFREM, Perpignan, France) and Prof. Francesco Latino Chiocci (Università di Roma "La Sapienza", Rome, Italy) for their very helpful observations, allowing a significant improvement of the paper. Finally, we thank Patricia Sclafani (CNR-IAMC, Naples, Italy) for reviewing the English.

REFERENCES

AIELLO, G., BRAVI, S., BUDILLON, F., CARUSO, A., D'ARGENIO, B., DE LAURO, M., FERRARO, L., MARSELLA, E., MOLISSO, F., PELOSI, N., SACCHI, M., TOSCANO, F., AND TRAMONTANO, M.A., 1994, La piattaforma continentale pugliese al largo di Brindisi e tra Capo d'Otranto e S. Maria di Leuca (Adriatico meridionale): Approccio ad uno studio integrato di geologia marina: Istituto di Ricerca "Geomare Sud", National Research Council, Napoli, Italy, Technical Report no. 2, 25 p.

AIELLO, G., BRAVI, S., BUDILLON, F., CARUSO, A., D'ARGENIO, B., DE LAURO, M., FERRARO, L., MARSELLA, E., MOLISSO, F., PELOSI, N., SACCHI, M., TOSCANO, F., AND TRAMONTANO, M.A., 1995, Marine Geology of the Salento shelf (Apulia, south Italy): preliminary results of a multidisciplinary study: Giornale di Geologia, v. 57 (1–2), p. 17–40.

ALLEN, P.A., AND ALLEN, J.R., 1990, Basin Analysis; Principles and Applications: Oxford, U.K., Blackwell Scientific Publications, 451 p.

ASHLEY, G.M., WELLNER, R.W., ESKER, D., AND SHERIDAN, R.E., 1991, Clastic sequences developed during late Quaternary glacio-eustatic sea-level fluctuations on a passive margin: Example from the inner continental shelf, Barnegat Inlet, New Jersey: Geological Society of America, Bulletin, v. 103, p. 1607–1621.

BARD, E., HAMELIN, B., AND FAIRBANKS, R.G., 1990a, U–Th ages obtained by mass spectrometry in corals from Barbados: sea-level during the past 130,000 years: Nature, v. 346, p. 456–458.

BARD, E., LABEYRIE, L.D., PICHON, J.J., LABRACHERIE, M., ARNOLD, J., DUPRAT, J., MOYES, J., AND DUPLESSY, J.C., 1990b, The last deglaciation in the southern and northern hemispheres: a comparison based on oxygen isotopes, sea surface temperatures estimates and accelerator ^{14}C dating from deep-sea sediments, *in* Bleil, U., and Thiede, J., eds., Geological History of the Polar Oceans: Arctic versus Antarctic: Boston, Kluwer Academic, p. 405–416.

BONIFAY, E., 1975, L'Ere Quaternaire: définition, limites and subdivision sur la base de la chronologie Méditerrannée: Société Géologique de France, Bulletin, v. 17, p. 380–393.

BORSETTI, A., LOIK, L., AND COLANTONI, P., 1984, Variazioni nella sedimentazione al passaggio glaciale-postglaciale e Olocene in alcuni bacini nord-tirrenici evidenziate dal contenuto microfaunistico: Società Geologica Italiana, Memorie, v. 27, p. 323–332.

BOSSIO, A., GUELFI, F., MAZZEI, R., MONTEFORTI, B., AND SALVATORINI, G., 1988, Studi sul Neogene ed il Quaternario della Penisola Salentina, V—Note geologiche sulla zona di Castro: Final Proceedings of the Conference "Conoscenze Geologiche della Penisola Salentina", Lecce, Italy, Facoltà di Ingegneria di Lecce, Quaderni Geotecnici, v. 11, p. 127–145.

BUDILLON, F., AND AIELLO, G., 1999, Evoluzione pleistocenica della piattaforma continentale del Salento orientale: fattori di controllo tettonici e/o eustatici: Il Quaternario, v. 12 (2), p. 149–160.

CATALANO, R., DI STEFANO, E., SULLI, A., VITALE, F.P., INFUSO, S., AND VAIL, P.R., 1998, Sequences and system tracts calibrated by high-resolution bio-chronostratigraphy: the central Mediterranean Plio-Pleistocene record, *in* de Graciansky, P.C., Hardenbol, J., Jacquin, T., and Vail, P.R., eds., Mesozoic and Cenozoic Sequence Stratigraphy of European Basins: SEPM, Special Publication 60, p. 155–177.

CHAPPELL, J., AND SHACKLETON, N.J., 1986, Oxygen isotopes and sea level: Nature, v. 324, p. 137–140.

CHIOCCI, F.L., D'ANGELO, S., ORLANDO, L., AND PANTALEONE, A., 1989, Evolution of the Holocene shelf sedimentation defined by high-resolution seismic stratigraphy and sequence analysis (Calabro-Tyrrhenian continental shelf): Società Geologica Italiana, Memorie, v. 48, p. 359–380.

CHIOCCI, F.L., ERCILLA, G., AND TORRES, J., 1997, Stratal architecture of western Mediterranean margin as the result of the stacking of Quaternary lowstand deposits below "glacio-eustatic fluctuation base-level": Sedimentary Geology, v. 112, p. 195–217.

CIARANFI, N., PIERI, P., AND RICCHETTI, G., 1992, Note alla carta geologica delle Murge e del Salento (Puglia centro-meridionale): Società Geologica Italiana, Memorie, v. 41, p. 449–460.

COLANTONI, P., GALLIGNANI, P., AND LENAZ, R., 1989: Late Pleistocene and Holocene evolution of the North Adriatic continental shelf (Italy): Marine Geology, v. 33, p. 41–50.

CORREGGIARI, A., ROVEIRI, M., AND TRINCARDI, F., 1992, Regressioni "forzate", regressioni deposizionali e fenomeni di instabilità in unità progradazionali tardo-quaternaire (Adriatico centrale): Giornale di Geolgia, v. 54, p. 19–36.

COSENTINO, D., AND GLIOZZI, E., 1992, Considerazioni sulla velocità di sollevamento dei depositi eutirreniani dell'Italia meridionale e della Sicilia: Società Geologica Italiana, Memorie, v. 41, p. 653–665.

D'ARGENIO, B., PESCATORE, T., AND SCANDONE, P., 1973, Schema geologico dell'Appennino meridionale (Campania e Lucania): Final Proceedings of the Conference "Moderne vedute sulla Geologia dell'Appennino", Accademia Nazionale dei Lincei, v. 183, p. 49–72.

DOGLIONI, C., MONGELLI, F., AND PIERI, P., 1994, The Puglia uplift: an anomaly of the foreland of the Apenninic subduction due to the buckling of a thick continental lithosphere: Tectonics, v. 13, p. 1309–1321.

FAIRBANKS, R.G., AND MATTHEWS, R.K., 1978, The marine oxygen isotope record in Pleistocene coral, Barbados: Quaternary Research, v. 10, p. 181–196.

FIELD, M.E., AND TRINCARDI, F., 1992, Regressive coastal deposits on Quaternary continental shelves: preservation and legacy, *in* Osborne, R.H., ed., From Shoreline to Abyss: Contributions in Marine Geology in Honor of Francis Parker Shepard: SEPM, Special Publication 46, p. 107–122.

GENSOUS, B., WILLIAMSON, D., AND TESSON, M., 1993, Late Quaternary transgressive and highstand deposits on a deltaic shelf (Rhone delta, France), *in* Posamentier, H.W., Summerhayes, C.P., Haq, B.U., and Allen, G.P., eds., Sequence Stratigraphy and Facies Associations: International Association of Sedimentologists, Special Publication 18, p. 197–211.

HEARTY, P.J., AND DAI PRA, G., 1992, The age and stratigraphy of middle Pleistocene and younger deposits along the Gulf of Taranto (south-east Italy): Journal of Coastal Research, v. 8, p. 882–905.

HUNT, D., AND TUCKER, M.E., 1992, Stranded parasequences and the forced regression wedge system tract: deposition during base-level fall: Sedimentary Geology, v. 81, p. 1–9.

MARANI, M., TAVIANI, M., TRINCARDI, F., ARGNANI, A., BORSETTI, A., AND ZITELLINI, N., 1986, Pleistocene progradation and postglacial events on the tyrrhenian continental shelf between the Tiber river delta and Capo Circeo: Società Geologica Italiana, Memorie, v. 36, p. 67–89.

MARTINSON, D.G., PISIAS, N.G., HAYS, J.D., IMBRIE, J., MOORE, T.C., AND SHACKLETON, N.J., 1987, Age dating and the orbital theory of the ice ages: development of a high-resolution 0 to 300,000 year chronostratigraphy: Quaternary Research, v. 27, p. 1–29.

MITCHUM, R.M., VAIL, P.R., AND SANGREE, J.M., 1977, Seismic stratigraphy and global change of sea level, Part 6: Stratigraphic interpretation of seismic reflection patterns in depositional sequences, *in* Payton, C.E., ed., Seismic Stratigraphy—Applications to Hydrocarbon Exploration: American Association of Petroleum Geologists, Memoir 26, p. 117–133.

MORETTI, I., AND ROYDEN, L., 1987, Deflection, gravity anomalies and tectonics of doubly subducted continental lithosphere: Adriatic and Ionian seas: Tectonics, v. 7, p. 875–893.

OKAMURA, Y., AND BLUM, P., 1993, Seismic stratigraphy of Quaternary stacked depositional sequences in the Southwest Japan forearc: an example of fourth order sequences of an active margin, *in* Posamentier, H.W., Summerhayes, C.P., Haq, B.U., and Allen, G.P., eds., Sequence Stratigraphy and Facies Associations: International Association of Sedimentologists, Special Publication 18, p. 213–232.

PIRAZZOLI, P.A., 1993, Global sea level changes and their measurement: Global and Planetary Change Letters, v. 8, p. 135–148.

POSAMENTIER, H.W., AND VAIL, P.R., 1988, Eustatic control on clastic deposition II—sequence and system tract models, *in* Wilgus, C.K., Hastings, B.S., Kendall, C.G.St.C., Posamentier, H.W., Ross, C.A., and Van Wagoner, J.C., Sea Level Changes: An Integrated Approach: SEPM, Special Publication 42, p. 125–154.

POSAMENTIER, H.W., AND JAMES, D.P., 1993, An overview of sequence stratigraphic concepts: uses and abuses, *in* Posamentier, H.W., Summerhayes, C.P., Haq, B.U., and Allen, G.P., eds., Sequence Stratigraphy and Facies Associations: International Association of Sedimentologists, Special Publication 18, p. 3–18.

POSAMENTIER, H.W., ALLEN, G.P., JAMES, D.P., AND TESSON, M., 1992, Forced regression in a sequence stratigraphic framework: concepts, examples, and exploration significance: American Association Petroleum Geologists, Bulletin, v. 76, p. 1687–1709.

RICCHETTI, G., CIARANFI, N., LUPERTO SINNI, E., MONGELLI, F., AND PIERI, P., 1992, Geodinamica ed evoluzione stratigrafico-tettonica dell'avampaese apulo: Società Geologica Italiana, Memorie, v. 42, p. 287–300.

RIDENTE, D., AND TRINCARDI, F., 2002, Late Pleistocene depositional cycles and syn-sedimentary tectonics on the central and south Adriatic shelf: Società Geologica Italiana, Memorie, v. 57, p. 517–526.

SAITO, Y., 1991, Sequence stratigraphy on the shelf and the upper slope in response to the latest Pleistocene–Holocene sea level changes off Sendai, northeast Japan, *in* McDonald, D.I.M., ed., Sedimentation, Tectonics, and Eustasy: Sea-Level Changes at Active Margins: International Association of Sedimentologists, Special Publication 12, p. 133–150.

SHACKLETON, N.J., AND OPDYKE, N.D., 1973, Oxygen isotope and paleomagnetic stratigraphy of equatorial Pacific core V28-238: oxygen isotope temperature and ice volume on a 10 year scale: Quaternary Research, v. 3, p. 39–55.

SUTER, J.R., AND BERRYHILL, H.L., 1985, Late Quaternary shelf margin deltas, northwest Gulf of Mexico. American Association of Petroleum Geologists, Bulletin, v. 69, p. 77–91.

SUTER, J.R., BERRYHILL, H.L., AND PENLAND, S., 1987, Late Quaternary sea-level fluctuations and depositional sequences, southwest Louisiana continental shelf, *in* Nummedal, D., Pilkey, O.H., and Howard, J.D., eds., Sea-Level Fluctuation and Coastal Evolution: SEPM, Special Publication 41, p. 199–219.

TESSON, M., GENSOUS, B., ALLEN, G.P., AND RAVENNE, C., 1990, Late Quaternary deltaic lowstand wedges on the Rhone continental shelf, France: Marine Geology, v. 91, p. 325–332.

TESSON, M., ALLEN, G.P., AND RAVENNE, C., 1993, Late Pleistocene shelf perched lowstand wedges on the Rhone continental shelf, *in* Posamentier, H.W., Summerhayes, C.P., Haq, B.U., and Allen, G.P., eds., Sequence Stratigraphy and Facies Associations: International Association of Sedimentologists, Special Publication 18, p. 183–196.

TRINCARDI, F., AND FIELD, M.E., 1991, Geometry, lateral variation and preservation of downlapping regressive shelf deposits: Eastern Tyrrhenian sea margin, Italy: Journal of Sedimentary Petrology, v. 61, p. 775–790.

TRINCARDI, F., CORREGGIARI, A., AND ROVERI, M., 1994, Late Quaternary transgressive erosion and deposition in a modern epiconental shelf: the Adriatic semienclosed basin: Geomarine Letters, v. 14, p. 41–51.

TRINCARDI, F., AND CORREGGIARI, A., 2000, Quaternary forced regression deposits in the Adriatic basin and the record of composite sea-level cycles, *in* Hunt, D., and Gawthorpe, R.L., eds., Sedimentary Responses to Forced Regressions: Geological Society of London, Special Publication 172, p. 245–269.

VAIL, P.R., AUDEMARD, F., BOWMAN, S.A., EISNER, P.N., AND PEREZ CRUZ, C., 1991, The stratigraphic signatures of tectonics, eustacy and sedimentology—an overview, *in* Einsele, G., Ricken, W., Seilacher, A., eds., Cycles and Events in Stratigraphy: Berlin, Springer-Verlag, p. 617–659.

ANTARCTIC SEDIMENT DRIFTS AND PLIO-PLEISTOCENE ORBITAL PERIODICITIES (ODP SITES 1095, 1096, AND 1101)

MARINA IORIO
Istituto per l'Ambiente Marino Costiero, Geomare, National Research Council, Calata Porta di Massa, Porto di Napoli, 80133 Napoli, Italy
e-mail: iorio@gms01.geomare.na.cnr.it
THOMAS WOLF-WELLING
GEOMAR Research Center for Marine Geoscience, Wischhofstr. 1–3, D-24148 Kiel, Germany
e-mail: j.welling-wolf@netsurf.de
AND
TOBIAS MOERZ
RCOM, University Bremen, Tabgebäude (ECO5), Am Fallturm 1, 28359, Bremen, Germany
e-mail: tmoerz@uni-bremen.de

ABSTRACT: Petrophysical datasets and related spectral analysis from Plio-Pleistocene sediments cored on the Western Antarctic continental rise during the Ocean Drilling Program, Leg 178, are discussed. It is shown that in different cores, nonharmonic wavelength peaks, when normalized, exhibit a very high correlation factor with predicted Earth's orbital variations. It is also found that both short (~ 95–125 ky) and long (~ 400 ky) eccentricity periodicities emerge clearly from the signal during the whole Pleistocene, without an evident switch to obliquity at mid-Pleistocene (~ 0.9 Ma), as reported in the literature. This suggests that the lithological parameters, a proxy for glacial cycles, are controlled, directly or indirectly, by astronomically forced processes (Milankovitch cycles). Moreover, the good correlatability among distant coring sites, based on systematic sedimentological variations at intervals of about 140 and 370 ky, allows extension of the results to regional scale.

INTRODUCTION

The dynamics of the Antarctic ice sheet is a significant component of the global climate system, so the understanding of its driving mechanisms, still incompletely known, is of great importance in the analysis of the Earth's climatic evolution. The onshore geological record of Antarctic glaciation is geographically restricted and difficult to date (Webb and Harwood, 1991; Webb et al., 1996, Denton et al., 1993 , Stroeven et al., 1998; Harwood and Webb, 1998). The main evidence derives from low-latitude proxy data, where ice-sheet volume has been estimated through measurement of stable oxygen-isotope variation in sediments (Miller et al., 1987). Additional evidence has come from estimates of sea-level oscillation and the ensuing influence on deep-ocean circulation patterns, variation of biogenic sedimentation, and distribution of ice-rafted sediments (Miller et al., 1987; Barker, 1992), even though some estimates from these proxies are in poor agreement among themselves. Other insights into this problem have come from studies of the glacial–interglacial sedimentation of the continental margins of Antarctica and on the deposits of its slope and rise, where sediment progradation on the continental margin occurs during glacial stages in response to the sediment supply at the grounded line of the ice shelf (Larter and Barker, 1989; Anderson et al., 1991; Cooper et al., 1991; Barker et al., 1999a). The fine-grained part of these sediments is conveyed via slumping, debris flows, and turbidity currents from the offshore prograding wedges to the continental slope and rise and to the abyssal plain, either in suspension or entrained in bottom currents. This is the origin of the large hemipelagic sediment drifts distributed on the continental rise all around the Antarctic continent (e.g., Rebesco et al., 1998). Drift sediments alternate from rapidly deposited, poorly fossiliferous silt and clay during glacial intervals, to slowly deposited biogenic muds (with only a minor terrigenous component) during interglacial times. Existing seismic and drilling data from Antarctica have demonstrated that foreset and topset deposits of the prograding wedges, as well as rise-proximal drift sediments, contain complementary geological records of the Antarctic glacial history and a likely gross to fine climate sensitivity (Barker et al., 1999b; Barker and Camerlenghi, 2002; O'Brien et al., 2001). In this paper we discuss results from spectral analysis and correlation of sedimentological and petrophysical data from Sites 1095, 1096, and 1101, which were drilled in the sediment drifts of the continental rise off the West Antarctic margin during ODP Leg 178.

GEOLOGY OF SEDIMENT DRIFTS

The sector off the Pacific margin of the Antarctica Peninsula considered in this study spans over ~ 1000 km, from Smith Island in the northeast to Charcot Island in the southwest. In this area the continental rise can be divided into two parts: the upper rise, with irregular bottom relief, and the lower rise, with a more subdued morphology.

The irregular relief of the upper rise (Fig. 1) is due to large sediment drifts, which are separated by nine channel systems formed by turbidity fluxes and trending northwest across the rise (Tomlinson et al., 1992). The drifts form thick successions and are located mostly between the lobes of the rise, elongated in a direction approximately orthogonal to its margin (Rebesco et al., 1998). They are asymmetric, generally with a steep southwest-facing slope and a gentle northeast-facing slope (Barker et al., 1999b, Barker and Camerlenghi, 2002).

Petrophysical and coarse-fraction sedimentological data from Sites 1095, 1096, and 1101, (Fig. 2A, B, C), drilled respectively from drifts 7 and 4 (Fig. 1) during the ODP Leg 178 (Barker et al., 1999b), are considered in this study.

Cyclostratigraphy: Approaches and Case Histories
SEPM Special Publication No. 81, Copyright © 2004
SEPM (Society for Sedimentary Geology), ISBN 1-56576-108-1, p. 231–244.

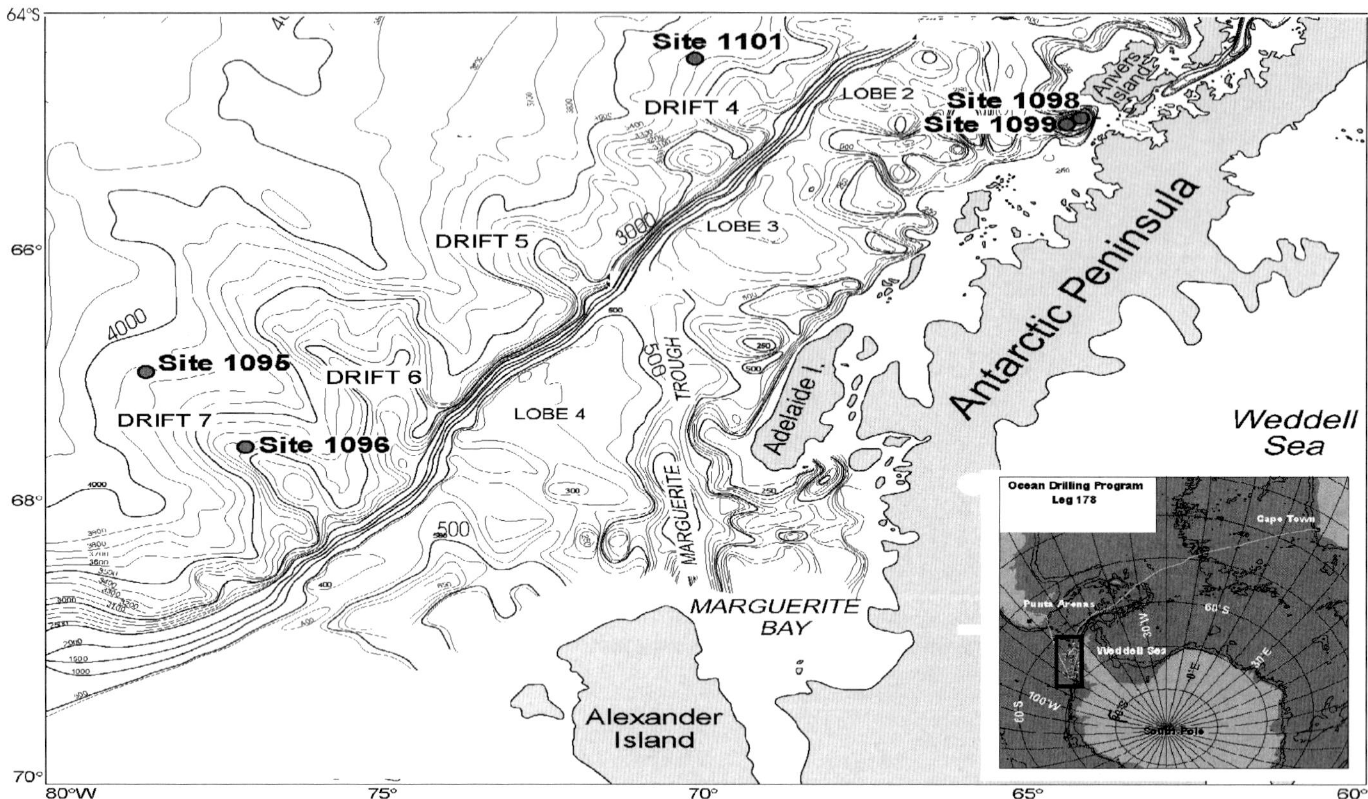

FIG. 1.—Location of ODP, Sites 1095, 1096, and 1101 (Leg 178) on the Antarctic continental rise and sediment drifts 4 to 7 (after Barker et al., 1999b; modified).

SITE SEDIMENTOLOGY

Sites 1095 and 1096 lie on the northwestern lower flank of sediment drift 7 and on its crest, respectively, and Site 1101 is located centrally within sediment drift 4, approximately 500 km northeast of Sites 1095 and 1096 (Fig. 1). The sequences used in this study (Fig. 2A, B, C) are characterized by continuous recovery, extending from 0 to 150 and from 0 to 220 mcd (meters composite depth; the mcd scale combines by correlation holes 1095A, 1095B, 1095D of Site 1095, and holes 1096A, 1096B, 1096C of Site 1096), and from 0 to 217.7 mbsf (meters below sea floor), for Site 1101 (Barker, 2001; Barker et al., 1999b). The age of the studied sediments spans from Holocene to Early Pliocene at Site 1095 and to Late Pliocene at Sites 1096 and 1101 (Iwai et al., 2002; Acton et al., 2002).

Site 1095

In the 150 mcd of the studied section of Site 1095, the uppermost ~ 50 mcd shows a marked cyclic pattern of alternating gray terrigenous, silty clays and brown biogenic-rich silty clays, interpreted as an expression of cyclic glacial and interglacial sedimentation, respectively (Barker et al., 1999b; Barker and Camerlenghi, 2002). The lower part of the section consists mainly of green laminated silts and muds, whose sediment supply was probably also controlled by glacial cycles (Pudsey, 2001). Between these two segments a minor hiatus has been supposed (Barker et al., 1999b). However, the presence of all expected magnetostratigraphic events and the lack of any noticeable change in sedimentation rate (Acton et al., 2002) suggest that the time gap evidenced by an erosional contact in Site 1095D and the related seismic unconformity is smaller than 200 ky (Iwai et al., 2002).

Site 1096

The 220 mcd of studied sequence encloses, from the top, about 33 mcd of laminated and massive, diatom-bearing silty clays, which have a well-defined alternation of biogenic-rich and biogenic-poor levels, again interpreted as an expression of glacial–interglacial cyclicity (Barker and Camerlenghi, 2002). Down to 220 mcd, an alternation of laminated deposits, interpreted as fine resedimented deposits and bioturbated hemipelagites (Barker et al., 1999b), records cyclic fluctuations in sediment supply. Some of these fluctuations may be related to glacial–interglacial cycles along the Antarctic Peninsula margin, similar to those identified within the upper 33 mcd. Moreover, lower-frequency cycles also exist (Barker et al., 1999b).

Site 1101

The studied sequence, consisting of 217.7 mbsf of predominantly hemipelagic clayey silt, is composed for the first 142.7 mbsf of alternating massive and laminated biogenic clayey silts interpreted, by analogy with the equivalent interval at Site 1096, as the sediment expression of interglacial and glacial periods (Barker et al., 1999b; Barker and Camerlenghi, 2002). In the interval from 142.7 to 198 mbsf, massive, barren clayey silts are present. They could have originated in turbidity plumes or sediment gravity flows from glacier fronts near the continental shelf edge; alternatively they could be thin ice-rafted deposits (Barker et al., 1999b). The remaining

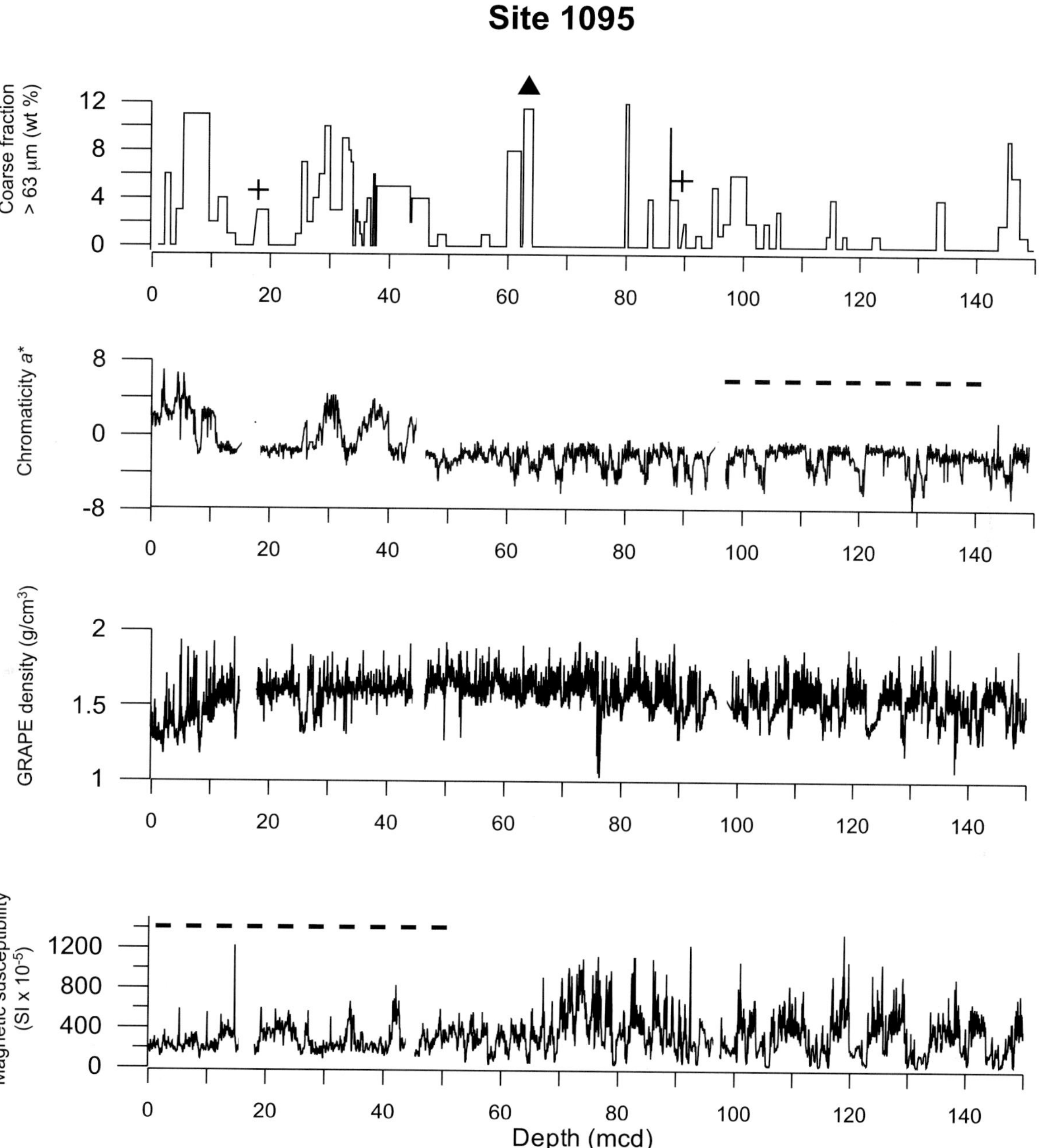

FIG. 2.—Physical properties and coarse-fraction granulometry (raw data) of the analyzed sediments. **A)** Site 1095; cross and triangle in the coarse-fraction graph indicate 52 and 27 weight % peak values, respectively.

interval (from 198 down to 217.7 mbsf) is characterized by diatom-bearing massive and laminated deposits, which may represent an interval of warmer climatic conditions (Barker et al., 1999b).

DATA SETS AND METHODS

The data sets used for the present study are based on the assemblage of four parameters (Fig. 2 A, B, C). Two of them are GRAPE (gamma ray attenuation porosity evaluator) bulk density and magnetic susceptibility. They are derived by sensors collecting data at 2 cm intervals on a multi sensor track (MST). The third parameter, chromaticity a^*, was routinely measured downhole at evenly spaced intervals of 5 cm using a Minolta CM 2002 spectrophotometer. The chromaticity measurements were taken as soon as possible after the cores were split, to minimize redox-associated color changes, which generally occur when deep-sea sediments are exposed to the atmosphere. Among the chromaticity parameters available, a^* was chosen because it is thought to give a maximum response to changes of lithology in the sediments (Wolf-Welling et al., 2001a). All these measurements were collected on board during Leg 178.

The fourth parameter, evaluation coarse fraction (>63 μm wt %), was obtained from discrete samples, collected at an average interval of 140 cm, during post-cruise analysis at Geomar, Kiel.

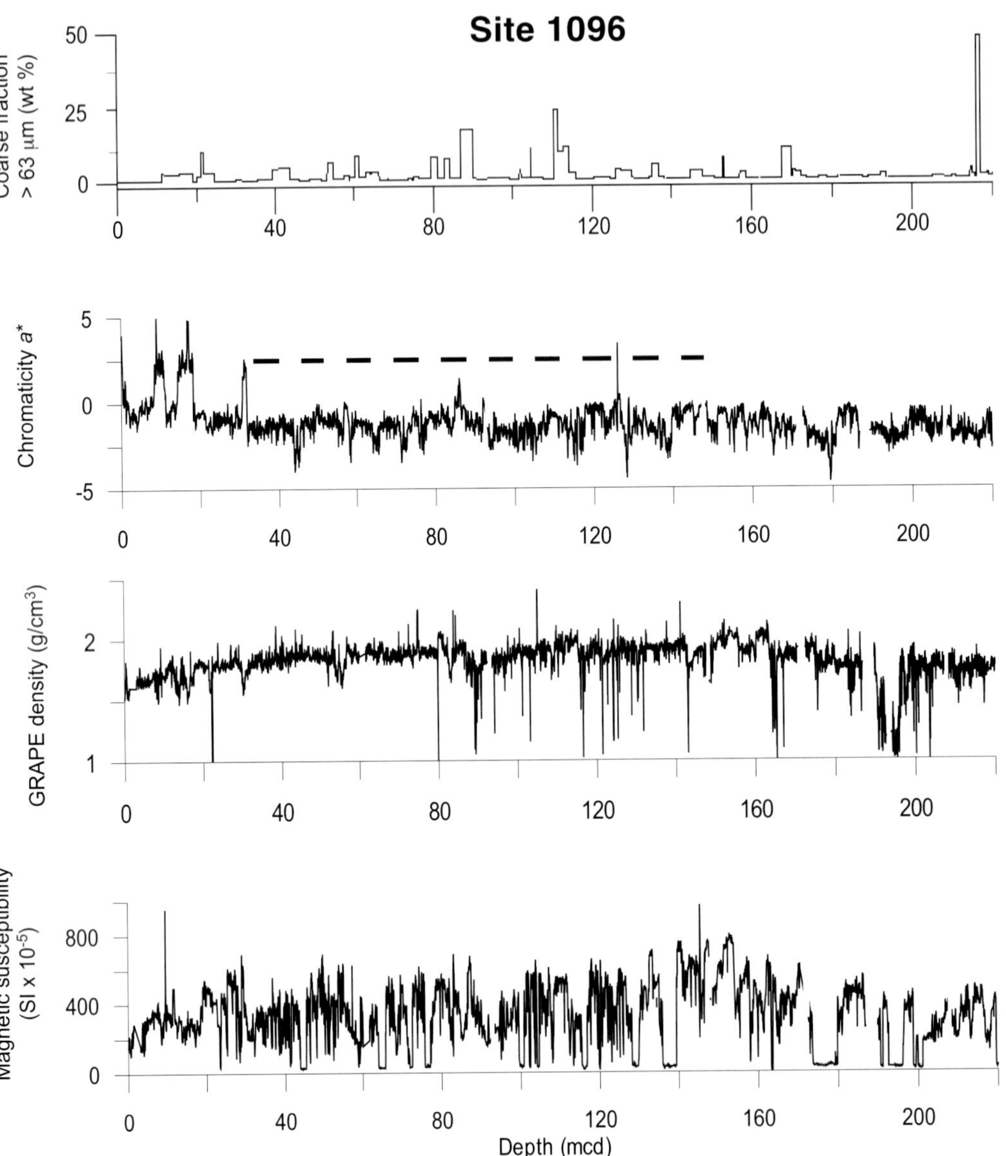

FIG. 2 (continued).—**B)** Site 1096.

These data are considered as a further proxy for ice-rafted debris (Wolf-Welling et al., 2001b).

FAST FOURIER TRANSFORM ANALYSIS

From the data sets of Fig. 2 A, B, C, several intervals (Table 1), showing consistent behavior between the four different parameters and characterized by continuous recovery and constant sedimentation rates (Barker et al., 1999b), were selected. They were processed mathematically by mean of Fast Fourier Transforms to search for petrophysical and/or sediment periodicities. The coarse-fraction data set of Site 1096 was excluded from this analysis because the sampling was discontinuous.

The spectral-analysis program used (Longo et al., 1994; Iorio et al., 1995; Brescia et al., 1995; Tagliaferri et al., 2001) originally was written to deal with unevenly spaced astronomical data sets, and it is based on a technique originally developed by Scargle (1982) and Horne and Baliunas (1986).

This technique aims at detecting the presence and significance of periodicities in unevenly sampled data series. In fact, as was found in many stratigraphic data sets, rebinding the unevenly sampled data to equally spaced bins and calculating them by means of a conventional periodogram often alters the perceived frequency and significance of a periodic signal. In addition this technique allows discrimination between meaningful frequencies and spurious ones on a physical basis (using the evaluation of the false-alarm probability) and renders the power spectra invariant to any shift in the signal origin (Brescia et al., 1995).

To tailor the specific needs of the stratigraphic data used, the signals were processed with a window function (tapering) before spectral analysis. The tapering is needed to eliminate spectral leakage due to the finite width of the sampled signal. Finally it was also necessary to exclude those peaks above the 90% false-alarm probability, which could have arisen from harmonics of lower frequencies.

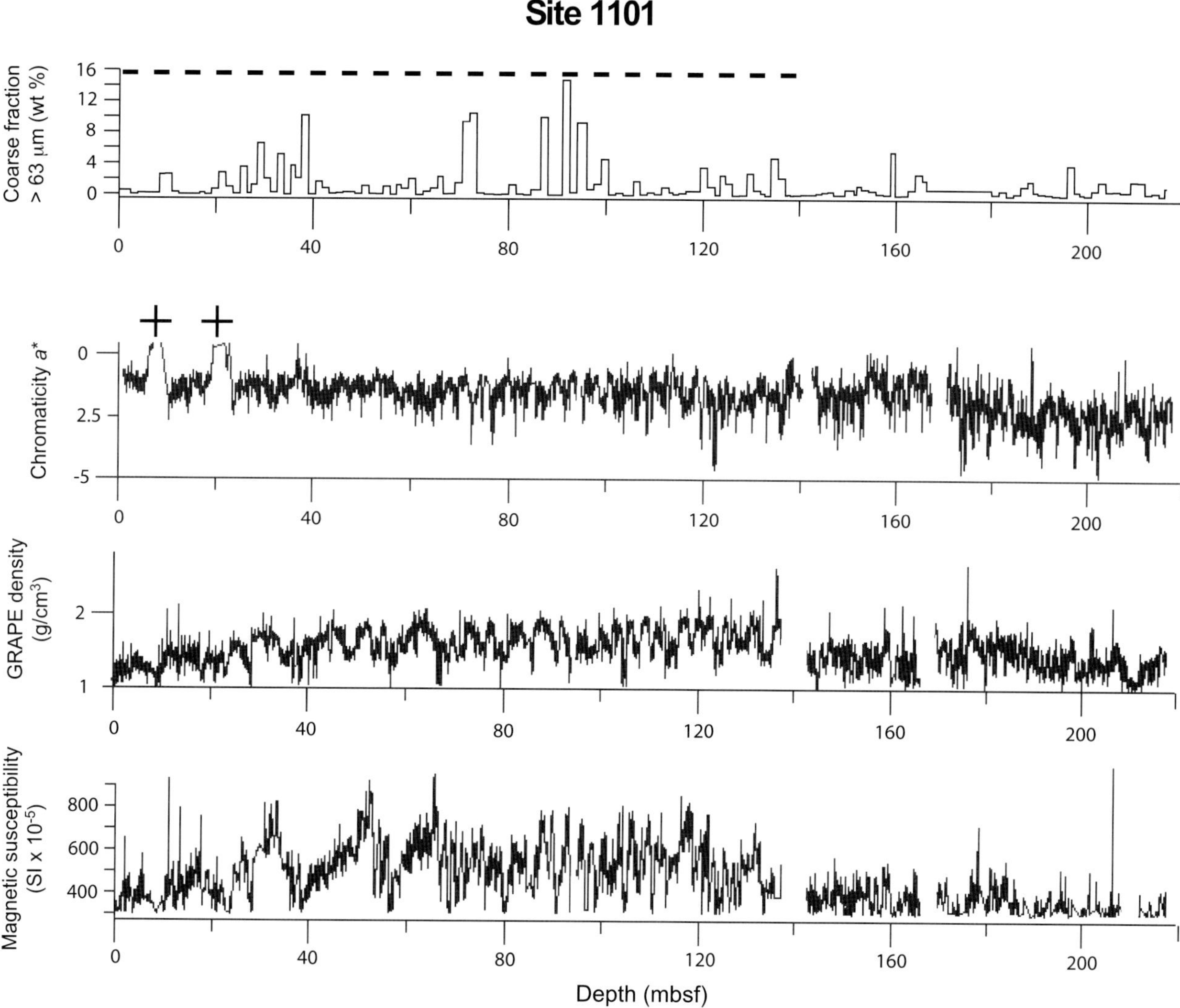

FIG. 2 (continued).—**C)** Site 1101; crosses in the graph of chromaticity a^* indicate peak values at 4.14 and 3.39, respectively, (dashed lines in A, B, C, indicate some selected intervals analyzed by means of Fast Fourier Transform and plotted in Part D for visual observation).

RESULTS

Among all parameters in the studied intervals, only those with well defined power spectra with more than two wavelengths were considered to record a true cyclic signal (Table 1).

Six examples of the resulting power spectra of the parameter analyzed, with well defined wavelengths at different amplitude, are shown in Fig. 3.

It is important to note that the magnetic susceptibility parameter is the most sensitive to cyclic variation. In fact in all sites and time intervals the power spectra are well defined (Table 1), and so they were chosen to test the spectral variability, analyzed together with and without the Upper Pleistocene interval (Site 1101, 0–53 m). In both power spectra (Fig. 3E, F) a positive persistence of peaks linked to all standard Earth's orbital periodicities was found (see next paragraph and Table 1c). For visual observation of the cycles identified by mathematical treatment, four FFT analyzed intervals (details of Fig. 2 A, B, C, shown by dashed lines) are plotted in Figure 2D at larger scale. In graphs A, C, D, arrows indicate the ~ 400 ky periodicity (Table 1A–C), and the circles in graphs B and C highlight a visual estimate of the ~ 100 ky cycles.

In order to compare the computed wavelengths expressed in sediment thickness (m) to orbital periods expressed in temporal units (ky), the wavelengths of each power spectrum were normalized to their highest well defined frequency (example in Table 2B), and each orbital calculation for the last 3 My computed by Berger and Loutre (1992) was normalized (Table 2A) with respect to the others, (Iorio et al., 1998; Brescia et al., 1995; D'Argenio et al., 1998).

The two tables of nondimensional ratios so obtained (Table 2A, B) were then cross-correlated, searching for the highest correlation coefficient. The contoured ratio set in Table 2A indicates the best match (correlation coefficient 0.984) between the two nondimensional ratio sets. This allows tying the sedimentary thickness wavelengths 2.35 m, 2.80 m, and 3.80 m to the obliquity periodicities (39.719 ky and 50.0 ky) and to a combined periodicity term (64.0 ky), respectively. In the same way it is possible to attach the 4.80 m, 6.80 m, 8.35 m, and 10.70 m wavelengths to the

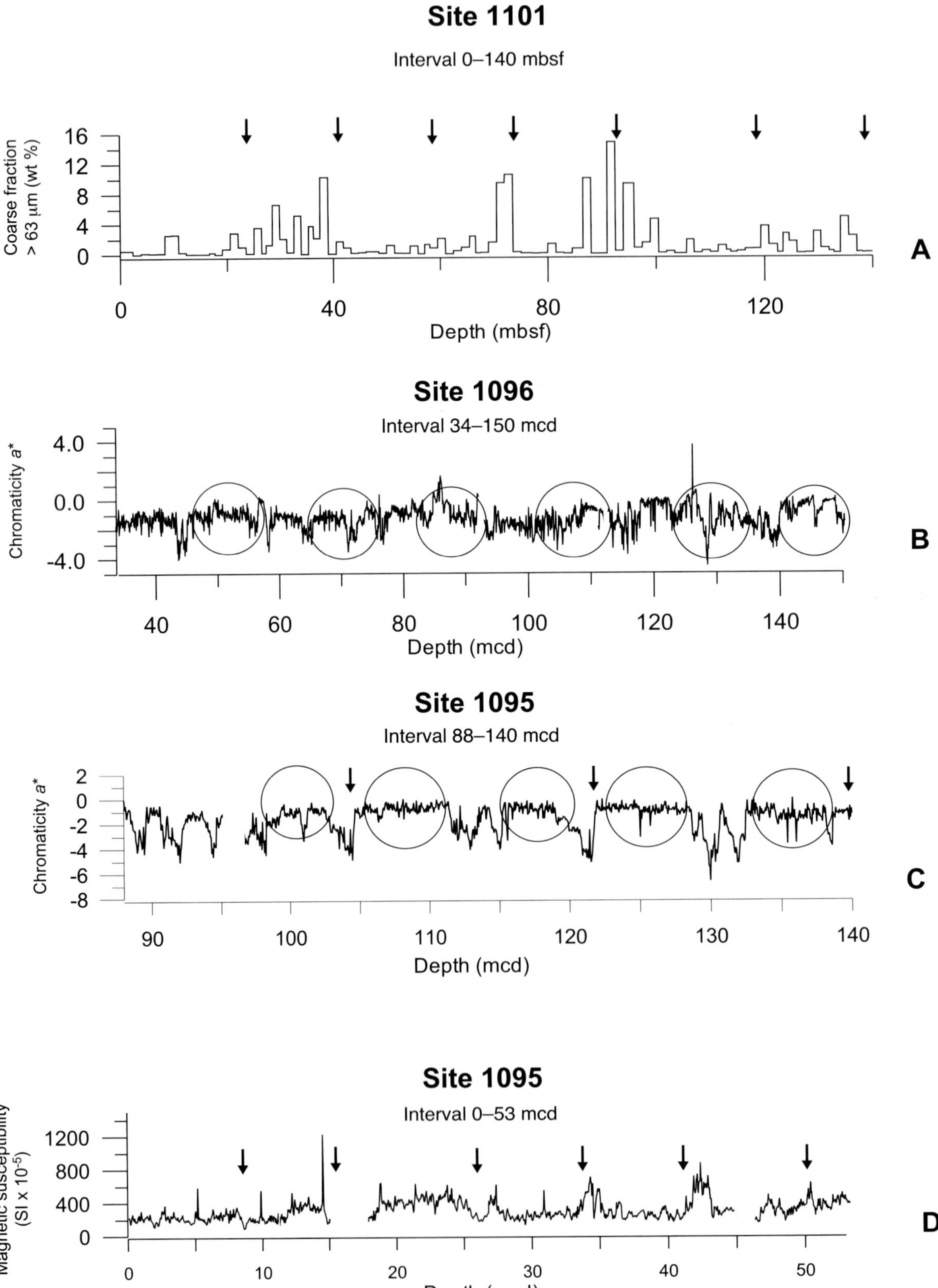

FIG. 2 (continued).—**D)** Selected data from Sites 1095,1096, and 1101. Arrows and circles indicate respectively reiterated peaks and troughs patterns, whose lengths, in agreement with the mathematical data analysis, appear linked to the long- and short-eccentricity patterns.

TABLE 1.—Depth intervals and data-set parameters used for Fast Fourier Transform analysis and computed wavelengths. **A)** Site 1095; **B)** Site 1096; **C)** Site 1101. MS, magnetic susceptibility; GRAPE (gamma ray attenuation porosity evaluator) density; Rfa*, chromaticity parameter *a**; Fr > 63, percentage of coarse fraction > 63 μm in weight %; Astr, computed astronomical periodicities (Berger, 1977; Berger and Loutre, 1992), expressed in years. The boldface figures indicate the highest-energy peaks of the lithological power spectra and the corresponding astronomical periodicities. Corr. is the correlation coefficient between the wavelength data set and the astronomical periodicity set. In Part A, the coarse fraction was analyzed from 88 to 150 mcd and is marked with +.

A

SITE 1095						
0–53 mcd		88–140 mcd				
Late Pliocene to Pleistocene		Early to Late Pliocene				
MS (m)	Astr. (years)	Ms (m)	GRAPE (m)	Rfa*(m)	+Fr > 63 μm (m)	Astr.(years)
		0.98				22394
1.12	39719	1.73			1.50	39719
		2.08	2.13	2.14	2.06	50000
1.60	53864					
		2.36		2.44	2.29	59000
1.99	64000	3.06	**2.74**	2.82	2.77	**64000**
					4.01	**98715**
3.09	**111000**					
4.15	123818	**6.06**	**6.83**	5.36	**5.76**	**123818**
8.24	**404178**			17.87		404178
Corr. 0.993		Corr. 0.996	Corr. 0.999	Corr. 0.998	Corr. 0.993	

B

SITE 1096				
0–33 mcd		34–150 mcd		
Middle to Late Pleistocene		Late Pliocene to Middle Pleistocene		
MS (m)	Astr. (years)	MS (m)	Rfa*(m)	Astr. (years)
1.05	15000			
1.45	18964			
1.79	**23708**	2.04	2.08	23708
		3.34	3.51	39719
3.39	50000	—	4.39	50000
		—	**5.34**	**64000**
6.10	**94782**	**8.06**	7.49	**94782**
		11.50	**11.18**	**123818**
		19.58	—	**227000**
		34.25	—	404178
Corr. 0.995		Corr. 0.999	Corr. 0.998	

C

SITE 1101							
0–143 mbsf			53–143 mbsf				
Late Pliocene to Pleistocene			Late Pliocene to Middle Pleistocene				
MS (m)	Fr > 63 μm (m)	Astr. (years)	Ms (m)	GRAPE (m)	Rfa*(m)	Fr > 63 μm (m)	Astr. (years)
1.0	—	15000	0.91				15000
1.5	—	22394	**1.53**	1.37	1.53		**22394**
1.8	—	23708	**1.76**				**23708**
2.4	2.35	39719	2.22	2.44	2.27		39719
			2.70			2.68	41090
—	2.8	50000		2.80			50000
—	—	53864		3.15	**3.36**		**53864**
						3.89	**59000**
3.7	3.8	64000	3.99	**4.21**			**64000**
	4.8	**94782**			4.67	**4.77**	**94782**
6.8	6.8	98715	6.30				98715
—	8.35	111000				**7.19**	**111000**
9.5	10.7	**123818**	9.45	10.80		10.88	123818
15.4	—	**231000**					
24.7	**26.0**	**404178**		25.21	28.04	21.42	404178
Corr.0.993	Corr. 0.984		Corr. 0.988	Corr. 0.995	Corr. 0.990	Corr. 0.975	

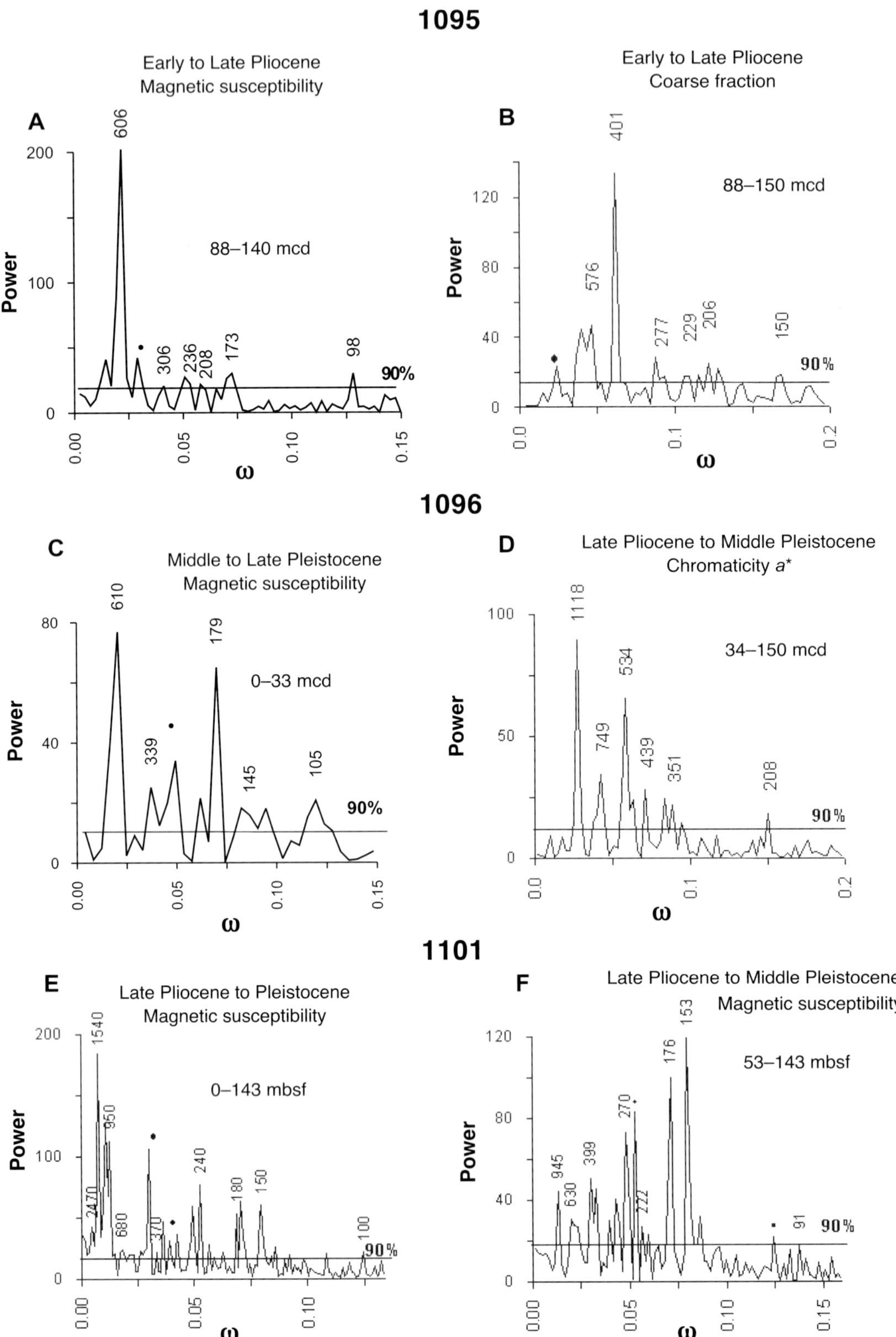

FIG. 3.—Power spectra of some lithological and petrophysical data sets. ω indicates frequency. The horizontal line marks the 90% probability that a given peak does not originate from the noise. Peaks above the 90% horizontal line, not arising from aliasing or harmonics (dots), are marked with the relative wavelengths computed in centimeters. **A, B)** magnetic susceptibility and coarse fraction, the Pliocene interval at Site 1095 (88–150 mcd). **C, D)** Middle to Late Pleistocene magnetic susceptibility (0–33 mcd) and Late Pliocene to Middle Pleistocene chromaticity *a** (34–150 mcd) at Site 1096. **E, F)** Magnetic susceptibility in partly overlapping intervals (0–143 and 53–143 mbsf) at Site 1101.

TABLE 2.—**A)** Relative ratio sets of the Plio-Pleistocene computed astronomical periodicities. All periods listed are associated with the four main largest-amplitude terms, except for 15.0 ky, 50.0 ky, and 111.0 ky. Instead 59.0 ky and 64.0 ky refer to the combined effects of both precession and obliquity (Berger, 1977; Berger and Loutre, 1992). The rectangular contour indicates the ratio sets of the astronomical periodicities showing the highest correlation coefficient with the ratio sets of the periodicities found in the coarse fraction at Site 1101 (Table 1C and Table 2B). **B)** Site 1101. Example of relative ratio sets in the Late Pliocene–Pleistocene computed wavelengths (m) of the coarse fraction (> 63 μm), correlated with the relative ratio sets of astronomical periodicities (rectangular contour from Part A). Note the very high correlation coefficient.

A

	15000	**18964**	**19116**	**22394**	**23708**	**39719**	**40392**	**41090**	**50000**	**53864**	**59000**	**64000**	**94782**	**98715**	**111000**	**123818**	**404178**	**Cor.**
15000	1	0.791	0.785	0.670	0.633	0.378	0.371	0.365	0.300	0.278	0.254	0.234	0.158	0.152	0.135	1.000	0.037	
18964	1.264	1	0.992	0.847	0.800	0.477	0.469	0.462	0.379	0.352	0.321	0.296	0.200	0.192	0.171	0.153	0.047	
19116	1.274	1.008	1	0.854	0.806	0.481	0.473	0.465	0.382	0.355	0.324	0.299	0.202	0.194	0.172	0.154	0.047	
22394	1.493	1.181	1.171	1	0.945	0.564	0.554	0.545	0.448	0.416	0.380	0.350	0.236	0.227	0.202	0.181	0.055	
23708	1.581	1.250	1.240	1.059	1	0.597	0.587	0.577	0.474	0.440	0.402	0.370	0.250	0.240	0.214	0.191	0.059	
39719	2.648	2.094	2.078	1.774	1.675	1	0.983	0.967	0.794	0.737	0.673	0.621	0.419	0.402	0.358	0.321	0.098	0.984
40392	2.693	2.130	2.113	1.804	1.704	1.017	1	0.983	0.808	0.750	0.685	0.631	0.426	0.409	0.364	0.326	0.100	
41090	2.739	2.167	2.150	1.835	1.733	1.035	1.017	1	0.822	0.763	0.696	0.642	0.434	0.416	0.370	0.332	0.102	
50000	3.333	2.637	2.616	2.233	2.109	1.259	1.238	1.217	1	0.928	0.847	0.781	0.528	0.507	0.450	0.404	0.124	
53864	3.591	2.840	2.818	2.405	2.272	1.356	1.334	1.311	1.077	1	0.913	0.842	0.568	0.546	0.485	0.435	0.133	
59000	3.933	3.111	3.086	2.635	2.489	1.485	1.461	1.436	1.180	1.095	1	0.922	0.622	0.598	0.532	0.477	0.146	
64000	4.267	3.375	3.348	2.858	2.700	1.611	1.584	1.558	1.280	1.188	1.085	1	0.675	0.648	0.577	0.517	0.158	
94782	6.319	4.998	4.958	4.232	3.998	2.386	2.347	2.307	1.896	1.760	1.606	1.481	1	0.960	0.854	0.765	0.235	
98715	6.581	0.000	5.164	4.408	4.164	2.485	2.444	2.402	1.974	1.833	1.673	1.542	1.041	1	0.889	0.797	0.244	
111000	7.400	5.853	5.807	4.957	4.682	2.795	2.748	2.701	2.220	2.061	1.881	1.734	1.171	1.124	1	0.896	0.275	
123818	8.255	6.529	6.477	5.529	5.223	3.117	3.065	3.013	2.476	2.299	2.099	1.935	1.306	1.254	1.115	1	0.306	
404178	26.945	21.313	21.143	18.048	17.048	10.176	10.006	9.836	8.084	7.504	6.850	6.315	4.264	4.094	3.641	3.264	1	

B

Relative ratio sets correlation								
1101 Fr > 63	**2.35**	**2.80**	**3.80**	**4.80**	**6.80**	**8.35**	**10.70**	**26.00**
2.35	1	0.839	0.618	0.490	0.346	0.281	0.220	0.090
			Correlation coefficient 0.984					
39719	1	0.794	0.621	0.419	0.402	0.358	0.321	0.098
Astron.	**39719**	**50000**	**64000**	**94782**	**98715**	**111000**	**123818**	**404178**

94.782 ky, 98.715 ky, 111.0 ky, and 123.818 ky values of the short eccentricity, respectively. Finally the 26.00 m peak matches the 404.178 ky periodicity of the long eccentricity.

Linear regression analysis of the relative ratios of all wavelengths and orbital periods give correlation factors that are above 0.98, for all power spectra analyzed (Table 1A, B, C), hence they can be considered strongly linked.

Moreover, the above data enable a precise computation of the average sediment accumulation rates in the studied intervals, which show also a close similarity (Table 3) with both biostratigraphically and paleomagnetically computed sedimentation rates (Iwai et al., 2002; Acton et al., 2002). Looking at the highest-energy peaks (marked in bold in Table 1), we find out that the short eccentricity (95 and 123 ky) strongly influences the Early and Late Pliocene (Site 1095) and the whole Pleistocene of Sites 1095, 1101, and 1096. The ~ 400 ky long eccentricity emerges strongly from the Plio-Pleistocene sedimentary record of Sites 1095 and 1101. Moreover the ~ 20 ky precession is evident in the Late Pliocene–Pleistocene of Sites 1096 and 1101. Finally the ~ 64 ky period characterizes the Early–Late Pliocene and Middle Pleistocene of Sites 1095, 1096, and 1101.

In order to allow a comparison of the FFT analysis results among the data sets from all three sites, all sedimentological and petrophysical data, which were collected at different intervals in each site, were smoothed with an average window of 140 cm (which is the longest sampling interval) in order to allow a comparison among all data sets. The smoothed data were calibrated in Ma (Fig. 4) using the astronomically computed as well as biostratigraphically and paleomagnetically constrained sedimentation rates (Table 3) (Iwai et al., 2002; Acton et al., 2002) in the intervals excluded from FFT analysis (i.e., 53–88 mcd and 150–220 mcd for Site 1095 and Site 1096 and 143–217.7 mbsf for Site 1101).

Subsequently, the variations of all smoothed and calibrated parameters, synchronously occurring in each site, were considered to be due to a change in the sedimentary record, and the relative age was marked accordingly (horizontal black lines in Fig. 4, and Table 4, Columns I–III).

Moreover, any lithological variation occurring at the same time (± 0.05 Ma) in at least two sites belonging to two distinct sediment drifts (drifts 7 and 4) was considered to be an expression of the same event, and the averaged age was computed (Table 4, Column IV). Finally, when the same event was recorded in all three sites (as at 0.39, 0.9, 1.07, 1.42, 2.01, 2.13, 2.61 Ma), it was considered an even stronger event, and its age was marked in bold (Table 4, Column IV, and stars in Fig. 4).

The time interval between two subsequent events during the last 2.6 Ma was computed and listed in Table 4, Column V, while in Column VI the time interval between the events recorded at all

TABLE 3.—Comparison of sedimentation rates computed on the base of biostratigraphic and paleomagnetic data (Acton et al., 2002; Iwai et al., 2002) with those obtained from astronomical periodicities (this study). Sedimentation rates based only on magnetostratigraphic analysis are indicated by *.

Comparison of sedimentation rates		
Site 1095 sedimentation rates		
	Late Pliocene to Pleistocene (0–53 mcd)	Early to Late Pliocene (88–140 mcd)
Astron. (this study)	2.8 cm/ky	4.3 cm/ky
Bio-paleomagnetic	*2.4 cm/ky	5.0 cm/ky
Site 1096 sedimentation rate		
	Middle to Late Pleistocene (0–33 mcd)	Late Pliocene to Middle Pleistocene (34–150 mcd)
Astron. (this study)	7.0 cm/ky	8.6 cm/ky
Bio-paleomagnetic	8.0 cm/ky	8.0 cm/ky
Site 1101 sedimentation rate		
	Late Pliocene to Pleistocene (0–143 mcd)	Late Pliocene to Middle Pleistocene (53–143 mcd)
Astron. (this study)	6.3 cm/ky	6.4 cm/ky
Bio-paleomagnetic	7.0 cm/ky	7.0 cm/ky

three sites are shown. The average time interval (Table 4) for Column V is 0.14 ± 0.04 My, and for column VI, 0.37 ± 0.17 My. So these time intervals are close to that computed for the Plio-Pleistocene power spectra of higher-energy wavelengths (Table 1A, B, C).

It is also important to note the absence of common events recurring with a frequency of about 100 ky in the sediments of the first ~ 2.1 My at Site 1095 (Table 4), whereas the FFT analysis in the same interval shows a strong ~ 100 ky periodicity, even if this discrepancy could be due to the 50 ky interval used for the smoothing window.

DISCUSSION

New insights into the behavior of the Plio-Pleistocene Antarctic Ice Sheet were gained from the analysis of three ODP wells. Even if given that a single region can record only a part of the Antarctic glacial history, the investigated area is among the most sensitive to the changes in ice-sheet volume in Antarctica (Barker et al., 1999b) .

The ODP Leg 178 results (Barker et al., 1999b; Barker and Camerlenghi, 2002) show that the release of glacial sediments at the drilled Sites was cyclic, and it was suggested by Barker et al. (1999b) that this reflects the orbital variation through much of the period examined.

However, Barker and Camerlenghi (2002) left open the question about the possibility of finding clear evidence of frequencies associated with orbital variations in the sedimentary parameters of the drift at Sites 1095 and 1096. Our results based on FFT analysis of lithological data and their correlation with the astronomical periodicities confirm this hypothesis at least for the Plio-Pleistocene. A comparison with other results of spectral analysis (Lauer-Leredde et al., 2002; Pudsey, 2001) for times in partial or total overlap with our intervals shows that astronomical cyclicity can be found in other intervals of the Early to Late Pliocene, considering the recurrence of bioturbated beds, and the magnetic susceptibility and chromaticity a^* parameters at Site 1095 (periods at 56 ky and 87 ky, from hole 1095B in Lauer-Laredde et al., 2002, periods at 19 ky, 63 ky, and 140 ky, from hole 1095 A in Pudsey, 2001). Some results have been obtained from the Plio-Pleistocene data logs (covering time interval 1.2–1.8 Ma) of Site 1096 (periodicities at 55 ky and, strong, at 100 ky, from hole 1096 C; Lauer-Laredde et al., 2002).

So, even though all the orbital periodicities appear reflected in the climatic signal of the studied intervals, the short eccentricity seems to be constant and stronger since the Early Pliocene, not showing in our analysis the claimed switch from a predominant 41 ky to a predominant 100 ky cycle at about 0.9 Ma (the mid-Pleistocene transition; e.g., Pisias and Moore, 1981). This finding gives a new insight into the so called "100 ky problem" (e.g., Raymo et al., 1997; see also the review papers by Hinnov, 2000, and Elkibbi and Rial, 2001). Several models try to explain how so small a variation in insolation forcing, like that acting at 100 ky, could be the direct cause of the great ice ages of the Pleistocene. Among these models are those featuring self-sustaining internal oscillations with no relation to insolation (e.g., Maasch and Saltzman, 1990) and those supporting the insolation as main control of the 100 ky oscillation, by some form of linear resonance or by nonlinear coupling of higher-frequency forcing (e.g., Hagelberg et al., 1991; Imbrie et al., 1993). Among recent models a temperature forcing has been proposed by Shackleton (2000). Shackleton shows that, in the later part of the Pleistocene, the 100 ky cycle in the oxygen isotope record of deep-sea sediments did not arise from ice-sheet dynamics but instead probably derived from fluctuation in atmospheric CO_2 concentration.

To interpret our results we need to take into account the different nature of the parameters analyzed. In fact, although variations in magnetic susceptibility and reflectance in deep-sea sediments may be sensitive indicators of both cyclic environmental changes and terrigenous influx, the coarse-fraction changes can be considered as mostly linked to ice-sheet growth and decay. So our findings, including significant spectral peaks at 413 ky over the past 1.2 My in the data from all three sites, support the Milankovitch climate theory in the studied area.

Spectral peaks, linked to 15.0 ky and 50.0 ky frequencies as predicted by Berger (1977) in the geological records, as well as to 59.0 ky and 64.0 ky frequencies (the combined effect of precession and obliquity), were found. In addition a non-orbital period at about 230 ky could be interpreted as combination tones of the forcing frequencies (Rial, 1999). It is also important to note that all three sites present synchronous events at 1.07, 2.01, and 2.61 Ma (black stars in Fig. 4), which correspond to time intervals interpreted as the expression of periods of intense ice rafting at Site 1101 (Cowan, 2001) and fall at relative sea-level highs in the sea-level oscillation curve of Haq et al. (1987), recalibrated by Berggren et al. (1995), suggesting a possible superimposed eustatic influence.

Finally, the constant presence of all the orbital periodicities in the signals since the Early Pleistocene and in part of the Early–Late Pliocene of Site 1095 confirms the relative stability of the Antarctic ice sheet since Late Neogene (Barker et al., 1999a; Barker and Camerlenghi, 2002; Hillenbrand and Ehrmann, 2001) and implies its sensitivity to the insolation changes during the studied time interval.

CONCLUSIONS

Spectral analysis of data sets from Plio-Pleistocene bore cores from ODP Sites 1096, 1101, and 1095 on the West Antarctic

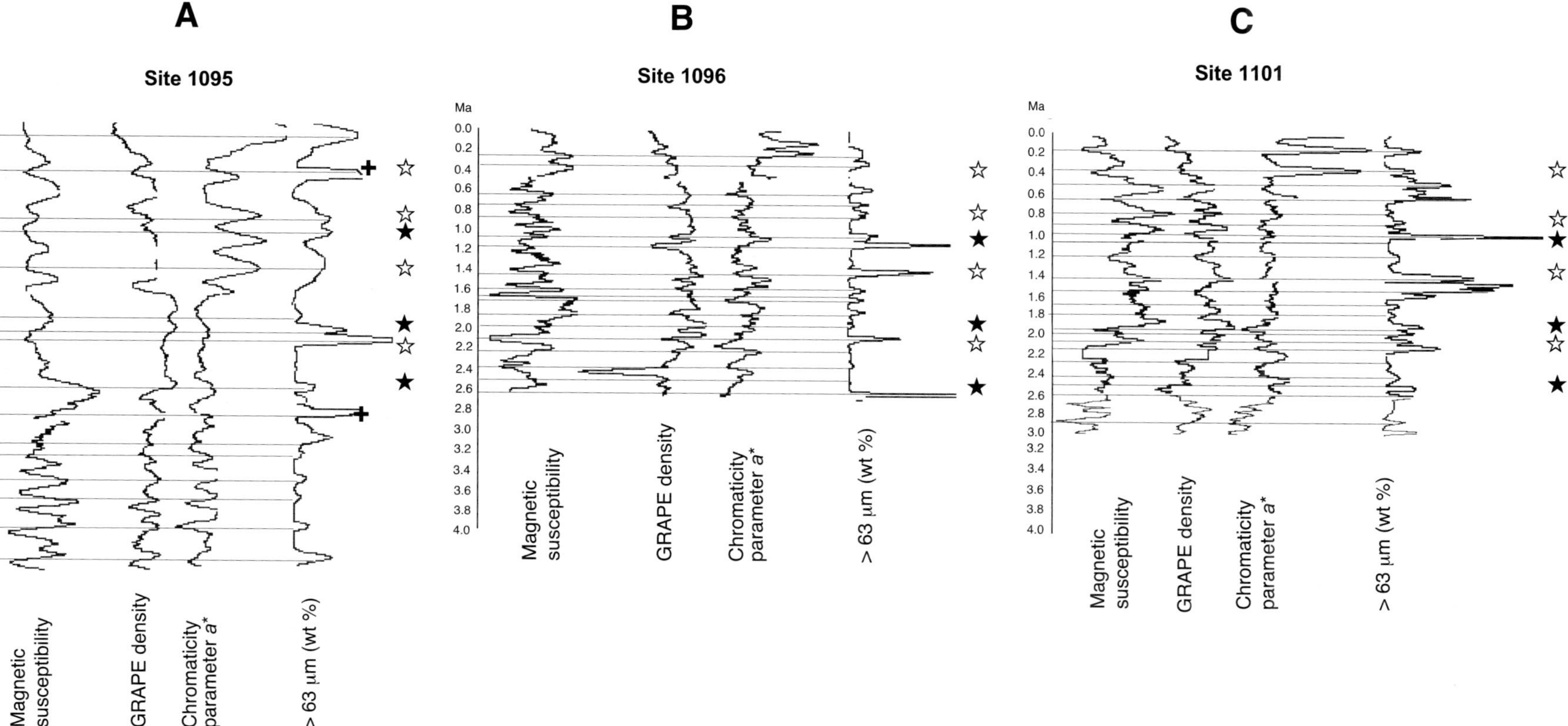

FIG. 4.—Smoothed curves of magnetic susceptibility, GRAPE (gamma ray attenuation porosity evaluator) density, chromaticity a^*, and coarse fraction (> 63 μm wt %). The horizontal black lines refer to the coeval variation of four parameters for Sites 1095, 1096, and 1101. Stars show synchronous changes occurring at all three sites. Black stars indicate time of particularly intense periods of ice rafting at Site 1101 (Cowan, 2001); the crosses at Site 1095 indicates higher values, omitted for scale clarity. See also caption of Fig. 2.

TABLE 4.—Timing, in Ma, of correlated events at all sites. Columns I, II, and III: correlated lithological variations in age for each site (see text and Fig. 4). Column IV: timing of averaged events corresponding in two and in three (marked in bold) sites. Column V: averaged time intervals of events occurring at two and three sites. Column VI: averaged time intervals of events occurring only at all three sites. (Aver = Average, Sd = Standard deviation).

	SITES				
1096	1101	1095			
I	II	III	IV	V	VI
		0.08			
0.24	0.19		0.21		
0.38	0.38	0.42	**0.39**	0.18	**0.39**
	0.5				
0.65	0.65		0.65	0.26	
0.77	0.8		0.78	0.13	
0.88	0.93	0.9	**0.9**	0.12	**0.51**
	0.96				
1.08	1.07	1.04	**1.07**	0.17	**0.17**
1.18	1.23		1.2	0.13	
1.42	1.45	1.4	**1.42**	0.22	**0.35**
1.6	1.55		1.57	0.15	
1.64					
1.7	1.72		1.71	0.14	
1.84	1.8		1.82	0.11	
	1.94	1.9	1.92	0.1	
1.99	2	2.04	**2.01**	0.09	**0.59**
	2.08				
2.11	2.16	2.13	**2.13**	0.12	**0.12**
2.28	2.3		2.29	0.16	
2.38	2.43		2.4	0.11	
2.5	2.51		2.5	0.1	
2.65	2.6	2.6	**2.61**	0.11	**0.48**
	2.9	2.9	2.9		
		3.18			
		3.32			
		3.55			
		3.73			
		4			
		4.31			
			Aver. ± Sd	0.14 ± 0.04	0.37 ± 0.17

continental rise show non-harmonic wavelength peaks, with probabilities > 90%. When the wavelength peaks are normalized, they show an extremely high correlation coefficient to each other and to *all* predicted orbital variations (Milankovitch cycles) for the same time interval, suggesting the influence of orbitally controlled insolation on the glacial signal recorded in the marine sediments.

In particular, looking at the highest-energy peaks, in the Early–Late Pliocene (Site 1095) and in the whole Quaternary intervals (Sites 1095, 1101, 1096) it appears that the short eccentricity (~ 95 ky and 123 ky) strongly influences the glacial signal, with no evidence of a switch at about 0.9 Ma, towards the 41 ky obliquity cycle (Pisias and Moore, 1981). Moreover, the ~ 400 ky long eccentricity emerges strongly from the Plio-Pleistocene sedimentary record of Sites 1095 and 1101, where the combined effect of both precession and obliquity (~ 64 ky) seems to characterize the Early–Late Pliocene and Early Pleistocene of Sites 1095, 1096, and 1101. Moreover the ~ 20 ky precession is evident in the Late Pliocene–Pleistocene of Sites 1096 and 1101.

As a consequence of the link between the orbital and sedimentary periodicities, an astronomical sedimentation rate was computed in the studied intervals and compared with that calculated from biostratigraphic–paleomagnetic analysis (Table 4) .

The constant presence of all orbital periods in the signal since the Early Pleistocene and the good correlation for the last 2.6 My among sites widely far apart allows us to extend the results at regional scale, confirming the hypothesis of the relative stability of the Antarctic ice sheet, at least since the Late Neogene (Barker et al., 1999a; Barker and Camerlenghi, 2002; Hillenbrand and Ehrmann, 2001) and suggesting that the ice sheet in the Early–Late Pliocene and in the Pleistocene could have been sensitive directly or indirectly to temperature changes.

Finally, an age correspondence emerges between the correlated events at 1.07, 2.01, and 2.61 Ma, corresponding to periods of particularly intense ice rafting (Cowan, 2001) and relative sea-level highs, suggesting a superimposed eustatic influence.

ACKNOWLEDGMENTS

The data used in the present paper were provided by the Ocean Drilling Program (ODP), sponsored by the U.S. National Science Foundation (NSF) and participating countries, under management of Joint Oceanographic Institutions (JOI). Funding support was also provided by the Italian Antarctic Research Program. We would like to thank B. D'Argenio, K. Strand, and an anonymous referee for their constructive comments.

REFERENCES

ACTON, D.G., GUYODO, Y., AND BRACHFELD, S.A., 2002, Magnetostratigraphy of sediment drift on the continental Rise of west Antarctica (ODP Leg 178, Sites 1095, 1096, and 1101), *in* Barker, P.F., Camerlenghi, A., Acton, G.D., and Ramsay, A.T.S., eds., Proceedings of the Ocean Drilling Program, Scientific Results, v. 178, p. 1–61. [Online] available from world wide web <http://www-odp.tamu.edu./publications/178_SR/Volume/Chapters/SR178_37.PDF >.

ANDERSON, J.B., BARTEK, R.L., THOMAS, M.A., AND ASHLEY, G.M., 1991, Seismic and sedimentologic record of glacial events on the Antarctic peninsula shelf, *in* Thomson, M.R.A., Crame, J.A., and Thomson, J.W., eds., Geological Evolution of Antarctica: Cambridge, U.K., Cambridge University Press, p. 678–691.

BARKER, P.F., 1992, The sedimentary record of Antarctic climate change: Royal Society (London), Philosophical Transactions, B, v. 338, p. 259–267.

BARKER, P.F., BARRET, P.J., COOPER, A.K., AND HUYBRECHTS, P., 1999a, Antarctic glacial history from numerical models and continental margin sediments: Palaeogeography, Palaeoclimatology, Palaeoecology, v. 150, p. 247–267.

BARKER, P.F., CAMERLENGHI, A., ACTON, G.D., ET AL., 1999b, Proceedings of the Ocean Drilling Program, Initial Reports, v. 178 [CD-ROM]. Available from: Ocean Drilling Program, Texas A&M University, College Station, TX 77845-9547, U.S.A.

BARKER, P.F., 2001, Data report: composite depths and spliced sections for Leg 178 Sites 1095 and 1096, Antarctic Peninsula Continental Rise, *in* Barker, P.F., Camerlenghi, A., Acton, G.D., and Ramsay, A.T.S., eds., Proceeding Ocean Drilling Program, Scientific Results, v. 178, p. 1–15. [Online] available from world wide web <http://www-odp.tamu.edu./publications/178_SR/Volume/Chapters/SR178_06.PDF >.

Barker, P.F., and Camerlenghi, A., 2002 Glacial history of the Antarctic Peninsula from Pacific margin sediments, *in* Barker, P. F., Camerlenghi, A., Acton, G. D., and Ramsay, A.T.S., Proceedings of the Ocean Drilling Program, Scientific Results, v. 178, p. 1–40 [Online]. available from world wide web <http: / / www-odp.tamu.edu. / publications / 178_SR / Volume / Synth / Synth.PDF >.

Berger, A.L., 1977, Support for the astronomical theory of climatic change: Nature, v. 269 p. 44–45.

Berger, A., and Loutre, M.F., 1992, Astronomical solutions for paleoclimate studies over the last 3 million years: Earth and Planetary Science Letters, v. 111, p. 369–382.

Berggren, W.A., Kent, D.V., Swisher, C., III, and Aubry, M.-P., 1995, A revised Cenozoic geochronology and chronostratigraphy, *in* Berggren, W.A., Kent, D.V., Aubry, M.-P, and Hardenbol, J., eds., Geochronology, Time Scales and Global Stratigraphic Correlation: SEPM, Special Publication 54, p. 129–212.

Brescia, M., D'Argenio, B., Ferreri, V., Longo, G., Pelosi, N., Rampone, S., and Tagliaferri, R., 1995, Neural net aided detection of astronomical periodicities in geologic records: Earth and Planetary Science Letters, v. 139, p. 33–45.

Cooper, A.K., Barret, P.J., Hinz, K., Traube, V., Leitchenkov, G., and Stagg, H.M.J, 1991, Cenozoic prograding sequences of the Antarctic continental margin: a record of glacio-eustatic and tectonic events: Marine Geology, v. 102, p. 175–213.

Cowan, E.A., 2001, Identification of the glacial signal from the Antarctica peninsula since 3.0 Ma at Site 1101 in a continental Rise sediment drift, *in* Barker, P.F., Camerlenghi, A., Acton, G.D., and Ramsay, A.T.S, eds., Ocean Drilling Program, Proceedings, Scientific Results, v. 178, p. 1–22. [Online]. available from world wide web <http: / / www-odp.tamu.edu. / publications / 178_SR / Volume / Chapters / SR178_10.PDF >.

D'Argenio, B., Fischer, A.G., Richter, G.M., Longo, G., Pelosi, N., Molisso, F., and Duarte Morais, M.L., 1998, Orbital cyclicity in the Eocene of Angola: visual and image-time-series analysis compared: Earth and Planetary Science Letters, v. 160, p. 147–161.

Denton, G.H., Sugden, D.E., Marchant, D.R., Hall, B.L., and Wilch, T.I., 1993, East Antarctic ice sheet sensitivity to Pliocene climate change from a dry Valleys perspective: Geografiska Annaler, v. 75A, p. 155–204.

Elkibbi, M., and Rial, J.A., 2001, An outsider's review of the astronomical theory of the climate: is the eccentricity-driven insolation the main driver of the ice ages?: Earth-Science Reviews, v. 56, p. 161–177.

Hagelberg, T.K., Pisias, N., and Elgar, S., 1991, Linear and nonlinear couplings between orbital forcing and the marine $\delta^{18}O$ record during the late Neogene: Paleoceanography, v. 6, p. 729–746.

Haq, B.U., Hardenbol, J., and Vail, P.R., 1987, Chronology of fluctuating sea level since Triassic (250 million years ago to present): Science, v. 235, p. 1156–1167.

Harwood, D.M., and Webb, P.-N., 1998, Glacial transport of diatoms in the Antarctic Sirius Group: Pliocene refrigerator: Geological Society of America, GSA Today, v. no. 8, p. 1–8.

Hillenbrand, C.-D. and Ehrmann, W., 2001, Distribution of clay minerals in drift sediments on the continental rise west of the Antarctic peninsula ODP leg 178, Sites 1095 and 1096, *in* Barker, P.F., Camerlenghi, A., Acton, G.D., and Ramsay, A.T.S. eds., Ocean Drilling Program, Proceedings, Scientific Results, v. 178, p. 1-29. [Online]. available from world wide web <http: / / www-odp.tamu.edu. / publications / 178_SR / Volume / Chapters / SR178_08.PDF >.

Hinnov, L.A., 2000, New perspectives on orbitally forced stratigraphy: Annual Review of Earth and Planetary Sciences, v. 28, p. 419–475.

Horne, J.H., and Baliunas, S. L., 1986, A prescription for period analysis of unevenly sampled time series: Astrophysical Journal, v. 302, p. 757–763.

Imbrie, J., Berger, A., Boyle, E.A., Clemens, S.C., Duffy, A., Howard, W.R., Kukla, G., Kutzbach, J., Martinson, D.G., McIntyre, A., Mix, A.C., Molfino, B., Morley, J.J., Peterson, L.C., Pisias, N.G., Prell, W.L., Raymo, M.E., Shackleton, N.J., and Toggweiler, J.R., 1993, On the structure and origin of major glaciation cycles: 2. The 100,000-year cycle: Paleoceanography, v. 8, p. 699–735.

Iorio, M., Tarling, D., D'Argenio, B., Nardi, G., and Hailwood, A.E., 1995, Milankovitch cyclicity of magnetic directions in Cretaceous shallow-water carbonate rocks, southern Italy: Bolletino di Geofisica Teorica ed Applicata, v. 37, p. 109–118.

Iorio, M., Tarling, D.H., and D'Argenio, B, 1998 Periodicity of magnetic intensities in magnetic anomaly profiles. The Cretaceous of the Central Pacific: Geophysical Journal International, v. 134, p. 13–24.

Iwai, M., Acton G.D., Lazarus D., Osterman L.E., and Williams, T., 2002, Magnetobiochronologic synthesis of ODP Leg 178 rise sediments from the Pacific sector of the Southern Ocean: Sites 1095, 1096, and 1101, *in* Barker P.F., Camerlenghi A., Acton G.D., and Ramsay A.T.S., eds., Proceedings of the Ocean Drilling Program, Scientific Results, v. 178, p. 1-40. [Online]. available from world wide web <http: / / www-odp.tamu.edu. / publications / 178_SR / Volume / Chapters / SR178_36.PDF >.

Larter, R.D., and Barker, P.F., 1989, Seismic stratigraphy of the Antarctic Peninsula Pacific margin: a record of Pliocene–Pleistocene ice volume and paleoclimate: Geology, v. 17, p. 731–734.

Lauer-Leredde, C., Briqueu, L., and Williams, T., 2002 A wavelet analysis of physical properties measured downhole and on core from Holes 1095B and 1096C (Antarctic Peninsula), *in* Barker, P.F., Camerlenghi, A., Acton, G.D., and Ramsay, A.T.S., eds., Proceedings of the Ocean Drilling Program, Scientific Results, v. 178, p. 1-43. [Online]. available from world wide web <http: / / www-odp.tamu.edu. / publications / 178_SR / Volume / Chapters / SR178_32.PDF >.

Longo, G., D'Argenio, B., Ferreri, V., and Iorio, M., 1994, Fourier evidence for high-frequency astronomical cycles recorded in Early Cretaceous carbonate platform strata, Monte Maggiore, Southern Apennines, Italy, *in* de Boer, P.L., and Smith, D.G., eds., Orbital Forcing and Cyclic Sequences: International Association of Sedimentologist, Special Publication 19, p. 77–85.

Maasch, K.A., and Saltzman, B., 1990, A low-order dynamical model of the global climatic variability over the full Pleistocene: Journal of Geophysical Research, v. 95, p. 1955–1963.

Miller, K.G., Fairbanks, R.G., and Mountain, G.S., 1987, Tertiary oxygen isotope synthesis, sea-level history and continental margin erosion: Paleoceanography, v. 2, p. 1–19.

O'Brien, P.E., Cooper, A.K. Richter, C., et al., 2001, Proceedings of the Ocean Drilling Program, Initial Reports, v. 188 [CD-ROM]. Available from: Ocean Drilling Program, Texas A&M University, College Station, TX 77845-9547, U.S.A.

Pisias, N.G., and Moore, T.C., Jr., 1981, The evolution of Pleistocene climate: a time series approach: Earth and Planetary Science Letters, v. 52, p. 450–458.

Pudsey, C.J., 2001, Neogene record of Antarctic peninsula glaciation in Continental Rise sediments: ODP Leg 178, Site 1095, *in* Barker, P.F., Camerlenghi, A., Acton, G.D., and Ramsay, A.T.S., eds., Proceedings of the Ocean Drilling Program, Scientific Results, v. 178, p. 1-25. [Online]. available from world wide web <http: / / www-odp.tamu.edu. / publications / 178_SR / Volume / Chapters / SR178_25.PDF >.

Raymo, M.E., Oppo, D.W., and Curry, W., 1997, The mid-Pleistocene climate transition: a deep sea carbon perspective: Paleoceanography, v. 12, p. 546–559.

Rebesco, M., Camerlenghi, A., and Zanolla, C., 1998, Bathymetry and morphogenesis of the continental margin west of the Antarctic Peninsula: Terra Antarctica, v. 5, p. 715–725.

Rial, J.A., 1999, Pacemaking the ice ages by frequency modulation of Earth's orbital eccentricity: Science, v. 285, p. 564–568.

Scargle, J.D., 1982, Studies in astronomical time series analysis. II. Statistical aspects of spectral analysis of unevenly spaced data: Astrophysical Journal, v. 263, p. 835–853.

SHACKLETON, N.J., 2000, The 100,000-year ice-age cycle identified and found to lag temperature, carbon dioxide, and orbital eccentricity: Science, v. 289, p. 1897–1902.

STROEVEN, A.P., BURCKLE, L.H., KLEMAN, J., AND PRENTICE, M.L., 1998, Atmospheric transport of diatoms in the Antarctic Sirius Group: Pliocene deep freeze: Geological Society of America, GSA Today, v. 8 (4), p. 1–5.

TAGLIAFERRI, R., PELOSI, N., CIARAMELLA, A., LONGO, G., MILANO, M., AND BARONE, F., 2001, Soft computing methodologies for spectral analysis in cyclostratigraphy: Computers & Geosciences, v. 27, p. 535–548.

TOMLINSON, J.S., PUDSEY, C.J., LIVERMORE, R.A., LARTER, R.D., AND BARKER, P.F., 1992, Long-range sidescan sonar (GLORIA) survey of the Antarctic Peninsula Pacific margin, *in* Yoshida, Y., Kaminuma, K., and Shiraishi, K., eds., Recent Progress in Antarctic Earth Science: Tokyo, Terra Scientific Publishing Company, p. 423–429.

WEBB, P.-N., AND HARWOOD, D.M., 1991, Late Cenozoic glacial history of the Ross Sea embayment, Antarctica: Quaternary Science Reviews, v. 10, p. 215–223.

WEBB, P.-N., HARWOOD, D.M., MABIN, M.C.G., AND MCKELVEY, B.C., 1996, A marine and terrestrial Sirius Group succession, middle Beardmore Glacier–Queen Alexandra Range, Transantarctic Mountains: Antarctica: Marine Micropalaeontology, v. 27, p. 273–298.

WOLF-WELLING, T.C.W., COWAN, E.A., DANIELS, J., EYLES, N., MALDONADO, A., AND PUDSEY, C.J., 2001a, Data Report: Diffuse spectral reflectance data from rise Sites 1095, 1096, and 1101 and Palmer Deep Sites 1098 and 1099 (Leg 178, Western Antarctic Peninsula), *in* Barker, P.F., Camerlenghi, A., Acton, G.D., and Ramsay, A.T.S., eds., Proceedings of the Ocean Drilling Program, Scientific Results, v. 178, p. 1–22 [Online]. available from world wide web <http:// www-odp.tamu.edu./publications/178_SR/Volume/Chapters/SR178_21.PDF >.

WOLF-WELLING, T.C.W., MOERZ, T., HILLENBRAND, C-D, PUDSEY, C.J., AND COWAN, E.A., 2001b, Data Report: Bulk sediment parameters ($CaCO_3$, TOC, and > 63 mm) of Sites 1095, 1096, and 1101 and Coarse Fraction analysis of Site 1095 (ODP Leg 178, Western Antarctic Peninsula), *in* Barker, P.F., Camerlenghi, A., Acton, G.D., and Ramsay, A.T.S., eds., Proceedings of the Ocean Drilling Program, Scientific Results, v. 178, p. 1–19, [Online]. available from world wide web <http:// www-odp.tamu.edu./publications/178_SR/Volume/Chapters/SR178_15.PDF >.

FACIES PATTERNS THAT DEFINE ORBITALLY FORCED THIRD-, FOURTH-, AND FIFTH-ORDER SEQUENCES OF SIXTH-ORDER CYCLES AND THEIR RELATIONSHIP TO OSTRACOD FAUNICYCLES: THE PURBECKIAN (BERRIASIAN) OF DORSET, ENGLAND

EDWIN J. ANDERSON
Department of Geology, Temple University, Philadelphia, Pennsylvania 19122, U.S.A.
e-mail: andy@temple.edu

ABSTRACT: The Purbeck Group (Berriasian) in Dorset, England, is divisible into meter-scale rock cycles that are in turn bundled into a multi-tiered hierarchy of cycle sets (sequences). These cycles and sequences are the product of "Croll–Milankovitch" orbital forcing, by which the smallest-scale cycles (sixth order) are the product of precession and three orders of cycle bundles (sequences) are produced by modulation of precession by variation in the degree of eccentricity of the Earth's orbit. The Purbeckian at the thickest section in Durlston Bay (over 100 m of marginal marine, brackish, and freshwater facies and paleosols) is divisible into four third-order (2 My) sequences that in turn comprise fourth-order (400 ky) and fifth-order (100 ky) sequences. All levels of this hierarchy are recognized by asymmetry in their patterns of facies distribution. Larger facies changes occur at the bases of cycles lower in sequences whereas smaller facies changes occur at cycle boundaries higher in sequences. In the Purbeck Group coarse-grained skeletal limestone is the dominant facies at the base of cycles and low in cyclic sequences, whereas fine carbonates, shale, and paleosols characterize the upper parts of cycles and sequences.

Clements (1969, 1993) described the biofacies and lithofacies of 245 beds that constitute the Purbeck Group at Durlston Bay. Morter (1984) described nine molluscan salinity (~ depth) zone assemblages and used them to interpret patterns of sea-level rise and fall through the Middle Purbeck stratigraphic interval. Anderson (1973) divided the Purbeck of southern England into four ostracod assemblage zones and later (1985) into 40 ostracod faunicycles. Each faunicycle consists of a lower more-marine **S-phase** and an upper more-freshwater **C-phase** assemblage of ostracods. These faunicycles are traceable laterally, and Anderson (1985) suggested that they might be the products of salinity fluctuations. Using Clements' (1993) beds as markers in association with the ostracod assemblage zones it is possible to integrate Anderson's ostracod faunicycles and Morter's salinity-based molluscan assemblages with the lithologically determined four-tiered hierarchy of allocycles developed in this paper. Anderson's faunicycles (1985) appear to coincide closely with fifth-order (100 ky) sequences, and the transgressive events interpreted by Morter (1984) occur at either the first or second fifth-order boundaries in fourth-order sequences. This independent paleobiological evidence corroborates the interpretation of a cyclic hierarchy determined by lithofacies analysis on the basis of assumptions of orbital forcing. In turn, recognition of the orbitally forced hierarchy of rock cycles provides an explanation for the published patterns of repeated (cyclic) faunal change.

INTRODUCTION

The "Croll–Milankovitch" orbital forcing hypothesis (Croll, 1875; Milankovitch, 1941) has been used to interpret the cyclic stratigraphy of Purbeckian facies in the French Jura (Strasser, 1994; Strasser and Hillgärtner, 1998) and at the Sierra del Pozo section in southern Spain (Jiménez de Cisneros and Vera, 1993; Anderson, 1999, 2000a). However, time-equivalent and similar facies at the type locality of the Purbeck in Dorset England have not previously been interpreted in terms of this mechanism (see Allen, 1998, p. 227). This is particularly surprising in that these rocks represent a classic set of localities that have been studied extensively. In this paper the hierarchic cyclic structure of one third-order (2 My) sequence at Durlston Bay is described in detail (the third of four third-order sequences that constitute the Purbeckian of Dorset).

In this study the term **cycle** (or sixth-order cycle) is used to describe the smallest-scale rock cycle in the Purbeck stratigraphic interval. This usage is the same as a "punctuated aggradational cycle" (PAC) defined by Goodwin and Anderson (1985), Anderson and Goodwin (1990), and an "elementary sequence" defined by Strasser (1994). A sixth-order cycle (PAC or elementary sequence) is defined as a meter-scale sedimentary rock package bounded by surfaces marked by abrupt change to deeper facies. These rock cycles typically shallow upward and may contain sea-level fall surfaces. These cycles are interpreted as the product of the 20 ky precessional signal (Anderson and Goodwin, 1992; Strasser, 1994). The term **sequence** is used in this paper to describe bundles or sets of sixth-order cycles. Periodic variation in eccentricity bundles sixth-order cycles into 100 ky sets (fifth-order sequences), 400 ky sets (fourth-order sequences) and 2 My sets (third-order sequences).

House (1993) has published an excellent summary of the geology of the Dorset coast in which all the key sections are located, described briefly, and set in a larger stratigraphic context. Two other notable publications include a detailed description of every bed at two key localities: Durlston Bay (Clements, 1969, 1993) and Worbarrow Tout (Ensom, 1985). Most significantly, Clements has described 245 beds in 105 m of section at Durlston Bay and summarized his results in a detailed stratigraphic log that can be compared directly to the logs depicting cyclic structure in this paper. Clements (1969, 1993) based his lithological description of these beds on field observations, and the relative abundance of biological components was determined from material picked from residues of disaggregated and dissolved samples. Ensom (1985), applying similar techniques, described 209 beds in 77.5 m of section at Worbarrow Tout. In the field analysis used to construct the cyclic stratigraphic logs that are the basis for this study, virtually all of Clements' beds were identified. Clements' bed numbers included on the new stratigraphic logs presented here provide reference points for locating all cycle and sequence boundaries in the cyclic hierarchy. In addition the lithological and

Cyclostratigraphy: Approaches and Case Histories
SEPM Special Publication No. 81, Copyright © 2004
SEPM (Society for Sedimentary Geology), ISBN 1-56576-108-1, p. 245–260.

biological descriptions (of Clements and Ensom) provide supporting data for the cyclostratigraphic analysis presented in this paper.

Paleobiological analyses by Anderson (1985) and Morter (1984) suggest frequent fluctuations in salinity (possibly related to sea-level fluctuations) during deposition of the Purbeckian sedimentary succession in southern England (Fig. 1). In particular, Anderson (1985) described 40 ostracod faunicycles in the Purbeck on the basis of data derived from cored boreholes in the Weald basin. Of these, 18–19 faunicycles were located in the Middle Purbeck (between 50 and 60 m thick in both the Weald basin and Durlston Bay). A faunicycle consists of alternating marine (**S-phase**) and freshwater (**C-phase**) ostracod associations (Anderson, 1985). A single faunicycle begins with an **S-phase** characterized by the dominance of non-cypridean ostracod genera that in turn is overlain by a nonmarine **C-phase** dominated by cypridean genera (see, e.g., Fig. 2). Anderson (1985) interpreted this rhythmic recurrence of marine and nonmarine ostracod associations as a response to basinwide salinity fluctuations. He located faunicycles with reference to ostracod assemblage zones and key beds in the Purbeck and suggests that they (faunicycles) provide a basis for refined stratigraphic correlation within the Purbeck Group.

On the basis of study of molluscan faunas also from cored boreholes through the Purbeck sequence in the Weald Basin, Morter (1984) recognized nine molluscan assemblages controlled by variation in salinity. These range from a nearly freshwater association (# 8, ***Viviparus–Unio***) to a euhaline association (# 1, containing a more cosmopolitan marine fauna including ***Trigonia***, ***Nanogyra***, and the echinoid ***Hemicidaris***). Morter (1984) used this evidence to interpret ten transgressive–regressive events in the Middle Purbeck stratigraphic interval and correlated these sea-level events to Anderson's ostracod faunicycles (see Fig. 1). It is the intention here to describe lithological variability that defines each order of the orbitally forced cyclic hierarchy in the Middle Purbeck section at Durlston Bay, Dorset, and then to use this cyclic structure to explain the molluscan and ostracod faunal patterns described by Morter (1984) and Anderson (1985).

Horne (1995) critically reviewed F.W. Anderson's analysis of ostracod assemblage zones and faunicycles. The assemblage zones (with minor modifications) have been widely used to correlate Purbeck sections in Britain (Horne, 1995; Westhead and Mather, 1996) and into other European basins (Allen and Wimbledon, 1991). Horne (1995) proposed a revised biozonation

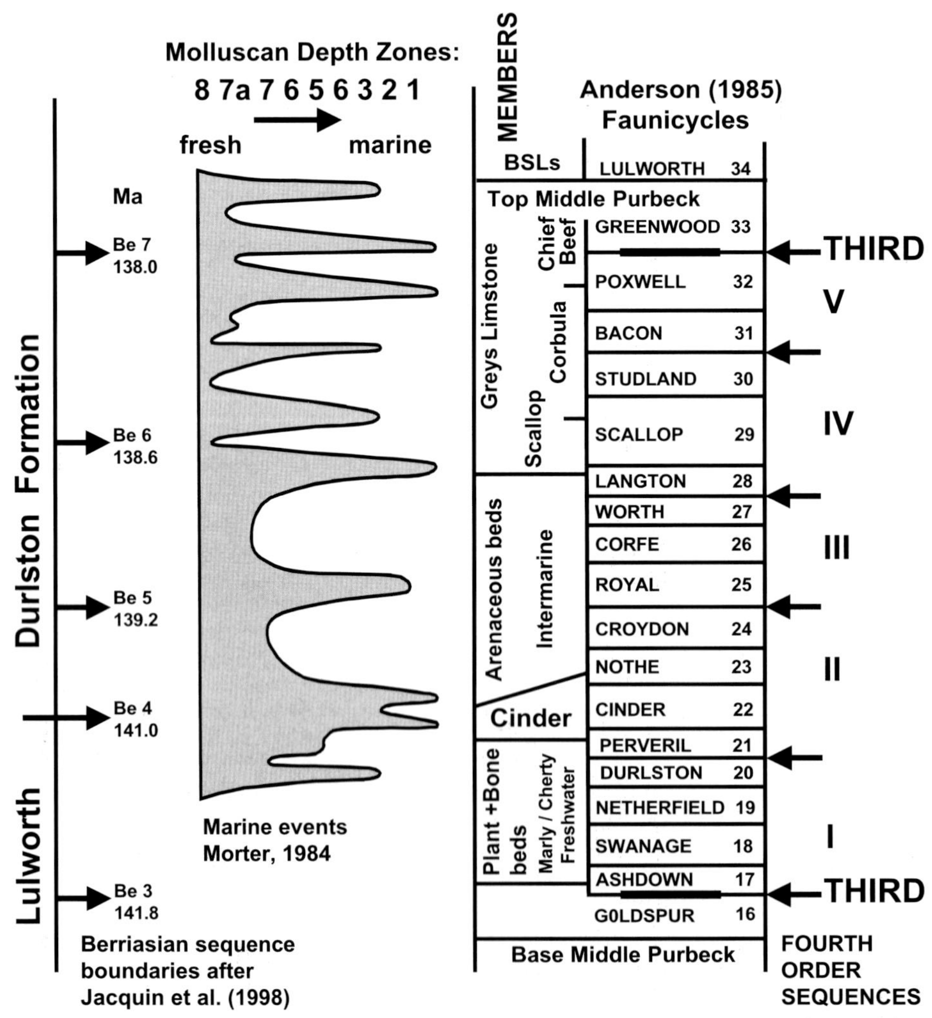

FIG. 1.—Correlation of transgressive events based on molluscan salinity-controlled associations (curve) and F.W. Anderson's ostracod faunicycles for the Middle Purbeck (adapted from Morter, 1984). The position of lithologically determined fourth-order sequence boundaries in the Middle Purbeck third-order sequence are shown with reference to members and faunicycles (arrows on right). Estimated position and ages of Berriasian sequence boundaries (arrows on left) from Jacquin et al. (1998).

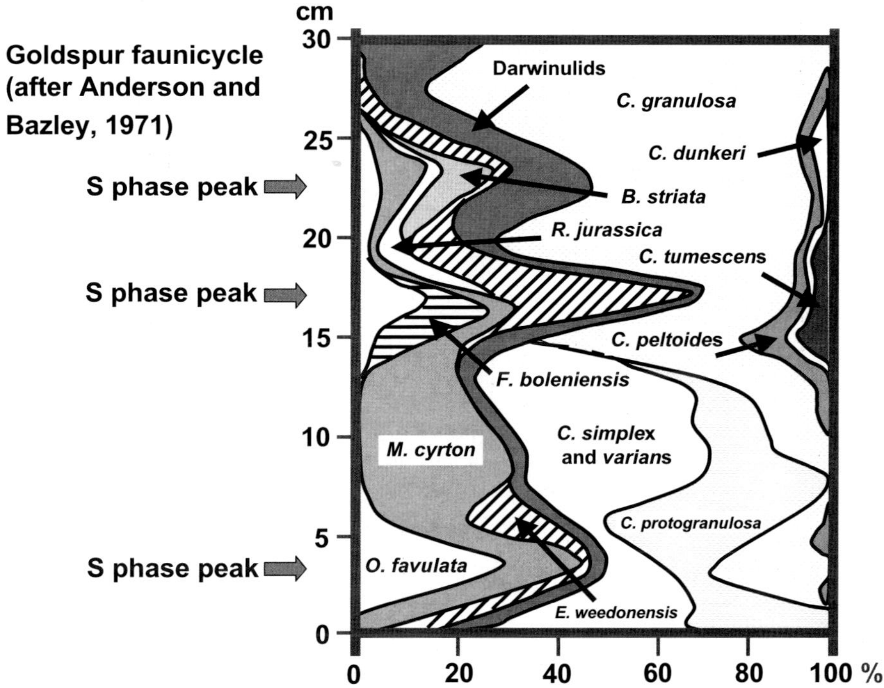

FIG. 2.—The distribution of non-cypridean, **S-phase**, and cypridean, **C-phase**, ostracod associations in the lower part of the Goldspur faunicycle (adapted from Anderson and Bazley, 1971). S-phase peaks are associated with sixth-order cycles in the lower part of a fifth-order sequence. The interpreted stratigraphic position of the Goldspur faunicycle is shown in Figure 10.

scheme and discussed shortcomings in the faunicycle concept. With regard to the latter, he points out that the Goldspur faunicycle contains three **S-phase** peaks (Fig. 2) and that it is unexplained why these strata represent a single faunicycle rather than three separate ones. Horne also pointed out that the evidence for true freshwater conditions is equivocal and that there are problems concerning the definition of faunicycle boundaries. Curiously, and without further comment, he quoted Anderson and Bazley (1971) as saying that the possibility that "Milankovitch cyclicity might have controlled the faunicycles deserves further consideration". In this paper this consideration is evaluated, and it is demonstrated that Anderson's and Bazley's suggestion has merit. In particular, it is demonstrated here that there is an underlying allogenic mechanism (i.e., orbital forcing) that accounts for ostracod faunicycles and patterns of change in molluscan assemblages in the Dorset Purbeck. Establishing this link between paleobiologic changes and periodic sea-level fluctuations provides a rational basis for using this paleoecologic evidence as a correlation tool.

Correlation of the Purbeckian strata within southern England (Boreal realm) and between England and central and southern Europe (Tethyan realm) has frequently been debated in the literature. The main problem is that key fossils for zonation of the Berriasian (i.e., Tethyan ammonites) are missing in the Boreal English sections and rare for paleoecological reasons in the Tethyan localities of the Jura and southern Spain (see Cope et al., 1980a, 1980b). In the sections in southern England ostracod zones have been established (and modified) by Anderson (1973, 1985), and Anderson and Bazley (1971), discussed by Morter (1984), and revised by Horne (1995). This evidence has been supplemented by palynological studies (see work on charophytes by Feist et al., 1995) and event stratigraphy (Wimbledon, 1987). Wimbledon recognized six lithologically determined transgressive events in the Purbeck Group and advised using these signals only in conjunction with biostratigraphical evidence; he showed the location of these events for the Isle of Purbeck associated with boreal ammonite zones (see figure 2 in Wimbledon, 1987). Wimbledon and Hunt (1983) and Allen and Wimbledon (1991) discussed issues of unconformity at the Portland Purbeck junction on the basis of the biostratigraphic distribution of dinoflagellates. They discussed problems in the correlation of Purbeckian sections between basins in southern England and from there to other European basins. They also assessed the use of ostracods, palynomorphs, and event stratigraphy for interbasinal correlation of nonmarine Purbeckian facies. Finally Allen (1998) presented detailed evidence for paleoclimatic fluctuations throughout deposition of the Dorset Purbeck and discussed these changes in relation to the diversity of fauna and flora associated with Purbeckian deposits in southern England. Allen (1998) concluded that Purbeck–Wealden climates were highly variable through time with a long-term trend from hot, dry conditions (with short wet spells) to hot but more humid conditions with increasing rainfall. However, many reversals of these climatic conditions occurred at widely ranging timescales. Allen also noted that except for ostracod faunicycles (of uncertain climatic significance) no regular stratigraphic periodicities are known which might be related to orbitally forced Milankovitch cycles like those reported from the Berriasian of France and Spain.

The Purbeckian stratigraphic section at Durlston Bay Dorset is correlative with the Berriasian and lowermost Valanginian stages (Jacquin et al., 1998). Sequence stratigraphic analysis of the Berriasian stage (Hardenbol et al., 1998; Gradstein et al., 1995) indicates that its duration is 7.2 My (144.2 ± 2.6 to 137.0 ± 2.2 Ma). This stratigraphic interval (the Berriasian stage) contains eight "third-order" sequence boundaries, Be 1 to Be 8 (Hardenbol et al., 1998). Whereas Allen and Wimbledon (1991) showed the entire type section of the Purbeck in Dorset to be equivalent to just the Berriasian stage (i.e., 7.2 My), Jacquin et al., (1998) revised this correlation, extending the top of the Purbeck Group to the first sequence boundary in the Valanginian (Va 1 at 136.5 Ma). In this interpretation the Purbeck Group in Dorset would represent 7.7 My with uncertainties of approximately ± 2 My on the boundary dates. In the French Jura this same interval comprises the Tidalites-de-Vouglans, Goldberg, Pierre-Châtel, Vions, and lower part of the Chambotte Formations (Strasser and Hillgärtner, 1998).

The stratigraphic sequence analyzed in this paper is approximately equivalent to the Middle Purbeck in Dorset (Allen and Wimbledon, 1991; Clements, 1993) and equivalent to the Pierre-Châtel and Vions Formations in the Jura, that is, the stratigraphic interval between sequence boundaries Be 4 and Be 7 (Strasser and Hillgärtner, 1998). Note that although the estimate of 7.7 My for the Dorset Purbeck Group by Jacquin et al. (1998) is consistent with an interpretation in this paper of four 2 My sequences, the estimated dates for sequence boundaries within the Berriasian stage are inconsistent with the cyclostratigraphic interpretation (in this paper) that divides the Purbeck Group of Dorset into four sequences of equal duration (i.e., 2 My each). However, Hardenbol et al. (1998) and Jacquin et al. (1998) give uncertainties only for the Berriasian boundary dates, not for its internal sequence boundaries. The ages of the sequence boundaries internal to the Berriasian stage are estimated by correlating these sequence boundaries to the estimated ages of magnetic polarity zones (Gradstein et al., 1995). Nevertheless the estimated positions and absolute ages of Berriasian sequence boundaries Be 3 to Be 7 (from Jacquin et al., 1998) are shown for reference in Figure 1.

Ogg et al. (1994) presented magnetostratigraphic data for the Purbeck Group at Durlston Bay, Dorset. They concluded that "the terrestrial to marginal-marine Purbeck Beds encompass polarity chrons M19r through M14r, implying deposition from latest Tithonian (Late Portlandian) to the beginning of the Valanginian stage. The Cinder Bed transgression of the Middle Purbeck Beds occurred at the beginning of chron M17n, corresponding to the *Berriasella privasensis* ammonite Subzone in the Tethyan realm." These observations permit correlation to the global magnetic time scale and biozones presented in Gradstein et al., (1995) and Hardenbol et al., (1998). This data in turn helps establish the time scale set out in Jacquin et al., (1998) for the Dorset Purbeck summarized in Fig. 1 and supports the conclusion that the Dorset Purbeck Beds comprise a stratigraphic record of approximately 8 Ma.

ASSUMPTIONS IN THE ANALYSIS OF CYCLES

Deductive Methodology

In this study the first and most fundamental assumption is adoption of a deductive methodology. Specifically the hypothesis is adopted (and tested by ongoing application) that "Croll–Milankovitch" type orbital-forcing processes impose a hierarchic, allocyclic structure on the stratigraphic record. This hypothesis is not derived from the Dorset Purbeck but rather from 25 years of experience in measuring and interpreting cyclic patterns, primarily in Silurian and Devonian strata in the Appalachian Basin. The model has subsequently been tested and confirmed by analysis of cyclic structure in the Carboniferous of Wales (Anderson and Goodwin, 1990), evaluation of the cyclic structure of the Sierra del Pozo section in southern Spain (Jiménez de Cisneros and Vera, 1993; Anderson, 1999, 2000a), and field checking the cyclic structure of Berriasian strata in the French Jura presented by Strasser (1994) and Strasser and Hillgärtner (1998).

The hypothesis is sustained and reinforced to the degree that strata of the Dorset Purbeck are consistent with the model. The hypothesis is to be adjusted, revised, or even rejected if the observations on newly analyzed sections (like the Dorset Purbeck) are in significant and inexplicable conflict with the model. This method is the same as that expressed in Goodwin and Anderson (1985, p. 515), who state "Our deductive approach is consistent with the hypothetico-deductive interpretation of the scientific method (Popper, 1959) and with the views of Kuhn (1962) that science advances principally through generation and testing of paradigms. In agreement with Medawar (1964), Eldredge and Gould (1972), and Laudan (1977) we think that ideas precede detailed observations so that all observations are to a large degree made under the influence of some previously formulated hypothesis."

Precession and Eccentricity

Beginning with the orbital-forcing variables summarized by Berger (1988), Fischer and Bottjer (1991), de Boer and Smith (1994), and House and Gale (1995), in this study it is assumed that the fundamental process in producing a cyclic stratigraphic record is the precessional signal (Anderson and Goodwin, 1990, 1992). The strength of the precessional signal is directly related to the degree of eccentricity of the Earth's orbit (Fig. 3). As eccentricity increases and summers occur near perihelion (in the precessional cycle), insolation at high latitudes reaches a maximum. If those summers are in a hemisphere with large areas of continent at high latitudes, then high insolation values may trigger the melting of significant amounts of accumulated continental glacial ice. When summers occur near aphelion with high eccentricity, the resulting long series of cool summers may trigger renewed buildup of high-latitude, continental glacial ice. While it has been suggested by many that the Mesozoic was a "greenhouse" world with no significant continental ice, recent work by Price (1999) provided new evidence of high-latitude continental glaciation at several times in the Mesozoic (including the Berriasian and Valanginian). Price suggested that the volume of continental glacial ice might have been as much as 30% of the volume estimated for the Pleistocene. If this evidence is accepted, it removes the "ice-free world" objection, and changes in sea level can be linked directly to fluctuations in global ice volume in the Early Cretaceous. Note that Read et al. (1986) stated that even "low amplitude sea-level oscillation(s) of a few meters" generate tidal-flat cycles in their modeling experiments. Thus variations in ice volumes only 5–10% of that demonstrated for the Pleistocene could be expected to be recorded in the stratigraphic record. Grotzinger (1986) discussed the role of low-amplitude (orbitally forced) sea-level fluctuations in the production of small-scale carbonate cycles and linked the origin of such cycles to decreased surface area of high-latitude land masses on which continental glaciation might develop.

If fluctuation in sea level is primarily a response to precession of the position of the summer solstice in the Earth's orbit, then the magnitude of sea-level rise and fall would be proportional to the degree of eccentricity of the orbit (Fig. 3). When eccentricity is high there would be large changes in high-latitude summer insolation between perihelion and aphelion and larger sea-level fluctuations. When the Earth's orbit approaches circular the magnitude of change in insolation through a precessional cycle would be minimized, resulting in more stable sea levels. The degree of eccentricity of the Earth's orbit varies periodically. It varies from nearly circular to as much as 5% eccentric and back to circular with a period of about 100 ky (see Berger, 1988, p. 634, fig. 9). Croll (1875) recognized the significance of this phenomenon and showed that at a maximum eccentricity (7.7%, later revised down) the difference in distance between the Earth and the Sun in the aphelion versus the perihelion positions would be some 22 million km. This number is closer to 14 million km at 5% eccentricity and 5.5 million km at the Earth's current eccentricity (~ 1.7%). In addition, the 100 ky eccentricity maxima in the Earth's orbit are more enhanced at 400 ky intervals and possibly 2 My intervals (Berger, 1988; Fischer, 1991). These periodic patterns in degree of eccentricity of the Earth's orbit modulate the strength of the precessional signal (and potentially changes in ice volume and associated sea-level fluctuations), thus producing 100 ky and 400 ky bundles of rock cycles, i.e., fifth-order and fourth-order sequences (see Fig. 3).

Obliquity

The role of obliquity (the 41 ky obliquity cycle) in the assumed orbital-forcing model is limited to modulating the magnitude of insolation changes through the precessional cycle. For example, if a maximum precessional sea-level rise is in phase with maximum tilt, the magnitude of this rise would be amplified, but 20 ky later the next precessional rise would occur at minimum tilt and be muted. This could lead to a cyclic stratigraphic record in which the first precessional rise (in a 100 ky sequence of rises) produced a marked facies change at the associated precessional (sixth-order) cycle boundary but where the next sixth-order boundary might be difficult to detect.

The Origin of Cycle Boundaries

A final working hypothesis is that all rock-cycle boundaries are surfaces produced by rates of (precessional) sea-level rise that exceed a critical value (Figs. 4, 5; Anderson, 2000b). At rates above the critical value of sea-level change, sedimentation is thought to effectively stop (see Tipper, 1997, 2000). Tipper (1997) predicted that the critical variable for the production of lag is colonization rate and that it produces the lag phenomenon during both sea-level rise and sea-level fall. In this manner, higher-amplitude sea-level rises produce discontinuities between lowstand and highstand facies at rock-cycle boundaries representing longer time intervals and displaying larger facies contrasts across boundary surfaces (Fig. 4, outer box). During lower-amplitude sea-level rises, the time interval of discontinuity would be shorter and the facies change across the cycle boundary (= depth difference) would be less, as indicated by the inner box on Figure 4. Notice that this concept of a mechanism for producing cycle boundaries is distinct from modeling methods that artificially generate lag time or lag depth (e.g., Read et al., 1986), and (the concept) explains the genesis of cycle boundaries in totally subtidal deposits. The critical value for rate of sea-level change may also be exceeded during (precessional) sea-level falls (also predicted by Tipper, 1997), producing sea-level-fall surfaces between highstand and lowstand facies within sixth-order cycles (Fig. 5). The computer-modeling-based concepts of lag time and lag depth do not explain sea-level-fall surfaces. Sea-level-rise surfaces are more marked and better preserved stratigraphically than sea-level-fall surfaces because rates of relative sea-level rise are amplified by subsidence whereas rates of relative sea-level fall are diminished.

In summary, in the assumed model, all orbitally forced sea-level rises, and therefore all surfaces at allocycle boundaries, are the product of the precessional signal (Anderson and Goodwin, 1992; Goodwin and Anderson 1997). The term **PAC** (punctuated aggradational cycle) is exclusively equated to precessionally forced sixth-order rock cycles (Goodwin and Anderson, 1985, 1988, 1997; Goodwin et al., 1986). These rock cycles appear to be equivalent to Strasser's (1994) elementary sequences. Variation in the degree of eccentricity bundles these sixth-order cycles (into fifth-order and fourth-order sequences or cycle sets) by periodically varying the magnitude of the precessional signal (Fig. 3).

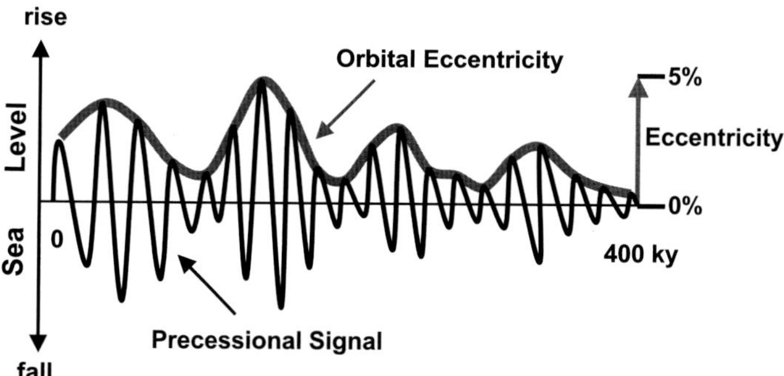

FIG. 3.—Model for control of sea level by the magnitude of the precessional signal where the degree of eccentricity of the Earth's orbit determines the intensity of high-latitude summer insolation producing fluctuation in global ice volume between perihelion and aphelion summer positions.

Genetic Hierarchy and Stacking Pattern

The hierarchy of cycles and sequences applied in this study is genetic (see Busch and West, 1987; D' Argenio et al., 1997) in the sense that each rank in the hierarchy is related to a specific stratigraphic process (Fig. 6). In particular, as described above, fourth-, fifth-, and sixth-order sequences (or cycles) are each related to a particular orbital-forcing process involving the interplay of precession and eccentricity (Goodwin et al., 1992). In this scheme the terms second-order and third-order sequences are adopted and modified from the usage of seismic and sequence stratigraphy in which the fundamental unit of sequence stratigraphy is the "third-order sequence" (a genetically related set of strata bounded by unconformities and their correlative conformities; Van Wagoner et al., 1988).

Although "third-order sequences" are not defined in terms of duration (Van Wagoner et al., 1988), in practice third-order sequences commonly are 1–2 My in duration and second-order (b) supersequences, tend toward 10 My in duration. For this reason the third-order rank in the genetic cyclostratigraphic hierarchy defined here is assigned to the 2 My sequence. Note that when sequences of less than two million years duration are identified by the practitioners of the sequence stratigraphic model (and called third-order), the boundaries of such shorter-duration sequences could have been placed at prominent 400 ky sequence boundaries within a 2 My sequence. However, the sequence-stratigraphic model does not recognize an orbitally forced hierarchic structure internal to third-order sequences. Furthermore, this structure (orbitally forced fourth-, fifth-, and sixth-order ranks used in this study) is conceptually distinct from both parasequences and parasequence sets and thus from the hierarchy of stratal units employed in the sequence-stratigraphy model (Van Wagoner et al., 1988; Kamola and Van Wagoner, 1995). The genetically specific ranks used in this paper are also distinct from the "orbitally related" hierarchy of orders used by Goldhammer et al. (1990). Goldhammer et al. used time ranges in their definition of ranks in a hierarchy (i.e., fifth order is 10–100 ky and fourth order is 100–1000 ky). This kind of definition of ranks precludes a specific genetic connection to each rank in their hierarchy.

If complete preservation of the orbitally forced hierarchy occurs, the predicted stacking pattern consists of five sixth-order

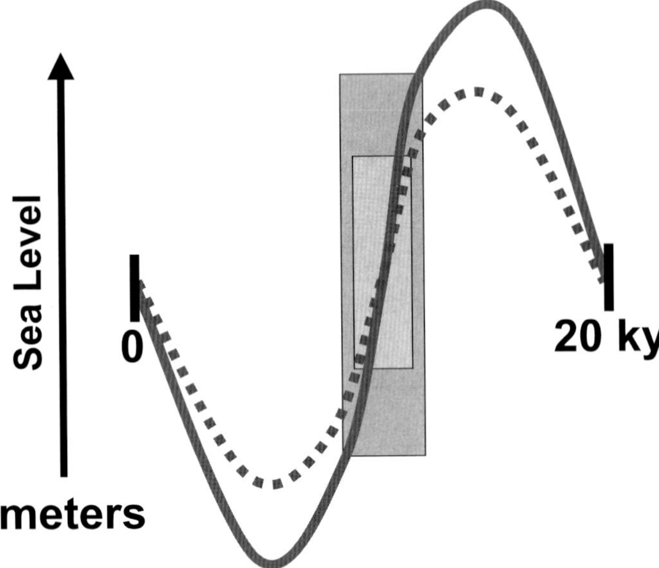

FIG. 4.—Model illustrating the relationship between the magnitude of precessionally forced sea-level rise and degree of disjunctness at sixth-order cycle (PAC) boundaries. Sedimentation ceases at some critical rate of sea-level rise. With higher-magnitude rises the critical rate is reached for a longer period of time and a greater facies change (difference in depth) occurs across the boundary (outer box). With lower-magnitude sea-level rises the cycle boundary surface represents less time and a smaller facies change (difference in depth) produced by the interval of nondeposition (inner box).

cycles in each fifth-order sequence. At the next level, four fifth-order sequences would occur in each fourth-order sequence, and at a larger scale, five fourth-order sequences are predicted in a third-order sequence (Fig. 7). The largest facies changes occur at the bases of sixth-order rock cycles and (by arbitrary definition) in the lower part of the second cycle or sequence in each of the larger-scale bundles in the hierarchy. Experience demonstrates that the basic cycle and the bundled sets of these cycles (sequences) are asymmetric in facies distribution, and it is then most parsimonious to have the bundled sets reach a shallowest point (or a minimal facies contrast across cycle boundaries) at or near the tops of cycles and larger-scale sequences.

THE CYCLIC STRUCTURE OF THE DORSET PURBECK

The Time Constraints

To apply the orbital-forcing model to the Dorset Purbeck requires that an assessment of the time interval represented by the Purbeckian Group in Dorset be consistent with the interpreted cyclic structure. The correlative strata (to the Dorset section) in the nearly complete stratigraphic section at Mt Salève in the French Jura encompasses the *Jacobi* to *Otopeta* ammonite zones and calpionellid zones B to E (Strasser and Hillgärtner, 1998). This correlation between Dorset and the Jura is on the basis of Allen and Wimbledon (1991), Feist et al. (1995), Jacquin et al. (1998), and measurement of the cyclic structure (cyclostratigraphy) in both regions by the writer. This 7.7 My stratigraphic interval (136.5 to 144.2 Ma), includes the entire Berriasian and the lower Valanginian from sequence boundaries Be 1 to Va 1 (Strasser and Hillgärtner, 1998; Gradstein et al., 1994, 1995; Hardenbol et al., 1998; Jacquin et al., 1998).

There is thus time for four third-order (2 My) sequences in the Purbeck Group of Dorset. The first third-order sequence begins at the Portland–Purbeck boundary, includes the fossil forest, the Great Dirt Bed, and other paleosols (Francis, 1986), the "Broken Beds", the "Cypris" Freestone and Hard Cockle members, and ends with the overlying massive gypsum beds. The second third-order sequence lies between the top of the massive gypsum beds and the "Mammal Bed" unconformity and includes the Soft Cockle member and the lower half of the Marly Freshwater Beds, ending at the top of Clements' bed 81. The third third-order sequence at Durlston Bay (to be described in detail in this paper) is the best exposed, thickest, and most complete. This sequence begins at the "Mammal Bed" unconformity and ends at the top of Clements' bed 199 in the middle of the Chief Beef member of the Durlston Formation (i.e., it is approximately equivalent to the Middle Purbeck). The top third-order sequence in the Purbeck at Durlston Bay includes the upper half of the Chief Beef, the Broken Shell Limestone, the "Unio" and Upper "Cypris" Clays and Shales members (member designations of Clements, 1993) and extends into facies of the lower Wealden (possibly to the early Valanginian transgressive event interpreted at the base of the Wadhurst Clay; Allen, 1998).

Facies Change and Sea-Level Fluctuation

The orbital-forcing mechanism is thought to cause both climate change and sea-level fluctuation. Allen (1998) presented extensive evidence of climate change during the deposition of the Purbeck Group in southern England. However, evidence of frequent and repeated development of paleosols at cycle and sequence boundaries and other emersion features such as karst development require that sea-level fluctuations (including sea-

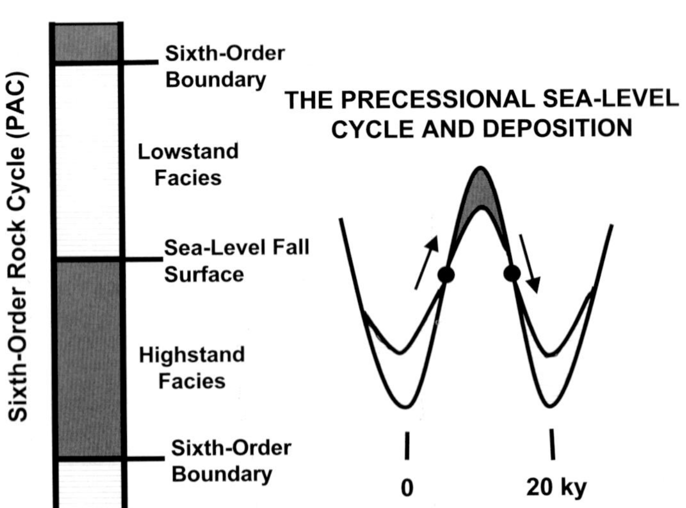

FIG. 5.—The boundaries of sixth-order rock cycles (PACs) are surfaces across which facies change from shallower to deeper (column). These surfaces are associated with intervals of maximum (precessionally forced) rates of sea-level rise, i.e., near the inflection point of the sea-level-rise curve (see text for explanation). A fully preserved PAC is represented by a highstand and lowstand facies between two sea-level-rise boundary surfaces. A sea-level-fall surface may be preserved internally (from Goodwin and Anderson, 1997).

Second-Order	**10 My**	**Tectonoeustasy**
Third-Order	**2 My**	**Eccentricity ?**
Fourth-Order	**400 ky**	**Long Eccentricity**
Fifth-Order	**100 ky**	**Short Eccentricity**
Sixth-Order	**20 ky**	**Precession**

FIG. 6.—The hierarchy of allocyclic orders applied in this paper (from Goodwin and Anderson, 1997). Each order is defined in terms of a specific orbital forcing process; therefore, each rank is genetic.

level falls) play a significant role in the development of the observed cyclic structure in the Dorset Purbeck (see argument by Strasser and Hillgärtner, 1998). It is thought that sea-level rises result in the trapping of clays in onshore nonmarine coastal-plain paleoenvironments. This effect would then result in contemporaneous carbonate deposition (at highstand) in marginal marine and shoreline paleoenvironments. Subsequent return to equilibrium between the coastal-plain aggradation and base-level and later sea-level fall would force a return to input of terrigenous clastic (clays and sometimes quartz sand) to marginal marine areas. Strasser and Hillgärtner (1998) make a similar argument for recurrent facies patterns in the Berriasian of the French Jura. This process is thought to vary as a function of the magnitude of precessionally forced sea-level rises an falls and, as a consequence, determines the asymmetric facies patterns observed at each scale (rank) of the cyclic hierarchy (see Figs. 10–21).

The final line of evidence supporting an interpretation of orbital forcing is the record in the Dorset Purbeck of a four-tiered hierarchy itself. At Durlston Bay the studied third-order sequence comprises five fourth-order sequences and is recognized by a distinctive facies asymmetry (see Fig. 7). A similar facies asymmetry characterizes each order in the hierarchy of sequences (third, fourth, and fifth orders). In particular, the facies in lower cycles or sequences in a bundled set are enriched in shelly carbonates (limestone) whereas cycles higher in bundled sets are enriched in shale and commonly are topped with paleosols (see Anderson et al., 2001; Stynchula and Anderson, 2001). For example, the second fourth-order sequence in a third-order bundle is more enriched in limestone than subsequent sequences. Likewise the second fifth-order sequence in a fourth-order bundle tends to be the most limestone-enriched sequence in a fourth-order set. In fifth-order (100 ky) sequences the most limestone-rich sixth-order cycle in the set occurs in the first or second position. Finally, at the smallest scale, individual sixth-order cycles (**PACs**) in the Purbeckian typically contain limestone at the base overlain by more marly or shale-rich facies at their tops. PAC boundaries are seen as surfaces marked by abrupt change from marl, shale, or soil to more carbonate-rich (typically calcarenite) facies.

The third (i.e., Middle Purbeck), third-order sequence is 45 m thick and is well exposed in continuous outcrop in Durlston Bay (Fig. 8) just south of Peveril Point (see the same stratigraphic interval in the log of El-Shahat and West, 1983). At this excellent exposure the typical facies asymmetry that defines sequences at each tier of the cyclic hierarchy is well expressed. In this paper a separate stratigraphic log is presented to depict each fourth-order sequence in the Middle Purbeck third-order sequence (e.g., Sequence I, Figs. 9, 10). On these logs fourth-order sequences are designated by Roman numerals I–V (see Figs. 1 and 7) and named after key beds that occur within them. In turn, fifth-order sequences are identified on the logs by the capital letters A, B, C, or D, and bold arrows mark their boundaries. PAC (or sixth-order) boundaries are indicated by tick marks (note that all third-, fourth-, and fifth-order boundaries are also sixth-order boundaries).

The "Mammal Bed" Fourth-Order Sequence

At Durlston Bay, the first fourth-order sequence (Sequence I in the Middle Purbeck third-order sequence) is named from the "Mammal Bed" that occurs in the lowstand phase of the first sixth-order cycle (Clements' bed 83, at 2.9 m on the log; see Figs. 9, 10). This fourth-order sequence begins above a meter-thick paleosol characterized by well-developed peds (Clements' bed 81; see Fig. 9) and is 5.7 m thick (Fig. 10). It comprises three moderately complete 100 ky sequences (A, B, and C) and a highly truncated D sequence. The first three 100 ky sequences (A, B, and C) all show the typical fining-up facies asymmetry and each contains four or five PACs. Each begins with a carbonate bed, sequences A and C with marly limestone (Figs. 11, 13) and sequence B with a coarse calcarenite bed (Clements' bed 87; Fig. 12). A well-developed paleosol is found at the top of sequence B; it contains extensively preserved root structures (i.e., the "fern bed" at 6.3 m; Clements' bed 93; Fig. 10). The facies at the top of sequence C, although not shale-rich or a paleosol, is a distinctive micritic, chert-bearing, freshwater limestone (Figs. 13, 14). Most of the well-preserved charophyte material from the Purbeck comes from this horizon (Feist et al., 1995). The C–D fifth-order sequence boundary has a half-meter of erosional relief (Fig. 14). Note that the overlying D sequence is very incomplete as a consequence of exposure and a period of erosion prior to the deposition of D and due to exposure and nondeposition at the fourth-order boundary above. A well developed paleosol 40–50 cm thick lying on an erosional surface with 10 cm of relief represents the truncated D sequence (depicted in Figures 10 and 14), less than 1 km away in the southern part of Durlston Bay (just south of the Zig-Zag path).

It is possible to correlate F.W. Anderson's ostracod faunicycles with the internal cyclic structure of this fourth-order sequence.

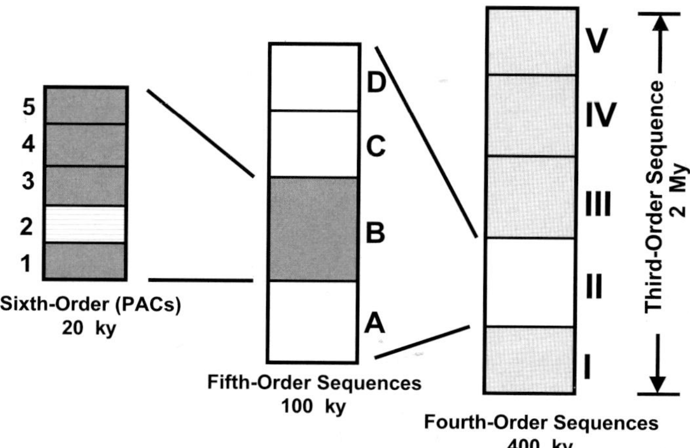

FIG. 7.—Predicted stacking patterns of cycles and sequences produced by orbital forcing assuming complete preservation of the stratigraphic record. Roman numerals designate fourth-order sequences and capital letters fifth-order sequences. The largest facies changes and/or deepest facies typically are associated with the second cycle or sequence in each set (i.e., **2**, **B**, or **II**). This asymmetry defines fifth-, fourth-, and third-order sequences, respectively.

Anderson (1985) showed six faunicycles in the Marly and Cherty Freshwater members (Fig. 1). The first, the Goldspur faunicycle, occurs below the "Mammal Bed" fourth-order sequence, and the sixth faunicycle, the Peveril, is coincident with the A fifth-order sequence in the next fourth-order sequence just under the Cinder member (Figs. 10, 15, 16). This leaves four faunicycles in the interval of the "Mammal Bed" fourth-order sequence (Ashdown, Swanage, Netherfield, and Durlston). The most reasonable hypothesis is that these four faunicycles are coincident with 100 ky sequences A, B, C, and D, which comprise the "Mammal Bed" fourth-order sequence. At this locality the Durlston faunicycle is not well represented because the D fifth-order sequence is truncated, as explained above.

Horne (1995) depicted the Goldspur faunicycle (Fig. 2) and showed that it contains three **S-phase** peaks within 30 cm at the bottom of the cycle; as mentioned earlier, he cited this as a problem with the faunicycle concept. The Goldspur faunicycle corresponds to the fifth-order sequence just below the "Mammal Bed" fourth-order sequence. The problem of three **S-phase** peaks is explained when one sees that these peaks (suggesting marine incursions) match three carbonate beds (Clements' beds 76, 78, and 80) that define the basal parts of three PACs in the lower part of a fifth-order sequence (Fig. 10).

On the basis of molluscan assemblages (Fig. 1; Morter, 1984) interpreted three transgressive events in the stratigraphic interval depicted in Figure 10. The first of these events is at the base of the Middle Purbeck (base of the Goldspur faunicycle), the second at the level of the Durlston–Peveril faunicycles, and the third at the base of the Cinder member. The positions of these events are shown in Figure 10 by gray block arrows. Wimbledon (1987) recognized five transgressive events in the Middle Purbeck; the first, his event 7, is placed at the base of Clements' bed 88 near the base of fifth-order sequence B, and the next, his event 8, is at the base of the Cinder member. In these ways it can be seen that paleoecological and lithostratigraphical evidence developed by others is consistent with the new interpretation of cyclic structure developed herein.

The "Cinder Bed" Fourth-Order Sequence

The second fourth-order sequence (Sequence II of the Middle Purbeck) is named for the very prominent Cinder member of the Durlston Formation. The "Cinder Bed" fourth-order sequence (Fig. 16) is 11.3 m thick and is the most carbonate-rich fourth-order in the Middle Purbeck third-order sequence. The entire fourth-order sequence shows marked asymmetry with the most marine and carbonate-rich fifth-order sequence, the Cinder member itself, in the second or B sequence position (Figs. 15, 16). The Cinder member is characterized by a nearly normal-marine biofacies association, including *Praeexogyra, Protocardia, Modiolus,* and *Neomiodion* (Clements, 1993; Morter, 1984). Sequence A contains massive calcarenite (the "New Vein") in the lower two sixth-

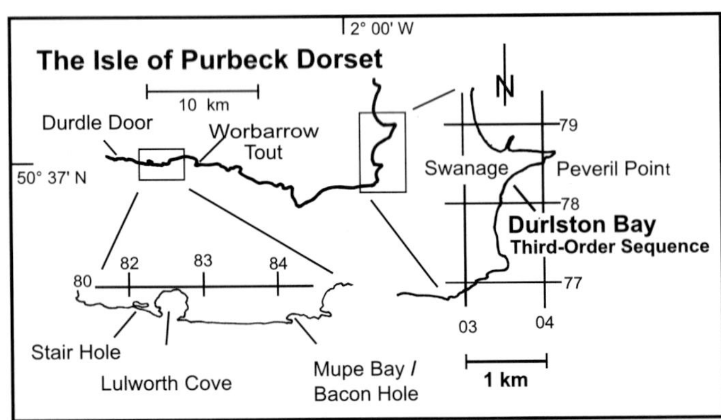

Fig. 8.—Location of the studied third-order sequence at Durlston Bay. The Purbeck section extends from Peveril Point to the southwest for a distance of 1 km. The top of the Purbeck Group and access to the section is at Peveril Point (from Ordinance Survey Sheet 195, Bournemouth, with British National Grid System).

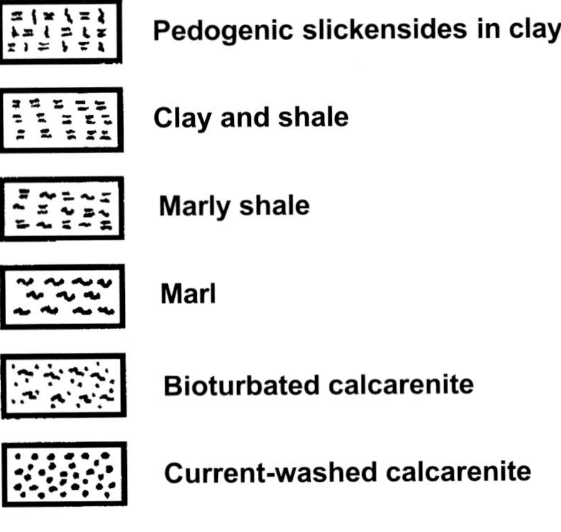

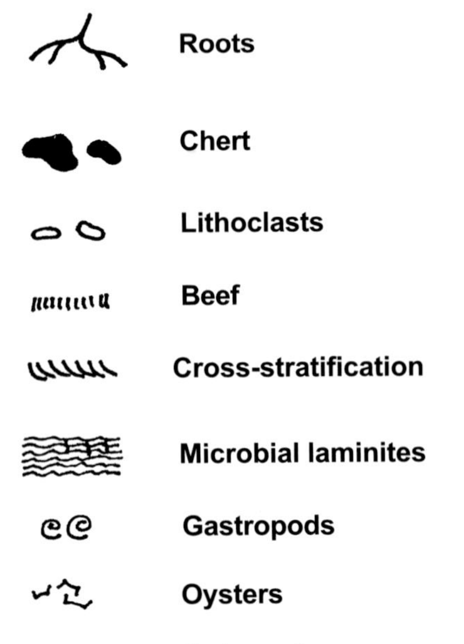

Fig. 9.—Legend for detailed stratigraphic columns in Figs. 10, 16, 17, 18, 20, and 21. Facies are arranged from most restricted (lowstand—clay) at the top to most open (highstand—calcarenite) at the bottom.

order cycles and thinner-bedded calcarenite with argillaceous partings and shale in the upper three sixth-order cycles (Fig. 16). Sequences C and D (above the Cinder member) are progressively thinner and finer grained (relative to sequence B), more argillaceous, and represent more restricted facies (see the top of Fig. 16). In turn, each fifth-order sequence is also asymmetric, with coarser carbonate facies dominating at the base of each sequence and finer, more argillaceous deposits towards the top of each. Sequences C and D are both topped by paleosols, and the carbonate bed in D at 19 m (Clements' bed 118) represents a very restricted facies, a microbial laminite.

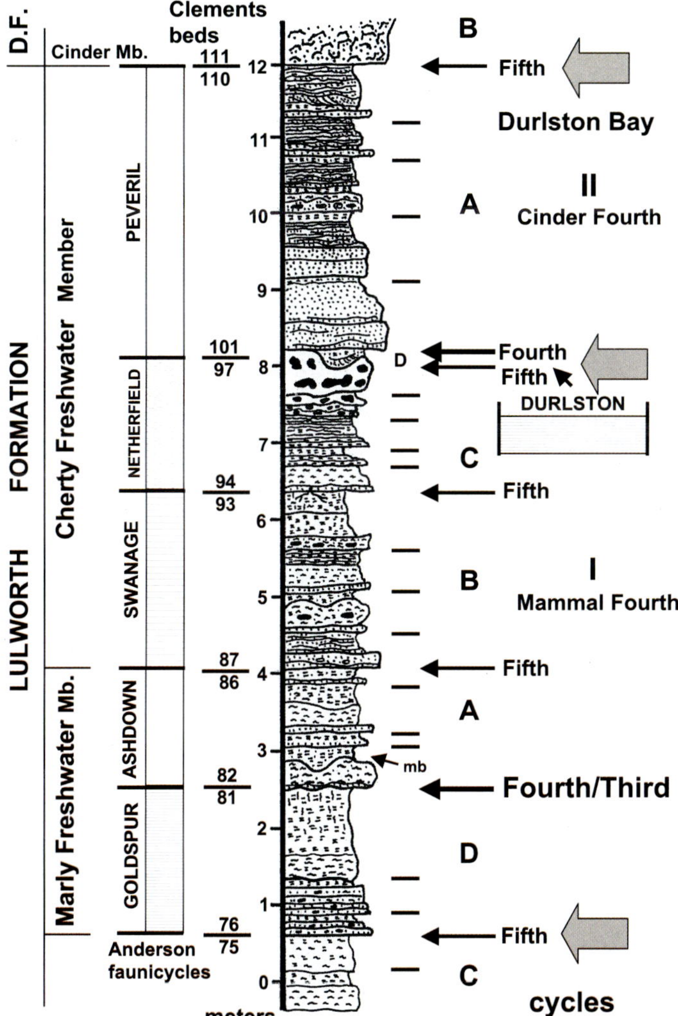

Fig. 10.—Stratigraphic log of the "Mammal Bed" fourth-order sequence. Labeled arrows and tick marks indicate the positions of sequence and cycle boundaries (note that all such boundaries are a product of precessional sea-level rises). Key-bed numbers from Clements are shown to the left of the column. Biofacies and lithofacies descriptions of all beds are available in Clements (1993). The interpreted positions of F.W. Anderson's faunicycles are shown in relationship to the cyclic hierarchy. Faunicycles (indicated with names and block patterns) appear to be coincident with fifth-order sequences. The D fifth-order sequence (correlated with the Durlston faunicycle) is highly truncated at this locality. All stratigraphic logs in this paper apply the sequence terminology defined in Figure 7. Weathering profiles reflect carbonate-rich versus shale-rich facies (dots, calcarenite; wavy dashes, micrite or marl; double dashes, shale; black blebs, chert; vertical marks at 2–2.5 m, peds; roots are drawn at 6.3 m; mb, mammal bed). Heavy gray arrows indicate the position of transgressive events determined by Morter (1984) from mapping the distribution nine molluscan associations (see Fig. 1).

Fig. 11.—Field photo of the base of the Middle Purbeck third-order sequence and the first fifth-order sequence (sequence A) of the "Mammal" fourth-order sequence. A paleosol represented by a meter-thick gray shale with well-developed ped structure is seen just below the third-order boundary. The dark brown dirt bed labeled (mb) is the "Mammal Bed". Sixth-order cycle boundaries, marked by abrupt superposition of disjunct facies, are well expressed at the bottom and top of the first PAC (at the third-order boundary and at the top of the "Mammal Bed"). In all field photos sequence boundaries are marked with yellow arrows and PAC boundaries with red arrows. The scale is one meter.

Four of F.W. Anderson's (1985) faunicycles, Peveril, Cinder, Nothe, and Croydon, are correlative with this fourth-order sequence (Figs. 1, 16). The most parsimonious explanation is that each fifth-order sequence in the Cinder fourth-order sequence is represented by an ostracod faunicycle. This assessment is on the basis of the relationship of faunicycles, key beds, and ostracod-assemblage zone boundaries in Figure 2 of Anderson (1985), and Clements' log (1993), and corroborated by Morter's (see Fig. 1, this paper) correlation of faunicycles and molluscan assemblages. In addition to the transgressive event at the basal Cinder fourth-order sequence boundary, Morter (1984, p. 233, Fig. 3) specifically located the Peveril–Cinder and the Cinder–Nothe faunicycle boundaries and interpreted transgressive events at each boundary. Also, Wimbledon's lithologically determined transgressive event 8 is located at the base of the Cinder Bed (Clements' bed 111).

The "Intermarine Beds" Fourth-Order Sequence

The third fourth-order sequence (Sequence III of the Middle Purbeck) is named for the Intermarine member of the Durlston Formation. Sequence III is nearly 9 m thick and contains four fifth-order sequences (Fig. 17). In comparison to the underlying Cinder fourth-order sequence, Sequence III is thinner, contains a higher ratio of shale to limestone, is characterized by more restricted facies, and its cyclic structure is less complete. This indicates that

FIG. 12.—Field photo of sequence B of the "Mammal" fourth-order sequence. A massive calcarenite is seen near the base (just above the A–B boundary) and a well-developed paleosol (the Fern Bed) with root structures occurs just below the upper fifth-order boundary (B–C). Thinner calcarenite beds mark the bases of sixth-order cycles (PACs). Compare the photo with the log of fifth-order sequence B in Figure 10.

in Sequence III more sixth-order cycles are missing at fifth-order sequence boundaries. Specifically, fifth-order sequences A, C, and D contain only two or three sixth-order cycles.

Anderson (1985) recognized only three faunicycles (Royal, Corfe, and Worth) in this stratigraphic interval (Fig. 17). It appears that the sea-level-rise events that initiated fifth-order sequences A and B were sufficiently large to produce a response that can be recorded in the ostracod faunas (the Royal and Corfe faunicycles). However, in the stratigraphic interval occupied by sequences C and D the paleoenvironment did not become sufficiently marine to establish an **S-phase** ostracod association at the C–D sequence boundary. A single faunicycle, the Worth, thus occupies the C–D interval. Morter's "Royal transgressive event" appears to be correlative with the base of this fourth-order sequence. Two massive shelly calcarenite beds (Clements' beds 131 and 133, the under rag and the red rag) define the bases of the first and third PACs in the C fifth-order sequence. This pattern might be an example of obliquity being in phase with the first and third precessional sea-level rise events, but out of phase with the second rise, muting the facies change at the base of the second sixth-order cycle (Fig. 17, 25–26 m). Wimbledon's (1987) transgressive event 9 is associated with the "upper intermarine freestones and rags", which associates this event with either the B or C fifth-order sequence boundaries.

The "Scallop Beds" Fourth-Order Sequence

The fourth fourth-order sequence is named for the beds that define the Scallop member, Clements' beds 145–153; this sequence is 9.3 m thick (Fig. 18). The base of Sequence IV is marked by the occurrence of a sandy, bedded calcarenite with shaly partings, overlain by a massive hard calcarenite bed (Clements' beds 139 and 140) that overlie dark selenite–limonite bearing shale of the top of the previous, Intermarine fourth-order sequence. These beds along with the massive "Laning vein" fossiliferous calcarenite (Clements' bed 144) comprise most of the A fifth-order sequence. Note that a fault occurs at or near this fourth-order boundary and leaves open the question of the true

FIG. 13.—Field photo of sequence C of the "Mammal Bed" fourth-order sequence. The gray marl (with a calcarenite 2 cm thick at its base) representing the first sixth-order cycle in sequence C lies just above the paleosol at the top of sequence B. Overall, sequence C is just less than two meters thick and fines up to a massive chert-bearing (freshwater) micrite. Internally thin calcarenite beds mark the bases (red arrows) of sixth-order cycles (compare to the log, Fig. 10).

FIG. 14.—Field photo of sequence C–D boundary showing up to 30 cm of erosional relief. The truncated D sequence overlies the massive micrite-bearing chert (Clements' bed 97) and is overlain by the thick massive calcarenite at the base of the next fourth-order sequence (Clements' bed 101, the "New Vein"). Less than 1 km away, at Durlston Bay south, this D sequence (with the chert bed below and the "New Vein" above) is represented by a paleosol 40 cm thick.

thickness of facies above and below the boundary (Clements, 1993, p. 193). The base of the most open-marine facies in Sequence IV (the Scallop member, Clements' beds 146–153) defines the base of fifth-order sequence B (Fig. 18). These beds contain a nearly normal-marine fauna including *Praeexogyra*, *Chlamas*, and *Modiolus* (Clements, 1993). Fine-grained carbonate (marl) or shale occurs in the upper parts of fifth-order sequences B and C and throughout sequence D. This asymmetrical distribution of facies defines the fourth-order sequence as well as its composite fifth-order sequences.

Just as in the Intermarine fourth-order sequence, the Scallop fourth-order sequence contains only three ostracod faunicycles. The Langton and Scallop faunicycles are correlated with fifth-order sequences A and B, respectively, but the third faunicycle (Studland) is coincident with the top two fifth-order sequences in the Scallop fourth-order sequence because again the C–D boundary event (sea-level rise) was not large enough to introduce **S-phase** ostracods. Three of Morter's transgressive events can be correlated to the cyclic hierarchy of this fourth-order sequence (the A–B, B–C, and upper fourth-order boundaries; Fig. 18). Wimbledon's transgressive event 10 is placed at the base of the Scallop member, which is the base of fifth-order sequence B.

The "Corbula Beds" Fourth-Order Sequence

The top fourth-order sequence (Sequence V) in the Middle Purbeck third-order sequence, the "Corbula Beds" fourth-order sequence, is also 9 m thick. It occurs in the upper half of the Corbula member and the lower half of the Chief Beef member (Clements' beds 167–199). This sequence is the most shale-rich fourth-order sequence in the Middle Purbeck. Nevertheless, it internally displays the facies asymmetry typical at each scale in the cyclic hierarchy, the second or B fifth-order sequence is most limestone-rich and the C and D fifth-order sequences are progressively finer-grained and more shale-rich. This enrichment in shale and the marked change to carbonate at the overlying third-order boundary are well expressed in the field photo and log (Figs. 19, 20).

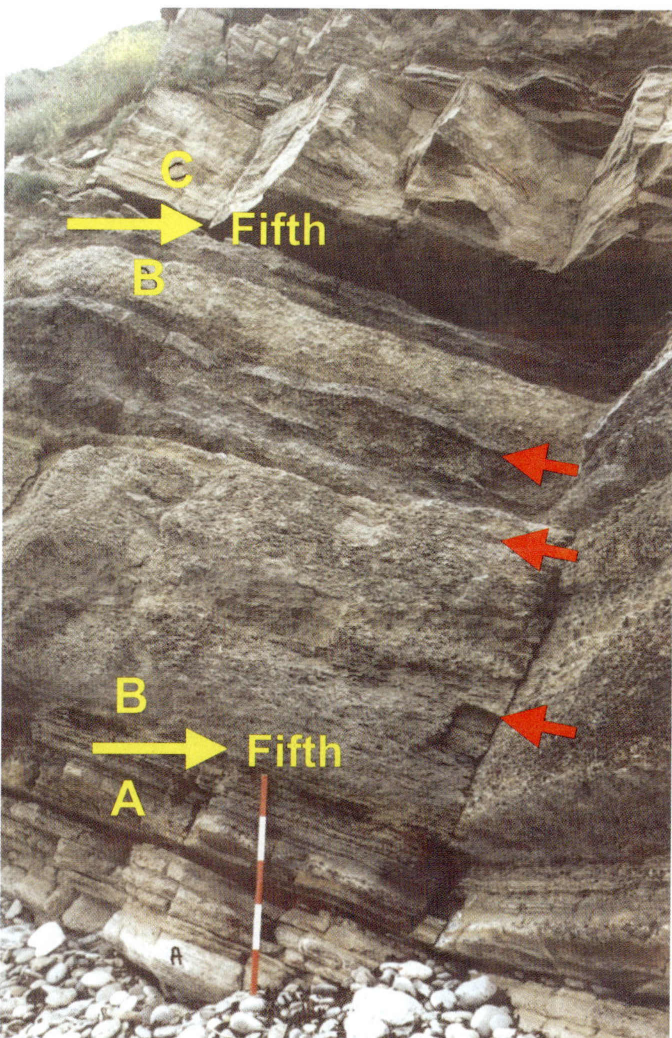

FIG. 15.—Field photo of sequences A, B, and C of the "Cinder" fourth-order sequence. The upper two sixth-order cycles of sequence A are behind the meter scale. The massive oyster-rich calcirudite between the fifth-order boundaries is sequence B (equals the Cinder member, Clements' beds 111 and 112). The massive calcarenite that constitutes the base of sequence C is seen above the upper fifth-order boundary (compare with the log in Fig. 16).

Three of Morter's molluscan-based transgressive events can be correlated to sequence boundaries in the "Corbula" fourth. These are here interpreted to occur at the lower fourth-order boundary, the A–B fifth-order boundary, and the overlying third-order boundary (Fig. 20). Because the upper part of Sequence V is associated with minimal eccentricity fluctuation, the B–C and C–D boundaries are formed by relatively small sea-level changes. These events then do not introduce **S-phase** ostracod associations into the stratigraphic section, thus Anderson (1985) recognized only one faunicycle (the Poxwell) in this stratigraphic interval (Fig. 20). As in previous fourth-order sequences the A fifth-order sequence does correspond to a faunicycle (the Bacon). Wimbledon placed his transgressive event 11 at the Chlamas bed (Clements' bed 175), which occurs at the A–B fifth-order sequence boundary in fourth-order sequence V.

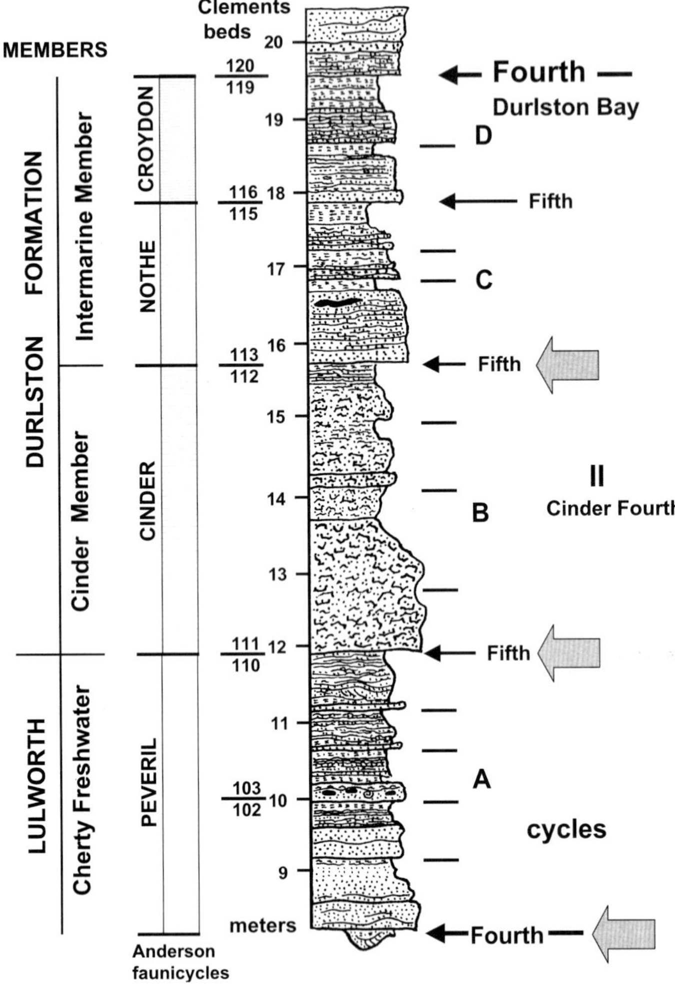

FIG. 16.—Stratigraphic log of the "Cinder Bed" fourth-order sequence. This sequence is named for the oyster-rich shell bed (Clements' bed 111), which represents the first member of the Durlston Formation. The positions of cycle and sequence boundaries are indicated by labeled arrows and tick marks (PACs). Each fifth-order sequence in the Cinder fourth-order sequence displays a well-developed facies asymmetry with thinner bedding, shales, and/or paleosols occurring toward the top of each sequence. F.W. Anderson's faunicycles (indicated with names and block patterns to the left side of the log) are coincident with fifth-order sequences.

The Upper Third-Order Boundary

The placement of the upper boundary of the Middle Purbeck third-order sequence is on the basis of a progressive but stepwise return to massive limestone facies beginning at 47 m on the log and culminating with the Broken Shell Limestone member at 52.5 m (Figs. 20 and 21). Just above the upper third-order boundary a thicker limestone bed occurs in the upper part of the Chief Beef member (e.g., Clements' bed 200). This bed marks the base of the overlying Upper Purbeck third-order sequence. The first fourth-order sequence (Fig. 19) is only 4 m thick and incomplete (note that it could be interpreted as containing only two fifth-order sequences, B and C, by recognizing only the boundary at 49.3 m as opposed to the three shown on the log). However, the overall pattern mimics the succession of facies changes at the base of the Middle Purbeck third-order sequence, where the marly, restricted, and incomplete "Mammal Bed" fourth-order sequence (with soils) is overlain by the coarse-grained, carbonate-rich, marine "Cinder Bed" fourth-order sequence. The second fourth-order sequence in the Upper Purbeck is 8.5 m thick, and the B fifth-order sequence is a massive calcarenite, the Broken Shell Limestone member. The last transgressive event shown by Morter (1984) is placed just below the Broken Shell Limestone. In F.W. Anderson's listing of faunicycles there is only one faunicycle, the Greenwood, in the Chief Beef fourth-order sequence (no **S-phase** ostracods are introduced at fifth-order sequence boundaries). However, the Lulworth faunicycle is coincident with a B fifth-order sequence (the Bro-

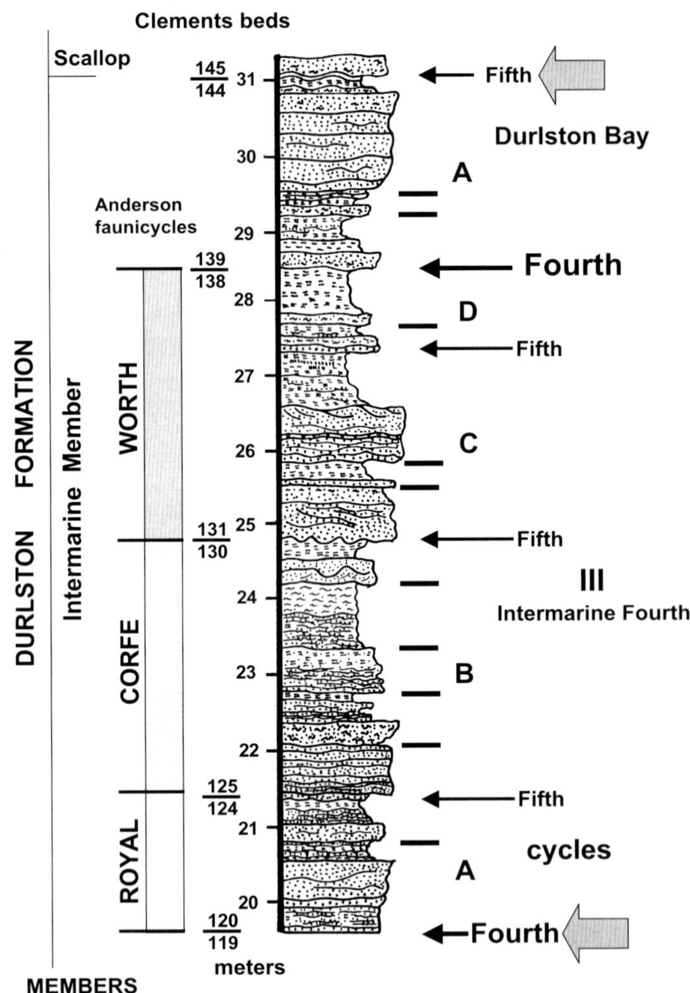

FIG. 17.—Stratigraphic log of the Intermarine fourth-order sequence. This sequence occupies most of the Intermarine member and is almost nine meters thick. Both the fourth-order and fifth-order sequences show well-developed facies asymmetry. The Royal and Corfe faunicycles are coincident with fifth-order sequences A and B, but the Worth faunicycle overlaps both the C and D sequences. The massive calcarenite beds in sequence C at 25 and 26 meters (Clements' beds 131 and 133) are at the bases of the first and third PACs in the sequence. See obliquity explanation in the text. Large gray arrows mark the positions of Morter's transgressive events.

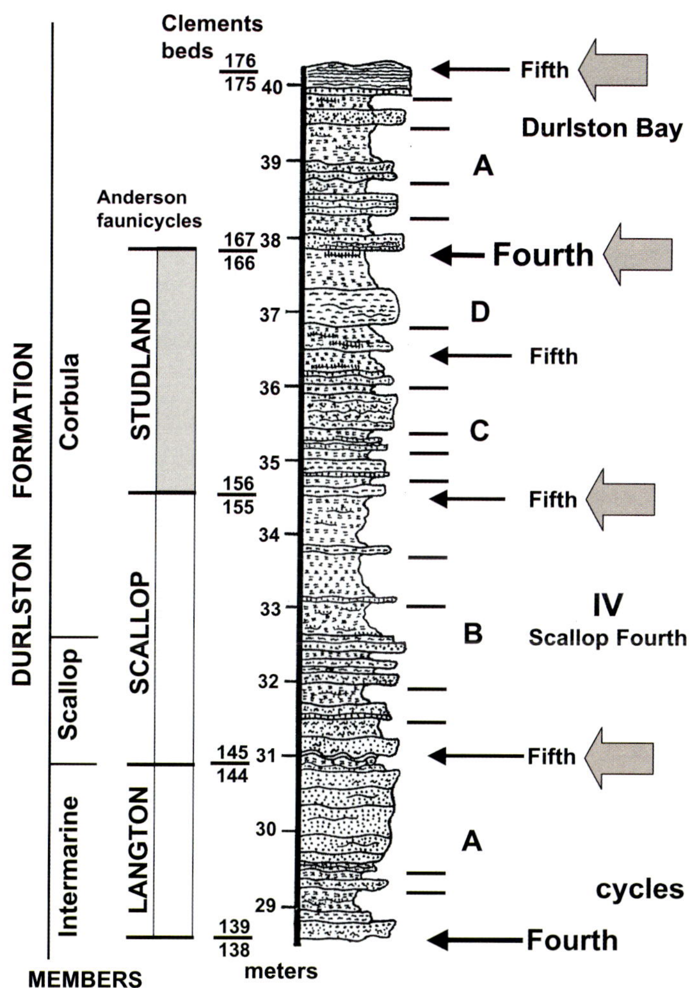

FIG. 18.—Stratigraphic log of the "Scallop Bed" fourth-order sequence. This sequence is close to nine meters thick and shows a marked facies asymmetry. Calcarenite is the dominant facies in fifth-order sequence A and the lower half of B, whereas shale and soils are more abundant throughout the rest of sequence IV. The most diverse marine fauna occurs in the Scallop member in the lower half of fifth-order sequence B (Clements' beds 145–153). Again the Langton and Scallop faunicycles are coincident with fifth-order sequences A and B, whereas the Studland faunicycle is correlated with sequences C and D.

FIG. 19.—Field photo of the upper boundary of the Middle Purbeck third-order sequence (compare with the log in Fig. 20). The facies below the boundary are predominantly selenite- and ostracod-rich shales with thin, rotted bivalve shells (the Chief Beef member); above the boundary there is a marked return to cycles and sequences with thick calcarenite beds at their bases. The Broken Shell Limestone is the thick calcarenite unit at the top of the photograph (above the fifth-order boundary).

ken Shell Limestone itself), the second fifth-order sequence, in the Broken Shell Limestone fourth-order sequence (Fig. 21).

CONCLUSIONS

1. The Middle Purbeck (from the base of Clements' bed 82 to the base of bed 200) is a 2 My, third-order sequence. This sequence is the third preserved third-order sequence (of four) in the Purbeckian of Dorset. It is recognized on the basis of gross facies asymmetry and its internal cyclic structure.

2. The Middle Purbeck third-order sequence comprises a multi-tiered cyclic hierarchy (i.e., fourth-order, fifth-order, and sixth-order subdivisions) each defined on analogous facies asymmetry. In general, limestone is the dominant facies at or near the bases of sequences or cycles at each scale in the hierarchy and shale is the predominant facies toward the tops.

3. This interpretation is on the basis of application of a strict assumption of an orbital-forcing model in which each tier of the hierarchy of rock sequences (or cycles) is related to a particular stratigraphic process (precession modulated by 100 ky, 400 ky, and 2 My eccentricity).

4. All sequence and cycle boundaries are interpreted as the product of precessionally forced sea-level rises due to cessation of sedimentation when a critical rate of rise is attained. The degree of facies change that occurs at cycle boundaries is a function of the magnitude of sea-level rise and is the basis for bundling rock cycles into sets (i.e., fifth-, fourth-, and third-order sequences).

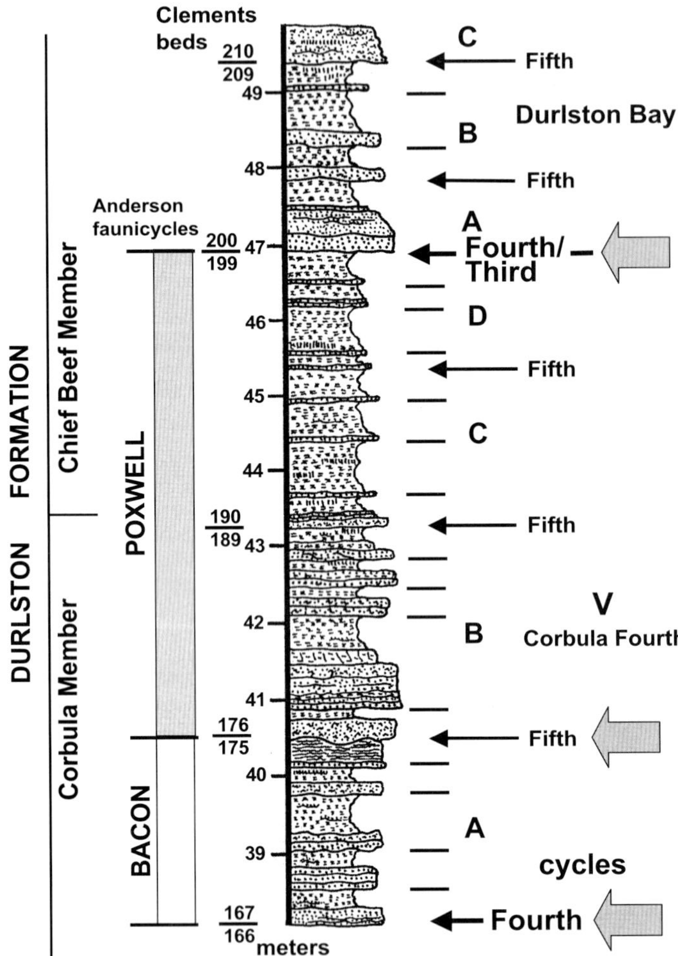

FIG. 20.—Stratigraphic log of the Corbula fourth-order sequence. This sequence overlaps parts of the Corbula and Chief Beef members and again is close to nine meters thick. It (Sequence V) constitutes the most restricted (and shale-rich) facies in the Middle Purbeck third-order sequence. Overall it displays a marked facies asymmetry and has the most marine and carbonate-rich facies in the B fifth-order sequence position. Ostracod faunicycles are initiated at the basal fourth-order boundary and at the A–B fifth-order boundary. Owing to restriction of marine conditions at the top of the third-order sequence, only one faunicycle (the Poxwell) occurs in the rest of Sequence V.

5. There is an excellent correspondence between individual ostracod faunicycles defined by Anderson (1985) and 100 ky sequences (fifth-order) of the cyclic hierarchy. The exceptions to this rule occur high in fourth-order sequences in the upper part of the Middle Purbeck third-order sequence where the magnitudes of marine incursions are minimal. At these stratigraphic intervals two or three fifth-order sequences may be coincident with one faunicycle.

6. Transgressive events in the English Purbeck defined on the basis of molluscan associations (Morter, 1984) and lithostratigraphy (Wimbledon, 1987) generally are correlative with the first or second fifth-order sequence boundaries within fourth-order sequences.

7. The hierarchical cyclic structure of the Dorset Purbeck primarily determined by lithofacies analysis provides an explanation for previously interpreted sea-level fluctuations (or salinity changes) derived from paleoecological analyses. Conversely the paleoecological data reinforces the interpretation of cyclic structure on the basis of applying the "Croll–Milankovitch" model of orbital forcing.

ACKNOWLEDGMENTS

Professors Ian West and André Strasser provided stimulating discussion of ideas and much helpful guidance at the field localities in Dorset and the French Jura, respectively. Conceptual models relating orbital forcing to cyclic stratigraphy have been developed over the past two decades with friend and colleague Peter W. Goodwin. This material is based on work supported by a Research Incentive Fund Grant from Temple University in 1998 and by the National Science Foundation under Grant No. EAR-9902680.

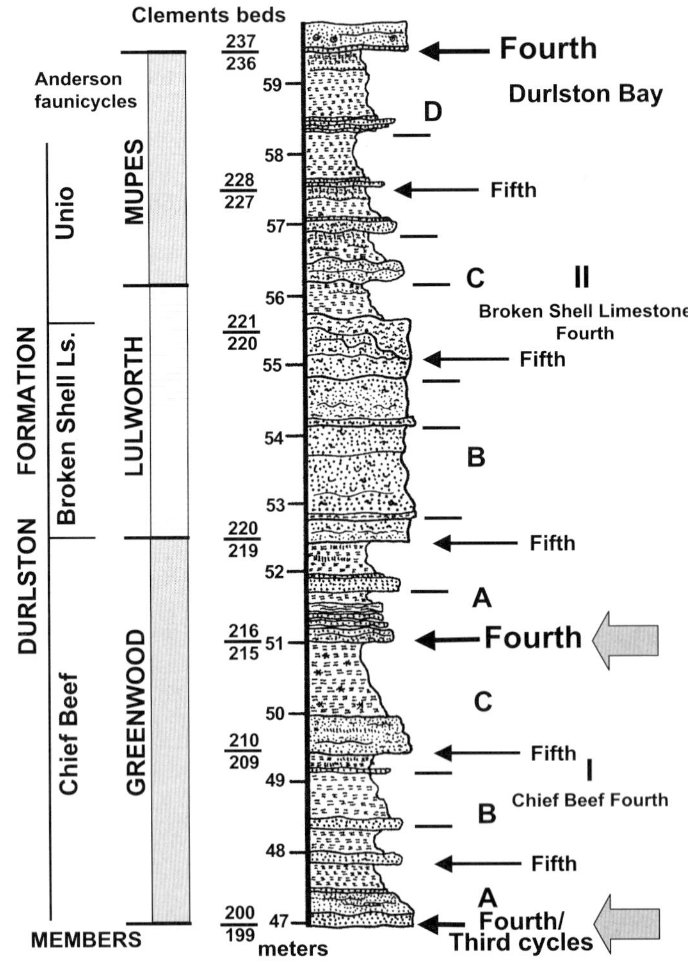

FIG. 21.—Stratigraphic log of the first two fourth-order sequences in the Upper Purbeck. This log shows the return to sequences marked by massive calcarenite units at their bases above the third-order sequence boundary (see photo, Fig. 19). The Broken Shell Limestone member is the B fifth-order sequence in the second fourth-order sequence and thus has a position in the cyclic structure analogous to that of the Cinder member near the base of the Middle Purbeck third-order sequence.

REFERENCES

ALLEN, P., 1998, Purbeck–Wealden (early Cretaceous) climates: Proceedings of the Geologists' Association, v. 109, p. 197–236.

ALLEN, P., AND WIMBLEDON, W.A., 1991, Correlation of NW European Purbeck–Wealden (nonmarine Lower Cretaceous) as seen from the English type-areas: Cretaceous Research, v. 12, p. 511–526.

ANDERSON, E.J., 1999, The cyclic structure of the 'Purbeckian' at the Classic Sierra del Pozo section of the Prebetic (Berriasian) Southern Spain (abstract): Geological Society of America, Abstracts with Programs, v. 33, p. 323.

ANDERSON, E.J., 2000a, Criteria for recognizing an orbitally forced cyclic hierarchy: the 'Purbeckian' at the Classic Sierra del Pozo Section (Berriasian) Southern Spain (abstract): Abstract Program, Sediment 2000, Leoben, Austria, v. 43, p. 20.

ANDERSON, E.J., 2000b, Cycle boundaries and the magnitude and frequency of sea-level rise events: examples from the Purbeckian of Dorset (abstract): International Association of Sedimentologists, 20th Regional Meeting, Dublin, p. 4.

ANDERSON, E.J., AND GOODWIN, P.W., 1990, The significance of meter-scale cycles in the quest for a fundamental stratigraphic unit: Geological Society of London, Journal, v. 147, p. 507–518.

ANDERSON, E.J., AND GOODWIN, P.W., 1992, The primacy of the precessional signal (abstract): Geological Society of America, Abstracts with Programs, v. 24, p. 110.

ANDERSON, E.J., PERRY, L.L., AND STYNCHULA, J.A., 2001, Lateral continuity and discontinuity of 100 ky eccentricity sequences within an orbitally forced cyclic hierarchy: The Purbeck Group, Lower Cretaceous, Dorset, England (abstract): Geological Society of America, Annual Meeting, Abstracts with Programs, v. 33, p. 323.

ANDERSON, F.W., 1973, The Jurassic–Cretaceous transition: the nonmarine ostracod faunas, *in* Casey, R., and Rawson, P.F., eds., The Boreal Lower Cretaceous: Geological Journal, Special Issue, v. 5, p. 101–110.

ANDERSON, F.W., 1985, Ostracod faunas in the Purbeck and Wealden of England: Journal of Micropalaeontology, v. 4, p. 1–68.

ANDERSON, F.W., AND BAZLEY, R.A.B., 1971, The Purbeck Beds of the Weald (England): Geological Survey of Great Britain, Bulletin, v. 34, 174 p.

BERGER, A., 1988, Milankovitch theory and climate: Reviews of Geophysics, v. 26, p. 624–657.

BUSCH, R.M., AND WEST, R.R., 1987, Hierarchal genetic stratigraphy: a framework for paleoceanography: Paleoceanography, v. 2, p. 141–164.

CLEMENTS, R.G., 1969, Annotated cumulative section of the Purbeck Beds between Peveril Point and the Zig-Zag Path, Durlston Bay, *in* Torrens, H.S., ed., International Field Symposium on the British Jurassic: Excursion No. 1, Guide for Dorset and South Somerset, University of Keele, p. A1–A77.

CLEMENTS, R.G., 1993, Type section of the Purbeck Limestone Group, Durlston Bay, Swanage, Dorset: Dorset Natural History and Archaeological Society, Proceedings, v. 114, p. 181–206.

COPE, J.C.W., DUFF, K.L., PARSONS, C.F., TORRENS, H.S., WIMBLEDON, W.A., AND WRIGHT, J.K., 1980a, Boreal British Ammonite zones: Geological Society of London, Special Report 14, p. 73–108.

COPE, J.C.W., DUFF, K.L., PARSONS, C.F., TORRENS, H.S., WIMBLEDON, W.A., AND WRIGHT, J.K., 1980b, A correlation of Jurassic rocks in the British Isles. Part 2: Middle & Upper Jurassic: Geological Society of London, Special Report 15, p. 1–109.

CROLL, J., 1875, Climate and Time in Their Geological Relations; A Theory of Secular Changes of the Earth's Climate: London, Daldy, Ibister & Co., 577 p.

D'ARGENIO, B., AMODIO, S., FERRERI, V., AND PELOSI, N., 1997, Hierarchy of high frequency orbital cycles in Cretaceous carbonate platform strata: Sedimentary Geology, v. 113, p. 169–193.

DE BOER, P.L., AND SMITH, D.G., EDS., 1994, Orbital Forcing and Cyclic Sequences: International Association of Sedimentologists, Special Publication 19, 559 p.

ELDREDGE, N., AND GOULD, S.J., 1972, Punctuated equilibria: an alternative to phyletic gradualism, *in* Schopf, T.J.M., ed., Models in Paleobiology: San Francisco, Freeman, p. 82–115.

EL-SHAHAT, A., AND WEST, I.M., 1983, Early and late lithification of aragonitic bivalve beds in the Purbeck Formation (Upper Jurassic–Lower Cretaceous) of southern England: Sedimentary Geology, v. 35, p. 15–41.

ENSOM, P.C., 1985, An annotated section of the Purbeck Limestone Formation at Worbarrow Tout, Dorset: Dorset Natural History and Archaeological Society, Proceedings, v. 106, p. 87–91.

FEIST, M., LAKE, R.D., AND WOOD, C.J., 1995, Charophyte biostratigraphy of the Purbeck and Wealden of southern England: Palaeontology, v. 38, p. 407–442.

FISCHER, A.G., 1991, Orbital cyclicity in Mesozoic strata, *in*, Einsele, G., Ricken, W., and Seilacher, A., eds., Cyclic and Event Stratification: Berlin, Springer-Verlag, p. 48–62.

FISCHER, A.G., AND BOTTJER, D.J., EDS., 1991, Orbital Forcing and Sedimentary Sequences: Journal of Sedimentary Petrology, Special Issue, v. 61, p. 1063–1252.

FRANCIS, J.E., 1986, The calcareous paleosols of the basal Purbeck Formation (Upper Jurassic) Southern England, *in* Wright, V.P., ed., Paleosols; Their Recognition and Interpretation: Oxford, U.K., Blackwell, p. 112–138.

GOLDHAMMER, R.K., DUNN, P.A., AND HARDIE, L.A., 1990, Depositional cycles, composite sea-level changes, cycle stacking patterns and the hierarchy of stratigraphic forcing: examples from the Alpine Triassic platform carbonates: Geological Society of America, Bulletin, v. 102, p. 535–562.

GOODWIN, P.W., AND ANDERSON, E.J., 1985, Punctuated aggradational cycles: a general hypothesis of episodic stratigraphic accumulation: Journal of Geology, v. 93, p. 515–533.

GOODWIN, P.W., AND ANDERSON, E.J., 1988, Episodic development of Helderbergian paleogeography, New York State, Appalachian Basin, *in*, McMillan, N.J., Embry, A.F., and Glass, D.J., eds., Devonian of the World: Canadian Society of Petroleum Geologists, Second International Symposium on the Devonian System, p. 553–568.

GOODWIN, P.W., AND ANDERSON, E.J., 1997, Stratigraphic incompleteness: Milankovitch in the Manlius at the margin, *in* Rayne, T.W., Bailey, D.G., and Tewksbury, B.J., eds., New York State Geological Association, 69th Annual Meeting, Field Trip Guide, p. 237–249.

GOODWIN, P.W., ANDERSON, E.J., GOODMAN, W.M., AND SARAKA, L.J., 1986, Punctuated aggradational cycles: implications for stratigraphic analysis: Paleoceanography, v. 1, p. 417–430.

GOODWIN, P.W., ANDERSON, E.J., AND ORZECHOWSKI, C., 1992, The case for a process-determined hierarchy of allocycles (abstract): Geological Society of America, Abstracts with Program, v. 24, p. 110.

GRADSTEIN, F.M., AGTERBERG, F.P., OGG, J.G., HARDENBOL, J., VAN VEEN, P., THIERRY, J., AND HUANG, Z., 1994, A Mesozoic Time Scale: Journal of Geophysical Research, v. 99, p. 24,051–24,074.

GRADSTEIN, F.M., AGTERBERG, F.P., OGG, J.G., HARDENBOL, J., VAN VEEN, P., THIERRY, J., AND HUANG, Z., 1995, A Triassic, Jurassic and Cretaceous time scale, *in* Berggren, W.A., Kent, D.V., Aubry, M.P., and Hardenbol, J., eds., Geochronology, Time Scales and Global Stratigraphic Correlation: SEPM, Special Publication 54, p. 95–126.

GROTZINGER, J.P., 1986, Upward shallowing platform cycles: a response to 2.2 billion years of low-amplitude, high-frequency (Milankovitch band) sea level oscillations: Paleoceanography, v. 1, p. 403–416.

HARDENBOL, J., THIERRY, J., FARLEY, M.B., JACQUIN, T., DE GRACIANSKY, P., AND VAIL, P.R., 1998, Mesozoic and Cenozoic sequence chronostratigraphic framework of European basins, *in*, de Graciansky, P., Hardenbol, J., Jacquin, T., and Vail, P.R., eds., Mesozoic and Cenozoic Sequence Stratigraphy of European Basins: SEPM, Special Publication 60, p. 3–13.

HORNE, D.J., 1995, A revised ostracod biostratigraphy for the Purbeck–Wealden of England: Cretaceous Research, v. 16, p. 639–663.

HOUSE, M.R., 1993, Geology of the Dorset Coast, Second Edition: Geologists' Association Guide, 164 p. and 32 plates,.

HOUSE, M.R., AND GALE, A.S., EDS., 1995, Orbital Forcing Timescales and Cyclostratigraphy: Geological Society of London, Special Publication 85, 210 p.

JACQUIN, T., RUSCIADELLI, G., DE GRACIANSKY, P., AND MAGNIEZ-JANNIN, F., 1998, The North Atlantic cycle: an overview of 2nd-order transgressive/regressive facies cycles in the Lower Cretaceous of Western Europe, *in* de Graciansky, P., Hardenbol, J., Jacquin, T., and Vail, P.R., eds., Mesozoic and Cenozoic Sequence Stratigraphy of European Basins: SEPM, Special Publication 60, p. 397–409.

JIMÉNEZ DE CISNEROS, C., AND VERA, J.A., 1993, Milankovitch cyclicity in Purbeck peritidal limestones of the Prebetic (Berriasian, southern Spain): Sedimentology, v. 40, p. 513–537.

KAMOLA, D.K., AND VAN WAGONER, J.C., 1995, Stratigraphy and facies architecture of parasequences with examples from the Spring Canyon Member, Black Hawk Formation, Utah, *in* Van Wagoner, J.C., and Bertram, G.T., eds., Sequence Stratigraphy of Foreland Basin Deposits: American Association of Petroleum Geologists, Memoir 64, p. 27–54.

KUHN, T.S., 1962, The Structure of Scientific Revolutions: The University of Chicago Press, 172 p. (Second Edition, 1970, 210 p.).

LAUDAN, L., 1977, Progress and Its Problems; Towards a Theory of Scientific Growth: Berkeley, University of California Press, 257 p.

MEDAWAR, P.B., 1964, Is the scientific paper fraudulent?: Saturday Review, August 1, p. 42–43.

MILANKOVITCH, M., 1941, Kanon der Erdbestrahlung und seine Anwendung auf das Eiszeitenproblem: Royal Serbian Academy, v. 132, Section of Mathematical and Natural Sciences, v. 33, 633 p., Belgrade (Canon of Insolation and the Ice-Age Problem). English translation by Israel Program for Scientific Translations, Jerusalem, 1969.

MORTER, A.A., 1984, Purbeck–Wealden beds Mollusca and their relationship to ostracod biostratigraphy, stratigraphical correlation and palaeoecology in the Weald and adjacent areas: Proceedings of the Geologists' Association, v. 95, p. 217–234.

OGG, J.G., HASENYAGER, R.W., II, AND WIMBLEDON, W.A., 1994, Jurassic–Cretaceous boundary: Portland–Purbeck magnetostratigraphy and possible correlation to the Tethyan faunal realm: Géobios, Mémoire Speciale no. 17, v. 2, p. 519–527.

POPPER, K., 1959, The Logic of Scientific Discovery: New York, Basic Books, 479 p.

Price, G.D., 1999, The evidence and implications of polar ice during the Mesozoic: Earth-Science Reviews, v. 48, p. 183–210.

READ, J.F., GROTZINGER, J.P., BOVA, J.A., AND KOERSCHNER, W.F., 1986, Models for generation of carbonate cycles: Geology, v. 14, p. 107–110.

STRASSER, A., 1994, Milankovitch cyclicity and high-resolution sequence stratigraphy in lagoonal–peritidal carbonates (Upper Tithonian–Lower Berriasian, French Jura Mountains), *in* de Boer, P.L., and Smith, D.G., eds., Orbital Forcing and Cyclic Sequences: International Association of Sedimentologists, Special Publication 19, p. 285–301.

STRASSER, A., AND HILLGÄRTNER, H., 1998, High-frequency sea-level fluctuations recorded on a shallow carbonate platform (Berriasian and Lower Valanginian of Mount Salève, French Jura): Eclogae Geologicae Helvetiae, v. 91, 375–390.

STYNCHULA, J.A., AND ANDERSON, E.J., 2001, The relative maturity of paleosols (chrono and toposequences) formed within an orbitally forced cyclic hierarchy of cycles: The Purbeck Group, Lower Cretaceous, Dorset, England (abstract): Geological Society of America, Abstracts with Programs, v. 33, p. 446.

TIPPER, J.C., 1997, Modeling carbonate platform sedimentation—Lag comes naturally: Geology, v. 25, p. 495–498.

TIPPER, J.C., 2000, Patterns of stratigraphic cyclicity: Journal of Sedimentary Research, v. 70, p. 1262–1279.

VAN WAGONER, J.C., POSAMENTIER, H.W., MITCHUM, R.M., VAIL, P.R., SARG, J.F., LOUTIT, T.S., AND HARDENBOL, J., 1988, An overview of the fundamentals of sequence stratigraphy and key definitions, *in* Wilgus, C.K., Hastings, B.S., Kendall, C.G. St. C., Posamentier, H.W., Ross, C.A., and Van Wagoner, J.C., eds., Sea-Level Changes: An Integrated Approach: SEPM, Special Publication 42, p. 39–45.

WESTHEAD, R.K., AND MATHER, A.E., 1996, An updated lithostratigraphy for the Purbeck Limestone Group in the Dorset type area: Proceedings of the Geologists' Association, v. 107, p. 117–128.

WIMBLEDON, W.A., 1987, Rhythmic sedimentation in the Late Jurassic–Early Cretaceous: Dorset Natural History and Archaeological Society, Proceedings, v. 108, p. 127–133.

WIMBLEDON, W.A., AND HUNT, C.O., 1983, The Portland–Purbeck junction (Portlandian–Berriasian) in the Weald, and correlation of latest Jurassic–early Cretaceous rocks in southern England: Geological Magazine, v. 120, p. 267–280.

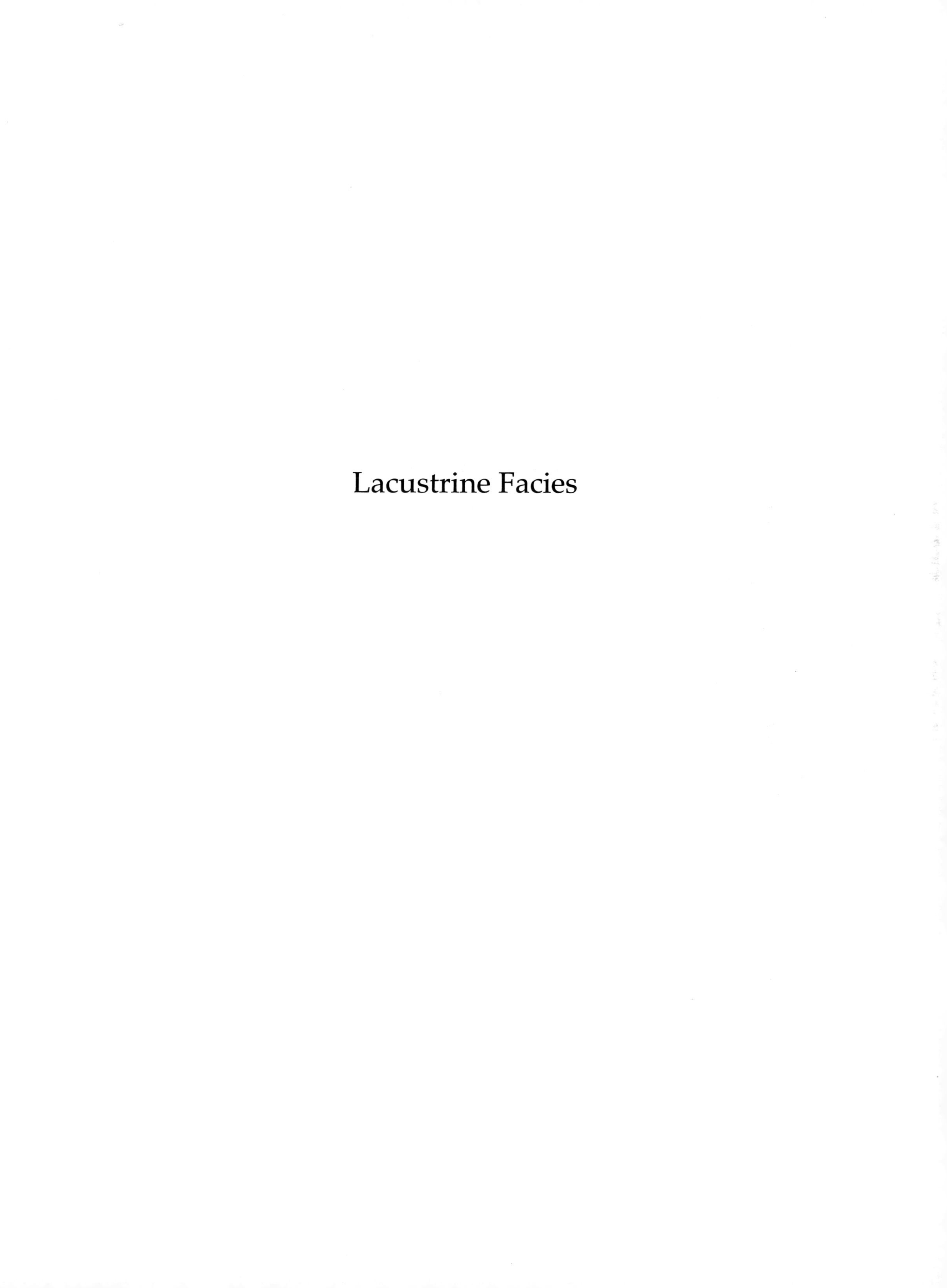

Lacustrine Facies

MAGNETIC SUSCEPTIBILITY CYCLES IN UPPER PLIOCENE LACUSTRINE DEPOSITS OF THE NORTHERN APENNINES, ITALY

GIOVANNI NAPOLEONE
Dipartimento di Scienze della Terra, and Museo di Storia Naturale, Università di Firenze, via La Pira 4, I-50121, Firenze FI, Italy
e-mail napo19@unifi.it
ANDREA ALBIANELLI
Dipartimento di Scienze della Terra, Università di Firenze, via La Pira 4, I-50121, Firenze FI, Italy
e-mail albix@geo.unifi.it
AND
ALFRED G. FISCHER
Department of Earth Sciences, University of Southern California, Los Angeles California 90089-0740, U.S.A.

ABSTRACT: Limnic deposits in Plio-Pleistocene graben basins in the Apennine Mountains provide detailed records of magnetostratigraphy leading into the Pleistocene. Their lacustrine clays retain both pollen that record the vegetation and, in magnetic signals, the history of magnetic reversals and a record of the orbital variations. The magnetically defined precessional units, with mean thicknesses of 3–8 m depending on accumulation rate, are modulated in amplitude by the eccentricity cycles, and are commonly associated with (and at times overshadowed by) obliquity cycles. The magnetic stratigraphy of the Upper Valdarno and Valtiberina basins serves to trace these cyclicities through the late Gauss and early Matuyama chrons, and the Olduvai. These cyclic records set a chronologic framework for the history of both basins. Varied accumulation rates, beginning of oscillations between moister and drier plant assemblages, and a rise in the strength of the obliquity signal coincide with the 400 ky eccentricity low at 2.84 Ma.

INTRODUCTION

In the development of the Apennine mountain system, a front of compressional deformation migrated eastward, creating thrusts and folds. Behind this followed a zone of tensional deformation, which superimposed a succession of grabens on this deformed and eroded terrane (Fig. 1A) (Martini and Sagri, 1993; Sagri et al., 1994). This typically accumulated some hundreds of meters of sediment. The northwestern grabens of late Miocene–early Pliocene age were largely filled with marine sediments, and the northeastern ones, here dealt with, by fluvio-lacustrine facies of the Plio-Pleistocene.

Exposures are generally poor owing to poor consolidation of the sediments and low dip. However, centuries of lignite mining and clay extraction for brick manufacture provided large exposures. Numerous fossil finds provided motivation for early and continued studies.

The alluvial deposits of the Plio-Pleistocene basins yielded the classical Villafranchian and younger mammalian faunas. Azzaroli in 1977 established the continental biochronology and in 1983 tied it to the marine record by correlating the Triversa faunal unit of Villafranca to the Piacenzian marine clays now assigned to the mid-Pliocene. The faunal record of the Upper Valdarno basin, starting with a Triversa-related fauna, thus seemed to roughly equate the Villafranchian ages to a mid-Pliocene to mid-Pleistocene time span (Torre et al., 1993; Azzaroli, 2001; Azzaroli et al., 1997).

The lacustrine clays (Fig. 1B, C) contain a pollen record which reveals a complex history in which subtropical forests gave way, in fluctuations, to the precursors to deciduous broadleaf forests, with dry spells bringing dominance of steppe pollen, and with times of montane spruce and fir pollen—a mixture of climatic and tectonic change (Bertini, 1994; Suc et al., 1995; Bertini and Roiron, 1997; Pontini, 1997).

Vital to the reconstruction of this history is the establishment of a high-resolution chronology, aided here by accumulation rates of 200–400 mm/ky. A first step in this direction was the delineation of a magnetic-reversal stratigraphy tied to the geomagnetic polarity time scale (GPTS; Berggren et al., 1995) and to the stratotype section at Vrica (Van Couvering, 1997). For the progressive understanding of magnetic polarity sequence in the Upper Valdarno basin, see Albianelli et al. (1995), Albianelli et al. 1997), and Albianelli et al. (2002). The record of short polarity events, the Reunion and the Olduvai with its internal split (Zijderveld et al., 1991), provided 2 ky resolution to the timing of faunal occurrences around the Plio-Pleistocene boundary (Napoleone et al., 2001; Napoleone and Azzaroli, 2002). For refinement of longer-polarity chrons we must turn to the record of orbital variations (Laskar, 1988, 1990; Lourens et al., 1996)

MAGNETIC DATA

The magnetic record of orbital variations has been used to good effect in marine sequences (Weedon et al., 1997; Napoleone and Ripepe, 1989; Napoleone et al., 1990); The lacustrine clay facies here examined also recorded variations in all magnetic parameters measured. Here, we demonstrate this cyclicity in the susceptibility of three successive profiles (Fig. 1) covering the interval from the late Gauss through the Olduvai. The Meleto Clay in the Upper Valdarno basin provided a record of Late Gauss age. For the early Matuyama, a sandy deltaic facies interrupts the record in the Upper Valdarno basin, and we turned to the clays of the Valtiberinea basin near Perugia, returning to the Upper Valdarno basin for a record of Olduvai time.

Sampling

Profiles were initially sampled with oriented hand samples at 1 m spacing, considered sufficient for resolving the 10 ky time span needed for a polarity change. Specimens of one-inch (~ 2.5

Cyclostratigraphy: Approaches and Case Histories
SEPM Special Publication No. 81, Copyright © 2004
SEPM (Society for Sedimentary Geology), ISBN 1-56576-108-1, p. 263–274.

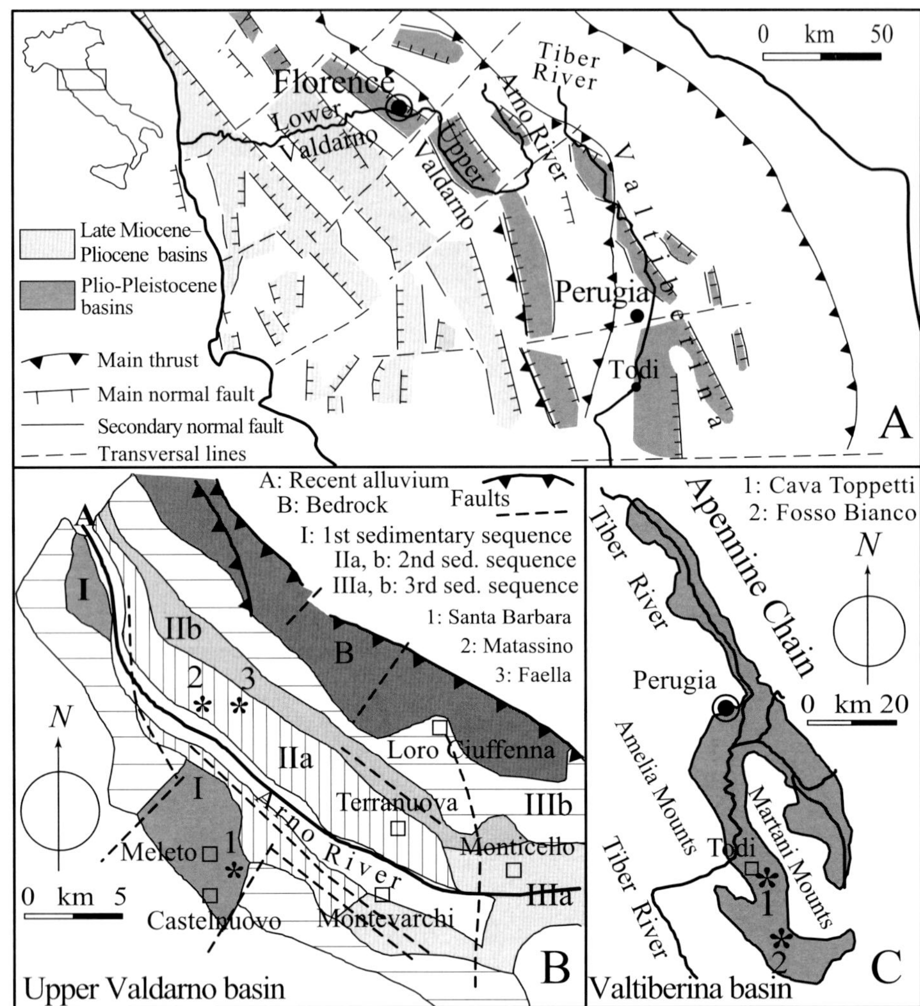

FIG. 1.—**A)** General setting of the Northern Apennine basins after the uplift of the mountain belts (after Martini and Sagri, 1993). **B)** Simplified geologic sketch of the Upper Valdarno basin, with its sedimentary sequences and main sections (after Sagri et al., 1994). **C)** Valtiberina basin (after Basilici, 1992).

cm) cubes were cut for the laboratory experiments. In most cases the bulk susceptibility was obtained from the same specimens, with a sampling density sufficient to reveal the 20 ky precessional cycles. The susceptibility profile for the Meleto Clay, however, was based on 50 cm sampling, that of the Olduvai interval on 20 cm spacing.

Laboratory Procedures

The rock-magnetic parameters vary with mineralogy, grain size, and diagenetic overprints. The natural remanent magnetization (NRM) is generally weak, on the order of 10^{-3} A m^{-1}. But both it and the susceptibility of ca. 10^{-4} SI units are stable, dispersed only in few cases of very low remanence.

Magnetic polarities were determined from the characteristic directional changes obtained from stepwise thermal demagnetization. Figure 2 shows the profiles of NRM, susceptibility, inclination, declination, the latitudinal changes of the virtual geomagnetic pole, and polarity, obtained from the Cava Toppetti section (Valtiberina). The parallelism of susceptibility and NRM profiles attests to ferromagnetic behavior of high integrity. This is substantiated in Figure 3 by the characters enhanced by different rock types. Saturation of a silty sample (Fig. 3A) reveals a low coercivity below 0.1 Tesla and a high one (typical of hematite) not yet saturated at 1.0 Tesla; its subsequent thermal demagnetization according to Lowrie's (1990) procedure shows a high blocking temperature with a minor kink at nearly 350°C, which is enlarged by the susceptibility curve when a change is produced in the magnetic mineralogy. The sample used for comparison (Fig. 3B) has a rapidly saturated mineral phase, typical of ferrimagnetic minerals like magnetite and sulfides, and a blocking temperature much lower typical of greigite, whose thermal alteration is evidenced by the sudden increase of susceptibility. This is even more evident in the Meleto Clay, where the sulfides content decreases with distance from the lignite seam (Albianelli, 1995).

The grain size, according to Oldfield et al. (1985), produces different behavior in the lacustrine sediments of Valtiberina and Valdarno basins. The NRM and susceptibility of the Fosso Bianco profile (Fig. 3C) closely reflect the behavior of Cava Toppetti (Fig. 2), testifying to a uniform fine-grained sediment.

The Santa Barbara section in the Meleto Clay (Fig. 3D) shows a large variability from bottom to top, on which the mineralogical

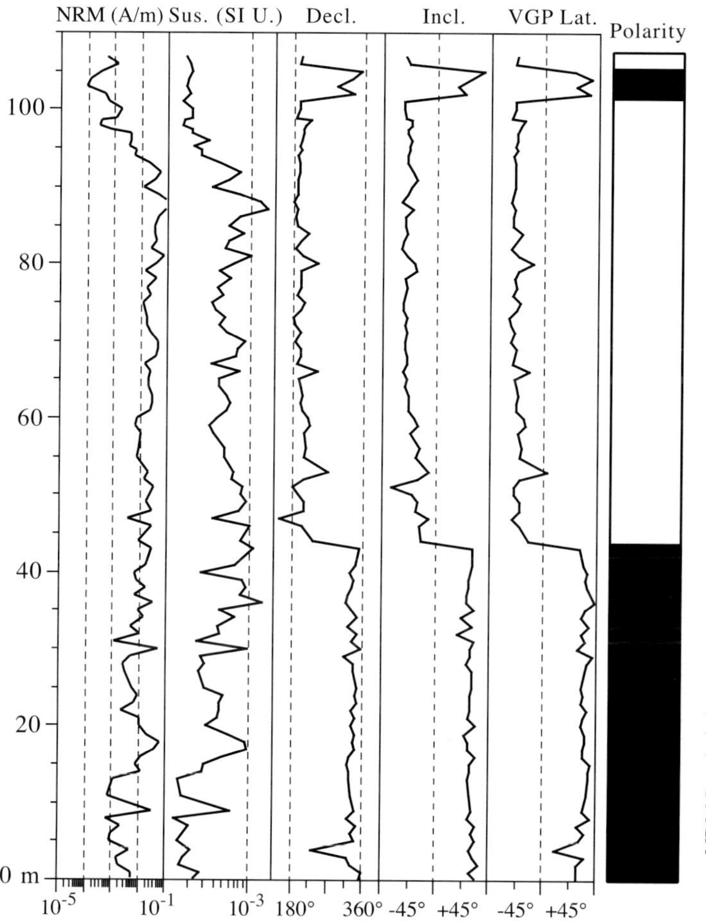

FIG. 2.—Profiles of magnetic parameters in the Cava Toppetti section at Todi (Valtiberina). Note similarity of NRM intensity and susceptibility.

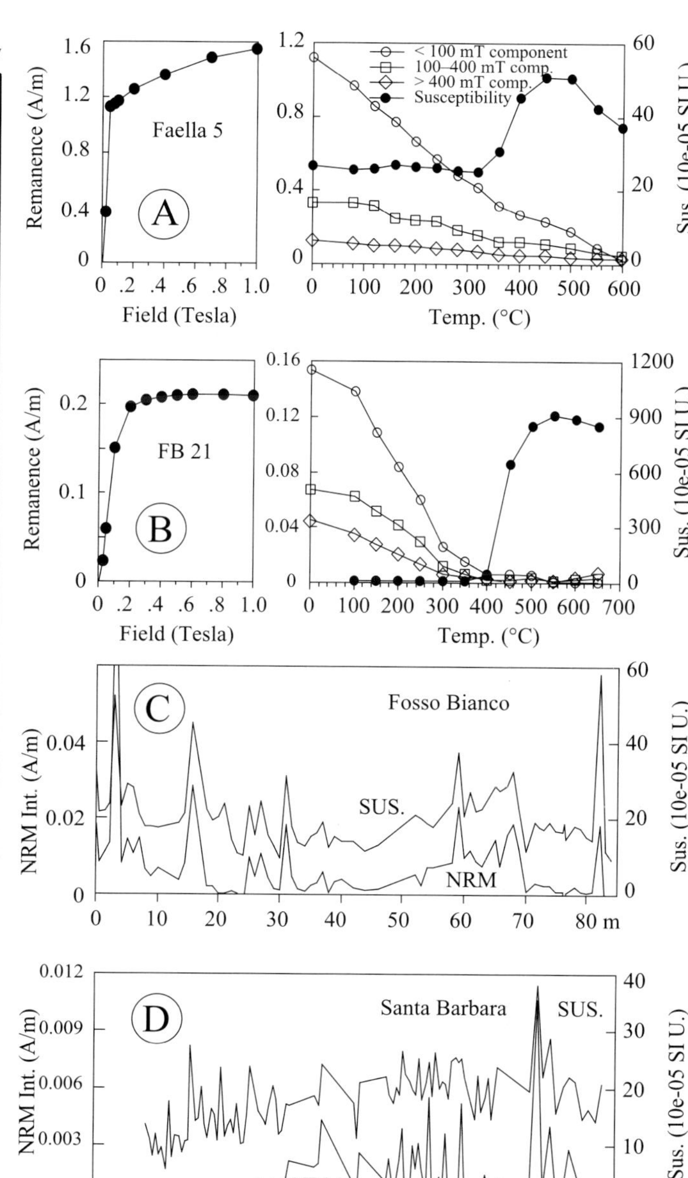

FIG. 3.—Evidence for stable primary magnetization. **A)** The coercivity spectrum for saturation of a sample (Faella 5), from the Olduvaian interval in the Upper Valdarno, under an applied field of up to 1 Tesla (diagram on left) and subsequent thermal demagnetization of IRM, according to Lowrie's (1990) procedure; the blocking temperature is close to 600°C. **B)** Same experiment from the Fosso Bianco profile, Valtiberina, sample (FB21), with lower blocking temperature and larger increase of susceptibility under the effect of higher content of greigite altered into magnetite. **C)** The profiles of susceptibility and NRM with their parallel trend demonstrate the ferromagnetic character of the susceptibility signal; the intensity of NRM is notably higher in the Valtiberina clays of the Fosso Bianco profile than in the next one. **D)** The profiles of susceptibility and NRM in the Upper Valdarno Gaussian sequence of Santa Barbara, whose discrepancies are explained in the text.

differences are superimposed. Weathering of greigite is responsible for the weak NRM in the lower profile because it predominates over the magnetite, and the increasing content of sand in the upper profile produces the discrepancy between the two curves.

MAGNETIC POLARITY FRAME

In the 1.5 My Valdarno profile (Albianelli et al., 1997; Napoleone et al., 2001), complemented by that of the Valtiberina (Fig. 4) (Albianelli et al., 2002), the magnetic polarity record provides the time frame for the geological and palynological events and establishes chronological ties to the GPTS numerical reference section (Berggren et al., 1995). At ca. 3.3 Ma, the basal gravel was deposited on the Oligocene basement. Lacustrine sedimentation started at 3.1 Ma and ended at 2.64 Ma with the beginning of a tectonic phase that tilted the lacustrine series by more than 30 degrees. A lacustrine and deltaic phase recurred between 2.0 Ma and 1.7 Ma. Alternations between warmer and cooler flora began near the middle of the late Gauss chron, with the decay trend of the *Taxodium* pollen percentage, and ended in the latest Olduvai, just before the Pliocene–Pleistocene boundary (Bertini, 1994). Other events were the polarity changes at the Mammoth and Kaena reversed chrons in the Gauss, lasting 3.33–3.22 Ma and 3.11–3.04 Ma, respectively, and the early Matuyama (2.58–2.15 Ma) with its Reunion (2.15–2.14 Ma) and Olduvai (1.95–1.77 Ma)

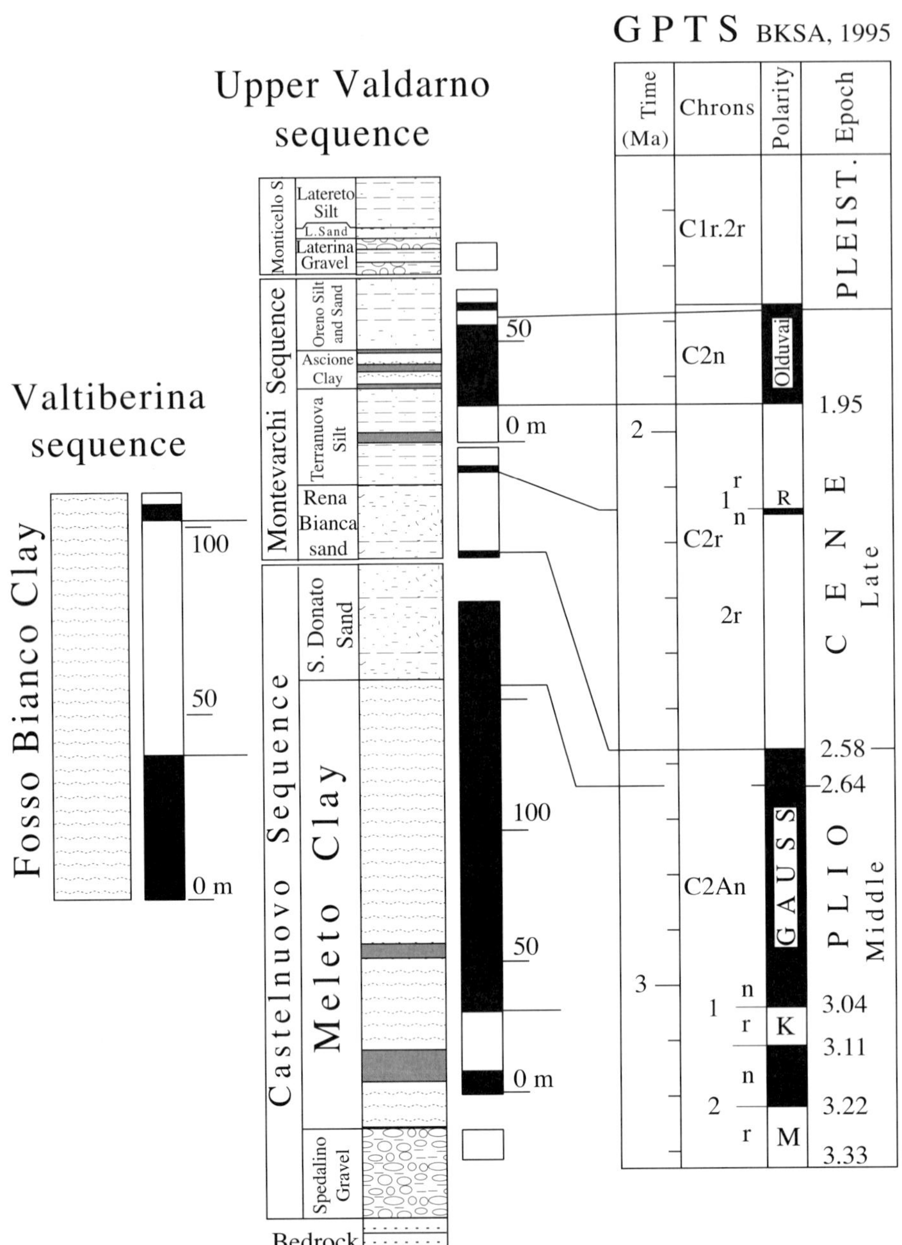

FIG. 4.—Magnetostratigraphy of the Upper Valdarno composite sequence and the Valtiberina sequence correlated to the Berggren et al. (1995) GPTS, extending from the Mammoth (M) and Kaena (K) through Reunion (R) and Olduvai.

intervals. The latter chron is split by a normal event from 1.815 to 1.785 Ma. These dates fix critical boundaries.

The polarity profile in the Valtiberina basin (Fig. 4) extends from the Upper Gauss into the Matuyama, barely beyond the Reunion event (Abbazzi et al., 1997).

Further refinement of chronology depends on extrapolations troubled by variations in accumulation rates. Such rates appeared likely to vary less in the continuous clay sequences than in alternating clay and sand such as shown in the middle parts of the Upper Valdarno sequence. For this reason we sampled the corresponding ages in the deeper Valtiberina basin, which remained in clay facies, to return to the Upper Valdarno basin for Olduvai time, recorded in the second lacustrine development.

ORBITAL CYCLICITY

Late Gauss: Meleto Clay, Upper Valdarno Basin

The Upper Valdarno basin was filled by a complex of shallow lake deposits variously constricted and interrupted by fan deltas (Billi et al., 1991; Magi, 1999; Magi and Sagri, 1996; Magi et al., 1992). The enormous open pit of the Santa Barbara lignite mine, now being filled, greatly facilitated the measurement and sampling of continuous long profiles extending from middle Gauss times through the Olduvai (Fig. 4). In the pit the sequence starts with the lignite bed, which here lies on older sandstones, but in deeper parts of the basin the lignite was preceded by lacustrine clay, sampled from cores. The composite sequence of 156 m was used to establish the magnetic-polarity stratigraphy. Early pollen studies are now being refined by Bertini and co-workers to take advantage of the new detailed chronology provided to patterns of orbital forcing.

The reported dates measured in this profile are based on the good magnetic behavior shown by the laboratory experiments for all the parameters leading to the polarity sequence. The intensity values of the NRM and susceptibility were measured together with the directions (declination and inclination), and compared to the other profiles in the Upper Valdarno and to those shown in Figure 2.

Late Gauss time is represented in the Upper Valdarno basin mainly by the Meleto Clay, but it extends slightly into the overlying San Donato Sand. Recent data (Albianelli et al., unpublished work) suggest that the boundary with the Matuyama chron lies in the subaerial Rena Bianca Sand (Fig. 4). The basal portion of the Meleto Clay includes a ca. 20 m lignite seam, within which lies the base of the Kaena event. A profile measured by Bertini and extending from the base of the lignite to the base of the S. Donato Sand was sampled at intervals not exceeding 50 cm and was used for our susceptibility study.

Toward ascertaining accumulation rates we considered first the entire sequence, extending from below the base of the Kaena to nearly the end of the Gauss chron, and measuring 156 m. The 460 ky allotted to the complete latest Gauss chron (Berggren et al., 1995) implies a mean accumulation rate of 340 mm/ky. But for the clays that figure is probably distorted by inclusion of the more rapidly deposited lignite.

The profile considered in Figure 5 therefore begins at the top of the lignite. The 137.5 m were stripped of periodicities below 4 m and above 70 m by band-pass filtering (Fig. 5A). A fast Fourier transform spectrum extracted by the Stratabase procedure (Ripepe, 1988) (Fig. 5B) shows a large series of peaks culminating in one at 7 m. This aggregation of peaks suggests the presence of several periodicities, multiplied by variations in accumulation rate. An accumulation rate of 300 mm/ky would assign values of 115 ky to the 34.5 m peak at far left and of 23 ky to the tallest. These values lie only 10% above those to be expected for the short-eccentricity cycle and the precessional mean, and we tentatively accept these identifications.

Figure 5C omits the lower 20 m of this time series, sampled irregularly, to start at the top of the Kaena, and shows the series decomposed into a sliding 48 m window moved in increments of 0.4 m. The spectra of these windows are arranged from oldest at top to youngest at bottom. The eccentricity peak is lost owing to the short time intervals captured by each window. Three spectral peaks are well aligned in the early portion and in the latest portion, and are complexly shifted in the middle part of the profile. It seems clear that variable accumulation rates have interfered seriously with the identification of periodicities, and that tuning is called for.

For this we turn to Figure 6, in which the time series has been decomposed into nine segments. In each of these the spectrum for each window has been stretched or condensed to bring the peaks into alignment. This yields a well-defined, simple spectrum for each segment, in which one peak ranges from 13 to 19 ky, one from 22 to 27 ky, one at 31 ky, and another at 44 ky. With exception of the 31 ky term, these correspond to the short and long mode of the precession and to the obliquity cycle. We translated this into simple changes in accumulation rate, of 240 mm/ky for the lowest 20 m, 200 mm/ky for the next 15 m, 290 mm/ky to the following 30 m, and 360 mm/ky for the final 55 m, rising to 390 mm/ky in the last few sandy meters. Tuning the time series to these values simplifies the spectrum (Fig. 5E) to three major peaks, those for obliquity (40 ky), precession (19 ky), and an intermediate one at ca. 31 ky. Its refinement is not increased by the stacked values in Figure 5F. The tuned and filtered time series (Fig. 5D) displays the individual precessions, with a taller peak at 5–6 precession intervals to mark the ca. 100 ky short-eccentricity cycle.

Breaking this time series into halves (Fig. 5G, H) shows that the lower half has lost its obliquity signal to the 32 ky peak of uncertain origin, whereas in the upper half a 40 ky obliquity signal overshadows the precession.

Late Gauss–Early Matuyama and Reunion: The Fosso Bianco Clay, Valtiberina Basin

The early Matuyaman sands of the Upper Valdarno basin in the San Donato and Rena Bianca units lack continuity of exposures and were not sampled thoroughly. We therefore now turn to the Valtiberina basin, near Perugia (Fig. 1C), where a deeper and long-persistent lake accumulated the thick Fosso Bianco Clay. Mid-Villafranchian mammals in the overlying fluvial deposits (Abbazzi et al., 1997) imply that the upper, magnetically reversed portion of this formation is Matuyama and the lower, magnetically normal portion Gauss. The clays are notably silty, but their magnetic behavior during demagnetization is exemplary. Two profiles were measured, and sampled. The Cava Toppetti profile was obtained in the large, active pit of the ceramic company of that name. A 110 m section begins in the Gauss and extends through the early Matuyama, slightly beyond the Reunion interval (Figs. 2, 4). A fault at the 80 m level implies that part of the Matuyama sequence is missing.

Exposures in the Fosso Bianco gorge yielded an 84 m profile, which is reversely magnetized throughout, missing the Gauss–Matuyama contact and falling short of the Reunion event. Thus the early Matuyama is incomplete in both profiles, the one for loss of a portion to a fault, the other also for lacking base and top. We are therefore obliged to approach the timing of cycles by trial and error.

The susceptibility profile for the Cava Toppetti section (Fig. 7A), processed like that of the Meleto Clay, yielded a spectrum of three major periodicities at 17.7, 8.8, and 6.3 m, and a minor one centered on 4.6 m (Fig. 7B). Assuming a mean accumulation rate independent of magnetostratigraphy at 200 mm/ky (Fig. 2), this leads to the durations of 89, 44, 31, and 23 ky, fairly matching the astronomical rhythms. The evolutive spectral plot (Fig. 7C) shows little variation in accumulation rate and stacked averages of 100 and 33 and 20 ky for the eccentricity, obliquity, and precessional periods. However, the upper profile (lower portion of the spectral lines) actually does not show the obliquity peak whereas the other two are dominant, and an adjustment by splitting the diagram would improve its resolution.

The length of the interval between the base of the Matuyama chron and the base of the Reunion was estimated at 430 ky by Berggren et al. (1995). This estimate yields a length of 86 m for the thickness of these sediments at the Cava Toppetti, in place of the 58 m measured, suggesting a stratigraphic displacement of 28 m by the fault encountered.

The Fosso Bianco profile (Fig. 8A) measured 84 m of magnetically reversed sediment, implying a sequence still 2 m shorter than that restored at the Cava Toppetti. The spectrum (Fig. 8B) shows a multiplicity of peaks, at a slightly higher accumulation rate (240 mm/ky). Considering the missing portion, the duration of this profile could be estimated close to 400 ky, which would yield 96 m at the above mentioned rate of 240 mm/ky, and a reduction of its thickness by 12 m. The disturbance in the mid section could still be an effect of the main fault.

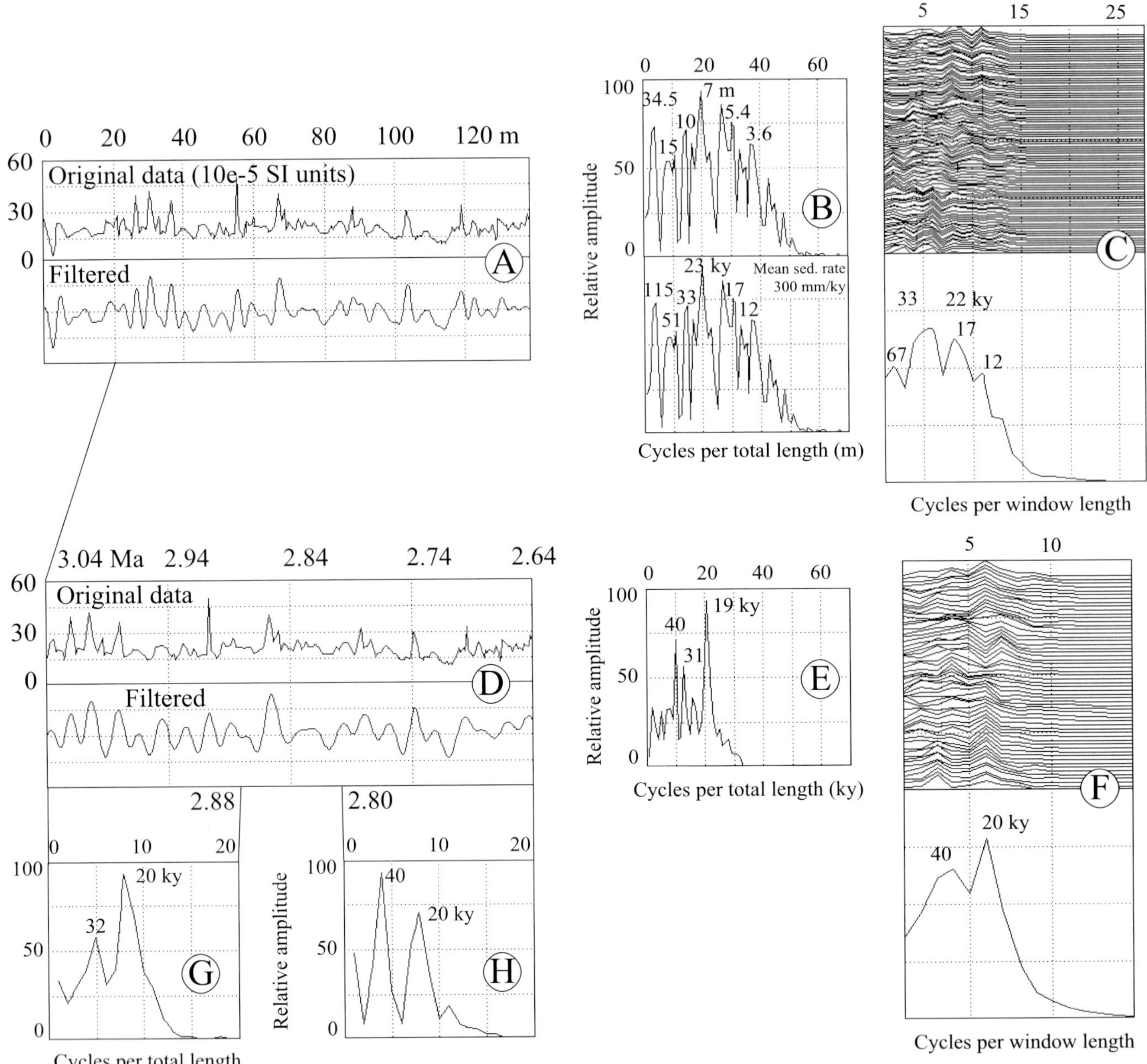

FIG. 5.—**A)** Magnetic susceptibility profile for the Upper Valdarno Meleto Clay in original values and after a low-grade bandpass filtering. **B)** Its overall spectral content with periods labeled in meters, and its conversion to time by applying an average accumulation rate of 300 mm/ky. **C)** The FFT spectrum and its evolutionary decomposition by a moving window, beginning at meter 20. Tuning to variations in accumulation rate (Fig. 6) yields the time series and spectra of **D, E,** and **F. G, H)** Spectra of lower and upper halves of this tuned time series.

In the evolutive spectral plot (Fig. 8C) the interval has the character of the early Cava Toppetti sequence: steady accumulation reflecting the obliquity and precession cycles. Noteworthy is the great strength of the obliquity signal in the upper part of the sequence (spectra in Fig. 8D, E). Chaos prevails in the intervening mid-section of Figures 8A, C. The absence of such chaos at the Cava Toppetti suggests that its origin is to be sought not in a regional disturbance of the sedimentary regime but in a localized one.

Despite the excellence of these spectra, the two susceptibility traces cannot be readily correlated. We suggest that this derives from two factors. Interference from a strong obliquity signal, and a sampling rate which, at ca. 5 ky intervals, is marginal for revealing a cyclicity as unstable as is that of the precession.

Olduvai: Montevarchi Silt, Upper Valdarno

In Olduvai time the Upper Valdarno basin again came to be occupied by a lake in which clays, silts, and sands recorded the orbital variations in fluctuations of magnetic susceptibility. Near Faella (Fig. 1B), the nearly 65 m of strata from the Terranuova Silt, the Ascione Clay, and the Oreno Silt-and-Sand unit record almost

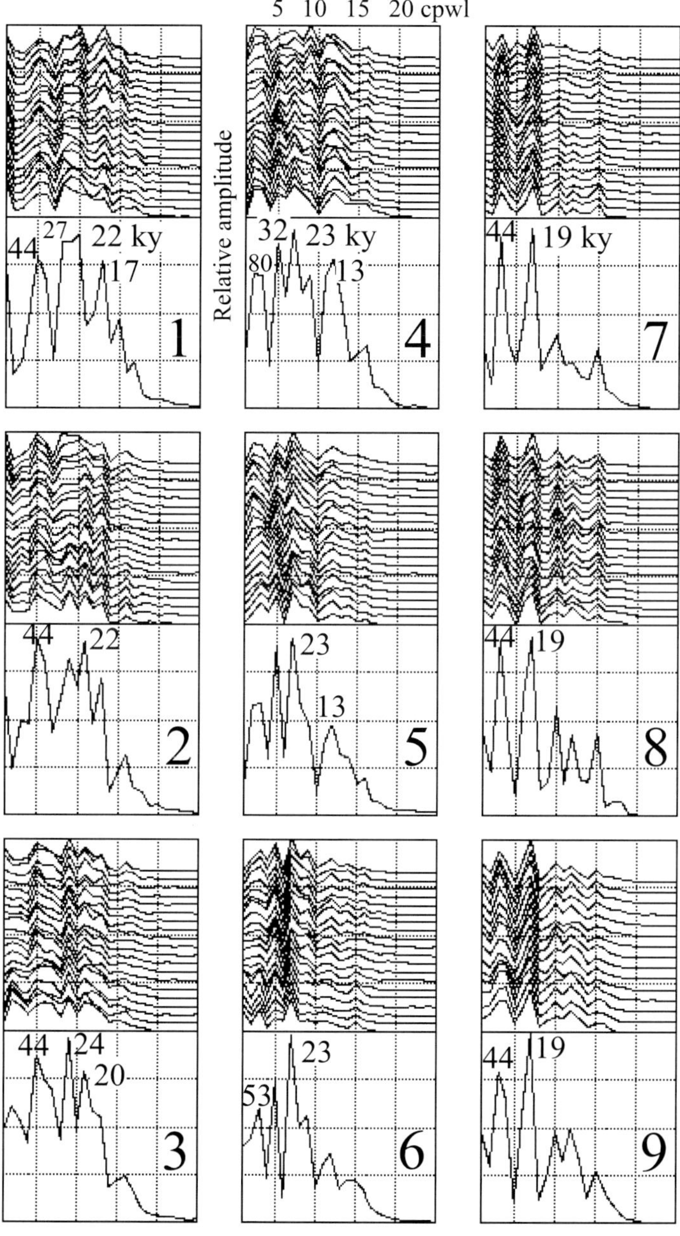

FIG. 6.—Tuning the Meleto Clay susceptibility time series (Fig. 5) by evolutive spectra. The evolutive spectrum, stepwise from bottom of the same time series as in Fig. 5C, is shown in nine steps. For each of these the spectra have been extended or shortened to bring corresponding peaks into line. Resulting time series and spectra are those of Figure 5D–H. Moving window spectra: window length is 48 m, shift is 0.4 m.

all of the Olduvai interval. Its full duration of 180 ky in the GPTS is represented by the composite section with the Tasso tunnel section (Albianelli et al., 2002). The lower Olduvai, represented by the interval before the split in the stratotype section of Vrica, lasted for 135 ky (Zijderveld et al., 1991). Its thickness at Faella is 47 m, with an apparent mean accumulation rate of 350 mm/ky. Samples at 20 cm intervals imply a sampling rate better than 600 years.

The susceptibility plot and its spectrum (Fig. 9A) show cyclicity peaks at 34 and 10 m, corresponding to 97 ky and 28 ky. The 135 ky duration of the lower Olduvai provides two tie points with Laskar's (1990) astronomical curve for the precessions over this 1.950 to 1.815 Ma interval. This implies about 6.5 precessional cycles, as shown on the target curve, with a mean thickness of 7.4 m. That is indeed apparent, but the ca. 600 year sampling revealed an oscillation of higher frequency, with ca. three oscillations per precessional cycle, for a period of ca. 7 ky. Matching sets of precessional peaks against the astronomic target curve requires only minor adjustments in accumulation rate, and yields the spectral correspondence shown in Figure 9B.

Correlation with the Astronomical Curves

The time series, reconstructed through the profiles of Santa Barbara, Cava Toppetti, and Faella pits and tightened to the magnetic constraints of the GPTS, served to test the correspondence of the calibrated susceptibility curves to the astronomical eccentricity and precession target curves of Laskar (1990) (Fig. 10). This correspondence is striking, considering that no direct tuning at all with the astronomical curves was applied. In turn, the Santa

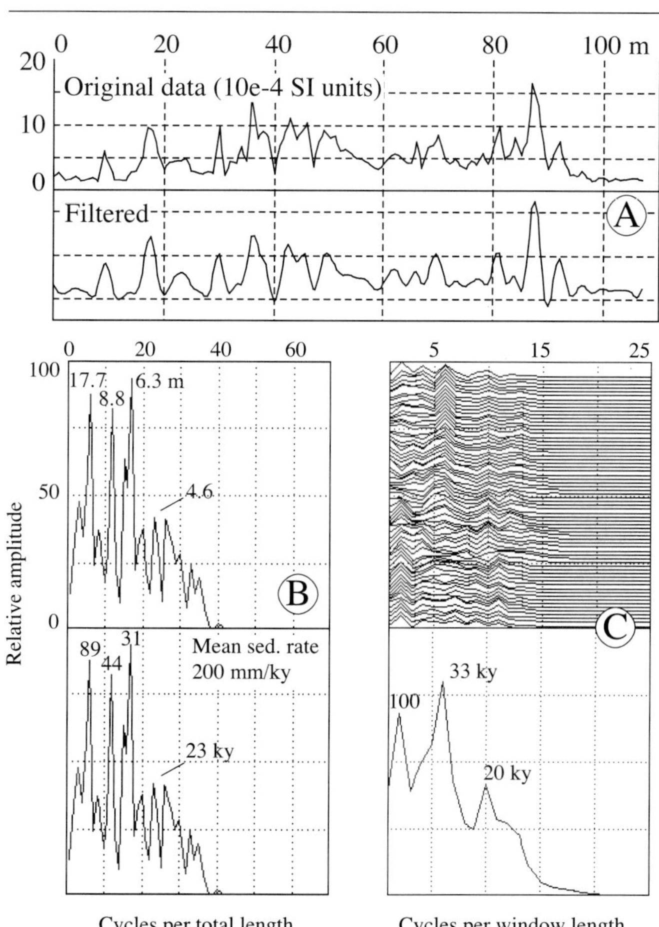

FIG. 7.—Magnetic susceptibility of Fosso Bianco Clay at Cava Toppetti (CT), processed in the manner of the Meleto Clay sequence (Fig. 5). **A)** Raw and filtered time series. **B)** Spectra in meters and in time for a constant accumulation rate of 200 mm/ky demonstrate a rather stable sedimentation rate. **C)** The evolutive spectrum enhances how the changes in the accumulation rate disturb the spectral distribution, which would be improved by a minor shift to higher rates in the upper section.

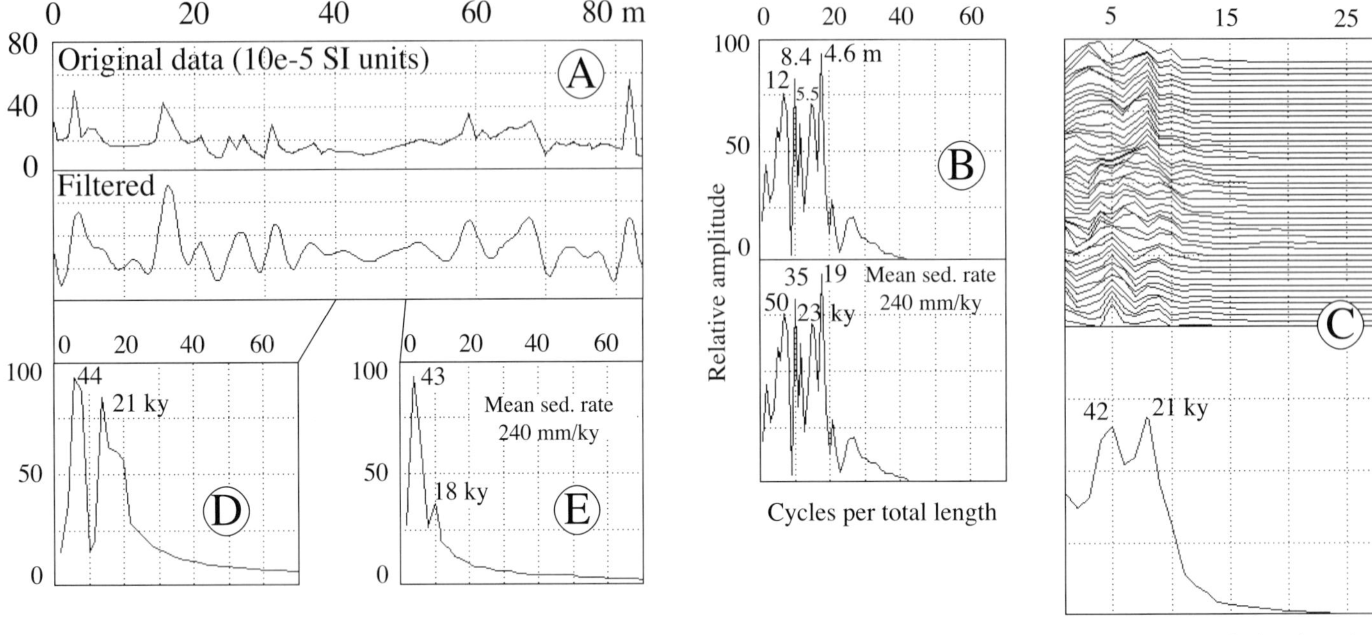

FIG. 8.—Susceptibility profile of Fosso Bianco sequence, Valtiberina basin, processed like Cava Toppetti sequence (Fig. 5). **A)** Susceptibility series, raw and filtered time series. **B)** Spectra in meters and converted to time assuming an accumulation rate of 240 mm/ky. **C)** Evolutive spectra by moving window reveals chaotic mid-portion, and similar shift to higher rates in the upper section. **D, E)** Spectra of lower and upper parts, showing subequal strength of precessional and obliquity cycles in lower part and predominant obliquity signal in the upper.

Barbara and Cava Toppetti profiles show an opposite correlation with the eccentricity curve. The sediment in the former lake were deposited under a higher supply from the catchment area during the drier conditions, which in the Upper Valdarno are produced under cooler conditions, as at the eccentricity minima (Shackleton et al., 1995). The erosion of the Oligocene sandstone yields a more abundant supply of the magnetic minerals. In contrast, the increase of erosion in the Umbrian limestone and marl produces a dilution of the magnetic particles in the Valtiberina basin. This results in a direct correlation between eccentricity and susceptibility, as shown in the pelagic limestones at Piobbico.

In this general frame, also two major events can be placed. One, at the eccentricity minimum at ca. 2.82 Ma, is close to the change when obliquity became a dominant susceptibility signal in the Santa Barbara section (Fig. 5H), in the interval 2.88–2.80 Ma, and the pollen record began to decline towards a cooler climate (Albianelli et al., 1999). The second minimum, at ca. 2.4 Ma, corresponds to the ingression of a notable amplitude in the short eccentricity at the Cava Toppetti section (Fig. 7C).

DISCUSSION

Lakes in general are ideal sediment traps, and they tend to show good continuity in accumulation. Interruptions result from filling by deltaic sands or from desiccation of the lake, both readily identified. Response to snowmelt and storms induces strong variations in accumulation rates over short time scales. These tend to average out in the Milankovitch frequency band, yet there the effects of climatic and tectonic changes on accumulation rate become notable. Raw spectra of long time series are therefore very complex, but evolutive spectra with moving windows generally serve well to identify the main frequencies, to which the accumulation rates can then be tuned, as shown in Figures 5, 6, 7, and 8.

Cyclicities Recorded

The profile of the Olduvaian deposits at Faella in the Upper Valdarno basin was sampled at intervals of ca. 600 years. Matched to the astronomical target curve (Fig. 9) this reveals a trans-Milankovitch oscillation with a period of ca. 7 ky. These cycles are arranged in triplets within the precessional oscillation.

Precessions are well resolved in time-series profiles as well as in spectra either in the Faella profile of the Olduvai chron or in the Gaussian Meleto clay, where, with less than 50 cm spacing, most of the sequence was measured at 1–2 ky intervals. In the Valtiberina sections, on the other hand, sampling ca. 5 ky intervals was adequate for resolving the precession in spectra but not in the time-series profiles.

An obliquity cycle is generally present, but at times it becomes dominant, as in the upper part of the Meleto Clay (Fig. 5) and the lower part of the Fosso Bianco Formation (Fig. 7).

The 100 ky short eccentricity, generally displayed prominently in geological time series (e.g., Fischer, 1986; Premoli Silva et al., 1989; Ripepe and Fischer, 1991; Schwarzacher, 1975, 1993), is not very conspicuous in the susceptibility time series here assembled, though it consistently appears in spectra.

The 400 ky eccentricity cycle, the most stable of the orbital variations, is generally not displayed as prominently in geological time series as the 100 ky short-eccentricity cycle. The Upper Valdarno profiles are not extended sufficiently to test for the 400 ky periodicity, but in the Valtiberina (Cava Toppetti, Fig. 10) record the 400 ky signal stands out clearly. High amplitudes of susceptibility match times of maximal eccentricity, when the

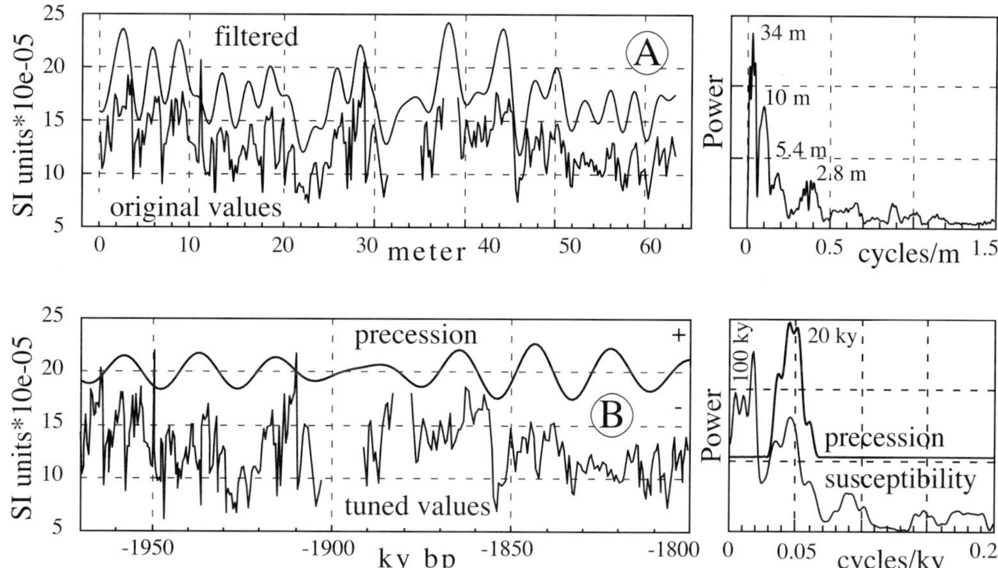

FIG. 9.—**A)** The profile of the 63 m section at Faella, sampled at 20 cm (ca. 600 years), spectrum alongside in meters. **B)** Same, tuned to the astronomical target curve for precession (Laskar, 1990), for the date fixed by magnetostratigraphy, and with respective spectra.

precessional alternations between low and high seasonality reach maximal contrast.

Relation to Other Studies

The chronology provided by this study will find application in coordinated studies of climatic changes. Palynological studies by Bertini and co-workers (Pontini and Bertini, 2000; Pontini et al., 2002), partly on the very samples here used, show lively fluctuations of a complex nature. These involve on the one hand the balance between the moist subtropical elements dominant prior to 2.84 Ma and the fluctuating climates that followed. This change coincided with a 400 ky minimum in eccentricity (Fig. 10) and with the change from precessional to obliquity dominance in cyclicity. The botanical change brought alternations of the subtropical vegetation with precursors to the present warm temperate broadleaf forest. They also brought episodes of steppe conditions recorded by floods of *Artemisia*, and of fir and spruce pollen related to the rise of the Apennines. The picture is far from simple, and the orbital variations recorded will serve not only as a chronological frame. The revelation of individual precessional and obliquity cycles in densely sampled time series implies opportunity to tie pollen profiles to specific precessional and obliquity events, with significance to biotic history and to climatology.

The Tie of Susceptibility to Orbital Cyclicity—A Problem

Various cyclostratigraphic studies on marine sediments (e.g., Weedon et al., 1997) have utilized magnetic susceptibility as an index to orbital forcing. In these cases fluctuations in magnetic susceptibility have been attributed to variations in the ratio of detrital matter to authigenic carbonate, and serves as a proxy for orbitally driven variations in carbonate productivity or supply of detritus. Such is not the case in these lacustrine clays. These lack an appreciable carbonate component yet show consistent fluctuations of susceptibility with orbital (Milankovitch) periods.

The magnetic susceptibility of the clay fraction, typically paramagnetic, is largely a measure of its magnetite content. Why this should have fluctuated with the orbital variations in precession, obliquity, and eccentricity remains unclear.

The association of susceptibility with clay is not altogether simple. Places associated with the lignites tend to yield poor data, whereas sands associated with the Olduvaian sequence studied yield good information. The best magnetic results were obtained from the silty clays in the Valtiberina, deposited somewhat slowly and in what appears to have been a deeper lake (Basilici, 1992, 1997).

As noted by Bloemendal (1980) and Thompson and Oldfield (1986), the amount of detrital magnetite supplied to lakes may be influenced by climatic fluctuations in two ways. Weathering eventually destroys all magnetite and other authigenic ferromagnetic minerals, and its higher rate in warm, moist climates can be expected to supply less magnetite in a given setting than would drier and cooler ones. It must also vary with erosion rate: times when soils were stripped deeply must increase the supply of magnetite.

Another factor may be a variable eolian supply derived from afar, richer or poorer in magnetite than the clay derived from within the drainage basin.

Yet another possible source of variation is the contribution by magnetotactic bacteria. These inhabit settings in which a redox gradient provides a good setting for chemolithotrophy, and they are widely distributed in lakes and ponds as well as marine settings (Blakemore, 1975; Kirshvink et al., 1985; Petersen et al., 1986). It seems likely that abundance of bacteria should vary with the intensity of redox gradients at the water–sediment interface, a matter largely dependent on the rate at which organic matter is supplied. That, in turn, is governed by many factors, including climate.

How these factors interacted in these Italian lakes remains to be discovered.

Iron Flux and Magnetic Signals

The magnetic record of orbitally forced climatic oscillations is here transmitted by way of variations in iron entrapment in lakes.

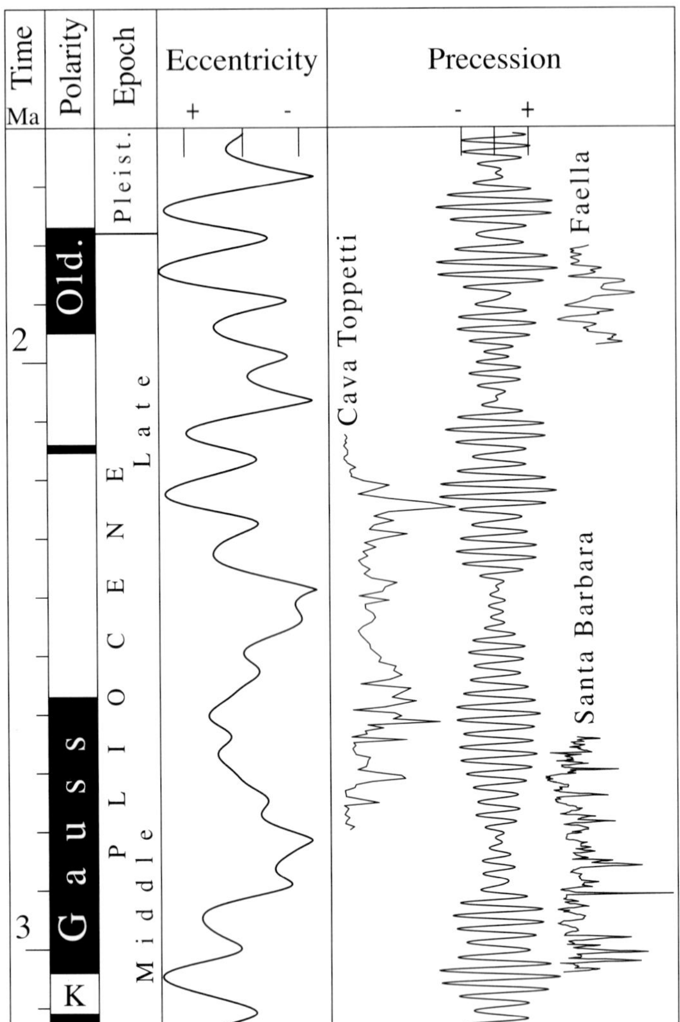

FIG. 10.—The Upper Valdarno and Valtiberina profiles from the Kaena through Olduvai chrons compared to the astronomical target curve (Laskar's 1990). Note the correlation of susceptibility amplitude with the 400 ky peaks in eccentricity.

Weathering in the source area, probably with involvement of iron bacteria, transforms ferrous and semi-ferric iron minerals into the ferric state, and although some of this ferric oxide or hydroxide forms nodules or crusts, much of it comes to be adsorbed by clay minerals to yield the familiar brown and red soils. Climatic factors may control the form of the iron delivered to a lacustrine depositional site, by varying the rate of weathering and rates and patterns of erosion, but these are hardly likely to yield the sharp response to Milankovitch forcing revealed in this study. This seems to call for direct response of the depositional system to orbital forcing.

Such a response is demonstrated in the Miocene sediments of Lake Pannon, described by Sacchi and Muller (this volume). In a two-million-year record of the late Miocene, median grain size fluctuated regularly between clay and fine sand, in 20 ky cycles representing the precession, roughly bundled into groups corresponding to the 100 ky eccentricity cycle. It would appear that the hydrodynamics of the limnic system oscillated with climate, to produce regular modulations in the clay–silt (sand) ratio in the sediments deposited at any one place. Diagenesis in turn altered the clay-adsorbed iron to greigite. Proof will have to await granulometric studies, but for the present this appears compelling.

CONCLUSIONS AND OUTLOOK

1. The lake clays in the Plio-Pleistocene intramontane basins of the Apennines show a well-developed magnetic stratigraphy. This is expressed on the one hand in a magnetic-reversal stratigraphy tied to geochronology, and on the other in regular fluctuations in magnetic intensity and susceptibility, resulting from variations in content of fine-grained magnetite.

2. The relations of this stratigraphy to the magnetic-reversal chronology shows that magnetite content fluctuated with orbital forcing. A 600-year spacing of samples revealed a sub-Milankovitch periodicity of ca. 7 ky. The 1–2 ky sampling revealed precession cycles in both time series and spectra, whereas 4–5 ky sampling served to reveal the precession in spectra only. The obliquity signal is variable but generally less powerful. Precessional signals are weakly modulated by the 100 ky eccentricity, and the longest of our profiles shows a strong reflection of the 400 ky eccentricity.

3. The persistence of this variation, in space and time, suggests that it is a normal feature of orbital forcing in lacustrine clays. Amongst the possible causes are climatically induced variations in the rapidity of weathering in source terranes, climatically driven variations in the stripping of the regolith in source terranes, and eolian introduction of magnetite-rich dust from outside the drainage basin. How these factors interacted here remains to be discovered.

4. This magnetic cyclostratigraphy is currently being put to work. On the one hand, it provides a highly refined chronology for palynological studies aimed at reconstructing the vegetational history of the region, in this time of change from the warmer, moister climates of the early Pliocene to the periglacial regimes of the Pleistocene. On the other hand, it links the vegetational–climatic record to orbital forcing, the process to which much of this change may be attributable.

5. Many questions remain to be resolved. We are only beginning to understand the connections between orbital forcing of climates, between climates and organisms, between climates and sediments, and specifically between climates and the magnetic record.

6. The chronologies being established here will on the one hand link to similar chronologies of Plio-Pleistocene climates and organic responses elsewhere, to provide a far more precise history of the changes that led up to the great Pleistocene glaciations. On the other hand, the nature of these magnetic, lithic, and biotic responses to orbital forcing will also come to provide insight into the ways in which all of these factors interact, in how the outer earth works.

ACKNOWLEDGMENTS

The paleomagnetic surveys in continental basins in peninsular Italy were supported by the Italian Ministry of Education to the University of Firenze Grant Program. Early measurements of the Valdarno sequences were supported by the Italian Consiglio Nazionale delle Ricerche and the Natural History Museum of Firenze. Maurizio Magi and Giorgio Basilici provided all the

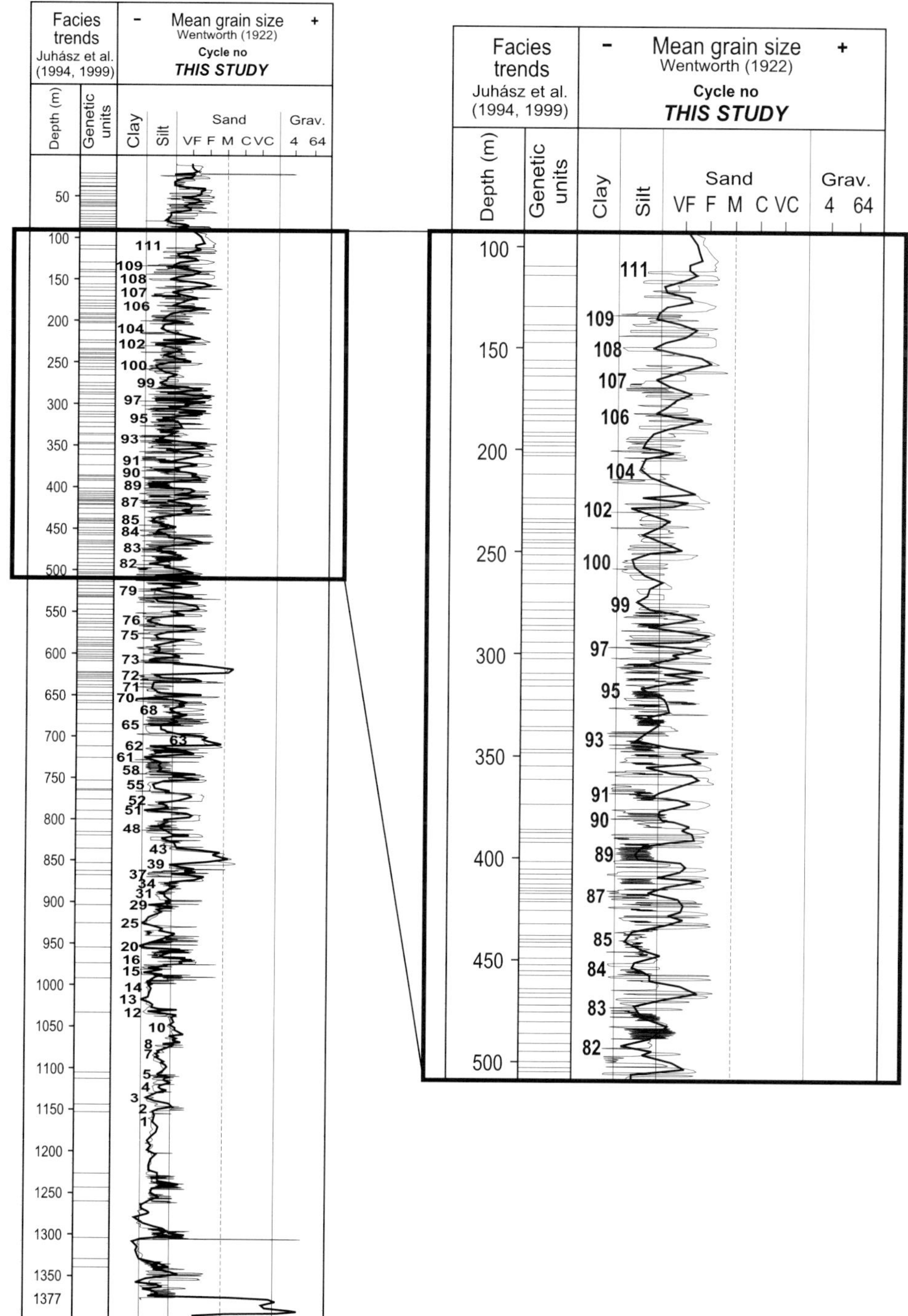

FIG. 13.—Recognition of stratigraphic cycles in the Iharosberény-I section based on variations of the (smoothed) mean grain size against depth. Horizontal scale of the diagram is expressed in terms of the classic Udden–Wentworth scale for clastic sediments (Wentworth, 1922). Correlation with facies trends (genetic units *sensu* Homewood et al., 1992) is also shown.

maxima. Moreover, individual coal (mostly lignite) seams or clusters and/or organic-rich silt/clay layers often mark minima of the curve of mean grain size in marginal lacustrine and delta-plain facies at depth of 750 m to 50 m beneath the surface.

Correlation of individual minima in mean grain size to precession minima and summer insolation maxima was substantially confirmed along almost the entire cored succession, with the exception of the interval between ca. 1000 and 640 m (between cycles 71 and 15), where the minima of the smoothed curve of mean grain size appear to correlate with short-eccentricity cycles (ca. 100 ky) instead. This apparent contradiction can be resolved by considering that within such an interval, which includes three long-eccentricity (ca. 400 ky) cycles (between ca. 7.85 Ma and 9.08 Ma), depositional rates are significantly lower than elsewhere in

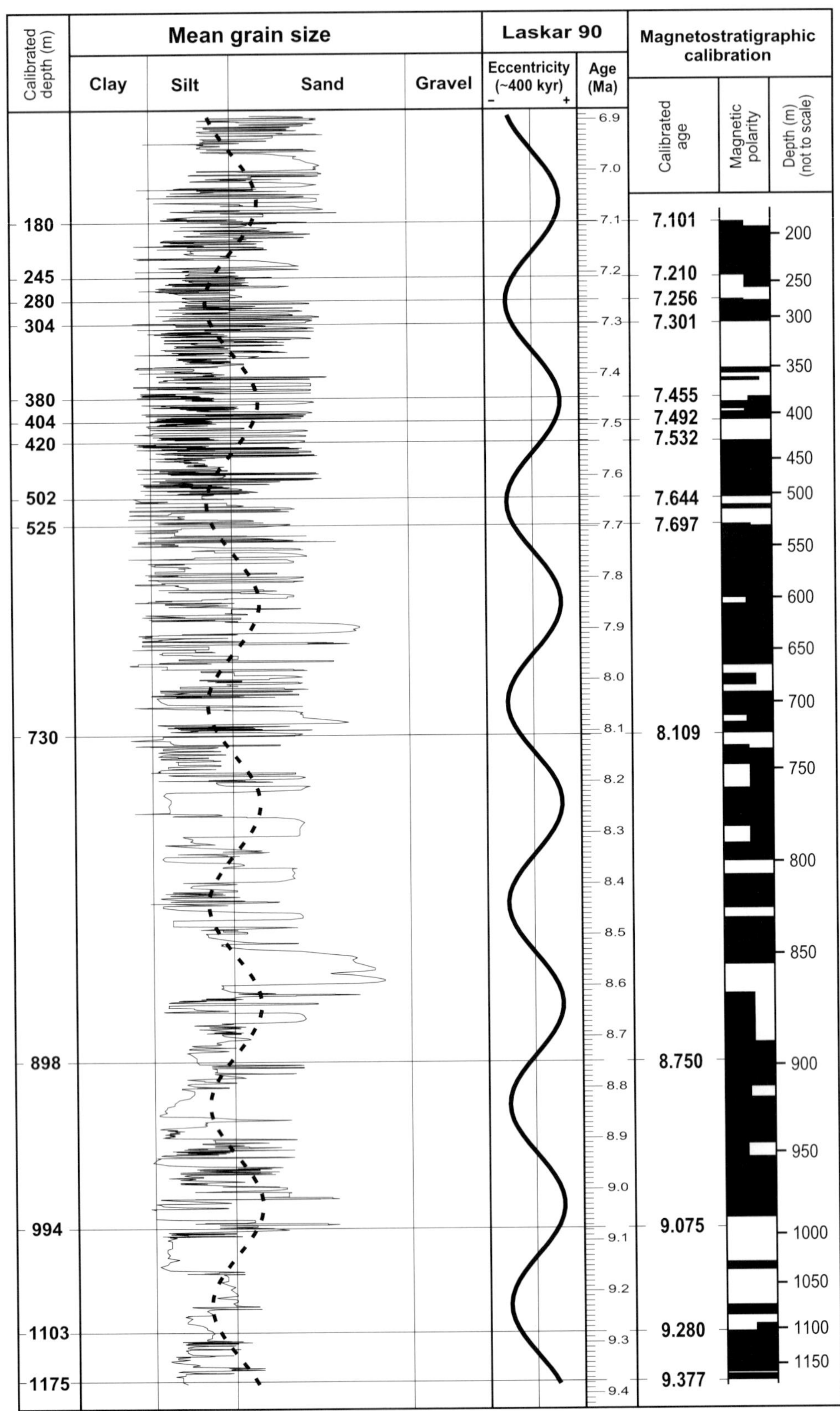

Fig. 14.—Magnetostratigraphic calibration and conversion of the vertical scale of the grain-size curve and magnetic-polarity profile of the Iharosberény-I well from depth to time. The phase relations between astronomical and sedimentary cycles ensure a first-order pattern matching of the grain-size curve to the long (400,000 year) eccentricity curve.

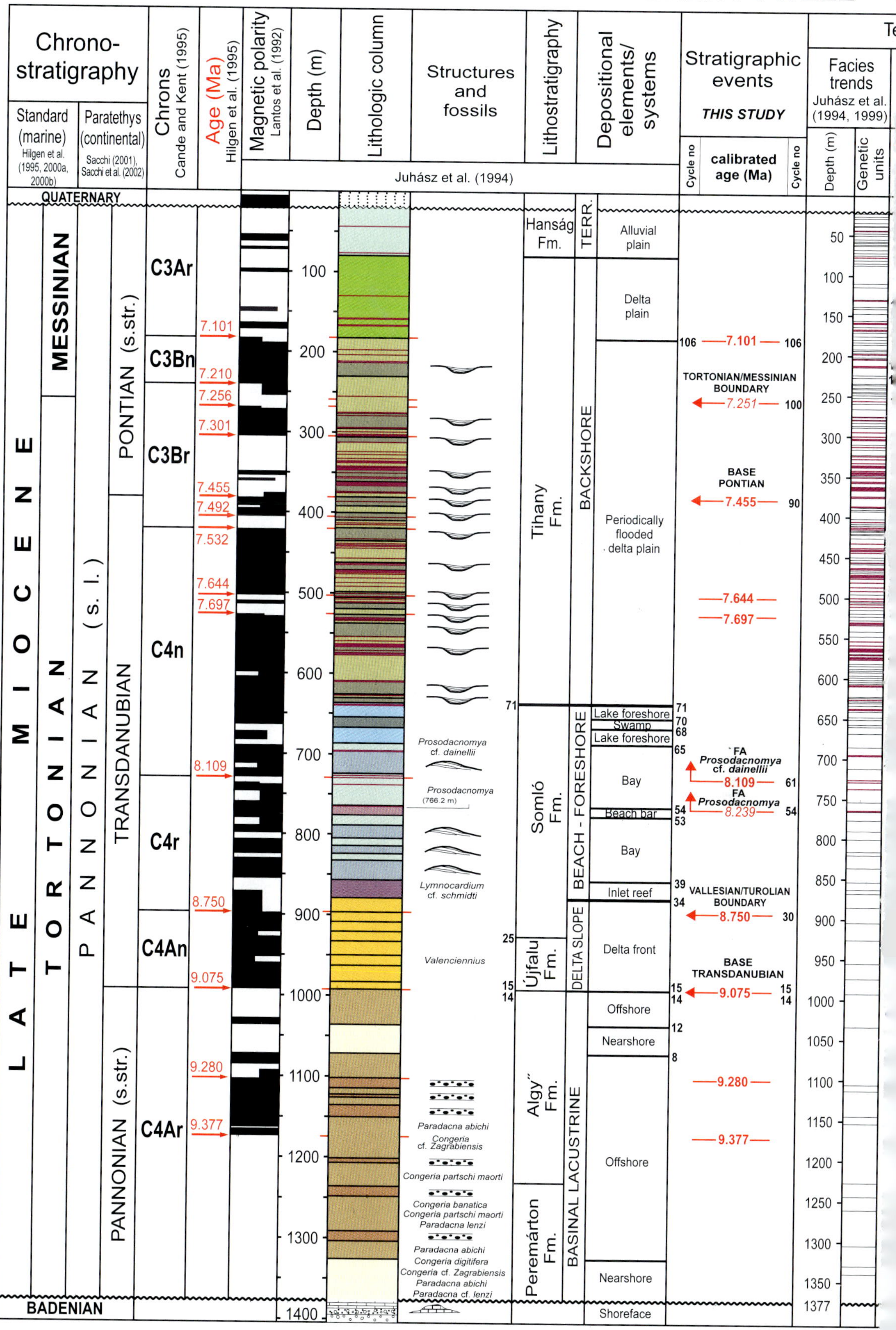

FIG. 15.—Integrated stratigraphy and astronomical calibration of the Upper Miocene (Tortor

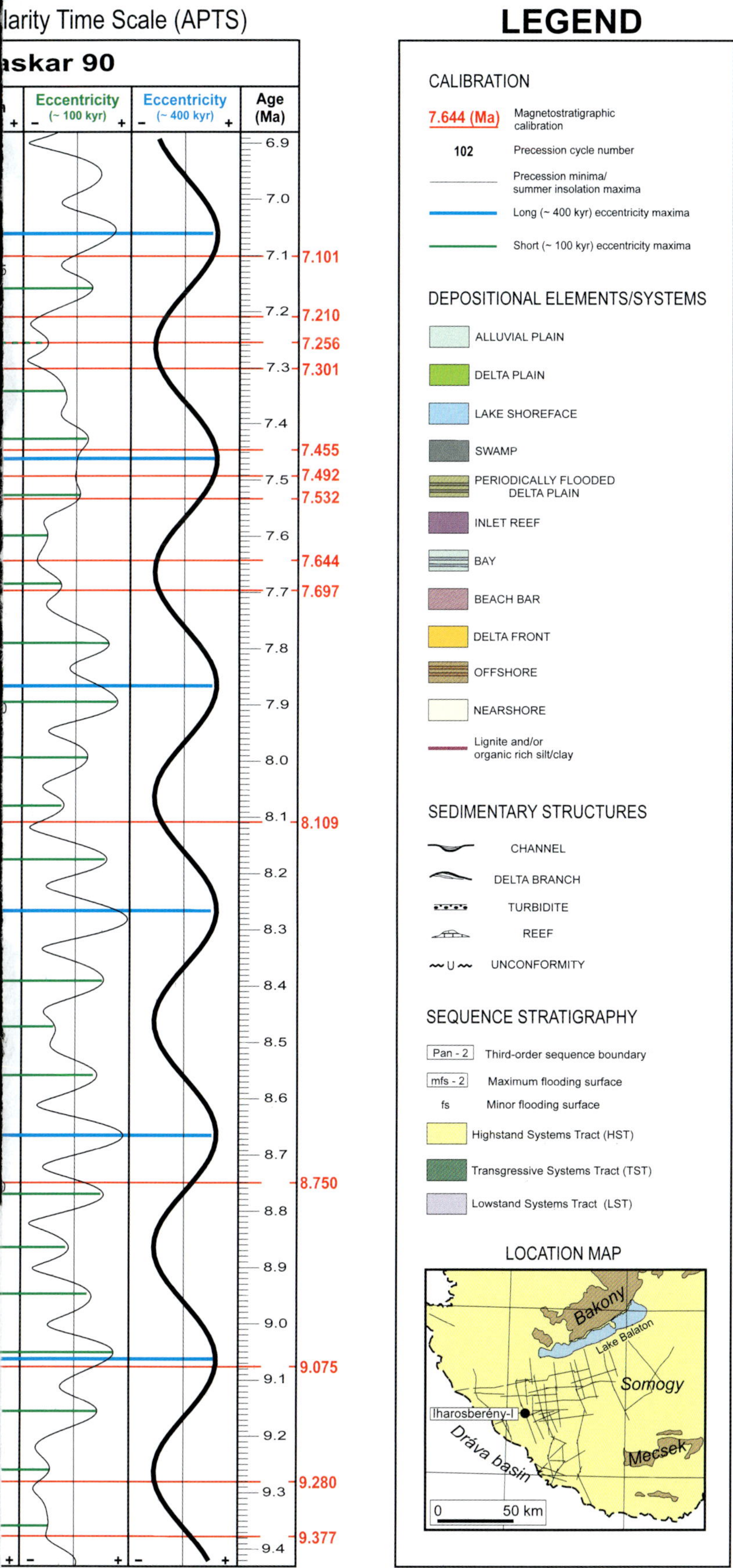

nnonian basin, Hungary (after Sacchi, 2001).

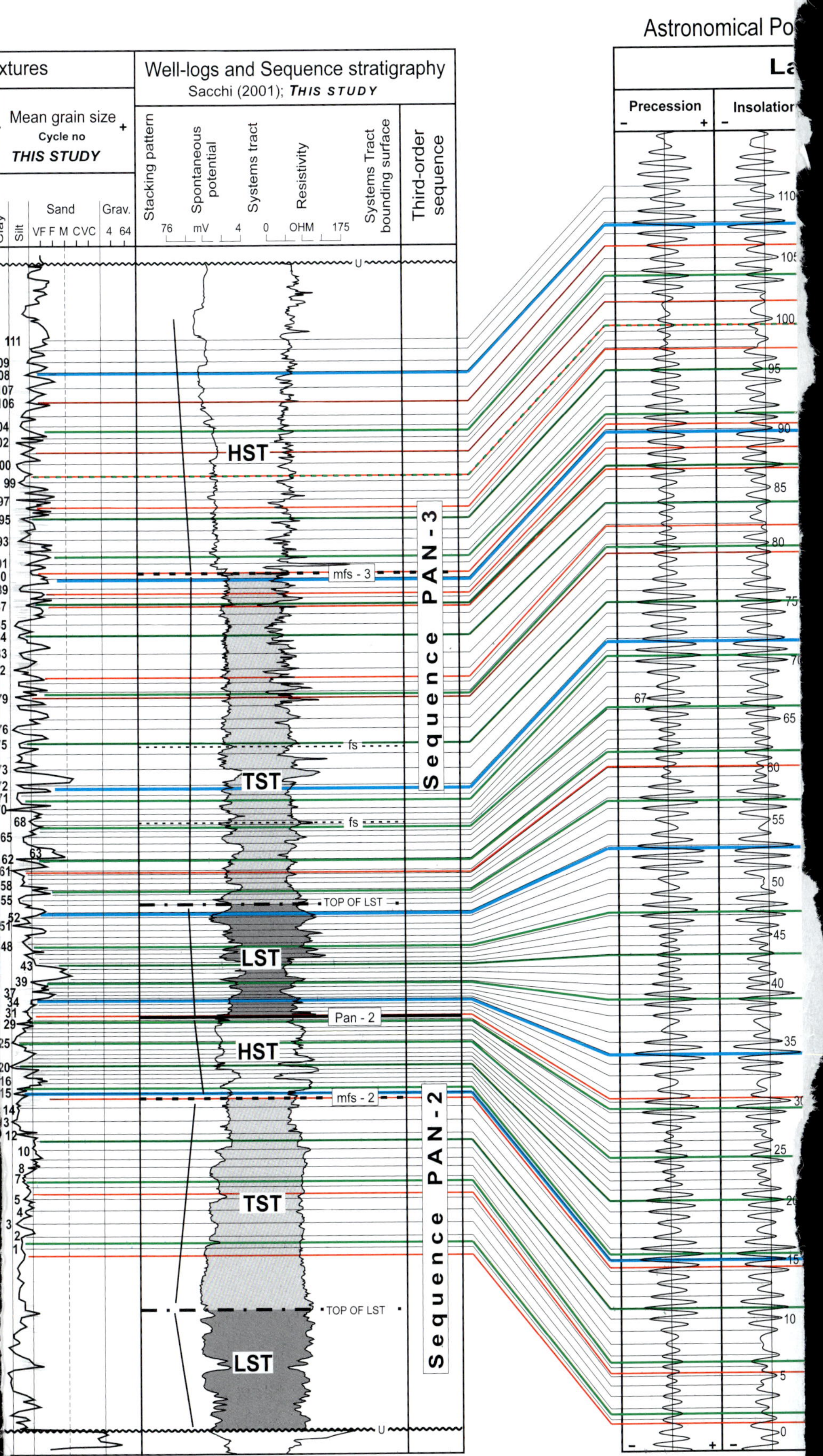

[T]an–Messinian) continental strata cored at the Iharosberény-I well site, western Pa

the stratigraphic column (Fig. 16). As a consequence the resolution of the smoothed curve of mean grain size in this specific interval falls beyond the precession-related frequency of the original (non-smoothed) curve of mean grain size. Nevertheless, minima of the curve of original mean grain size display a fair correlation with precession minima and summer insolation maxima (Fig. 15).

The genetic units described by Juhász et al. (1994, Juhász et al. (1997), and Juhász et al. (1999) display a very wide range of thicknesses through the borehole section, varying from several tens of meters (max. 150 m), mostly in the basinal lacustrine, delta slope, and beach foreshore settings at depth between 1377 m and 750 m, to a few meters (min. 4 m), particularly in the delta-plain settings, between 750 m and 24 m (Fig. 15). Astronomical calibration showed that the duration of each genetic unit can vary significantly, ranging from ca. 140 ky to a few thousand years. Therefore the oscillations in facies trends defined on the basis of genetic units are unlikely to be directly dictated by Milankovitch-type cyclicity.

Astronomical calibration also revealed substantial correlation between groups of depositional elements and systems, third-order sequence stratigraphic surfaces, and eccentricity periodicities. The delta-front depositional system that lies between 990 m and 880 m correlates with an individual long-eccentricity (400 ky) cycle (9.06–8.66 Ma). The beach–foreshore system that follows upwards between 880 m and 640 m correlates instead with two consecutive long-eccentricity cycles (7.86–8.66 Ma) (Fig. 15).

The boundary between stratigraphic sequences PAN-2 and PAN-3, namely Pan-2 SB, corresponds to cycle 30 and coincides fairly well with the short-eccentricity (100 ky) maximum dated at ca. 8.75 Ma, which is also the age of the boundary between mammal zones MN 10/MN11 (Vallesian–Turolian), dated astronomically by Krijgsman et al. (1996) in Spain.

According to Sacchi et al. (1999a) and Sacchi (2001), Pan-2 SB (ca. 8.75 Ma) developed in response to a significant lowering of the base level in Lake Pannon.

Both maximum flooding surface mfs-2 (cycle 14/15) and mfs-3 (cycle 90) correlate with distinct long-eccentricity maxima and can be dated at 9.075 Ma and 7.455 Ma, respectively. The stratigraphic unit bounded by mfs-2 and mfs-3 includes four long-eccentricity cycles (1.6 My). This unit substantially correlates with the Transdanubian stage, or substage (Sacchi et al, 1999a; Sacchi et al, 1999b; Sacchi and Horváth, 2002) we adopt in this study as a chronostratigraphic unit intermediate between the Pannonian *s. str.* (Stevanović, 1951) and the Pontian *s. str.* (Andrussov, 1887) of the Paratethys stage system (Figs. 3, 6, 11, and 15).

CONCLUSIONS

The cyclic Upper Miocene continental (mostly brackish lacustrine) succession of the western Pannonian basin, cored in western Hungary at the Iharosberény-I (Ib-I) well site, has been correlated with the global polarity time scale (GPTS) of Cande and Kent (1995) and Hilgen et al. (1995) and calibrated to the astronomical record (Laskar, 1990). Astrochronology of this upper Tortonian–lower Messinian succession of the central Paratethys was obtained by tuning the cyclostratigraphic patterns of the Ib-I well log to target plots of the 65°N summer insolation curve of Laskar (1990), solved by assuming present-day values for the dynamical ellipticity of the Earth and tidal dissipation by the Sun and the Moon.

Sedimentary cycles were defined in the Ib-I well section on the basis of cyclic variations of the smoothed curve of mean grain size (Fig. 13). These cycles display an average thickness on the order of 13 m and show good to excellent correlation with precession cycles (ca. 21 ky). Namely, minima of the curve of mean grain size correlate with precession minima and insolation maxima (Fig. 15).

Individual lignite layers or clusters and/or organic-rich silt–clay layers often correspond to minima of the curve of mean grain size in marginal lacustrine and delta-plain facies of the cored section between 750 m and 50 m (Fig. 15). As a consequence they also correlate with precession minima and insolation maxima. Hence we suggest, as a working hypothesis, that these organic-rich layers may represent a continental equivalent of the marine sapropels of the Mediterranean (Hilgen, 1991b; Nijenhuis et al., 1996; Negri et al., 1999; Schenau et al., 1999).

In the stratigraphic interval between ca. 1000 and 640 m (cycles 71 to 15), which corresponds to delta slope and beach–foreshore setting and includes three long-eccentricity (ca. 400 ky) cycles between ca. 7.85 Ma and 9.05 Ma, individual minima of the curve of mean grain size can be correlated with short-eccentricity cycles (ca. 100 ky). This can be explained by considering the relatively low accumulation rate in this stratigraphic interval (Figs. 15–16), which is the cause for the lack of an adequate resolution of the smoothed curve of mean grain size compared to the precession-related frequency of the original (non-smoothed) curve of mean grain size.

The genetic units described by Juhász et al. (1994), Juhász et al. (1997), Juhász et al. (1999) (Fig. 12) represent a time interval varying from ca. 140 ky to a few thousand years. Therefore the oscillations in facies trends defined on the basis of genetic units are likely to reflect autocyclic control or at least interference between astronomical forcing and autocyclicity.

The astronomical calibration of the Pannonian *s. l.* section at the Ib-I well site yields accurate ages for individual sedimentary cycles and other stratigraphic events discussed in this study (Fig. 15). The base of Pontian (cycle 90) correlates with a third-order maximum flooding surface (mfs-3) and is dated at 7.455 Ma. The base of the Transdanubian (cycles 14–15) again correlates with a third-order maximum flooding surface (mfs-2) and is dated at 9.075 Ma.

The FAD of endemic bivalves, *Prosodacnomya* (gen.) (cycle 54) and *Prosodacnomya* cf. *dainellii* (cycle 61) are dated at 8.293 Ma and 8.109 Ma, respectively.

The Tortonian–Messinian boundary, astronomically calibrated on Mediterranean sections and dated by Hilgen et al.

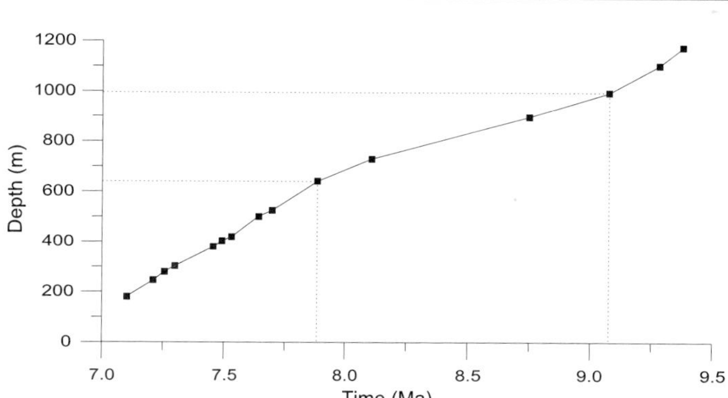

FIG. 16.—Non-decompacted time–depth curve of the Pannonian *s. l.* succession cored at the Iharosberény-I well site. Note the decrease in the sedimentation rate between ca. 640 m (7.87 Ma) and 1000 m (9.07 Ma). Also see Fig. 15 and text for discussion.

(2000b) at 7.251 Ma, is correlated with cycle 100 of the Ib-I borehole.

Sequence boundary Pan-2, which has been detected on a regional scale on seismic profiles across the southern Transdanubia (western Hungary) and cored at depth of ca. 900 m at the Ib-I well site, appears to correlate with the boundary between mammal zones MN10/MN11 (Vallesian–Turolian), astronomically dated at 8.75 Ma by Krijgsman et al. (1996) at the La Gloria section in Spain. This sequence boundary corresponds to a significant base-level drop (several tens of meters) that occurred at the western margin of Lake Pannon during the Tortonian (cycle 30).

The results of this study suggest that there is a high potential for new research in the field of orbital cyclostratigraphy applied to the Neogene sedimentary record of the Paratethys and other vast continental realms.

ACKNOWLEDGMENTS

This research work was made possible thanks to the Geological Institute of Hungary (MÁFI), which kindly provided the original data set from the Iharosberény-I well. Seismic profiles used in this study were selected from the database of the Department of Geophysics of the Eötvös University (ELTE), Budapest.

The authors gratefully acknowledge Annamaria Nádor and Silvia Iaccarino for their critical reviews which greatly improved the manuscript.

Sincere thanks are also due to Larry Phillips for his comments on an early version of the paper, to Bruno D'Argenio, Vittoria Ferreri, and Mario Sprovieri for helpful discussion at various stages, and to Mary Sclafani for the proofreading of the English text.

This research was carried out in the frame of the Italian–Hungarian cooperation agreement (CNR–MTA) for the period 2001–2003. Financial support was also provided by the National Science Foundation of Hungary and the Geological Institute of Hungary, Budapest.

REFERENCES

Abdul Aziz, H., Hilgen, F.J., Krijgsman, W., Sanz, E., and Calvo, J.P., 2000, Astronomical forcing of sedimentary cycles in the middle to late Miocene continental Calatayud basin (NE Spain): Earth and Planetary Science Letters, v. 177, p. 9–22.

Allen, P., and Collison, J.D., 1986, Lakes, *in* Reading, H.G., ed., Sedimentary Environments and Facies: Oxford, U.K., Blackwell Science, p. 63–94.

Andrussov, N., 1887, Geologische Untersuchungen in der westlichen Hälfte der Halbinsel Kertsch: Société Impériale des Naturalistes de Moscou (Russia), Bulletin, v. 2, p. 69–147.

Báldi, T., 1980, The early history of the Paratethys: Földtani Közlöny, v. 110/3–4, p. 456–472.

Báldi, T., 1989, Tethys and Paratethys through Oligocene times. Remarks to a comment: Geologica Carpathica, v. 41, p. 85–99.

Balogh, K., 1995, K/Ar study of the Tihany Volcano, Balaton Highland, Hungary. Report on the work supported by the European Community in the frame of program Integrated Basin Studies: Institute of Nuclear Research, Hungarian Academy of Sciences. Debrecen, 1995, 12 p.

Balogh, K., Árva-Sós, E., Pécskay, Z., and Ravasz-Baranyai, L., 1986, K/Ar dating of Post-Sarmatian alkali basaltic rocks in Hungary: Acta Mineralogica-Petrographica, v. 28, p. 75–93.

Bartha, F., 1971, Biostratigraphic investigation of the Pannonian layers, Hungary, *in* Goczán, F., and Benkő, J., eds., A magyarországi pannonkori képzQdmények kutatásai: Akadémiai Kiadó Budapest, 11–172 p.

Basic Drillings, 1987, The geological stratotypes of Hungary: Budapest, Hungarian Geological Institute 125 p.

Berggren, W.A., Hilgen, F.J., Langereis, C.G., Kent D.V., Obradovich, J.D., Raffi, I., Raymo, M.E., and Shackleton, N.J., 1995, Late Neogene chronology: New perspectives in high-resolution stratigraphy: Geological Society of America, Bulletin, v. 107, p. 1272–1287.

Cande, S.C., and Kent, D.V., 1995, Revised calibration of the geomagnetic polarity timescale for the Late Cretaceous and Cenozoic: Journal of Geophysical Research, v. 100, p. 6093–6095.

Channell, J.E.T., Rio, D., and Tunnel, R.C., 1988, Miocene/Pliocene boundary magnetostratigraphy at Capo Spartivento, Calabria, Italy: Geology, v. 16, p. 1096–1099.

Csató, I., 1993, Neogene sequences in the Pannonian Basin, Hungary: Tectonophysics, v. 226, p. 377–400.

Dam, G., and Surlyk, F., 1992, Forced regression in a large wave-and storm-dominated anoxic lake, Rhaetian–Sinemurian Kap Stewart Formation, East Greenland: Geology, v. 20, p. 749–752.

Elston, D., Lantos, M., and Hámor, T., 1990, Magnetostratigraphic and seismic stratigraphic correlations of Pannonian (s. l.) deposits in the Great Hungarian: Hungarian Geological Institute, Annual Report, 1988 v. I, p. 110–134.

Elston, D.P., Lantos, M., and Hámor, T., 1994, High-resolution polarity records and the stratigraphic and magnetostratigraphic correlation of the Late Miocene and Pliocene (Pannonian s. l.) deposits of Hungary, *in* Teleki, P.G., Mattick, R.E., and Kókai, J., eds., Basin Analysis in Petroleum Exploration. A Case Study from the Békés Basin, Hungary: Dordrecht, The Netherlands, Kluwer Academic Publishers, 111–142 p.

Garcés, M., Agustí, J., Cabrera, L., and Parés, J.M., 1996, Magnetostratigraphy of the Vallesian (Late Miocene) in the Vallès–Penedès basin (northeast Spain): Earth and Planetary Science Letters, v. 142, p. 381–396.

Hilgen, F.J., 1991a, Astronomical calibration of Gauss to Matuyama sapropels in the Mediterranean and implication for the Geomagnetic Polarity Time Scale: Earth and Planetary Science Letters, v. 104, p. 226–244.

Hilgen F.J., 1991b, Extension of the astronomically calibrated (polarity) time scale to the Miocene/Pliocene boundary: Earth and Planetary Science Letters, v. 107, p. 349–368.

Hilgen, F.J., and Langereis, C.G., 1988, The age of the Miocene–Pliocene boundary in the Capo Rossello area (Sicily): Earth and Planetary Science Letters, v. 91, p. 214–222.

Hilgen, F.J., Krijgsman, W., Langereis, C.G., and Lourens, L.J., 1997, Breakthrough made in dating of the geological record: Eos, Transactions, American Geophysical Union, v. 78, no. 28, p. 288–289.

Hilgen, F.J., Krijgsman, W., Raffi, I., Turco, E., and Zachariasse, W.J., 2000a, Integrated stratigraphy and astronomical calibration of the Serravallian/Tortonian boundary section at Monte Gibliscemi (Sicily, Italy): Marine Micropaleontology, v. 38, p. 181–211.

Hilgen, F.J., Krijgsman, W., Langereis, C.G., Lourens, L.J., Santarelli, A., and Zachariasse, W.J., 1995, Extending the astronomical (polarity) time scale into the Miocene: Earth and Planetary Science Letters, v. 136, p. 495–510.

Hilgen, F.J., Bissoli, L., Iaccarino, S., Krijgsman, W., Meijer, R., Negri, A., and Villa, G., 2000b, Integrated stratigraphy and astrochronology of the Messinian GSSP at Oued Akrech (Atlantic Morocco): Earth and Planetary Science Letters, v. 182, p. 237–251.

Homewood, P., Guillocheau, F., Eschard, R., and Cross, T.A., 1992, Corrélations haute résolution et stratigraphie génétique: Une démarche intégrée. (High-resolution correlations and genetic stratigraphy: an integrated approach): Centres de Recherche Exploration-Production Elf-Aquitaine, Bulletin, v. 16, p. 357–381.

Horváth, F., 1993, Towards a mechanical model for the formation of Pannonian Basin: Tectonophysics, v. 226, p. 333–357.

Horváth, F., and Tari, G., 1999, IBS Pannonian Basin project: a review of the main results and their bearings on hydrocarbon exploration, *in*

Durand, B., Jolivet, L., Horváth, F., and Séranne, M., eds., The Mediterranean Basins: Tertiary Extension within the Alpine Orogen: Geological Society of London, Special Publications 156, p. 195–213.

Jámbor, Á., 1980, A Dunántúli-középhegység pannóniai képzQdményei: Annales Instituti Geologici Publici Hungarici, v. 62, p. 1–259.

Juhász, E., Kovács, L.Ó., Müller, P., Tóth-Makk, Á., Phillips, L., and Lantos, M., 1997, Climatically driven sedimentary cycles in the Late Miocene sediments of the Pannonian basin, Hungary: Tectonophysics, v. 282, p. 257–276.

Juhász, E., Budai, T., Farkas-Bulla, J., Hámor, T., Lantos, M., Müller, P., and Tóth-Makk, Á., 1994, A Pannon-medence Neogén üledékeinek vészletes sedimentológiai fácies-elemzése és értékelése. Jelentés az IBS project kekretében végzett munkákról: Eötvös University of Budapest, Department of Geophysics könyvtár, 48 p.

Juhász, E., Phillips, L., Müller, P., Ricketts, B., Tóth-Makk, Á., Lantos, M., and Kovács, L.Ó., 1999, Late Neogene sedimentary facies and sequences in the Pannonian Basin, Hungary, *in* Durand, B. Jolivet, L., Horváth, F., and Séranne, M., eds., The Mediterranean Basins: Tertiary Extension within the Alpine Orogen: Geological Society of London, Special Publication 156, p. 335–356.

Juhász, Gy., 1993, Sedimentological and stratigraphical evidences of water-level fluctuations in the Pannonian Lake: Földtani Közlöny, v. 123 (4), p. 379–398.

Kázmér, M., 1990, Birth, life and death of the Pannonian Lake: Palaeogeography, Palaeoclimatology, Palaeoecology, v. 79, p. 171–188.

Kokay, J., Hamor, T., Lantos, M., and Müller, P., 1991, The paleomagnetic and geological study of borehole section Berhida 3: Hungarian Geological Institute, Annual Report, 1989, p. 45–63.

Korpás-Hódi, M., and Pogácsás, Gy., 1992, Paleogeographic outline of the Pannonian s.l. of the southern Danube–Tisza interfluve: Acta Geologica Hungarica, v. 35, p. 145–163.

Körössy, L., 1990, Hydrocarbon geology of SE Transdanubia, Hungary: Általános Földtani Szemle, v. 25, p. 3–53.

Krijgsman, W., 1996, Miocene magnetostratigraphy and cyclostratigraphy in the Mediterranean: extension of the astronomical polarity time scale: Ph.D. Dissertation, Aardwetenschappe, Universiteit Utrecht, Geologica Ultraiectina, 207 p.

Krijgsman, W., Hilgen, F.J., Langereis, C.G., Santarelli, A., and Zachariasse, W.J., 1995, Late Miocene magnetostratigraphy, biostratigraphy and cyclostratigraphy from the Mediterranean: Earth and Planetary Science Letters, v. 136, p. 475–494.

Krijgsman, W., Garcés, M., Langereis, C.G., Daams, R., van der Meulen, A.J., Agustí, J., and Cabrera, L., 1996, A new chronology for the middle to late Miocene continental record in Spain: Earth and Planetary Science Letters, v. 142, p. 367–380.

Langereis, C.G., and Hilgen, F.J., 1991, The Rossello composite: A Mediterranean and global reference section for the early to late Late Pliocene: Earth and Planetary Science Letters, v. 104, p. 211–225.

Lantos, M., and Elston, D.P., 1995, Low to high-amplitude oscillations and secular variation in a 1.2 km Late Miocene inclination record: Physics of the Earth and Planetary Interiors, v. 90, p. 37–53.

Lantos, M., Hámor, T., and Pogácsás, G., 1992, Magneto- and seismostratigraphic correlations of Pannonian s. l. (late Miocene and Pliocene) deposits in Hungary: Paleontologia i Evolució, v. 24–25, p. 35–46.

Laskar, J., 1990, The chaotic motion of the solar system: a numerical estimate of the size of the chaotic zones: Icarus, v. 88, p. 266–291.

Laskar, J., Joutel, F., and Boudin, F., 1993, Orbital precession and insolation quantities for the Earth from −20 Myr to +10 Myr: Astronomy and Astrophysics, v. 270, p. 522–533.

Laskarev, V., 1924, Sur les équivalents du Sarmatien supérieur en Serbie. Recueil des travaux offert a M. Jovan Cvijic, Beograd, v. 151–165.

Lőrenthey, I., 1900, Foraminifera der pannonischen Stufe Ungarns: Neues Jahrbuch für Mineralogie, Geologie und Paläontologie, v. 2, p. 99–107.

Magyar, I., Geary, D.H., and Müller, P., 1999a, Paleogeographic evolution of the Late Miocene Lake Pannon in Central Europe: Palaeogeography, Palaeoclimatology, Palaeoecology, v. 147, p. 151–167.

Magyar, I., Geary D.H., Sütő-Szentai, M., Lantos, M., and Müller, P., 1999b, Integrated bio-, magneto- and chronostratigraphic correlations of the Late Miocene Lake Pannon deposits, *in* Magyar, I., and Geary, D.H., eds., Fossils and Strata of Lake Pannon, a Long-Lived Lake from the Upper Miocene of Hungary: Acta Geologica Hungarica, v. 42, p. 5–31.

Mitchum, R.M., and Van Wagoner, J.C., 1991, High-frequency sequences and their stacking patterns: sequence stratigraphic evidence of high-frequency eustatic cycles: Sedimentary Geology, v. 70, p. 131–160.

Müller, P., Geary, D.H., and Magyar, I., 1999, The endemic molluscs of the Late Miocene Lake Pannon: their origin, evolution and family-level taxonomy: Lethaia., v. 32, p. 47–60.

Nagymarosy, A., 1990, From Tethys to Paratethys, a way of survival: Acta Geodaetica, Geophysica et Montanistica Hungarica, v. 25 (3–4), p. 373–385.

Nagymarosy, A., and Müller, P., 1988, Some aspects of Neogene biostratigraphy in the Pannonian Basin, *in* Royden, L.H., and Horváth, F., eds., The Pannonian Basin; A Study in Basin Evolution: American Association of Petroleum Geologists, Memoir 45, p. 69–78.

Negri, A., Hilgen, F.J., Giunta, S., Krjigsman, W., and Vai, G.B., 1999, Calcareous nannofossil biostratigraphy of the M. del Casino section (northern Apennines, Italy) and paleoceanographic conditions at times of Late Miocene sapropel formation: Marine Micropaleontology, v. 36, p. 13–30.

Nevesskaya, L.A., 1990, Definition of the Pontian Stage according to N. P. Barbot de marny (1869) and N. I., Andrussov (1917), *in* Stevanović, P.M., Nevesskaya, L.A., Marinescu, F., Sokać, A., and Jámbor, Á., eds., Chronostratigraphie und Neostratotypen, Neogen der Westlichen ("Zentrale") Paratethys 8, P11 Pontien: Zagreb–Belgrade, Jazu–Sanu, p. 39–40.

Nijenhuis, I.A., Schenau, S.J., Van Der Weijden, C.H., Hilgen, F.J., Lourens, L.J., and Zachariasse, W.J., 1996, On the origin of upper Miocene sapropelites: A case study from the Faneromeni section, Crete (Greece): Paleoceanography, v. 11, p. 633–645.

Papp, A., Jámbor, Á., and Steininger, F.F., 1985, Chronostratigraphie und Neostratotypen, Miozän der Zentralen Paratethys 7, Pannoniaen: Akadémiai Kiadó, Budapest, 636 p.

Pécskay, Z., Lexa, J., Szakács, A., Balogh, K., Seghedi, I., Konečny, V., Kovács, M., Márton, E., Kaličiak, M., Széky-Fux, V., Póka, T., Gyarmati, P., Edelstein, O., Rosu, E., and Žec, B., 1995, Space and time distribution of Neogene–Quaternary volcanism rocks in the Carpatho–Pannonian Region, *in* Downes, H., and Vaselli, O., eds., Neogene and Related Magmatism in the Carpatho–Pannonian Region: Acta Vulcanologica, v. 7, p. 15–28.

Picard, M.D., and High, L.R., Jr., 1972, Criteria for recognizing lacustrine rocks, *in* Rigby, J.K., and Hamblin, W.K., eds., Recognition of Ancient Sedimentary Environments: Society of Economic Paleontologists and Mineralogists, Special Publication 16, p. 108–145.

Pogácsás, Gy., Lakatos, L., Újszászi, K., Vakarcs, G., Várkonyi, L., Várnai, P., and Révész, I., 1988, Seismic facies, electro facies and Neogene sequence chronology of the Pannonian basin: Acta Geologica Hungarica, v. 31 (3–4), p. 175–207.

Posamentier, H.W., and Vail, P.R., 1988, Eustatic controls on clastic deposition II—sequence and systems tract models, *in* Wilgus, C.K., Hastings, B.S., Kendall, C.G.St.C., Posamentier, H.W., Ross, C.A., and Van Wagoner, J.C., eds., Sea Level Changes: An Integrated Approach: SEPM, Special Publication 42, p. 125–154.

Posamentier, H.W., Jervey, M.T., and Vail, P.R., 1988, Eustatic controls on clastic deposition I—conceptual framework, *in* Wilgus, C.K., Hastings, B.S., Kendall, C.G.St.C., Posamentier, H.W., Ross, C.A., and Van Wagoner, J.C., eds., Sea Level Changes: An Integrated Approach: SEPM, Special Publication 42, p. 125–154.

RCMNS, 1975, VI Congress of the Regional Committee of Mediterranean Neogene Stratigraphy, Bratislava, 1975.

Roth, L., 1879, Geologische Skizze des Kroisbach-Ruster Bergzuges und des südlichen Teiles des Leita-Gerbiges: Földtani Közlöny, Budapest, v. 9, p. 139–150,

Royden, L.H., and Horváth, F., 1988, The Pannonian Basin; A Study in Basin Evolution: American Association of Petroleum Geologists, Memoir 45, 394 p.

Rögl, F., 1996, Stratigraphic correlation of the Paratethys Oligocene and Miocene: Gesellschaft der Geologie und Bergbaustudenten in Österreich, Mitteilungen, v. 41, p. 65–73.

Rögl, F., 1998, Paleogeographic considerations for Mediterranean and Paratethys seaways (Oligocene and Miocene): Naturhistorisches Museum in Wien, Annalen, v. 99 A, p. 279–310.

Rögl, F., and Steininger, F.F., 1983, Vom Zerfall de Tethys zu Mediterran und Paratethys. Die neogene Paläogeographie und palinspastik des zirkum-mediterranen Raum: Naturhistorisches Museum in Wien, Annalen, v. 85 / A, p. 135–163.

Sacchi, M., 2001, Late Miocene evolution of the western Pannonian basin, Hungary. Ph.D. Dissertation, Eötvös Loránd University, Budapest, Hungary, 193 p.

Sacchi, M., and Horváth F., 2002, Towards a new time scale for the Upper Miocene continental series of the Pannonian basin (Central Paratethys): European Geosciences Union, Stephan Mueller Special Publication Series, v. 3, p. 1–16.

Sacchi, M., Horváth, F., and Magyari, O., 1999a, Role of unconformity-bounded units in the stratigraphy of the continental record: a case study from the Late Miocene of the western Pannonian basin, Hungary, *in* Durand, B., Jolivet, L., Horváth F., and Séranne, M., eds., The Mediterranean Basins: Tertiary Extension within the Alpine Orogen: Geological Society of London, Special Publication 156, p. 357–390.

Sacchi, M., Horváth, F., Magyar, I., and Müller, P., 1997, Problems and progress in establishing a Late Neogene Chronostratigraphy for the Central Paratethys: Neogene Newsletter, Padova, v. 4, p. 37–46.

Sacchi, M., Horváth, F., Magyar, I., and Müller, P., 1999b, Problems and progress in establishing a Late Neogene Chronostratigraphy for the Central Paratethys. Comments and Replies: Neogene Newsletter, v. 6, p. 25–59.

Schenau, S.J., Antonarakau, A., Hilgen, F.J., Lourens, L.J., Nijenhuis, I.A., Van Der Weijden, C.H., and Zachariasse, W.J., 1999, Organich-rich layers in the Metochia section (Gavdos, Greece): evidence for a single sapropel formation mechanism in the eastern Mediterranean during the last 10 Myr: Marine Geology, v. 153, p. 117–135.

Seneš, J., 1969, Unsere Kentnisse über die Paläogeographie der Zentralparatethys: Geologicke Prace, Spravy, v. 55, p. 83–108.

Shackleton, N.J., and Crowhurst, S., 1997, Sediment fluxes based on an orbitally tuned time scale 5 Ma to 14 Ma, site 926: Application to ODP leg 154 sites: Proceedings of the Ocean Drilling Program, Scientific Results, v. 154, p. 69–82.

Shackleton, N.J. Berger, A., and Peltier, W.R., 1990, An alternative astronomical calibration of the lower Pleistocene time scale based on ODP site 66: Royal Society of Edinburgh, Transactions, Earth Sciences, v. 81, p. 251–261.

Shackleton, N.J., Crowurst, S., Hagelberg, T., Pisias, N.G., and Shneider, D.A, 1995, A new late Neogene time scale: Application to ODP leg 138 sites, *in* Pisias, N.G., Mayer, L.A., Janececk, T.R., Palmer-Julson, A., and Van Andel, T.H., eds, Proceedings of the Ocean Drilling Program, Scientific Results, v. 138, p. 73–101.

Shanley, K.W., and McCabe, P.J., 1994, Perspectives on the sequence stratigraphy of continental strata: American Association of Petroleum Geologists, Bulletin, v. 78, p. 544–568.

Sprovieri, M., Sacchi, M., and Rohling, E.J., 2003, Climatically influenced interactions between the Mediterranean and the Paratethys during the Tortonian: Paleoceanography, v. 18, p. 1034–1044.

Steininger, F.F., and Nevesskaya, L.A., 1975, Stratotypes of Mediterranean Neogene stages: Committee on Mediterranean Neogene Stratigraphy, v. 2, p. 1–364.

Steininger, F.F., Bernor, R.L., Fahlbusch, V., and Mein, P., 1990, European Neogene marine / continental chronologic relations, *in* Lindsay, E.H., Fahlbusch, V., and Mein, P., eds., European Neogene Mammal Chronology: New York, Plenum Press, p. 15–46.

Stevanović. P.M., 1951, Pontische Stufe im engeren sinne—Obere Congerienschichten Serbiens und der angrenzenden Gebiete: Serbische Akademie der Wissenschaften, Sonderausgabe 187, Mathematisch-Naturwissenschaftliche Klasse no. 2, Beograd, 351 p.

Stevanović. P.M., Nevesskaya. L.A., Marinescu. F., Sokać, A., and Jámbor. Á., 1990, Chronostratigraphie und Neostratotypen, Neogen der Westlichen ("Zentrale") Paratethys 8, Pontien: Jazu-Sanu, Zagreb-Beograd, 952 p.

Surlyk, F., Noe-Nygaard, N., and Dam, G., 1993, High and low resolution sequence stratigraphy in lithological prediction—examples from the Mesozoic around the Northern Atlantic, *in* Parker, J.R., ed., Petroleum Geology of Northwest Europe: Proceedings of the 4th Conference: Geological Society of London, p. 199–214.

Swann, D.H., 1964, Late Mississippian rhythmic sediments of the Mississippi Valley: American Association of Petroleum Geologists, Bulletin, v. 48, p. 637–658.

Tari, G., Horváth, F., and Rumpler, J., 1992, Styles of extension in the Pannonian Basin: Tectonophysics, v. 208, p. 203–219.

Ujszászi, K., and Vakarcs, G., 1993, Sequence stratigraphic analysis in the South Transdanubian region, Hungary: Geophysical Transactions, v. 38, 2–3, p. 69–87.

Vail, P.R., 1987, Seismic Stratigraphy Interpretation Procedure: American Association of Petroleum Geologists, Studies in Geology no. 27, v. 1, p. 1–10.

Vakarcs, G., 1997, Sequence stratigraphy of the Cenozoic Pannonian basin, Hungary: Ph.D. Dissertation, Rice University, Houston, Texas. 514 p.

Vakarcs, G., and Várnai, P., 1991, Seismostratigraphic model of the Derecske trough: Magyar Geofizika, Budapest, v. 32, p. 38–50.

Vakarcs, G., Vail, P.R., Tari, G., Pogácsás, Gy., Mattick, R.E., and Szabó, A., 1994, Third-order Miocene–Pliocene depositional sequences in the prograding delta complex of the Pannonian Basin: Tectonophysics, v. 240, p. 81–106.

Van Wagoner, J.C., Mitchum, R.M., Campion, K.M., and Rahmanian, V.D., 1990, Siliciclastic sequence stratigraphy in well-logs, core and outcrops: concepts for high-resolution correlation of time and facies: American Association of Petroleum Geologists, Methods in Exploration Series, v. 7, 55 p.

Várnai, P. 1992, Seismic stratigraphy, paleogeographic reconstruction, and petroleum potential of Pannonian strata (Upper Miocene–Lower Miocene), south-western Hungary (abstract): American Association of Petroleum Geologists, Program with Abstracts, p. 134.

Wentworth, C.K., 1922, A scale of grade and class terms for clastic sediments: Journal of Geology, v. 30, p. 377–392.

Wilson, D.S., 1993, Confirmation of the astronomical calibration of the magnetic polarity timescale from sea-floor spreading rate: Nature, v. 364, p. 788–790.

Zijderveld, J.D.A., Hilgen, F.J., Langereis, C.G., Verhallen, P.J.J.M., and Zachariasse, W.J., 1991, Integrated magnetostratigraphy and biostratigraphy of the upper Pliocene–lower Pleistocene from the Monte Singa and Crotone areas in Calabria, Italy: Earth and Planetary Science Letters, v. 107, p. 697–714.

Appendix

OBTAINING TIMESCALES FOR CYCLOSTRATIGRAPHIC STUDIES

WALTHER SCHWARZACHER
School of Geography, The Queen's University of Belfast, Belfast, Northern Ireland BT7 1NN, U.K.
e-mail: w.schwarzacher@qub.ac.uk

ABSTRACT: Time in stratigraphy is derived from measuring the thickness of accumulated sediment. The quality of a time scale depends on the uniformity of the sedimentation. By subdividing a section using markers such as beds or cycles the uniformity of a section can be evaluated. By assuming that similar beds represent similar time intervals accumulation rates can be estimated. A cumulative plot of bed thickness against number tends towards a straight line if the sedimentation is uniform. Deviations from the straight line are due to both changes in accumulation rates and changes in the length of the time "unit". Examples from Pliocene varves and Cenomanian limestones are used to demonstrate the use of beds and cycles as markers. The inaccuracies of time scales should be remembered when interpreting cyclic records.

INTRODUCTION

A complete understanding of cyclicity without knowing the times involved in its formation is impossible. The pattern of repetition, which is the basic geological information, must be developed as a function of time, and this involves knowledge of a time scale. It is almost universal practice that time is deduced from the amount of deposited sediment that determines the stratigraphic position of the cycle. The stratigraphic position is measured in units of length above a base line in a logged section.

In order to obtain a valid time scale, it is necessary firstly to know the time interval represented by the section and, secondly, to have some information on the accumulation rates throughout the section. Only if the accumulation rate is constant can each point of the profile be dated by interpolation, using the stratigraphic scale of thickness. Constant accumulation rates throughout sections of any length are unlikely: to study such sections in detail, they must be subdivided into shorter intervals which can be represented by constant average accumulation rates. Bedding planes or cycle boundaries can provide the markers for such intervals; the intervals can however also be created artificially by a series of (often equidistant) sample points. The accumulation rates can of course be calculated only if the time represented by the thickness between markers is known.

One practical solution, which is often taken, is that the most prominent cycle of a sequence is used as a marker of equal time and the remaining cyclicity is expressed in units of the basic cycle.

There are, however, a large number of geological processes that can lead to cycle formation, and different processes may represent some quite different time intervals, which, if they are not well defined, introduce errors into the time scales. Scales based on cycles can further incorporate errors from the misidentification of the "unit" cycle or by missing a cycle boundary, either because of observational errors or because of a "missed beat" in the section. For this reason, it is often advisable to record successive cycles not only by numbering but also by their position in the stratigraphic profile. The two methods of measuring "time", either in centimeters or in cycle units, have been called the stratigraphic and metronomic approaches (Herbert, 1994).

It is often convenient to use the two terms *cycles* and *beds* when describing and subdividing sections. The bed (however defined) is usually the smallest unit, and in a sequence does not necessarily represent the same definite time interval, whereas the cycle (which may consist of one or more beds) is believed to represent a definite time interval. One can argue, however, that similar types of frequently repeated beds represent similar time intervals, and the numbering of such markers can therefore provide an approximate time scale.

THE UNIFORMITY OF SECTIONS

The degree of uniformity of a section reflects the uniformity in the deposition processes. The first step in any cyclostratigraphic study should concern the uniformity of the markers (which could be either cycles or beds), and which determines the quality of the time scale. The thickness variations of the time intervals between the markers involve, with only one exception, two unknown variables: the time interval between the markers, and the accumulation rate within the time span indicated by the markers. This time span includes the time spent in the actual deposition of the sediment between the markers and the time spent in either nondeposition or erosion at the cycle boundaries or bedding planes. The one exception is varved sediments, where the time interval is the constant year, the official unit of geological time. The ratio of time spent in deposition and nondeposition is also important in determining the quality of time markers. For example, event stratification, where the time spent in deposition can be regarded as instantaneous, cannot be used as a time marker.

A first step in examining the regularity of markers is to look at the distribution of bed thickness or cycle thickness. If beds or cycles of the same order have been used, the distributions should be unimodal and provide a mean μ_m and a standard deviation of σ_m. The position of the nth unit is the sum of consecutive cycles, plotted against the number of beds or cycles. This is the cumulative curve of the marker units.

In the limiting case when the cumulative curve is a straight line, all intervals must have been the same length but also both the time intervals and the rates must have been constant, because it is extremely unlikely that rate and time change in such a way that they result, as it were by accident, in units of equal sediment thickness (Sander, 1936).

To obtain a better picture of the random fluctuations of the cumulative curve, it is useful to subtract a trend corresponding to an average accumulation rate. One can either fit a straight line to the data or one can connect the two endpoints of the section by a straight line. The latter is the more appropriate method if the age of the endpoints is known.

The cumulative curve constitutes a random walk which is a self-affine fractal process. The Hurst exponent of a normal ran-

Cyclostratigraphy: Approaches and Case Histories
SEPM Special Publication No. 81, Copyright © 2004
SEPM (Society for Sedimentary Geology), ISBN 1-56576-108-1, p. 297–302.

dom walk is 0.5, giving it a fractal dimension of 3/2. The length of the random walk that corresponds to the total length of a section is a random variable that tends towards a normal distribution with a mean of $n\ \mu_m$ and a standard deviation of $\sigma_m \sqrt{n}$. Knowing the statistical parameters of the markers μ_m and σ_m together with the expected position after n cycles permits the calculation of confidence limits within which the expected cumulative curve should fall.

Two examples illustrate how the cumulative curve can be used to examine the uniformity of a section, which should reflect the constancy of sedimentation conditions. The example in Figure 1 shows the cumulative curve of "cycle" thicknesses from the Triassic carbonate platform of the Latemar (Italy). This section was measured by Goldhammer (1987), and the same data were used by Hinnov and Goldhammer (1991).

The cycle thickness in this section is far from uniform, and therefore an average cycle thickness was subtracted from successive cumulative values. The curve shows that relatively uniform sedimentation occurred in two intervals from cycle 1 to approximately cycle 160 and again from cycle 160 to the top of the section.

The lower two thirds of the cycles in the section have an average thickness of 38 cm, which increases to 116 cm in the upper third of the section. The linear metric scale, which is given together with the cycle number scale, gives the position of a bed only after it has been corrected by the corresponding deviation. Identifying a bed by the linear position scale can be wrong by up to 20 m. The deviations of the cumulative curve from the straight line would represent accumulation rates, if it can be assumed that the cycles represent equal time intervals; however, with the available data it cannot be assessed how far this is the case.

Figure 2 gives the cumulative curve of Pliocene lacustrine varves from the Villarroya Basin (Spain) measured by Muñoz et al. (2001). The cumulative curve shows the central part of the profile with varves 500 to 1200 in the Muñoz profile. The trend that was subtracted from the cumulative curve represents the accumulation rate for the interval and is the straight line that connects the end points of the section. The confidence limits shown assume zero error at the endpoints and a maximum error in the middle of the section. Because the time interval of the markers is known and constant, the trend-free cumulative curve represents accurate accumulation rates as deviations from the average rate. Again, the rates fall outside the confidence limits and are higher than would be expected from random variations. The interval therefore shows non-uniform development.

The power spectrum of the cumulative curve plotted using a logarithmic scale for both the frequency and power shows an almost straight-line decrease of power with increase of frequency (Fig. 3). The slope of this line is –2.177, giving a Hurst exponent of 0.588, which is close to the expected exponent of 0.5 for the Gaussian random walk. The analysis of the varve series by Muñoz et al. (2001) has shown that a considerable amount of variability in the accumulation rate can be attributed to climatic factors, and in particular the effect of the sunspot cycle can be demonstrated. A periodicity of about 12 years is visible in the spectrum of the cumulative curve, but its main feature is the linear increase of logarithmic power with reciprocal frequency, which indicates the much wider complexity of the accumulation process.

BEDDING AND CYCLES IN THE CENOMANIAN OF THE GUBBIO DISTRICT

The Cenomanian (Scaglia Bianca) of the Gubbio district in the Umbria Apennines, central Italy, is a well bedded limestone marl sequence. Data from two measured sections at M. Petrano and Contessa highway (Schwarzacher, 1994) have been used to illustrate the use of beds as stratigraphic markers. Both sections represent probably the complete Cenomanian, and they can be correlated very accurately by the well-established Bonarelli anoxic event, which occurs at the top of the sections. The distance between the two localities is approximately 20 km.

Sedimentary beds in limestone–marl sequences are often referred to as limestone–marl couplets, but this is very much a simplification for the Gubbio successions. Marl layers can range in thickness from 20 cm to a few millimeters, and limestones can be subdivided by bedding planes without any visible marl or clay seams. Cherts frequently occur in the limestone beds and near limestone–marl boundaries. To obtain a relatively objective definition of a bed, the following rule was followed: any change on top of a limestone with a thickness of one centimeter or more was called a bedding plane. A bed was taken as any sediment between two successive bedding planes. The most common bed consists of 1 cm to 2 cm of marl followed by about 10 cm of limestone. Cherts for this purpose were treated as limestone. Using this definition, cumulative bed thicknesses are shown in Figure 4, where deviations from the average trend are plotted against bed numbers. The horizontal thickness scale is linear, therefore does not give the precise position of the bed for which the deviation is plotted.

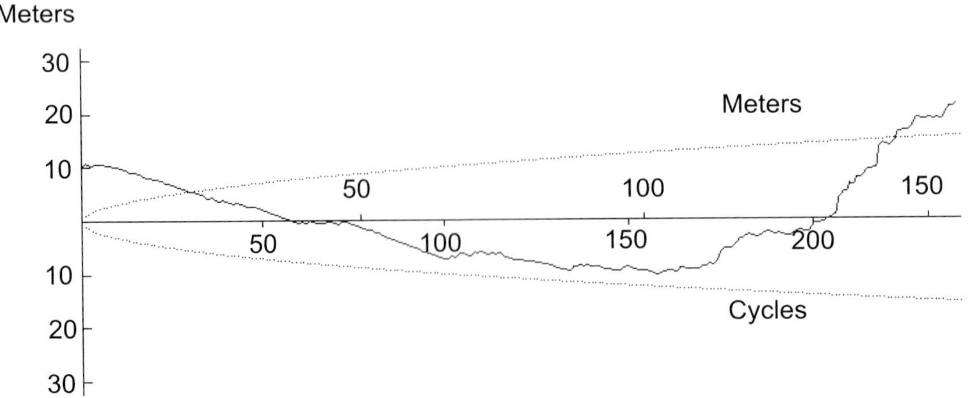

FIG. 1.—Cumulative thickness of carbonate platform cycles in the Triassic Latemar limestone. An average cycle thickness of 64 cm has been subtracted as a cumulative trend. The upper horizontal scale gives the stratigraphic positions in meters measured from the base of the section, and the lower scale gives the positions in number of cycles from the base. The 95% confidence limits for a Gaussian random walk are given by dotted lines. Data from Goldhammer (1987).

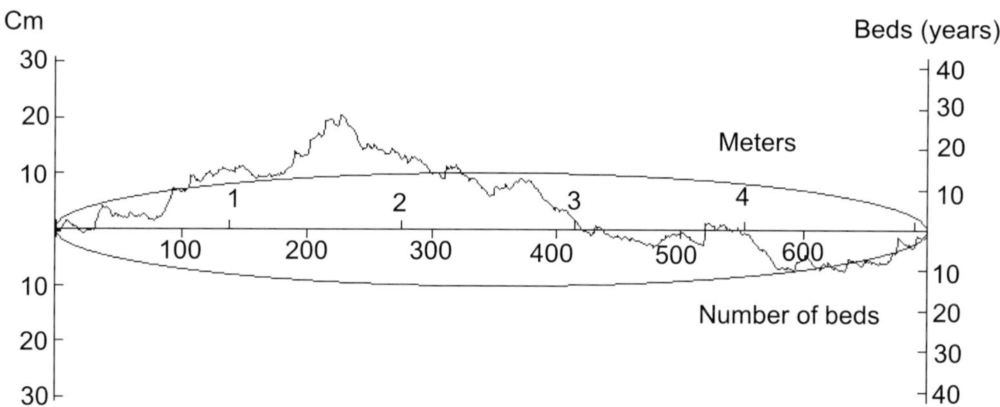

FIG. 2.—Trend-free cumulative curve of 700 lacustrine varves from the Vilarroya Basin (Spain). The upper horizontal scale gives the stratigraphic positions in meters measured from the base of the section, and the lower scale gives the positions in varves (years). The 95% confidence limits for a Gaussian random walk are given by dotted lines. Data from Muñoz et al. (2001).

To obtain this, the corresponding deviation has to be added to or subtracted from the linear scale.

The mean values of thicknesses in the two sections are very similar, 17.6 cm for the M. Petrano section and 16.2 cm for the Contessa highway section. The standard deviations are 10.8 and 9.1, respectively. Both sections are almost within the confidence limits for a normal random walk, and the somewhat higher deviations of the M. Petrano section could in part be due to misinterpreting some slight tectonic disturbance. The similarity between the two curves would be good enough to use them for correlation and indicates that the variation in bed thickness is regionally important for the history of sedimentation. Nevertheless, the closeness of the curves to the Gaussian random walk suggests that care must be taken when trying to interpret the long-term trend of the cumulative curves.

The beds in both sections are grouped into well-developed bundles, and the bundle boundaries can readily be observed in the field. The bundles have been interpreted as orbitally controlled Milankovitch cycles (Schwarzacher and Fischer, 1982). The time interval represented by the cycles is believed to be 100 ky, and it is therefore possible to determine accumulation rates at approximately regularly spaced intervals.

The curves of cumulative cycle thickness (Fig. 5) again show great similarity with each other, but also some similarity to the bed-thickness curves. In particular, all curves show a distinct change of deviations at about 38 m above the bases of the sections. Comparing the two types of data, the average cycle is found to be 103.8 cm in the M. Petrano section and 102.9 cm in the Contessa section, and on average should contain 5.8 and 6.3 beds.

The power spectrum of bed thickness as a function of position in the M. Petrano section (Fig. 6) has a strong maximum at a cycle length of 95.3 cm, which corresponds to 5.4 beds. The spectra of the Contessa sections are very similar.

It should be noted that a spectral analysis based on the Lomb–Scargle method can deal with unevenly spaced values like the positions of successive beds (Schulz, 1996), and the results from such analyses come very close to spectra based on the evenly spaced bed-number series. Therefore, using the terminology of Herbert, the stratigraphic method becomes equivalent to the metronomic method, and differentiating between the two be-

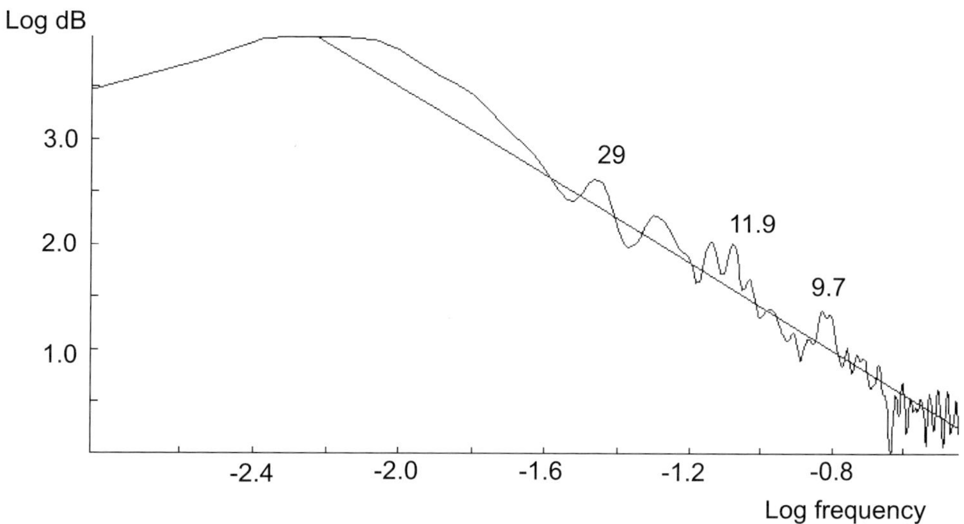

FIG. 3.—Power spectrum of the varve data. The slope of the fitted straight line indicates a fractional random walk.

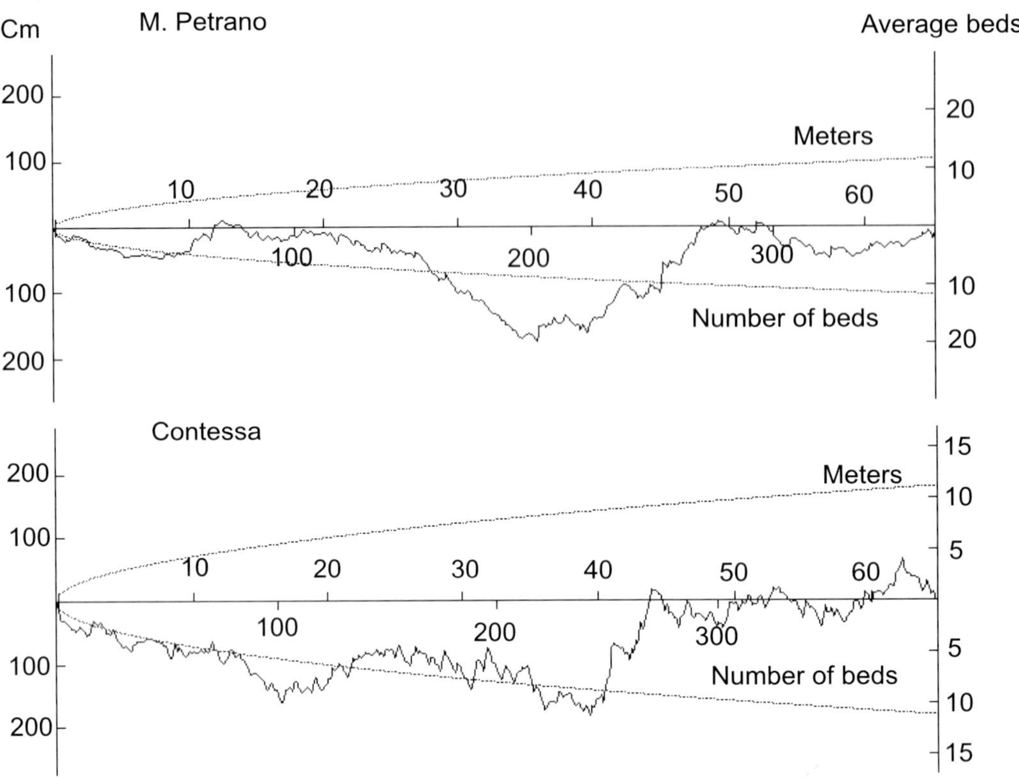

FIG. 4.—Cumulative curves of limestone–marl from the Cenomanian in the Gubbio district (Italy). The 95% confidence limits for Gaussian random walks are given by dotted lines. Deviations are expressed in centimeters and units of average bed thickness.

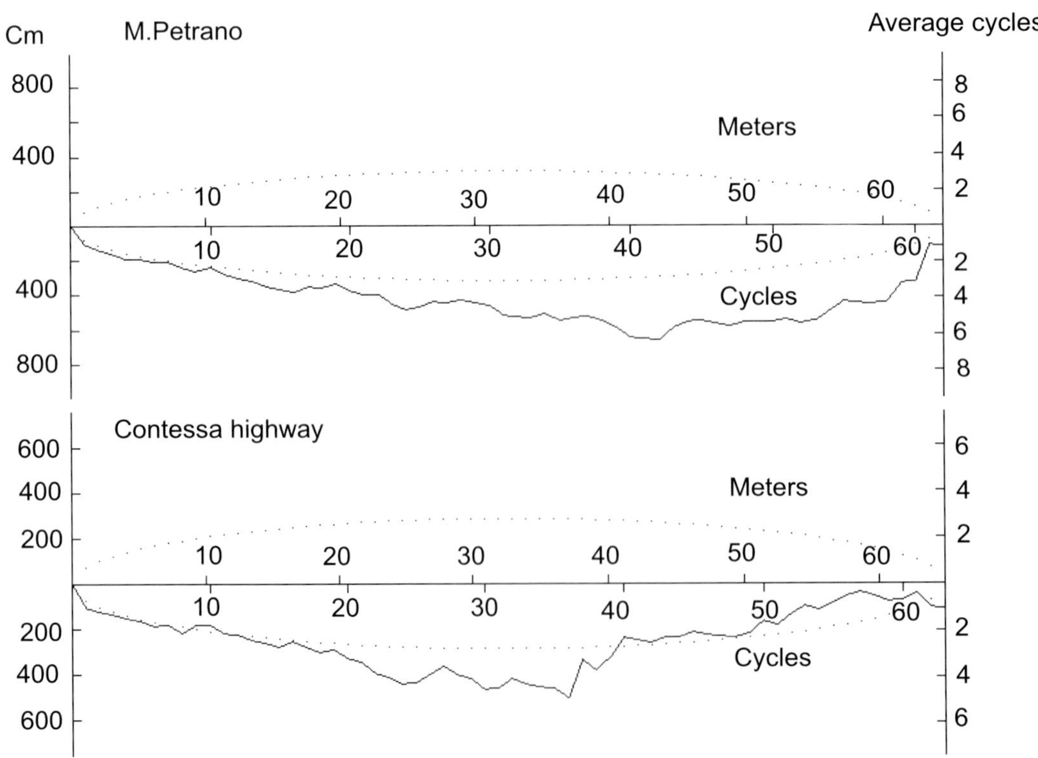

FIG. 5.—Cumulative curve of cycles from the Gubbio localities. The 95% confidence limits for a Gaussian random walk are given by dotted lines. Deviations are expressed in centimeters and units of average cycle thickness.

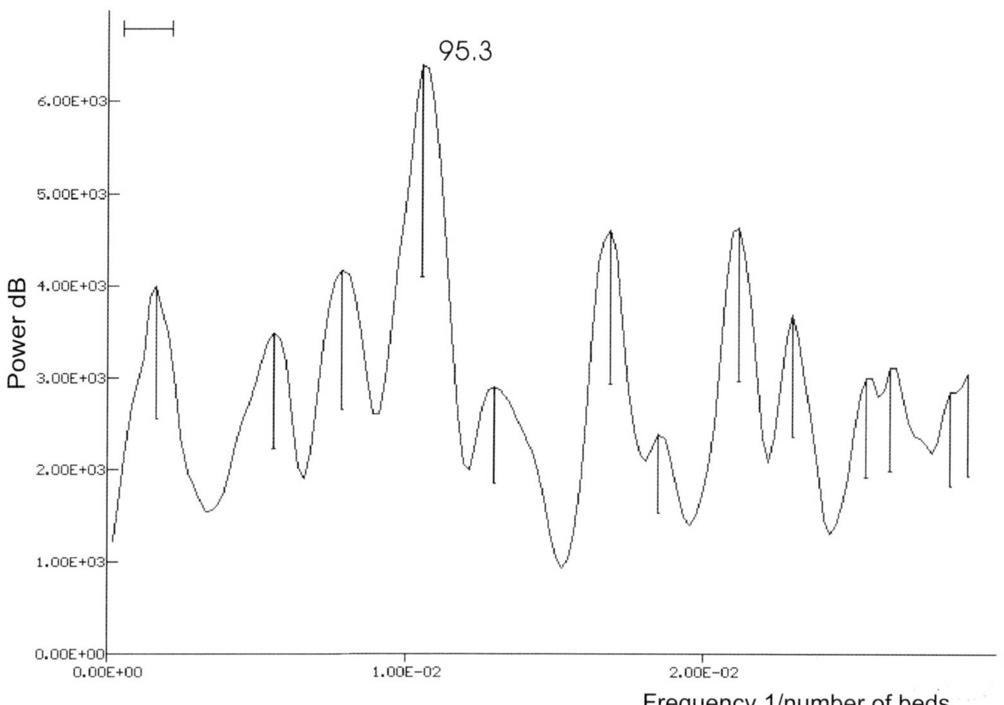

FIG. 6.—Power spectrum of the M. Petrano data. Frequency is expressed in numbers of beds per unit. The spectrum has a maximum at a frequency corresponding to a wavelength of 95.3 cm.

comes relevant only when one is dealing with a genuine metronomic series, in other words with varve analysis.

THE TIME SCALE

To calibrate the stratigraphic scale as a time scale, the ages of the end points of a section or its average rate of sedimentation must be known. In the case of the Gubbio sections, one is dealing approximately with the complete Cenomanian. The duration of this stage is given as 5.4 My (Gradstein and Ogg, 1996) and as 6.6 My by the slightly older timescale of Harland et al. (1989). Assuming that the 100 cm cycle in the sections represents the 100 ky eccentricity period, then one obtains additionally an estimate of 6.2 to 6.3 My for the interval.

Using the last estimate, one can estimate the time equivalent of the measured beds as being between 17.9 ky and 13.5 ky, which would suggest that the average 21 ky precession cycle is composed of 1.5 to 1.1 beds. These figures do not provide compelling evidence for Milankovitch cyclicity, but it is very likely that bedding is controlled largely by the precession cycle and that the additional beds are due to lithological changes not necessarily related to orbital control.

The cumulative curves can now be used to assess the error that is made in timing when interpolating from the total length or using the linear scales of either bed numbers or positions. Assuming the total duration to be 6 My, then the maximum deviations from the linear trend are 320 ky to 150 ky when bed thickness is used and 600 ky to 500 ky when the cycle scale is used. This is about 5–10% of the admittedly uncertain duration of 6 My representing the total section. The higher error is always in the M. Petrano section, which contains some small faults. Errors are only apparently smaller in the more detailed scale based on bed thickness because the average thickness is probably a less accurate estimate of time intervals than the cycle mean.

DISCUSSION AND CONCLUSIONS

The estimates of time error rely on the assumption that the average bed and cycle are a valid measure of time. The thickness distribution of beds (Fig. 7A) shows a wide range, and the average bed, which is used as time estimator, is therefore not very efficient. Furthermore, if bedding is determined largely by the precession cycle, then one is dealing with quite variable time intervals. Figure 7B, for example, shows the distribution of intervals between maxima in a calculated insolation curve. Although the mean for this distribution is 21 ky, the range of values is from 12 ky to 27 ky.

The construction of a time scale therefore faces two difficulties. Even constant time intervals produce variable thicknesses of sediments, as is clearly demonstrated by varve records. This variability can be called the recording error of sedimentation. The source of this error is very complex. Some components, like the climatic variations in the varve record, can be identified, and others may be long-range changes in the sedimentary basin or local variations, but all combine to create a noise component.

Secondly, orbital signals and the even more complicated insolation changes are quasi-periodic functions in which the local periods and amplitudes are not constant. Because we are here concerned with the formation of repeated markers such as bedding planes or cycle boundaries, we are interested in repeated features of the signal, such as the extrema or maxima or zero crossings of the signal. It is the time interval between such points that determines the local frequency of the orbital signals.

A Thickness of beds, M.Petrano

0 20 40 cm

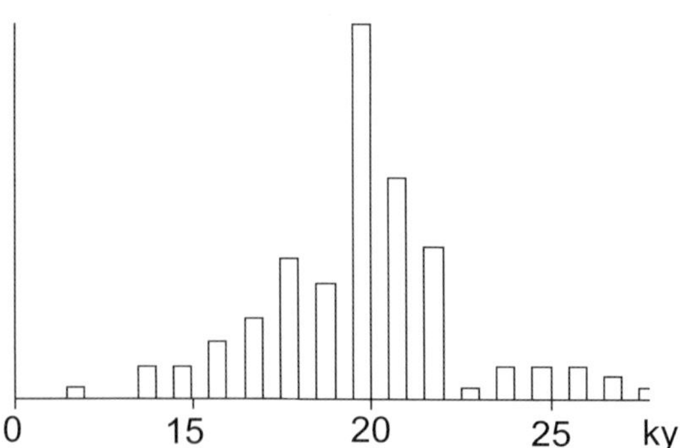

FIG. 7.—**A)** Frequency distribution of bed thicknesses in the M. Petrano section. **B)** Frequency distribution of length of time between maxima in a calculated insolation curve

Hinnov and Park (1999) examined the local frequency variation of both precession and obliquity by finding the distances between zero crossings of calculated orbital data. Their method, which they call metronomic FM analysis, is in fact a simplified demodulation of the signal. The term is misleading, because frequency modulation is normally the modulation of a constant carrier wave. Neither precession nor obliquity can be regarded as constant waves. The terminology is also hardly appropriate for the analysis of bed thickness or cycle thickness.

To transform the time intervals from the local frequency analysis into corresponding sediment thicknesses, the time intervals have to be multiplied by the accumulation rates, which not only are unknown but also contain unknown errors. To proceed any further, either one has to ignore the variation of accumulation rates or one can ignore the local frequency variations and assume an average length for the cycles of the signal. The model of a constant accumulation rate is quite unrealistic because it would, for example, imply that different intensities (amplitudes) of insolation produce identical amounts of sediment. Indeed there would not be any record of cycles. Ignoring the local frequency changes also produces errors, but at least one can obtain an estimate of the error involved.

For the practical analysis of sedimentary cycles, it is obvious that one of the first requirements for a useful time scale is the uniformity of the sedimentation. Any change in the deviation pattern of the cumulative curves indicates either a change of accumulation rate or a change of the time values represented by the markers. It may be useful, under such circumstances, to examine different segments of a section separately.

Whether one uses the so-called metronomic or cycle-by-cycle approach, or the stratigraphic approach, which locates markers by a metric scale, depends on knowledge of the time interval represented by the markers. If there are any uncertainties about the time, the method of locating by stratigraphic position is probably more convenient. In any case, it is very important that an accurate record of the stratigraphic position of any observed feature of the section be kept.

Finally, it should be recognized that any stratigraphic analysis is affected by the inherent irregularities of sedimentation and that it is meaningless to apply sophisticated statistical tests and high-resolution methods of analysis unless the quality of data justify such methods.

REFERENCES

GOLDHAMMER, R.K., 1987, Platform carbonate cycles. Middle Triassic of Northern Italy: the interplay of local tectonics and global eustasy: Ph.D. Dissertation, The Johns Hopkins University , 468 p.

GRADSTEIN, F.M., AND OGG, J.G., 1996, Geologic time scale for the Phanerozoic: Episodes, v. 19, p. 1–2

HARLAND, W.B., ARMSTRONG, R.L., COX, A.V., CRAIG, L.E., AND SMITH, D.G., 1990, A Geologic Time Scale 1989: Cambridge, U.K., Cambridge University Press, 263 p.

HERBERT, T.D., 1994, Reading orbital signals by sedimentation: models and examples, *in* de Boer, P.L., and Smith, D.G., eds., Orbital Forcing and Cyclic Sequences: International Association of Sedimentologists, Special Publication 19, p. 439–457.

HINNOV, L.A., AND GOLDHAMMER, R.K., 1991, Spectral analysis of the Middle Triassic Latemar Limestone: Journal of Sedimentary Petrology, v. 61, p. 1173–1193.

HINNOV, L.A., AND PARK, J.J., 1999, Strategies for assessing Early–Middle (Pliensbachian–Aalenian) Jurassic cyclochronologies: Royal Society (London), Philosophical Transactions, series A, v. 357, p. 1831–1859.

MUÑOZ, A., OJEDA, J., AND SANCHEZ-VALVERDE, B., 2001, Sunspot-like and ENSO/NAO-like periodicities in lacustrine laminated sediments of the Pliocene Vilarroya Basin (La Rioja, Spain): Paleoclimatology, v. 27, p. 453–463.

SANDER, B., 1936, Beiträge zur Kenntnis der Anlagerungsgefüge: Mineralogische und Petrographische Mitteilungen, v. 48, p. 27–139.

SCHWARZACHER, W., 1994, Cyclostratigraphy of the Cenomanian in the Gubbio district, Italy: a field study, *in* de Boer, P.L., and Smith, D.G., eds., Orbital Forcing and Cyclic Sequences: International Association of Sedimentologists, Special Publication 19, p. 87–97.

SCHWARZACHER, W., AND FISCHER, A.G., 1982, Limestone–shale bedding and perturbations in the Earth's orbit, *in* Einsele, G., and Seilacher, A., eds., Cyclic and Event Stratification: Berlin, Springer-Verlag, p. 72–95.

SCHULZ, M., 1996, Spectrum und Envelope, Computer programme zur Spektralanalyse nicht aequidister Palaeoklimatischer Zeitreihen: University of Kiel, Berichte aus dem Sonderforschungsbereich 313, no. 65, p. 1–131.

APPENDIX:

CONCEPT AND DEFINITIONS IN CYCLOSTRATIGRAPHY (SECOND REPORT OF THE CYCLOSTRATIGRAPHY WORKING GROUP)

International Subcommission on Stratigraphic Nomenclature of the IUGS Commission on Stratigraphy

FREDRIK HILGEN
Faculty of Earth Sciences, University of Utrecht, Budapestlaan 4, 3584 CD Utrecht, The Netherlands
e-mail: fhilgen@geo.uu.nl
WALTHER SCHWARZACHER
School of Geography, The Queen's University, BT7 1NN Belfast, Northern Ireland, U.K.
e-mail: w.schwarzacher@qub.ac.uk
AND
ANDRÉ STRASSER
Department of Geosciences, University of Fribourg, Pérolles, CH-1700 Fribourg, Switzerland
e-mail: andreas.strasser@unifr.ch

... to correlate these with an astronomical cycle of known period, and to deduce from this correlation an estimate in years of a portion of Cretaceous time (G.K. Gilbert, 1895)

INTRODUCTION

The application of sedimentary cycles in geochronology goes back into the 19th century, when Gilbert (1895) correctly linked *repetitive limestone–shale alternations* to the astronomical cycle of precession to determine the absolute duration of part of the Cretaceous. However, because of the often poor age control, geologists remained reluctant to link such repetitive changes to cyclic processes with identifiable time periods and often considered them to result from stochastic processes. As a consequence, the term sedimentary cycle was often defined in a purely descriptive sense to indicate "... recurrent sequences of strata each consisting of several lithologically distinctive members arranged in the same order" (Weller, 1960).

However, owing to improved age control and breakthrough studies of Pleistocene glacial cyclicity in marine cores, we now can use sedimentary cycles and cyclic changes in climatic proxy records with identifiable time periods for improving the resolution and accuracy of the geological time scale and for understanding natural climate variability. The cyclic changes in the sedimentary record referred to above are related to climate variations that are ultimately controlled by variations in the shape of the Earth's orbit and the inclination of its rotational axis. They are often called Milankovitch cycles, after the Serbian astronomer Milutan Milankovitch for his milestone contribution in linking orbital variation to the climatic changes of the Earth (Milankovitch, 1941).

At about the same time as the revival of Milankovitch cyclicity, the increasing concern about man-induced global warming spurred a host of detailed paleoclimatic and paleoceanographic studies of the last 10,000 to 150,000 years. These studies in particular aimed to understand natural climate variability on much shorter time scales at the extremely young end of Earth's history. For this purpose, ultra-high-resolution age models were needed and developed. The chronologies are based on counts of annual layers in corals, ice cores, and sediment cores from lakes and marine settings, thereby following the construction of a dendrochronology for the Holocene.

It is clear that accurate and high-resolution age models based on coherent cyclostratigraphic frameworks, and if possible integrated with biostratigraphic, magnetostratigraphic, chemostratigraphic, and tephrostratigraphic data, are an indispensable tool if we are to better understand natural climate variability and changes in sedimentary and ecological systems on different time scales. In view of the recent developments outlined above and the overwhelming evidence that cyclic variations are ubiquitously present in the geological record, the time is ripe to formally define a concept of cyclostratigraphy. The aim of our Working Group, which was installed under the auspices of the International Subcommission on Stratigraphic Classification (ISSC), is to formulate such a concept, to clarify its relationship with other schemes of stratigraphic classification, and to provide some definitions that are based on a broad consensus within our community. For this purpose, a first draft of the report was sent to all members of the ISSC as well as to a large number of specialists from various disciplines for review. Furthermore, the draft was discussed at the SEPM (Society for Sedimentary Geology) International Workshop "Multidisciplinary Approach to Cyclostratigraphy", held on May 26 to 28, 2001, in Sorrento, Italy. The critical comments have been taken into account during the preparation of this revised version of the report.

PROPOSED CONCEPT AND TERMINOLOGY

The term **cyclostratigraphy** was probably first publicly launched at the a meeting held in Perugia, Italy, and Digne, France (Fischer et al., 1988), to indicate the branch of stratigraphy that deals with sedimentary cycles and other cyclic variations in

Cyclostratigraphy: Approaches and Case Histories
SEPM Special Publication No. 81, Copyright © 2004
SEPM (Society for Sedimentary Geology), ISBN 1-56576-108-1, p. 303–305.

the stratigraphic record of any description but with identifiable time periods and, in particular, with their application in geochronology by improving the resolution of time-stratigraphic frameworks.

At present, the term cycle is employed in so many different ways that is has become almost meaningless. In its broadest sense, it is used to describe any regular or even irregular repetition of lithofacies in sedimentary successions, irrespective of the time involved. The members of the WG unanimously are of the opinion that **the term sedimentary cycle (as used in cyclostratigraphy) should be restricted to these repetitive changes in the stratigraphic record that have, or are inferred to have, a time significance.** Thus, individual cycles represent at least approximately equal time intervals.

The frequency stability of cyclic processes generating sedimentary cycles differs considerably. Relatively stable and therefore most useful oscillations are invariably directly connected with planetary and lunar motions. Therefore, astronomically controlled cycles are the main tool of cyclostratigraphy. They fall into different frequency bands: the calendar band with diurnal, monthly (lunar), and annual variations, the solar band with multi-annual to centennial variations, the sub-Milankovitch band with millennial variations, and the Milankovitch band with the precession, obliquity, and eccentricity orbital variations.

Some oscillations, like solar activity, climatic cycles such as Dansgaard–Oeschger cycles, Heinrich events, or ENSO cycles, are not yet fully explained and are less stable as far as their frequencies are concerned. Such cycles are therefore less suitable for establishing chronostratigraphic timescales, also because they cannot be predicted. However, they can be of considerable use in stratigraphy. Longer-term cycles with periods of tens to hundreds of millions of years are not included here because evidence for their presence in the geological record is weak and an underlying cause has not been convincingly demonstrated.

The translation of any cyclic or nearly cyclic variations of the environment into the sedimentary record involves nonlinearities as well as random disturbances. It is therefore not to be expected that stratigraphic series measured by the geologist will always produce precisely predictable periods. Clearly, this can make the identification of astronomically controlled cycles difficult and presents one of the main challenges for the stratigrapher. The sedimentary succession may contain cycles of several orders, indicating that the system in which it formed was governed by more than one periodic process, or by a combination of periodic and aperiodic processes. In the case of Milankovitch forcing, ratios of the superimposed cycles are known and can in principle be used to extract time information irrespective of the availability of other dating tools.

In this report, it is not our intention to provide a detailed and complete overview of cyclic forcing and how this forcing is eventually recorded in the widely different depositional environments that are found on Earth. Good reviews of the different forcing mechanisms and many examples of applying cyclostratigraphic methods are found in de Boer and Smith (1994) and House and Gale (1995). A general text is given by Schwarzacher (1993).

METHODOLOGY

The primary data are extracted from the sedimentary record in its broadest sense. This includes detailed analysis of (e.g.) lithology, facies, paleontology, mineralogy, stacking pattern of beds, geochemical proxies including stable isotopes, and/or physical properties. Very dense sampling is needed in order to obtain several data points within the smallest discernible sedimentary cycle.

Various methods of spectral analysis can be employed to detect cyclicity in such records (e.g., Weedon, 2003). Without time control the records can be analyzed in the depth domain and the resulting frequencies of cyclic components are given in number of cycles per meter. The depth series can be transformed into a time series using widely different methods (e.g., biochronology, radioisotopic dating, magnetostratigraphic calibration, astronomical tuning) to date the records as accurately as possible and thus to obtain the necessary age calibration points. The resulting time series of the original depth record can then be analyzed by the same spectral methods. In all cases confidence levels can be added to indicate the significance of the cyclicities.

LINKS TO OTHER BRANCHES IN STRATIGRAPHY

To Sequence Stratigraphy

The term "sedimentary cycle" as used in cyclostratigraphy differs from the terminology usually applied in sequence stratigraphy, where some authors use "cycle" to describe repetitive changes that are inferred to have time significance on at least a regional scale but that are not necessarily periodic. Evidently, the use of the term "cycle" in sequence stratigraphy is not consistent with cycles as used in cyclostratigraphy. However, sequences that represent equal time periods, such as many of the higher-order sequences according to the cycle hierarchy of Vail et al. (1991), are also "sedimentary cycles" as defined in cyclostratigraphy. On the other hand, the term "sedimentary cycle" includes many cyclic repetitions that are not sequences according to the sequence-stratigraphic concept (see definitions proposed by the ISSC Working Group on Sequence Stratigraphy).

The situation that some cyclic repetitions in sedimentary successions should be considered both as a sedimentary cycle and as a sequence, and thus are the subject of both cyclostratigraphy and sequence stratigraphy, is unavoidable but is not considered a problem by the members of the WG. In fact, the integration of high-resolution sequence stratigraphy and cyclostratigraphy can lead to a better understanding of the sedimentary systems and of how they react to sea-level, hydrodynamic, climatic, physical, chemical, and biological changes.

To Event Stratigraphy

Event stratigraphy deals with the identification and application of beds in the stratigraphic record that are caused by sudden events. Typical examples are turbidites and tempestites. As a consequence, there is no strong link between cyclostratigraphy and event stratigraphy because events are episodic rather than periodic in nature. However, individual event-stratigraphic units may occur in clusters that are related to cyclic processes.

To Geochronology

The absolute ages of sedimentary cycles and other cyclic variations in the Milankovitch frequency band can be determined by calibrating them to target curves, which are calculated from astronomical solutions for the Solar System. Such calibrations have been used to establish absolute time scales for the Paleocene to Recent and already underlie the standard geological time scale for the Neogene (Lourens et al., 2004). They can be considered "anchored" because they are tied to the Recent. "Floating" astrochronologies have been established for older parts of the geological record by multiplying the number of, for example, precession-induced cycles with the expected precessional periodicity at that time. Floating time scales can be tied to biostratigraphy, chemostratigraphy, or magnetostratigraphy, which are themselves calibrated by radiometric dates.

The standard geological time scale is now based on two independent absolute dating methods. The astronomical method is used for establishing the youngest part of the time scale, while the radiometric method is used for the older parts. Serious attempts are now being made to rigorously intercalibrate both dating methods (Kuiper, 2003). The most recent numerical astronomical solution permits a reliable reconstruction of the orbital parameters for the last 35 million years (Laskar, 1999; Varadi et al., 2003; Laskar et al., manuscript submitted).

Working Definitions

Cyclostratigraphy. – The subdiscipline of stratigraphy that deals with the identification, characterization, correlation, and interpretation of cyclic (periodic or nearly periodic) variations in the stratigraphic record and, in particular, with their application in geochronology by improving the accuracy and resolution of time-stratigraphic frameworks.

Sedimentary cycle (as used in cyclostratigraphy). – One succession of lithofacies that repeats itself many times in the sedimentary record and that is, or is inferred to be, causally linked to an oscillating system and, as a consequence, is (nearly) periodical and has a time significance. Different cycles can be described by their period, for example as a 100-ky cycle or, if this is not precisely known, as a cycle on the order of 100 ky.

Astronomical time scale (ATS). – A geological time scale with absolute ages derived from the calibration of sedimentary cycles and other cyclic variations in sedimentary successions to astronomical time series. Chron boundaries and biostratigraphic events are directly tied to such a time scale via first-order calibrations, i.e., they have been located in the same astronomically dated sections that have been used to construct the time scale. This approach is already common practice for the Neogene part of the geological time scale.

Suggestion for a Formal Codification of Milankovitch Cycles

At present, different codification schemes exist in the literature for sedimentary cycles and cyclic variations in oxygen-isotope records that occur in the Milankovitch frequency band of the spectrum and that have been tuned to the astronomical record. Oxygen-isotope stages have been numbered back from the Recent or have been given a numbering linked to the chron nomenclature via magnetostratigraphy. Sedimentary cycles in the Mediterranean Plio-Pleistocene have been coded after the correlative peak in precession or insolation time series numbered back from the Recent.

However, probably the most suitable cycle for establishing a formal codification scheme for astronomically calibrated variations is the 400-ky eccentricity cycle, because it is the most stable longer-term orbital cycle over prolonged intervals of time. The 400-ky cycles can be numbered back from the Recent and subdivided into individual precession cycles. In addition the 2.3 My eccentricity cycle might be of considerable use as well. Deeper in the geologic past, floating time scales must be tied to well-dated stratigraphic intervals. Type sections should be defined where these 400-ky and 2.3-My cycles are particularly well developed and linked to other stratigraphical scales. Astronomically calibrated time-stratigraphic units can be incorporated in the standard geological time scale. This is particularly favorable if the standard time scale is based on these units itself.

ACKNOWLEDGMENTS

We thank our colleagues who have responded to the questionnaire accompanying the first version of this report, and/or who have commented on the draft: E.J. Anderson, T. Bechstädt, Ki-Hong Chang, I. Chlupac, M.B. Cita, R. Cooper, B. D'Argenio, C.N. Drummond, C.H. Holland, A.G. Fischer, J. Haas, Wang Hongzhen, M. Kominz, A.D. Miall, M.A. Perlmutter, B. Pittet, W.A. Read, A. Salvador, Zhang Shouxin, J.B. Waterhouse, G.P. Weedon, and H. de la R. Winter. We also thank B. D'Argenio for having initiated the Workshop on Cyclostratigraphy in Sorrento, which gave us the opportunity to present and discuss our propositions. We have tried to reach a compromise that considers as many of the comments received as possible.

REFERENCES

de Boer, P.L., and Smith, D.G., eds., 1994, Orbital Forcing and Cyclic Sequences: International Association of Sedimentologists, Special Publication 19, 559 p.

Fischer, A.G., de Boer, P.L., and Premoli Silva, I., 1988, Cyclostratigraphy, *in* Beaudoin, B., and Ginsburg, R.N., eds., Global Sedimentary Geology Program: Cretaceous Resources, Events, and Rhythms: NATO ASI series, Dordrecht, The Netherlands, Kluwer Academic Publishers, p. 139–172.

Gilbert, G.K., 1895, Sedimentary measurement of Cretaceous time: Journal of Geology, v. 3, p. 121–127.

House, M.R., and Gale, A.S., eds., 1995, Orbital Forcing Timescales and Cyclostratigraphy: Geological Society of London, Special Publication 85, 210 p.

Kuiper, K.F., 2003, Direct intercalibration of radioisotopic and astronomical time in the Mediterranean Neogene: Geologica Ultraiectina, 235, 223 p. (http://www.geo.vu.nl/users/kuik).

Laskar, J., 1999, The limits of Earth orbital calculations for geological time scale use: Royal Society (London), Philosophical Transactions, v. A357, p. 1735–1759.

Lourens, L.J., Hilgen, F.J., Laskar, J., Shackleton, N.J., and Wilson, D., 2004, The Neogene Period, *in* Gradstein, F., Ogg, J., and Smith, A.G., eds., A Geologic Time Scale 2004: Cambridge, U.K., Cambridge University Press, in press.

Milankovitch, M., 1941, Kanon der Erdbestrahlung und seine Anwendung auf das Eiszeitenproblem: Akademie Royale Serbe, v. 133, 633 p.

Schwarzacher, W., 1993, Cyclostratigraphy and the Milankovitch Theory: Amsterdam, Elsevier, Developments in Sedimentology, v. 52, 225 p.

Vail, P.R, Audemard, F., Bowman, S.A., Eisner, P.N., and Perez-Cruz, C., 1991, The stratigraphic signatures of tectonics, eustasy and sedimentology—an overview, *in* Einsele, G., Ricken, W., and Seilacher, A., eds., Cycles and Events in Stratigraphy: Berlin, Springer-Verlag, p. 617–659.

Varadi, F., Runnegar, B., and Ghil, M., 2003, Successive refinements in long-term integrations of planetary orbits: Astrophysical Journal, v. 592, p. 620–630.

Weedon, G.P., 2003, Time-Series Analysis and Cyclostratigraphy: Cambridge, U.K., Cambridge University Press, 274 p.

Weller, J.M., 1960, Stratigraphic Principles and Practice: New York, Harper & Brothers, 725 p.

Index

A

B

C

Cyclostratigraphy: Approaches and Case Histories
SEPM Special Publication No. 81, Copyright © 2004
SEPM (Society for Sedimentary Geology), ISBN 1-56576-108-1, p. 307–311.

D

E

F

G

H

I

J

L

M

N

O

P

R

S

T

U

V

W

Z